Math

The Exponential Distribution

Theory, Methods and Applications

Edited by

N. Balakrishnan

McMaster University
Hamilton, Ontario, Canada

and

Asit P. Basu

University of Missouri–Columbia
Columbia, Missouri, U.S.A.

Gordon and Breach Publishers

Australia China France Germany India Japan Luxembourg Malaysia
The Netherlands Russia Singapore Switzerland Thailand
United Kingdom United States

Emmaplein 5
1075 AW Amsterdam
The Netherlands

British Library Cataloguing in Publication Data

Exponential Distribution:Theory, Methods
and Applications
I. Balakrishnan, N. II. Basu, Asit P.
519.24

ISBN 2-88449-192-9

To my brother Ramachandran
and sister-in-law Vijayalakshmi

To my wife Sandra Basu
and sons Amit and Shumit

Contents

Tables

Figures

Preface

The exponential distribution is one of the key distributions in the theory and practice of statistics. It possesses several significant statistical properties – most notably, its characterization through the lack of memory property. Yet it exhibits great mathematical tractability. Consequently, there is a vast body of literature on the theory and applications of the exponential distribution. In this volume, we have attempted to provide a systematic and comprehensive synthesis of this literature.

Over the last few decades, many handbooks have appeared consolidating the literature on various specific distributions. Initially, we were surprised that, given its importance, the exponential distribution had not received a similar special treatment. However, we quickly realized that the sheer volume of the literature on the exponential distribution may have discouraged many potential authors/editors. Our handbook comprises thirty-three distinct chapters discussing different aspects of the theory and applications of the exponential distribution. We express our deep appreciation and gratitude to all the authors for accepting our invitation and making valuable contributions, and for extending support and constant encouragement throughout the preparation of this volume.

We sincerely thank Debbie Iscoe for her excellent typesetting of the entire volume. We are grateful to the Natural Sciences and Engineering Research Council of Canada and the U.S. Air Force Office of Scientific Research for providing research grants (nos. F49620-95-1-0094 and F49620-94-1-0355).

Contributors

M. Ahsanullah, Department of Management Sciences, Rider University, 2083 Lawrenceville Road, Lawrenceville, NJ 08648-3099, U.S.A.
Ph: (609)895-5530, Fax: (609)896-5304, E-mail: *ahsan@enigma.rider.edu*

Barry C. Arnold, Department of Statistics, University of California, Riverside, CA 92521-0138, U.S.A.
Ph: (909)787-5939, Fax: (909)787-3286, E-mail: *barnold@ucrstat2.ucr.edu*

N. Balakrishnan, Department of Mathematics and Statistics, McMaster University, Hamilton, Ontario, Canada L8S 4K1
Ph: (905)525-4120, ext. 23420, Fax: (905)522-1676,
E-mail: *bala@mcmail.cis. mcmaster.ca*

Asit P. Basu, Department of Statistics, University of Missouri-Columbia, 222 Math Sciences Building, Columbia, MO 65211-0001, U.S.A.
Ph: (314)882-8283, Fax: (314)882-9676, E-mail: *basu@statserv1.cs.missouri.edu*

Gouri K. Bhattacharyya, Department of Statistics, University of Wisconsin-Madison, 1210 West Dayton Street, Madison, WI 53706-1685, U.S.A.
Ph: (608)262-3851

Rick Chappell, Department of Statistics, University of Wisconsin-Madison, 1210 West Dayton Street, Madison, WI 53706-1685, U.S.A.
Ph: (608)263-5572, Fax: (608)263-1059, E-mail: *chappell@stat.wisc.edu*

Aaron Childs, Department of Mathematics and Statistics, McMaster University, Hamilton, Ontario, Canada L8S 4K1
Ph: (905)525-9140

A. Clifford Cohen, Department of Statistics, University of Georgia, Athens, GA 30602-1952, U.S.A.
Ph: (706)542-3312, Fax: (706)542-3391, E-mail: *cliff@rolf.stat.uga.edu*

Nader Ebrahimi, Department of Mathematical Sciences, Division of Statistics, Northern Illinois University, DeKalb, IL 60115-2854, U.S.A.
Ph: (815)753-6884

Max Engelhardt, Principal Scientist, Statistics, Reliability and Analysis, Idaho National Engineering Laboratory, 2065 Ironwood Drive, Idaho Falls, ID 83402-5560, U.S.A.
Ph: (208)526-2100, Fax: (208)526-5647, E-mail: *mee@inel.gov*

Ursula Gather, Fachbereich Statistik, Universität Dortmund, Vogelpothsweg 87, D-44221 Dortmund, Germany
Ph: 49(231)755-3110, Fax: 49(231)755-3110,
E-mail: *gather@omega.statistik.uni-dortmund.de*

Jayanta K. Ghosh, Indian Statistical Institute, 203 Barrackpore Trunk Road, Calcutta-700 035, India
Fax: 91(33)556-6925, E-mail: *jkg@isical.ernet.in*

Alan J. Gross, Department of Biometry and Epidemiology, Medical University of South Carolina, 171 Ashley Avenue, Charleston, SC 29425-2503, U.S.A.
Ph: (803)792-7578, Fax: (803)792-0539

John E. Hewett, Department of Statistics, University of Missouri-Columbia, Math Sciences Building, Columbia, MO 65211-0001, U.S.A.
Ph: (314)882-6376, Fax: (314)884-5524, E-mail: *statjh3@mizzou1.missouri.edu*

J.S. Huang, Department of Mathematics and Statistics, University of Guelph, Guelph, Ontario, Canada N1G 2W1
Ph: (519)824-4120, Fax: (519)837-0221,
E-mail: *jhuang@msnet.mathstat.uoguelph.ca*

Norman L. Johnson, Department of Statistics, Phillips Hall, University of North Carolina, Chapel Hill, NC 27599-3260, U.S.A.
Ph: (919)962-0234

Richard A. Johnson, Department of Statistics, University of Wisconsin-Madison, 1210 West Dayton Street, Madison, WI 53706-1685, U.S.A.
Ph: (608)262-2357, Fax: (608)262-0032, E-mail: *rich@stat.wisc.edu*

John P. Klein, Department of Biostatistics, Medical College of Wisconsin, 8701 West Watertown Plank Road, Milwaukee, WI 53226-4801, U.S.A.
Ph: (414)456-8379, Fax: (414)266-8481, E-mail: *klein@hp03.biostat.mcw.edu*

Samuel Kotz, Department of Management Sciences and Statistics, University of Maryland at College Park, College of Business and Management, University of Maryland, College Park, MD 20742-0001, U.S.A.
Ph: (301)405-2254, Fax: (301)314-9157

Geoff J. McLachlan, Department of Mathematics, The University of Queensland, St. Lucia, Queensland 4072, Australia
Ph: 61(7)365-2150, Fax: 61(7)365-1477, E-mail: *gjm@maths.uq.oz.au*

Nitis Mukhopadhyay, Department of Statistics, University of Connecticut-Storrs, Box U-120, CT 06269-3120, U.S.A.
Ph: (203)486-6144, Fax: (203)486-4113, E-mail: *mukhop@uconnvm.uconn.edu*

H.N. Nagaraja, Department of Statistics, Ohio State University, 141 Cockins Hall, 1958 Neil Avenue, Columbus, OH 43210-1247, U.S.A.
Ph: (614)292-6072, Fax: (614)292-2092, E-mail: *hnn@stat.mps.ohio-state.edu*

S. Panchapakesan, Department of Mathematics, Southern Illinois University at Carbondale, Carbondale, IL 62901-4408, U.S.A.
Ph: (618)453-5302, Fax: (618)453-5300, E-mail: *GA3694@siucvmb.siu.edu*

R.A. Sandhu, Department of Mathematics and Statistics, McMaster University, Hamilton, Ontario, Canada L8S 4K1
Ph: (905)525-9140

Pranab K. Sen, Departments of Biostatistics and Statistics, University of North Carolina, Chapel Hill, NC 27599-7400, U.S.A.
Ph: (919)966-7274, Fax: (919)966-3804, E-mail: *pksen@uncmvs.oit.unc.edu*

Ashis SenGupta, Indian Statistical Institute, 203 Barrackpore Trunk Road, Calcutta-700 035, India
E-mail: *ashis@isical.ernet.in*

J. George Shanthikumar, Walter A. Haas School of Business Administration, University of California at Berkeley, Berkeley, CA 94720, U.S.A.
Ph: (510)642-2571, Fax: (510)642-2826, E-mail: *sjgsha@garnet.berkeley.edu*

Samuel S. Shapiro, Department of Statistics, Florida International University, Miami, FL 33199-0001, U.S.A.
Ph: (305)348-2741, Fax: (305)348-4172

Bimal K. Sinha, Department of Mathematics and Statistics, University of Maryland-Baltimore County, Baltimore, MD 21228, U.S.A.
Ph: (410)455-2347, Fax: (410)455-1066, E-mail: *sinha@umbc2.umbc.edu*

Pandu R. Tadikamalla, Graduate School of Business, University of Pittsburgh, 258 Mervis Hall, Pittsburgh, PA 15260, U.S.A.
Ph: (412)648-1596, Fax: (412)648-1693, E-Mail: *pandu@vms.cis.pitt.edu*

Román Viveros, Department of Mathematics and Statistics, McMaster University, Hamilton, Ontario, Canada L8S 4K1
Ph: (905)525-9140, ext. 23425, Fax: (905)522-0935,
E-mail: *rviveros@icarus.math.mcmaster.ca*

Lam Yeh, Department of Statistics, The Chinese University of Hong Kong, Shatin, New Territories, Hong Kong

S. Zacks, Department of Mathematical Sciences, Binghamton University, Binghamton, NY 13902-6000, U.S.A.
Ph: (607)777-6035, Fax: (607)777-2450, E-mail: *bg0525@bingvmb.bitnet*

Qiqing Zhang, Department of Mathematics and Statistics, McMaster University, Hamilton, Ontario, Canada L8S 4K1
Ph: (905)525-9140

CHAPTER 1

Genesis

Norman L. Johnson,[1] Samuel Kotz,[2] and N. Balakrishnan[3]

1.1 Preliminaries

Exponential distributions are encountered as life-time distributions with *constant hazard rate*. If the cumulative distribution function of X, a random variable representing lifetime, is $F_X(x)$ and the (constant) hazard rate is λ (not depending on x), then

$$-\frac{d\log(1-F_X(x))}{dx}=\lambda \qquad (x>0)\ , \tag{1.1.1}$$

or, in terms of the survival function $\overline{F}_X(x)=1-F_X(x)$,

$$-\frac{d\log\overline{F}_X(x)}{dx}=\lambda \qquad (x>0)\ . \tag{1.1.2}$$

Solving the differential equation in (1.1.2) upon using the condition $\overline{F}_X(0)=1$, we get

$$\overline{F}_X(x)=e^{-\lambda x} \qquad (x\ge 0)\ , \tag{1.1.3}$$

and the corresponding probability density function of X as

$$f_X(x)=\lambda\, e^{-\lambda x} \qquad (x\ge 0)\ . \tag{1.1.4}$$

If $\lambda=1$, we have the standard exponential distribution with the survival function

$$\overline{F}_X(x)=e^{-x} \qquad (x\ge 0) \tag{1.1.5}$$

and the probability density function

$$f_X(x)=e^{-x} \qquad (x\ge 0)\ . \tag{1.1.6}$$

[1] University of North Carolina, Chapel Hill, North Carolina
[2] University of Maryland, College Park, Maryland
[3] McMaster University, Hamilton, Ontario, Canada

Note that for $y > 0$ and $x > 0$

$$P[X > x + y \mid X > x] = P[X > y] , \tag{1.1.7}$$

that is, the future lifetime distribution, given survival to age x, does not depend on x. This property, termed *lack of memory* - or, equivalently in this context, *lack of aging*, characterizes exponential distributions, as can easily be seen. Eq. (1.1.7) can also be written equivalently as

$$S_X(x + y)/S_X(x) = S_X(y) , \tag{1.1.8}$$

i.e.,

$$\log S_X(x + y) = \log S_X(x) + \log S_X(y) . \tag{1.1.9}$$

Differentiating (1.1.9) with respect to y, we get

$$-\frac{d \log S_X(x+y)}{dy} = -\frac{d \log S_X(y)}{dy} \quad [= \lambda_X(y)] . \tag{1.1.10}$$

Since

$$-\frac{d \log S_X(x+y)}{dy} = -\frac{d \log S_X(x+y)}{d(x+y)} \quad [= \lambda_X(x+y)] , \tag{1.1.11}$$

the hazard rate at y is the same as at $(x+y)$ for *any* x *and any* y - that is, it is constant.

This lack of memory (or lack of aging) property is a critical part of many of the statistical analyses based on the exponential distribution as will be seen throughout the rest of this volume.

Exponential distributions are commonly employed in the formation of models of lifetime distributions, and stochastic processes in general. Even when the simple mathematical form of the distribution is inadequate to describe real-life complexity, it often serves as a bench-mark with reference to which effects of departures to allow for specific types of disturbance can be assessed.

Exponential distributions also appear as distributions of squared aiming errors in two dimensions when the error is the resultant of independent homoscedastic and unbiased errors in horizontal and vertical directions. If each component has a normal distribution with expected value zero (for unbiasedness) and the same variance σ^2, then the square of the combined aiming error is distributed as $2\sigma^2 \times$ (standard exponential variable).

1.2 Historical Remarks

Despite being a special case of the gamma family of distributions (Type III) discussed by Karl Pearson as early as 1895, it took another three-and-a-half-decades for the exponential distribution to appear on its own in the statistical literature. Kondo (1931) referred to the exponential distribution, while discussing the sampling distribution of standard deviation, as Pearson's Type X distribution. Applications of the exponential distribution in actuarial, biological and engineering problems were demonstrated subsequently by Steffensen (1930), Teissier (1934) and Weibull (1939), respectively.

Though Poisson distribution appeared (as an approximation to binomial) in Section 81 of the well-known *Récherches sur la Probabilité des Jugements* by S.D. Poisson in 1837 and later in the monograph *Das Gesetz der kleinen Zahlen* by L. Bortkiewicz in 1898, it did not receive much recognition till later. The relationship between Poisson and exponential distributions was noted in a physical context in *Studien über das Gleichgewicht der lebendigen Kraft zwischen bewegten materiellen Punkten* by L. Boltzmann in 1868, and then in *Choice and Chance* by W.A. Whitworth in 1886. Yet, it took about another 50 years for the general acceptance of the exponential distribution in statistical circles.

In a remarkable paper in 1937, P.V. Sukhatme indicated that the exponential distribution could serve, in the case of problems where the form of variation is far removed from the normal, as a suitable alternative to the normal distribution. He then went on to study properties of exponential order statistics quite extensively, and established, in particular, that the exponential spacings are themselves independently exponentially distributed. As pointed out by Galambos and Kotz (1978), this property of exponential spacings has been rediscovered by many authors; for example, see Malmquist (1950), Rényi (1953), and Epstein and Sobel (1953). Though these papers also contributed significantly to the development of the theory of exponential distributions, the last mentioned paper was primarily responsible for a flurry of activity that followed soon after.

W. Weibull, in a pioneering paper in 1951, considered an extension of the exponential distribution which is now referred to as the *Weibull distributions*. This family of distributions includes the exponential distribution as a special case when the shape parameter equals one. In the following year, D.J. Davis (1952) discussed the analysis of failure data using the exponential distribution and compared the analysis with that based on the normal distribution.

This late recognition of the exponential distribution by the statistical community explains why characterizations of the exponential distribution started so late too. The basic characterization of the exponential distribution based on lack of memory is simply the logarithmic equivalent of the functional equation $f(x+y) = f(x)f(y)$, which is due to Cauchy (1821) and Lobachevskii (1829); see also Aczél (1966). A complete solution of the logarithmic equivalent of this functional equation for both continuous and discrete cases was provided by Wilson (1899) and Hamel (1905). However, Galambos and Kotz (1978) conjecture that the first explicit characterization of the exponential distribution may have appeared sometime in the 19th century in the physical science literature in connection with some earlier theories of atoms since atoms do exhibit the lack of memory property. These have been discussed by Feller in 1950. Current work on characterizations of the exponential distribution seems to have originated from European mathematicians such as Rényi (1956) (who characterized exponential through Poisson processes and renewal processes), Fisz (1958) (who characterized through order statistics), and Rossberg (1960) (who characterized through range and ratios of order statistics). Independently, Ghurye (1960) and Teicher (1961) derived characterizations of the exponential distribution which are modifications of characterizations of the normal distribution through the distribution of t-type statistics and the maximum likelihood property. Since then, characterization results for the exponential distribution have enjoyed great prominence as is clearly evident from the monograph by Galambos and Kotz (1978) and the recent volume by Rao and Shanbhag (1994), among others.

Feller (1966), while discussing various properties of the exponential distribution, has also highlighted the following natural applications in addition to few other problems:

Tensile strength. Let $\overline{F}(t)$ be the probability that a thread of length t of a given material can sustain a certain fixed load. A thread of length $t+s$ does not snap if and only if the two segments individually sustain the given load. If there is no interaction, the two events must be independent in which case the functional equation $\overline{F}(t+s) = \overline{F}(t)\overline{F}(s)$ must be satisfied. Hence, the length at which the thread will break is an exponentially distributed random variable.

Free-path problem. Consider an ensemble of stars in space (a star is a ball of fixed radius $\rho > 0$) and choose for the origin a point not contained in any star. Taking the x-axis as representative of an arbitrary direction, one may be interested in the longest interval $(0, x)$ not intersecting any star representing the "visibility in the x-direction." Suppose we describe the ensemble by Poisson distributions which serve as a model of "perfect randomness". Then, the probability that a segment of length $t + s$ intersects no star must be equal to the product of the corresponding probabilities for the two segments of length t and s. This, as in the first example, means that the visibility in a given direction must be an exponentially distributed random variable. This has been used by Larsen and Marx (1981), in a popular textbook, in the framework of kinetic theory of gases wherein the distance that a molecule travels before colliding with another molecule is described by an exponential distribution.

In the 60's, an important paper that appeared on the exponential distribution was that of Zelen and Dannemiller (1961) in which the authors demonstrated that inference procedures based on the exponential assumption may be quite non-robust, i.e., highly sensitive to departures from exponentiality. Another ground breaking paper was that of Feigl and Zelen (1965), in which estimation of exponential survival probabilities with concomitant information was discussed.

Methods based on exponential distribution have flourished since then, and this volume with its 33 chapters provides an ample testimony to that. A lucid article by Galambos (1982), and the books of Galambos and Kotz (1978) and Johnson, Kotz, and Balakrishnan (1994) can provide interested readers with further details on the history of exponential distributions.

References

Aczél, J. (1966). *Lectures on Functional Equations and Their Applications*, Academic Press, New York.

Cauchy, A.L. (1821). *Cours d'analyse de l'École Polytechnique*, Vol. I, Analyse algébrique, V. Paris.

Davis, D.J. (1952). An analysis of some failure data, *Journal of the American Statistical Association*, **47**, 113-150.

Epstein, B. and Sobel, M. (1953). Life testing, *Journal of the American Statistical Association*, **48**, 486-502.

Feigl, P. and Zelen, M. (1965). Estimating exponential survival probabilities with concomitant information, *Biometrics*, **21**, 826-838.

Feller, W. (1966). *An Introduction to Probability Theory and Its Applications - Vol. II*, Second edition, John Wiley & Sons, New York.

Fisz, M. (1958). Characterization of some probability distributions, *Skandinavisk Aktuarietidskrift*, **41**, 65-70.

Galambos, J. (1982). Exponential distribution, In *Encyclopedia of Statistical Sciences*, Vol. 2 (Eds., S. Kotz and N.L. Johnson), pp. 582-587, John Wiley & Sons, New York.

Galambos, J. and Kotz, S. (1978). *Characterizations of Probability Distributions*, Lecture Notes in Mathematics No. 675, Springer-Verlag, New York.

Ghurye, S.G. (1960). Characterization of some location and scale parameter families of distributions, In *Contributions to Probability and Statistics*, pp. 202-215, Stanford University Press, Stanford, California.

Hamel, G. (1905). Eine Basis aller Zahlen und die unstetigen Lösungen der Funktionalgleichung $f(x+y) = f(x) + f(y)$, *Mathematische Annalen*, **60**, 459-462.

Johnson, N.L., Kotz, S., and Balakrishnan, N. (1994). *Continuous Univariate Distributions - Vol. 1*, Second edition, John Wiley & Sons, New York.

Kondo, T. (1931). Theory of sampling distribution of standard deviations, *Biometrika*, **22**, 31-64.

Larsen, R.J. and Marx, M.L. (1981). *An Introduction to Mathematical Statistics and Its Applications*, Prentice-Hall, Englewood Cliffs, New Jersey.

Lobachevskii, N.L. (1829). On the foundation of geometry, III, Section 12, *Kasaner Bote*, **27**, 227-243 (in Russian).

Malmquist, S. (1950). On a property of order statistics from a rectangular distribution, *Skandinavisk Aktuarietidskrift*, **33**, 214-222.

Pearson, K. (1895). Contributions to the mathematical theory of evolution. II. Skew variations in homogeneous material, *Philosophical Transactions of the Royal Society of London, Series A*, **186**, 343-414.

Rao, C.R. and Shanbhag, D.N. (1994). *Choquet-Deny Type Functional Equations with Applications to Stochastic Models*, John Wiley & Sons, Chichester, England.

Rényi, A. (1953). On the theory of order statistics, *Acta Mathematica Academiae Scientarium Hungaricae*, **4**, 191-232.

Rényi, A. (1956). A characterization of the Poisson process, *Magyar Tud. Akad. Mat. Kutato Int. Kozl.*, **1**, 519-527 (in Hungarian). Translated into English in *Selected Papers of Alfréd Rényi*, Vol. 1, Akademiai Kiadó, Budapest, 1976.

Rossberg, H.J. (1960). Über die Verteilungsfunktionen der Differenzen und Quotienten von Ranggrössen, *Mathematische Nachrichten*, **21**, 37-79.

Steffensen, J.F. (1930). *Some Recent Research in the Theory of Statistics and Actuarial Science*, Cambridge University Press, Cambridge, England.

Sukhatme, P.V. (1937). Tests of significance for samples of the χ^2 population with two degrees of freedom, *Annals of Eugenics*, **8**, 52-56.

Teicher, H. (1961). Maximum likelihood characterization of distributions, *Annals of Mathematical Statistics*, **32**, 1214-1222.

Teissier, G. (1934). Recherches sur le vieillissement et sur les lois de mortalité, *Annals of Physics, Biology and Physical Chemistry*, **10**, 237-264.

Weibull, W. (1939). The phenomenon of rupture in solids, *Ingen. Vetensk. Akad. Handl.*, **153**.

Weibull, W. (1951). A statistical distribution function of wide applicability, *Journal of Applied Mechanics*, **18**, 293-297.

Wilson, E.B. (1899). Note on the function satisfying the functional relation $\phi(u)\phi(v) = \phi(u+v)$, *Annals of Mathematics*, **1**, 47-48.

Zelen, M. and Dannemiller, M.C. (1961). The robustness of life testing procedures derived from the exponential distribution, *Technometrics*, **3**, 29-49.

CHAPTER 2

Basic Distributional Results and Properties

Asit P. Basu[1] and Nader Ebrahimi[2]

2.1 Introduction

In this chapter we consider the basic properties of the exponential distribution. Other relevant results will be derived in subsequent chapters. Consider the negative exponential distribution having the probability density function (p.d.f.)

$$f(t;\mu,\lambda) = \lambda \exp\{-\lambda(t-\mu)\} I(t>\mu) \ , \tag{2.1.1}$$

where $I(\cdot)$ is the indicator function of $(\cdot)$. That is,

$$I(t>\mu) = \begin{cases} 1 & \text{if } t>\mu \\ 0 & \text{otherwise} \ . \end{cases}$$

Here μ is the location parameter (guarantee period) and λ is the scale parameter measuring the failure rate.

The survival function corresponding to (2.1.1) is given by

$$P(T>t) = \overline{F}(t;\mu,\lambda) = \exp\{-\lambda(t-\mu)\} I(t>\mu) + I(t \le \mu) \ , \tag{2.1.2}$$

and the failure rate

$$\rho(t) = \frac{f(t;\mu,\lambda)}{\overline{F}(t;\mu,\lambda)} = \lambda \text{ for } \overline{F}(t;\mu,\lambda) > 0 \ . \tag{2.1.3}$$

Also, the mean residual life time function is

$$\mu(t) = E[T-t|T>t] = \frac{\int_t^\infty \overline{F}(x)\,dx}{\overline{F}(t)} = \begin{cases} (\mu - t) + \frac{1}{\lambda} \ , & t \le \mu \\ \frac{1}{\lambda} \ , & t > \mu \end{cases} \ . \tag{2.1.4}$$

The moment generating function is

$$M(\tau) = E[\exp(\tau X)] = \int_\mu^\infty (e^{\tau x}) \lambda\, e^{-\lambda(x-\mu)} dx = \left(\frac{\lambda}{\lambda-\tau}\right) \exp(\tau\mu) \ . \tag{2.1.5}$$

For simplicity throughout this chapter we will assume that $\mu = 0$.

[1]University of Missouri-Columbia, Columbia, Missouri
[2]Northern Illinois University, Dekalb, Illinois

2.2 Basic Results and Properties

First we list several basic properties of the exponential distribution.

(i) Let $X_1, X_2, \ldots, X_n$ be independent and identically distributed random variables with common probability density function $f(x;\lambda)$, that is, $f(x,\lambda) = \lambda \exp(-\lambda x)$. Then $\sum_{i=1}^{n} X_i$, a convolution of $X_1, \ldots, X_n$, is gamma with parameters n and λ.

(ii) If $X \sim e(\lambda)$, then $\frac{2X}{\theta}$ is chi-squared with two-degrees of freedom and $\frac{2\sum_{i=1}^{n} X_i}{\theta}$ is chi-squared with $2n$-degrees of freedom. Here $\theta = \frac{1}{\lambda}$ and X_i's are independent and identically distributed as $e(\lambda)$.

Consider a system which does not age stochastically, that is, its survival probability over an additional period of duration t is the same regardless of the age of the system. More specifically,

$$\overline{F}(t+x \mid x) = \frac{\overline{F}(t+x)}{\overline{F}(x)} = \overline{F}(t)$$

or equivalently,

$$\overline{F}(t+x) = \overline{F}(t)\overline{F}(x) \text{ for all } x \geq 0,\ t \geq 0\ . \tag{2.2.1}$$

Then we have the following theorem.

Theorem 2.1 *T is a non-negative non-degenerate random variable satisfying (2.2.1) if and only if $\overline{F}(t) = \exp(-\lambda t)$, for some $\lambda > 0$ and all $t > 0$.*

Proof: First we assume that (2.2.1) holds. Now let $c > 0$, and let m and n be positive integers. Using (2.2.1) we get

$$\overline{F}(nc) = (\overline{F}(c))^n\ , \tag{2.2.2}$$

and

$$\overline{F}(c) = \left(\overline{F}\left(\frac{c}{m}\right)\right)^m\ . \tag{2.2.3}$$

Putting $c = \frac{1}{m}$ in (2.2.2) and $c = 1$ in (2.2.3) we obtain

$$\overline{F}\left(\frac{n}{m}\right) = \left(\overline{F}\left(\frac{1}{m}\right)\right)^n = (\overline{F}(1))^{n/m}\ . \tag{2.2.4}$$

We claim that $0 < \overline{F}(1) < 1$. Because, if $\overline{F}(1) = 1$, then by substituting $c = 1$ in (2.2.2), $\overline{F}(n) = \overline{F}^n(1) = 1$ for all positive integers n which contradicts $\overline{F}(+\infty) = 0$. On the other hand if $\overline{F}(1) = 0$, then by (2.2.3) $\overline{F}\left(\frac{1}{m}\right) = (\overline{F}(1))^{1/m} = 0$, and hence using right continuity of $\overline{F}$, $\overline{F}(0) = 0$ which contradicts the non-degeneracy of T. Thus, $0 < \overline{F}(1) < 1$. Write $\overline{F}(1) = \exp(-\lambda)$, $0 < \lambda < \infty$. Then, it is clear that $\overline{F}\left(\frac{n}{m}\right) = \exp\left(-\lambda\, \frac{n}{m}\right)$ for all positive rationals $\frac{n}{m}$. By right continuity it follows that $\overline{F}(t) = \exp(-\lambda t)$ for all $t > 0$. Using right continuity one more time we get $\overline{F}(0) = 1$.

Now if $\overline{F}(t) = \exp(-\lambda t)$, then it is clear that $\overline{F}(t+x) = \overline{F}(t)\overline{F}(x)$ for all $t,\ x > 0$. This completes the proof. □

Property (2.2.1) is referred to as the *loss of memory* or *memoryless property* (LMP) of the exponential distribution. Thus, it follows from Theorem 2.1 that

(a) in continuous case the property (2.2.1) characterizes an exponential distribution;

(b) since a used component with exponentially distributed life time is as good as new, there is no advantage of following a planned replacement policy of the used components still functioning;

(c) in statistical estimation of mean life, percentiles, survival function and so on, data may be collected consisting only of the number of hours of the observed life and number of observed failures. The ages of components under observation are irrelevant.

Another property of the exponential distribution is that it is preserved under the formation of a series system of independent components.

Theorem 2.2 *Consider a series system with n independent components. (A series system works only if all its components work.) Let T be the life-time of the system and T_i be the life-time of the component i, $i = 1, \ldots, n$, then T has the survival function $\overline{F}(t) = \exp(-\lambda t)$, $\lambda = \sum_{i=1}^{n} \lambda_i$. Here $T_i \sim e(\lambda_i)$.*

Proof:

$$\begin{aligned} \overline{F}(t) &= P(T > t) = P\left(\min_{1\le i\le n} T_i > t\right) = P(T_1 > t, \ldots, T_n > t) \\ &= \prod_{i=1}^{n} P(T_i > t) = \exp\left\{-\left(\sum_{i=1}^{n}\lambda_i\right)t\right\} = \exp(-\lambda t) \, . \end{aligned}$$

This completes the proof. □

Let $\mu = \int_0^\infty \overline{F}(x)dx < \infty$; the total time on test transform (TTT-transform) is defined by

$$\Psi(t) = H_F^{-1}(t) \text{ for } 0 \le t \le 1 \, , \tag{2.2.5}$$

where $H_F^{-1}(t) = \int_0^{F^{-1}(t)} \overline{F}(s)ds$, and $F^{-1}(t)$ is the inverse function of F. This concept, due to Barlow and Campos (1975) and discussed in more detail by Barlow (1979), has been found useful in many statistical analyses of life testing problems.

Assume that $t_{(1)} \le t_{(2)} \le \cdots \le t_{(n)}$ is an ordered sample from a life distribution F (and let $t_{(0)} = 0$). The n spacings are defined by $D_k = t_{(k)} - t_{(k-1)}$, $k = 1, \ldots, n$. The total time on test $Y_i(t)$, for $t_{(i-1)} \le t \le t_{(i)}$, is given by

$$\begin{aligned} Y_i(t) &= nt_{(1)} + (n-1)(t_{(2)} - t_{(1)}) + \cdots + (n-i+2)(t_{(i-1)} - t_{(i-2)}) \\ &\quad +(n-i+1)(t - t_{(i-1)}) \, . \end{aligned} \tag{2.2.6}$$

Therefore,

$$Y_i(t_{(i)}) = \sum_{k=1}^{i} (n-k+1)\, D_k \, . \tag{2.2.7}$$

A natural choice of estimator of TTT-transform is the empirical TTT-transform

$$H_{F_n}^{-1}(t) = \int_0^{F_n^{-1}(t)} \overline{F}_n(s)ds = Y_i(t), \qquad t_{(i-1)} \le t \le t_{(i)}, \; i = 1, \ldots, n \, .$$

In life testing from an exponential distribution, a statistic that plays a central role is $Y_i(t_{(i)})$, $i = 1, \ldots, n$. For more details see Bain (1978).

Theorem 2.3 *Let $T_1, \ldots, T_n$ be independent with common probability density function $f(t) = \lambda \exp(-\lambda t)$, $\lambda > 0$, $t > 0$. Then, $(n-k+1)D_k$, $k = 1, \ldots, n$, are independent having common probability density function $f(t)$.*

Proof : The joint probability density function of $t_{(1)}, \ldots, t_{(n)}$ is (see Chapter 3)

$$f(t_{(1)}, \ldots, t_{(n)}) = n!\lambda^n \exp\left(-\lambda \sum_{i=1}^{n} t_{(i)}\right) .$$

Let $D_k = t_{(k)} - t_{(k-1)}$, $k = 1, 2, \ldots, n$. Then the Jacobian of transformation is 1 and the joint probability density function of $D_1, \ldots, D_n$ is

$$f(d_1, \ldots, d_n) = n!\lambda^n \exp\left(-\lambda \sum_{i=1}^{n} \sum_{j=1}^{i} d_j\right) .$$

Since

$$\sum_{i=1}^{n} \sum_{j=1}^{i} d_j = \sum_{j=1}^{n} \left(\sum_{i=j}^{n} d_i\right) = \sum_{j=1}^{n} (n-j+1) d_j ,$$

$$\begin{aligned} f(d_1, \ldots, d_n) &= \prod_{j=1}^{n} (n-j+1)\lambda^n \exp\left(-\lambda \sum_{j=1}^{n} (n-j+1) d_j\right) \\ &= \prod_{j=1}^{n} \left\{(n-j+1)\lambda\, e^{-\lambda(n-j+1)d_j}\right\}, \ d_j \geq 0, \ j = 1, \ldots, n . \end{aligned}$$

This completes the proof. □

From Theorem 2.3 it follows that

(a) $E[D_k] = \frac{1}{\lambda(n-k+1)}$, $k = 1, \ldots, n$;

(b) $\mathrm{Var}(D_k) = \frac{1}{\lambda^2(n-k+1)^2}$, $k = 1, \ldots, n$;

(c) $Y(t_{(i)}) = \sum_{k=1}^{i} (n-k+1) D_k$ is gamma with parameters i and λ.

Theorem 2.4 *Let $X_1, \ldots, X_n$ be independent and identically distributed random variables with common cumulative distribution function F. Then the limiting distribution of $U_n = nF(X_{(1)})$ is exponentially distributed with failure rate 1.*

Proof : Define $Y = F(X_{(1)})$. Then it is clear that the probability density of Y is $f_Y(y) = n(1-y)^{n-1}$, $0 \leq y \leq 1$ and zero elsewhere. Therefore,

$$f_{U_n}(u) = \left(1 - \frac{u}{n}\right)^{n-1} ,$$

and as $n \to \infty$, $f_{U_n}(u) = e^{-u}$. □

To obtain additional results, we first consider the following definition.

Definition 2.1 A counting process is an integer valued process $\{N(t);\ t \geq 0\}$ which counts the number of events occurring in an interval following a random mechanism, so that the time to occurrence of each event is a random variable. $M(t) = E[N(t)]$ is the expected number of events in $[0,t]$ and $M'(t)$, if is exists, it called the *intensity* or *mean rate*.

As an example, $T_1, T_2, \ldots, T_5$ are the arrival times in the picture given below:

$$\overline{0 \qquad T_1 \qquad T_2 \qquad T_3 \qquad T_4 \qquad T_5}\ .$$

Here $X_1 = T_1,\ X_2 = T_2 - T_1,\ X_3 = T_3 - T_2,\ X_4 = T_4 - T_3,\ X_5 = T_5 - T_4$ are the interarrival times.

Definition 2.2 A counting process $\{N(t);\ t \geq 0\}$ is a Poisson process with mean rate (or intensity) λ, if the following assumptions are fulfilled:

(i) $N(0) = 0$;

(ii) $\{N(t);\ t \geq 0\}$ has stationary independent increments. That is, for all choices of indices $t_0 < t_1 < \cdots < t_n$, the n random variables $N(t_i) - N(t_{i-1}),\ i = 1, \ldots, n$, are independent and $N(s_2 + h) - N(s_1 + h)$ and $N(s_2) - N(s_1)$ have the same distribution for all choices of indices $s_1,\ s_2$ and for every $h > 0$;

(iii) For any pair of times s, t such that $s < t$, the number of counts of occurrence of an event of $N(t) - N(s)$ in the interval $[s,t]$ has a Poisson distribution with mean $\lambda(t-s)$. That is,

$$P(N(t) - N(s) = k) = \exp\{-\lambda(t-s)\}\ \frac{\{\lambda(t-s)\}^k}{k!}\ ,\quad k = 0, 1, 2, \ldots$$

Theorem 2.5 *Let $\{N(t);\ t \geq 0\}$ be a Poisson process with mean rate λ. Then the length of the interval from some fixed time to the next event has the probability density function $f(x) = \lambda \exp(-\lambda x)$.*

Proof: Let T denote the time to occurrence of the first event from some fixed time T_0. Then,

$$\begin{aligned} P(T > t) &= P\ (\text{first event from } T_0 \text{ occurs after } t) \\ &= P(N(T_0 + t) - N(T_0) = 0) \\ &= \exp(-\lambda t)\ . \end{aligned}$$

The last equality comes from (iii) of Definition 2.2. □

Theorem 2.6 *The counting process $\{N(t);\ t \geq 0\}$ is a Poisson process with intensity rate λ if and only if*

(i) $N(0) = 0$;

(ii) the process has stationary and independent increments;

(iii) $0 < P(N(t) = 0) < 1$;

(iv) $\lim_{h\to 0} \frac{P(N(h)\geq 2)}{P(N(h)=1)} = 0$.

Proof: First we will assume that the assumptions (i) - (iv) hold. The characteristic function of $N(t+h)$ is

$$\begin{aligned}\phi_{N(t+h)}(u) &= E[\exp(iuN(t+h))] \\ &= E[\exp\{iu(N(t+h)-N(t))+iuN(t)\}] \\ &= \phi_{N(t+h)-N(t)}(u)\,\phi_{N(t)}(u) \\ &= \phi_{N(h)}(u)\,\phi_{N(t)}(u)\,. \qquad (2.2.8)\end{aligned}$$

The last result is obtained by using (ii) and then (i). Using 2.2.8

$$\begin{aligned}\frac{d}{dt}\,\phi_{N(t)}(u) &= \lim_{h\to 0}\frac{\phi_{N(t+h)}(u)-\phi_{N(t)}(u)}{h} = \frac{\phi_{N(t)}(u)\phi_{N(h)}-\phi_{N(t)}(u)}{h} \\ &= \lim_{h\to 0}\phi_{N(t)}(u)\,\frac{\phi_{N(h)}(u)-1}{h}\,.\end{aligned}$$

That is,

$$\frac{\frac{d}{dt}\,\phi_{N(t)}(u)}{\phi_{N(t)}(u)} = \lim_{h\to 0}\frac{\phi_{N(h)}(u)-1}{h}\,.$$

It is now sufficient to show that

$$\lim_{h\to 0}\frac{\phi_{N(h)}(u)-1}{h} = \lambda\{\exp(iu)-1\}$$

is the characteristic function of the Poisson distribution. Assuming $P(N(t)=0)=\alpha(t)$, $P(N(t+s)=0)=\alpha(t+s)$ we obtain, using (ii)

$$P(N(t+s)-N(t)=0,\ N(t)=0) = P(N(s)=0)P(N(t)=0) = \alpha(t)\alpha(s)\,.$$

Therefore by Theorem 2.1, $\alpha(t)=\exp(-\lambda t)$ or $\alpha(t)=0$, where $\lambda>0$. Also,

$$P(N(h)=0)=\exp(-\lambda h) \text{ and } \lim_{h\to 0}\frac{P(N(h)=0)-1}{h} = -\lambda\,.$$

But we know that

$$1 = P(N(h)=0)+P(N(h)=1)+P(N(h)\geq 2)$$

and therefore

$$\frac{1-P(N(h)=0)}{h} = \left(\frac{P(N(h)=1)}{h}\right)\left(1+\frac{P(N(h)\geq 2)}{P(N(h)=1)}\right)\,.$$

Using (iv) we get $\lim_{h\to 0}\frac{P(N(h)=1)}{h} = \lambda$. Now

$$\begin{aligned}\lim_{h\to 0}\frac{\phi_{N(h)}(u)-1}{h} &= \lim_{h\to 0}\frac{P(N(h)=0)-1}{h} + \lim_{h\to 0}\frac{P(N(h)=1)\,e^{iu}}{h} \\ &\quad + \lim_{h\to 0}\frac{\sum_{m=2}^{\infty}\exp(ium)P(N(h)=m)}{h} \\ &= \lambda\{\exp(iu)-1\}\,.\end{aligned}$$

Assume that $N(t)$ is a Poisson process. Then

$$P(N(h) = k) = \frac{e^{-\lambda h}(\lambda h)^k}{k!} \; .$$

Now, $0 < P(N(h) = 0) = e^{-\lambda h} < 1$. Moreover,

$$\begin{aligned}\lim_{h\to 0} \frac{P(N(h) \geq 2)}{P(N(h) = 1)} &= \lim_{h\to 0} \frac{1 - e^{-\lambda h} - \lambda h\, e^{-\lambda h}}{\lambda h e^{-\lambda h}} = \lim_{h\to 0} \frac{e^{\lambda h} - 1 - \lambda h}{\lambda h} \\ &= \lim_{h\to 0} \frac{\lambda\, e^{\lambda h} - \lambda}{\lambda} = 0 \; .\end{aligned}$$

This completes the proof. □

Theorem 2.7 *Let $\{N(t);\ t \geq\}$ be a Poisson process with intensity rate λ and let $X_1, X_2, \ldots$ be inter-arrival times. Then, X_n, $n = 1, 2, \ldots$ are independent identically distributed exponential random variables having the failure rate λ.*

Proof: It is clear that

$$P(X_1 > t) = P(N(t) = 0) = \exp(-\lambda t) \; .$$

Hence, X_1 is an exponential random variable with failure rate λ. Now,

$$P(X_2 > t | X_1 = s) = P(N(t + s) - N(s) = 0) = \exp(-\lambda t) \; .$$

Therefore, X_2 is also an exponential random variable with failure rate λ which is independent of X_1. Repeating the same argument we get the result. Using arguments similar to those in Theorem 2.7, one can easily show that if the inter-arrival times are independent with common exponential distribution then $\{N(t);\ t \geq 0\}$ is a Poisson process. □

Theorem 2.8 *If a Poisson process is observed up to a fixed time T and if r events occur in $[0, T]$ at times $T_1 \leq T_2 \leq \cdots \leq T_r$, then these times, after being subjected to a random permutation, can be considered as r independent observations on a random variable uniformly distributed over $[0, T]$.*

Proof:

$$\begin{aligned}&P(t_i < T_i \leq t_i + dt_i,\ i = 1, \ldots, r \mid r \text{ events occurred in } [0, T]) \\ &= \frac{\lambda^r \exp(-\lambda T)}{\{\exp(-\lambda T)\}(\lambda T) r / r!} dt_1 dt_2 \ldots dt_r \\ &= \frac{r!}{T^r} dt_1 \ldots dt_r \; .\end{aligned}$$

Therefore, the joint probability density function of $T_1, T_2, \ldots, T_r$ is

$$f(t_1, \ldots, t_r) = \frac{r!}{T^r},\ 0 < t_1 < t_2 < \cdots < t_r < T.$$

This completes the proof. □

In Theorem 2.8 note that

(i) $\frac{T_i}{T}$, $i = 1, \ldots, r$, when unordered, is a random sample from the uniform distribution $(0, 1)$;

(ii) Let $T_1 \leq T_2 \leq \cdots \leq T_r$ be the first r-ordered observations from an exponential distribution. Then, the conditional distribution of $T_1, \ldots, T_{r-1}$ given T_r is the distribution of $(r - 1)$ ordered observations from the uniform distribution $(0, T_r)$.

The basic uncertainty measure for distribution F is differential entropy

$$H(f) = -\int_0^\infty f(x) \log f(x) dx \ .$$

$H(f)$ is commonly referred to as the *Shannon information measure* [see Shannon (1948)]. Motivated by the fact that highly uncertain components are inherently unreliable, Ebrahimi (1993) proposed a new approach to measure uncertainty. Given that a system has survived up to time t, he proposed to measure uncertainty by

$$\begin{aligned} H(f;t) &= -\int_t^\infty \frac{f(x)}{\overline{F}(t)} \log \frac{f(x)}{\overline{F}(t)} \, dx \\ &= 1 - \frac{1}{\overline{F}(t)} \int_t^\infty \{\log \rho(x)\} f(x) \, dx \ , \end{aligned} \tag{2.2.9}$$

where $\rho(x)$ is defined in (2.1.3). Since

$$\frac{f(x)}{\overline{F}(t)} = \rho(x) \frac{\overline{F}(x)}{\overline{F}(t)} \ ,$$

the second equality in (2.2.9) follows, after some simplification.

Theorem 2.9 *$H(f;t)$ remains constant if and only if $\overline{F}(t) = \exp(-\lambda t)$.*

Proof: If $\overline{F}(t) = \exp(-\lambda t)$, then $H(f;t) = 1 - \log \lambda$, which is constant. Now suppose $H(f;t)$ remains constant. Then, differentiating (2.2.9), we get

$$\overline{F}(t) \log \overline{F}(t) - \int_t^\infty f(x) \log \rho(x) \, dx = 0 \ .$$

Differentiating a second time, we obtain $\rho'(x) = 0$. That is, $\rho(x) = \lambda$, and $\overline{F}(t) = \exp(-\lambda t)$, where λ is some positive constant.

References

Bain, L.J. (1978). *Statistical Analysis of Reliability and Life-Testing Models: Theory and Methods*, Marcel Dekker, New York.

Barlow, R.E. (1979). Geometry of the total time on test transform, *Naval Research Logistics Quarterly*, **26**, 393-402.

Barlow, R.E. and Campos, R. (1975). Total time on test processes and applications to failure data analysis, In *Reliability and Fault Tree Analysis* (Eds., R.E. Barlow, J. Fussel and N. Singpurwalla), pp. 451-481, SIAM, Philadelphia.

Ebrahimi, N. (1993). How to measure uncertainty in the residual life time distribution, *Unpublished report.*

Shannon, C.E. (1948). The mathematical theory of communication, *Bell System Technical Journal*, 279-423, 623-656.

CHAPTER 3

Order Statistics and their Properties

N. Balakrishnan[1] **and Bimal K. Sinha**[2]

3.1 Introduction

In this chapter, we present various results for order statistics based on a random sample of size n from a standard exponential distribution. As mentioned already in Chapters 1 and 2, the exponential distribution plays a prominent role in analyzing many data sets obtained from life-tests, and the use of order statistics in this connection arises quite naturally. It is thus not surprising that over the last fifty years or so, this particular area of research dealing with properties of exponential order statistics has drawn the attention of numerous researchers. Here, we will attempt to provide a comprehensive summary of various developments that have taken place in this area.

To fix the notations, let $X_1, X_2, \ldots, X_n$ be independent and identically distributed standard exponential, $e(1)$, random variables with a density function

$$f(x) = e^{-x}, \qquad x \geq 0 . \tag{3.1.1}$$

Let $X_{1:n} \leq X_{2:n} \leq \cdots \leq X_{n:n}$ be the order statistics obtained by arranging the X_i's in increasing order of magnitude. Let us denote the single moment of order statistics $E[X_{i:n}^k]$ by $\alpha_{i:n}^{(k)}$ for $1 \leq i \leq n$ and $k = 0, 1, 2, \ldots$, the product moment of order statistics $E[X_{i:n}X_{j:n}]$ by $\alpha_{i,j:n}$ for $1 \leq i < j \leq n$, and the covariance of order statistics $\text{Cov}(X_{i:n}, X_{j:n})$ by $\beta_{i,j:n}$ for $1 \leq i < j \leq n$. For convenience, we shall also use $\alpha_{i:n}$ for $\alpha_{i:n}^{(1)}$ and $\beta_{i,i:n}$ for $\text{Var}(X_{i:n})$.

In Section 3.2, we present some basic distributional results for order statistics and establish some interesting properties of the exponential order statistics. In Section 3.3, we present some recurrence relations satisfied by single and product moments of order statistics. The Lorenz ordering results of exponential order statistics are provided in Section 3.4. A brief discussion on the asymptotic distributions of exponential order statistics is made in Section 3.5. Sections 3.6-3.8 deal with relationships of exponential order statistics to order statistics from uniform, geometric, and double exponential

[1] McMaster University, Hamilton, Ontario, Canada
[2] University of Maryland, Baltimore County Campus, Maryland

distributions, respectively. In Section 3.9, some interesting distributional results are presented for the case when the order statistics arise from a Type-II progressively censored sample. Sections 3.10 and 3.11 are devoted to results on order statistics from right-truncated and doubly-truncated exponential distributions, respectively.

To conclude this section, we note that results for order statistics from $e(\theta)$ distribution, $Y_{i:n}$ (say), may be obtained from the results presented here on standard exponential order statistics through the relationship $Y_{i:n} \stackrel{D}{=} \theta X_{i:n}$. Similarly, the relationship $Y_{i:n} \stackrel{D}{=} \mu + \theta X_{i:n}$ will enable one to derive results for order statistics from $e(\mu, \theta)$ distribution.

3.2 Basic Distributional Results and Properties

The density function of $X_{i:n}$, $1 \le i \le n$, is given by [see, for example, David (1981, p. 9) and Arnold, Balakrishnan, and Nagaraja (1992, p. 10)]

$$\begin{aligned} f_{i:n}(x) &= \frac{n!}{(i-1)!(n-i)!}\{F(x)\}^{i-1}\{1-F(x)\}^{n-i}f(x) \\ &= \frac{n!}{(i-1)!(n-i)!}(1-e^{-x})^{i-1}e^{-(n-i+1)x}, \qquad 0 \le x < \infty\,. \end{aligned} \tag{3.2.1}$$

For the case $i = 1$, it is clear from (3.2.1) that the density function of the smallest order statistic $X_{1:n}$ is

$$f_{1:n}(x) = n\,e^{-nx}, \qquad 0 \le x < \infty\,; \tag{3.2.2}$$

from (3.2.2), it readily follows that $X_{1:n}$ is distributed exactly as $e(\theta = 1/n)$ as has already been noted in Chapter 2.

Similarly, the joint density function of $X_{i:n}$ and $X_{j:n}$, $1 \le i < j \le n$, is given by [see, for example, David (1981, p. 10) and Arnold, Balakrishnan, and Nagaraja (1992, p. 16)]

$$\begin{aligned} f_{i,j:n}(x,y) &= \frac{n!}{(i-1)!(j-i-1)!(n-j)!}(1-e^{-x})^{i-1}(e^{-x}-e^{-y})^{j-i-1} \\ &\quad \cdot e^{-x}\,e^{-(n-j+1)y}, \qquad 0 \le x < y < \infty\,. \end{aligned} \tag{3.2.3}$$

The joint density function of all n order statistics is similarly given by

$$\begin{aligned} f_{1,2,\ldots,n:n}(x_1,x_2,\ldots,x_n) &= n!\prod_{i=1}^{n} f(x_i) = n!\,e^{-\sum_{i=1}^{n} x_i}, \\ &\quad 0 \le x_1 < x_2 < \cdots < x_n < \infty\,. \end{aligned} \tag{3.2.4}$$

From (3.2.4), it is easily seen that the *spacings* $D_i = X_{i:n} - X_{i-1:n}$ (with $X_{0:n} \equiv 0$), $i = 1, 2, \ldots, n$, are independently distributed as already noted in Chapter 2. Further, by considering the transformation

$$Z_i = (n-i+1)D_i = (n-i+1)(X_{i:n} - X_{i-1:n}), \qquad i = 1, 2, \ldots, n\,, \tag{3.2.5}$$

and noting that the Jacobian of this transformation is $1/n!$, we readily obtain the result that the *normalized spacings* Z_i's are distributed as i.i.d. $e(1)$. This result, originally due to Sukhatme (1937), has played a key role in many problems concerning the exponential distribution (as shall be seen in subsequent chapters).

In view of (3.2.5), we have

$$X_{i:n} \stackrel{D}{=} \sum_{r=1}^{i} Z_r/(n-r+1), \qquad i = 1, 2, \ldots, n\ , \tag{3.2.6}$$

which expresses the ith standard exponential order statistic as a linear combination of i i.i.d. $e(1)$ variables. As a by-product, we observe the interesting phenomenon that the standard exponential order statistics form an *additive Markov chain* [Rényi (1953), Karlin and Taylor (1975)]. From (3.2.6), we also observe the interesting property of the standard exponential order statistics that

$$X_{i+r:n} - X_{i:n} \stackrel{D}{=} X_{r:n-i}, \qquad 1 \le r \le n-i\ . \tag{3.2.7}$$

It also follows from (3.2.6) that

$$\alpha_{i:n} = \sum_{r=1}^{i} E[Z_r]/(n-r+1) = \sum_{r=1}^{i} 1/(n-r+1), \qquad 1 \le i \le n\ , \tag{3.2.8}$$

$$\beta_{i,i:n} = \sum_{r=1}^{i} \mathrm{Var}(Z_r)/(n-r+1)^2 = \sum_{r=1}^{i} 1/(n-r+1)^2,\ 1 \le i \le n\ , \tag{3.2.9}$$

and

$$\beta_{i,j:n} = \beta_{i,i:n} = \sum_{r=1}^{i} 1/(n-r+1)^2, \qquad 1 \le i < j \le n\ . \tag{3.2.10}$$

Higher order moments of standard exponential order statistics may also be similarly derived from (3.2.6). Note the interesting structure of the variance-covariance matrix in (3.2.9). This feature, as will be seen later in Chapter 5, will help in the exact explicit derivation of the best linear unbiased estimators of the parameters of both $e(\theta)$ and $e(\mu, \theta)$ populations.

3.3 Recurrence Relations

Even though standard exponential order statistics admit such simple explicit expressions for their moments, Joshi (1978) established some recurrence relations satisfied by single and product moments of order statistics. The primary advantage of this recursive method is that it easily extends to the truncated exponential distribution case (as will be seen in Sections 3.10 and 3.11).

By starting with the characterizing differential equation $f(x) = 1 - F(x)$, $x \ge 0$, and employing integration by parts, Joshi (1978) established the following recurrence relations:

$$\alpha_{1:n}^{(k)} = \frac{k}{n}\, \alpha_{1:n}^{(k-1)}, \qquad n \ge 1,\ k \ge 1\ , \tag{3.3.1}$$

$$\alpha_{i:n}^{(k)} = \alpha_{i-1:n-1}^{(k)} + \frac{k}{n}\, \alpha_{i:n}^{(k-1)}, \qquad 2 \le i \le n,\ k \ge 1\ , \tag{3.3.2}$$

$$\alpha_{i,i+1:n} = \alpha_{i:n}^{(2)} + \frac{1}{n-i}\, \alpha_{i:n}, \qquad 1 \le i \le n-1\ , \tag{3.3.3}$$

and

$$\alpha_{i,j:n} = \alpha_{i,j-1:n} + \frac{1}{n-j+1}\,\alpha_{i:n}, \qquad 1 \le i < j \le n,\ j-i \ge 2\ . \tag{3.3.4}$$

These recurrence relations are *complete* in the sense that they may be used in a simple recursive manner in order to compute *all* the single and *all* the product moments of order statistics. Extensions of these results for moments of orders up to four have been given by Balakrishnan and Gupta (1992) who have then used those results to develop inference based on doubly Type-II censored samples (see Chapter 5).

By considering the independent and non-identical case, i.e., X_i's are independent but are distributed as $e(\theta_i)$ for $i = 1, 2, \ldots, n$, Balakrishnan (1994a) has recently generalized the recurrence relations in (3.3.1) - (3.3.4). He then showed that these recurrence relations can be used in a simple systematic manner in order to compute all the single and all the product moments of order statistics arising from a multiple-outlier model (with a slippage of p observations). Applications of these results to the robust estimation of the parameter of $e(\theta)$ distribution have also been illustrated by Balakrishnan (1994a) (see also Chapter 14).

3.4 Lorenz Ordering

In this section, we present some results on the Lorenz ordering of standard exponential order statistics established recently by Arnold and Nagaraja (1991).

First of all, let us recall that the *Lorenz curve* associated with a non-negative random variable X (with distribution function $F(x)$ and a finite mean) is given by

$$L_X(u) = \frac{\int_0^u F^{-1}(y)dy}{\int_0^1 F^{-1}(y)dy}\,, \qquad 0 \le u \le 1\ , \tag{3.4.1}$$

where $F^{-1}(y)$ is the inverse distribution function defined as $\sup\{x : F(x) \le y\}$.

Now given two non-negative random variables X and Y, *X is said to exhibit less inequality (variability) than Y in the Lorenz sense* if

$$L_X(u) \ge L_Y(u) \qquad \text{for every } u \in [0,1]\ , \tag{3.4.2}$$

and write $X <_L Y$. Of course, if the inequality in (3.4.2) is in fact an equality, we say that *X and Y have the same amount of variability in the Lorenz sense* and write $X =_L Y$. On the other hand, if the two Lorenz curves $L_X(u)$ and $L_Y(u)$ cross each other, then we say that *X and Y are not comparable in the Lorenz sense.* Books by Marshall and Olkin (1979) and Arnold (1987) will provide the interested readers with great details on the Lorenz ordering.

We now are in a position to state some key results on the Lorenz ordering of the standard exponential order statistics established by Arnold and Nagaraja (1991):

(a) $X_{i:n+1} <_L X_{i:n}$ (3.4.3)

(b) $X_{i+1:n+1} <_L X_{i:n}$ (3.4.4)

(c) $X_{i+1:n} <_L X_{i:n}$ iff $\alpha_{i:n} \le 1$ (3.4.5)

(d) $X_{i+1:n}$ and $X_{i:n}$ are not Lorenz ordered if $\alpha_{i:n} > 1$. (3.4.6)

Realize that $\alpha_{i:n} > 1$ when $i > n(1 - e^{-1})$.

Interestingly, Arnold and Villaseñor (1991) proved that in the case of uniform order statistics $X_{i:n} <_L X_{i:n+1}$ for all i and n. It is then clear from (a) that the Lorenz ordering could go in opposite directions for different populations. Furthermore, as pointed out by Arnold and Nagaraja (1991), (c) and (d) disprove the conjecture made by Arnold and Villaseñor (1991) that $X_{i+1:n} <_L X_{i:n}$ for all i for the exponential distribution.

Due to (3.2.7) that

$$D_{i,r,n} = X_{i+r:n} - X_{i:n} \stackrel{D}{=} X_{r:n-i} ,$$

we can also readily claim the following:

$$\text{(e)} \qquad D_{i,r,n+1} <_L D_{i,r,n} <_L D_{i+1,r,n} \tag{3.4.7}$$

$$\text{(f)} \qquad D_{i,r+1,n+1} <_L D_{i,r,n} <_L D_{i+1,r-1,n} \tag{3.4.8}$$

$$\text{(g)} \qquad D_{i,r+1,n+1} <_L D_{i,r,n} \text{ iff } \sum_{k=n-i-r+1}^{n-i} 1/k \leq 1 . \tag{3.4.9}$$

3.5 Asymptotic Distributions

In this section, we present the asymptotic joint and marginal distributions of various order statistics, suitably normalized, as $n \to \infty$. In general there are two distinct types of results for the asymptotic distribution of $X_{i:n}$, depending on whether $i \to \infty$ as $n \to \infty$ such that $i/n \to \xi$ $(0 < \xi < 1)$ or i is fixed. Examples of the former are the sample quantiles and the latter are the sample extremes such as $X_{1:n}$ and $X_{n:n}$. These asymptotic distributions are quite useful in developing asymptotically optimal inference procedures for the parameter of $e(\theta)$; see, for example, Chapter 5.

Writing $i_j = [n\xi_j] + 1$, $j = 1, 2, \ldots, k$ where $0 < \xi_1 < \xi_2 < \cdots < \xi_k < 1$, quite generally the asymptotic joint distribution of the k normalized sample quantiles $\sqrt{n}(X_{i_1:n} - \eta_{\xi_1}), \ldots, \sqrt{n}(X_{i_k:n} - \eta_{\xi_k})$ can be shown to be k-dimensional normal with zero mean vector and covariance matrix given by [Mosteller (1946)]

$$\boldsymbol{B} = \left(\left(\frac{\xi_j(1-\xi_\ell)}{f(\eta_{\xi_j})f(\eta_{\xi_\ell})} \right) \right), \qquad 1 \leq j \leq \ell \leq k , \tag{3.5.1}$$

where η_{ξ_j} is the ξ_jth quantile of the population $(j = 1, 2, \ldots, k)$.

In the case of the standard exponential population with $f(x) = e^{-x}$ and $F(x) = 1 - e^{-x}$, it is easy to verify that $\eta_{\xi_j} = -\log(1 - \xi_j)$ and consequently $f(\eta_{\xi_j}) = 1 - \xi_j$. Hence, the variance-covariance matrix in (3.5.1) assumes in this case the form

$$\boldsymbol{B} = \begin{bmatrix} \frac{\xi_1}{1-\xi_1} & \frac{\xi_1}{1-\xi_1} & \cdots & \frac{\xi_1}{1-\xi_1} \\ & \frac{\xi_2}{1-\xi_2} & \cdots & \frac{\xi_2}{1-\xi_2} \\ & & \ddots & \vdots \\ & & & \frac{\xi_k}{1-\xi_k} \end{bmatrix} . \tag{3.5.2}$$

Observe once again the special structure of the asymptotic variance-covariance matrix in (3.5.2).

For the smallest order statistic $X_{1:n}$, we have already seen in Section 3.2 [see Eq. (3.2.5)] that $nX_{1:n}$ is always distributed as standard exponential. In order to derive the limiting distribution of the largest order statistic $X_{n:n}$, let us consider

$$\begin{aligned} P(X_{n:n} - \log n \le x) &= \begin{cases} 0 &, \quad x \le -\log n \\ [1 - e^{-(x+\log n)}]^n &, \quad x > -\log n \end{cases} \\ &= (1 - e^{-x}/n)^n, \qquad x > -\log n \\ &\to e^{-e^{-x}}, \qquad -\infty < x < \infty . \end{aligned} \tag{3.5.3}$$

From (3.2.6), since we have

$$X_{i:n} \stackrel{D}{=} \frac{Z_1}{n} + \frac{Z_2}{n-1} + \cdots + \frac{Z_i}{n-i+1} ,$$

we have when $n \to \infty$ and i held fixed

$$nX_{i:n} \stackrel{D}{=} Z_1 + Z_2 + \cdots + Z_i \stackrel{D}{=} \text{Gamma}(i) . \tag{3.5.4}$$

That is, for any fixed i, $nX_{i:n}$ is distributed as Gamma(i) as $n \to \infty$.

3.6 Relationship to Uniform Order Statistics

It is well-known that the probability integral transformation $U = F(X)$ transforms the order statistic $X_{i:n}$ from the distribution $F(\cdot)$ to the order statistic $U_{i:n}$ from the uniform $U(0,1)$ distribution. In the case of the exponential distribution, this means $1 - e^{-X_{i:n}} \stackrel{D}{=} U_{i:n}$, or

$$X_{i:n} \stackrel{D}{=} -\log U_{n-i+1:n} \tag{3.6.1}$$

since $1 - U_{i:n} \stackrel{D}{=} U_{n-i+1:n}$. Using (3.6.1), we have

$$\begin{aligned} U_i^* = \left(\frac{U_{i:n}}{U_{i+1:n}}\right)^i &\stackrel{D}{=} \left(\frac{e^{-X_{n-i+1:n}}}{e^{-X_{n-i:n}}}\right)^i = e^{-i(X_{n-i+1:n} - X_{n-i:n})} \\ &\stackrel{D}{=} e^{-Z_{n-i+1}} \end{aligned} \tag{3.6.2}$$

due to (3.2.5), where Z_i's are independent standard exponential random variables. Since e^{-Z_i}'s are distributed independently as uniform $U(0,1)$, we simply have the result that the variables

$$U_i^* = \left(\frac{U_{i:n}}{U_{i+1:n}}\right)^i, \qquad i = 1, 2, \ldots, n \tag{3.6.3}$$

are all independent uniform $U(0,1)$ random variables. This result, due to Malmquist (1950), is quite useful in generating order statistics from the uniform distribution without using any sorting algorithms. This simulational algorithm has recently been extended by Balakrishnan and Sandhu (1995) in order to assist in the generation of order statistics from a progressive Type-II censored sample from the uniform distribution.

3.7 Relationship to Geometric Order Statistics

For the standard exponential random variable X, it is known that $Y = [X]$, the integer part of X, is distributed as Geometric(p) with $p = 1-e^{-1}$; further, the random variables $[X]$ and $\langle X \rangle = X - [X] = X - Y$, the fractional part of X, are statistically independent (see Chapter 18 for details). Then, since $nX_{1:n} \stackrel{D}{=} Z_1$ [see Eq. (3.2.5)], we readily have from Eq. (3.2.6) that

$$X_{i:n} \stackrel{D}{=} \sum_{r=1}^{i} Z_r/(n-r+1) \stackrel{D}{=} \sum_{r=1}^{i} Z_{1:n-r+1} \,, \tag{3.7.1}$$

where $Z_{1:n-r+1}$'s are independent.

Steutel and Thiemann (1989) established a parallel result for the geometric order statistics using the relationship between exponential and geometric random variables mentioned above. To this end, let us use $Y_{i:n}$ to denote the ith order statistic in a random sample of size n from the Geometric(p) distribution. Since $nY_{1:n}$ is also distributed as Geometric(p), we readily have the result that

$$Y_{1:m} \stackrel{D}{=} Y_1/m \stackrel{D}{=} [Z_1/m] \,. \tag{3.7.2}$$

Next, we have

$$Y_{i:n} \stackrel{D}{=} [X]_{i:n} \stackrel{D}{=} [X_{i:n}] \stackrel{D}{=} \left[\sum_{r=1}^{i} Z_r/(n-r+1)\right] \tag{3.7.3}$$

from (3.2.6). Now since

$$\left[\sum_{r=1}^{i} Z_r/(n-r+1)\right] = \sum_{r=1}^{i} [Z_r/(n-r+1)] + \left[\sum_{r=1}^{i} \langle Z_r/(n-r+1)\rangle\right] \,, \tag{3.7.4}$$

we obtain from (3.7.3) that

$$\begin{aligned} Y_{i:n} &\stackrel{D}{=} [Z_r/(n-r+1)] + \left[\sum_{r=1}^{i} \langle Z_r/(n-r+1)\rangle\right] \\ &\stackrel{D}{=} \sum_{r=1}^{i} Y_{1:n-r+1} + \left[\sum_{r=1}^{i} \langle Z_r/(n-r+1)\rangle\right] \,, \end{aligned} \tag{3.7.5}$$

where all the variables on the right hand side are independent due to the independence of $[Z]$ and $\langle Z \rangle$, and due to (3.7.1) and (3.7.2).

Note that this distributional property of geometric order statistics is similar to that of the exponential order statistics in (3.7.1) except for the presence of the last term on the right hand side of (3.7.5).

3.8 Relationship to Double Exponential Order Statistics

As in the preceding sections, let $X_{i:n}$ denote the ith order statistic in a sample of size n from the standard exponential distribution with density function $f(x) = e^{-x}$,

$0 \leq x < \infty$, and cumulative distribution function $F(x) = 1 - e^{-x}$, $0 \leq x < \infty$. Next, let $Y_{i:n}$ denote the ith order statistic in a sample of size n from the double exponential distribution with density function

$$g(y) = \frac{1}{2}\, e^{-|y|}\,, \qquad -\infty < y < \infty\,, \tag{3.8.1}$$

and cumulative distribution function

$$G(y) = \begin{cases} \frac{1}{2}\, e^{y} & , \quad y \leq 0 \\ 1 - \frac{1}{2}\, e^{-y} & , \quad y > 0 \end{cases} . \tag{3.8.2}$$

It is clear that $X \stackrel{D}{=} |Y|$ and

$$F(x) = 2G(x) - 1 \text{ and } f(x) = 2g(x) \text{ for } x \geq 0\,. \tag{3.8.3}$$

Govindarajulu (1963) used the relations in (3.8.3) to express the moments of order statistics from a symmetric distribution in terms of moments of order statistics from its folded distribution and, in particular, those of double exponential distribution in terms of the moments of exponential order statistics; see Govindarajulu (1966). Specifically, Govindarajulu (1963) established the relations

$$E[Y_{i:n}^{k}] = 2^{-n}\left\{\sum_{r=0}^{i-1}\binom{n}{r}E[X_{i-r:n-r}^{k}] + (-1)^{k}\sum_{r=i}^{n}\binom{n}{r}E[X_{r-i+1:r}^{k}]\right\}, \quad 1 \leq i \leq n\,, \tag{3.8.4}$$

and

$$\begin{aligned} E[Y_{i:n}Y_{j:n}] = 2^{-n}\Bigg\{&\sum_{r=0}^{i-1}\binom{n}{r}E[X_{i-r:n-r}X_{j-r:n-r}] \\ &-\sum_{r=i}^{j-1}\binom{n}{r}E[X_{r-i+1:r}]E[X_{j-r:n-r}] \\ &+\sum_{r=j}^{n}\binom{n}{r}E[X_{r-j+1:r}X_{r-i+1:r}]\Bigg\}, \\ &1 \leq i < j \leq n\,. \end{aligned} \tag{3.8.5}$$

Balakrishnan (1989) generalized these relations to the independent non-identical case which, in particular, express the moments of order statistics from independent non-identically distributed double exponential random variables in terms of the moments of order statistics from independent non-identically distributed exponential random variables.

By using probability arguments, Balakrishnan, Govindarajulu, and Balasubramanian (1993) established that for $Y_{i:n} > 0$, conditional on exactly r of the n Y's negative,

$$Y_{i:n} \stackrel{D}{=} X_{i-r:n-r} \qquad \text{for } r = 0, 1, \ldots, i-1\,, \tag{3.8.6}$$

and for $Y_{i:n} \leq 0$, conditional on exactly r of the n Y's negative,

$$Y_{i:n} \stackrel{D}{=} -X_{r-i+1:r} \qquad \text{for } r = i, i+1, \ldots, n\,. \tag{3.8.7}$$

Eqs. (3.8.6) and (3.8.7), along with the fact that the probability that exactly r of the n Y's are negative is equal to the binomial probability $\binom{n}{r}/2^n$, readily yields the relation in (3.8.4). A similar proof can be provided for the relation in (3.8.5). Finally, we should mention about the generalizations provided by Balakrishnan, Govindarajulu, and Balasubramanian (1993) to the non-independent non-identical case by using this probabilistic approach.

3.9 Results for a Progressive Type-II Censored Sample

A Type-II progressively censored (on the right) life test is performed in practice as follows: A random sample of n items are subjected to a life test under identical conditions. All n items are placed in test at once and the test can be terminated at the time of any individual failure. Further, one or more surviving items may be censored from the test at the time of each failure occurring before the termination of the experiment.

Let us now denote by $X_{1:m:n} \le X_{2:m:n} \le \cdots \le X_{m:m:n}$ the observed failure times and by $R_1, R_2, \ldots, R_m$ (each ≥ 0) the corresponding number of units censored from the test. Obviously, $m + R_1 + R_2 + \cdots + R_m = n$. Observe that the complete sample case and the Type-II right-censored sample case are special cases of this general progressive Type-II censored sample when $R_1 = R_2 = \cdots = R_m = 0$ $(m = n)$ and $R_1 = R_2 = \cdots = R_{m-1} = 0$, $R_m = n - m$, respectively.

Inference methods under such progressive Type-II censored samples are discussed in Chapters 4 and 5. In this section, we present some properties of the order statistics arising from such a progressive Type-II censored sample from a standard exponential distribution. These results have been established recently by Viveros and Balakrishnan (1994), and they generalize the corresponding results for the standard exponential order statistics presented in Section 3.2.

The joint density function of $X_{1:m:n}, X_{2:m:n}, \ldots, X_{m:m:n}$ is given by

$$\begin{aligned}
&f_{X_{1:m:n},\ldots,X_{m:m:n}}(x_1, \ldots, x_m)\\
&\quad = n\, f(x_1)\{\overline{F}(x_1)\}^{R_1} \times (n - R_1 - 1) f(x_2)\{\overline{F}(x_2)\}^{R_2} \times \cdots\\
&\qquad \times \left(n - \sum_{i=1}^{m-1} R_i - (m-1)\right) f(x_m)\{\overline{F}(x_m)\}^{R_m}\\
&\quad = n(n - R_1 - 1)\cdots\left(n - \sum_{i=1}^{m-1} R_i - (m-1)\right) \exp\left\{-\sum_{i=1}^{m}(R_i + 1)x_i\right\},\\
&\qquad\qquad 0 \le x_1 < x_2 < \cdots < x_m < \infty . \qquad (3.9.1)
\end{aligned}$$

Let us now define the ith normalized spacing as

$$Z_i^* = \left(n - \sum_{j=1}^{i-1}(R_j + 1)\right)(X_{i:m:n} - X_{i-1:m:n}) \ , \ i = 1, 2, \ldots, m \ , \qquad (3.9.2)$$

with $X_{0:m:n} \equiv 0$. Then from (3.9.1) and (3.9.2), we obtain the joint density function of

$Z_1^*, Z_2^*, \ldots, Z_m^*$ as

$$f_{Z_1^*,\ldots,Z_m^*}(z_1, z_2, \ldots, z_m) = \exp\left\{-\sum_{i=1}^{m} z_i\right\}, \qquad 0 \le z_1, \ldots, z_m < \infty, \tag{3.9.3}$$

which simply established that the normalized spacings $Z_1^*, Z_2^*, \ldots, Z_m^*$ are all independently distributed as standard exponential. From this, it readily follows that $2Z_1^*$ and $2\sum_{i=2}^{m} Z_i^*$ are independently distributed as χ_2^2 and χ_{2m-2}^2, respectively.

These distributional results have been used successfully by Viveros and Balakrishnan (1994) to develop interval estimation for the parameters of the one- and two-parameter exponential distributions based on progressive Type-II censored samples.

In view of (3.9.2), we also have

$$X_{i:m:n} \stackrel{D}{=} \sum_{r=1}^{i} Z_r^* \Big/ \left(n - \sum_{j=1}^{r-1}(R_j + 1)\right), \qquad i = 1, 2, \ldots, m, \tag{3.9.4}$$

which expresses the ith progressively censored order statistic as a linear combination of i i.i.d. $e(1)$ variables. This extends the additive Markov chain property of exponential order statistics described in Section 3.2 to the case when the order statistics arise from an exponential progressive Type-II censored sample. It also follows from (3.9.4) that

$$E[X_{i:m:n}] = \sum_{r=1}^{i} 1 \Big/ \left(n - \sum_{j=1}^{r-1}(R_j + 1)\right), \qquad i = 1, 2, \ldots, m, \tag{3.9.5}$$

$$\mathrm{Var}(X_{i:m:n}) = \sum_{r=1}^{i} 1 \Big/ \left(n - \sum_{j=1}^{r-1}(R_j + 1)\right)^2, \qquad i = 1, 2, \ldots, m, \tag{3.9.6}$$

and

$$\mathrm{Cov}(X_{i:m:n}, X_{j:m:n}) = \mathrm{Var}(X_{i:m:n}), \qquad 1 \le i < j \le m. \tag{3.9.7}$$

Note the interesting structure of the variance-covariance matrix in (3.9.7), similar to the variance-covariance matrix of the standard exponential order statistics seen earlier in Section 3.2. This will be helpful in deriving the best linear unbiased estimators of the parameters of $e(\theta)$ and $e(\mu, \theta)$ populations as shall be seen later in Chapter 5.

3.10 Results for Right-Truncated Exponential Distribution

Let us consider the right-truncated exponential distribution with density function

$$\begin{aligned} f(x) &= e^{-x}/P, \quad 0 \le x \le P_1 \\ &= 0, \quad \text{otherwise,} \end{aligned} \tag{3.10.1}$$

where $1 - P$ $(0 < P < 1)$ is the proportion of truncation on the right of the standard exponential distribution, and $P_1 = -\log(1 - P)$. The corresponding cumulative distribution function is

$$F(x) = \frac{1}{P}\,(1 - e^{-x})\;, \qquad 0 \le x \le P_1\;. \tag{3.10.2}$$

From (3.10.1) and (3.10.2), we observe the characterizing differential equation

$$f(x) = \{1 - F(x)\} + P_2\;, \qquad 0 \le x \le P_1\;, \tag{3.10.3}$$

where $P_2 = (1-P)/P$. Joshi (1978, 1982) made use of (3.10.3) to establish the following recurrence relations for single and product moments of order statistics from the right-truncated exponential distribution in (3.10.1):

$$\alpha_{1:n}^{(k)} = \frac{k}{n}\,\alpha_{1:n}^{(k-1)} - P_2\,\alpha_{1:n-1}^{(k)}, \qquad n \ge 2,\; k \ge 1, \tag{3.10.4}$$

$$\alpha_{i:n}^{(k)} = \frac{1}{P}\,\alpha_{i-1:n-1}^{(k)} + \frac{k}{n}\,\alpha_{i:n}^{(k-1)} - P_2\,\alpha_{i:n-1}^{(k)}, \quad 2 \le i \le n-1,\; k \ge 1, \tag{3.10.5}$$

$$\alpha_{n:n}^{(k)} = \frac{1}{P}\,\alpha_{n-1:n-1}^{(k)} + \frac{k}{n}\,\alpha_{n:n}^{(k-1)} - P_2 P_1^k, \qquad n \ge 2,\; k \ge 1, \tag{3.10.6}$$

$$\alpha_{n-1,n:n} = \alpha_{n-1:n}^{(2)} + \alpha_{n-1:n} - n\,P_2\{P_1\alpha_{n-1:n-1} - \alpha_{n-1:n-1}^{(2)}\}, \quad n \ge 2, \tag{3.10.7}$$

$$\alpha_{i,i+1:n} = \alpha_{i:n}^{(2)} + \frac{1}{n-i}\,[\alpha_{i:n} - n\,P_2\{\alpha_{i,i+1:n-1} - \alpha_{i:n-1}^{(2)}\}], \quad 1 \le i \le n-2, \tag{3.10.8}$$

$$\alpha_{i,j:n} = \alpha_{i,j-1:n} + \frac{1}{n-j+1}\,[\alpha_{i:n} - n\,P_2\{\alpha_{i,j:n-1} - \alpha_{i,j-1:n-1}\}], \quad 1 \le i < j \le n-1,\; j-i \ge 2, \tag{3.10.9}$$

and

$$\alpha_{i,n:n} = \alpha_{i,n-1:n} + \alpha_{i:n} - n\,P_2\{P_1\alpha_{i:n-1} - \alpha_{i,n-1:n-1}\}, \quad 1 \le i \le n-2\;. \tag{3.10.10}$$

These recurrence relations generalize those presented in Section 3.3 which may be deduced from here by letting $P \to 1$. The recurrence relations in (3.10.4)-(3.10.10) are also complete and will enable the computation of all single moments and all product moments of order statistics for all sample sizes in a simple recursive manner for any choice of the truncation parameter P.

Saleh, Scott, and Junkins (1975) earlier derived explicit series expressions for means, variances and covariances of order statistics. Balakrishnan and Gupta (1992) extended the results of Joshi (1978, 1982) and established recurrence relations for moments of orders up to four. By considering the independent and non-identical case, i.e., X_i's are independent but are distributed as right-truncated $e(\theta_i)$ with the same truncation point, Balakrishnan (1994b) has recently generalized the recurrence relations in (3.10.4)-(3.10.10). He then showed that these recurrence relations can be used in a systematic manner in order to compute all the single and the product moments of order statistics arising from a multiple-outlier model (with a slippage of p observations).

3.11 Results for Doubly-Truncated Exponential Distribution

Now, let us consider the doubly-truncated exponential distribution with density function

$$\begin{aligned} f(x) &= e^{-x}/(P-Q) \quad , \quad Q_1 \le x \le P_1 \\ &= 0 \quad , \quad \text{otherwise,} \end{aligned} \tag{3.11.1}$$

where Q and $1-P$ $(0 < Q < P < 1)$ are the proportions of truncation on the left and right of the standard exponential distribution, respectively, and $Q_1 = -\log(1-Q)$ and $P_1 = -\log(1-P)$. The corresponding cumulative distribution function is

$$F(x) = \frac{1-Q-e^{-x}}{P-Q} , \qquad Q_1 \le x \le P_1 . \tag{3.11.2}$$

From (3.11.1) and (3.11.2), we observe the characterizing differential equation

$$f(x) = \{1 - F(x)\} + P_2, \qquad Q_1 \le x \le P_1 , \tag{3.11.3}$$

where $P_2 = (1-P)/(P-Q)$. Let $Q_2 = (1-Q)/(P-Q)$. Then, by making use of the differential equation in (3.11.3), Joshi (1979) established the following recurrence relations for single moments:

$$\alpha_{1:n}^{(k)} = Q_2 Q_1^k + \frac{k}{n}\, \alpha_{1:n}^{(k-1)} - P_2 \alpha_{1:n-1}^{(k)}, \qquad n \ge 2,\ k \ge 1 , \tag{3.11.4}$$

$$\alpha_{i:n}^{(k)} = Q_2 \alpha_{i-1:n-1}^{(k)} + \frac{k}{n}\, \alpha_{i:n}^{(k-1)} - P_2 \alpha_{i:n-1}^{(k)}, \ 2 \le i \le n-1,\ k \ge 1 , \tag{3.11.5}$$

and

$$\alpha_{n:n}^{(k)} = Q_2 \alpha_{n-1:n-1}^{(k)} + \frac{k}{n}\, \alpha_{n:n}^{(k-1)} - P_2 P_1^k, \qquad n \ge 2,\ k \ge 1 . \tag{3.11.6}$$

Proceeding on the lines of Joshi (1979) and exploiting the differential equation in (3.11.3), Balakrishnan and Joshi (1984) established the following results:

$$\begin{aligned} \alpha_{i:n}^{(k)} &= \alpha_{i-1:n}^{(k)} + \binom{n}{i-1} \Big[k \sum_{r=i}^{n} (1-Q_2)^{n-r} B(i, r-i+1) \alpha_{i:r}^{(k-1)} \\ &\quad + (1-Q_2)^{n-i+1} \{P_1^k - \alpha_{i-1:i-1}^{(k)}\} \Big], \ 2 \le i \le n,\ k \ge 1 , \end{aligned} \tag{3.11.7}$$

$$\alpha_{i:n}^{(k)} = \alpha_{i-1:n}^{(k)} + \binom{n}{i-1} \Big[Q_2^{i-1} \{\alpha_{1:n-i+1}^{(k)} - Q_1^k\}$$

$$-k \sum_{r=2}^{i} Q_2^{i-r} B(r-1, n-i+2) \alpha_{r-1:n-i+r}^{(k-1)} \Bigg] ,$$
$$2 \le i \le n, \; k \ge 1 , \qquad (3.11.8)$$

$$\alpha_{n-1,n:n} = \alpha_{n-1:n}^{(2)} + \alpha_{n-1:n} - n P_2 \{ P_1 \alpha_{n-1:n-1} - \alpha_{n-1:n-1}^{(2)} \}, \quad n \ge 2 , \qquad (3.11.9)$$

$$\alpha_{i,i+1:n} = \alpha_{i:n}^{(2)} + \frac{1}{n-i} \left[\alpha_{i:n} - n P_2 \{ \alpha_{i,i+1:n-1} - \alpha_{i:n-1}^{(2)} \} \right] , \quad 1 \le i \le n-2 , \qquad (3.11.10)$$

$$\alpha_{i,j:n} = \alpha_{i,j-1:n} + \frac{1}{n-j+1} [\alpha_{i:n} - n P_2 \{ \alpha_{i,j:n-1} - \alpha_{i,j-1:n-1} \}] , \quad 1 \le i < j \le n-1, \; j-i \ge 2 , \qquad (3.11.11)$$

and

$$\alpha_{i,n:n} = \alpha_{i,n-1:n} + \alpha_{i:n} - n P_2 \{ P_1 \alpha_{i:n-1} - \alpha_{i,n-1:n-1} \} , \quad 1 \le i \le n-2 . \qquad (3.11.12)$$

These results generalize those presented in Section 3.10 which may be deduced from here by letting $Q \to 0$. The recurrence relations in (3.11.4)-(3.11.6) and (3.11.9)-(3.11.12) are also complete and will enble one to determine all single moments and all product moments of order statistics for all sample sizes in a simple recursive manner for any choice of the truncation parameters Q and P.

Joshi and Balakrishnan (1984) have derived the distributions of range and quasi-range from the doubly truncated exponential distribution.

References

Arnold, B.C. (1987). *Majorization and the Lorenz Order: A Brief Introduction*, Lecture Notes in Statistics No. 43, Springer-Verlag, New York.

Arnold, B.C., Balakrishnan, N., and Nagaraja, H.N. (1992). *A First Course in Order Statistics*, John Wiley & Sons, New York.

Arnold, B.C. and Nagaraja, H.N. (1991). Lorenz ordering of exponential order statistics, *Statistics & Probability Letters*, **11**, 485-490.

Arnold, B.C. and Villaseñor, J.A. (1991). Lorenz ordering of order statistics, In *Stochastic Order and Decisions under Risk*, IMS Lecture Notes - Monograph Series, **19**, 38-47.

Balakrishnan, N. (1989). Recurrence relations among moments of order statistics from two related sets of independent and non-identically distributed random variables, *Annals of the Institute of Statistical Mathematics*, **41**, 323-329.

Balakrishnan, N. (1994a). Order statistics from non-identical exponential random variables and some applications (with discussion), *Computational Statistics & Data Analysis*, **18**, 203-253.

Balakrishnan, N. (1994b). On order statistics from non-identical right-truncated exponential random variables and some applications, *Communications in Statistics - Theory and Methods*, **23**, 3373-3393.

Balakrishnan, N., Govindarajulu, Z., and Balasubramanian, K. (1993). Relationships between moments of two related sets of order statistics and some extensions, *Annals of the Institute of Statistical Mathematics*, **45**, 243-247.

Balakrishnan, N. and Gupta, S.S. (1992). Higher order moments of order statistics from exponential and right-truncated exponential distributions and applications to life-testing problems, *Technical Report No. 92-07C*, Department of Statistics, Purdue University, West Lafayette, IN.

Balakrishnan, N. and Joshi, P.C. (1984). Product moments of order statistics from doubly truncated exponential distribution, *Naval Research Logistics Quarterly*, **31**, 27-31.

Balakrishnan, N. and Sandhu, R.A. (1995). A simple simulational algorithm for generating progressive Type-II censored samples, *The American Statistician* (To appear).

David, H.A. (1981). *Order Statistics*, Second edition, John Wiley & Sons, New York.

Govindarajulu, Z. (1963). Relationships among moments of order statistics in samples from two related populations, *Technometrics*, **5**, 514-518.

Govindarajulu, Z. (1966). Best linear estimates under symmetric censoring of the parameters of a double exponential population, *Journal of the American Statistical Association*, **61**, 248-258.

Joshi, P.C. (1978). Recurrence relations between moments of order statistics from exponential and truncated exponential distributions, *Sankhyā, Series B*, **39**, 362-371.

Joshi, P.C. (1979). A note on the moments of order statistics from doubly truncated exponential distribution, *Annals of the Institute of Statistical Mathematics*, **31**, 321-324.

Joshi, P.C. (1982). A note on the mixed moments of order statistics from exponential and truncated exponential distributions, *Journal of Statistical Planning and Inference*, **6**, 13-16.

Joshi, P.C. and Balakrishnan, N. (1984). Distribution of range and quasi-range from doubly truncated exponential distribution, *Trabajos de Estadistica y de Investigaciones Operationes*, **35**, 231-236.

Karlin, S. and Taylor, H.M. (1975). *A First Course in Stochastic Processes*, Second edition, Academic Press, New York.

Malmquist, S. (1950). On a property of order statistics from a rectangular distribution, *Skandinavisk Aktuarietidskrift*, **33**, 214-222.

Marshall, A.W. and Olkin, I. (1979). *Inequalities: Theory of Majorization and Its Applications*, Academic Press, New York.

Mosteller, F. (1946). On some useful "inefficient" statistics, *Annals of Mathematical Statistics*, **17**, 377-408.

Rényi, A. (1953). On the theory of order statistics, *Acta Mathematica Academiae Scientarium Hungaricae*, **4**, 191-231.

Saleh, A.K.Md.E., Scott, C., and Junkins, D.B. (1975). Exact first and second order moments of order statistics from the truncated exponential distribution, *Naval Research Logistics Quarterly*, **22**, 65-77.

Steutel, F.W. and Thiemann, J.G.F. (1989). On the independence of integer and fractional parts, *Statistica Neerlandica*, **43**, 53-59.

Sukhatme, P.V. (1937). Tests of significance for samples of χ^2-population with two degrees of freedom, *Annals of Eugenics*, **8**, 52-56.

Viveros, R. and Balakrishnan, N. (1994). Interval estimation of parameters of life from progressively censored data, *Technometrics*, **36**, 84-91.

CHAPTER 4

MLEs Under Censoring and Truncation and Inference

A. Clifford Cohen[1]

4.1 Introduction

As the title indicates, this chapter is concerned with truncation and censoring in the exponential distribution. This distribution is an important model in life time and reaction time studies. Truncation arises on the left when observations less than a truncation point, which may be known or unknown, are eliminated from the population. The number of observations eliminated from any given sample is thus an unknown quantity. Truncation on the left is of little importance in life time studies, although it might be of interest in other applications. Censoring on the right occurs in life testing and related studies when sample items are removed from a test before all have reacted. Survivors removed from a test are said to be censored at the time (point) of removal. Censoring may occur in a single stage or it may be progressive and occur in multiple stages. These problems have been previously considered by numerous authors including Epstein and Sobel (1953, 1954), Sarhan (1954, 1955), Epstein (1954, 1960), Bartholomew (1957), Sarhan and Greenberg (1962), Freireich et al. (1963), Cohen (1963, 1966, 1991), Barlow, Madansky, and Proschan (1968), Nelson, W. (1968, 1969, 1972, 1982), David (1970, 1981), Grubbs (1971), Cohen and Helm (1973), Mann, Schafer, and Singpurwalla (1974), Gross and Clark (1975), Engelhardt and Bain (1978), Kambo (1978), Lawless (1982), Cohen and Whitten (1982, 1988), Cohen, Whitten, and Ding (1984), Chen and Bhattacharyya (1988), Gehan (1990), Balakrishnan (1990), Bain and Engelhardt (1991), Balakrishnan and Cohen (1991), Arnold, Balakrishnan, and Nagaraja (1992), Balasubramanian and Balakrishnan (1992), Viveros and Balakrishnan (1994), Johnson, Kotz, and Balakrishnan (1994), and many others. Apologies are offered to any whose contributions might have been overlooked.

[1]The University of Georgia, Athens, Georgia

4.1.1 The Exponential Distribution and Its Characteristics

The exponential distribution, sometimes referred to as the negative exponential distribution, is a positively skewed reverse J-shaped distribution. It is a special case of the Weibull and also of the gamma distribution. In the notation of Chapter 2, the pdf and the cdf of the two-parameter distribution are

$$\begin{aligned} f(x;\mu,\theta) &= \tfrac{1}{\theta}\, e^{-(x-\mu)/\theta} \quad , \; \mu < x < \infty \, , \\ &= 0 \quad , \; \text{elsewhere,} \end{aligned} \tag{4.1.1}$$

and

$$F(x;\mu,\theta) = \int_\mu^x f(t)\,dt = 1 - e^{-(x-\mu)/\theta} \, , \tag{4.1.2}$$

where μ is the origin (threshold parameter) and θ is the scale parameter. In many applications, the origin is at zero, that is; $\mu = 0$, and in these cases the resulting one-parameter distribution is completely characterized by the single parameter θ. Basic characteristics of this distribution are

$$\begin{aligned} E[X] &= \mu + \theta \, , \qquad V(X) = \theta^2 \, , \\ \text{Med}(X) &= \mu + \theta \ln 2 \, , \end{aligned} \tag{4.1.3}$$

and the coefficients of skewness and kurtosis are

$$\alpha_3(X) = 2, \qquad \alpha_4(X) = 9 \, .$$

The hazard function (i.e., the instantaneous failure rate) is (see Chapter 2)

$$\rho(x) = f(x) \,/\, \bar{F}(x) = 1/\theta \, , \tag{4.1.4}$$

and the cumulative hazard function is

$$H(x) = \int_\mu^x \rho(t)\,dt = (x-\mu)/\theta \, , \tag{4.1.5}$$

which is a linear function of x. Wayne Nelson (1969, 1972, 1982) employed this function in a graphical procedure for estimating parameters of the exponential distribution. It is to be noted that the distribution is completely and uniquely characterized by either the pdf, the cdf or the hazard function.

Since the hazard function of the exponential distribution is constant for all values of the random variable, this distribution is a suitable model for lifetime data where used items are considered to be as good as new ones as already mentioned in Chapter 2.

4.1.2 Types of Censored Samples

Censored samples are classified as progressively censored, doubly censored, singly left and right censored, and hybrid censored. In all censored samples, let n designate the total number of randomly selected items in the sample. In a progressively censored sample, let $T_1 < T_2 < \cdots < T_j < \cdots < T_k$ designate points (times) at which censoring occurs. At $x = T_j$, c_j sample items are randomly removed (censored) from further observation, and for these it is known only that $x > T_j$. The total number of censored observations is $\sum_{j=1}^{k} c_j$, and the number of complete (fully measured) observations is $m = n - \sum_{j=1}^{k} c_j$. For Type I censoring, the T_j's are fixed (known) constants. For Type II censoring the T_j's are random variables. In hybrid censoring, the points (times) of censoring are either fixed times, T_j, or times at which numbers of survivors attain fixed levels, whichever occurs first.

Singly right censored samples may be considered as a special case of progressively censored samples in which $k = 1$. In these samples, n is the total sample size, m (in the context of a life test) is the number of failures, and c is the number of survivors when the test is terminated at time T. Of course, $n = m + c$. For a Type I censored sample, T is a fixed constant, and c is a random variable. In a Type II censored sample, c is a fixed constant and T is a random variable. More specifically, in this case T is the mth order statistic in a sample of size n.

Doubly censored samples are censored both on the left and on the right. The c_1 smallest observations are censored on the left at T_1 and the c_2 largest observations are censored on the right at T_2. In a sample of total size n, the number of uncensored (fully measured) observations is $m = n - c_1 - c_2$. In Type I samples, T_1 and T_2 are fixed and c_1 and c_2 are random variables. In Type II samples, c_1 and c_2 are fixed while T_1 and T_2 are random variables. More specifically, T_1 is the c_1-th order statistic in a sample of size n, and T_2 is the $(m + c_1)$th order statistic in a sample of size n. Singly censored samples become special cases of doubly censored samples. If $c_1 = 0$, the sample is singly censored on the right. As previously mentioned, this case also results as a special case of progressively censored samples. If $c_2 = 0$, the sample is singly censored on the left. Progressively and singly right censored samples occur frequently in life span and reaction time studies. Doubly censored and singly left censored samples have other applications.

4.2 Parameter Estimation from Censored Samples

4.2.1 Estimation in Progressively and Singly Censored Samples

The log-likelihood function of a k-stage progressively censored sample as described above is

$$\ln L = -m \ln \theta - \sum_{i=1}^{m} \left(\frac{x_i - \mu}{\theta} \right) - \sum_{j=1}^{k} c_j \left(\frac{T_j - \mu}{\theta} \right) + \text{ constant} . \tag{4.2.1}$$

This is an increasing function of μ, and the maximum likelihood estimate of μ is therefore the largest permissible value. Since $\mu \leq x_{1:n}$, where $x_{1:n}$ is the sample first order statistic in a sample of size n, it follows that the MLE of μ is

$$\hat{\mu} = x_{1:n} . \tag{4.2.2}$$

The estimating equation for θ is $\partial \ln L / \partial \theta = 0$, and this follows from (4.2.1) as

$$\frac{\partial \ln L}{\partial \theta} = -\frac{m}{\theta} + \frac{1}{\theta^2} \left[\sum_{i=1}^{m} (x_i - \mu) + \sum_{j=1}^{k} c_j (T_j - \mu) \right] = 0 . \tag{4.2.3}$$

The maximum likelihood estimators can thus be written as

$$\begin{aligned} \hat{\mu} &= x_{1:n} , \\ \hat{\theta} &= [ST - nx_{1:n}]/m , \end{aligned} \tag{4.2.4}$$

where

$$ST = \sum_{i=1}^{m} x_i + \sum_{j=1}^{k} c_j T_j ,$$

is the "Sum Total" of all observations, both complete and censored. In the *one-parameter exponential* distribution, maximum likelihood estimating equation (4.2.3), with $\mu = 0$, becomes

$$\hat{\theta} = ST/m . \tag{4.2.5}$$

As frequently happens with maximum likelihood estimators, the MLE in this instance is biased. However, the first order statistic $x_{1:n}$ is well known to be a sufficient statistic for the estimation of μ, and the bias can be readily removed if we require that $E[X_{1:n}] = x_{1:n}$. The pdf of the first order statistic in a random sample of size n from an exponential distribution with pdf (4.1.1) is (see Chapter 3)

$$\begin{aligned} f(x_{1:n}; \mu, \theta) &= (n/\theta)e^{-n(x_{1:n}-\mu)/\theta} &&, \ \mu < x_{1:n} , \\ &= 0 &&, \ \text{elsewhere} . \end{aligned} \tag{4.2.6}$$

Note that (4.2.6) differs from (4.1.1) only in that θ has been replaced by θ/n, and the expected value of the first order statistic is

$$E[X_{1:n}] = \mu + \theta/n . \tag{4.2.7}$$

The first equation of (4.2.4) is replaced with $E[X_{1:n}]$, as given above, equated to $x_{1:n}$, and the resulting estimating equations become

$$\begin{aligned} \hat{\mu} &= x_{1:n} - \hat{\theta}/n , \\ \hat{\theta} &= [ST - n\hat{\mu}]/m . \end{aligned} \tag{4.2.8}$$

Minimum variance unbiased estimators (MVUEs) then follow as

$$\begin{aligned} \hat{\mu} &= [mx_{1:n} - ST/n]/(m-1) , \\ \hat{\theta} &= [ST - nx_{1:n}]/(m-1) . \end{aligned} \tag{4.2.9}$$

The estimators in (4.2.9) are applicable for both singly right censored and for progressively censored samples. For singly censored samples, $ST = \sum_{i=1}^{m} x_i + cT$.

For complete (uncensored) samples $c = 0$, $m = n$, and the estimators of (4.2.9) become

$$\hat{\mu} = \frac{nx_{1:n} - \bar{x}}{n-1} , \text{ and } \hat{\theta} = \frac{n(\bar{x} - x_{1:n})}{n-1} . \tag{4.2.10}$$

The estimators in (4.2.10) were originally given as best linear unbiased estimators (BLUEs) by Sarhan (1954), who derived them by employing a somewhat laborious least-square technique of Lloyd (1952) (see Chapter 5). At approximately the same time, Epstein and Sobel (1954) independently obtained these estimators as functions of the maximum likelihood estimators and then pointed out that since the MLE are both sufficient and complete, and since the estimators of (4.2.10) are unbiased, they are then both BLUE and minimum variance unbiased (MVUE). Cohen and Helm (1973) derived these estimators from the estimating equations, $E[X] = \bar{x}$ and $E[X_{1:n}] = x_{1:n}$. Exact variances and covariance of these estimators obtained by standard expected value procedures are

$$\begin{aligned} V(\hat{\mu}) &= \theta^2/\{n(n-1)\}\ , \qquad V(\hat{\theta}) = \theta^2/(n-1)\ , \\ \mathrm{Cov}(\hat{\mu}, \hat{\theta}) &= -\theta^2/\{n(n-1)\}\ . \end{aligned} \tag{4.2.11}$$

In the one-parameter exponential distribution with $\mu = 0$, the estimator follows from the second equation of (4.2.8) as

$$\hat{\theta} = ST/m\ . \tag{4.2.12}$$

This estimator is applicable for progressively censored samples, for singly censored samples, and for complete samples. Of course, in complete samples, $ST = n\bar{x}$.

4.2.2 Estimation in Doubly Censored Samples

The loglikelihood function for a doubly censored sample from the two-parameter exponential distribution with pdf (4.2.1) is

$$\ln L = c_1 \ln F(T_1) - m \ln\theta - \sum_{i=1}^{m}\{(x_i - \mu)/\theta\} + c_2 \ln \bar{F}(T_2) + \text{ constant}\ , \tag{4.2.13}$$

where $F(x)$ is given by (4.1.2).

The resulting estimating equation, when the partial derivative of $\ln L$ with respect to θ is equated to zero, becomes

$$\begin{aligned} \frac{\partial \ln L}{\partial \theta} &= -c_1\left(\frac{T_1 - \mu}{\theta^2}\right) J(T_1) - \frac{m}{\theta^2} + \frac{1}{\theta^2}\sum_{i=1}^{m}(x_i - \mu) \\ &\quad + c_2\left(\frac{T_2 - \mu}{\theta^2}\right) = 0\ , \end{aligned} \tag{4.2.14}$$

where

$$J(T) = [e^{(T-\mu)/\theta} - 1]^{-1}\ . \tag{4.2.15}$$

Equation (4.2.14) can be rewritten as

$$\theta = \frac{1}{m}\left[\sum_{i=1}^{m}(x_i - \mu) - c_1(T_1 - \mu)J(T_1) + c_2(T_2 - \mu)\right]\ . \tag{4.2.16}$$

Although Eq. (4.2.16) gives the appearance of being an explicit estimator of θ, this is not the case since $J(T_1)$ is a function of θ. When the threshold parameter μ is known,

(4.2.16) can be solved for the maximum likelihood estimate of θ by employing iterative procedures. In most instances, a trial and error technique will suffice.

Sarhan (1954) derived best linear unbiased estimators (BLUE) for a doubly censored sample, which in a notation consistent with that employed throughout this chapter, may be written as (see Chapter 5)

$$\begin{aligned} \hat{\theta} &= \frac{ST - nT_1}{m-1}, \\ \hat{\mu} &= \left[T_1 - \left(\frac{ST - nT_1}{m-1} \right) \sum_{i=1}^{c_1+1} \frac{1}{n-i+1} \right], \end{aligned} \tag{4.2.17}$$

where $\bar{x}$ is the mean of the m fully measured observations (i.e., $\bar{x} = \sum_{i=1}^{m} x_i/m$) and $ST = m\bar{x} + c_1 T_1 + c_2 T_2$. In Type I samples, the terminals T_1 and T_2 are known (fixed) points whereas c_1 and c_2 are random variables. In Type II samples, c_1 and c_2 are fixed while T_1 and T_2 are the order statistics, $T_1 = x_{c_1+1:n}$ and $T_2 = x_{n-c_2:n}$.

Estimating Eqs. (4.2.9) and (4.2.10) for singly right censored and for complete samples follow from (4.2.17) as special cases when we set $c_1 = 0$ for singly right censored samples, and $c_1 = c_2 = 0$ for complete samples. Accordingly, these estimators are both MVUE and BLUE. Additional details concerning linear estimators can be found in Chapter 5.

4.2.3 Hybrid Censoring

Hybrid censoring is a mixture of Type I and Type II censoring. Consider a life testing experiment in which n items are placed on test, successive failure times are recorded, and the test is terminated either with a specified number of survivors, c, or at a specified time T, whichever occurs first. Chen and Bhattacharyya (1988) considered this problem for the one-parameter exponential distribution and constructed an exact lower confidence bound for the parameter θ.

For any given sample of size n of this type, with T and the maximum value of m specified, the data consist of m ordered observations x_i of the m failed items, plus the value of T^*, the terminus, where T^* is the smaller of T or $x_{m:n}$. Note that in all cases, $x_{m:n} \le T$, and the number of censored observations is $n - m$. Furthermore, $(n-m) \ge c$. Estimates may then be calculated by using the appropriate estimators given in Section 4.2 with T^* as the terminus or censoring point. However, calculation of confidence intervals as derived by Chen and Bhattacharyya (1988) becomes more complex in this case. Approximate confidence intervals may be calculated from (4.4.1), if the distribution is a one-parameter exponential or from (4.4.2) and (4.4.3) if it is a two-parameter distribution.

As an illustration of the practical application of estimates based on hybrid censoring, an example originally given by Barlow et al. (1968) and also used by Chen and Bhattacharyya (1988) is reexamined as Example 4.5 in Section 4.5.

4.3 Estimation in Truncated Distributions

In this section, samples from truncated distributions are referred to as truncated samples. They differ from censored samples in that the number of missing observations is an unknown quantity. Samples consist of m randomly selected observations from a population from which certain small and/or large observations have been removed (i.e., truncated). We consider singly left truncated samples, singly right truncated samples, and doubly truncated samples.

4.3.1 Singly Truncated Samples

Singly Left Truncation

The pdf of an exponential distribution that is singly truncated on the left at $x = T_1$ may be written as

$$f_{LT}(x) = \frac{1}{\theta} e^{-(x-\mu)/\theta}[\bar{F}(T_1)]^{-1} . \tag{4.3.1}$$

After simplification this becomes

$$f_{LT}(x) = \frac{1}{\theta} e^{-(x-T_1)/\theta} , \qquad T_1 < x < \infty , \tag{4.3.2}$$

which is the pdf of another two-parameter exponential distribution with the origin at T_1. As a consequence of this property, the exponential distribution is a suitable model for life spans when used items are considered to be as good as new ones (see Chapter 2). The log-likelihood function of a random sample of size m from the singly left truncated sample with pdf (4.3.2) then follows as

$$\ln L = -m \ln \theta - \sum_{i=1}^{m}(x_i - T_1)/\theta , \tag{4.3.3}$$

which differs from the corresponding function for a sample from the original complete (untruncated) distribution only in that μ has been replaced by T_1. The maximum likelihood estimating equations become

$$\begin{aligned} \hat{T}_1 &= x_{1:m} , \\ \hat{\theta} &= \bar{x} - x_{1:m} . \end{aligned} \tag{4.3.4}$$

When T_1 is a fixed (known) truncation point, we simply have

$$\hat{\theta} = \bar{x} - T_1 . \tag{4.3.5}$$

When T_1 is unknown, it then becomes a new threshold parameter, and the estimators of (4.2.10) which are both BLUE and MVUE are applicable where T_1 is estimated in place of μ. In this case,

$$\hat{T}_1 = \frac{m x_{1:m} - \bar{x}}{m-1} \qquad \text{and} \qquad \hat{\theta} = \frac{m(\bar{x} - x_{1:m})}{m-1} . \tag{4.3.6}$$

In a sample of this type, no information is provided for estimating the value of μ other than that $\mu \le T_1$.

Singly Right Truncation

The pdf of a two-parameter exponential distribution that is singly truncated on the right at $x = T_2$ is

$$\begin{aligned} f_{RT}(x) &= \tfrac{1}{\theta}\, e^{-(x-\mu)/\theta}[F(T_2)]^{-1} \quad , \ \mu < x \le T_2 \\ &= 0 \quad , \ \text{otherwise} . \end{aligned} \tag{4.3.7}$$

The log-likelihood function of a random sample of size m from this distribution is

$$\begin{aligned} \ln L &= -m \ln \theta - \frac{1}{\theta} \sum_{i=1}^{m} (x_i - \mu) \\ &\quad - m \ln[1 - e^{-(T_2-\mu)/\theta}] + \text{ constant} . \end{aligned} \tag{4.3.8}$$

Maximum likelihood estimating equations in this case are $\hat{\mu} = x_{1:m}$ and $\frac{\partial \ln L}{\partial \theta} = 0$. Accordingly, it follows that

$$\begin{aligned} \hat{\mu} &= x_{1:m} , \\ \hat{\theta} &= (\bar{x} - \hat{\mu}) + (T_2 - \hat{\mu}) J(T_2; \hat{\mu}, \hat{\theta}) , \end{aligned} \tag{4.3.9}$$

where $J(T)$ is defined by (4.2.15).

With $\hat{\mu} = x_{1:m}$, the second equation in (4.3.9) can be solved by employing a 'trial and error' iterative procedure. The first order statistic is a biased estimator of μ, and in small samples, the bias can be substantial. We are therefore led to replace the first equation in (4.3.9) with $E[X_{1:n}] = x_{1:n}$, as was done in the corresponding censored sample. However, in truncated samples, the number of missing observations and thus n is unknown. Only m is known. However, we can estimate n as

$$E[n] = m/F(T_2) = m/[1 - e^{-(T_2-\hat{\mu})/\hat{\theta}}] , \tag{4.3.10}$$

and estimates $\hat{\mu}$ and $\hat{\theta}$ can now be found as the simultaneous solution of estimating equations

$$\begin{aligned} &\mu + [1 - e^{-(T-\mu)/\theta}]\theta/m = x_{1:n} , \\ &\theta = (\bar{x} - \mu) + (T - \mu) J(T_2; \mu, \theta) . \end{aligned} \tag{4.3.11}$$

These two equations can be solved for the required estimates by employing standard iterative methods.

4.3.2 Doubly Truncated Samples

The log-likelihood function of a doubly truncated sample of size m from the one-parameter exponential distribution with fixed (known) truncation points T_1 and T_2 such that $T_1 < x < T_2$ is

$$\ln L = -m \ln \theta - \sum_{i=1}^{m} x_i/\theta - m \ln[e^{-T_1/\theta} - e^{-T_2/\theta}] . \tag{4.3.12}$$

When the partial derivative of ln L with respect to θ is equated to zero, the resulting estimating equation becomes

$$\begin{aligned}\frac{\partial \ln L}{\partial \theta} &= -\frac{m}{\theta} + \frac{1}{\theta^2}\sum_{i=1}^{m} x_i \\ &\quad - \frac{m}{\theta^2}\left[\frac{T_1\, e^{-T_1/\theta} - T_2\, e^{-T_2/\theta}}{e^{-T_1/\theta} - e^{-T_2/\theta}}\right] = 0 \ . \end{aligned} \tag{4.3.13}$$

This can be simplified to

$$\theta + \left[\frac{T_2\, e^{-T_2/\theta} - T_1\, e^{-T_1/\theta}}{e^{-T_2/\theta} - e^{-T_1/\theta}}\right] = \bar{x} \ , \tag{4.3.14}$$

which can be solved by employing 'trial and error' iterative procedures for the estimate $\hat{\theta}$.

Estimators for θ when samples are singly truncated can be obtained as special cases from (4.3.14). When $T_2 \to \infty$ in a doubly truncated distribution, the result is *singly left truncation*. When $T_1 = \mu$, (0 in the one-parameter exponential) the result is *singly right truncation*. It can be shown that $\lim_{T_2 \to \infty} T_2 \exp(-T_2/\theta) = 0$, and thus for a *singly left truncated sample*, estimating equation (4.3.14) becomes

$$\hat{\theta} = \bar{x} - T_1 \ , \tag{4.3.15}$$

and for a singly right truncated sample, it becomes

$$\hat{\theta} = \bar{x} + T_2 J(T_2) \ . \tag{4.3.16}$$

Note that estimating Eqs. (4.3.15) and (4.3.16) are in agreement with estimating Eqs. (4.3.5) and (4.3.9) previously obtained directly from corresponding singly truncated samples.

4.4 Errors of Estimates

Epstein and Sobel (1954) noted relationships that exist between estimates of exponential parameters from censored samples and the chi-square distribution. These were then employed to develop confidence intervals on the distribution parameters.

For the one-parameter exponential distribution, it was shown that $2m\hat{\theta}/\theta$ is distributed as chi-square with $2m$ degrees of freedom. In this case, a two-sided $100(1-\alpha)\%$ confidence interval on θ is

$$\frac{2m\hat{\theta}}{\chi^2_{2m,1-\frac{\alpha}{2}}} < \theta < \frac{2m\hat{\theta}}{\chi^2_{2m,\frac{\alpha}{2}}} \ . \tag{4.4.1}$$

For the two-parameter exponential, it was shown that $2(m-1)\hat{\theta}/\theta$ is distributed as chi-square with $2(m-1)$ degrees of freedom. Accordingly, for this case, a two-sided $100(1-\alpha)\%$ confidence interval on θ is

$$\frac{2(m-1)\hat{\theta}}{\chi^2_{2(m-1),1-\frac{\alpha}{2}}} < \theta < \frac{2(m-1)\hat{\theta}}{\chi^2_{2(m-1),\frac{\alpha}{2}}} \ . \tag{4.4.2}$$

A confidence interval on μ can be obtained by using the fact that $2n(x_{1:n} - \mu)/\theta$ and $2(m-1)\hat{\theta}/\theta$ are independently distributed and are distributed as chi-square with 2 and $2(m-1)$ degrees of freedom, respectively. It then follows that the ratio $W = n(x_{1:n} - \mu)/\hat{\theta}$ has an F distribution with 2 and $2(m-1)$ degrees of freedom. Thus a lower $100(1-\alpha)\%$ confidence bound on μ is given as

$$LCB = x_{1:n} - \hat{\theta}\, F_{2,2(m-1),1-\alpha} \,/\, n \ , \tag{4.4.3}$$

where $F_{2,2(m-1),1-\alpha}$ is the $100(1-\alpha)$th percentile of the F distribution with 2 and $2(m-1)$ degrees of freedom. With probability one, the smallest sample observation (i.e., the observed value of the first order statistic in a sample of size n) is, of course, an upper bound on μ.

The need for tables of the F distribution in calculating lower confidence bounds as given by (4.4.3) can be eliminated by employing the following result due to Epstein and Sobel (1954) for calculating the $100p$th percentile:

$$F_{2,2(m-1),p} = (m-1)[(1-p)^{-1/(m-1)} - 1] \ . \tag{4.4.4}$$

It is to be noted that confidence intervals (4.4.2) and (4.4.3), given above for censored samples are also applicable when samples are complete, in which case, $n = m$.

In most circumstances, the above confidence intervals provide more information concerning the reliability of estimates than is given by estimate variances. However, it is deemed useful to have the variances presented in Table 4.1 available.

It seems appropriate at this time to examine interpretations to be placed on the threshold parameter μ. Epstein and Sobel (1954) mention two possible interpretations:

1. In the context of life distributions, μ is the minimum life. It is assumed that life is measured from time zero, and the mean life then is $\mu + \theta$. If we begin with a one-parameter exponential distribution and then truncate it on the left at $x = T_1$, which is assumed to be unknown, the truncated distribution then becomes a two-parameter exponential with T_1 as the minimum life to be estimated from sample data. Additional details are included in Section 4.3.

2. Life is assumed to begin at time μ, which is known. Under this interpretation, θ which is to be estimated from sample data, is the expected (mean) life span.

4.5 Illustrative Examples

Example 4.1 This example was originally given by Herd (1956), and has also been used as an illustration by Sarhan and Greenberg (1962). It consists of test results on the time to failure of eleven gyroscopes in a life test. The test engineer decided to withdraw three units at the time of the first failure. At the time of the second failure, he withdrew two additional units and at the third failure, he again withdrew two units. The remaining unit was allowed to remain on test until failure. The test results are summarized in Table 4.2.

This is a progressively censored sample with $n = 11$, $m = 4$, $T = x_{1:11} = 34$, $\sum_{i=1}^{4} x_i = (34+133+169+237) = 533$, $\sum_{j=1}^{3} c_j T_j = 666$, and $ST = 553+666 = 1219$. We assume this to be a sample from a two-parameter exponential distribution with a

Table 4.1: Variances of parameter estimates

One-Parameter Distributions	Two-Parameter Distributions
Singly Right Censored Samples	
$V(\hat{\theta}) = \theta^2/m$	$V(\hat{\theta}) \doteq [m/\theta^2 - 2ST/\theta^3]^{-1}$ $V(\hat{\mu}) = \theta^2/n^2$
Progressively Censored Samples	
$V(\hat{\theta}) = \theta^2/m$	$V(\hat{\theta}) = [m/\theta^2 - 2ST/\theta^3]^{-1}$ $V(\hat{\mu}) = \theta^2/n^2$
Two-Sided Censored Samples	
$V(\hat{\theta}) = \theta^2/K$	$V(\hat{\theta}) = \theta^2/K$ $V(\hat{T}_1) = \theta^2/n^2$
$K = \left[\left\{\left(\sum_{i=1}^{c_1+1} \frac{1}{n-i+1}\right)^2 \Big/ \sum_{i=1}^{c_1+1} \frac{1}{(n-i+1)^2}\right\} + (m-1)\right]$.	
Singly Right Truncated Samples	
$V(\hat{\theta}) \doteq -\frac{\partial^2 \ln L}{\partial \theta^2}\Big\vert_{\substack{\mu=0 \\ \theta=\hat{\theta}}}$	$V(\hat{\theta}) \doteq -\frac{\partial^2 \ln L}{\partial \theta^2}\Big\vert_{\substack{\mu=\hat{\mu} \\ \theta=\hat{\theta}}}$ $V(\hat{\mu}) \doteq (\theta^2/m^2)[1 - e^{-(T_2-\mu)/\theta}]^2$
Two-Sided Truncated Samples	
$V(\hat{\theta}) = \theta^2/(m-1)$	$V(\hat{\theta}) = \theta^2/(m-1)$ $V(\hat{T}_1) \doteq \frac{\theta^2}{m^2}\left[1 - e^{-(T_2-T_1)/\theta}\right]^2$
Complete (Uncensored) Samples	
$V(\hat{\theta}) = \theta^2/n$	$V(\hat{\theta}) = \theta^2/(n-1)$ $V(\hat{\mu}) = \theta^2/\{n(n-1)\}$

$$\frac{\partial^2 \ln L}{\partial \theta^2} = \frac{m}{\theta^2} - \frac{2}{\theta^3}\sum_{i=1}^{m}(x_i - \mu) + \frac{m}{\theta^2}(T-\mu)J'(\theta) - \frac{2m(T-\mu)}{\theta^3}J(\theta)$$

where $J(T;\mu,\theta) = [e^{(T-\mu)/\theta} - 1]^{-1}$, and $J'(T;\mu,\theta) = \left(\frac{T-\mu}{\theta^3}\right)e^{(T-\mu)/\theta}J^2(\theta)$.

Table 4.2: Time to failure of gyroscopes

Time in Hours, x_i	Number of Failures	Number Withdrawn (Censored), c_j
34	1	3
113	1	2
169	1	2
237	1	0
Totals	4	7

threshold parameter (origin) μ and a scale parameter θ. Substitution in Eqs. (4.2.9) yields the estimates

$$\hat{\mu} = [4(34) - 1219/11](4 - 1) = 8.39,$$

and

$$\hat{\theta} = [1219 - 11(34)]/(4 - 1) = 281.67,$$

which are both MVUE and BLUE. The distribution mean then is estimated as $\widehat{E[X]} = \hat{\mu} + \hat{\theta} = 8.39 + 281.67 = 290.06$. 95% confidence intervals on μ and θ, calculated by substitution in (4.4.2) and (4.4.3), are

$$117 < \theta < 1363 \ , \qquad \text{and} \qquad -97.6 < \mu < 34 \ .$$

Herd and also Sarhan and Greenberg assumed this sample to be from a one-parameter exponential distribution with origin at zero. In this case, with $\mu = 0$, the applicable estimator for θ is Eq. (4.2.12), and it follows that $\hat{\theta} = ST/m = 1219/4 = 304.75$ hours, which in this instance is the mean and is to be compared with the two-parameter estimate of 290.06. A 95% confidence interval on θ based on the one-parameter distribution as calculated by substitution in (4.4.1) becomes

$$139 < \theta < 1118 \ .$$

Example 4.2 Data for this example consist of life span observations of 70 generator fans originally given by W. Nelson (1969) to illustrate estimation of exponential parameters from a progressively censored sample by using a graphical procedure based on the cumulative hazard function. This example is used here to illustrate the practical application of estimators presented in Section 4.2.1 for progressively censored samples. Sample data are tabulated in Table 4.3.

Substitution in Eqs. (4.2.9) yields the BLU and MVU estimates

$$\begin{aligned} \hat{\mu} &= [12(4500) - 3,444,400/70]/(12 - 1) = 435.8 \ , \\ \hat{\theta} &= [3,444,400 - 70(4500)]/(12 - 1) = 284,490.9 \ , \end{aligned}$$

and the mean life is estimated as

$$\widehat{E[X]} = \hat{\mu} + \hat{\theta} = 435.8 + 284,490.9 = 284,926.7 \text{ hours} \ .$$

Table 4.3: Life span observations of 70 generator fans

Hours to Failure, x_i	Censoring Times, T_j	Number Censored, c_j	Hours to Failure, x_i	Censoring Times, T_j	Number Censored, c_j
4500			46000		
	4600	1		48500	4
11500				50000	3
11500			61000		
	15600	1		61000	3
1600				63000	1
	16600	1		64500	2
	18500	5		67000	1
	20300	3		74500	1
20700				78000	2
20700				81000	2
20800				82000	1
	22000	1		85000	3
	30000	4	87500		
31000				87500	2
	32000	1		94000	1
34500				99000	1
	37500	2		101000	3
	41500	4		115000	1
	43000	4			

$$n = 70,\ m = 12,\ \sum_{j=1}^{27} c_j = 58,\ T_1 = x_{1:70} = 4500,\ \sum_{i=1}^{12} x_i = 365,700,$$

$$\sum_{j=1}^{27} c_j T_j = 3,078,700, \qquad ST = 365,700 + 3,078,700 = 3,444,400.$$

The maximum likelihood estimate of mean life based on a one-parameter exponential distribution is obtained by substitution in (4.3.5) as

$$\widehat{E[X]} = \hat{\theta} = ST/m = 3,444,400/12 = 287,033 \text{ hours} .$$

Substitution in Eq. (4.4.1) yields a 95% confidence interval on θ that is applicable if the distribution is a one-parameter exponential as

$$174,842 < \theta < 555,548 .$$

For the two-parameter distribution, Eqs. (4.4.2) and (4.4.3) yield

$$170,076 < \theta < 568,982, \qquad \text{and} \qquad -9481 < \mu < 4500 .$$

The negative lower bound on μ suggests that the one-parameter exponential distribution with origin at zero, is a reasonable model in this case.

According to Nelson (1982), the fan data were accumulated in order to consider the possibility of providing a warranty that would guarantee a minimum fan life of 80,000 hours. The probability that this life would not be realized (based on the one-parameter distribution) is

$$F(80,000) = 1 - \exp[-(80,000/287,033)] = 0.243 .$$

Accordingly, the manufacturer could expect to service 24.3 percent of all fans of the present design and quality to be sold with the proposed 80,000 hour warranty. In order to achieve a more acceptable quality level (i.e. mean life) and a corresponding reduction in the percentage of failures prior to the warranted life, either a redesign or a reduction in the warranty life would be required.

Example 4.3 McCool (1974) gave test results on the fatigue life in hours for ten bearings of an unspecified type. As originally given, the sample was complete, but for this illustration the test is assumed to have been terminated on failure of the ninth unit. Life spans of the first nine sample items are 152.7, 172.0, 172.5, 173.3, 193.0, 204.7, 216.5, 234.9, and 262.6. The sample is thus assumed to be singly right censored at $T = x_{9:10} = 262.6$. The sample is further assumed to be from a two-parameter exponential distribution, and in summary, $n = 10$, $m = 9$, $c = 1$, $x_{1:10} = 152.7$, $\sum_{i=1}^{9} x_i = 1782.2$, $cT = 262.6$, and $ST = 1782.2 + 262.6 = 2044.8$. When these values are substituted into Eqs. (4.2.9), we calculate

$$\begin{aligned} \hat{\mu} &= [9(152.7) - 2044.8/10]/(9-1) = 146.23 , \\ \hat{\theta} &= [2044.8 - 10(152.7)]/(9-1) = 64.73 , \end{aligned}$$

and the estimate of the population mean is

$$\widehat{E[X]} = 146.23 + 64.73 = 210.96 .$$

These estimates are to be compared with complete sample estimates, $\hat{\mu} = 145.17$, $\hat{\theta} = 75.31$, and $\widehat{E[X]} = 220.48$. The tenth observation, censored in this illustration, was originally recorded at 422.6.

95% confidence intervals calculated by substitution in (4.4.2) and (4.4.3) are

$$35.96 < \theta < 149.9 \qquad \text{and} \qquad 129 < \mu < 152.7 .$$

Table 4.4: Time to remission of leukemia patients

Remission Time in Weeks, x_i	Censoring Times T_j	Number Censored c_j	Remission Time in Weeks, x_i	Censoring Times T_j	Number Censored c_j
6			16		
6				17	1
6				19	1
	6	1		20	1
7			22		
	9	1	23		
10				25	1
	10	1		32	2
	11	1		34	1
13				35	1

$$n = 21,\ m = 9,\ T_1 = x_{1:21} = 6,\ \sum_{i=1}^{9} x_i = 109,$$

$$\sum_{j=1}^{11} c_j T_j = 250, \qquad ST = 109 + 250 = 359.$$

Example 4.4 This sample of survival data was originally presented by Freireich et al. (1963). It was subsequently considered by Gehan (1970), and it was used as an illustration in a graphical hazard function analysis by Gross and Clark (1975). The data consist of ordered times to remission in weeks of twenty-one leukemia patients who had undergone a chemo-therapy treatment to maintain their previously induced remission. Sample observations are presented in Table 4.4.

Substitution in Eqs. (4.2.9) yields MVU and BLU estimates to be

$$\begin{aligned}\hat{\mu} &= [9(6) - 359/21]/(9-1) = 4.61\ ,\\ \hat{\theta} &= [359 - 21(6)]/(9-1) = 29.13\ .\end{aligned}$$

The mean remission time based on a two-parameter exponential distribution is then estimated as

$$\widehat{E[X]} = \hat{\mu} + \hat{\theta} = 4.613 + 29.125 = 33.74 \text{ weeks}\ .$$

If we assume the origin to be at zero as did Gross and Clark, the maximum likelihood estimated of the mean, based on a one-parameter exponential distribution, follow from (4.2.12) as

$$\widehat{E[X]} = \hat{\theta} = ST/m = 359/9 = 39.89 \text{ weeks}\ .$$

With μ assumed to be zero, Gross and Clark obtained 36.2 weeks from a cumulative hazard plot as the estimated mean remission time.

A 95% confidence interval on θ in the one-parameter case is calculated from (4.4.1) as

$$22.8 < \theta < 87.2 \ .$$

If we assume the distribution to be a two-parameter exponential, 95% confidence intervals follow from (4.4.2) and (4.4.3) to be

$$16.2 < \theta < 67.5 \qquad \text{and} \qquad 0.96 < \mu < 6 \ .$$

Example 4.5 This example was employed by Chen and Bhattacharyya (1988) to illustrate estimation in the one-parameter exponential distribution under hybrid censoring. $n = 10$ unspecified items were placed on test and failures were observed at times 4, 9, 11, 18, 27, and 38 preceding a fixed termination time of $T = 50$. For these data, the observed value of the first order statistic is $x_{1:10} = 4$ and $m = 6$. The number of censored observations is $n - m = 4$. The maximum number of censored observations is $c = 4$. For this example, the test was terminated at $T^* = x_{6:10} = 38$ since this occurred prior to $T = 50$. The sum of the life spans of the six items which failed is $4+9+11+18+27+38 = 107$, and $ST = 107+4(38) = 259$. Accordingly, the estimate of θ, which in this case is the mean, is calculated from the second equation of (4.2.8) with $\mu = 0$, as

$$\hat{\theta} = ST/m = 259/6 = 43.167 \ .$$

Chen and Bhattacharyya calculated an exact lower 95% confidence bound on θ with due regard for the hybrid censoring scheme as $LB = 19.34$. The corresponding approximate lower 95% confidence bound calculated from (4.4.1) without regard for hybrid censoring is 22.2.

If we regard this sample as coming from a two-parameter exponential distribution, the estimates would be calculated from Eqs. (4.2.9) as

$$\hat{\mu} = [6(4) - 259/10]/5 = -0.38 \ ,$$

and

$$\hat{\theta} = [259 - 10(4)]/5 = 43.8 \ ,$$

and the mean would then be estimated as $\widehat{E[X]} = -0.38 + 43.8 = 43.42$.

For the one-parameter distribution, the 95% confidence interval on θ is calculated from (4.4.1) as

$$22.2 < \theta < 117.7 \ .$$

For a two-parameter distribution, 95% confidence intervals calculated by substitution in (4.4.2) and (4.4.3) are

$$21.4 < \theta < 134.8 \qquad \text{and} \qquad -14 < \mu < 4 \ .$$

The negative lower bound on μ suggests that the one-parameter exponential with $\mu = 0$ is a reasonable model for these data.

References

Arnold, B.C., Balakrishnan, N., and Nagaraja, H.N. (1992). *A First Course in Order Statistics*, John Wiley & Sons, New York.

Bain, L.J. and Engelhardt, M. (1991). *Statistical Analysis of Reliability and Life-Testing Models*, Second edition, Marcel Dekker, New York.

Balakrishnan, N. (1990). On the maximum likelihood estimation of the location and scale parameters of exponential distribution based on multiply Type II censored samples, *Journal of Applied Statistics*, **17**, 55-61.

Balakrishnan, N. and Cohen, A.C. (1991). *Order Statistics and Inference: Estimation Methods*, Academic Press, San Diego.

Balasubramanian, K. and Balakrishnan, N. (1992). Estimation for one- and two-parameter exponential distributions under multiple type-II censoring, *Statistische Hefte*, **33**, 203-216.

Barlow, R.E., Madansky, A., Proschan, F., and Schever, F. (1968). Statistical estimation procedures for the "burn-in" process, *Technometrics*, **10**, 51-62.

Bartholomew, D.J. (1957). A problem in life testing, *Journal of the American Statistical Association*, **52**, 350-355.

Chen, S.-M. and Bhattacharyya, G.K. (1988). Exact confidence bounds for an exponential parameter under hybrid censoring, *Communications in Statistics - Theory and Methods*, **17**, 1857-1870.

Cohen, A.C. (1963). Progressively censored samples in life testing, *Technometrics*, **5**, 327-339.

Cohen, A.C. (1966). Life testing and early failure, *Technometrics*, **8**, 539-549.

Cohen, A.C. (1991). *Truncated and Censored Samples*, Marcel Dekker, New York.

Cohen, A.C. and Helm, R. (1973). Estimation in the exponential distribution, *Technometrics*, **14**, 841-846.

Cohen, A.C. and Whitten, B.J. (1982). Modified moment and maximum likelihood estimators for parameters of the three-parameter gamma distribution, *Communications in Statistics - Simulation and Computation*, **11**, 197-216.

Cohen, A.C. and Whitten, B.J. (1988). *Parameter Estimation In Reliability and Life Span Models*, Marcel Dekker, New York.

Cohen, A.C., Whitten, B.J., and Ding, Y. (1984). Modified moment estimation for the three-parameter Weibull distribution, *Journal of Quality Technology*, **16**, 159-167.

David, H.A. (1981). *Order Statistics*, Second edition, John Wiley & Sons, New York.

Engelhardt, M. and Bain, L.J. (1978). Construction of optimal unbiased inference procedures for the parameters of the gamma distribution, *Technometrics*, **20**, 485-489.

Epstein, B. (1954). Truncated life tests in the exponential case, *Annals of Mathematical Statistics*, **25**, 555-564.

Epstein, B. (1960). Estimation from life test data, *Technometrics*, **2**, 447-454.

Epstein, B. and Sobel, M. (1953). Life testing, *Journal of the American Statistical Association*, **48**, 485-502.

Epstein, B. and Sobel, M. (1954). Some theorems relevant to life testing from an exponential distribution, *Annals of Mathematical Statistics*, **25**, 373-381.

Freireich, E.J., et al. (1963). The effect of 6-mercaptoporine on the duration of steroid-induced remission in acute leukemia, *Blood*, **21**, 699-716.

Gehan, E.A. (1990). *Unpublished Notes On Survivability Theory*, Univ. of Texas, M.D. Anderson Hospital and Tumor Institute, Houston, Texas.

Gross, A.J. and Clark, V.A. (1975). *Survival Distributions: Reliability Applications in the Biomedical Sciences*, John Wiley & Sons, New York.

Grubbs, F.E. (1971). Fiducial bounds on reliability for the two-parameter negative exponential distribution, *Technometrics*, **13**, 873-876.

Herd, G.R. (1956). *Estimation of the Parameters From a Multicensored Sample*, Ph.D. dissertation, Iowa State College, Iowa.

Johnson, N.L., Kotz, S., and Balakrishnan, N. (1994). *Continuous Univariate Distributions - I*, Second edition, John Wiley & Sons, New York.

Kambo, N.S. (1978). Maximum likelihood estimators of the location and scale parameters of the exponential distribution from a censored sample, *Communications in Statistics - Theory and Methods*, **7**, 1129-1132.

Lawless, J.F. (1982). *Statistical Models & Methods for Lifetime Data*, John Wiley & Sons, New York.

Lloyd, E.H. (1952). Least-squares estimation of location and scale parameters using order statistics, *Biometrika*, **39**, 88-95.

Mann, N.R., Schafer, R.E., and Singpurwalla, N.D. (1974). *Methods for Statistical Analysis of Reliability and Life Data*, John Wiley & Sons, New York.

McCool, J.I. (1974). Inferential techniques for Weibull populations, *Aerospace Research Laboratories Report ARL TR 74-0180*, Wright-Patterson AFB, Ohio.

Nelson, W. (1968). A method for statistical hazard plotting of incomplete failure data that are arbitrarily censored, *TIS Report 68-C-007*, General Electric Research and Development Center, Schenectady, New York.

Nelson, W. (1969). Hazard plotting for incomplete failure data, *Journal of Quality Technology*, **1**, 27-52.

Nelson, W. (1972). Theory and applications of hazard plotting for censored failure data, *Technometrics*, **14**, 945-966.

Nelson, W. (1982). *Applied Life Data Analysis*, John Wiley & Sons, New York.

Sarhan, A.E. (1954). Estimation of the mean and standard deviation by order statistics, *Annals of Mathematical Statistics*, **25**, 317-328.

Sarhan, A.E. (1955). Estimation of the mean and standard deviation by order statistics, Part III, *Annals of Mathematical Statistics*, **26**, 576-592.

Sarhan, A.E. and Greenberg, B.G. (Eds.) (1962). *Contributions to Order Statistics*, John Wiley & Sons, New York.

Viveros, R. and Balakrishnan, N. (1994). Interval estimation of parameters of life from progressively censored data, *Technometrics*, **36**, 84-91.

CHAPTER 5

Linear Estimation Under Censoring and Inference

N. Balakrishnan and R.A. Sandhu[1]

5.1 Introduction

Exponential order statistics and their properties were discussed in detail earlier in Chapter 3. In this chapter, we will use those results to derive the best linear unbiased estimators (BLUEs) of the parameters of one- and two-parameter exponential distributions as explicit linear functions of the observed order statistics. These are derived for the Type-II right censored, doubly censored, multiply censored, progressively right censored, and general progressively censored situations. In most of these cases, it will be further shown that these estimators are the maximum likelihood estimators, seen in Chapter 4, corrected for their bias. The asymptotic best linear unbiased estimation of the exponential mean based on k optimally selected order statistics is also briefly discussed. Finally, some examples are presented to illustrate the method of estimation described in this chapter.

5.2 Type-II Right Censored Samples

Consider a life-testing experiment in which n identical components are placed on test simultaneously. Suppose the experiment was terminated as soon as the $(n - s)$th component failed, thus censoring the s remaining components. Such a sample is called a *Type-II right censored sample.*

Let $X_{1:n} \le X_{2:n} \le \cdots \le X_{n-s:n}$ denote such a Type-II right censored sample. Let $\alpha_{i:n}$ $(1 \le i \le n - s)$ and $\beta_{i,j:n}$ $(1 \le i \le j \le n)$ denote the means and covariances of order statistics from the standard exponential distribution. We have already seen in

[1] McMaster University, Hamilton, Ontario, Canada

Chapter 3 that

$$\alpha_{i:n} = \sum_{r=1}^{i} \frac{1}{(n-r+1)} \text{ and } \beta_{i,j:n} = \beta_{i,i:n} = \sum_{r=1}^{i} \frac{1}{(n-r+1)^2} \ . \tag{5.2.1}$$

Now, let us denote $\boldsymbol{\alpha} = (\alpha_{1:n}, \ldots, \alpha_{n-s:n})^T$, $\mathbf{1} = (1, \ldots, 1)^T$, $\boldsymbol{X} = (X_{1:n}, \ldots, X_{n-s:n})^T$ and $\boldsymbol{B} = ((\beta_{i,j:n};\ 1 \le i,j \le n-s))$.

As pointed out already in Chapter 3, it is the special structure of the variance-covariance matrix of the order statistics in (5.2.1) that enables the derivation of the BLUEs in explicit forms. For this purpose, we present the following lemma.

Lemma 5.1 *For a symmetric matrix $\boldsymbol{V}_{k\times k}$ with $v_{ij} = v_i$, $1 \le i \le j \le k$, the inverse is a symmetric tri-diagonal matrix given by $\boldsymbol{V}_{k\times k}^{-1} = ((v^{ij}))$ where*

$$\begin{aligned} v^{ii} &= w_i + w_{i+1} \text{ for } i = 1, 2, \ldots, k-1, \ v^{kk} = w_k, \\ v^{i,i+1} &= -w_{i+1} \text{ for } i = 1, 2, \ldots, k-1, \\ v^{i,j} &= 0 \text{ for } j - i \ge 2, \end{aligned}$$

with $w_1 = 1/v_1$ and $w_i = 1/(v_i - v_{i-1})$, $i = 2, 3, \ldots, k$.

One-parameter exponential case

Suppose the Type-II right censored sample arises from the one-parameter exponential, $e(0, \theta)$, distribution. Then, by using

$$E(X_{i:n}) = \theta\, \alpha_{i:n} \text{ and } \mathrm{Cov}(X_{i:n}, X_{j:n}) = \theta^2 \beta_{i,j:n}$$

and minimizing the generalized variance with respect to θ, we obtain the BLUE of θ to be [Lloyd (1952) and Balakrishnan and Cohen (1991, p. 74)]

$$\theta^* = (\boldsymbol{\alpha}^T \boldsymbol{B}^{-1} \boldsymbol{X}) / (\boldsymbol{\alpha}^T \boldsymbol{B}^{-1} \boldsymbol{\alpha}) \tag{5.2.2}$$

and its variance to be

$$\mathrm{Var}(\theta^*) = \theta^2 / (\boldsymbol{\alpha}^T \boldsymbol{B}^{-1} \boldsymbol{\alpha}). \tag{5.2.3}$$

By noting in this case that $w_i = (n-i+1)^2$, $i = 1, 2, \ldots, n-s$, and using Lemma 5.1, we obtain the BLUE of θ in (5.2.2) to be

$$\theta^* = \frac{1}{n-s} \left\{ \sum_{i=1}^{n-s} X_{i:n} + s\, X_{n-s:n} \right\} \tag{5.2.4}$$

and its variance to be

$$\mathrm{Var}(\theta^*) = \theta^2 / (n-s) \ . \tag{5.2.5}$$

We note that the BLUE θ^* in (5.2.4) is exactly the same as the maximum likelihood estimator of θ based on the Type-II right censored sample.

Two-parameter exponential case

Suppose the Type-II right censored sample arises from the two-parameter exponential, $e(\mu, \theta)$, distribution. Then, by using

$$E(X_{i:n}) = \mu + \theta\,\alpha_{i:n} \text{ and } \mathrm{Cov}(X_{i:n}, X_{j:n}) = \theta^2 \beta_{i,j:n}$$

and minimizing the generalized variance with respect to both μ and θ, we obtain the BLUEs of μ and θ to be [Lloyd (1952), David (1981, p. 130) and Balakrishnan and Cohen (1991, pp. 80-81)]

$$\mu^* = \left\{ \frac{\boldsymbol{\alpha}^T \boldsymbol{B}^{-1} \boldsymbol{\alpha} \mathbf{1}^T \boldsymbol{B}^{-1} - \boldsymbol{\alpha}^T \boldsymbol{B}^{-1} \mathbf{1} \boldsymbol{\alpha}^T \boldsymbol{B}^{-1}}{(\boldsymbol{\alpha}^T \boldsymbol{B}^{-1} \boldsymbol{\alpha})(\mathbf{1}^T \boldsymbol{B}^{-1} \mathbf{1}) - (\boldsymbol{\alpha}^T \boldsymbol{B}^{-1} \mathbf{1})^2} \right\} \boldsymbol{X} \tag{5.2.6}$$

and

$$\theta^* = \left\{ \frac{\mathbf{1}^T \boldsymbol{B}^{-1} \mathbf{1} \boldsymbol{\alpha}^T \boldsymbol{B}^{-1} - \mathbf{1}^T \boldsymbol{B}^{-1} \boldsymbol{\alpha} \mathbf{1}^T \boldsymbol{B}^{-1}}{(\boldsymbol{\alpha}^T \boldsymbol{B}^{-1} \boldsymbol{\alpha})(\mathbf{1}^T \boldsymbol{B}^{-1} \mathbf{1}) - (\boldsymbol{\alpha}^T \boldsymbol{B}^{-1} \mathbf{1})^2} \right\} \boldsymbol{X} ; \tag{5.2.7}$$

the variances and covariance of these BLUEs are given by

$$\mathrm{Var}(\mu^*) = \theta^2 (\boldsymbol{\alpha}^T \boldsymbol{B}^{-1} \boldsymbol{\alpha}) / \{(\boldsymbol{\alpha}^T \boldsymbol{B}^{-1} \boldsymbol{\alpha})(\mathbf{1}^T \boldsymbol{B}^{-1} \mathbf{1}) - (\boldsymbol{\alpha}^T \boldsymbol{B}^{-1} \mathbf{1})^2\} , \tag{5.2.8}$$

$$\mathrm{Var}(\theta^*) = \theta^2 (\mathbf{1}^T \boldsymbol{B}^{-1} \mathbf{1}) / \{(\boldsymbol{\alpha}^T \boldsymbol{B}^{-1} \boldsymbol{\alpha})(\mathbf{1}^T \boldsymbol{B}^{-1} \mathbf{1}) - (\boldsymbol{\alpha}^T \boldsymbol{B}^{-1} \mathbf{1})^2\} , \tag{5.2.9}$$

and

$$\mathrm{Cov}(\mu^*, \theta^*) = -\theta^2 (\boldsymbol{\alpha}^T \boldsymbol{B}^{-1} \mathbf{1}) / \{(\boldsymbol{\alpha}^T \boldsymbol{B}^{-1} \boldsymbol{\alpha})(\mathbf{1}^T \boldsymbol{B}^{-1} \mathbf{1}) - (\boldsymbol{\alpha}^T \boldsymbol{B}^{-1} \mathbf{1})^2\} . \tag{5.2.10}$$

By noting in this case once again that $w_i = (n-i+1)^2$, $i = 1, 2, \ldots, n-s$, using Lemma 5.1 and simplifying the expressions in (5.2.6) and (5.2.7), we obtain the BLUEs of μ and θ to be

$$\mu^* = X_{1:n} - \frac{1}{n(n-s-1)} \left\{ \sum_{i=1}^{n-s} X_{i:n} + s\,X_{n-s:n} - n\,X_{1:n} \right\} \tag{5.2.11}$$

and

$$\theta^* = \frac{1}{(n-s-1)} \left\{ \sum_{i=1}^{n-s} X_{i:n} + s\,X_{n-s:n} - n\,X_{1:n} \right\} . \tag{5.2.12}$$

Further, the variances and covariance in (5.2.8)-(5.2.10) simplify to

$$\mathrm{Var}(\mu^*) = \frac{(n-s)\theta^2}{n^2(n-s-1)}, \qquad \mathrm{Var}(\theta^*) = \frac{\theta^2}{n-s-1} \text{ and}$$

$$\mathrm{Cov}(\mu^*, \theta^*) = -\frac{\theta^2}{n(n-s-1)} . \tag{5.2.13}$$

It is of interest to note here that the BLUEs μ^* and θ^* in (5.2.11) and (5.2.12) are the same as the maximum likelihood estimators of μ and θ (seen earlier in Chapter 4) corrected for their bias.

5.3 Type-II Doubly Censored Samples

Consider a life-testing experiment in which n identical components are placed on test simultaneously. Suppose the lifetimes of the first r components failed were not observed and that the experiment was terminated as soon as the $(n-s)$th component failed. Such a sample is called a *Type-II doubly censored sample.*

Let $X_{r+1:n} \le X_{r+2:n} \le \cdots \le X_{n-s:n}$ denote such a Type-II doubly censored sample. Let us denote $\boldsymbol{\alpha} = (\alpha_{r+1:n}, \ldots, \alpha_{n-s:n})^T$, $\mathbf{1} = (1, \ldots, 1)^T$, $\boldsymbol{X} = (X_{r+1:n}, \ldots, X_{n-s:n})^T$, and $\boldsymbol{B} = ((\beta_{i,j:n};\ r+1 \le i, j \le n-s))$.

One-parameter exponential case

Suppose the Type-II doubly censored sample arises from the one-parameter exponential, $e(0,\theta)$, distribution. By noting in this case that $w_1 = \left\{\sum_{i=1}^{r+1} 1/(n-i+1)^2\right\}^{-1}$ and $w_i = (n-i+1)^2$, $i = 2, \ldots, n-r-s$, and using Lemma 5.1, we obtain the BLUE of θ from (5.2.3) to be

$$\theta^* = \frac{1}{K}\left[\{K_1 - (n-r)\}X_{r+1:n} + \sum_{i=r+1}^{n-s} X_{i:n} + s\,X_{n-s:n}\right], \qquad (5.3.1)$$

where

$$K_1 = \left(\sum_{i=n-r}^{n} 1/i\right) \Big/ \left(\sum_{i=n-r}^{n} 1/i^2\right) \qquad (5.3.2)$$

and

$$K = (n-r-s-1) + \left(\sum_{i=n-r}^{n} 1/i\right)^2 \Big/ \left(\sum_{i=n-r}^{n} 1/i^2\right). \qquad (5.3.3)$$

Further, the variance of the BLUE in (5.2.4) simplifies to

$$\mathrm{Var}(\theta^*) = \theta^2/K \qquad (5.3.4)$$

where K is as given in (5.3.3).

By using the formulas in Eqs. (5.3.1) and (5.3.4), Sarhan and Greenberg (1957) tabulated the coefficients of the order statistics and the value of $\mathrm{Var}(\theta^*)/\theta^2 = 1/K$ for doubly censored samples in samples of size up to ten. Epstein (1956, 1962), Epstein and Sobel (1953, 1954) and Herd (1956) proposed some simplified linear estimators of θ in the case of doubly and multiply censored samples.

Two-parameter exponential case

Suppose the Type-II doubly censored sample arises from the two-parameter exponential, $e(\mu,\theta)$, distribution. Upon using Lemma 5.1 with $w_1 = \left\{\sum_{i=1}^{r+1} 1/(n-i+1)^2\right\}^{-1}$ and

$w_i = (n-i+1)^2$, $i = 2, 3, \ldots, n-r-s$, and simplifying the expressions in (5.2.7) and (5.2.8), we obtain the BLUEs of μ and θ in this case to be

$$\mu^* = X_{r+1:n} - \frac{\sum_{i=n-r}^{n} 1/i}{n-r-s-1} \left\{ \sum_{i=r+1}^{n-s} X_{i:n} + s\, X_{n-s:n} - (n-r) X_{r+1:n} \right\} \tag{5.3.5}$$

and

$$\theta^* = \frac{1}{n-r-s-1} \left\{ \sum_{i=r+1}^{n-s} X_{i:n} + s\, X_{n-s:n} - (n-r) X_{r+1:n} \right\} . \tag{5.3.6}$$

Further, the variances and covariance in (5.2.9)-(5.2.11) simplify in this case to

$$\mathrm{Var}(\mu^*) = \theta^2 \left\{ \frac{1}{n-r-s-1} \left(\sum_{i=n-r}^{n} 1/i \right)^2 + \sum_{i=n-r}^{n} 1/i^2 \right\} , \tag{5.3.7}$$

$$\mathrm{Var}(\theta^*) = \theta^2/(n-r-s-1) , \tag{5.3.8}$$

and

$$\mathrm{Cov}(\mu^*, \theta^*) = -\theta^2 \left(\sum_{i=n-r}^{n} 1/i \right) \Big/ (n-r-s-1) . \tag{5.3.9}$$

It needs to be mentioned here that the BLUEs μ^* and θ^* in (5.3.5) and (5.3.6) are once again the same as the maximum likelihood estimators of μ and θ corrected for their bias.

From (5.3.5) and (5.3.6), we also obtain the BLUE of the expected lifetime $\xi = \mu + \theta$ to be

$$\xi^* = X_{r+1:n} + \frac{1 - \sum_{i=n-r}^{n} 1/i}{n-r-s-1} \left\{ \sum_{i=r+1}^{n-s} X_{i:n} + s\, X_{n-s:n} - (n-r) X_{r+1:n} \right\} ; \tag{5.3.10}$$

Eqs. (5.3.7)-(5.3.9) readily yield the variance of the BLUE ξ^* to be

$$\mathrm{Var}(\xi^*) = \theta^2 \left\{ \sum_{i=n-r}^{n} \frac{1}{i^2} + \frac{1}{n-r-s-1} \left(1 - \sum_{i=n-r}^{n} \frac{1}{i} \right)^2 \right\} . \tag{5.3.11}$$

For the case when the available sample is complete (i.e., $r = s = 0$), the BLUEs of μ and θ in (5.3.5) and (5.3.6) reduce to

$$\mu^* = (n\, X_{1:n} - \overline{X})/(n-1) \tag{5.3.12}$$

and

$$\theta^* = n(\overline{X} - X_{1:n})/(n-1) ; \tag{5.3.13}$$

their variances and covariance in (5.3.7)-(5.3.9) reduce to

$$\mathrm{Var}(\mu^*) = \frac{\theta^2}{n(n-1)}, \quad \mathrm{Var}(\theta^*) = \frac{\theta^2}{n-1} \text{ and } \mathrm{Cov}(\mu^*, \theta^*) = -\frac{\theta^2}{n(n-1)} . \tag{5.3.14}$$

The coefficients of order statistics in the BLUEs μ^*, θ^* and ξ^* in (5.3.5), (5.3.6) and (5.3.10), the variances and covariance of these estimators, and their relative efficiencies (compared to the BLUEs based on the complete sample) were tabulated by Sarhan and Greenberg (1957) for sample sizes up to 10 and all possible choices of r and s. Epstein (1956, 1962) and Epstein and Sobel (1953, 1954) proposed some simplified linear estimators for μ, θ and ξ in the case of doubly and multiply censored samples. Epstein (1962) also described the construction of confidence intervals for μ and θ.

5.4 Type-II Multiply Censored Samples

Suppose n identical components are placed on a life-testing experiment and that the r_1th, r_2th, ..., r_kth failure times are only available. Such a sample is called a *Type-II multiply censored sample*. Best linear unbiased estimators of exponential parameters based on such a multiply censored sample were derived recently by Balasubramanian and Balakrishnan (1992).

Let $X_{r_1:n} \le X_{r_2:n} \le \cdots \le X_{r_k:n}$ denote such a Type-II multiply censored sample. Let us denote $\boldsymbol{\alpha} = (\alpha_{r_1:n}, \ldots, \alpha_{r_k:n})^T$, $\mathbf{1} = (1, \ldots, 1)^T$, $\boldsymbol{X} = (X_{r_1:n}, \ldots, X_{r_k:n})^T$, and $\boldsymbol{B} = ((\beta_{r_i,r_j:n};\ 1 \le i, j \le k))$. From Chapter 3, we have

$$\alpha_{r_i:n} = \sum_{\ell=n-r_i+1}^{n} 1/\ell = T_1(n - r_i, n) \tag{5.4.1}$$

and

$$\beta_{r_i,r_j:n} = \beta_{r_i,r_i:n} = \sum_{\ell=n-r_i+1}^{n} 1/\ell^2 = T_2(n - r_i, n) . \tag{5.4.2}$$

One-parameter exponential case

Suppose the Type-II multiply censored sample arises from the one-parameter exponential, $e(0, \theta)$, distribution. By noting in this case that $w_1 = 1/T_2(n - r_1, n)$ and $w_i = 1/\{T_2(n - r_i, n) - T_2(n - r_{i-1}, n)\} = 1/T_2(n - r_i, n - r_{i-1})$, $i = 2, 3, \ldots, k$, using Lemma 5.1 and then simplifying the expression in (5.2.3), we obtain the BLUE of θ in this case to be

$$\theta^* = \frac{1}{K} \sum_{i=1}^{k} (a_i - a_{i+1}) X_{r_i:n} , \tag{5.4.3}$$

where

$$a_i = w_i T_1(n - r_i, n - r_{i-1}), \qquad i = 1, 2, \ldots, k ,$$

with $r_0 = 0$ and $a_{k+1} = 0$, and

$$K = \sum_{i=1}^{k} T_1^2(n - r_i, n - r_{i-1}) / T_2(n - r_i, n - r_{i-1}) . \tag{5.4.4}$$

Further, upon simplifying the expression in (5.2.4), we obtain the variance of θ^* in (5.4.3) to be

$$\text{Var}(\theta^*) = \theta^2/K \ . \tag{5.4.5}$$

Two-parameter exponential case

Suppose the Type-II multiply censored sample arises from the two-parameter exponential, $e(\mu,\theta)$, distribution. By noting once again that $w_1 = 1/T_2(n-r_1,n)$ and $w_i = 1/T_2(n-r_i,n-r_{i-1})$, $i=2,\ldots,k$, using Lemma 5.1 and then simplifying the expressions in (5.2.7) and (5.2.8), we obtain the BLUEs of μ and θ in this case to be

$$\mu^* = \frac{1}{K^*}\sum_{i=1}^{k} b_i\, X_{r_i:n} \tag{5.4.6}$$

and

$$\theta^* = \frac{w_1}{K^*}\sum_{i=2}^{k} a_i(X_{r_i:n} - X_{r_{i-1}:n}) \ , \tag{5.4.7}$$

where

$$K^* = \frac{1}{T_2(n-r_1,n)}\sum_{i=2}^{k} T_1^2(n-r_i,n-r_{i-1})\,/\,T_2(n-r_i,n-r_{i-1}) \tag{5.4.8}$$

and

$$b_1 = K^* + a_1a_2 \text{ and } b_i = a_1(a_{i+1}-a_i),\ i=2,\ldots,k \ . \tag{5.4.9}$$

Furthermore, the variances and covariance of these estimators reduce to

$$\text{Var}(\mu^*) = \frac{\theta^2 K}{K^*},\ \text{Var}(\theta^*) = \frac{\theta^2}{K^*T_2(n-r_1,n)} \text{ and } \text{Cov}(\mu^*,\theta^*) = -\frac{\theta^2 a_1}{K^*} \ . \tag{5.4.10}$$

From Eqs. (5.4.6) and (5.4.7), we also obtain the BLUE of the expected lifetime $\xi = \mu + \theta$ to be

$$\xi^* = X_{r_1:n} + \frac{a_1 - w_1}{K^*}\left\{a_2X_{r_1:n} + \sum_{i=2}^{k}(a_{i+1}-a_i)X_{r_i:n}\right\} \tag{5.4.11}$$

with its variance given by

$$\text{Var}(\xi^*) = \theta^2\left\{K + \frac{1}{T_2(n-r_1,n)} - 2a_1\right\}\Big/ K^* \ . \tag{5.4.12}$$

5.5 Type-II Progressively Right Censored Samples

Suppose n identical components are placed in a life-test at once and upon observing the first failure, R_1 randomly selected survivors are withdrawn from the test, and at the time of the second failure, R_2 survivors are randomly withdrawn from the test, and so on. At the final stage of the mth failure, the remaining R_m survivors are withdrawn from the experiment. Such a sample is called a *Type-II progressively right censored sample.* Balakrishnan and Sandhu (1995a) have recently presented a simple simulational algorithm for the generation of such progressively censored samples.

Let $X_{1:m:n} \le X_{2:m:n} \le \cdots \le X_{m:m:n}$ denote such a Type-II progressively right censored sample with $R_1, R_2, \ldots, R_m$ (each ≥ 0) denoting the corresponding number of units censored. Obviously, we have $m + \sum_{i=1}^{m} R_i = n$. We have already seen in Section 3.9 that, in the case of the standard exponential distribution,

$$\alpha_{i:m:n} = E[X_{i:m:n}] = \sum_{r=1}^{i} 1 \Big/ \left(n - \sum_{s=1}^{r-1}(R_s + 1)\right), \quad i = 1, 2, \ldots, m, \tag{5.5.1}$$

$$\beta_{i,i:m:n} = \mathrm{Var}(X_{i:m:n}) = \sum_{r=1}^{i} 1 \Big/ \left(n - \sum_{s=1}^{r-1}(R_s + 1)\right)^2, \quad i = 1, 2, \ldots, m, \tag{5.5.2}$$

and

$$\beta_{i,j:m:n} = \mathrm{Cov}(X_{i:m:n}, X_{j:m:n}) = \beta_{i,i:m:n}, \qquad 1 \le i < j \le m . \tag{5.5.3}$$

Now, let us denote $\boldsymbol{\alpha} = (\alpha_{1:m:n}, \ldots, \alpha_{m:m:n})^T$, $\mathbf{1} = (1, \ldots, 1)^T$, $\boldsymbol{X} = (X_{1:m:n}, \ldots, X_{m:m:n})^T$, and $\boldsymbol{B} = ((\beta_{i,j:m:n};\ 1 \le i, j \le m))$. Once again, the special structure of the variance-covariance matrix $\boldsymbol{B}$ [due to Eq. (5.5.3)] will facilitate the explicit derivation of the BLUEs.

One-parameter exponential case

Suppose the Type-II progressively right censored sample arises from the one-parameter exponential, $e(0, \theta)$, distribution. By noting in this case that $w_i = \left(n - \sum_{j=1}^{i-1} R_j - i + 1\right)^2$, $i = 1, 2, \ldots, m$, and using Lemma 5.1, we obtain the BLUE of θ in (5.2.3) to be

$$\theta^* = \frac{1}{m} \sum_{i=1}^{m} (R_i + 1) X_{i:m:n} \tag{5.5.4}$$

and its variance to be

$$\mathrm{Var}(\theta^*) = \theta^2 / m . \tag{5.5.5}$$

We note that the BLUE of θ^* in (5.5.4) is exactly the same as the maximum likelihood estimator of θ based on the Type-II progressively right censored sample seen already in the last chapter. Further, the result that the normalized spacings are i.i.d. $e(0, \theta)$

variables (see Section 3.9 for details) can be used to prove that $2m\theta^*/\theta$ has a central chi-square distribution with $2m$ degrees of freedom which may be used to develop confidence intervals or carry out tests of hypotheses for θ. Finally, it is important to note from (5.5.5) that the precision of the BLUE θ^* depends only on m and not on the Type-II progressive right censoring scheme $(R_1, R_2, \ldots, R_m)$.

Two-parameter exponential case

Suppose the Type-II progressively right censored sample arises from the two-parameter exponential, $e(\mu, \theta)$, distribution. By noting once again that $w_i = \left(n - \sum_{j=1}^{i-1} R_j - i + 1\right)^2$, $i = 1, 2, \ldots, m$, using Lemma 5.1 and simplifying the expressions in (5.2.7) and (5.2.8), we obtain the BLUEs of μ and θ in this case to be

$$\mu^* = X_{1:m:n} - \frac{1}{n(m-1)} \sum_{i=2}^{m} (R_i + 1)(X_{i:m:n} - X_{i-1:m:n}) \tag{5.5.6}$$

and

$$\theta^* = \frac{1}{m-1} \sum_{i=2}^{m} (R_i + 1)(X_{i:m:n} - X_{i-1:m:n}) \ . \tag{5.5.7}$$

Further, the variances and covariance in (5.2.9)-(5.2.11) simplify in this case to

$$\text{Var}(\mu^*) = \frac{m\,\theta^2}{n^2(m-1)}, \quad \text{Var}(\theta^*) = \frac{\theta^2}{m-1} \text{ and } \text{Cov}(\mu^*, \theta^*) = -\frac{\theta^2}{n(m-1)} \ . \tag{5.5.8}$$

These results were derived by Viveros and Balakrishnan (1994). They also noted that the BLUEs μ^* and θ^* in (5.5.6) and (5.5.7) are the same as the maximum likelihood estimators based on the Type-II progressively censored sample (seen in the last chapter) having been corrected for their bias.

Once again, by using the result that the normalized spacings are i.i.d. $e(0, \theta)$ variables (see Section 3.9), one may show that $2(m-1)\theta^*/\theta$ and $2(X_{1:m:n} - \mu)/\theta$ are independently distributed as central chi-square variables with $2(m-1)$ and 2 degrees of freedom, respectively. Consequently, as shown by Viveros and Balakrishnan (1994), confidence intervals or tests of hypotheses concerning the parameters θ and μ may be based on χ^2_{2m-2} and $F_{2,2m-2}$ distributions, respectively. It is also important to observe from (5.5.8) that the precision of the BLUEs μ^* and θ^* depend only on m and n and not on the Type-II progressive right censoring scheme $(R_1, R_2, \ldots, R_m)$.

5.6 General Type-II Progressively Censored Samples

Consider a life-testing experiment in which both left censoring as well as progressive censoring on the right takes place. In other words, suppose the lifetimes of the first r components failed were not observed and at the time of the $(r+1)$th failure R_{r+1} randomly selected survivors are withdrawn from the experiment, at the time of the

$(r+2)$th failure R_{r+2} randomly selected survivors are withdrawn, and so on. At the final stage of the mth failure, all the remaining R_m survivors are withdrawn. Such a sample is called a *General Type-II progressively censored sample.*

Let $X_{r+1:m:n} \le X_{r+2:m:n} \le \cdots \le X_{m:m:n}$ denote such a general Type-II progressively censored sample with $R_{r+1}, R_{r+2}, \ldots, R_m$ (each ≥ 0) denoting the corresponding number of units censored. Obviously, $m + \sum_{i=r+1}^{m} R_i = n$. Then, by using the results presented in Chapter 3, we have in the case of the standard exponential distribution

$$\alpha_{r+1:m:n} = E[X_{r+1:m:n}] = \sum_{i=1}^{r+1} 1/(n-i+1) , \tag{5.6.1}$$

$$\alpha_{j:m:n} = E[X_{j:m:n}] = \alpha_{r+1:m:n} + \sum_{i=r+2}^{j} 1\Big/\Big\{n - \sum_{k=r+1}^{i-1} R_k - i + 1\Big\}, \quad j = r+2, \ldots, m, \tag{5.6.2}$$

$$\beta_{r+1,r+1:m:n} = \mathrm{Var}(X_{r+1:m:n}) = \sum_{i=1}^{r+1} 1/(n-i+1)^2 , \tag{5.6.3}$$

$$\beta_{j,j:m:n} = \mathrm{Var}(X_{j:m:n}) = \beta_{r+1,r+1:m:n} + \sum_{i=r+2}^{j} 1\Big/\Big\{n - \sum_{k=r+1}^{i-1} R_k - i + 1\Big\}^2 , \quad j = r+2, \ldots, m, \tag{5.6.4}$$

and

$$\beta_{i,j:m:n} = \mathrm{Cov}(X_{i:m:n}, X_{j:m:n}) = \beta_{i,i:m:n} \text{ for } r+1 \le i < j \le m . \tag{5.6.5}$$

Once again, the special structure of the variance-covariance matrix seen in (5.6.5) will enable the explicit derivation of the BLUEs of the parameters in this case too.

Now, let us denote $\boldsymbol{\alpha} = (\alpha_{r+1:m:n}, \ldots, \alpha_{m:m:n})^T$, $\mathbf{1} = (1, \ldots, 1)^T$, $\boldsymbol{X} = (X_{r+1:m:n}, \ldots, X_{m:m:n})^T$, and $\boldsymbol{B} = ((\beta_{i,j:m:n};\ r+1 \le i, j \le m))$. Then, Balakrishnan and Sandhu (1995b) have recently derived the BLUEs for the one- and two-parameter exponential distributions.

One-parameter exponential case

Suppose the general Type-II progressively censored sample arises from the one-parameter exponential, $e(0,\theta)$, distribution. By noting in this case that $w_1 = \left\{\sum_{i=1}^{r+1} 1/(n-i+1)^2\right\}^{-1}$ and $w_i = \left(n - \sum_{k=r+1}^{i-1} R_k - i + 1\right)^2$, $i = 2, \ldots, m-r$, and using Lemma 5.1, we obtain the BLUE of θ from (5.2.3) to be

$$\theta^* = \frac{1}{m-r-1+\frac{\alpha^2_{r+1:m:n}}{\beta_{r+1,r+1:m:n}}} \left\{ \sum_{i=r+2}^{m} (R_i+1)(X_{i:m:n} - X_{r+1:m:n}) + \frac{\alpha_{r+1:m:n}}{\beta_{r+1,r+1:m:n}} X_{r+1:m:n} \right\} . \quad (5.6.6)$$

Further, the variance of the BLUE in (5.2.4) simplifies to

$$\text{Var}(\theta^*) = \theta^2 \Big/ \left\{ m - r + 1 + \frac{\alpha^2_{r+1:m:n}}{\beta_{r+1,r+1:m:n}} \right\} . \quad (5.6.7)$$

We note in this general case too that the precision of the BLUE in (5.6.7) depends only on r, m and n and not on the progressive censoring scheme $(R_{r+1}, \ldots, R_m)$. Also, Balakrishnan and Sandhu (1995b) have observed that the maximum likelihood estimator of θ does not exist in an explicit form and has to be determined numerically. However this is not the situation in the case of two-parameter exponential distribution to be considered next.

Two-parameter exponential case

Suppose the general Type-II progressively censored sample arises from the two-parameter exponential, $e(\mu, \theta)$, distribution. By noting once again that $w_1 = \left\{ \sum_{i=1}^{r+1} 1/(n-i+1)^2 \right\}^{-1}$ and $w_i = \left(n - \sum_{k=r+1}^{i-1} R_k - i + 1 \right)^2$, $i = 2, \ldots, m-r$, using Lemma 5.1, and simplifying the expressions in (5.2.7) and (5.2.8), we obtain the BLUEs of μ and θ in this case to be

$$\mu^* = X_{r+1:m:n} - \frac{\alpha_{r+1:m:n}}{m-r-1} \sum_{i=r+2}^{m} (R_i+1)(X_{i:m:n} - X_{r+1:m:n}) \quad (5.6.8)$$

and

$$\theta^* = \frac{1}{m-r-1} \sum_{i=r+2}^{m} (R_i+1)(X_{i:m:n} - X_{r+1:m:n}) . \quad (5.6.9)$$

Further, the variances and covariance in (5.2.9)-(5.2.11) simplify in this case to

$$\text{Var}(\mu^*) = \theta^2 \left\{ \frac{\alpha^2_{r+1:m:n}}{m-r-1} + \beta_{r+1,r+1:m:n} \right\} , \quad (5.6.10)$$

$$\text{Var}(\theta^*) = \theta^2/(m-r-1) \quad (5.6.11)$$

and

$$\text{Cov}(\mu^*, \theta^*) = -\theta^2 \, \alpha_{r+1:m:n}/(m-r-1) . \quad (5.6.12)$$

It is important to mention here that Balakrishnan and Sandhu (1995b) have shown that the BLUEs μ^* and θ^* in (5.6.8) and (5.6.9) are same as the maximum likelihood estimators of μ and θ corrected for their bias. These authors have also derived the maximum likelihood estimators of μ and θ to be

$$\widehat{\mu} = X_{r+1:m:n} + \frac{1}{m-r} \log\left(1 - \frac{r}{n}\right) \sum_{i=r+2}^{m} (R_i + 1)(X_{i:m:n} - X_{r+1:m:n}) \qquad (5.6.13)$$

and

$$\widehat{\theta} = \frac{1}{m-r} \sum_{i=r+2}^{m} (R_i + 1)(X_{i:m:n} - X_{r+1:m:n}) \,. \qquad (5.6.14)$$

Further, the precision of the BLUEs [see Eqs (5.6.10)-(5.6.12)] depend once again only on r, m and n and not on the progressive censoring scheme $(R_{r+1}, \ldots, R_m)$. Finally, as in the case of Type-II progressively right censored samples, confidence intervals and tests of hypotheses may be carried out for θ based on chi-square distribution with $2(m-r-1)$ degrees of freedom.

5.7 Asymptotic Best Linear Unbiased Estimation Based On Optimally Selected Order Statistics

Let $0 < \xi_1 < \xi_2 < \cdots < \xi_k < 1$ denote a spacing of k proportions, and $\xi_0 = 0$ and $\xi_{k+1} = 1$. Corresponding to these proportions, let us determine the ranks of the order statistics as $n_i = [n\,\xi_i] + 1$. $X_{n_i:n}$ is called as the *sample quantile of order* ξ_i; thus, $X_{n_1:n}, \ldots, X_{n_k:n}$ will be the k selected order statistics from a sample of size n.

Then, as already seen in Section 3.5, we have in the case of the standard exponential distribution

$$\alpha_{n_i:n} = -\log(1-\xi_i) + O(n^{-1}), \qquad 1 \le i \le k \,, \qquad (5.7.1)$$

and

$$\beta_{n_i,n_j:n} = \frac{\xi_i}{n(1-\xi_i)} + O(n^{-2}), \qquad 1 \le i \le j \le k \,. \qquad (5.7.2)$$

Now, by making use of the approximations in (5.7.1) and (5.7.2) in Eqs. (5.2.3) and (5.2.4), we derive the asymptotic BLUE of θ based on the k selected order statistics (say, θ^{**}) with its variance as [Ogawa (1951, 1952)]

$$\mathrm{Var}(\theta^{**}) = \theta^2/(nV) \,, \qquad (5.7.3)$$

where

$$V = \sum_{i=1}^{k+1} \{u_i e^{-u_i} - u_{i-1} e^{-u_{i-1}}\}^2 \,/\, (e^{-u_{i-1}} - e^{-u_i}) \,; \qquad (5.7.4)$$

here, $u_i = -\log(1-\xi_i)$, $i = 1, 2, \ldots, k$. The asymptotic BLUE of θ is given by

$$\theta^{**} = \frac{1}{V}\sum_{i=1}^{k} a_i X_{n_i:n} = \sum_{i=1}^{k} b_i X_{n_i:n} , \tag{5.7.5}$$

where

$$a_i = e^{-u_i}\left\{\frac{u_i e^{-u_i} - u_{i-1}e^{-u_{i-1}}}{e^{-u_{i-1}} - e^{-u_i}} - \frac{u_{i+1}e^{-u_{i+1}} - u_i e^{-u_i}}{e^{-u_i} - e^{-u_{i+1}}}\right\} , \tag{5.7.6}$$

$$b_i = a_i/V , \tag{5.7.7}$$

and

$$n_i = [n\,\xi_i] + 1 \qquad \text{for } i = 1, 2, \ldots, k .$$

The required optimal spacing is the set $(\xi_1^0, \ldots, \xi_k^0)$, or $(u_1^0, \ldots, u_k^0)$, that minimizes the variance in (5.7.3) or equivalently maximizes V in (5.7.4).

Fortunately, the determination of these optimal spacings may be done easily in a step by step fashion in the case of the exponential distribution. For example, for $k = 1$ we have $V = u_1^2/(e^{u_1} - 1)$ which attains its maximum value of 0.647610 at $u_1^0 = 1.593624$ which readily yields $\xi_1^0 = 0.796812$. This means that the 79th sample quantile is the optimal single quantile estimator for θ with a relative efficiency of 79.68%.

For the case when $k = 2$, by letting $u_2 = u_1 + v$, we may write the expression of V in (5.7.4) as

$$V = e^{-u_1}\left\{\frac{u_1^2}{e^{u_1} - 1} + u_1^2 + \frac{v^2}{e^v - 1}\right\} .$$

As already noted (in the case of $k = 1$), we have $v^0 = 1.593624$ and then maximization with respect to u_1 yields $u_1^0 = 1.017578$, with the corresponding maximum value of V being 0.820263. From these values, we get the optimal values of u_1 and u_2 to be 1.017578 and 2.611202 which readily yields $\xi_1^0 = 0.638531$ and $\xi_2^0 = 0.926554$. This means that the 64th and 93rd sample quantiles are the two optimal sample quantiles for the estimation of θ and the relative efficiency of this estimator is 82.02%. One may similarly continue to determine the optimal quantiles for $k = 3, 4, 5, \ldots$. Balakrishnan and Cohen (1991) have presented tables of these optimal quantiles for $k = 1(1)25$, the corresponding coefficients b_i $(i = 1, 2, \ldots, k)$, and the relative efficiency values. Greenberg and Sarhan (1958) had prepared such tables earlier and had commented on the usefulness of such estimators in studies about the average age at death for infants particularly in countries where ages of early deaths are not properly recorded.

Several other issues relating to the asymptotic BLUE based on k optimally selected order statistics, such as the estimation of population quantiles and tests of hypotheses, have been discussed by numerous authors. Due to the special nature of these discussions and lack of space to detail these, we refer the readers to the following papers: Ali, Umbach, and Hassanein (1981), Ali, Umbach, and Saleh (1982, 1985), Chan and Cheng (1988), Harter (1961, 1970), Kaminsky (1973, 1974), Kulldorff (1963), Saleh (1966, 1967, 1981), Saleh and Ali (1966), and Umbach, Ali, and Hassanein (1981).

5.8 Illustrative Examples

Example 5.1: 12 identical components were placed on a lifetest and the failure times (in hours) of the first 8 components to fail were observed as follows [Lawless (1982)]:

31, 58, 157, 185, 300, 470, 497, 673

By assuming a one-parameter exponential distribution for the data we find the BLUE of θ, from (5.2.4), to be

$$\theta^* = \frac{1}{8}\left\{\sum_{i=1}^{8} X_{i:12} + 4\,X_{8:12}\right\} = 632.875 \text{ hours},$$

and its standard error, from (5.2.5), to be

$$SE(\theta^*) = \theta^*/\sqrt{8} = 223.755 \text{ hours}.$$

Example 5.2: Let us consider the data obtained from an experiment on insulating fluid breakdowns [Nelson (1982, Table 2.1)]. Of the 12 specimens tested at 45 kV, 3 failed before 1 second and the times to breakdown (in seconds) of the remaining 9 specimens were as follows:

2, 2, 3, 9, 13, 47, 50, 55, 71.

Realize in this case that the sample observed is strictly not a Type-II censored sample. However, if we consider it to be a Type-II censored sample from the one-parameter exponential distribution (that is, the first three times to breakdown were not observed), we have $n = 12$, $r = 3$ and $s = 0$. Then, we find from Section 5.3 that

$$\begin{aligned} K_1 &= 10.26115786, \\ K &= 11.95417346, \end{aligned}$$

and the BLUE of θ to be

$$\theta^* = 21.292 \text{ seconds}$$

and its standard error as

$$SE(\theta^*) = 6.158 \text{ seconds}.$$

This particular example has also been used in Chapter 11 to illustrate the conditional inference approach.

Example 5.3: Consider the following data which represent failure times (in minutes) for a type of electrical insulation in a life-testing experiment in which the insulation was subjected to a continuously increasing voltage stress [Lawless (1982)]; in this experiment, 12 insulations were tested and the experiment was stopped as soon as the 11th insulation failed:

12.3, 21.8, 24.4, 28.6, 43.2, 46.9, 70.7, 75.3, 95.5, 98.1, 138.6

By assuming a two-parameter exponential distribution for this data we find the BLUEs of μ and θ, from (5.2.11) and (5.2.12), to be

$$\mu^* = 6.91 \text{ minutes and } \theta^* = 64.64 \text{ minutes.}$$

The BLUE of the expected lifetime ξ is then

$$\xi^* = \mu^* + \theta^* = 71.55 \text{ minutes,}$$

and its standard error is obtained from (5.2.13) to be

$$SE(\xi^*) = 19.50 \text{ minutes.}$$

Example 5.4: The following data, presented by Sarhan and Greenberg (1957), is part of an experiment in which 10 rabbits were inoculated with 0.2 ml of graded inoculum containing varying numbers of *Treponema pallidum*. Each rabbit received six injections from solutions containing 10^1, 10^2, 10^3, 10^4, 10^5, and 10^6 spirochetes per milliliter, and each was then observed for a period of 90 days to observe whether a syphilitic lesion developed at the site of injection.

The incubation time required for a lesion to appear is an index of the amount and potency of the inoculation, as well as the susceptibility of the individual rabbit. The distribution of incubation periods is assumed to be a two-parameter exponential distribution.

Knowledge of the reaction mechanism in rabbits has indicated that censoring the observation at 90 days after inoculation is desirable because only a infinitesimal proportion of rabbits will have a incubation period beyond that. In fact, the present data has been considered at one-half that period, viz., 45 days.

During the experiment, the rabbits were examined about twice a week for lesions. Those lesions that developed in the interim between examinations were undetected until the next period. Lesions that were one or two days old at time of first observation could be distinguished by their greater size. The first examination was performed approximately one week following inoculation. This meant that certain observations could be considered censored from the left if the size of the lesion was large at the first examination period.

The data from the above-described experiment are presented in Table 5.1, along with the BLUEs of μ, θ and ξ calculated from the censored samples and the standard error of the estimates of ξ. There is, however, a slight deviation from the original data in one respect. For illustrative purposes of censoring from the left, Sarhan and Greenberg (1957) have assumed that some of the rabbits had lesions large enough at the time of first examination to allow the presumption that the true incubation period ended a few days earlier. For example, we may note that when 10^6 inoculum was used, rabbits 7 and 9 were considered to have lesions large enough that it could be presumed that the incubation was less than seven days.

Table 5.1: Best linear unbiased estimates of parameters for Sarhan and Greenberg's (1957) data on days of incubation among 10 rabbits following inoculation with graded amounts of *Treponema pallidum*

	Inoculum					
Rabbit number	10^6	10^5	10^4	10^3	10^2	10^1
7	< 7	< 11	18	< 18	< 25	> 45
8	11	11	18	18	40	> 45
9	< 7	< 11	14	> 45	> 45	> 45
10	7	11	18	< 18	< 25	< 25
11	11	14	18	25	25	25
12	14	14	18	21	29	25
13	7	11	18	18	29	32
14	> 45	35	40	25	> 45	> 45
15	7	14	18	25	29	40
16	11	14	18	21	29	> 45
μ^*	5.544	9.271	13.356	16.096	20.899	18.174
θ^*	4.333	5.143	6.444	5.667	12.200	32.333
ξ^*	9.877	14.414	19.800	21.763	33.099	50.507
SE(ξ^*)	1.446	1.634	2.038	1.892	4.333	15.500

Example 5.5: Thirty units were placed on a life-testing experiment and their times to failure (in hours) were recorded as [Balasubramanian and Balakrishnan (1992)]

> 0.961, 0.990, 1.565, 2.031, 2.204, 2.340, 3.642, 6.008, 6.538, 7.145, –, –, –,
> 11.937, 15.433, 18.234, 18.307, 22.096, –, –, –, 28.799, 30.692, 30.737,
> 33.702, 34.245, –, –, –, –

Some middle observations were not recorded due to experimental difficulties and the last four failure times were censored since the experiment was terminated as soon as the 26th item failed.

For the above given Type-II multiply censored sample, by assuming a one-parameter exponential distribution we obtain the BLUE of θ, from Eq. (5.4.3), to be

$$\theta^* = 19.943 \text{ hours}$$

and its standard error, from (5.4.5), to be

$$SE(\theta^*) = 3.916 \text{ hours}.$$

If we assume a two-parameter exponential distribution for the above Type-II multiply censored sample, we obtain the BLUEs of μ, θ and ξ, from Eqs. (5.4.6), (5.4.7) and (5.4.9), to be

$$\mu^* = 0.308 \text{ hours}, \ \theta^* = 19.587 \text{ hours and } \xi^* = 19.894 \text{ hours}.$$

The standard error of ξ^* is found from (5.4.12) to be

$$SE(\xi^*) = 3.847 \text{ hours}.$$

Example 5.6: Let us consider Nelson's data (1982, Table 6.1) which gives times to breakdown of an insulating fluid in an accelerated test conducted at various test voltages. In analyzing the complete data, Nelson assumed a scaled Weibull distribution for the times to breakdown and showed by the 90% confidence interval for the shape parameter that an exponential distribution is appropriate for this data. For the purpose of illustration, Viveros and Balakrishnan (1994) generated the following Type-II progressively right censored sample from this data ($n = 19$, $m = 8$):

i	1	2	3	4	5	6	7	8
$X_{i:m:n}$	0.19	0.78	0.96	1.31	2.78	4.85	6.50	7.35
R_i	0	0	3	0	3	0	0	5

By assuming a one-parameter exponential distribution for this data, we obtain from (5.5.4) the BLUE of θ to be

$$\theta^* = 9.09$$

and its standard error is obtained from (5.5.5) to be

$$SE(\theta^*) = 3.21.$$

Example 5.7: Suppose in the above given Type-II progressively censored data, the smallest breakdown time was also not observed. Then, we have $n = 19$, $m = 8$, $r = 1$, resulting in the following general Type-II progressively censored sample:

i	1	2	3	4	5	6	7	8
$X_{i:m:n}$	-	0.78	0.96	1.31	2.78	4.85	6.50	7.35
R_i	-	0	3	0	3	0	0	5

By assuming a one-parameter exponential distribution for this data we obtain, from (5.6.6), the BLUE of θ to be [Balakrishnan and Sandhu (1995b)]

$$\theta^* = 9.11$$

and its standard error, from (5.6.7), to be

$$SE(\theta^*) = 3.22.$$

A comparison of these with the corresponding results in Example 5.6 reveals that the BLUE of θ does not face any significant loss in precision due to the censoring on the left.

References

Ali, M. Masoom, Umbach, D., and Hassanein, K.M. (1981). Estimation of quantiles of exponential and double exponential distributions based on two order statistics, *Communications in Statistics - Theory and Methods*, **10**, 1921-1932.

Ali, M. Masoom, Umbach, D., and Saleh, A.K.Md.E. (1982). Small sample quantile estimation of the exponential distribution using optimal spacings, *Sankhyā, Series B*, **44**, 135-142.

Ali, M. Masoom, Umbach, D., and Saleh, A.K.Md.E. (1985). Tests of significance for the exponential distribution based on selected quantiles, *Sankhyā, Series B*, **47**, 310-318.

Balakrishnan, N. and Cohen, A.C. (1991). *Order Statistics and Inference: Estimation Methods*, Academic Press, San Diego.

Balakrishnan, N. and Sandhu, R.A. (1995a). A simple simulational algorithm for generating progressive Type-II censored samples, *The American Statistician* (To appear).

Balakrishnan, N. and Sandhu, R.A. (1995b). Best linear unbiased and maximum likelihood estimation for exponential distributions under general progressive Type-II censored samples, *Sankhyā, Series B* (To appear).

Balasubramanian, K. and Balakrishnan, N. (1992). Estimation for one- and two-parameter exponential distributions under multiple type-II censoring, *Statistische Hefte*, **33**, 203-216.

Chan, L.K. and Cheng, S.W. (1988). Linear estimation of the location and scale parameters based on selected order statistics, *Communications in Statistics - Theory and Methods*, **17**, 2259-2278.

David, H.A. (1981). *Order Statistics*, Second edition, John Wiley & Sons, New York.

Epstein, B. (1956). Simple estimators of the parameters of exponential distributions when samples are censored, *Annals of the Institute of Statistical Mathematics*, **8**, 15-26.

Epstein, B. (1962). Simple estimates of the parameters of exponential distributions, In *Contributions to Order Statistics* (Eds., A.E. Sarhan and B.G. Greenberg), pp. 361-371, John Wiley & Sons, New York.

Epstein, B. and Sobel, M. (1953). Life testing, *Journal of the American Statistical Association*, **48**, 486-502.

Epstein, B. and Sobel, M. (1954). Some theorems relevant to life testing from an exponential distribution, *Annals of Mathematical Statistics*, **25**, 373-381.

Greenberg, B.G. and Sarhan, A.E. (1958). Applications of order statistics to health data, *American Journal of Public Health*, **48**, 1388-1394.

Harter, H.L. (1961). Estimating the parameters of negative exponential populations from one or two order statistics, *Annals of Mathematical Statistics*, **32**, 1078-1090.

Harter, H.L. (1970). *Order Statistics and Their Use in Testing and Estimation, Vol. 2*, U.S. Government Printing Office, Washington, D.C.

Herd, R.G. (1956). Estimation of the parameters of a population from a multi-censored sample, *Ph.D. Thesis*, Iowa State College, Ames, Iowa.

Kaminsky, K.S. (1972). Confidence intervals for the exponential scale parameter using optimally selected order statistics, *Technometrics*, **14**, 371-383.

Kaminsky, K.S. (1973). Comparison of approximate confidence intervals for the exponential scale parameter from sample quantiles, *Technometrics*, **15**, 483-487.

Kaminsky, K.S. (1974). Confidence intervals and tests for two exponential scale parameters based on order statistics in compressed samples, *Technometrics*, **16**, 251-254.

Kulldorff, G. (1963). Estimation of one or two parameters of the exponential distribution on the basis of suitably chosen order statistics, *Annals of Mathematical Statistics*, **34**, 1419-1431.

Lawless, J.F. (1982). *Statistical Models & Methods for Lifetime Data*, John Wiley & Sons, New York.

Lloyd, E.H. (1952). Least-squares estimation of location and scale parameters using order statistics, *Biometrika*, **39**, 88-95.

Nelson, W. (1982). *Applied Life Data Analysis*, John Wiley & Sons, New York.

Ogawa, J. (1951). Contributions to the theory of systematic statistics, I, *Osaka Mathematical Journal*, **3**, 175-213.

Ogawa, J. (1952). Contributions to the theory of systematic statistics, II, *Osaka Mathematical Journal*, **4**, 41-61.

Saleh, A.K.Md.E. (1966). Estimation of the parameters of the exponential distribution based on order statistics in censored samples, *Annals of Mathematical Statistics*, **37**, 1717-1735.

Saleh, A.K.Md.E. (1967). Determination of the exact optimum order statistics for estimating the parameters of the exponential distribution from censored samples, *Technometrics*, **9**, 279-292.

Saleh, A.K.Md.E. (1981). Estimating quantiles of exponential distribution, In *Statistics and Related Topics* (Eds., M. Csörgö, D.A. Dawson, J.N.K. Rao and A.K.Md.E. Saleh), pp. 279-283, North-Holland, Amsterdam.

Saleh, A.K.Md.E. and Ali, M.M. (1966). Asymptotic optimum quantiles for the estimation of the parameters of the negative exponential distribution, *Annals of Mathematical Statistics*, **37**, 143-151.

Sarhan, A.E. and Greenberg, B.G. (1957). Tables for best linear estimates by order statistics of the parameters of single exponential distributions from singly and doubly censored samples, *Journal of the American Statistical Association*, **52**, 58-87.

Umbach, D., Ali, M. Masoom, and Hassanein, K.M. (1981). Small sample estimation of exponential quantiles with two order statistics, *Aligarh Journal of Statistics*, **1**, 113-120.

Viveros, R. and Balakrishnan, N. (1994). Interval estimation of parameters of life from progressively censored data, *Technometrics*, **36**, 84-91.

CHAPTER 6

Reliability Estimation and Applications

Max Engelhardt[1]

6.1 Introduction

The exponential distribution plays an important role in the field of reliability. Justifications for using the exponential assumption in reliability applications can be found in the early works of Davis (1952), and Epstein and Sobel (1953). Further justification, in the form of theoretical arguments supporting the use of the exponential distribution as the failure law of complex equipment, can be found in books by Barlow and Proschan (1965, 1975).

If the failure-time X of a component or system is exponentially distributed with scale parameter θ and location parameter μ, then the *reliability function* is

$$R(t) = \bar{F}(t) = P[X > t] \begin{array}{ll} = e^{-(t-\mu)/\theta} & \text{if } t > \mu \\ = 0 & \text{otherwise.} \end{array} \tag{6.1.1}$$

An important property of the one-parameter exponential distribution which is often cited in connection with reliability applications is the *no memory* property (see Chapter 2),

$$P[X > s + t \mid X > s] = P[X > t] \tag{6.1.2}$$

for all $s > 0$ and $t > 0$. In reliability terms, this property says that an operating component or system which has not yet failed is as reliable as a new one. The no memory property uniquely characterizes the one-parameter exponential model within the class of continuous life-models, and it provides a rationale for its use in situations where wearout is not a factor. One such example is an electronic component which does not wear out, but fails due to a random shock such as a power surge. While providing a justification for the exponential assumption for some types of components, this property also precludes its use with many other types of components where wearout is a dominant cause of failure, such as bearings or other moving parts in a machine. In other words, an old component is as good as new, provided that failure has not yet occurred.

[1] Idaho National Engineering Laboratory, EG&G Idaho, Inc.

Another property which is often used as a rationale in choosing the one-parameter exponential model is the *constant hazard* property (see Chapter 2),

$$\rho(x) = \frac{f(x)}{\bar{F}(x)} = \lambda \tag{6.1.3}$$

for all $x > 0$. The *hazard function* $\rho(x)$, as defined by the left side of (6.1.3), is also called the *force of mortality* in actuarial science and the *intensity function* in extreme-value theory. It is used in reliability to express the likelihood of failure in the next instant of time, say $(x, x + \Delta x]$, given it is still working at time x. The usual interpretation of the constant hazard property is that the component or system is not affected by aging. Although the constant hazard property appears to be a less global property than the no memory property, it also uniquely characterizes the one-parameter exponential model within the class of continuous life-models. This reiterates, in a slightly different way, the message expressed earlier by the no memory property that an old component or system which has not yet failed is as good as new.

6.2 Estimation With the One-Parameter Model

It is easily shown that for the one-parameter exponential model the quantity on the right side of (6.1.3), called the *hazard rate*, is the reciprocal of the mean, $\lambda = 1/\theta$. Because of the special importance of the hazard rate in reliability applications, the one-parameter model will be reparametrized in terms of λ for the remainder of the section. Thus, the density function will be written as

$$f(x; \lambda) = \lambda\, e^{-\lambda x} \tag{6.2.1}$$

for $x > 0$ and $\lambda > 0$, with the notation $X \sim e(\lambda)$.

Most of the results in the remainder of the chapter will assume that the data were obtained from a random sample. Ideally, the complete random sample can be observed. However, it is often the case, in reliability applications, that an experiment or study will be terminated early, before all units on test have failed, resulting in right-censored data. The term *Type I censoring* is used when a test is terminated after a fixed time, and the term *Type II censoring* applies when a test is terminated after a fixed number of failures have occurred (see Chapter 4 for more details).

6.2.1 Complete Samples

We will restrict attention in this section to point estimates, confidence intervals, and tests of hypotheses for complete samples. We first consider point estimation. The maximum likelihood estimate (MLE), and the uniformly minimum variance unbiased (UMVU) estimate for λ and $R(t)$ are derived in the following theorem. Unless otherwise specified, the indices on all summations in this subsection will range from 1 to n.

Theorem 6.1 *Assuming a complete random sample of size n from $e(\lambda)$, if $s = \Sigma x_i$,*

1. *The MLE of λ is $\hat{\lambda} = n/s = 1/\bar{x}$.*

2. *The MLE of $R(t)$ is $\hat{R}(t) = e^{-\hat{\lambda} t}$.*

3. *The UMVU estimate of* λ *is* $\tilde{\lambda} = (n-1)/s$.

4. *The UMVU estimate of* $R(t)$ *is* $\tilde{R}(t) = \begin{cases} (1-t/s)^{n-1} & \text{if } s > t \\ 0 & \text{if } s \le t \ . \end{cases}$

Proof: Denote by $X_1, X_2, \ldots, X_n$ a random sample of size n from $e(\lambda)$, so that the joint density function is

$$f(x_1, x_2, \ldots, x_n; \lambda) = \lambda^n \, e^{-\lambda \Sigma x_i} \tag{6.2.2}$$

for all $x_i > 0$. Eq. (6.2.2) also provides the likelihood function for λ, say $L(\lambda)$. The derivative of the log-likelihood is

$$\frac{d}{d\lambda} \ln L(\lambda) = \frac{n}{\lambda} - \Sigma x_i \ . \tag{6.2.3}$$

Part 1 follows by equating (6.2.3) to zero and solving for $\lambda = \hat{\lambda}$ (also see Chapter 4). Part 2 follows from Part 1 and the invariance property of MLE's. In order to show Parts 3 and 4, we first note, based on well-known results about the regular exponential family of distributions, that $S = \Sigma X_i$ is a complete sufficient statistic for λ. It follows from results found in Bain and Engelhardt (1992, Chapter 8) that S has a gamma distribution with scale parameter $\theta = 1/\lambda$, and shape parameter n, which also implies that $Y = 2\lambda S \sim \chi^2_{2n}$. It also follows that $E[Y^{-1}] = 2^{-1}\Gamma(2n/2-1)/\Gamma(2n/2) = [2(n-1)]^{-1}$, and $E[(n-1)/S] = \lambda$. Consequently, from completeness and sufficiency of S, $\tilde{\lambda} = (n-1)/s$ is the UMVU estimate of λ, which is Part 3. Part 4 follows by a similar, though more involved argument. We consider first the conditional density function of X_1 given $S = s$

$$\begin{aligned} f_{X_1|S}(x_1 \mid s) &= \frac{f_{X_1,S}(x_1, s)}{f_S(s)} \\ &= \frac{f_{S|X_1}(s \mid x_1) f_{X_1}(x_1)}{f_S(s)} \\ &= \frac{\frac{\lambda^{n-1}}{\Gamma(n-1)} (s - x_1)^{n-2} \, e^{-\lambda(s-x_1)} \lambda \, e^{-\lambda x_1}}{\frac{\lambda^n}{\Gamma(n)} \, s^{n-1} \, e^{-\lambda s}} \\ &= \frac{n-1}{s} \left(1 - \frac{x_1}{s}\right)^{n-2} \end{aligned} \tag{6.2.4}$$

if $s > x_1$, and zero otherwise. We then define a Bernoulli random variable U to be 1 if $X_1 > t$, and 0 otherwise. The random variable U is an unbiased statistic for $R(t)$, since $E[U] = 0 \times P[X \le t] + 1 \times P[X > t] = R(t)$. Although U is not a function of the complete sufficient statistic S, the conditional expectation $E[U \mid S]$ is a statistic which is a function of S, and also unbiased for $R(t)$. Thus, $E[U \mid S]$ is the UMVU estimate of $R(t)$. The specific form of the UMVU estimate is obtained by integration. Specifically,

$$\begin{aligned} E[U \mid s] &= \int_t^s \frac{n-1}{s} \left(1 - \frac{x_1}{s}\right)^{n-2} dx_1 \\ &= (n-1) \int_{t/s}^1 (1-w)^{n-2} dw \\ &= \left(1 - \frac{t}{s}\right)^{n-1} \end{aligned}$$

if $s > t$, and zero otherwise, and Part 4 follows, thus completing the proof of the theorem. □

The chi-square distribution of $2\lambda S$ can also be used to derive confidence limits for λ and $R(t)$. These limits are given in the following theorem.

Theorem 6.2 *Assuming a complete random sample of size n from $e(\lambda)$, if $s = \Sigma x_i$,*

1. *A $100(1-\alpha)\%$ confidence interval for λ is given by*

$$[\lambda_L, \lambda_U] = \left[\hat{\lambda}\ \frac{\chi^2_{2n,\alpha/2}}{2n}\ ,\ \hat{\lambda}\ \frac{\chi^2_{2n,1-\alpha/2}}{2n}\right] .$$

2. *A $100(1-\alpha)\%$ confidence interval for $R(t)$ is given by*

$$[R_L(t), R_U(t)] = \left[e^{-\lambda_U t},\ e^{-\lambda_L t}\right] .$$

3. *A test of size λ for the null hypothesis $H_0 : \lambda = \lambda_0$ versus $H_1 : \lambda \neq \lambda_0$ rejects H_0 if $y \leq \chi^2_{2n,\alpha/2}$ or $y \geq \chi^2_{2n,1-\alpha/2}$ where $y = \frac{2n\lambda_0}{\hat{\lambda}}$. This provides a test of the null hypothesis $H_0 : R(t) = R_0$ versus $H_1 : R(t) \neq R_0$ by setting $\lambda_0 = -\frac{1}{t}\, R_0$.*

Proof: Since $Y = 2\lambda S = 2n\lambda/\hat{\lambda} \sim \chi^2_{2n}$, Part 1 follows from the expression

$$P\left[\chi^2_{2n,\alpha/2} < \frac{2n\lambda}{\hat{\lambda}} < \chi^2_{2n,1-\alpha/2}\right] = 1 - \alpha . \qquad (6.2.5)$$

Part 2 follows from the confidence limits in Part 1, and the fact that $R(t) = e^{-\lambda t}$ is a decreasing function of λ. Part 3 also follows immediately from (6.2.5). This completes the proof as stated. However, it is also worth mentioning that one-tailed tests of size α are obtained if either of the comparisons in Part 3 are used separately with $\alpha/2$ replaced by α, and these one-tailed tests are Uniformly Most Powerful (UMP) tests. It is also possible to construct two-tailed UMP unbiased tests, but it is necessary to allocate the total α differently than the equal-tails allocation of Part 3. In particular, the UMP unbiased test rejects H_0 if $y \leq c_1$ or $y \geq c_2$, where c_1 and c_2 are constants such that both $P[c_1 < \chi^2_{2n} < c_2] = 1 - \alpha$ and $P[c_1 < \chi^2_{2n+2} < c_2] = 1 - \alpha$. Since these equations are rather difficult to solve simultaneously, the equal-tails solution of Part 3 is usually preferred in practice. It is also the case that the UMP unbiased test and the equal-tails test differ only slightly for large n. □

Since many of the results for Type II censored samples parallel those of complete samples, these results will be considered next.

6.2.2 Type II Censored Samples

Suppose the data consists only of the smallest r out of n ordered observations from $e(\lambda)$, $x_{1:n} < x_{2:n} < \cdots < x_{r:n}$. Such data are called Type II right censored data. We first consider point estimation. The maximum likelihood estimate (MLE), and the UMVU estimate for λ and $R(t)$ for Type II censored data are given in the following theorem.

Theorem 6.3 *Denote by $x_{1:n} < x_{2:n} < \cdots < x_{r:n}$ the smallest r out of n ordered observations from $e(\lambda)$.*

1. *The MLE of* λ *is* $\hat{\lambda} = r / \left[\sum_{i=1}^{r} x_{i:n} + (n-r)x_{r:n}\right]$.

2. *The MLE of* $R(t)$ *is* $\hat{R}(t) = e^{-\hat{\lambda}t}$.

3. *The UMVU estimate of* λ *is* $\tilde{\lambda} = (r-1) / \left[\sum_{i=1}^{r} x_{i:n} + (n-r)x_{r:n}\right]$.

4. *The UMVU estimate of* $R(t)$ *is* $\tilde{R}(t) = \begin{cases} (1 - \hat{\lambda}t/r)^{r-1} & \text{if } r > \hat{\lambda}t \\ 0 & \text{if } r \leq \hat{\lambda}t \end{cases}$.

Proof: Denote by $X_{1:n} < X_{2:n} < \cdots < X_{r:n}$ the smallest r out of n order statistics from $e(\lambda)$. The joint density function is

$$f(x_{1:n}, x_{2:n}, \ldots, x_{r:n}; \lambda) = \frac{n!}{(n-r)!} \lambda^r e^{-\lambda\left[\sum_{i=1}^{r} x_{i:n} + (n-r)x_{r:n}\right]} \tag{6.2.6}$$

for all $0 < x_{1:n} < x_{2:n} < \cdots < x_{r:n}$. Equation (6.2.6) also provides the likelihood function for λ, say $L(\lambda)$. The derivative of the log-likelihood is

$$\frac{d}{d\lambda} \ln L(\lambda) = \frac{r}{\lambda} - \left[\sum_{i=1}^{r} x_{i:n} + (n-r)x_{r:n}\right] . \tag{6.2.7}$$

Part 1 follows by equating (6.2.7) to zero and solving for $\lambda = \hat{\lambda}$ (see also Chapter 4). Part 2 follows from Part 1 and the invariance property of MLE's. In order to show Parts 3 and 4, we first note, based on well-known results about the regular exponential family of distributions, that $S_r = \sum_{i=1}^{r} X_{i:n} + (n-r)X_{r:n}$ is a complete sufficient statistic for λ. However, it is important to note that the sufficiency in this case is relative to other statistics which are functions only of the first r out of n observations. The statistic S_r defined here clearly can not be sufficient relative to functions of the complete random sample, since information is lost when the last $n-r$ observations are censored. Similar to the case of a complete random sample, the sufficient statistic has a gamma distribution with scale parameter $\theta = 1/\lambda$, but the shape parameter is r in this case. This also implies that $Y = 2\lambda S_r \sim \chi^2_{2r}$. It follows from an argument similar to that found in Section 6.2.1 that $E[(r-1)/S_r] = \lambda$, and by completeness and sufficiency of S_r that $\tilde{\lambda} = (r-1)/s_r$ is the UMVU estimate of λ. This verifies Part 3. The proof of Part 4 of Theorem 6.1 can be modified to yield Part 4 of the present theorem, but the derivation is somewhat more complicated since the terms in S_r are not independent, which was the case with S. Instead, some special properties of order statistics must be used. However, rather than going through the same parallel argument, instead we will simply show that the statistic defined in Part 4 is unbiased for $R(t)$. In particular, we note that $\tilde{R}(t) = (1 - \hat{\lambda}t/r)^{r-1} = (1 - 2\lambda t/y)^{r-1}$ if $y > 2\lambda t$, with the relationships $y = 2\lambda s_r$ and $\hat{\lambda} = r/s_r$. Thus,

$$\begin{aligned}
E[\tilde{R}(t)] &= \int_{2\lambda t}^{\infty} \left(1 - \frac{2\lambda t}{y}\right)^{r-1} \frac{1}{2^{2r/2}\Gamma(2r/2)} y^{(2r/2)-1} e^{-y/2} \, dy \\
&= \frac{1}{2^{2r/2}\Gamma(2r/2)} \int_{2\lambda t}^{\infty} (y - 2\lambda t)^{r-1} e^{-y/2} \, dy \\
&= \frac{1}{2^{2r/2}\Gamma(2r/2)} \int_{0}^{\infty} w^{r-1} e^{-(\lambda t + w/2)} \, dw
\end{aligned}$$

$$= e^{-\lambda t} \frac{1}{2^{2r/2}\Gamma(2r/2)} \int_0^{\infty} w^{r-1} e^{-w/2} dw$$
$$= e^{-\lambda t} .$$

Since S_r is complete and sufficient relative to statistics which are functions of the first r out of n observations, the statistic defined in Part 4 must be the UMVU estimate of $R(t)$, thus completing the proof of the theorem. □

The chi-square distribution of $2\lambda S_r$ can also be used to derive confidence limits for λ and $R(t)$. The confidence intervals and tests of hypotheses stated in Theorem 6.2 are valid for the case of a Type II censored sample if the MLE $\hat{\lambda}$ for the complete sample case is replaced by the MLE $\hat{\lambda}$ of Theorem 6.3, and if n is replaced by the number of uncensored observations, r.

Theorem 6.4 *Assume that the data consists of the smallest r out of n ordered observations $x_{1:n} < x_{2:n} < \cdots < x_{r:n}$ from $e(\lambda)$, and that $\hat{\lambda}$ is the MLE from Part 1 of Theorem 6.3.*

1. *A $100(1-\alpha)\%$ confidence interval for λ is given by*

$$[\lambda_L, \lambda_U] = \left[\hat{\lambda}\,\frac{\chi^2_{2r,\alpha/2}}{2r}\ ,\ \hat{\lambda}\,\frac{\chi^2_{2r,1-\alpha/2}}{2r}\right] .$$

2. *A $100(1-\alpha)\%$ confidence interval for $R(t)$ is given by*

$$[R_L(t), R_U(t)] = \left[e^{-\lambda_U t},\ e^{-\lambda_L t}\right] .$$

3. *A test of size α for the null hypothesis $H_0 : \lambda = \lambda_0$ versus $H_1 : \lambda \neq \lambda_0$ rejects H_0 if $y \leq \chi^2_{2r,\alpha/2}$ or $y \geq \chi^2_{2r,1-\alpha/2}$ where $y = \frac{2r\lambda_0}{\hat{\lambda}}$. This provides a test of the null hypothesis $H_0 : R(t) = R_0$ versus $H_1 : R(t) \neq R_0$ by setting $\lambda_0 = -\frac{1}{t}\,R_0$.*

6.2.3 Time Truncation and Type I Censored Samples

Suppose there is a time τ such that failure times greater than this time are not observed. If a random sample of units, with lifetimes distributed according to $e(\lambda)$, are placed on test, and only r failure times $x_i \leq \tau$ are recorded, then there are two possible situations. Either the number of unobserved failure times is not known, in which case the sample is said to be *time truncated* above τ, or the number of unobserved failure times is known, in which case the sample is said to be *time censored* above τ. In the latter case, it is more common to say that the sample is Type I censored at τ, with data consisting of the number of failures r and the smallest r out of n ordered failure times (see Chapter 4).

We will first consider some results relating to the case of time truncation. Suppose a component has a exponential failure time, $X \sim e(\lambda)$. Suppose also that the measuring process or device is not capable of recording the failure time if it exceeds a fixed time τ. If the failure time occurs before τ, then the observed time has a truncated exponential distribution with density function

$$f_\tau(x;\lambda) = \frac{\lambda\, e^{-\lambda x}}{1 - e^{-\lambda\tau}} \tag{6.2.8}$$

if $0 < x < \tau$, and zero otherwise. There are a number of possible situations which might lead to data which are distributed according to (6.2.8). First, suppose an unknown number of units with exponential lifetimes are in operation. When failures occur, the times are recorded, but only until a specified time τ. Conditional on the observed number r, the ordered observations are distributed as r order statistics of a random sample of size r from (6.2.8). Another sort of application would involve particles whose diameters, which are exponentially distributed, are to be measured, but the particles are first sifted through a screen which only allows particles to pass with diameters less than τ. In this example, the sifting is done only until a predetermined number r have passed through the screen. A third possibility is simply that lifetimes of individuals in the original population are distributed according to (6.2.8), and a random sample of size r is taken. We will study the problem in this latter situation, although, as noted above, the results also have a bearing on the problems of time truncation and Type I censoring.

If lifetimes are distributed according to a truncated exponential distribution, then the reliability function, which is obtained by integrating (6.2.8), is

$$R_\tau(t;\lambda) = \frac{e^{-\lambda t} - e^{-\lambda\tau}}{1 - e^{-\lambda\tau}} \tag{6.2.9}$$

for $0 < t < \tau$. Methods for estimating λ and the reliability are given in the following theorem.

Theorem 6.5 *Assume a random sample $x_1, x_2, \ldots, x_r$ of size r from $f_\tau(x;\lambda)$ and let $s = \sum_{i=1}^r x_i = r\bar{x}$.*

1. *The MLE of λ is the unique solution $\hat{\lambda}$ of the equation $\frac{1}{\hat{\lambda}\tau} - \frac{1}{\exp(\hat{\lambda}\tau)-1} = \frac{\bar{x}}{\tau}$ if $0 < \bar{x}/\tau < 1/2$, with no solution if $1/2 \leq \bar{x}/\tau$.*

2. *The MLE of $R_\tau(t;\lambda)$ is $R_\tau(t;\hat{\lambda})$.*

3. *The UMVU estimate of $R_\tau(t;\lambda)$ is*

$$\tilde{R}_\tau(t;\lambda) = 1 - \frac{\sum_{i=0}^{m_1} (-1)^i \binom{r-1}{i} \left(1 - \frac{i\tau}{s}\right)^{r-1} - \sum_{i=0}^{m_2} (-1)^i \binom{r-1}{i} \left(1 - \frac{i\tau+t}{s}\right)^{r-1}}{\sum_{i=0}^{m_1} (-1)^i \binom{r}{i} \left(1 - \frac{i\tau}{s}\right)^{r-1}}$$

where $m_1 = [s/\tau]$ and $m_2 = [(s-t)/\tau]$, and $[\cdot]$ is the greatest integer function.

Proof: If $X_1, X_2, \ldots, X_r$ is a random sample of size r from (6.2.8), then the joint density function is

$$f_\tau(x_1, x_2, \ldots, x_r;\lambda) = \frac{\lambda^r \exp\{-\lambda \sum_{i=1}^r x_i\}}{(1 - e^{-\lambda\tau})^r} \tag{6.2.10}$$

if all $0 < x_i < \tau$, and zero otherwise. The sum $S = \sum_{i=1}^r X_i = r\bar{X}$ is a complete sufficient statistic for λ in this situation. Since S is the sum of all r observations, it does not matter whether the observations are ordered.

Part 1 follows easily by equating to zero the derivative with respect to λ of the log-likelihood, based on (6.2.10). The problem of maximum likelihood estimation for the parameter $\theta = 1/\lambda$ was also considered by Deemer and Votaw (1955). In terms of the parameter λ, the MLE of λ is the solution $\hat{\lambda}$ of the equation in Part 1 of the theorem. Part 2 follows by the invariance property of MLE's. The UMVU estimate of $R_\tau(t;\lambda)$ can be derived using the density function of S, as derived by Bain and Weeks (1964). In the present notation, this density function is

$$f_S(s;\lambda) = \frac{\lambda^r e^{-\lambda s}}{(1-e^{-\lambda\tau})^r (r-1)!} \sum_{i=0}^{m} \binom{r}{i} (-1)^i (s - i\tau)^{r-1} \tag{6.2.11}$$

if $0 < s \le r\tau$ with $m = m(s/\tau) = [s/\tau]$. The UMVU estimate can be derived using a modification of the approach of Theorem 6.1. In particular, the density function for S which was gamma in (6.2.4), would be replaced with (6.2.11). The resulting conditional density of X_1 given $S = s$ is somewhat more complicated than (6.2.4). The conditional expectation of the Bernoulli variable U given $S = s$, could then be derived. A different approach, used by Sathe and Varde (1969), involves a derivation based on Laplace transform. By either method, it follows that the UMVU estimate of $R_\tau(t;\lambda)$ is given in Part 3 of the theorem. □

The distribution given by (6.2.11) can also be used to construct tests of hypothesis and confidence intervals for λ. Denote by $s_\gamma(\lambda)$ the 100γth percentile of the sum S, which can be obtained by integration of (6.2.11). Tests and confidence limits are given in the following theorem.

Theorem 6.6 *Assume a random sample $x_1, x_2, \ldots, x_r$ of size r from $f_\tau(x;\lambda)$, and denote the sample sum by $s = \sum_{i=1}^{r} x_i = r\bar{x}$.*

1. *A UMP test of size α for the null hypothesis $H_0 : \lambda \le \lambda_0$ versus $H_1 : \lambda > \lambda_0$ rejects H_0 if $s \le s_\alpha(\lambda_0)$.*

2. *A UMP test of size α for the null hypothesis $H_0 : \lambda \ge \lambda_0$ versus $H_1 : \lambda < \lambda_0$ rejects H_0 if $s \ge s_{1-\alpha}(\lambda_0)$.*

3. *A size α test for the null hypothesis $H_0 : \lambda = \lambda_0$ versus $H_1 : \lambda \ne \lambda_0$ rejects H_0 if $s \le s_{\alpha/2}(\lambda_0)$ or $s \ge s_{1-\alpha/2}(\lambda_0)$.*

4. *Limits for a $100(1-\alpha)\%$ confidence interval for λ are obtained as solutions to the equations $s = s_{\alpha/2}(\lambda)$ and $s = s_{1-\alpha/2}(\lambda)$.*

Proof: Parts 1 and 2 follow from the fact that (6.2.8) is a member of the regular exponential class, and the monotone likelihood property. Part 3 follows from Parts 1 and 2, replacing α with $\alpha/2$. Part 4 is obtained by inverting the two-tailed test of Part 3. Specifically, the acceptance set consists of all values of λ such that $s_{\alpha/2}(\lambda) \le s \le s_{1-\alpha/2}(\lambda)$, and the limits in Part 4 are the minimum and maximum values of this acceptance set. It should be noted that the test of Part 3 is not UMP, but if the tail probabilities are chosen properly it is possible to construct a two-tailed UMP unbiased test for λ. We will not pursue this point further, but the method for constructing such tests is given in Lehmann (1959). □

One problem with applying Theorem 6.6 is solving for the percentiles $s_\gamma(\lambda)$. However, Bain, Engelhardt, and Wright (1977) found that the distribution of the related variable $S/(r\tau) = \bar{X}/\tau$ can be approximated by a Beta distribution with parameters

$$a = \frac{r\mu[\mu(1-\mu) - \sigma^2/r]}{\sigma^2}, \quad b = \frac{r(1-\mu)[\mu(1-\mu) - \sigma^2/r]}{\sigma^2} \tag{6.2.12}$$

where μ and σ^2 are the mean and variance of X/τ, namely

$$\mu = \frac{1}{\lambda\tau} - \frac{e^{-\lambda\tau}}{1-e^{-\lambda\tau}}, \quad \sigma^2 = \left(\frac{1}{\lambda\tau}\right)^2 - \frac{e^{-\lambda\tau}}{[1-e^{-\lambda\tau}]^2} . \tag{6.2.13}$$

The well-known relationship between the Beta and F-distributions provides the following convenient approximation:

$$Y = (a/b)(\tau\,/\,\bar{X} - 1) \sim F_{2b,2a} . \tag{6.2.14}$$

Thus, using the values $a = a(\lambda)$ and $b = b(\lambda)$ which are computed from (6.2.12) and (6.2.13), it follows that $s_\gamma(\lambda) \doteq r\tau/[1 + (b/a)F_{2b,2a,1-\gamma}]$. Perhaps the most convenient way to run such a test is to base it directly on the statistic $y = (a/b)(\tau/\bar{x} - 1)$. For example, an approximate size α test for the null hypothesis $H_0 : \lambda \leq \lambda_0$ versus $H_1 : \lambda > \lambda_0$ would reject H_0 if $y \geq F_{2b,2a,1-\alpha}$ with $a = a(\lambda_0)$ and $b = b(\lambda_0)$. Similarly, an approximate size α test for the null hypothesis $H_0 : \lambda \geq \lambda_0$ versus $H_1 : \lambda < \lambda_0$ would reject H_0 if $y \leq F_{2b,2a,\alpha}$. Also, limits for an approximate $100(1-\alpha)\%$ confidence interval for λ are solutions to the equations $\tau/\bar{x} - 1 = \frac{b}{a}\, F_{2b,2a,\alpha/2}$ and $\tau/\bar{x} - 1 = \frac{b}{a}\, F_{2b,2a,1-\alpha/2}$ with $a = a(\lambda)$ and $b = b(\lambda)$ defined by (6.2.12) and (6.2.13).

We now consider the case of Type I censoring from $e(\lambda)$. Suppose it is known that n independent units are tested for a fixed length of time τ. The data consist only of r, the number of failures prior to τ, and the observed failure times. Such data are called Type I right censored data. The ML estimates for λ and $R(t)$ for Type I censored data are given in the following theorem.

Theorem 6.7 *Suppose $r \geq 1$ and denote by $x_{1:n} < x_{2:n} < \cdots < x_{r:n} < \tau$ the r observed failure times from $e(\lambda)$.*

1. *The MLE of λ is $\hat{\lambda} = r/\left[\sum_{i=1}^{r} x_{i:n} + (n-r)\tau\right]$.*

2. *The MLE of $R(t)$ is $\hat{R}(t) = e^{-\hat{\lambda}\,t}$.*

Proof: The joint density function for a Type I censored sample is (see Chapter 4)

$$f(x_{1:n}, x_{2:n}, \ldots, x_{r:n}; \lambda) = \frac{n!}{(n-r)!}\,\lambda^r\, e^{-\lambda\left[\sum_{i=1}^{r} x_{i:n} + (n-r)\tau\right]} \tag{6.2.15}$$

for all $0 < x_{1:n} < x_{2:n} < \cdots < x_{r:n} < \tau$; $r = 1, \ldots, n$,

$$P[X_{1:n} > \tau] = e^{-n\lambda\tau}$$

if $r = 0$, and zero otherwise. In the case of Type I censoring, there is also a possibility that no failures will be observed prior to τ, and the probability that the smallest ordered observation, which is given above, accounts for this case. The number of observed failures, say R, is a random variable which has the binomial distribution with parameters n and $p = p(\lambda) = 1 - e^{-\lambda\tau}$. Consequently, the probability that no failures will be observed out of n is $P[R = 0] = e^{-n\lambda\tau}$. Thus, the likelihood function for λ, say $L(\lambda)$, is given by (6.2.15) if $r > 0$, and $e^{-n\lambda\tau}$ if $r = 0$. In this latter case when $r = 0$, the likelihood is maximized by $\lambda = 0$. However, when $r > 0$, the maximum occurs at a positive value of λ, and it can be obtained by equating the derivative of the log-likelihood to zero and solving for $\lambda = \hat{\lambda}$. In particular, by solving the equation

$$\frac{d}{d\lambda} \ln L(\lambda) = \frac{r}{\lambda} - \left[\sum_{i=1}^{r} x_{i:n} + (n-r)\tau\right] = 0$$

for $\lambda = \hat{\lambda}$, Part 1 follows. Part 2 follows from Part 1 and the invariance property of MLE's. This completes the proof. □

It is not possible, in the case of Type I censoring, to derive a UMVU estimate of λ. The difficulty lies in the fact that there is not a single minimal sufficient statistic, but rather a pair, R and $\sum_{i=1}^{R} X_{i:n}$ which are jointly sufficient. However, there is only one unknown parameter, namely λ, so that a pair of minimal sufficient statistics will not be complete, and the usual method for constructing UMVU estimates fails.

For inferences on λ, in the case of Type I censored data, there are essentially two possible approaches, one based on the MLE $\hat{\lambda}$, and the other based only on the number of observed failures, $R = r$, but not on the exact failure times.

Since R is binomial with parameters n and $p = 1 - e^{-\lambda\tau}$, a number of common inference procedures for the binomial parameter p are available for analyzing λ. Note that a small value of λ corresponds to a small value of p, in which case r, the observed value of R, will tend to be small. Thus, for example, with Type I censored data, a point estimate of λ is $\tilde{\lambda} = -(1/\tau)\ln(1-\tilde{p})$ where $\tilde{p} = r/n$ is the proportion of observed failure times prior to τ. Also, a point estimate of $R(t)$ is $\tilde{R}(t) = e^{-\tilde{\lambda}t}$. It should be noted that although these are biased estimates, and they are less efficient than the MLE's which are given in Theorem 6.7, they are asymptotically unbiased.

It is also possible to construct tests of hypotheses and confidence intervals for λ. For example, a conservative test of size α for the null hypothesis $H_0 : \lambda \leq \lambda_0$ versus $H_1 : \lambda > \lambda_0$ would reject H_0 if $\text{Bin}(r, n, p_0) \leq \alpha$ with $p_0 = 1 - e^{-\lambda_0\tau}$. Similarly, a conservative size α test of the null hypothesis $H_0 : \lambda \geq \lambda_0$ versus $H_1 : \lambda < \lambda_0$ would reject H_0 if $\text{Bin}(r, n, p_0) \geq 1 - \alpha$ with $p_0 = 1 - e^{-\lambda_0\tau}$.

Standard methods [see, Bain and Engelhardt (1992, p. 374)] for constructing conservative confidence limits for the parameter p can also be used to derive confidence limits for λ. A convenient method for constructing a conservative confidence interval for both λ and $R(t)$ can be based on the well-known relationships between the binomial distribution, the Beta distribution, and the F-distribution. We notice first that limits for a conservative two-sided $100(1-\alpha)\%$ confidence interval for reliability at τ, $R(\tau) = e^{-\lambda\tau}$, are

$$R_L(\tau) = \frac{1}{1 + \frac{r+1}{n-r} F_{2(r+1),2(n-r),1-\alpha/2}} \tag{6.2.16}$$

and

$$R_U(\tau) = \frac{1}{1 + \frac{r}{n-r+1} F_{2r,2(n-r+1),\alpha/2}} . \tag{6.2.17}$$

Limits for a conservative $100(1-\alpha)\%$ confidence interval for $R(t)$, can then be obtained from the relationship, $R(t) = [R(\tau)]^{t/\tau}$; specifically, $R_L(t) = [R_L(\tau)]^{t/\tau}$ and $R_U(t) = [R_U(\tau)]^{t/\tau}$. Limits for a conservative two-sided $100(1-\alpha)\%$ confidence interval for λ are

$\lambda_L = -(1/\tau)\ln R_U(\tau)$ and $\lambda_U = -(1/\tau)\ln R_L(\tau)$. Note that it is possible to compute a conservative lower one-sided confidence limit for $R(t)$, and a lower one-sided confidence limit for λ when no failures have occurred (i.e. with $r = 0$). In this case, we replace $\alpha/2$ with α, and the degrees of freedom in (6.2.16) are 2 and $2n$. The resulting lower confidence limit for $R(\tau)$ has the simple form $R_L(\tau) = \alpha^{1/n}$, and the upper confidence limit for λ is $\lambda_U = -\frac{1}{n\tau}\ln\alpha$.

Although the procedures described above are relatively simple and convenient to apply, an obvious matter of concern is the amount of efficiency lost by not using the actual failure times. In other words, the question is how much efficiency is lost by using the estimate $\tilde{\lambda}$ which is based solely on count data versus using the MLE $\hat{\lambda}$. This problem was studied by Bartholomew (1963). By comparing asymptotic variances, an asymptotic efficiency of about 96 percent was found for $\tilde{\lambda}$ when $p \leq 0.5$, which corresponds to $\lambda \leq \ln 2/\tau$. For large λ, there is an approximate normal test, also due to Bartholomew, for the exponential problem which uses all of the data. The test statistic is

$$Z = \frac{V\sqrt{np}}{\sqrt{1 - 2(q\ln q)V/p + qV^2}} \tag{6.2.18}$$

where $V = \lambda/\hat{\lambda} - 1$, $p = 1 - e^{-\lambda\tau}$ and $q = 1 - p = e^{-\lambda\tau}$. Conditional on $R > 0$, the variable defined by (6.2.18) has an approximate standard normal distribution, with the best approximation occurring when $p \geq 0.5$ (or when $\lambda \geq \ln 2/\tau$), which is the range where the test based only on $\tilde{\lambda}$ is least efficient. Thus, it seems reasonable to use the MLE $\hat{\lambda}$ and procedures based on (6.2.18) for testing large values of λ, and procedures based on $\tilde{\lambda}$ for small values of λ.

Tests of hypotheses about λ based on (6.2.18) are relatively easy to perform, but confidence limits would be rather tedious to calculate. It should also be noted that (6.2.18) is only defined when $r \geq 1$, but the case $r = 0$ should not be totally ignored in the construction of tests in which the alternative hypotheses includes small values of λ. This point is addressed in the following theorem.

Theorem 6.8 *Consider data based on a random sample with Type I censoring at τ from $e(\lambda)$. Suppose $\lambda_0 \geq \ln 2/\tau$, and if $r \geq 1$, denote by z_0 the value of (6.2.18) based on an observed value of $\hat{\lambda}$ with $\lambda = \lambda_0$.*

1. *A size α test of the null hypothesis $H_0 : \lambda \leq \lambda_0$ versus $H_1 : \lambda > \lambda_0$ would reject H_0 if $\Phi(z_0) \leq \alpha$.*

2. *A size α test of the null hypothesis $H_0 : \lambda \geq \lambda_0$ versus $H_1 : \lambda < \lambda_0$ (assuming $\alpha \geq e^{-n\lambda_0\tau}$) rejects H_0 if $r = 0$, or if $r \geq 1$ and $\Phi(z_0) \geq 1 - (\alpha - e^{-n\lambda_0\tau})$.*

3. *A size α test for the null hypothesis $H_0 : \lambda = \lambda_0$ versus $H_1 : \lambda \neq \lambda_0$ rejects H_0 if $\Phi(z_0) \leq \alpha/2$ or (assuming $\alpha/2 \geq e^{-n\lambda_0\tau}$) if $r = 0$, or if $r \geq 1$ and $\Phi(z_0) \geq 1 - (\alpha/2 - e^{-n\lambda_0\tau})$.*

6.3 Estimation With the Two-Parameter Model

This section deals with reliability analysis based on the two-parameter model with a scale parameter θ and a threshold parameter μ. Thus, the density function will be written as

$$f(x;\mu,\theta) = (1/\theta)\, e^{-(x-\mu)/\theta} \tag{6.3.1}$$

for $x > \mu$ and $\theta > 0$, with the notation $X \sim e(\mu,\theta)$. If μ is known, then the transformation $Y = X - \mu$ allows the use of results for the one-parameter model, as discussed in Section 6.2. Thus, in the remainder of this section, we assume that μ is unknown.

6.3.1 Complete and Type II Censored Samples

Suppose the data consists of the smallest r out of n ordered observations, $\mu \le x_{1:n} < x_{2:n} < \cdots < x_{r:n}$ from $e(\mu,\theta)$. Such data are called Type II right censored data. Again, we will emphasize point and interval estimation. We first consider point estimation. The maximum likelihood estimates (MLE) of θ, μ and $R(t)$, and the UMVU estimates of θ and μ are given in the following theorem.

Theorem 6.9 *Denote by $\mu \le x_{1:n} < x_{2:n} < \cdots < x_{r:n}$ the smallest r out of n ordered observations from $e(\mu,\theta)$.*

1. *The MLE of θ is $\hat\theta = [\sum_{i=1}^r x_{i:n} + (n-r)x_{r:n} - nx_{1:n}]/r$.*
2. *The MLE of μ is $\hat\mu = x_{1:n}$.*
3. *The MLE of $R(t)$ is $\hat R(t) = e^{-(t-\hat\mu)/\hat\theta}$.*
4. *The UMVU estimate of θ is $\tilde\theta = [\sum_{i=1}^r x_{i:n} + (n-r)x_{r:n} - nx_{1:n}]/(r-1)$.*
5. *The UMVU estimate of μ is $\tilde\mu = x_{1:n} - \tilde\theta/n$.*

Proof: Denote by $\mu \le X_{1:n} < X_{2:n} < \cdots < X_{r:n}$ the smallest r out of n order statistics from $e(\mu,\theta)$. The joint density function is (see Chapter 4)

$$f(x_{1:n}, x_{2:n}, \ldots, x_{r:n}) = \frac{n!}{(n-r)!}\,(1/\theta)^r\, e^{-(1/\theta)[\sum_{i=1}^r (x_{i:n}-\mu)+(n-r)(x_{r:n}-\mu)]}$$

for all $\mu \le x_{1:n} < x_{2:n} < \cdots < x_{r:n}$, and zero otherwise. This also provides the likelihood function, say $L(\mu,\theta)$. We first note for each fixed $\theta > 0$ that $L(\mu,\theta)$ is an increasing function of μ, provided $\mu \le x_{1:n}$, but $L(\mu,\theta) = 0$ if $\mu > x_{1:n}$. Consequently, the MLE of μ is $\hat\mu = x_{1:n}$. The rest of the derivation involves the usual approach, based on setting the derivative of the log-likelihood to zero (with μ replaced by $\hat\mu$), and solving for $\hat\theta$. The result is

$$\begin{aligned}\hat\theta &= \left[\sum_{i=1}^r (x_{i:n} - \hat\mu) + (n-r)(x_{r:n} - \hat\mu)\right] \Big/ r \\ &= \left[\sum_{i=1}^r x_{i:n} + (n-r)x_{r:n} - nx_{1:n}\right] \Big/ r\,.\end{aligned}$$

The MLE of $R(t)$ follows from the invariance property of MLE's.

In order to derive the UMVU estimates, it is convenient to relate the order statistics to another set of ordered variables. In particular, an important property of exponential order statistics is that they are distributed as a linear combination of independent exponential random variables. In particular, for $1 \leq j \leq n$

$$Y_{j:n} - \mu \sim \sum_{i=1}^{j} \frac{Z_{n-i+1}}{n-i+1} \tag{6.3.2}$$

where $Z_1, Z_2, \ldots$ are independent one-parameter exponential, $e(\theta)$ (see Chapter 3). We now define a new set of related variables as follows:

$$\begin{aligned} X^*_{j:n-1} &= X_{j+1:n} - X_{1:n} \\ &= (X_{j+1:n} - \mu) - (X_{1:n} - \mu) \sim \sum_{i=1}^{j-1} \frac{Z_{(n-1)-i+1}}{(n-1)-i+1} \end{aligned}$$

for $1 \leq j \leq n-1$. Thus, for any $2 \leq r \leq n$, $0 < X^*_{1:n-1} < X^*_{2:n-1} < X^*_{r-1:n-1}$ are distributed as the first $r-1$ out of $n-1$ order statistics from a one-parameter exponential distribution, $e(\theta)$. These new variables, $X^*_{j:n-1}$, are free of the threshold parameter μ, their joint distribution is free of μ, and thus, many of the results from Section 6.2.2 can be adapted for inferences on θ. For example, it follows that a suitably scaled version of the estimator $\tilde{\theta}$ defined in Part 4 is chi-square distributed. In particular, if $2 \leq r \leq n$, then

$$\begin{aligned} 2(r-1)\tilde{\theta}/\theta &= 2\left[\sum_{i=1}^{r} X_{i:n} + (n-r)X_{r:n} - nX_{1:n}\right]\Big/\theta \\ &= 2\left[\sum_{i=2}^{r}(X_{i:n} - X_{1:n}) + (n-r)(X_{r:n} - X_{1:n}) - nX_{1:n}\right]\Big/\theta \\ &= 2\left[\sum_{i=1}^{r-1} X^*_{i:n-1} + [(n-1)-(r-1)]X^*_{r-1:n-1}\right]\Big/\theta \\ &\sim \chi^2_{2(r-1)}\ . \end{aligned} \tag{6.3.3}$$

Note also that $X_{1:n}$ is a complete sufficient statistic in the case where θ is known. Also, since the distribution of $\tilde{\theta}$ does not depend on μ, it follows that $X_{1:n}$ and $\tilde{\theta}$ are independent random variables [see Bain and Engelhardt (1992, p. 352)]. It can also be shown that these are jointly complete sufficient statistics. Since the mean of a chi-square distribution is also the degrees of freedom, $E[\tilde{\theta}] = \theta$, and $\tilde{\theta}$ must be the UMVU estimator of θ. Note also, for the estimator of Part 5

$$\begin{aligned} E[\tilde{\mu}] &= \mu + E[X_{1:n} - \mu] - E[\tilde{\theta}]/n \\ &= \mu + \theta/n - \theta/n = \mu\ , \end{aligned}$$

and $\tilde{\mu}$ must be the UMVU estimator of μ. This concludes the proof of the theorem. □

The fact that $2(r-1)\tilde{\theta}/\theta \sim \chi^2_{2(r-1)}$ also has obvious implications for inferences about θ with μ an unknown nuisance parameter. Confidence intervals for the parameters θ and μ each with the other as an unknown nuisance parameter are given in the following theorem.

Theorem 6.10 *Denote by $\mu \leq x_{1:n} < x_{2:n} < \cdots < x_{r:n}$ the smallest r out of n ordered observations from $e(\mu, \theta)$.*

1. Limits for a $100(1-\alpha)\%$ *confidence interval for* θ *is given by*

$$\theta_L = \frac{2(r-1)\tilde{\theta}}{\chi^2_{2(r-1),1-\alpha/2}},$$
$$\theta_U = \frac{2(r-1)\tilde{\theta}}{\chi^2_{2(r-1),\alpha/2}}.$$

2. Limits for a $100(1-\alpha)\%$ *confidence interval for* μ *is given by*

$$\mu_L = x_{1:n} - \frac{(r-1)\tilde{\theta}}{n}\left[(\alpha/2)^{-1/(r-1)} - 1\right],$$
$$\mu_U = x_{1:n} - \frac{(r-1)\tilde{\theta}}{n}\left[(1-\alpha/2)^{-1/(r-1)} - 1\right].$$

Proof: Part 1 follows immediately from the chi-square results in (6.3.3). Part 2 requires a bit more effort. First, we recall the variable $S_r = \sum_{i=1}^{r} X_{i:n} + (n-r)X_{r:n}$ which was encountered in Section 6.2.2. It is easily verified that S_r is sufficient for θ when μ is known, and thus the conditional distribution of $X_{1:n}$ given $S_r = s_r$ does not depend on θ, which suggests basing inferences on μ with θ an unknown nuisance parameter on this conditional distribution. The rest of the derivation, which we will only sketch, is based on the relationship $S_r = (r-1)\tilde{\theta} + nX_{1:n}$, and the facts that $X_{1:n} \sim e(\mu, \theta/n)$, that $2(r-1)\tilde{\theta}/\theta \sim \chi^2_{2(r-1)}$, and that $\tilde{\theta}$ and $X_{1:n}$ are independent. From these results, it is possible to obtain the conditional density of $X_{1:n}$ given $S_r = s_r$ as

$$\begin{aligned}
f_{X_{1:n}|S_r}(x_{1:n}|s_r) &= \frac{f_{X_{1:n},S_r}(x_{1:n}, s_r)}{f_{S_r}(s_r)} \\
&= \frac{f_{S_r|X_{1:n}}(s_r|x_{1:n}) f_{X_{1:n}}(x_{1:n})}{f_{S_r}(s_r)} \\
&= \frac{n(r-1)(s_r - nx_{1:n})^{r-2}}{(s_r - n\mu)^{r-1}}
\end{aligned}$$

if $\mu \le x_{1:n} \le s_r/n$, and zero otherwise. Thus, the conditional cumulative distribution function of $X_{1:n}$ given $S_r = s_r$ is

$$F_{X_{1:n}|S_r}(x_{1:n}|s_r) = 1 - \left(\frac{s_r - nx_{1:n}}{s_r - n\mu}\right)^{r-1} \tag{6.3.4}$$

if $\mu \le x_{1:n} \le s_r/n$, 0 if $x_{1:n} < \mu$ and 1 if $s_r/n < x_{1:n}$. The limits of Part 2 are obtained by equating (6.3.4) to $\alpha/2$ and $1-\alpha/2$, and solving for μ_L and μ_U, respectively. This concludes the proof. □

Confidence limits for reliability can be based on the MLE. In particular, the MLE of $R = R(t)$ is $\hat{R} = \hat{R}(t) = e^{-(t-\hat{\mu})/\hat{\theta}}$, which means that

$$\begin{aligned}
\ln \hat{R} &= -\frac{t-\hat{\mu}}{\hat{\theta}} \\
&= -\frac{(t-\mu)/\theta - (\hat{\mu}-\mu)/\theta}{\hat{\theta}/\theta}
\end{aligned}$$

$$= -\frac{\ln R - (\hat{\mu} - \mu)/\theta}{\hat{\theta}/\theta} .$$

It follows that the distribution of $\hat{R}$ depends only on θ, μ, and t only through R and not separately. Of course, this is true in general for location-scale models, and $e(\mu, \theta)$ is such a model. The inferences can also be made in terms of the related quantities $Y = n \ln(\hat{R})$ and $\xi = n \ln(R)$. This has the advantage that the distribution Y depends only on r, the number of uncensored observations, and not on the sample size n.

If we define $U = n(\hat{\mu} - \mu)/\theta$, and $W = 2r\hat{\theta}/\theta = 2(r-1)\tilde{\theta}/\theta$, so that $2U \sim \chi^2_2$, and $W \sim \chi^2_{2(r-1)}$, with U and W independent, it follows that $Y = 2r(U + \xi)/W$. The cumulative distribution function of Y can be written as

$$\begin{aligned} F_Y(y) &= P[Y \leq y] \\ &= 1 - P[2r(U + \xi)/y \geq W] \\ &= 1 - \int_{A_y} h(u; 2) h(w; 2r-2)\, du\, dw \end{aligned}$$

where $h(\cdot; \nu)$ is the density function of a chi-square distribution with ν degrees of freedom, and $A_y = \{(u, w) | 2r(u + \xi)/y \geq w\}$. The ease with which this integration can be done depends on the value of y. In the case $y > 0$ (or $\hat{R} > 1$), which occurs when $x_{1:n} > t$, the result is

$$F_Y(y) = 1 - e^{\xi} \left(1 + \frac{y}{r}\right)^{-(r-1)} . \tag{6.3.5}$$

For all other values of y the integral is more complicated and it can only be evaluated by numerical methods. Such an evaluation was done by Guenther, Patil, and Uppuluri (1976) for the related tolerance limit problem. However, expression (6.3.5) provides a very good approximation in many cases of practical interest in reliability. A table with ranges of r, n and $\hat{R}$ for which this approximation is adequate is given by Engelhardt and Bain (1978), and an asymptotic normal approximation which can be used in the other cases is also provided. The corresponding confidence limits are given in the following theorem.

Theorem 6.11 *Denote by $\mu \leq x_{1:n} < x_{2:n} < \cdots < x_{r:n}$ the smallest r out of n ordered observations from $e(\mu, \theta)$.*

1. *A lower $100(1-\alpha)\%$ confidence limit for $R = R(t)$ is given by*

$$R_L = \alpha^{1/n} \left[1 + \frac{n}{r} \ln(\hat{R})\right]^{(r-1)/n}$$

which is exact if $x_{1:n} > t$, and approximate otherwise.

2. *An approximate lower $100(1-\alpha)\%$ confidence limit for $R = R(t)$ is given by*

$$R_L = \exp\left[-\frac{1}{n} + \frac{r(r-5/2)}{na} \left(y - \frac{z_{1-\alpha}}{r} \sqrt{ry^2 + a}\right)\right]$$

where $y = n \ln(\hat{R})$ and $a = r\left(r - z^2_{1-\alpha}\right)$.

A table giving ranges of n and r for which each approximation is best is provided by Engelhardt and Bain (1978).

6.3.2 Type I Censored Samples

We now consider the case of Type I censoring from $e(\mu,\theta)$. Suppose it is known that n independent units are tested for a fixed length of time τ. The data consist only of r, the number of failures prior to τ, and the observed failure times. Such data are called Type I right censored data. The ML estimates for θ and $R(t)$ for Type I censored data are given in the following theorem (see Chapter 4).

Theorem 6.12 *Suppose $r \geq 1$ and denote by $x_{1:n} < x_{2:n} < \cdots < x_{r:n} < \tau$ the r observed failure times from $e(\mu,\theta)$.*

1. *The MLE of μ is $\hat{\mu} = x_{1:n}$.*
2. *The MLE of θ is $\hat{\theta} = [\sum_{i=1}^{r} x_{i:n} + (n-r)\tau - nx_{1:n}]/r$.*
3. *The MLE of $R(t)$ is $\hat{R}(t) = e^{-(t-\hat{\mu})/\hat{\theta}}$.*

Proof: The joint density function for a Type I censored sample is (see Chapter 4)

$$f(x_{1:n}, x_{2:n}, \ldots, x_{r:n}, r; \theta, \mu) = \frac{n!}{(n-r)!\theta^r} e^{-[\sum_{i=1}^{r} x_{i:n} + (n-r)\tau - n\mu]/\theta}$$

for all $\mu \leq x_{1:n} < x_{2:n} < \cdots < x_{r:n} < \tau$; $r = 1, \ldots, n$,

$$P[X_{1:n} > \tau] = e^{-n(\tau-\mu)/\theta}$$

if $r = 0$, and zero otherwise. As in the case of Type I censoring from a one-parameter exponential distribution, there is also a possibility that no failures will be observed prior to τ. Also, as in the case of the one-parameter model, the number of observed failures, say R, is a random variable which has the binomial distribution with parameters n and $p = p(\mu,\theta) = 1 - e^{-(\tau-\mu)/\theta}$. Consequently, the probability that no failures will be observed out of n is $P[R = 0] = e^{-n(\tau-\mu)/\theta}$. Thus, the joint likelihood function for μ and θ, say $L(\mu,\theta)$, is given by the above joint density function for any value $r = 0, 1, \ldots, n$. We note that for any fixed $\theta > 0$, $L(\mu,\theta)$ is an increasing function of μ, provided that $\mu \leq x_{1:n}$, but zero if $\mu > x_{1:n}$. Thus, the likelihood is maximized by $\hat{\mu} = x_{1:n}$, yielding Part 1 of the theorem. When $r > 0$, the value of θ which maximizes $L(\mu,\theta)$, with $\mu = \hat{\mu} = x_{1:n}$, can be obtained by equating the derivative of the log-likelihood to zero and solving for $\theta = \hat{\theta}$. In particular, Part 2 of the theorem follows by solving the equation

$$\frac{d}{d\theta} \ln L(\theta,\hat{\mu}) = -r/\theta + \left[\sum_{i=1}^{r} x_{i:n} + (n-r)\tau - nx_{1:n}\right] \Big/ \theta^2 = 0 .$$

Part 3 follows from Parts 1 and 2, and the invariance property of MLE's. This completes the proof. □

It can be shown, by inspection of the joint density function that $\sum_{i=1}^{R} X_{i:n}$, R and $X_{1:n}$ are joint minimal sufficient statistics. However, since there are three minimal sufficient statistics and only two parameters, no set of statistics can be complete. However, inference procedures can be derived, using an approach similar to that used with Type II censoring.

We first note that the marginal density function of $X_{1:n}$ is a mixture with both discrete and continuous parts,

$$\begin{aligned} f(x_{1:n}) &= \tfrac{n}{\theta}\, e^{-n(x_{1:n}-\mu)/\theta} && \text{if} \quad \mu \le x_{1:n} < \tau\,, \\ &= e^{-n(\tau-\mu)/\theta} && \text{if} \quad \tau \le x_{1:n}\,, \end{aligned}$$

and zero if $x_{1:n} < \mu$. Finally, as noted above, the number of observed failures R is binomial with $p = 1 - e^{-(\tau-\mu)/\theta}$. Thus, if $r \ge 2$, the conditional density function of $X_{2:n}, \ldots, X_{r:n}$ and $R = r$ given $X_{1:n} = x_{1:n}$ is

$$\begin{aligned} f(x_{2:n}, \ldots, x_{r:n}, r | x_{1:n}) &= \frac{f(x_{1:n}, x_{2:n}, \ldots, x_{r:n}, r)}{f(x_{1:n})} \\ &= \frac{\frac{n!}{(n-r)!\theta^r}\, e^{-\left[\sum_{i=1}^{r} x_{i:n} + (n-r)\tau - n\mu\right]/\theta}}{\frac{n}{\theta}\, e^{-n(x_{1:n}-\mu)/\theta}} \\ &= \frac{(n-1)!}{(n-r)!\theta^{r-1}}\, e^{-\left[\sum_{i=1}^{r} x_{i:n} + (n-r)\tau - n x_{1:n}\right]/\theta} \end{aligned}$$

if $x_{1:n} < x_{2:n} < \cdots < x_{r:n} < \tau$, with $r \ge 2$, which does not depend on μ. It also follows that, the conditional density function of the differences $X^*_{j:n-1} = X_{j+1:n} - X_{1:n}$ given $X_{1:n} = x_{1:n}$ is

$$\begin{aligned} & f(x^*_{1:n-1}, x^*_{2:n-1}, \ldots, x^*_{r-1:n-1}; \theta) \\ &= \frac{(n-1)!}{(n-r)!\theta^{r-1}}\, e^{-\left[\sum_{j=1}^{r-1} x^*_{j:n} + (n-r)(\tau - x_{1:n})\right]/\theta} \end{aligned}$$

for $0 < x^*_{1:n-1} < x^*_{2:n-1} < \cdots < x^*_{r-1:n-1} < \tau - x_{1:n}$, and zero otherwise. This has the same form as (6.2.15), the joint density function encountered in Section 6.2.3, in connection with Type I censored data from a one-parameter exponential distribution, replacing r with $r^* = r-1$, n with $n^* = n-1$, and τ with $\tau^* = \tau - x_{1:n}$. Consequently, the inference procedures for θ which were developed in Section 6.2.3 for use with Type I censored data from a one-parameter exponential distribution, can be applied with r, n and τ replaced with the corresponding starred quantities. When μ is an unknown nuisance parameter, and the above modification of the earlier methods is used, in effect, the loss is a single observation.

When θ is an unknown nuisance parameter, it is possible to construct inference procedures for μ. In this case, since R and $S = \sum_{i=1}^{r} X_{i:n}$ are jointly sufficient for θ when μ is known, it follows that θ can be eliminated by considering the conditional density function of $X_{1:n}$ given $R = r$ and $S = s$. This density function which was derived by Wright, Engelhardt, and Bain (1978), and a table of percentiles for the distribution of $Y = rX_{1:n}/\tau$ given $R = r$ and $S = s$ for the case $\mu = 0$ was given.

We want to be able to test $H_0 : \mu = 0$ versus $H_1 : \mu > 0$, and it is intuitively clear that H_0 should be rejected for large values of $y = rx_{1:n}/\tau$. If we wish to test the more general hypothesis $H_0 : \mu = \mu_0$ versus $H_1 : \mu > \mu_0$ for some $\mu_0 \ne 0$, then the above test can be applied after subtracting μ_0 from all the failure times and also from τ. Since the distribution of Y is quite complicated, the following approximation, based on a heuristic argument, was also proposed by Wright, Engelhardt, and Bain. We define a function

$$m(u) = u + 1 - \frac{1}{1 - e^{-1/u}}\,. \tag{6.3.6}$$

It can be shown that when $\mu = 0$, $S/(R\tau)$ converges with probability one to $m(\theta/\tau)$. Note that this is the mean of a truncated exponential distribution (6.2.8) with $\lambda = 1/\theta$. Also, (6.3.6) is an increasing function of u, which converges to zero as $u \to 0$, and converges to $1/2$ as $u \to \infty$. We now define the variable

$$V = \frac{rX_{1:n} - \mu}{\tau - \mu} . \tag{6.3.7}$$

The conditional distribution of V given $R = r$ and $S = s$ is, approximately, one-parameter exponential, $e(\lambda)$, with parameter $\lambda = [(\tau - \mu)/\theta^*]/\{1 - \exp[-(\tau - \mu)/\theta^*]\}$ where $\theta = \theta^*$ is the solution to the equation

$$m\left(\frac{\theta}{\tau - \mu}\right) = \frac{s - r\mu}{r(\tau - \mu)} , \tag{6.3.8}$$

which admits a solution if the right side is less than 1/2. This approximation, is somewhat simpler to use than the exact solution, and it is easily carried out with the use of a PC.

References

Bain, L.J. and Engelhardt, M. (1992). *Introduction to Probability and Mathematical Statistics*, Second edition, PWS-Kent, Boston.

Bain, L.J., Engelhardt, M., and Wright, F.T. (1977). Inferential procedures for the truncated exponential distribution, *Communications in Statistics*, **A2**, 103-112.

Bain, L.J. and Weeks, D.L. (1964). A note on the truncated exponential distribution, *Annals of Mathematical Statistics*, **35**, 1366-1367.

Barlow, R.E. and Proschan, F. (1975). *Statistical Theory of Reliability and Life Testing*, Holt, Reinhart and Winston, New York.

Davis, D.J. (1952). An analysis of some failure data, *Journal of the American Statistical Association*, **47**, 113-150.

Deemer, W.L. and Votaw, D.F. (1955). Estimation of parameters of truncated or censored exponential distributions, *Annals of Mathematical Statistics*, **26**, 498-504.

Engelhardt, M. and Bain, L.J. (1978). Tolerance limits and confidence limits on reliability for the two-parameter exponential distribution, *Technometrics*, **20**, 37-39.

Epstein, B. and Sobel, M. (1953). Life testing, *Journal of the American Statistical Association*, **48**, 486-502.

Guenther, W.C., Patil, S.A., and Uppuluri, V.R.R. (1976). One-sided β-content tolerance factors for the two parameter exponential distribution, *Technometrics*, **18**, 333-340.

Lehmann, E.L. (1959). *Testing of Statistical Hypotheses*, John Wiley & Sons, New York.

Sathe, Y.S. and Varde, S.D. (1969). Minimum variance unbiased estimation of reliability for the truncated exponential distribution, *Technometrics*, **11**, 609-612.

Wright, F.T., Engelhardt, M., and Bain, L.J. (1978). Inferences for the two-parameter exponential distribution under type I censored sampling, *Journal of the American Statistical Association*, **73**, 650-655.

CHAPTER 7

Inferences Under Two-Sample and Multi-Sample Situations

Gouri K. Bhattacharyya[1]

7.1 Introduction

Comparative life test experiments are of paramount importance when the object of a study is to ascertain the relative merits of two or more competing products in regard to the duration of their service life. This chapter presents methods for comparing the parameters of k (≥ 2) exponential distributions based on complete or censored samples from each population. The relevant backgrounds of this chapter such as the distributional properties of exponential order statistics and inferences for a single exponential population including parameter estimation, hypotheses tests and confidence intervals have been discussed in Chapters 2-5. With samples from two or several exponential populations, the inference procedures concerning single sample analysis are of course useful in drawing inferences separately for each of the populations. Our focus here is solely on comparisons which, in the setting of two-sample problems, involve testing the null hypothesis that there are no differences between the parameters of the two exponential distributions which model the life lengths of the two products under study. The related issues of setting confidence intervals for the ratio of the scale parameters as well as for the difference between the location parameters are also briefly discussed.

For expediency of making statistical inferences, life test experiments are often terminated before all units under test fail. A Type II (right) censored sample consists of the collection of the first r order statistics of a random sample of size n where r $(< n)$ is a prespecified number. In reliability contexts such a sample arises when n prototypes of a product are simultaneously put on a life test, and successive failure times are recorded till the occurrence of the rth failure. It was apparent in the preceding chapters that, for the exponential model, the structure of the inference procedures and the forms of the underlying sampling distributions are quite analogous between the situations of fully observed samples and Type II right-censored samples. A major portion of this chapter,

[1]University of Wisconsin, Madison, Wisconsin

namely, Sections 7.2, 7.3 and 7.4, is devoted to the treatment of inference procedures when each sample is either fully observed or is Type II censored. In fact, it is this sampling scheme that has led to the development of the bulk of the literature on two- and multi-sample inferences under the exponential model. The reason for this basically lies in the property of independence of the spacings of order statistics (seen earlier in Chapter 3) that lead to analytically tractable results for the sampling distributions of the estimators or the test statistics.

Section 7.2 introduces the basic notation in the context of k (≥ 2) independent Type II censored samples, and presents the fundamental results of sampling distributions that come for repeated use while developing the inference procedures in Sections 7.3 and 7.4. Section 7.3 is confined to the discussion of two-sample inference problems. It presents the details of hypotheses tests for equality of the parameters of two exponential populations. For generality, the setting of two-parameter exponential model involving both location parameter (μ) and scale parameter (θ) is considered. The hypotheses test problems are organized into several subsections according to the type of the parameter on which hypotheses are formed, and the assumptions made about the other parameter. Although emphasis is laid on hypotheses tests, confidence interval constructions readily follow from the test procedures. Even in the setting of two-sample problems, it turns out that in some cases, the sampling distribution of the likelihood ratio (LR) statistic is intractable. Along with the large-sample approximation of the LR statistics, some simple test procedures are also presented.

Section 7.4 forms a natural extension from the two-sample problems of Section 7.3 to the problems of comparison of k (≥ 3) exponential populations. Here the problem types are classified exactly as in Section 7.3. For most of the cases in Section 7.3, the exact sampling distributions of the relevant test statistics are available in simple closed forms, and standard tables such as χ^2 or F can be used to perform the tests. These nice features are generally lacking in the multi-sample testing problems. Consequently, one has to turn to approximations based on simulation or asymptotic expansions in order to determine the critical values for the LR tests. In some cases, alternative test procedures have been proposed for reasons of their simplicity.

Section 7.5 deals with two-sample inferences when Type II censoring is implemented on the two samples in a combined manner. Under this scheme, prototypes of the two products under study are put on life test at the same time and successive failure times and the corresponding product types are recorded. The experiment is terminated at a specified total number of failures. Although this scheme essentially amounts to Type II censoring of each sample, a major distinction is that the resulting samples become dependent. Consequently, the likelihood function, its properties, and the forms of the relevant sampling distributions are substantially different from, and in fact, considerably more complex than those for the case of independent censoring treated in Section 7.3. The joint censoring scheme is of practical significance in conducting comparative life tests of two products at the same facilities. While this sampling scheme has been the focal point in the development of optimal rank tests for censored data, it has received little attention in the area of parametric inferences. The available results concern only with two samples, and these are presented in Section 7.5. With other schemes of censoring, for instance, double Type II censoring or Type I censoring, the exact sampling distributions for two or multi-sample problems are by and large intractable. The relevant literature is referenced in Section 7.6.

7.2 Notation and Preliminaries

A two-parameter exponential model with probability density function

$$f(x;\mu,\theta) = \frac{1}{\theta}\, e^{-(x-\mu)/\theta}, \qquad \mu < x < \infty,\ \theta > 0,\ \mu > 0 \tag{7.2.1}$$

will be denoted by $e(\mu,\theta)$. In life testing and reliability contexts, the location parameter μ represents the threshold or the minimum assured life, while θ is a scale parameter. When $\mu = 0$, the model will be denoted simply as $e(\theta)$.

A Type II right censored sample, abbreviated as an (n,r)-sample, consists of the first r order statistics $X_{(1)} \leq X_{(2)} \leq \cdots \leq X_{(r)}$ of a random sample of size n from $e(\mu,\theta)$. For a real number a we denote

$$S(a) = \sum_{i=1}^{r}(X_{(i)} - a) + (n-r)(X_{(r)} - a), \qquad S = S(X_{(1)}) \tag{7.2.2}$$

so $S(\mu) = S + n(X_{(1)} - \mu)$.

The likelihood function for $\boldsymbol{X} = (X_{(1)},\ldots,X_{(r)})$ is then given by

$$L(\mu,\theta;\boldsymbol{X}) = \frac{n!}{(n-r)!}\,\theta^{-r}\exp[-\theta^{-1}\{S + n(X_{(1)} - \mu)\}] \tag{7.2.3}$$

for which $S(X_{(1)})$ and $X_{(1)}$ constitute complete sufficient statistics. The maximum likelihood estimates (MLE's) of θ and μ are $\hat{\theta} = S/r$ and $\hat{\mu} = X_{(1)}$, respectively, while the minimum variance unbiased estimates (MVUE's) are given by $\tilde{\theta} = S/(r-1)$ and $\tilde{\mu} = X_{(1)} - \tilde{\theta}/n$; see Chapters 4 and 5. The basic distributional results for order statistics from $e(\mu,\theta)$ entail that $X_{(1)}$ and $S(X_{(1)})$ are independent, $2n(X_{(1)} - \mu)/\theta \sim \chi^2_2$, and $2S(X_{(1)})/\theta \sim \chi^2_{2r-2}$.

If the location parameter μ is known, the model $e(\mu,\theta)$ can be reduced to $e(\theta)$ in terms of the observable random variable $X - \mu$ so μ can be taken to be 0 without loss of generality. In this case, a complete sufficient statistic for θ is $S(0) = \sum_{i=1}^{r} X_{(i)} + (n-r)X_{(r)}$, and $2S(0)/\theta \sim \chi^2_{2r}$.

In the context of independently Type II censored samples from k (≥ 2) exponential populations, we denote the observable data by $\boldsymbol{X} = (\boldsymbol{X}_1,\ldots,\boldsymbol{X}_k)$ where $\boldsymbol{X}_j = (X_{j(1)},\ldots,X_{j(r_j)})$ represents an (n_j,r_j) sample from $e(\mu_j,\theta_j)$, $j = 1,\ldots,k$. In analogy with (7.2.1), the relevant random variables in this case are:

$$\begin{aligned} S_j(\mu_j) &= \sum_{i=1}^{r_j}(X_{j(i)} - \mu_j) + (n_j - r_j)(X_{j(r_j)} - \mu_j)\,, \\ S_j &= S_j(X_{j(1)}), \qquad j = 1,\ldots,k\,. \end{aligned} \tag{7.2.4}$$

When both sets of parameters $\boldsymbol{\theta} = (\theta_1,\ldots,\theta_k)$ and $\boldsymbol{\mu} = (\mu_1,\ldots,\mu_k)$ are unknown and no constraints are imposed on them, the aforementioned one-sample results readily yield the following.

Theorem 7.1 *With independent (n_j,r_j) samples from $e(\mu_j,\theta_j)$, $j = 1,\ldots,k$,*

(a) The statistics S_j, $X_{j(1)}$, $j = 1,\ldots,k$, constitute complete sufficient statistics which are all independent.

(b) $2n_j(X_{j(1)} - \mu_j) \sim \chi_2^2$, $2S_j/\theta_j \sim \chi^2_{2r_j-2}$, *and* $2S_j(\mu_j)/\theta_j \sim \chi^2_{2r_j}$.

(c) *The MLE's are* $\hat{\theta}_j = S_j/r_j$, $\hat{\mu}_j = X_{j(1)}$ *while the MVUE's are* $\tilde{\theta}_j = S_j/(r_j - 1)$, $\tilde{\mu}_j = X_{j(1)} - \tilde{\theta}_j/n_j$, $j = 1, \ldots, k$.

For the case of equal location parameters, a few other distributional results will be needed for two- or multi-sample comparison problems. These are provided in Theorem 7.2 with some additional notation which we first introduce. Let

$$\begin{aligned} X_\bullet &= \min(X_{j(1)},\ j = 1, \ldots, k),\ A_j = [X_{j(1)} = X_\bullet],\ p_j = P(A_j)\ , \\ W_j &= n_j(X_{j(1)} - X_\bullet),\ V_j = 2W_j/\theta_j,\ j = 1, \ldots, k\ . \end{aligned} \tag{7.2.5}$$

Note that $X_\bullet$ is the combined sample minimum, $V_j = 0$ on the set A_j, and $V_j > 0$ on A_j^c.

Theorem 7.2 *Under the assumption that* $\mu_1 = \cdots = \mu_k$, *the following results hold:*

(a)
$$p_t = (n_t/\theta_t) / \sum_{j=1}^{k} (n_j/\theta_j),\ t = 1, \ldots, k\ . \tag{7.2.6}$$

(b) *Conditionally given* A_t, *the* $k-1$ *random variables* $\{V_j,\ 1 \le j \le k,\ j \ne t\}$ *are independent, and each is distributed as* χ_2^2.

(c) *Conditionally given* A_t, *the* $k-1$ *random variables*

$$-2(r_j - 1)\log[S_j/(S_j + W_j)], \qquad 1 \le j \le k,\ j \ne t \tag{7.2.7}$$

are independent, and each is distributed as χ_2^2.

Proof: For notational simplicity, we consider the case $t = k$, and assume that the common $\mu = 0$.

(a) Letting $Y = \min(X_{j(1)},\ j = 1, \ldots, k-1)$, we note that $A_k = [X_{k(1)} < Y]$. Because Y is the minimum of $k-1$ independent exponentials with scale parameters θ_j/n_j, $j = 1, \ldots, k-1$, we have that

$$Y \sim e\left(1 \Big/ \sum_{j=1}^{k-1} (n_j/\theta_j)\right)\ .$$

Also, $X_{k(1)} \sim e(\theta_k/n_k)$ and $X_{k(1)}$ is independent of Y. The stated expression for $p_k = P[X_{k(1)} < Y]$ is now readily obtained by integration.

(b) Let the conditional joint distribution function of $V_1, \ldots, V_{k-1}$, given A_k, be denoted by $G(v_1, \ldots, v_{k-1}|A_k)$, and let the corresponding density be $g(v_1, \ldots, v_{k-1}|A_k)$. Furthermore, let H_j and h_j respectively denote the distribution and density of $X_{j(1)}$, that is,

$$\begin{aligned} H_j(y) &= 1 - \exp(-n_j y/\theta_j)\ , \\ h_j(y) &= (n_j/\theta_j)\exp(-n_j y/\theta_j), \qquad 0 < y < \infty\ . \end{aligned}$$

We then have

$$\begin{aligned} & G(w_1, \ldots, w_{k-1}|A_k) \\ &= P[2n_j\theta_j^{-1}(X_{j(1)} - X_{k(1)}) \le w_j,\ 1 \le j \le k-1, \text{ and } A_k]\ p_k^{-1} \\ &= p_k^{-1} P[0 \le X_{k(1)} \le X_{j(1)} \le X_{k(1)} + \theta_j w_j/(2n_j),\ 1 \le j \le k-1] . \end{aligned} \tag{7.2.8}$$

By means of conditioning on the variable $X_{k(1)}$, the RHS of (7.2.8) can be written as

$$p_k^{-1} \int_0^\infty \left[\prod_{j=1}^{k-1} \{H_j(y + \theta_j w_j)/(2n_j)) - H_j(y)\} \right] h_k(y)\, dy .$$

Differentiating with respect to $w_1, \ldots, w_{k-1}$ and using the expressions for $h_j(y)$'s we obtain the density

$$\begin{aligned} & g(w_1, \ldots, w_{k-1}|A_k) \\ &= p_k^{-1} \left[\prod_{j=1}^{k-1} \frac{\theta_j}{2n_j} \right] \int_0^\infty \left[\prod_{j=1}^{k-1} h_j \left(y + \frac{\theta_j w_j}{2n_j} \right) \right] h_k(y)\, dy \\ &= p_k^{-1} \frac{1}{2^{k-1}} \exp\left[-\sum_{j=1}^{k-1} w_j/2 \right] \int_0^\infty \exp\left[-y \sum_{j=1}^{k} n_j/\theta_j \right] dy \\ &= \frac{1}{2^{k-1}} \exp\left[-\sum_{j=1}^{k-1} w_j/2 \right], \qquad 0 \le w_j < \infty , \end{aligned}$$

where the last equality obtains from the result in Part (a).

(c) Since W_j's and S_j's are independent, they are also conditionally independent given A_k. From Part (b) we have that, given A_k, $2W_j/\theta_j$, $j = 1, \ldots, k-1$ are i.i.d. χ_2^2. Also $2S_j/\theta_j \sim \chi^2_{2r_j-2}$, $j = 1, \ldots, k-1$, and these are independent of the W_j's. Consequently, given A_k, we have

$$U_j \equiv S_j/(S_j + W_j) \sim \text{beta}(r_j - 1, 1)$$

so $-2(r_j - 1) \log U_j \sim \chi_2^2$.

□

For the special case $k = 2$, the results (a) and (b) in Theorem 7.2 were given by Epstein and Tsao (1953). The multisample extensions as well as the result (c) are due to Hsieh (1986).

7.3 Two-Sample Inferences Under Independent Type II Censoring

When comparing two exponential populations $e(\mu_j, \theta_j)$, $j = 1, 2$, various hypotheses testing problems concerning the two sets of parameters $\boldsymbol{\theta} = (\theta_1, \theta_2)$ and $\boldsymbol{\mu} = (\mu_1, \mu_2)$

are of interest. In a seminal paper, Epstein and Tsao (1953) laid out a comprehensive list of these formulations, and provided the structures of the likelihood ratio (LR) test. Subsequently, other authors investigated further on the sampling distributions and power properties of the LR tests and advanced other procedures in some cases where the LR test is difficult to implement.

Concerned with Type II censored data, we denote an (n_j, r_j) sample from an $e(\mu_j, \theta_j)$ population by the order statistics vector $\boldsymbol{X}_j = (X_{j(1)}, \ldots, X_{j(r_j)})$, $j = 1, 2$. Following the notation of Section 7.2, the likelihood function of $\boldsymbol{X} = (\boldsymbol{X}_1, \boldsymbol{X}_2)$ is given by

$$L(\boldsymbol{\xi}, \boldsymbol{X}) \propto \theta_1^{-r_1}\theta_2^{-r_2} \exp\left[-\sum_{j=1}^{2} \theta_j^{-1}\{S_j + n_j(X_{j(1)} - \mu_j)\}\right] \tag{7.3.1}$$

where $\boldsymbol{\xi} = (\boldsymbol{\mu}, \boldsymbol{\theta})$. For testing $H_0 : \boldsymbol{\xi} \in \Omega_0$, a proper subset of an assumed parameter space Ω, the likelihood ratio test rejects H_0 for small values of

$$\lambda = \sup_{\Omega_0} L(\boldsymbol{\xi}; \boldsymbol{X}) \,/\, \sup_{\Omega} L(\boldsymbol{\xi}; \boldsymbol{X}) = L(\hat{\boldsymbol{\xi}}_0)/L(\hat{\boldsymbol{\xi}}) \tag{7.3.2}$$

where $\hat{\boldsymbol{\xi}}_0$ and $\hat{\boldsymbol{\xi}}$ denote the MLE's of $\boldsymbol{\xi}$ over Ω_0 and Ω, respectively. In the following subsections, the testing problems are classified according to the assumed Ω, and the nature of the null hypothesis that is to be tested. The test procedures are described in the context of two-sided alternatives unless specifically stated otherwise. The derivations are straightforward and their details are therefore omitted.

7.3.1 Tests Concerning the Scale Parameters

Here we consider the problem of testing $H_0 : \theta_1 = \theta_2$ when the location parameters μ_1 and μ_2 are known, and when they are unknown. In the first case, one can take $\mu_1 = \mu_2 = 0$ without loss of generality by simply subtracting μ_j from each X_{ji}.

Case (a): $\mu_1 = \mu_2 = 0$

The likelihood (7.3.1) reduces to

$$\theta_1^{-r_1}\theta_2^{-r_2} \exp[-\theta_1^{-1}S_{10} - \theta_2^{-1}S_{20}] \tag{7.3.3}$$

where

$$S_{j0} = \sum_{i=1}^{r_j} X_{j(i)} + (n_j - r_j)X_{j(r_j)}, \qquad j = 1, 2\ .$$

The unrestricted MLE's are $\hat{\theta}_j = S_{j0}/r_j$, $j = 1, 2$ while the MLE of the common θ under H_0 is given by $\hat{\theta}_0 = (S_{10} + S_{20})/(r_1 + r_2)$. It follows that the LR statistic λ is proportional to

$$S_{10}^{r_1} S_{20}^{r_2} [S_{10} + S_{20}]^{-(r_1+r_2)}$$

and, therefore, a function of

$$T_1 = (S_{10}/r_1)(S_{20}/r_2)^{-1}$$

which has an $F_{2r_1, 2r_2}$ distribution under H_0. Consequently, the LR test reduces to a two-sided F-test based on the test statistic T_1. The critical values are to be determined from the condition of size (α) and the equality of λ at the two critical values. An equal-tail F-test is usually performed for convenience.

In the present case, the underlying model conforms to a two-parameter exponential family. A uniformly most powerful unbiased test exists and is based on the same test statistic T_1. The conditions for determining the critical values can be obtained by an application of the generalized Neyman-Pearson lemma. Also, the fact that $(\theta_2/\theta_1)T_1 \sim F_{2r_1,2r_2}$, a $100(1-\alpha)\%$ confidence interval for θ_1/θ_2 can be easily obtained in the form

$$\frac{T_1}{c_2} \le \frac{\theta_1}{\theta_2} \le \frac{T_1}{c_1}$$

where c_1 and c_2 are the lower α_1-point and the upper α_2-point of the $F_{2r_1,2r_2}$ distribution, and $\alpha_1 + \alpha_2 = \alpha$. Based on the general theory for exponential families, one can determine the split of α into α_1 and α_2 in order to arrive at a uniformly most accurate unbiased confidence interval. However, the equal split $\alpha_1 = \alpha_2 = \alpha/2$ is a convenient approximation that is most commonly used in practice.

Case (b): μ_1 and μ_2 are unknown

Referring to the likelihood (7.3.1), the MLE's of μ_1 and μ_2 are $X_{1(1)}$ and $X_{2(1)}$, respectively, both under Ω and Ω_0 while the respective MLE's of θ_j are $\hat{\theta}_j = S_j/r_j$ and $\hat{\theta}_0 = (S_1+S_2)/(r_1+r_2)$ where $S_j = \sum_{i=1}^{r_j}(X_{ji} - X_{j(1)}) + (n_j - r_j)(X_{j(r_j)} - X_{j(1)})$, $j = 1, 2$ as defined in (7.2.4). Here the LR statistic is proportional to

$$S_1^{r_1} S_2^{r_2} (S_1+S_2)^{-(r_1+r_2)} ,$$

and a rejection region of the form $\lambda \le c$ is equivalent to a two-sided rejection region based on the statistic

$$T_2 = [S_1/(r_1-1)][S_2/(r_2-1)]^{-1}$$

which has an $F_{2r_1-2,2r_2-2}$ distribution under H_0. Confidence intervals for θ_1/θ_2 can be obtained by using the pivotal $(\theta_2/\theta_1)T_2 \sim F_{2r_1-2,2r_2-2}$.

7.3.2 Tests Concerning the Location Parameters

Here we consider the problem of testing $H_0 : \mu_1 = \mu_2$ under various assumptions about the scale parameters θ_1 and θ_2.

Case (a): θ_1 and θ_2 are known

The likelihood (7.3.1), in this case, is a function of μ_1 and μ_2 while θ_1 and θ_2 are known constants. The unrestricted MLE's of the location parameters are $\hat{\mu}_j = X_{j(1)}$, $j = 1, 2$ while, under H_0, the common μ has the MLE $\hat{\mu}_0 = X_\bullet = \min(X_{1(1)}, X_{2(1)})$. The LR statistic reduces to $\lambda = \exp(-\frac{1}{2}\, V)$ where

$$\begin{aligned} V &= 2n_1(X_{1(1)} - X_{2(1)})/\theta_1 && \text{if } X_{1(1)} > X_{2(1)} \\ &= 2n_2(X_{2(1)} - X_{1(1)})/\theta_2 && \text{if } X_{2(1)} > X_{1(1)}\,. \end{aligned} \tag{7.3.4}$$

Theorem 7.2 entails that under H_0, the conditional distribution of V given the event $X_{1(1)} > X_{2(1)}$, is χ_2^2, and given the event $X_{2(1)} > X_{1(1)}$ we also have $V \sim \chi_2^2$. Also, noting that the upper α-point of χ_2^2 equals $-2\log\alpha$, the LR test of $H_0 : \mu_1 = \mu_2$, in this case, becomes equivalent to the rule: reject H_0 if either V_1 or $V_2 \geq -2\log\alpha$ where $V_1 = 2n_1(X_{1(1)} - X_{2(1)})/\theta_1$ and $V_2 = 2n_2(X_{2(1)} - X_{1(1)})/\theta_2$.

In the case of one-sided alternatives, say $H_1 : \mu_1 > \mu_2$, the two-sided procedure described above is to be modified as follows: The rejection region of a level α test is set as $V_1 \geq c$ with c determined from the condition $P[V_1 \geq c] = \alpha$. From Theorem 7.2, we have (for $c > 0$)

$$\begin{aligned} P[V_1 \geq c] &= P[V_1 \geq c | A_2]P(A_2) \\ &= p_2(\boldsymbol{\theta})\exp(-c/2) \end{aligned}$$

where $p_2(\boldsymbol{\theta}) = (n_2/\theta_2)[(n_1/\theta_1) + (n_2/\theta_2)]^{-1}$. Consequently, the critical value for a right sided level α test is given by $c = -2\log[\alpha/p_2(\boldsymbol{\theta})]$ provided, of course, $p_2(\boldsymbol{\theta}) > \alpha$ which would ordinarily be the case. Likewise, the critical value $(-c')$ for a left-sided level α test appropriate for $H_1 : \mu_1 < \mu_2$ derives from the condition

$$P[V_1 \leq -c'] = p_1(\boldsymbol{\theta})\exp[-\theta_1 n_2 c'/(2\theta_2 n_1)] = \alpha \ .$$

Combining the two one-sided test procedures, a $100(1-\alpha)\%$ confidence interval for $(\mu_1 - \mu_2)$ can be obtained as

$$(X_{1(1)} - X_{2(1)}) - c_1(\alpha_1) < \mu_1 - \mu_2 < (X_{1(1)} - X_{2(1)}) + c_2(\alpha_2)$$

where $\alpha_1 + \alpha_2 = \alpha$ and

$$\begin{aligned} c_1(\alpha_1) &= -(\theta_1/n_1)\log[\alpha_1/p_2(\boldsymbol{\theta})] \ , \\ c_2(\alpha_2) &= -(\theta_2/n_2)\log[\alpha_2/p_1(\boldsymbol{\theta})] \ . \end{aligned} \tag{7.3.5}$$

Alternatively, one can also invert the acceptance region of the LR test to arrive at a confidence interval for $\mu_1 - \mu_2$. For further details, see Bain and Engelhardt (1991).

Case (b): θ_1 and θ_2 are unknown

The likelihood (7.3.1) in this case is to regarded as a function of all four parameters. The unrestricted MLE's of the parameters are

$$\hat{\mu}_j = X_{j(1)}, \ \hat{\theta}_j = S_j/r_j, \qquad j = 1, 2 \ ,$$

while the MLE's under H_0 are given by

$$\hat{\mu}_0 = X_\bullet, \ \hat{\theta}_{j0} = (S_j + W_j)/r_j, \qquad j = 1, 2$$

where $W_j = n_j(X_{j(1)} - X_\bullet)$. The LR statistic has the form

$$\lambda = \prod_{j=1}^{2}[S_j/(S_j + W_j)]^{r_j} \ . \tag{7.3.6}$$

Theorem 7.2 entails that, under $H_0 : \mu_1 = \mu_2$,

$$P[-2\log\lambda \leq y] = p_2(\boldsymbol{\theta})G[(r_1 - 1)y/r_1] + p_1(\boldsymbol{\theta})G[(r_2 - 1)y/r_2] \tag{7.3.7}$$

where $G(\cdot)$ denotes the cumulative distribution of χ_2^2. If $r_1 = r_2$, the right side of (7.3.7) simplifies to $G[(r_1-1)y/r_1]$ because $p_1(\boldsymbol{\theta}) + p_2(\boldsymbol{\theta}) = 1$. Since this distribution is free of the nuisance parameter $\boldsymbol{\theta}$, an exact level α test obtains from the LR criterion.

If $r_1 \neq r_2$, the LR test cannot be used as the null distribution of the test statistic depends on $\boldsymbol{\theta}$. Epstein and Tsao (1953) proposed a simple modification by replacing the power r_j by $(r_j - 1)$ in expression (7.3.6). The resulting test is equivalent to one based on the test statistic

$$M = -2\sum_{j=1}^{2}(r_j - 1)\log[S_j/(S_j + W_j)] \, . \tag{7.3.8}$$

It has an exact χ_2^2 distribution under H_0 so the rejection region of a level α test is given by $M \geq \chi_{2,\alpha}^2$.

The non-null distribution of this modified LR statistic M was investigated by Hsieh (1986) who established that the power function of the test depends on $(\boldsymbol{\mu}, \boldsymbol{\theta})$ only through the two ratios $n_1\theta_2/(n_2\theta_1)$ and $n_2(\mu_2 - \mu_1)/\theta_2$. Also, the test is unbiased and asymptotically optimal in Bahadur sense.

Case (c): $\theta_1 = \theta_2$, the common value θ unknown

Here the LR statistic takes the form

$$\lambda = \left[1 + \frac{RT}{d}\right]^{-(r_1+r_2)}$$

where

$$\begin{aligned} T &= \frac{d|X_{1(1)} - X_{2(1)}|}{S_1 + S_2}, \qquad d = r_1 + r_2 \, , \\ R &= n_1 \text{ if } X_{1(1)} > X_{2(1)} \\ &= n_2 \text{ if } X_{1(1)} < X_{2(1)} \, . \end{aligned} \tag{7.3.9}$$

From Theorem 7.2, it follows that, under H_0, $P[R = n_1] = n_2/(n_1+n_2)$ and that (RT) has the $F_{2,2d}$ distribution. The resulting level α test is given by the rejection rule:

$$\begin{array}{rlll} \text{Reject } H_0 & \text{if} & X_{1(1)} > X_{2(1)} & \text{and} \quad n_1 T \geq F_{2,2d,\alpha} \\ \text{or} & \text{if} & X_{1(1)} < X_{2(1)} & \text{and} \quad n_2 T \geq F_{2,2d,\alpha} \, . \end{array} \tag{7.3.10}$$

Kumar and Patel (1971) proposed a test based on the statistic T alone by dropping the multiplier R which is random. The null distribution of T is given by the density

$$f_T(t) = \frac{n_1 n_2}{n_1 + n_2}\left[\left(1 + \frac{n_2 t}{d}\right)^{-(d+1)} + \left(1 + \frac{n_1 t}{d}\right)^{-(d+1)}\right], \quad 0 < t < \infty$$

and if $n_1 = n_2$, then $n_1 T \sim F_{2,2d}$. Tables of critical values of this test were provided by Kumar and Patel (1971). In the case of equal sample sizes $n_1 = n_2$, the KP test is equivalent to the LR test. Weinman et al. (1973) derived an expression for the power function of the KP test and compared the powers of this test and the LR test. They found that, for two-sided H_1, the KP test is generally more (less) powerful than the LR test when $\mu_1 < \mu_2$ ($\mu_1 > \mu_2$), and the differences in the powers are much more prominent when $\mu_1 > \mu_2$ in which case the LR test is considerably better. In the case of a two-sided test, it would be appropriate to use the LR test to guard against a serious loss of power.

7.3.3 Tests Concerning Both Location and Scale Parameters

Here we consider the problem of testing $H_0 : \theta_1 = \theta_2,\ \mu_1 = \mu_2$ against the alternative $H_1 : \theta_1 \neq \theta_2$ or $\mu_1 \neq \mu_2$. Referring to the likelihood (7.3.1), we have the MLE's

$$\begin{aligned} \hat{\mu}_j &= X_{j(1)},\ \hat{\theta}_j = S_j/r_j,\ j = 1,2\ , \\ \hat{\mu}_0 &= X_\bullet,\ \hat{\theta}_0 = (S_1 + S_2 + W) \end{aligned}$$

where $W = W_1 + W_2$. Note that W equals W_1 or W_2 according as $X_{1(1)} >$ or $< X_{2(1)}$. The LR statistic is given by

$$\lambda = U \prod_{j=1}^{2} (r/r_j)^{r_j}$$

where

$$U = \frac{S_1^{r_1} S_2^{r_2}}{(S_1 + S_2 + W)^r}\ , \qquad r = r_1 + r_2$$

and, consequently, the LR rejection region is of the form $U \leq c$.

For the case of complete samples, Sukhatme (1936) derived the LR statistic and provided a rather complicated expression for its density under H_0. In the more general case of Type II censoring, Hseih (1981) established that, under H_0,

$$\begin{aligned} P[U \leq u] &= 1 + I_a(r_1 - 1, r_2 - 1) - I_b(r_1 - 1, r_2 - 1) \\ &\quad + [B(r_1 - 1, r_2 - 1)]^{-1} u^{(r-2)/r} \int_a^b x^{(2r_1/r)-2}(1 - x)^{(2r_2/r)-2}\, dx\ , \\ &\qquad \text{for } 0 \leq u \leq (r_1/r)^{r_1}(r_2/r)^{r_2} \end{aligned} \tag{7.3.11}$$

where $a \leq b$ are the roots of $x^{r_1}(1 - x)^{r_2} = u$, $I_c(\cdot, \cdot)$ denotes the incomplete beta integral, and $B(\cdot, \cdot)$ denotes the beta function. Critical values of the LR test can be obtained from this distribution function by using numerical methods.

Hsieh (1981) showed that a simple yet quite accurate approximation to the critical value of the LR test can be obtained by adapting Box's χ^2-approximation to the present case. With this approximation, the LR test is given by the rejection region $-2 \log \lambda \geq \rho^{-1} \chi^2_{3,\alpha}$, where

$$\rho = 1 - \frac{13}{18} \left[\frac{1}{r_1} + \frac{1}{r_2} - \frac{1}{r} \right] .$$

It was found that for the nominal levels of $\alpha = .10,\ .05$ or $.01$, and moderate or large sample sizes, this procedure yields fairly accurate results even when r_1 and r_2 are small or extremely unbalanced. Also, the accuracy improves when r_1 and r_2 are close to each other.

In view of the difficulties with an exact determination of the critical values of the LR test, two other test procedures have been proposed in the literature. They are both based on the principle of decomposing the global null hypothesis into subhypotheses that are easy to test by standard methods. However, they differ according to the manner in which the subhypotheses tests are eventually integrated. The method due to Hogg and Tanis (1963) separates the null hypothesis $H_0 : \theta_1 = \theta_2,\ \mu_1 = \mu_2$ into two nested hypotheses

$$\begin{aligned} H_{01} &: \quad \theta_1 = \theta_2 \\ H_{02} &: \quad \theta_1 = \theta_2,\ \mu_1 = \mu_2\ . \end{aligned}$$

First, the hypothesis H_{01} is tested by the statistic [cf. Section 7.3.1, case (b)]

$$T_2 = [S_1/(r_1 - 1)][S_2/(r_2 - 1)]^{-1}$$

at a level of significance α_1. If H_{01} is not rejected, the assumption $\theta_1 = \theta_2$ is incorporated into the model, and the null hypothesis of $\mu_1 = \mu_2$ is tested at a level α_2 by means of [cf. Section 7.3.2, case (c)] the test statistic RT which has an $F_{2,2d}$ distribution under H_0. Theorem 7.2 entails that, under H_0, these test statistics are independent so a level $\alpha = 1 - (1 - \alpha_1)(1 - \alpha_2)$ test is obtained by rejecting H_0 if and only if a rejection results at either of these steps. With α initially specified, a reasonable choice of the subhypotheses levels would be $\alpha_1 = \alpha_2 = 1 - (1 - \alpha)^{1/2}$.

The other test, due to Perng (1978), employs Fisher's method of combination of independent P-values obtained from the statistics T_2 and RT. Specifically, letting G_1 and G_2 denote the distribution functions of $F_{2r_1-2,2r_2-2}$ and $F_{2,2d}$ distributions, respectively, the proposed test rejects H_0 if

$$-2\log[1 - G_2(RT)] - 2\log[1 - G_1(T_2)] \geq \chi^2_{4,\alpha} \ .$$

The LR test as well as Perng's test are asymptotically optimal in the sense that they both attain the exact Bahadur slope; the exact slope for the iterated procedure of Hogg and Tanis remains unknown. Hsieh (1981) conducted a simulation study to compare the finite-sample power properties of these tests. His results indicate that none of the three tests emerges as the uniformly best. The iterated procedure has the highest power when the difference between location parameters is small and the difference between the scale parameters is large. Otherwise, the LR test generally does better than the other two procedures.

7.4 Comparison of Several Populations Under Independent Type II Censoring

In this section, extensions of the preceding results are discussed in the context of comparison of more than two exponential populations. For the setting of independent Type II censored sample from k (≥ 3) exponential populations, we denote the observations as $\boldsymbol{X} = (\boldsymbol{X}_1, \ldots, \boldsymbol{X}_k)$ where $\boldsymbol{X}_j = (X_{j(1)}, \ldots, X_{j(r_j)})$ is the vector of the first r_j order statistics of a random sample of size n_j from an $e(\mu_j, \theta_j)$ distribution, $j = 1, \ldots, k$, and the $\boldsymbol{X}_j$'s are independent. Following the notation of Section 7.1, the likelihood function is given by

$$L(\boldsymbol{\xi}, \boldsymbol{X}) \propto \left[\prod_{j=1}^{k} \theta_j^{-r_j}\right] \exp\left[-\sum_{j=1}^{k} \theta_j^{-1}\{S_j + n_j(X_{j(1)} - \mu_j)\}\right] \tag{7.4.1}$$

where $\boldsymbol{\xi} = (\boldsymbol{\mu}, \boldsymbol{\theta})$, $\boldsymbol{\theta} = (\theta_1, \ldots, \theta_k)$, and $\boldsymbol{\mu} = (\mu_1, \ldots, \mu_k)$.

Several hypotheses testing problems are addressed here, and these are classified along the lines of Section 7.3. A general difficulty with most of the multi-sample testing problems is that although the derivation of the LR statistics is fairly straightforward, their null distributions do not simplify to allow the use of standard tables. Consequently, numerical approximations, asymptotic expansions, modification of the likelihood function, or construction of alternative test procedures have been explored in the various test settings.

7.4.1 Tests Concerning the Scale Parameters

Here we consider the problem of testing

$$H_0 : \theta_1 = \cdots = \theta_k \textit{ versus } H_1 : \text{ not all } \theta_j\text{'s are equal}$$

when the location parameters μ_j's are either assumed known or are considered unknown. If the μ_j's are known, they can be taken to be all zero without loss of generality.

Case (a): Assumption $\mu_1 = \cdots = \mu_k = 0$

As in Case (a) of Section 7.3.1, the likelihood (7.4.1) reduces to

$$L(\boldsymbol{\theta}; \boldsymbol{X}) \propto \left[\prod_{j=1}^{k} \theta_j^{-r_j}\right] \exp\left[-\sum_{j=1}^{k} \theta_j^{-1} S_{j0}\right] \tag{7.4.2}$$

for which the MLE's are $\hat{\theta}_j = S_{j0}/r_j$, $j = 1, \ldots, k$. The MLE of the common θ under H_0 is given by $\hat{\theta}_0 = S_0/r$ where

$$S_0 = \sum_{j=1}^{k} S_{j0}, \qquad r = \sum_{j=1}^{k} r_j \ .$$

The LR statistic λ is then given by

$$\lambda = \left[\prod_{j=1}^{k} \hat{\theta}_j^{r_j}\right] \left[\sum_{j=1}^{k} r_j \hat{\theta}_j / r\right]^{-r} \propto \prod_{j=1}^{k} [S_{j0}/S_o]^{r_j} \ . \tag{7.4.3}$$

The rejection region $-2\log\lambda \geq \chi^2_{k-1,\alpha}$ will provide an approximate level α test when the r_j's are large. However, a modification can be made to improve the accuracy of the approximation. Note that $S_{j0}/\theta_j \sim \chi^2_{2r_j}$, $j = 1, \ldots, k$, and they are independent so the structure of λ has a direct analogy with that of the LR statistic when testing equality of the variances of k normal populations. In the latter case, the role of Bartlett's correction in improving the accuracy of χ^2 approximation is well known. The same adjustment in the present context leads to the rejection region $-2\log\lambda \geq c\chi^2_{k-1,\alpha}$, where

$$c = 1 + \frac{1}{3(k-1)} \left[\sum_{j=1}^{k} \frac{1}{2r_j} - \frac{1}{2r}\right] .$$

For the special case of equal n_j's and no censoring, Nagarsenkar (1980) obtained a structural representation of the distribution of λ in a computational form, and also developed an asymptotic expansion justifying a beta approximation. Similar developments for Type II censored samples are reported in Nagarsenkar and Nagarsenkar (1988).

Case (b): The μ_j's are unknown

As in Case (b) Section 7.3.1, the LR statistic in the k-sample case is proportional to $\prod_{j=1}^{k}[S_j/S]^{r_j}$, where the components S_j are independent and $2S_j/\theta_j \sim \chi^2_{2r_j-2}$. In order to justify use of the Bartlett correction to $-2\log\lambda$ by drawing a correspondence with the LR statistic for testing homogeneity of variances in normal samples, we need to replace the power r_j by $r_j - 1$ and take the modified LR statistic in the form

$$\lambda' = \prod_{j=1}^{k} \left(\frac{S_j}{r_j - 1} \right)^{r_j - 1} \Big/ \left(\frac{S}{r - k} \right)^{r-k} . \tag{7.4.4}$$

This corresponds to maximization of the marginal likelihood of $\boldsymbol{\theta}$:

$$L' = \left[\prod_{j=1}^{k} \theta_j^{-(r_j - 1)} \right] \exp\left[- \sum_{j=1}^{k} \theta_j^{-1} S_j \right] .$$

Noting that the null distribution of λ' has a direct correspondence with that of the LR statistic for normal model variance test, the marginal likelihood ratio test with Barlett correction (MLB) will have a rejection region of the form $-2 \log \lambda' \geq c' \chi^2_{k-1,\alpha}$, where

$$c' = 1 + \frac{1}{3(k-1)} \left[\sum_{j=1}^{k} \frac{1}{2(r_j - 1)} - \frac{1}{2(r-k)} \right] . \tag{7.4.5}$$

Besides the MLB test, three other alternative test procedures were proposed by Thiagaraja and Paul (1990). One is based on the quadratic statistic

$$Q = \sum_{j=1}^{k} (r_j - 1)[(\tilde{\theta}_j - \tilde{\theta})^2 / \tilde{\theta}]$$

where $\tilde{\theta}_j = S_j/(r_j - 1)$ and $\tilde{\theta} = S/(r-k)$ are the bias-corrected MLE's. Under H_0, Q is asymptotically χ^2_{k-1}. Another test statistic was constructed by taking the ratio of the extremal scale parameter estimates:

$$ESP = \max_{1 \leq j \leq k} \tilde{\theta}_j / \min_{1 \leq j \leq k} \tilde{\theta}_j \ ,$$

and a third was based on the $C(\alpha)$ statistic corresponding to the profile likelihood function. With an extensive simulation study of these four competing tests, Thiagaraja and Paul (1990) report that by and large, the MLB test attains the nominal level of significance most accurately for a wide range of sample sizes and censoring combinations. Also, in terms of power, it performs best or as good as others in most situations.

7.4.2 Tests Concerning the Location Parameters

Here we consider the problem of testing

$$H_0 : \mu_1 = \cdots = \mu_k \ \textit{versus} \ H_1 : \text{ not all } \mu_j\text{'s are equal}$$

under various assumptions on the scale parameters.

Case (a): $\theta_j,\ j = 1, \ldots, k$ are known

Treating the likelihood (7.4.1) as a function of $\boldsymbol{\mu}$ alone with $\boldsymbol{\theta}$ known, we obtain the MLE's $\hat{\mu}_j = X_{j(1)},\ j = 1, \ldots, k$, while, under H_0, the common μ has the MLE $\hat{\mu} = X_\bullet = \min\{X_{j(1)},\ j = 1, \ldots, k\}$. The *LR* criterion yields the expression

$$-2\log\lambda = \sum_{j=1}^{k} 2n_j(X_{j(1)} - X_\bullet)/\theta_j \ . \tag{7.4.6}$$

From Theorem 7.2 it follows that, conditionally given the event $[X_{t(1)} = X_\bullet]$, the $k-1$ random variables $2n_j(X_{j(1)} - X_\bullet)/\theta_j,\ 1 \le j \le k,\ j \ne t$, and i.i.d. χ_2^2 under H_0. Consequently, the null distribution of $-2\log\lambda$ is χ^2_{2k-2}, and this can be the basis for an exact level α test.

Case (b): θ_j's are unknown

Referring to the likelihood (7.4.1), the unrestricted MLE $(\hat{\boldsymbol{\mu}}, \hat{\boldsymbol{\theta}})$ and the MLE $(\hat{\boldsymbol{\mu}}_0, \hat{\boldsymbol{\theta}}_0)$ under H_0 are respectively given by

$$\begin{aligned} &\hat{\mu}_j = X_{j(1)}, \quad \hat{\theta}_j = S_j/r_j \ , \\ &\hat{\mu}_0 = X_\bullet, \qquad \hat{\theta}_{j0} = (S_j + W_j)/r_j,\ j = 1, \ldots, k, \end{aligned} \tag{7.4.7}$$

where $W_j = n_j(X_{j(1)} - X_\bullet)$. These lead to the *LR* statistic

$$\lambda = \prod_{j=1}^{k} [S_j/(S_j + W_j)]^{r_j} \ . \tag{7.4.8}$$

For the special case of $k = 2$, the null distribution of $-2\log\lambda$ has been given in (7.3.7) followed by a discussion of why an exact level α test cannot be based on this statistic. The same situation prevails in the multisample case. From Theorem 7.2 it follows that

$$P[-2\log\lambda \le y] = \sum_{t=1}^{k} p_t(\boldsymbol{\theta}) P\left[\left\{\sum_{\substack{j=1 \\ j\ne t}}^{k} r_j Z_j/(r_j - 1)\right\} \le y\right] \tag{7.4.9}$$

where $Z_1, \ldots, Z_k$ are i.i.d. χ_2^2, and $p_t(\boldsymbol{\theta})$ is as given in (7.2.6). If the r_j's are all equal, the right hand side of (7.4.9) simplifies to the cumulative distribution of χ^2_{2k-2} evaluated at $(r_0 - 1)y/r_0$ with r_0 denoting the common value of the r_j's. Consequently, $-2[(r_0 - 1)/r_0]\log\lambda \sim \chi^2_{2k-2}$ so an exact level α test is given by the rejection region

$$-2\log\lambda \ge [r_0/(r_0 - 1)]\chi^2_{2k-2,\alpha} \ .$$

In the case that the r_j's are not all equal, Hsieh (1986) proposed the modified *LR* statistic λ' by replacing the power r_j in (7.4.8) by $r_j - 1$, thus generalizing the two-sample modified *LR* test due to Epstein and Tsao (1953) to k samples. Specifically, the test statistic in this general case can be taken as

$$-2\log\lambda' = -2\sum_{j=1}^{k}(r_j - 1)\log[S_j/(S_j + W_j)] \ , \tag{7.4.10}$$

and the rejection region $-2\log\lambda' \geq \chi^2_{2k-2,\alpha}$ provides an exact level α test. This last statement follows from Theorem 7.2, Part (c) by using the relation

$$P[-2\log\lambda' \leq y] = \sum_{t=1}^{k} P(A_t)P[-2\log\lambda' \leq y|A_t]$$

and the fact that conditionally given A_t, $-2\log\lambda'$ is distributed as χ^2_{2k-2}. Hsieh (1986) further established that this modified LR test is consistent and asymptotically optimal in Bahadur sense under the assumptions that as the total sample size $n = \sum_{j=1}^{k} n_j \to \infty$, we have $0 < \lim(n_j/n) < 1$ and $0 < \lim(r_j/r) < 1$ for all $j = 1, \ldots, k$.

Case (c): $\theta_1 = \cdots = \theta_k$, the common θ unknown

By replacing the θ_j's in (7.4.1) by a common θ, we note that the likelihood function in the present case is proportional to

$$\theta^{-r}\exp[-S/\theta - \sum_{j=1}^{k} n_j(X_{j(1)} - \mu_j)/\theta] \,.$$

The unrestricted MLE's and the MLE's under H_0 are respectively given by

$$\begin{aligned} \hat{\mu}_j &= X_{j(1)}, \quad \hat{\theta} = S/r, \; j = 1, \ldots, k, \\ \hat{\mu}_0 &= X_{\bullet}, \quad\;\; \hat{\theta}_0 = \tfrac{1}{r}\textstyle\sum_{j=1}^{k}(S_j + W_j). \end{aligned} \tag{7.4.11}$$

These lead to the LR statistic

$$\lambda = [\hat{\theta}_0/\hat{\theta}]^{-r} = [1 + W/S]^{-r} \tag{7.4.12}$$

where $W = \sum_{j=1}^{k} W_j$. Employing the relevant results of Theorem 7.1 and 7.2, it follows that, under H_0, $2W/\theta \sim \chi^2_{2k-2}$, $2S/\theta \sim \chi^2_{2r-2k}$, and that W and S are independent. Consequently, the LR test reduces to an F-test based on the statistic

$$\frac{(r-k)W}{(k-1)S} \sim F_{2k-2,2r-2k} \,. \tag{7.4.13}$$

Incidentally, in the case of no censoring, the likelihood ratio F-test with degrees of freedom $(2k-2, 2n-2k)$ was derived by Sukhatme (1937). The results for the censored case were obtained by Khatri (1974) and Singh (1983).

7.4.3 Tests Concerning Both Location and Scale Parameters

Based on Type II censored samples from two-parameter exponential populations $e(\mu_j, \theta_j)$, $j = 1, \ldots, k$, we consider here the problem of testing the global null hypothesis

$$H_0 : \theta_1 = \cdots = \theta_k, \; \mu_1 = \cdots = \mu_k$$

against the alternative that either the θ_j's or the μ_j's are not all equal.

In analogy with the development in Section 7.3.3, the LR statistic has the form

$$\lambda = U\prod_{j=1}^{k}(r/r_j)^{r_j} \,,$$

where

$$U = \prod_{j=1}^{k} \left[\frac{S_j}{S+W} \right]^{r_j}$$

and small values of λ constitute the rejection region. For the special case of $k = 2$, the exact distribution of U was given in (7.3.11) and a χ^2 approximation for $-2\log\lambda$ was found to be effective. However, for $k \geq 3$, the exact distribution of λ turns out to be intractable. Singh and Narayan (1983) developed a simple and quite accurate approximation in terms of an F-distribution by employing an asymptotic expansion, due to Box (1949), concerning the distribution of a random variable whose moments involve gamma functions. In the present case, the genesis of the approximation rests on the fact that the random variables

$$Y_j = \left[\sum_{t=1}^{j} S_t \right] \Big/ \left[\sum_{t=1}^{j+1} S_t \right], \; j = 1, \ldots, k \; ,$$

where $S_{k+1} = S + W$, are independent with $Y_j \sim \text{beta}\left(\sum_{t=1}^{j} (r_t - 1), r_{j+1} - 1 \right)$. Moments of the LR statistic λ are then readily obtained in terms of ratios of gamma functions by using the relation

$$U = \prod_{j=1}^{k} \left\{ (1 - Y_{j-1}) \prod_{t=j}^{k} Y_t \right\} .$$

The resulting F-distribution approximation applies either to the statistic $M \equiv -2\log\lambda$ or to a ratio of the form $aM(b-M)^{-1}$ depending on the sample sizes. For details, see Singh and Narayan (1983).

In Section 7.3.3, we described some simple test procedures that are based on a decomposition of the global null hypothesis into sub-hypotheses for which standard tests as well as exact null distributions are available. Although our discussion there was in the confine of a two-sample comparison problem, the underlying ideas are general enough to permit an extension of these procedures to the k-sample problem.

The applicability of the Hogg and Tanis (1963) approach is restricted to the situation where there is an ordering of the k populations such that the investigator would be interested in testing the equality of the first two populations, then the equality of the first three populations, and so on, in a sequential manner. With both sets of parameters $\boldsymbol{\theta}$ and $\boldsymbol{\mu}$ unknown, this sequencing needs to be further refined in terms of including an additional θ_j or μ_j in the testing problem at the jth stage. Specifically, the nested sub-hypotheses are considered in the following order:

$$\begin{aligned}
H_{2,1} &: \; \theta_1 = \theta_2 \\
H_{2,2} &: \; \theta_1 = \theta_2, \; \mu_1 = \mu_2 \\
H_{3,2} &: \; \theta_1 = \theta_2 = \theta_3, \; \mu_1 = \mu_2 \\
H_{3,3} &: \; \theta_1 = \theta_2 = \theta_3, \; \mu_1 = \mu_2 = \mu_3 \\
\cdots & \quad \cdots \\
H_{k,k} &: \; \theta_1 = \cdots = \theta_k, \; \mu_1 = \cdots = \mu_k \; .
\end{aligned} \tag{7.4.14}$$

Letting

$$\begin{aligned}
&Z_\alpha = \min\{X_{1(1)}, \ldots, X_{\alpha(1)}\}, \quad W_{j\alpha} = n_j(X_{j(1)} - Z_\alpha), \; N_\alpha = \textstyle\sum_{j=1}^{\alpha} n_j \; , \\
&T_\alpha = \textstyle\sum_{j=1}^{\alpha} (S_j + W_{j\alpha}), \qquad U_\alpha = N_{\alpha-1} Z_{\alpha-1} + n_\alpha X_{\alpha(1)} - N_\alpha Z_\alpha \; ,
\end{aligned} \tag{7.4.15}$$

the test statistics employed in testing these subhypotheses are $K_2, J_2, K_3, J_3, \ldots, K_k, J_k$, respectively, where

$$\begin{aligned} K_\alpha &= \frac{S_\alpha/(2r_\alpha - 2)}{\sum_{j=1}^{\alpha-1}(S_j + W_{j\alpha}) \Big/ \left[\sum_{j=1}^{\alpha-1}(2r_j) - 2\right]} , \\ J_\alpha &= \frac{U_\alpha/2}{\left[\sum_{j=1}^{\alpha-1}(S_j + W_{j\alpha}) + S_\alpha\right] \Big/ \left(2\sum_{j=1}^{\alpha} r_j - 4\right)} , \quad \alpha = 2, 3, \ldots, k . \end{aligned} \tag{7.4.16}$$

Under the global null hypothesis, these test statistics have mutually independent F distributions so the target level of significance α can be prorated to the components by using the product rule. This sequential procedure is terminated as soon as a subhypothesis is rejected, otherwise the decision is not to reject the global null hypothesis. Evidently, in situations where a natural ordering of the populations exist, this method can result in a saving in the sample size which could be substantial when a rejection occurs at an early stage. On the other hand, if no ordering can be justified a-priori, the method would seem rather arbitrary and also prone to substantial loss of power compared to a single-stage procedure.

Another simple test of the global null hypothesis, advanced by Perng (1978), in the two-sample context, rests on Fisher's method of combining the P-values that result from the testing of $H_0 : \theta_1 = \cdots = \theta_k$ [cf. Section 7.4.1 Case (b)] and the testing of $H_0 : \mu_1 = \cdots = \mu_k$ assuming that $\theta_1 = \cdots = \theta_k$ [cf. Section 7.4.2 Case (c)]. It is asymptotically optimal in the Bahadur sense [see Singh (1985)] but its finite sample performance vis-a-vis the other two procedures has not been studied.

In Subsections 7.4.1-7.4.3 our focus was primarily on the development of the LR test for specific hypotheses testing problems regarding location and/or scale parameters. In some cases, the exact sampling distribution of the test statistic is tractable while in other cases it is not, and consequently, approximations drawn from asymptotic theory are used to determine the critical region. For some problems, other competing test procedures were also described. Their motivation draws from considerations of simplicity of the method and control of the Type I error probability. In all cases, there are no finite sample optimality results that could identify any test to be superior to its competitors. However, a theorem due to Kourouklis (1988) establishes that for a variety of testing problems under the exponential model, the LR test is asymptotically optimal in the Bahadur sense under very general conditions. In order to describe this optimality result, we consider the general form of a k-sample testing problem

$$H_0 : \boldsymbol{\xi} \in \Omega_0 \ \textit{versus} \ H_1 : \boldsymbol{\xi} \in \Omega - \Omega_0$$

where $\boldsymbol{\xi} = (\boldsymbol{\mu}, \boldsymbol{\theta})$, Ω is a subset of the positive orthant of the $2k$-dimensional Euclidean space, and Ω_0 is a proper subset of Ω. Let the likelihood function (7.4.1) corresponding to k independent Type II censored samples be denoted by $L_n(\boldsymbol{\xi})$, and let

$$\delta_n = \sup_{\boldsymbol{\xi} \in \Omega} L_n(\boldsymbol{\xi}) \Big/ \sup_{\boldsymbol{\xi} \in \Omega_0} L_n(\boldsymbol{\xi})$$

where $n = \sum_{j=1}^{k} n_j$. Assume that as $n \to \infty$, the following limits hold:

$$n_j/n \to \alpha_j, \qquad r_j/r \to \beta_j, \qquad r_j/n \to \gamma_j$$

with $0 < \alpha_j,\ \beta_j < 1,\ 0 < \gamma_j \le 1$ for $j = 1, \ldots, k$, and $\Sigma\alpha_j = \Sigma\beta_j = 1$. Then we have the following result.

Theorem 7.3 [Kourouklis (1988)]. *Under the assumption that*

$$\liminf n^{-1} \log \delta_n \ge J(\theta) \text{ with probability } 1$$

when θ obtains in $\Omega - \Omega_0$, and $2J(\theta)$ is the optimal Bahadur slope at θ, the LR test based on δ_n is optimal at all $\theta \in \Omega - \Omega_0$ with slope $2J(\theta)$.

In some individual testing problems, this optimality property of the LR test was also demonstrated by Samanta (1986).

7.5 Two-Sample Inferences Under Joint Type II Censoring

The preceding sections were concerned with inferences for two or more exponential populations based on independent samples which are either fully observed or Type II censored independently of each other. When comparing the life lengths of two products A and B, the scheme of independent Type II censoring refers to a life test experiment in which m prototypes of product A are simultaneously put on a life test and successive failure times are observed until a prespecified number r_1 of failures occur $(1 \le r_1 \le m)$. Likewise, n prototypes of product B are tested simultaneously, and a prespecified number r_2 $(1 \le r_2 \le n)$ of successive failure times are recorded. In this process, the two sets of observations are independent and the number of observations is predetermined for each group.

Another important design of comparative life test experiments is based on the following sampling scheme: m prototypes of product A and n prototypes of product B are simultaneously put on a life test; successive failure times as well as the type of the product that fails on each occasion are recorded, and observation is terminated at the occurrence of the rth failure where r is a predetermined number $1 \le r \le m + n$. In the case of no censoring, that is $r = m + n$, the data can be separated into two sets representing independent samples of size m for A and n for B so the methods described in Section 7.3 would readily apply. However, in the presence of censoring, the data can still be separated as censored samples of A-failures and B-failures but these censored samples would no longer be independent. In the course of the developments that follow, it will be transparent that the structure of the likelihood as well as the relevant sampling distributions are considerably different between the cases of independent censoring and joint censoring. Accordingly, the methodology of Section 7.3 need to be considerably modified. In order to avoid confusion, it will be convenient to introduce new notation rather than maintain similarities with the preceding sections.

Suppose the lifetimes $X_1, \ldots, X_m$ of m prototypes of product A are i.i.d. random variables with distribution $F(x)$ and density $f(x)$, and the lifetimes $Y_1, \ldots, Y_n$ of n prototypes of product B are i.i.d. random variables with distribution $G(x)$ and density $g(x)$. Let $W_1 \le \cdots \le W_N$ denote the order statistics of the $N = m+n$ random variables $\{X_1, \ldots, X_m, Y_1, \ldots, Y_n\}$. Under the joint Type II censoring scheme, the observable data consist of $(\boldsymbol{Z}, \boldsymbol{W})$ where $\boldsymbol{W} = (W_1, \ldots, W_r)$, r is a fixed integer $< N$, and $Z_i = 1$

or 0 according as W_i is an X or an Y. Letting M_r denote the number of X's in $\boldsymbol{W}$ and $N_r = N - M_r$, the likelihood of $(\boldsymbol{Z}, \boldsymbol{W})$ is

$$m!n![(m - M_r)!(n - N_r)!]^{-1} \prod_{i=1}^{r} \{f^{Z_i}(W_i) g^{1-Z_i}(W_i)\}$$
$$\times [\overline{F}(W_r)]^{m-M_r} [\overline{G}(W_r)]^{n-N_r} , \qquad (7.5.1)$$

where $\overline{F} = 1 - F$ and $\overline{G} = 1 - G$. Incidentally, this likelihood has been the key element in the development of rank tests due to Rao, Savage, and Sobel (1960), Johnson and Mehrotra (1972), and Basu (1968). Our concern here is with parametric procedures under the exponential model $\overline{F}(x) = \exp(-x/\theta_1)$, $\overline{G}(x) = \exp(-x/\theta_2)$, $0 < x < \infty$, for which the likelihood (7.5.1) reduces to

$$L(\boldsymbol{\theta}; \boldsymbol{Z}, \boldsymbol{W}) \propto \theta_1^{-M_r} \theta_2^{-N_r} \exp[-U_1/\theta_1 - U_2/\theta_2] , \qquad (7.5.2)$$

where

$$U_1 = \sum_{i=1}^{r} Z_i W_i + (m - M_r) W_r = \sum_{i=1}^{M_r} X_{(i)} + (m - M_r) W_r ,$$
$$U_2 = \sum_{i=1}^{r} (1 - Z_i) W_i + (n - N_r) W_r = \sum_{j=1}^{N_r} Y_{(j)} + (n - N_r) W_r , \qquad (7.5.3)$$

and $X_{(\cdot)}$ and $Y_{(\cdot)}$ denote the order statistics of the X- and Y-sample, respectively. Note that U_1 represents the total time on test ($TTOT$) of the A-items and U_2 that of the B-items, when observation is terminated at the random time W_r.

The likelihood (7.5.2) constitutes an exponential family with three sufficient statistics (M_r, U_1, U_2) whereas the parameter space is only two dimensional. Consequently, in contrast with the case of separate Type II censoring, the standard theory of optimal tests and associated confidence sets does not apply. Inference procedures developed by Bhattacharyya and Mehrotra (1981) and Mehrotra and Bhattacharyya (1982) are summarized in this section.

First we address the problem of testing the null hypothesis $H_0 : \theta_1 = \theta_2$. Since this problem is invariant under common scale transformations to the X's and Y's, the principle of invariant tests requires that a test be based on the maximal invariants M_r and U_1/U_2 or equivalently on

$$T_1 = M_r/r, \qquad T_2 = U_1/(U_1 + U_2) . \qquad (7.5.4)$$

Observe that T_1 represents the proportion of failure counts and T_2 the proportion of $TTOT$ of the X-sample to the pooled X- and Y-samples. Either T_1 or T_2 individually, or some function of T_1 and T_2 can be employed as a test statistic. In the case of no censoring, T_2 provides the UMP unbiased as well as UMP invariant test for the exponential model. On the other hand, the statistic T_1 has the property of robustness that T_2 lacks. In fact, it provides a simple and distribution-free test for the general model of Lehmann alternatives $\overline{G}(x) = [\overline{F}(x)]^{\Delta}$ of which our exponential model with $\Delta = \theta_1/\theta_2$ is just a special case. The basic distributional properties of T_1 and T_2 are listed here separately for the *null case*, the case of *fixed alternatives*, and the case of *local alternatives*.

(a) The null case: $\Delta = \theta_1/\theta_2 = 1$

The distribution of $M_r = rT_1$ is hypergeometric with the probability mass function

$$
\begin{aligned}
P[M_r = m_r] &= \binom{m}{m_r}\binom{n}{n_r}\Big/\binom{N}{r} \\
&\qquad\qquad \text{for } \max(r-n,0) \le m_r \le \min(m,r) \\
E[T_1] &= \frac{m}{N}\,, \qquad \mathrm{Var}(T_1) = \frac{mn(N-r)}{rN^2(N-1)}\,. \qquad (7.5.5)
\end{aligned}
$$

Under the scheme of separate censoring, the statistic corresponding to T_2 has a beta distribution but that does not hold for joint censoring. A characterization of the null distribution of T_2 is available in terms of a linear function of Dirichlet random variables, and from that characterization it follows that

$$
\begin{aligned}
E[T_2] &= \frac{m}{N}\,, \\
\mathrm{Var}(T_2) &= \frac{mn}{N(N-1)(r+1)}\left[\frac{2N-r-1}{N} - \frac{2(N-r)}{r}\sum_{i=1}^{r}(N-i+1)^{-1}\right], \\
\mathrm{Cov}(T_1,T_2) &= \frac{mn(N-r)}{N^2r(n-1)}\left[1-\frac{N}{r}\sum_{i=1}^{r}(N-i+1)^{-1}\right]. \qquad (7.5.6)
\end{aligned}
$$

(b) The non-null case - fixed alternatives

With $\Delta = \theta_1/\theta_2$ fixed, the probability mass function of M_r is

$$
\begin{aligned}
P_\Delta[M_r = m_r] &= \binom{m}{m_r}\binom{n}{n_r}[m-m_r+\Delta(n-n_r)] \\
&\quad \times \int_0^1 u^{m-m_r+\Delta(n-n_r)-1}(1-u)^{m_r}(1-u^{\Delta})^{n_r}\,du\,. \qquad (7.5.7)
\end{aligned}
$$

The distribution of T_2 in this case turns out to be intractable.

(c) The non-null case - local alternatives

Assuming that as $N \to \infty$, $m/N \to \lambda$, $0 < \lambda < 1$, and $r/N \to p$, $0 < p < 1$, the asymptotic joint distribution of $\sqrt{N}[(T_1-\lambda),(T_2-\lambda)]$ under the sequence of local alternatives $\Delta_N = 1 + h/\sqrt{N}$, is bivariate normal with the mean vector $-\lambda(1-\lambda)h\boldsymbol{\xi}$ and covariance matrix $\lambda(1-\lambda)\boldsymbol{\tau}$, where

$$
\begin{aligned}
\boldsymbol{\xi}' &= p^{-1}(\zeta q,\ \zeta q - p), \qquad \zeta = -\ln q, \qquad q = 1-p\,, \\
\boldsymbol{\tau} &= p^{-2}\begin{pmatrix} pq & q(p-\zeta) \\ q(p-\zeta) & p(2-p)-2\zeta q \end{pmatrix}. \qquad (7.5.8)
\end{aligned}
$$

The test statistic T_1 has several advantages: its null distribution is simple, and the availability of its exact non-null distribution permits a calculation of the power of the test thus facilitating a determination of the required sample sizes prior to conducting the comparative life test experiment. Furthermore, the test is robust in the sense that it is a distribution-free test, and a calculation of its power based on expression (7.5.7) holds for general Lehmann alternatives. Also, it requires a simple data base (the failure counts) rather than the actual failure times that are required by T_2.

Using the limiting distributions of T_1 and T_2 under local alternatives, the Pitman asymptotic relative efficiency of T_1 with respect to T_2 turns out to be

$$ARE(T_1 : T_2) = \zeta^2 q[p(2-p) - 2\zeta q][p(\zeta q - p)^2]^{-1} \ . \tag{7.5.9}$$

As a function of p, the proportion of uncensored observations, the ARE (7.5.9) is monotone decreasing, goes to zero as $p \to 1$ and to ∞ as $p \to 0$. The value of the ARE exceeds 1 for $p \leq .7$. This means that for censoring proportions of 30% or higher, the test T_1 is asymptotically more power-efficient than T_2.

In view of the complexity of the distribution of T_2, there is little justification for the construction of a test based on T_2 alone. This is in sharp contrast with the situation of separate Type II censoring where T_2 not only has a simple distribution but it also provides the UMP unbiased test. In the case of joint censoring, the simple test T_1 should be used in the presence of moderate to heavy censoring. With light censoring, an optimal combination of T_1 and T_2 will be no more difficult to use than the use of T_2 alone. From the aforementioned joint limiting distribution, it follows that the combination of T_1 and T_2 that maximizes the Pitman ARE among all invariant tests is given by $T = T_1 - T_2$. Moreover, it is found that the test based on T is asymptotically power-equivalent to the likelihood ratio test which is based on the test statistic

$$\frac{\hat{L}_{H_0}}{\hat{L}} = \hat{\theta}_1^{M_r} \hat{\theta}_2^{N_r} \hat{\theta}_0^{-r}$$

where $\hat{\theta}_1 = U_1/M_r$, $\hat{\theta}_2 = U_2/N_r$, and $\hat{\theta}_0 = (U_1 + U_2)/r$. Consequently, the two-sided test based on the limiting normal distribution of T is equivalent to the likelihood ratio test in large samples.

We now turn to the problem of testing the more general null hypothesis $H_0 : \Delta = \Delta_0$ where Δ_0 is not necessarily 1, and also discuss the related problem of setting confidence intervals for $\Delta = \theta_1/\theta_2$. A reduction of the problem of testing $H_0 : \Delta = \Delta_0$ to that of testing $H_0 : \Delta = 1$ by rescaling one of the samples is possible when we are dealing with either separate Type II censoring or no censoring at all. This is not applicable in the situation of joint censoring because such a rescaling alters the observable vector $(\boldsymbol{Z}, \boldsymbol{W})$ in an undetermined manner. Formulation of tests and associated confidence procedures are to be again based on the likelihood function (7.5.2) which involves the sufficient statistics (M_r, U_1, U_2). However, unlike the case of $\Delta_0 = 1$, invariance considerations do not permit any further reduction of the data.

An exact confidence procedure for Δ is available in terms of the failure count statistic M_r. Referring to the distribution (7.5.7), the corresponding distribution function is given by

$$\begin{aligned} a(i|\Delta) &= P_\Delta[M_r \leq i] \\ &= \Delta \sum_{j=0}^{i} \binom{m}{j} \binom{n}{r-i} (r-i) \\ &\quad \cdot \int_0^1 u^{q(j,i,\Delta)} (1-u)^j (1-u^\Delta)^{r-i-1} \, du \ , \\ &\quad \max\{0, r-n\} \leq i \leq \min\{r, m\} \ , \end{aligned} \tag{7.5.10}$$

where

$$q(j, i, \Delta) = m - j + \Delta(n - r + i + 1) - 1 .$$

For a given observed value $M_r = m_r$, a $100(1-\alpha)\%$ lower confidence bound for Δ is the solution $\underline{\Delta}(m_r)$ of the equation

$$a(m_r|\Delta) = \alpha .$$

This follows from the fact that for fixed i, the function $a(i|\Delta)$ is monotone strictly increasing in Δ. The lower confidence bound $\underline{\Delta}(m_r)$ can be computed by means of numerical integration using expression (7.5.9). Also, because of an inherent symmetry of our problem, the same program can be used to determine an upper confidence bound $\overline{\Delta}(m_r)$. Specifically, if we denote the lower confidence bound $\underline{\Delta}(m_r) = \underline{\Delta}(m_r; m, n)$ explicitly as a function of the sample sizes m and n, then a $100(1-\alpha)\%$ upper confidence bound for Δ satisfies the relation

$$\overline{\Delta}(m_r; m, n) = [\underline{\Delta}(r - m_r; n, m]^{-1} .$$

These confidence bounds are conservative in the sense that the actual confidence is no less than the nominal probability.

With increasing sample sizes computation of the exact confidence bounds or two-sided confidence intervals becomes more involved. An approximation can be readily obtained from the limiting normal distribution of M_r/N appropriately centered and scaled. In particular, a large-sample approximate $100(1-\alpha)\%$ confidence interval for Δ based on the failure count M_r is of the form $h(M_r/N) \pm z_{\alpha/2}\tilde{\tau}/\sqrt{N}$ where $z_{\alpha/2}$ denotes the upper $\alpha/2$ point of the $N(0,1)$ distribution, $h(\cdot)$ is a function such that $\sqrt{N}[h(M_r/N) - \Delta]$ is asymptotically $N(0, \tau^2)$, and $\tilde{\tau}$ is a consistent estimate of τ.

The aforementioned confidence procedures based only on the failure count M_r are to be recommended when the amount of censoring is moderate to heavy. In cases of light censoring, asymptotic results suggest that shorter confidence intervals would typically result from the use of the approximate normal distribution of $\hat{\Delta} = \hat{\theta}_1/\hat{\theta}_2$, the ratio of the MLE's. In particular, a $100(1-\alpha)\%$ large-sample confidence interval for Δ is given by

$$\hat{\Delta}[1 + z_{\alpha/2}(M_r N_r/r)^{-1/2}]^{-1} < \Delta < \hat{\Delta}[1 - z_{\alpha/2}(M_r N_r/r)^{-1/2}]^{-1} .$$

Numerical comparisons of the asymptotic lengths of the confidence intervals suggest that the procedure based on M_r incurs very little loss of efficiency in situations where censoring is heavy. Even with light censoring, high relative efficiency of M_r is realized when the true mean lifetimes are considerably different and a larger size sample is taken from the population with a smaller mean. Note that in this latter situation, the joint censoring scheme is likely to lead to a gross imbalance in the numbers of observations from the two populations.

7.6 Miscellaneous Other Results

The sampling distributions required for inference under the exponential model are relatively simple and tractable for the scheme of Type II right censoring applied independently to the individual samples. Some complications arising from joint censoring were discussed in Section 7.5. Another form of censoring that has been considered in the context of two- and multi-sample inferences is called "double censoring" under which predetermined numbers of order statistics are censored at both ends. Specifically, letting $X_{(1)} \leq \cdots \leq X_{(n)}$ denote the order statistics of a random sample of size n, a doubly Type II censored sample, henceforth referred to as an (n, r, s) sample, consist of the observations $(X_{(r+1)}, \ldots, X_{(n-s)})$.

Several authors have considered the problem of testing equality of the location parameters of two or more exponential populations assuming that the scale parameters are equal but unknown. Based on independent (n_j, r_j, s_j) samples

$$(X_{j(r_j+1)}, \ldots, X_{j(n_j-s_j)})$$

from $e(\mu_j, \theta)$, $j = 1, 2$, Shetty and Joshi (1987, 1989) developed the LR test for testing $H_0 : \mu_1 = \mu_2$. Let

$$\begin{aligned}
Y_1 &= X_{1(r_1+1)} , \qquad Y_2 = X_{2(r_2+1)} , \qquad m_j = n_j - r_j - s_j , \\
U_j &= \sum_{i=r_j+1}^{n_j-s_j} X_{j(i)} + s_j X_{j(n_j-s_j)} - (n_j - r_j) X_{j(r_j+1)} , \qquad j = 1, 2 , \\
U &= U_1 + U_2 , \qquad m = m_1 + m_2 .
\end{aligned}$$

Then the MLE's of the parameters are

$$\begin{aligned}
\hat{\mu}_j &= Y_j - \hat{\theta} \log[n_j/(n_j - r_j)], & j = 1, 2 , \\
\hat{\theta} &= [U - (n_2 - r_2)(Y_2 - Y_1)]/m & \text{if } Y_2 > Y_1 \\
&= [U - (n_1 - r_1)(Y_1 - Y_2)]/m & \text{if } Y_2 < Y_1 .
\end{aligned}$$

Under H_0 however, the MLE's $\hat{\mu}_0$ and $\hat{\theta}_0$ are not obtainable in closed form. One of the two estimating equations has be to solved iteratively. Consequently, the LR statistic λ takes on a rather complex form. For some special cases, the critical values and power of λ were obtained by Shetty and Joshi (1989) by means of simulation.

Tiku (1981) proposed a simple test statistic

$$(m - 2)|Y_1 - Y_2|/U$$

and derived its null distribution in an easily computable form. Khatri (1981) derived the non-null distribution for computation of power. The corresponding distributions for one-sided tests were obtained by Shetty and Joshi (1986). Based on an extensive simulation study, Shetty and Joshi (1989) reported that neither of these two tests uniformly dominates the other, and overall, the LR test fares better. Kambo and Awad (1985) constructed a k-sample extension of Tiku's test, obtained the null distribution of the test statistic, and tabled some critical values. For the k-sample double-censoring situation, Shetty and Joshi (1987) derived the best linear unbiased estimators and maximum likelihood estimators of the location parameters $\mu_j, j = 1, \ldots, k$, and the common scale parameter θ, and studied their relative efficiencies.

Type I or fixed time censoring is another important censoring scheme for which a fair amount of literature is available for single-sample inference problems under the exponential model [see for instance, Spurrier and Wei (1980)]. For two- or multi-sample inference problems, no exact sampling distributions are available. With large samples, inferences can be handled by means of the usual asymptotic distributions of the maximum likelihood estimate and the likelihood ratio test statistic; see Chapter 3 of Lawless (1982).

Besides formal hypotheses tests for comparing exponential populations, graphical methods via data plots on special probability papers are also useful for visual comparisons and also for detecting possible violations of the model assumptions. These are discussed and illustrated in Chapter 10 of Nelson (1982).

Concerned with estimation of the ratio of the scale parameters of two exponential distributions with unknown scale parameters, Madi and Tsui (1990a) showed that the best affine equivariant estimator is inadmissible under a large class of loss functions, and they constructed some improved estimators. For other work on estimation, see Madi and Tsui (1990b) and Chiou (1990).

References

Bain, L.J. and Engelhardt, M. (1991). *Statistical Analysis of Reliability and Life Testing Models*, Second edition, Marcel Dekker, New York.

Basu, A.P. (1968). On a generalized Savage statistic with applications to life testing, *Annals of Mathematical Statistics*, **39**, 1591-1604.

Bhattacharyya, G.K. and Mehrotra, K.G. (1981). On testing equality of two exponential distributions under combined Type II censoring, *Journal of the American Statistical Association*, **76**, 886-894.

Box, G.E.P. (1949). A general distribution theory for a class of likelihood criteria, *Biometrika*, **36**, 317-346.

Chen, H.J. (1982). A new range statistic for comparison of several exponential location parameters, *Biometrika*, **69**, 257-260.

Chiou, P. (1990). Estimation of scale parameters of two exponential distributions, *IEEE Transactions on Reliability*, **39**, 106-109.

Epstein, B. and Tsao, C.K. (1953). Some tests based on ordered observations from two exponential populations, *Annals of Mathematical Statistics*, **24**, 456-466.

Hogg, R.V. and Tanis, E.A. (1963). An iterated procedure for testing the equality of several exponential distributions, *Journal of the American Statistical Association*, **58**, 435-443.

Hsieh, H.K. (1981). On testing the equality of two exponential distributions, *Technometrics*, **23**, 265-269.

Hsieh, H.K. (1986). An exact test for comparing location parameters of k exponential distributions with unequal scales based on Type II censored data, *Technometrics*, **28**, 157-164.

Johnson, R.A. and Mehrotra, K.G. (1972). Locally most powerful rank tests for the two-sample problem with censored data, *Annals of Mathematical Statistics*, **43**, 823-831.

Kambo, N.S. and Awad, A.M. (1985). Testing equality of location parameters of k exponential distributions, *Communications in Statistics - Theory and Methods*, **14**, 567-583.

Khatri, C.G. (1974). On testing the equality of location parameters in k censored exponential distributions, *Australian Journal of Statistics*, **16**, 1-10.

Khatri, C.G. (1981). Power of a test for location parameters of two exponential distributions, *Aligarh Journal of Statistics*, **1**, 8-12.

Kourouklis, S. (1988). Asymptotic optimality of likelihood ratio tests for exponential distributions under Type II censoring, *Australian Journal of Statistics*, **30**, 111-114.

Kumar, S. and Patel, H.I. (1971). A test for the comparison of two exponential distributions, *Technometrics*, **13**, 183-189.

Lawless, J.F. (1982). *Statistical Models & Methods for Lifetime Data*, John Wiley & Sons, New York.

Madi, M. and Tsui, K.-W. (1990a). Estimation of the ratio of the scale parameters of two exponential distributions with unknown location parameters, *Annals of the Institute of Statistical Mathematics*, **42**, 77-87.

Madi, M. and Tsui, K.-W. (1990b). Estimation of the common scale of several exponential distributions with unknown locations, *Communications in Statistics - Theory and Methods*, **19**, 2295-2313.

Mehrotra, K.G. and Bhattacharyya, G.K. (1982). Confidence intervals with jointly Type-II censored samples from two exponential distributions, *Journal of the American Statistical Association*, **77**, 441-446.

Nagarsenkar, B.N. and Nagarsenkar, P.B. (1988). Non-null distribution of a modified likelihood ratio criterion associated with exponential distribution, *ASA Proceedings of Statistical Computing Section*, 288-290.

Nagarsenkar, P.B. (1980). On a test of equality of several exponential survival distributions, *Biometrika*, **67**, 475-478.

Nagarsenkar, P.B. and Nagarsenkar, B.N. (1988). Asymptotic distribution of a test of equality of exponential distributions, *ASA Proceedings of Statistical Computing Section*, 284-287.

Perng, S.K. (1978). A test for the equality of two exponential distributions, *Statistica Neerlandica*, **32**, 93-102.

Rao, V., Savage, I.R., and Sobel, M. (1960). Contribution to the theory of rank order statistics: The two sample censored case, *Annals of Mathematical Statistics*, **31**, 415-426.

Samanta, M. (1986). On asymptotic optimality of some tests for exponential distribution, *Australian Journal of Statistics*, **28**, 164-172.

Shetty, B.N. and Joshi, P.C. (1986). Testing equality of location parameters of two exponential distributions, *Aligarh Journal of Statistics*, **6**, 11-25.

Shetty, B.N. and Joshi, P.C. (1987). Estimation of parameters of k exponential distributions in doubly censored samples, *Communications in Statistics - Theory and Methods*, **16**, 2115-2123.

Shetty, B.N. and Joshi, P.C. (1989). Likelihood ratio test for testing equality of location parameters of two exponential distributions from doubly censored samples, *Communications in Statistics - Theory and Methods*, **18**, 2063-2072.

Singh, N. (1983). The likelihood ratio test for the equality of location parameters of k (≥ 2) exponential populations based on Type II censored samples, *Technometrics*, **25**, 193-195.

Singh, N. (1985). A simple and asymptotically optimal test for the equality of k (≥ 2) exponential distributions based on Type II censored samples, *Communications in Statistics - Theory and Methods*, **14**, 1615-1625.

Singh, N. and Narayan, P. (1983). The likelihood ratio test for the equality of k (≥ 3) two-parameter exponential distributions based on Type II censored samples, *Journal of Statistical Computation and Simulation*, **18**, 287-297.

Spurrier, J. and Wei, L.J. (1980). A test for the parameter of the exponential distribution in the Type I censoring case, *Journal of the American Statistical Association*, **75**, 405-409.

Sukhatme, P.V. (1936). On the analysis of k samples from exponential populations with especial reference to the problem of random intervals, *Statistical Research Memoirs*, **1**, 94-112.

Sukhatme, P.V. (1937). Tests of significance for samples of the χ^2 population with two degrees of freedom, *Annals of Eugenics*, **8**, 52-56.

Thiagaraja, K. and Paul, S.R. (1990). Testing for the equality of scale parameters of K (≥ 2) exponential populations based on complete and Type II censored samples, *Communications in Statistics - Simulation and Computation*, **19**, 891-902.

Tiku, M.L. (1981). Testing equality of location parameters of two exponential distributions, *Aligarh Journal of Statistics*, **1**, 1-7.

Weinman, D.G., Dugger, G., Franck, W.E., and Hewett, J.E. (1973). On a test for the equality of two exponential distributions, *Technometrics*, **15**, 177-182.

CHAPTER 8

Tolerance Limits and Acceptance Sampling Plans

Lam Yeh[1] **and N. Balakrishnan**[2]

8.1 Introduction

An important topic in Quality Control is the study of tolerance limits. For example, in a mass production process, assume that the quality of an item is measured by a continuous random variable X. It is known by the national standard or contract that an item with measurement X less than L or greater than U is defective. If the manufacturer would like to know how successfully the production process is performing, he/she should study the probability $P(L \leq X \leq U)$. Let the density function of X be $f(x;\boldsymbol{\theta})$, where $\boldsymbol{\theta}$ is a parameter of one or more dimension(s). Suppose it is known to the manufacturer that unless 90% of his/her production is acceptable, he/she will be incurring a loss. Thus, an interval $[L, U]$ satisfying

$$P\left[\int_L^U f(x;\boldsymbol{\theta})\,dx \geq 0.90\right] \geq 0.95$$

needs to be constructed. In other words, the probability that the manufacturer will incur a loss will be less than 0.05. Hence, the interval $[L, U]$ is called a 0.90 content tolerance interval at level 0.95. Research in tolerance limits has mainly concentrated on the normal distribution case [see Guttman (1970, 1988) for details]. Guenther, Patil, and Uppuluri (1976) and Engelhardt and Bain (1978) discussed one-sided tolerance limits for the two-parameter exponential distribution. They constructed many tables through Monte Carlo procedures. We will discuss their work on the one-sided tolerance limits in Section 8.2.

In Quality Control, another important topic which is closely related to tolerance limits is the acceptance sampling plans. Assume that we are given a batch of products for inspection, and we want to make a decision, to either accept or reject the batch.

[1]The Chinese University of Hong Kong, Hong Kong
[2]McMaster University, Hamilton, Ontario, Canada

This is the problem of acceptance sampling plans, and sometimes it is called the sampling inspection problem. There are two kinds of acceptance sampling plans: attribute sampling plans and variable sampling plans. For an attribute sampling plan, the quality of an item is measured by the attribute of the item, defective and nondefective say. Otherwise, if the quality of an item is measured by a random variable, it is a variable sampling plan. To design a variable sampling plan is to determine the sample size and one or more specification limit(s) of the sampling plan. Suppose that the sample size n and the specification limit(s) of a variable sampling plan are given. To implement the sampling plan, we can take a random sample of size n from the batch, and compare the measurements from the sample with the specification limit(s). Then, based on the comparison, we can make a decision whether we should accept the batch or not. Of course, the natural question that arises here is what criterion should be used for making the decision of acceptance or rejection of the lot, or how to compare the measurements from the sample with the given specification limit(s)? For variable sampling plans, if the variable which measures the quality of an item has a normal distribution, this is an easy problem. So far, only little attention has been paid to the non-normal case. Kocherlakota and Balakrishnan (1986) studied one- and two-sided sampling plans based on the exponential distribution, and they also presented some useful tables. Their approach and results will be described in Section 8.3.

Perhaps, a more crucial problem in sampling inspection is the design of a sampling plan, i.e. the determination of the sample size and the specification limit(s). Many schemes have been studied for the design of a sampling plan. There are: the producer's and consumer's risk point schemes, the defence sampling schemes, Dodge and Romig's schemes, the decision theory schemes, etc. [see Wetherill (1977) for details]. The decision theory schemes, including Bayesian approach as the special case, are more scientific and hence are widely employed by many statisticians. For the normal distribution case, the problem of the design of a variable sampling plan has received considerable attention. Although the exponential distribution is very important not only in theory but also in applications like life testing and reliability, very little work has been done on the design of a variable sampling plan for the exponential distribution. Recently, by using Bayesian approach, Lam (1990) developed a model of the acceptance sampling plan for an exponentially distributed random variable with Type II censoring. He introduced a polynomial loss function and evaluated the Bayes risk explicitly. Lam (1990) suggested a simple algorithm to determine an optimal sampling plan for minimizing the Bayes risk. The algorithm is finite in the sense that an optimal sampling plan can be found after a finite-step of searching. Thereafter, Lam (1994) generalized these results to the case of Type I censoring and then Lam and Choy (1994) to the case of random censoring. This work on the Bayesian sampling plans for the exponential distribution with censoring will be discussed in Section 8.4.

8.2 One-Sided Tolerance Limits

Let $X_1, X_2, \ldots, X_n$ be a random sample from a continuous distribution with density function $f(x; \boldsymbol{\theta})$, where $\boldsymbol{\theta}$ is a parameter of one or more dimension(s). Let $L = L(X_1, X_2, \ldots, X_n)$, $U = U(X_1, X_2, \ldots, X_n)$ be two statistics. Then the observed values ℓ, u of L and U are called the lower and upper one-sided β-content tolerance limits at level γ, respectively, if

$$P\left[\int_L^{\infty} f(x;\boldsymbol{\theta})\,dx \geq \beta\right] = \gamma \tag{8.2.1}$$

holds for L, and

$$P\left[\int_{-\infty}^{U} f(x;\boldsymbol{\theta})\,dx \geq \beta\right] = \gamma \tag{8.2.2}$$

holds for U. Note that the lower and upper tolerance limits are sometimes defined by the following inequalities:

$$P\left[\int_L^{\infty} f(x;\boldsymbol{\theta})\,dx \geq \beta\right] \geq \gamma$$

and

$$P\left[\int_{-\infty}^{U} f(x;\boldsymbol{\theta})\,dx \geq \beta\right] \geq \gamma$$

[see Guttman (1970, 1988)].

Now, we assume further that X has a two-parameter exponential distribution with density

$$f(x;\mu,\theta) = \frac{1}{\theta}\, e^{-(x-\mu)/\theta}, \qquad x \geq \mu,\ -\infty < \mu < \infty,\ \theta > 0\ . \tag{8.2.3}$$

Here, μ is the location parameter and θ is the scale parameter. In practice, if X is the lifetime of an item in a batch, then μ will represent the minimum lifetime and θ will be the expected additional lifetime. If μ and θ are known, with $L = \mu - \theta \ln \beta$, then

$$\int_L^{\infty} f(x;\mu,\theta)\,dx = \beta\ . \tag{8.2.4}$$

L is a lower one-sided β-content tolerance limit at level 1. In the case when one or both of μ and θ are unknown, we should replace $\mu - \theta \ln \beta$ by $\hat{\mu} - \hat{\theta} \ln \beta$, where $\hat{\mu}$ and $\hat{\theta}$ are the estimators of μ and θ respectively (if one of μ and θ is known, use true value in place of its estimator). However, in general (8.2.4) now is no longer true. Therefore, instead of $\hat{\mu} - \hat{\theta} \ln \beta$, we should look for a number c so that $\hat{\mu} + c\hat{\theta}$ is a one-sided β-content tolerance limit at level γ, i.e.

$$P\left[\int_{\hat{\mu}+c\hat{\theta}}^{\infty} f(x;\mu,\theta)\,dx \geq \beta\right] = \gamma\ , \tag{8.2.5}$$

c is sometimes called the *tolerance factor.* Let $X_{(1)} \leq X_{(2)} \leq \cdots \leq X_{(n)}$ be the order statistics of the random sample $X_1, X_2, \ldots, X_n$. Guenther (1971) showed that the "best" one-sided β-content tolerance limit is of the form $L = \hat{\mu} + c\hat{\theta}$, where

$$\hat{\mu} = X_{(1)} \tag{8.2.6}$$

and

$$\hat{\theta} = \frac{1}{n}\sum_{i=2}^{n}(n-i+1)(X_{(i)} - X_{(i-1)}) \tag{8.2.7}$$

are the sufficient statistics and also the MLEs of μ and θ, respectively. From Chapters 3 and 4, it is known that

$$T = 2n(\hat{\mu} - \mu)/\theta \sim \chi_2^2 \tag{8.2.8}$$

and

$$S = 2n\hat{\theta}/\theta \sim \chi_{2n-2}^2 \tag{8.2.9}$$

independently. Note that (8.2.5) is then equivalent to

$$\gamma = P[T + cS \le -2n \ln \beta] \ . \tag{8.2.10}$$

To determine a β-content tolerance limit from (8.2.10), we consider the following cases:

Case 1: $c \le 0$. Now, (8.2.10) becomes

$$\begin{aligned}\gamma &= \int_0^\infty P[T + cS \le -2n\ln\beta \,|\, S = s]\, \frac{1}{2^{n-1}\Gamma(n-1)}\, s^{n-2}e^{-s/2}ds \\ &= \int_0^\infty [1 - \beta^n\, e^{cs/2}]\, \frac{1}{2^{n-1}\Gamma(n-1)}\, s^{n-2}e^{-s/2}ds \\ &= 1 - \beta^n/(1-c)^{n-1} \ .\end{aligned}$$

Hence,

$$c = 1 - [\beta^n/(1-\gamma)]^{1/(n-1)} \ . \tag{8.2.11}$$

Case 2: $0 < c < 1$. Similarly, (8.2.10) gives

$$\begin{aligned}\gamma &= \int_0^{-2n\ln\beta/c} P[T + cS \le -2n\ln\beta \,|\, S = s]\, \frac{1}{2^{n-1}\Gamma(n-1)}\, s^{n-2}e^{-s/2}ds \\ &= P[\chi_{2n-2}^2 \le -2n\ln\beta/c] - \frac{\beta^n}{(1-c)^{n-1}}\, P[\chi_{2n-2}^2 \le -2n(1-c)\ln\beta/c] \ .\end{aligned} \tag{8.2.12}$$

Case 3: $c = 1$. Since $T + S \rightsquigarrow \chi_{2n}^2$, (8.2.10) yields

$$\gamma = P[\chi_{2n}^2 \le -2n\ln\beta] \ . \tag{8.2.13}$$

Case 4: $c > 1$. Then

$$\begin{aligned}\gamma &= P[\chi_{2n-2}^2 \le -2n\ln\beta/c] \\ &\quad - \frac{\beta^n}{2^{n-1}}\left\{(-2n\ln\beta/c)^{n-2}\beta^{-n(c-1)/c}\sum_{i=0}^{n-2}\frac{(-1)^i}{[-n(c-1)\ln\beta/c]^i(n-2-i)!}\right. \\ &\quad \left. + (-1)^{n-1}\,\frac{1}{[(c-1)/2]^{n-1}}\right\} \ .\end{aligned} \tag{8.2.14}$$

(8.2.14) is because of the integration equality

$$\int_0^\alpha s^n e^{as}\, ds = \frac{n!\alpha^n e^{a\alpha}}{a} \sum_{i=0}^{n} \frac{(-1)^i}{(a\alpha)^i (n-i)!} + (-1)^{n+1} \frac{n!}{a^{n+1}} . \tag{8.2.15}$$

Clearly, c depends on the values of n, β and γ. From (8.2.11)–(8.2.14), depending on the range of c, we can evaluate c numerically. Obviously, it is convenient for the evaluation of c if we can locate the range of c in advance. Let $\chi^2_{2n,p}$ be the pth quantile of χ^2 distribution with $2n$ degrees of freedom, i.e.

$$P[\chi^2_{2n} \le \chi^2_{2n,p}] = \int_0^{\chi^2_{2n,p}} \frac{1}{2^n \Gamma(n)} x^{n-1} e^{-x/2} dx = p .$$

Thus, if $c = 0$, (8.2.11) gives

$$n = \ln(1-\gamma)/\ln\beta = \chi^2_{2,\gamma}/\chi^2_{2,1-\beta} .$$

If $c = 1$, it follows from (8.2.13) that

$$n = \chi^2_{2n,\gamma}/(-2\ln\beta) = \chi^2_{2n,\gamma}/\chi^2_{2,1-\beta} .$$

Because the value of $\chi^2_{2n,p}$ is increasing in n, we then have the following theorem.

Theorem 8.1 [Guenther, Patil, and Uppuluri (1976)]

1. *$c < 0$ if and only if $n < \chi^2_{2,\gamma}/\chi^2_{2,1-\beta}$;*
2. *$0 \le c < 1$ if and only if $\chi^2_{2,\gamma}/\chi^2_{2,1-\beta} \le n < \chi^2_{2n,\gamma}/\chi^2_{2,1-\beta}$;*
3. *$c = 1$ if and only if $n = \chi^2_{2n,\gamma}/\chi^2_{2,1-\beta}$;*
4. *$c > 1$ if and only if $n > \chi^2_{2n,\gamma}/\chi^2_{2,1-\beta}$.*

Clearly, Theorem 8.1 is very useful for the specification of the range of c.

By a similar argument, an upper one-sided β-content tolerance limit $\hat\mu + c\hat\theta$ can be determined by

$$P\left[\int_\mu^{\hat\mu + c\hat\theta} f(x;\mu,\theta)\, dx \ge \beta\right] = \gamma . \tag{8.2.16}$$

Obviously (8.2.16) is equivalent to

$$P\left[\int_{\hat\mu + c\hat\theta}^\infty f(x;\mu,\theta)\, dx \ge 1-\beta\right] = 1-\gamma .$$

Therefore, in comparison with (8.2.5), the upper tolerance limit $\hat\mu + c\hat\theta$ satisfying (8.2.16) can be obtained by replacing β by $1-\beta$, and γ by $1-\gamma$ in formulas (8.2.11)–(8.2.14). Accordingly, Theorem 8.1 will be modified to have the following form.

Theorem 8.2 [Guenther, Patil, and Uppuluri (1976)]

1. *$0 \le c < 1$ if and only if $\chi^2_{2,1-\gamma}/\chi^2_{2,\beta} \le n < \chi^2_{2n,1-\gamma}/\chi^2_{2,\beta}$;*

2. *$c = 1$ if and only if $n = \chi^2_{2n,1-\gamma}/\chi^2_{2,\beta}$;*

3. *$c > 1$ if and only if $n > \chi^2_{2n,1-\gamma}/\chi^2_{2,\beta}$.*

This is because, in the present case, the condition

$$n \le \chi^2_{2,1-\gamma}/\chi^2_{2,\beta}$$

is no longer satisfied for conventional choice of β and γ as both of them are close to 1.

Since the tolerance factor c depends on n, β and γ only, Guenther, Patil, and Uppuluri (1976) constructed many tables for both lower and upper tolerance limits for different values of n, β and γ.

If one of the parameters μ or θ is known, the determination of c is much simpler. In fact, for the lower tolerance limit, if θ is known, then from

$$\gamma = P\left[\int_{\hat{\mu}+c\theta}^{\infty} f(x;\mu,\theta)\,dx \ge \beta\right] = P[T \le -2n(c + \ln\beta)] ,$$

we have

$$c = -\ln\beta - \chi^2_{2,\gamma}/2n . \tag{8.2.17}$$

On the other hand, if μ is known, then

$$\gamma = P\left[\int_{\mu+c\hat{\theta}}^{\infty} f(x;\mu,\theta)\,dx \ge \beta\right] = P[S \le -2n\ln\beta/c] ,$$

Hence,

$$c = -2n\ln\beta/\chi^2_{2n-2,\gamma} . \tag{8.2.18}$$

Similarly, in the case when θ or μ is known, one can obtain an upper β-content tolerance limit at level γ by replacing β by $1-\beta$, and γ by $1-\gamma$ in (8.2.17) and (8.2.18) [see Kocherlakota and Balakrishnan (1986) for details].

Now, assume that the random sample is subject to Type II censoring at the time of the rth failure. Then the full ordered sample is $\boldsymbol{Y} = (Y_1, Y_2, \ldots, Y_n)$, where

$$Y_k = \begin{cases} X_{(k)} & k = 1, \ldots, r , \\ X_{(r)} & k = r+1, \ldots, n . \end{cases} \tag{8.2.19}$$

On similar lines, we can derive the tolerance limits based on sufficient statistics

$$\hat{\mu} = X_{(1)} \tag{8.2.20}$$

and

$$\hat{\theta} = \frac{1}{r}\sum_{i=2}^{r}(n-i+1)(X_{(i)} - X_{(i-1)}) . \tag{8.2.21}$$

It is known from Chapters 3, 4, and 5 that

$$T = 2n(\hat{\mu} - \mu)/\theta \sim \chi_2^2 \tag{8.2.22}$$

and

$$S = 2r\hat{\theta}/\theta \sim \chi_{2r-2}^2 \tag{8.2.23}$$

independently. It then follows from (8.2.5) that

$$\gamma = P[T/(2n) + c\,S/(2r) \leq -\ln\beta] \ . \tag{8.2.24}$$

The one-sided β-content tolerance limits can be constructed in a similar way. Actually, for $c \leq 0$, analogous to (8.2.11), (8.2.24) yields

$$c = \frac{r}{n}\left\{1 - [\beta^n/(1-\gamma)]^{1/(r-1)}\right\} \ . \tag{8.2.25}$$

From (8.2.25), one can see that $c \leq 0$ is equivalent to $n \leq \ln(1-\gamma)/\ln\beta$. Engelhardt and Bain (1978) also derived an approximate expression of c for larger sample size n as follows:

$$c = \frac{r}{n}\left\{-m(\beta) - z_\gamma[m(\beta)^2/r + 1/r^2]^{1/2}\right\} \tag{8.2.26}$$

where $m(\beta) = (1 + n\ln\beta)/(r - 2.5)$ and z_γ is the γth quantile of the standard normal distribution. Engelhardt and Bain (1978) recommended that (8.2.25) be used for small sample sizes and (8.2.26) be employed for large sample sizes. Of course, (8.2.25) and (8.2.26) are also applicable for the uncensored case ($r = n$).

8.3 One- and Two-Sided Sampling Plans

In an acceptance sampling process, assume that the quality of an item in a large batch is measured by its lifetime X. Assume further that X has a two-parameter exponential distribution with density (8.2.3). If μ and θ are known, then we will accept the batch if $\mu + \theta$ is large, $\mu + \theta \geq L$ say, and reject the batch otherwise. In practice, μ or θ or both of them are usually unknown. Then, we should take a random sample $\boldsymbol{X} = (X_1, X_2, \ldots, X_n)$ of size n from the batch. Let $\hat{\mu}$ and $\hat{\theta}$ be the MLEs of μ and θ given by (8.2.6) and (8.2.7), respectively. Analogous to the tolerance limits in Section 8.2, we will accept the batch if $\hat{\mu} + c\hat{\theta} \geq L$ and reject the batch otherwise, where c is chosen by the constraint

$$P[\hat{\mu} + c\hat{\theta} \geq L] = \gamma \ ,$$

i.e.,

$$P[T + c\,S \leq 2n(L - \mu)/\theta] = 1 - \gamma \ . \tag{8.3.1}$$

Here, T and S are as defined in (8.2.8) and (8.2.9), respectively. Note that (8.3.1) is equivalent to (8.2.10) with $\beta = \exp[-(L - \mu)/\theta]$ and γ is replaced by $1 - \gamma$. Thus, determining c satisfying (8.3.1) is exactly the same as determining c in a lower one-sided β-content tolerance limit at level $1 - \gamma$. Therefore, the problem of one-sided sampling plans is essentially the same as the problem of one-sided tolerance limits. The latter has already been discussed in Section 8.2.

Now, let us turn our attention to the problem of two-sided sampling plans. Kocherlakota and Balakrishnan (1986) studied the following decision function: accept the batch if $L \le \hat{\mu} + c\hat{\theta} \le U$ and reject the batch otherwise. Here, c is chosen to satisfy the condition

$$P[L \le \hat{\mu} + c\hat{\theta} \le U] = \gamma \ . \tag{8.3.2}$$

(8.3.2) is equivalent to

$$P[d_1 \le T + cS \le d_2] = \gamma \ , \tag{8.3.3}$$

where $d_1 = 2n(L-\mu)/\theta$ and $d_2 = 2n(U-\mu)/\theta$. To solve (8.3.3) for c, as in Section 8.2, according to different values of c, Kocherlakota and Balakrishnan (1986) recognized the following four cases.

Case 1: $c \le 0$. Then (8.3.3) yields

$$\begin{aligned}\gamma &= \int_0^\infty P[d_1 \le T + c\,S \le d_2 \,|\, S = s]\ \frac{1}{2^{n-1}\Gamma(n-1)}\ s^{n-2}e^{-s/2}ds \\ &= [e^{-d_1/2} - e^{-d_2/2}]/(1-c)^{n-1} \ .\end{aligned}$$

Thus, the explicit solution for c is given by

$$c = 1 - \{[e^{-d_1/2} - e^{-d_2/2}]/\gamma\}^{1/(n-1)} \ . \tag{8.3.4}$$

Case 2: $0 < c < 1$. It follows from (8.3.3) that

$$\begin{aligned}\gamma &= P[T + c\,S \le d_2] - P[T + c\,S \le d_1] \\ &= P[d_1/c \le \chi^2_{2n-2} \le d_2/c] + e^{-d_1/2}\int_0^{d_1/c} \frac{1}{2^{n-1}\Gamma(n-1)}\ s^{n-2}e^{-s(1-c)/2}ds \\ &\quad -e^{-d_2/2}\int_0^{d_2/c} \frac{1}{2^{n-1}\Gamma(n-1)}\ s^{n-2}e^{-s(1-c)/2}ds \\ &= P[d_1/c \le \chi^2_{2n-2} \le d_2/c] - (1-c)^{-(n-1)} \\ &\quad \cdot\{e^{-d_1/2}P[\chi^2_{2n-2} \le d_1(1-c)/c] \\ &\qquad -e^{-d_2/2}P[\chi^2_{2n-2} \le d_2(1-c)/c]\} \ .\end{aligned} \tag{8.3.5}$$

Case 3: $c = 1$. Since $T + S \sim \chi^2_{2n}$, (8.3.3) becomes

$$\gamma = P[d_1 \le \chi^2_{2n} \le d_2] \ . \tag{8.3.6}$$

Case 4: $c > 1$. Note that $c - 1 > 0$; from (8.3.3) with the help of (8.2.15), an argument similar to Case 2 gives

$$\begin{aligned}\gamma &= P[d_1/c \le \chi^2_{2n-2} \le d_2/c] \\ &\quad + \frac{(-1)^n}{(c-1)^{n-1}}\left\{e^{-d_2/2}\left[1 - \sum_{i=0}^{n-2}\frac{(-1)^i}{i!}\ \alpha_2^i e^{\alpha_2}\right]\right. \\ &\qquad \left. -e^{-d_1/2}\left[1 - \sum_{i=0}^{n-2}\frac{(-1)^i}{i!}\ \alpha_1^i e^{\alpha_1}\right]\right\} ,\end{aligned} \tag{8.3.7}$$

where $\alpha_1 = d_1(c-1)/(2c)$ and $\alpha_2 = d_2(c-1)/(2c)$.

If one of the parameters is known, as in the one-sided case, it is quite easy to determine c. For example, if θ is known, then (8.3.2) will take the form

$$P[L \leq \hat{\mu} + c\theta \leq U] = \gamma ,$$

i.e.,

$$P[d_1 - 2nc \leq T \leq d_2 - 2nc] = \gamma .$$

Because $T \sim X_2^2$, it follows that

$$\gamma = e^{nc}[e^{-d_1/2} - e^{-d_2/2}] .$$

Hence

$$c = \frac{1}{n} \ln\{\gamma/[e^{-d_1/2} - e^{-d_2/2}]\} . \tag{8.3.8}$$

If μ is known, then (8.3.2) will be replaced by

$$P[L \leq \mu + c\hat{\theta} \leq U] = \gamma$$

or

$$P[d_1/c \leq S \leq d_2/c] = \gamma .$$

Since $S \sim \chi^2_{2n-2}$, we have

$$\gamma = P[d_1/c \leq \chi^2_{2n-2} \leq d_2/c] . \tag{8.3.9}$$

To determine c, one can see from (8.3.4)–(8.3.9) that the first thing we need to do is to specify the values of d_1 and d_2. However, d_1 and d_2 themselves depend on μ and θ which are unknown. To overcome this difficulty, instead of knowing L and U, we can start from two constraints:

$$P[X \leq L] = p_1 \qquad \text{and} \qquad P[X \geq U] = p_2 \tag{8.3.10}$$

with known p_1 and p_2. Then, it follows from (8.2.3) that $d_1 = -2n\ln(1-p_1)$ and $d_2 = -2n \ln p_2$ are completely determined.

Under the conditions that $p_1 = p_2$ and $\gamma = 0.90$, Kocherlakota and Balakrishnan (1986) have evaluated a large number of values c. They have tabulated the values of c for the cases when either μ or θ is unknown, and also for the case when both of them are unknown.

In the case of Type II censoring, the methodology can be developed in exactly the same way. For example, assume that both parameters μ and θ are unknown, and we are looking for the sampling plan: accept the batch if $L \leq \hat{\mu} + c'\hat{\theta} \leq U$ with c' being chosen to satisfy the condition

$$P[L \leq \hat{\mu} + c'\hat{\theta} \leq U] = \gamma , \tag{8.3.11}$$

where $\hat{\mu}$ and $\hat{\theta}$ are the MLEs of μ and θ given by (8.2.20) and (8.2.21), respectively. Moreover, instead of (8.3.3), we now have

$$P[d_1 \leq T + (nc'/r)S \leq d_2] = \gamma , \tag{8.3.12}$$

where T and S are given by (8.2.22) and (8.2.23) and follow, respectively, the distributions χ_2^2 and χ_{2r-2}^2 independently. Thus, the procedure discussed for complete sample case applies equally well in the present instance. Actually, in the determination of c', formulas (8.3.4)–(8.3.7) are still applicable except we should replace c by nc'/r and n by r.

8.4 Bayesian Sampling Plans

In Section 8.3, we assumed that the sample size n and the specification limit(s) of a sampling plan are given. To conduct the sampling plan, we should take a random sample of size n from the batch and compare the measurements from the sample with the specification limit(s), and then make the decision. The problem here is how to compare the measurements with the given specification limit(s). For the one-sided case, the problem has been solved in Section 8.3.

Perhaps, a more important problem is the design of a sampling plan. As mentioned already in Section 8.1, the decision theory schemes may be better than the others for this purpose. Through Bayesian approach, Lam (1990, 1994) suggested a model to study this problem for the exponential distribution with Type II and Type I censoring.

To start with, assume that a batch of products is presented for acceptance inspection. Let X be again the lifetime to failure of an item in the batch. Assume that X has the one-parameter exponential distribution $e(\lambda)$ with density function:

$$f(x;\lambda) = \lambda\, e^{-\lambda x}, \qquad x \geq 0,\ \lambda > 0\ . \tag{8.4.1}$$

Moreover, λ has a conjugate prior distribution which is the Gamma distribution Gamma(α, β) with density

$$g(\lambda) = \begin{cases} \beta^\alpha \lambda^{\alpha-1} e^{-\beta\lambda}/\Gamma(\alpha) & , \quad \lambda \geq 0\ , \\ 0 & , \quad \lambda < 0\ , \end{cases} \tag{8.4.2}$$

where α and β are both known.

Through Bayesian approach, the problem is to design a sampling plan so that the Bayes risk is minimized.

To do this, assume that a random sample $\boldsymbol{X} = (X_1, X_2, \ldots, X_n)$ of size n is taken from the batch. Let $X_{(1)} \leq X_{(2)} \leq \cdots \leq X_{(n)}$ be the order statistics of $X_1, X_2, \ldots, X_n$. The sample is censored after the rth failure. The full ordered observed sample is $\boldsymbol{Y} = (Y_1, Y_2, \ldots, Y_n)$, where

$$Y_i = \begin{cases} X_{(i)} & i = 1, \ldots, r\ , \\ X_{(r)} & i = r+1, \ldots, n\ . \end{cases} \tag{8.4.3}$$

It is known that the MLE of the average lifetime $\theta = 1/\lambda$ is given by (see Chapter 4)

$$\hat{\theta}_r = \frac{1}{r} \sum_{i=1}^{r} (n-i+1)(X_{(i)} - X_{(i-1)})\ . \tag{8.4.4}$$

It is known from Chapters 3, 4, and 5 that $\hat{\theta}_r$ has the Gamma distribution Gamma$(r, r\lambda)$.

A decision function $\delta(\boldsymbol{X})$ is introduced by Lam (1990) on the basis of $\hat{\theta}_r$. It is

$$\delta(\boldsymbol{X}) = \begin{cases} d_0 \quad , & \hat{\theta}_r \geq T \ , \\ d_1 \quad , & \text{otherwise} \ , \end{cases} \tag{8.4.5}$$

where d_0 and d_1 denote the decisions of accepting and rejecting the batch, respectively. T is called the minimum acceptance time. The use of decision function (8.4.5) is reasonable. In fact, if the quality of an item is measured by its lifetime X, the decision function $\delta(\boldsymbol{X})$ should be one-sided as in (8.4.5). Suppose the quality of an item is measured by the reliability $\bar{F}(t_0) = P[X > t_0] = e^{-t_0/\theta}$. From Chapter 6, the MLE of $\bar{F}(t_0)$ is known to be $\widehat{\bar{F}(t_0)} = e^{-t_0/\hat{\theta}_r}$. Hence, in virtue of the fact that the larger the average lifetime $\hat{\theta}_r$ the greater the reliability $\widehat{\bar{F}(t_0)}$, the decision function $\delta(\boldsymbol{X})$ is once again of the form (8.4.5).

A polynomial loss function is considered to be

$$L[\lambda, \delta(\boldsymbol{X})] = \begin{cases} nC_s + C_0 + C_1\lambda + \cdots + C_k\lambda^k & , \quad \delta(\boldsymbol{X}) = d_0 \ , \\ nC_s + C_r & , \quad \delta(\boldsymbol{X}) = d_1 \ , \end{cases} \tag{8.4.6}$$

where $C_0, \ldots, C_k$, C_r, and C_s are constants with a natural constraint

$$C_0 + C_1\lambda + \cdots + C_k\lambda^k \geq 0, \qquad \lambda > 0 \ .$$

Here, C_s refers to the cost of sampling per item and C_r is the cost of rejecting a batch.

The polynomial loss function has been considered earlier by Hald (1967, 1981), Lam (1988a, b) and then by Lam and Lau (1993) in the sampling plan problem for the normal distribution case. For the exponential distribution case also, Lam (1990) commented that a polynomial loss function is reasonable. In fact, suppose that a batch of N items is presented for acceptance sampling. Assume that the quality of an item is measured by its lifetime X. Let μ_0 be the critical value specified by national standard or the contract. It is known that if $X \geq \mu_0$, it can be sold at normal price; this means that we can accept the item without any extra charge. Otherwise, if $X < \mu_0$, it might be sold at a reduced price, and a loss is incurred. Assume that the loss is proportional to $\mu_0 - X$. Then the item will be accepted with cost $C(\mu_0 - X)$ where C is a proportional constant. Thus the cost function when the batch is accepted will be given by

$$N \int_0^{\mu_0} C(\mu_0 - x)\lambda e^{-\lambda x} dx = NC\{\mu_0 - [1 - e^{-\lambda\mu_0}]/\lambda\} \ . \tag{8.4.7}$$

Sometimes, the quality of an item is measured by the reliability of the item at t_0, i.e., the probability of the item surviving beyond time t_0. For example, in medicine, an index of the success of a cancer therapy is measured by the probability that a patient survives at least five years. This is, $\bar{F}(t_0) = P[X > t_0]$. Let the critical value of $\bar{F}(t_0)$ be p_0. The values of t_0 and p_0 are both specified by the national standard or the contract. Then if $\bar{F}(t_0) \geq p_0$, we can sell the item at normal price and hence accept it without additional cost; if $\bar{F}(t_0) < p_0$, by a similar argument as above, we may accept the item with cost $C'[p_0 - \bar{F}(t_0)]$, where C' is a proportional constant. Thus the cost function when the batch is accepted is given by

$$NC'[p_0 - \bar{F}(t_0)] = NC'[p_0 - e^{-\lambda t_0}] \ . \tag{8.4.8}$$

From (8.4.7) and (8.4.8), one can see that for both cases the exact cost functions are general analytic functions of λ. Therefore, as in the normal distribution case, a polynomial cost function should be a better choice than a linear cost function, since the former has a better approximation to (8.4.7) and (8.4.8).

In applications, the choice of the parameters α and β and the coefficients $C_0, \ldots, C_k$, C_r, and C_s is important. Some suggestions regarding such choice have been made by Lam (1990).

In the present situation, a sampling plan is denoted by (n, r, T), where the sample size is n, the sample is censored at the time of the rth failure, and the minimum acceptance time is T. By Bayesian approach, an optimal sampling plan is a sampling plan which minimizes the Bayes risk $R(n, r, T) = E[L(\lambda, \delta(\boldsymbol{X}))]$.

It is not difficult to evaluate the Bayes risk explicitly. Without loss of generality, let us examine here in detail the case of quadratic loss function, i.e. $k = 2$. Conditional on λ, the Bayes risk is given by

$$\begin{aligned}
R(n,r,T) &= E[L(\lambda,\delta(\boldsymbol{X}))] = E\{E[L(\lambda,\delta(\boldsymbol{X}))|\lambda]\} \\
&= E\{(C_0 + C_1\lambda + C_2\lambda^2)P[\hat{\theta}_r \geq T] + (nC_s + C_r)P[\hat{\theta}_r < T]\} \\
&= nC_s + C_0 + C_1\alpha/\beta + C_2\alpha(\alpha+1)/\beta^2 \\
&\quad - \int_0^\infty \int_0^T (C_0 - C_r + C_1\lambda + C_2\lambda^2)\, \frac{(r\lambda)^r}{\Gamma(r)}\, t^{r-1} e^{-r\lambda t} \\
&\quad \cdot \frac{\beta^\alpha}{\Gamma(\alpha)}\, \lambda^{\alpha-1} e^{-\beta\lambda}\, dt\, d\lambda \\
&= nC_s + C_0 + C_1\alpha/\beta + C_2\alpha(\alpha+1)/\beta^2 \\
&\quad - \frac{r^r\beta^\alpha}{\Gamma(r)\Gamma(\alpha)} \left[\int_0^T (C_0 - C_r)\Gamma(r+\alpha)\, \frac{t^{r-1}}{(rt+\beta)^{r+\alpha}}\, dt \right. \\
&\quad + \int_0^T C_1\Gamma(r+\alpha+1)\, \frac{t^{r-1}}{(rt+\beta)^{r+\alpha+1}}\, dt \\
&\quad \left. + \int_0^T C_2\Gamma(r+\alpha+2)\, \frac{t^{r-1}}{(rt+\beta)^{r+\alpha+2}}\, dt \right]. \qquad (8.4.9)
\end{aligned}$$

Denote the beta distribution function or the incomplete beta distribution by

$$I_x(a,b) = B_x(a,b)/B(a,b)\ ,$$

where

$$B(a,b) = \int_0^1 t^{a-1}(1-t)^{b-1} dt$$

is the beta function and

$$B_x(a,b) = \int_0^x t^{a-1}(1-t)^{b-1} dt \qquad (8.4.10)$$

is the incomplete beta function. Then we can reduce (8.4.9) to

$$\begin{aligned}
R(n,r,T) &= nC_s + C_0 + C_1\alpha/\beta + C_2\alpha(\alpha+1)/\beta^2 - (C_0 - C_r)I_s(r,\alpha) \\
&\quad - C_1\alpha I_s(r,\alpha+1)/\beta - C_2\alpha(\alpha+1)I_s(r,\alpha+2)/\beta^2\ , \qquad (8.4.11)
\end{aligned}$$

where $s = rT/(rT+\beta)$. Note that $I_x(a,b)$ can be computed by using numerical subroutines [e.g., IMSL Libraries (1991)]. Thus, the value of Bayes risk $R(n,r,T)$ can be evaluated directly from (8.4.11).

The problem of minimizing $R(n,r,T)$ seems to be a difficult problem, because there are three variables in which two are discrete and one is continuous. However, Lam (1990) suggested the following algorithm for searching an optimal sampling plan:

1. Fix sample size $n = 0$ first;

2. For fixed n and $r = 0, \ldots, n$, find the corresponding minimum acceptance time T by minimizing $R(n,r,T)$; let

$$R(n, r_n, T_n) = \min_{r \leq n, T \geq 0} R(n,r,T) \ ;$$

3. Move n to $n+1$ and go back to Step 2;

4. By comparison, an optimal sampling plan (n^*, r^*, T^*) is chosen which corresponds to the smallest value of the Bayes risk, i.e.

$$R(n^*, r^*, T^*) = \min_n R(n, r_n, T_n) \ .$$

Obviously, Step 2 is the key step of the algorithm. By classical method, we can solve the equation $\partial R/\partial T = 0$ which, after simplification, reduces to a quadratic equation

$$(C_0 - C_r)(rT+\beta)^2 + C_1(r+\alpha)(rT+\beta) + C_2(r+\alpha+1)(r+\alpha) = 0 \ . \quad (8.4.12)$$

Therefore, this algorithm is a simple algorithm in view of the simple quadratic equation in (8.4.12). Furthermore, the algorithm is finite, i.e. we can find an optimal sampling plan after a finite-step of searching. This is due to the following theorem.

Theorem 8.3 [Lam (1990)]
The optimal sample size n^* satisfies the inequality

$$n^* \leq \min\{C_r/C_s,\ [C_0\beta^2 + C_1\alpha\beta + C_2\alpha(\alpha+1)]/(C_s\beta^2)\} \ .$$

Proof: Let $(0,0,\infty)$ and $(0,0,0)$ be the sampling plans of rejecting the batch and of accepting the batch without sampling, respectively. Clearly, $R(0,0,\infty) = C_r$ and

$$R(0,0,0) = [C_0\beta^2 + C_1\alpha\beta + C_2\alpha(\alpha+1)]/\beta^2 \ .$$

Then from

$$R(n^*, r^*, T^*) \leq \min\{R(0,0,\infty),\ R(0,0,0)\}$$

and

$$R(n^*, r^*, T^*) \geq n^* C_s \ ,$$

Theorem 8.3 follows directly. □

By using exactly the same method, Lam (1994) showed that the optimal sample size n^* should satisfy

$$n^* \leq \min\{R(0,0,\infty)/C_s, R(0,0,0)/C_s, R(n, r_n, T_n)/C_s\} \ . \quad (8.4.13)$$

(8.4.13) gives an adaptive upper bound for the optimal sample size n^*, and hence use of (8.4.13) may save computer time considerably.

In general, if the loss function in (8.4.6) is a polynomial of general degree k, a similar argument shows that the algorithm above is still applicable. It is also a finite algorithm. Although the algorithm relies on the solution of an algebraic equation of kth order, it is again a simple algorithm.

Lam (1990) has also considered the sampling plans for the two-parameter exponential distribution with Type II censoring. For convenience, assume that the lifetime X of an item in a batch follows a two-parameter exponential distribution with density function

$$f(x;\mu,\lambda) = \lambda\, e^{-\lambda(x-\mu)}, \qquad x \geq \mu > 0,\ \lambda > 0\ , \tag{8.4.14}$$

where $1/\lambda$ is the parameter θ defined in (8.2.3). Note that the domain of definition of μ in (8.4.14) is different from (8.2.3). Assume further that λ and μ are both unknown. Actually, if μ is known, a simple transformation $X' = X - \mu$ reduces the present problem to the problem for one-parameter exponential distribution case which has already been discussed.

A four-parameter conjugate prior distribution for (λ, μ) is considered here. The density function of the prior distribution is defined by

$$p(\lambda,\mu) = \frac{1}{A}\,\lambda^{\alpha-1}e^{-\lambda(\beta-\gamma\mu)}, \qquad \lambda > 0,\ 0 < \mu \leq \eta\ , \tag{8.4.15}$$

where $\alpha > 0,\ \beta > 0,\ \alpha \leq \gamma < \beta/\eta$ and

$$A = \begin{cases} \Gamma(\alpha-1)[1/(\beta-\gamma\eta)^{\alpha-1} - 1/\beta^{\alpha-1}]/\gamma & ,\ \alpha \neq 1\ , \\ [\ln\beta - \ln(\beta-\gamma\eta)]/\gamma & ,\ \alpha = 1 \end{cases}$$

with known α, β, γ, and η. This conjugate prior distribution was suggested by Varde (1969), where he restricted α and γ to be positive integers only, but this is not necessary.

Assume that a random sample $\boldsymbol{X} = (X_1, X_2, \ldots, X_n)$ is taken from a batch. Suppose that the sample is censored at the time of the rth failure. Since μ and λ are both unknown, the MLEs of μ and $\theta = 1/\lambda$ are given by (8.2.20) and (8.2.21) respectively. Hence, the MLE $\hat{\Omega}_r$ of the average lifetime Ω is given by

$$\hat{\Omega}_r = \begin{cases} \hat{\mu} & ,\quad r = 1\ , \\ \hat{\mu} + \hat{\theta} & ,\quad r \geq 2\ . \end{cases} \tag{8.4.16}$$

From (8.2.22) and (8.2.23), $\hat{\mu}$ has a two-parameter exponential distribution with density

$$g(u) = n\lambda\, e^{-n\lambda(u-\mu)}, \qquad u \geq \mu > 0\ , \tag{8.4.17}$$

while $\hat{\theta}$ has a Gamma distribution Gamma$(r-1, r\lambda)$ with density

$$h(v) = \frac{(r\lambda)^{r-1}}{\Gamma(r-1)}\, v^{r-2}e^{-r\lambda v}, \qquad v \geq 0\ . \tag{8.4.18}$$

Furthermore, they are independent. As in the one-parameter exponential distribution case, the decision function should also be one-sided:

$$\delta(\boldsymbol{X}) = \begin{cases} d_0 & ,\quad \hat{\Omega}_r \geq T\ , \\ d_1 & ,\quad \text{otherwise}\ , \end{cases} \tag{8.4.19}$$

where T is again the minimum acceptance time. The loss function now is also a polynomial function of λ and μ:

$$L[\lambda, \mu; \delta(\boldsymbol{X})] = \begin{cases} nC_s + \sum_{0 \le i+j \le k} C_{ij} \lambda^i \mu^j & , \ \delta(\boldsymbol{X}) = d_0 \ , \\ nC_s + C_r & , \ \delta(\boldsymbol{X}) = d_1 \ . \end{cases} \tag{8.4.20}$$

Then, we can evaluate the Bayes risk $R(n, r, T) = E[L(\lambda, \mu; \delta(\boldsymbol{X}))]$ in a similar way. To obtain an explicit expression for the Bayes risk, without loss of generality, we will once again consider the quadratic loss function, i.e. $k = 2$. Then, if $r = 1$, it follows from (8.4.20) with the help of (8.4.17) and (8.4.18) that

$$\begin{aligned} R(n, 1, T) &= E[L(\lambda, \mu; \delta(\boldsymbol{X}))] = E\{E[L(\lambda, \mu; \delta(\boldsymbol{X}))|\lambda, \mu]\} \\ &= E[nC_s + C_r + (C_{00} - C_r + C_{10}\lambda + C_{01}\mu + C_{20}\lambda^2 \\ &\quad + C_{11}\lambda\mu + C_{02}\mu^2)e^{-n\lambda(T-\mu)}] \\ &= nC_s + C_r + \frac{\Gamma(\alpha)}{A} \int_0^\eta [(C_{00} - C_r + C_{01}\mu + C_{02}\mu^2)/a^\alpha \\ &\quad + (C_{10} + C_{11}\mu)\alpha/a^{\alpha+1} + C_{20}\alpha(\alpha+1)/a^{\alpha+2}]\, d\mu \ , \end{aligned} \tag{8.4.21}$$

where $a = nT + \beta - (n+\gamma)\eta$. If $r \ge 2$, the density function $q(\omega)$ of $\hat{\Omega}_r$ is the convolution of g and h. Hence

$$\begin{aligned} q(\omega) &= \int_0^{\omega-\mu} g(\omega - s)h(s)\, ds \\ &= \frac{n\lambda(r\lambda)^{r-1}}{\Gamma(r-1)} \int_0^{\omega-\mu} s^{r-2} \exp\{-\lambda[n(\omega-\mu) - (n-r)s]\}\, ds, \qquad \omega \ge \mu \ . \end{aligned}$$

Thus, conditional on λ and μ, the Bayes risk now is given by

$$\begin{aligned} R(n, r, T) &= nC_s + \sum_{0 \le i+j \le 2} C_{ij} E[\lambda^i \mu^j] - \frac{nr^{r-1}\Gamma(r+\alpha)}{A\Gamma(r-1)} \\ &\quad \cdot \int_0^T \int_0^\eta \int_0^{\omega-\mu} s^{r-2}[(C_{00} - C_r + C_{01}\mu + C_{02}\mu^2)/b^{r+\alpha} \\ &\quad + (C_{10} + C_{11}\mu)(r+\alpha)/b^{r+\alpha+1} \\ &\quad + C_{20}(r+\alpha)(r+\alpha+1)/b^{r+\alpha+2}]\, ds\, d\mu\, d\omega \ , \end{aligned} \tag{8.4.22}$$

where $b = n\omega + \beta - (n+\gamma)\mu - (n-r)s$. Besides, it easily follows that

$$E[\lambda] = \frac{(\alpha-1)[\beta^\alpha - (\beta - \gamma\eta)^\alpha]}{\beta^\alpha(\beta - \gamma\eta) - \beta(\beta - \gamma\eta)^\alpha} \ ,$$

$$E[\mu] = \frac{(\alpha-2)\gamma\eta\beta^{\alpha-1} - (\beta - \gamma\eta)\beta^{\alpha-1} + \beta(\beta - \gamma\eta)^{\alpha-1}}{(\alpha-2)\gamma[\beta^{\alpha-1} - (\beta - \gamma\eta)^{\alpha-1}]} \ ,$$

$$E[\lambda^2] = \frac{\alpha(\alpha-1)[\beta^{\alpha+1} - (\beta - \gamma\eta)^{\alpha+1}]}{\beta^{\alpha+1}(\beta - \gamma\eta)^2 - \beta^2(\beta - \gamma\eta)^{\alpha+1}} \ ,$$

$$E[\lambda\mu] = \frac{(\alpha-1)\eta\beta^{\alpha-1}}{\beta^{\alpha-1}(\beta - \gamma\eta) - (\beta - \gamma\eta)^\alpha} - \frac{1}{\gamma} \ ,$$

$$
\begin{aligned}
E[\mu^2] &= \frac{(\alpha-2)(\alpha-3)\gamma^2\eta^2\beta^{\alpha-1} - 2(\alpha-3)\gamma\eta\beta^{\alpha-1}(\beta-\gamma\eta)}{(\alpha-2)(\alpha-3)\gamma^2[\beta^{\alpha-1}-(\beta-\gamma\eta)^{\alpha-1}]} \\
&\quad + \frac{2\beta^{\alpha-1}(\beta-\gamma\eta)^2 - 2\beta^2(\beta-\gamma\eta)^{\alpha-1}}{(\alpha-2)(\alpha-3)\gamma^2[\beta^{\alpha-1}-(\beta-\gamma\eta)^{\alpha-1}]} .
\end{aligned}
$$

Note that for $\alpha = 1$ (or 2, 3), values of the moments $E[\lambda]$, $E[\mu]$, $E[\lambda^2]$, $E[\lambda\mu]$, and $E[\mu^2]$ can all be obtained by L'Hospital's rule.

Analogous to Theorem 8.3, we may then establish the following result.

Theorem 8.4 [Lam (1990)]
The optimal sample size n_0 satisfies the inequality

$$n_0 \leq \min\{R(0,0,\infty)/C_s,\ R(0,0,0)/C_s\} ,$$

where $R(0,0,\infty) = C_r$ and

$$R(0,0,0) = \sum_{0 \leq i+j \leq 2} C_{ij} E[\lambda^i \mu^j] .$$

Of course, (8.4.13) is still applicable, and it provides an adaptive upper bound for the optimal sample size.

Therefore, the algorithm suggested for the one-parameter exponential distribution case is again a simple and finite algorithm for the present instance. However, for fixed n and r, either we can solve the equation $\partial R/\partial T = 0$ for the minimum of the Bayes risk, or we can minimize $R(n,r,T)$ by a numerical method.

In acceptance sampling plans, if the items of a batch are sophisticated and of high cost, we may put n items on inspection and terminate the experiment when a pre-assigned number of items, say r ($\leq n$) have failed. This is Type II censoring. However, if the cost of inspection increases heavily with time, we may put n items on inspection and terminate it at a pre-assigned time t. This is Type I censoring. In practice, Type I censoring is also commonly used and hence should be considered for the problem at hand. Recently, Lam (1994) has studied this same problem in detail. To start with, we assume again that the lifetime of an item in a batch has a one-parameter exponential distribution $e(\lambda)$ with the density given by (8.4.1). Suppose that λ has the Gamma prior distribution Gamma(α,β) with density (8.4.2). A random sample $\boldsymbol{X} = (X_1, X_2, \ldots, X_n)$ taken from the batch is subject to Type I censoring at time t. Let $M = \max\{i; X_{(i)} \leq t\}$ be the number of failures by time t. Then from Chapter 4, the MLE of the average lifetime $\theta = 1/\lambda$ is given by

$$\hat{\theta}_t = \begin{cases} \left\{\sum_{i=1}^{M} X_{(i)} + (n-M)t\right\}/M & , M > 0 , \\ nt & , M = 0 . \end{cases} \tag{8.4.23}$$

The decision function $\delta(\boldsymbol{X})$ in the present situation is also given by (8.4.5) in which $\hat{\theta}_r$ is replaced by $\hat{\theta}_t$. The loss function in (8.4.6) is investigated again. For quadratic loss function, let $[a]$ represent the integer part of a, and write $m_1 = \left[\frac{nt}{T+t}\right] + 1$, $m_2 = \left[\frac{nt}{T}\right]$ and $m_3 = m_2 + 1$. By using Fourier transform and contour integral, Lam (1994) showed that the Bayes risk is given by

$$R(n,t,T) = R_1 + R_2 + R_3 .$$

The first term is

$$R_1 = nC_s + C_0 + C_1\alpha/\beta + C_2\alpha(\alpha+1)/\beta^2 \ .$$

The second term is quite complicated, and is given by

$$\begin{aligned} R_2 &= \sum_{m=m_1}^{m_2} \sum_{j=0}^{[a_m/t]} \binom{n}{m}\binom{m}{j} (-1)^j \frac{1}{(m-1)!} \frac{\beta^\alpha}{\Gamma(\alpha)} \\ &\quad \cdot [(C_r - C_0)\Gamma(m+\alpha)(d+jt)^2 B_\gamma(m,\alpha) \\ &\quad - C_1\Gamma(m+\alpha+1)(d+jt)B_\gamma(m,\alpha+1) \\ &\quad - C_2\Gamma(m+\alpha+2)B_\gamma(m,\alpha+2)]/(d+jt)^{\alpha+2} \end{aligned}$$

where $a_m = mT - (n-m)t$, $d = (n-m)t + \beta$, $\gamma = (a_m - jt)/(a_m + d)$ and $B_x(a,b)$ is the incomplete beta function (8.4.10). Finally, the last term is

$$\begin{aligned} R_3 &= \sum_{m=m_3}^{n} \sum_{j=0}^{m} \binom{n}{m}\binom{m}{j} (-1)^j \beta^\alpha \{(C_r - C_0)(d+jt)^2 \\ &\quad - C_1\alpha(d+jt) - C_2\alpha(\alpha+1)\}/(d+jt)^{\alpha+2} \ . \end{aligned}$$

For Type I censoring case, an algorithm similar to the case of Type II censoring is still applicable. It is a finite algorithm because Theorem 8.3 or (8.4.13) holds in this instance as well, so that we can find an optimal sampling plan after a finite-step of searching. Since the equations $\partial R/\partial t = 0$ and $\partial R/\partial T = 0$ are no longer algebraic equations, a numerical method was introduced; see Lam (1994) for details.

Sensitivity analysis investigates the behaviour of the optimal solution due to changes in parameters or coefficients in the model. In practice, usually we can not determine or estimate the parameters and coefficients precisely. Therefore, as the stationarity study in the solution to a differential equation, the sensitivity analysis in the optimal solution of sampling plans is also very important. Lam (1994) has considered a large number of numerical examples to examine the variation in the optimal sampling plans and in the corresponding minimum Bayes risks due to changes in parameters α, β and changes in coefficients C_0, C_1, C_2, C_s and C_r. From the numerical results of Lam (1994), one can see that the model is insensitive to the parameters and the coefficients. This is an advantage of the proposed sampling plan.

In life-testing, the lifetime of an item in a batch is quite often simultaneously affected by an extraneous factor, which may require to introduce random censoring. For example, a group of patients who suffer from both lung cancer and heart disease are administered with a new drug which is claimed to be beneficial to lung cancer patients. Lifetime of each participated patient is reported and recorded. For each patient, a censoring time is assigned which corresponds to the survival time of heart disease for the patient. Therefore, a patient who survives beyond this time is no longer observed and a censored data is collected. These censoring times are assumed to be iid with known distribution and hence random censoring is used. Lam and Choy (1994) considered the problem of optimal sampling plans for the one-parameter exponential distribution in (8.4.1) with random censoring. They developed a model for this problem and derived an explicit expression for the Bayes risk. On the basis of Theorem 8.3 or (8.4.13), a finite algorithm parallel to Type I or II cases can be developed in this case as well. For details, see Lam and Choy (1994).

Although in this section, we mainly studied the case of quadratic loss function, the model and the algorithm are all applicable to the case of a general polynomial loss function.

References

Engelhardt, M. and Bain, L.J. (1978). Tolerance limits and confidence limits on reliability for the two-parameter exponential distribution, *Technometrics*, **20**, 37-39.

Guenther, W.C. (1971). On the use of best tests to obtain best β-content tolerance intervals, *Statistica Neerlandica*, **25**, 191-202.

Guenther, W.C., Patil, S.A., and Uppuluri, V.R.R. (1976). One-sided β-content tolerance factors for the two-parameter exponential distribution, *Technometrics*, **18**, 333-340.

Guttman, I. (1970). *Statistical Tolerance Regions: Classical and Bayesian*, Griffin, London.

Guttman, I. (1988). Tolerance regions, statistical, In *Encyclopedia of Statistical Sciences* (Eds., S. Kotz, N.L. Johnson and C.B. Read), pp. 272-287, John Wiley & Sons, New York.

Hald, A. (1967). Asymptotic properties of Bayesian single sampling plans, *Journal of the Royal Statistical Society, Series B*, **29**, 162-173.

Hald, A. (1981). *Statistical Theory of Sampling Inspection by Attributes*, Academic Press, New York.

IMSL (1991). STAT/LIBRARY, Version 2.0. IMSL, Inc. Houston.

Kocherlakota, S. and Balakrishnan, N. (1986). One- and two-sided sampling plans based on the exponential distribution, *Naval Research Logistics Quarterly*, **33**, 513-522.

Lam, Y. (1988a). A decision theory approach to variable sampling plans, *Scientia Sinica, Series A*, **31**, 120-140.

Lam, Y. (1988b). Bayesian approach to single variable sampling plans, *Biometrika*, **75**, 387-391.

Lam, Y. (1990). An optimal single variable sampling plan with censoring, *The Statistician*, **39**, 53-67.

Lam, Y. (1994). Bayesian variable sampling plans for the exponential distribution with Type I censoring, *Annals of Statistics*, **22** (to appear).

Lam, Y. and Choy, S.T.B. (1994). Bayesian variable sampling plans for the exponential distribution with random censoring, *Journal of Statistical Planning and Inference* (revised).

Lam, Y. and Lau, L.C. (1993). Optimal single variable sampling plans, *Communications in Statistics - Simulation and Computation*, **22**, 371-386.

Varde, S.D. (1969). Life testing and reliability estimation for the two-parameter exponential distribution, *Journal of the American Statistical Association*, **64**, 621-631.

Wetherill, G.B. (1977). *Sampling Inspection and Quality Control*, Second edition, Chapman and Hall, London.

CHAPTER 9

Prediction Problems

H.N. Nagaraja[1]

9.1 Introduction

For a frequentist statistician, prediction means guessing the value of an unobserved random variable (rv) of interest, while estimation corresponds to guessing the value of an unknown, but fixed, parameter. Thus, the former is conceptually different from the latter, and hence different, though similar, optimality criteria have to be developed to determine best predictors and prediction regions. These subtle differences between the problem of estimation and of prediction in a general linear model context are highlighted in Bibby and Toutenberg (1977). In contrast, for a Bayesian, both are essentially the same. In any case, prediction procedures form an important component of inferential procedures associated with the exponential distribution. Geisser (1986) points to the long history of prediction problems by noting that Bayes had considered such a problem in 1763. In this chapter, we describe several scenarios where prediction is *the* problem of interest, discuss various predictors proposed in the literature, and study their properties. We will present a survey of the area and hopefully cover all major contributions in our effort to do so.

In most of the prediction problems related to the exponential distribution, the random variable of interest is an order statistic or a function of order statistics from the same or a future sample. These variables, themselves, lead to system reliability and quality assurance variables and thus are important to reliability engineers and manufacturers. Since the memoryless nature of the exponential distribution makes the distribution theory for order statistics nice and simple (see Chapters 2 and 3), the resulting predictors have simple form and have tractable distributions. These factors have contributed to the presence of a huge literature on prediction of order statistics and related statistics from the exponential distribution. The golden decade of research on this topic seems to have been the 1970's. Most of the results we discuss in this chapter were obtained during that time period.

As in the case of an estimator, a predictor can be either a point predictor or a

[1]The Ohio State University, Columbus, Ohio

prediction region. Earlier, in Chapters 4 and 5, MLE's and best linear unbiased estimators (BLUE's) of the location (μ) and scale (θ) parameters were discussed. On similar lines, one can consider maximum likelihood predictors (MLP's) and best linear unbiased predictors (BLUP's) of variables of interest. We will discuss these and other prediction procedures. Our focus will be on two-parameter exponential distribution; this is a member of location-scale family of distributions. Some of the earlier results on prediction were obtained for the one-parameter exponential distribution which belongs to the scale family.

Prediction problems associated with the exponential distribution can be classified as either one-sample or multi-sample problems. The key difference is in the fact that the variable to be predicted and the observed data are dependent in the former case, whereas in the latter case, they are independent (conditioned on the value of the parameters, in the Bayesian sense). These problems will be introduced and described in Section 9.2. Next we will embark on the presentation of various prediction criteria and techniques. In Sections 9.3 and 9.4 we describe the general methods of linear prediction and maximum likelihood prediction respectively, and apply the techniques to exponential prediction problems. In Section 9.5 we describe the prediction intervals for various variables of interest. Section 9.6 presents a brief overview of the various Bayes procedures suggested in the literature. In Section 9.7 we look at other prediction criteria like probability of nearness (PN), probability of concentration (PC), and median unbiased prediction. We compare the previously proposed predictors in terms of these criteria. Section 9.8 describes some interesting miscellaneous results and the chapter concludes with some remarks in Section 9.9. We will not cover sequential prediction procedures in our review.

In any area of research, consolidation and synthesis of the literature occurs from time to time and the topic of prediction is no exception. Bibby and Toutenberg (1977) is a good source for frequentist approach to prediction especially in the context of linear models. Aitchison and Dunsmore (1975) is an excellent reference on Bayesian approach to prediction. Kalbfleisch (1971) discusses general likelihood methods of prediction and Hinkley (1979) introduces a definition of predictive likelihood with close resemblance to Bayesian ideas. There are the survey papers on prediction intervals by Hahn and Nelson (1973) and Patel (1989), and the recent book on statistical intervals by Hahn and Meeker (1991). David (1981) and Nelson (1982) also contain discussion on prediction problems. However, there appears to be no survey of the work on prediction problems associated with the exponential distributions thus far, even though there has been plenty of papers on this topic.

We now introduce our notations and conventions. In our general discussion $\boldsymbol{X}$ represents the random vector being observed and Y is the variable being predicted. $X_{i:n}$ represents the ith order statistic from a random sample from an absolutely continuous distribution with probability density function (pdf) $f(x)$. In a two-sample problem, $Y_{j:m}$ represents the j-th order statistic and $\bar{Y}_m$ is the mean of an independent random sample of size m from the pdf f. The variables $Z_1, \ldots, Z_j$ represent independent identically distributed standard exponential, $e(1)$, random variables and $Z_{i:j}$'s represent their order statistics. Median and mode of a variable W are denoted by Med(W) and Mode(W), respectively. Gam$(\alpha, 1/\theta)$ represents the gamma distribution with shape parameter α and scale parameter θ. Thus, Gam$(j, 1/\theta)$ represents the sum of j i.i.d. $e(0, \theta)$ variables. The mean and variance of $Z_{j:n} - Z_{i:n} \stackrel{d}{=} Z_{j-i:n-i}$ are given by $\alpha_1(i, j; n)$ and $\alpha_2(i, j; n)$, respectively, where, when $i = 0$, we define $Z_{0:n} \equiv 0$. Thus,

$$\alpha_1(i,j;n) = \sum_{k=i+1}^{j} 1/(n-k+1) \text{ and } \alpha_2(i,j;n) = \sum_{k=i+1}^{j} 1/(n-k+1)^2 \qquad (9.1.1)$$

(see Chapter 3).

9.2 Types of Prediction Problems

Prediction problems come up naturally in several real life situations. As pointed out earlier, they can be broadly classified under two categories: (i) the variable to be predicted comes from the same experiment or sample so that it may be correlated with the observed data, (ii) it comes from an independent future experiment. Both of these situations do arise in the context of reliability and lifetesting. Formally we call the problems in the first category as one-sample problems and those in the latter constitute two- or multi-sample problems.

Suppose a machine consists of n components and fails whenever k of these components fail. Such a machine is referred to as $(n-k+1)$-out-of-n system. Suppose our observations consist of the first r failure times, and the goal is to predict the failure time of the machine. Assuming the components' life lengths are i.i.d. we have a prediction problem involving a Type II censored sample, and it falls into category (i). Formally, the problem is to predict $X_{k:n}$ having observed $X_{1:n}, \ldots, X_{r:n}$. A popular choice for the lifelength of the components is the exponential distribution. Thus, one can think of a point predictor or an interval predictor for the lifelength of the system as one-sample problem. Let us label this problem as P1 for future reference.

In another situation, one might be interested in predicting the average strength of all the items in a lot or the average strength of survivors having observed the first r failures. This may well be the case for the batches of certain electrical components. Suppose all the items in a batch are put to test and the voltage was gradually increased till r of the items failed. In this (destructive) experiment, the average strength of the surviving items will give us an idea about the reliability of the components that will be used. This again is a one-sample problem.

Consider the third situation where a manufacturer of a certain equipment is interested in setting up a warranty for the equipment in a lot being sent out to the market. Using the information based on a small sample, possibly censored, the goal is to predict and set a lower prediction limit for the weakest item in a future sample. Typical assumption here is that the two samples are independent. This falls into category (ii), and we call it a two-sample problem. One may be interested in the lifelength of the k-th weakest item and the average lifelength of the lot simultaneously. Then our focus will be on a prediction region for the two variables of interest. We will label the problem of predicting $Y_{k:m}$ based on $X_{1:n}, \ldots, X_{r:n}$ as P2 for future discussion.

While several researchers have concentrated on the above problems there are some papers dealing with special prediction situations involving several independent random samples. These belong to category (ii) and will be introduced later.

9.3 Best Linear Unbiased and Invariant Predictors

9.3.1 General Results for Location-Scale Family Distributions

Let us now consider the problem P1 more closely when f belongs to a location-scale family. The best unbiased predictor (BUP) of $X_{k:n}$ is $E[X_{k:n}|X_{1:n}, \ldots, X_{r:n}] = E[X_{k:n}|X_{r:n}]$ by the Markov property of order statistics. This will be a linear function of the parameters. By replacing them by their BLUE's will not produce the BLUP of $X_{k:n}$! So the question is what is the adjustment factor? To answer this and other problems we recall some general results that are relevant.

Let $\boldsymbol{X} = (X_1, \ldots, X_r)'$ be a vector of observed data and the goal is to predict a random variable Y. We say a linear predictor of Y based on $\boldsymbol{X}$, say $\boldsymbol{c}'\boldsymbol{X}$, is unbiased if $E[Y] = E[\boldsymbol{c}'\boldsymbol{X}]$ for all values of the parameters involved, where $\boldsymbol{c} = (c_1, \ldots, c_r)$. It will be a BLUP of Y if $\mathrm{Var}(Y - \boldsymbol{c}'\boldsymbol{X})$ is the smallest among linear unbiased predictors of Y.

Kaminsky and Nelson (1975a) applied the results of Goldberger (1962) for general linear model to develop the theory, and obtained expressions for BLUPs of order statistics in location-scale families. It is closely related to the theory of BLUE developed by Lloyd (1952) and reported in Chapter 5. Recently, in a discussion paper, Robinson (1991) describes the virtues and uses of BLUP's in the framework of a general linear model.

We will now give a brief description of the general results. Let $(X_1, \ldots, X_r, Y)$ be $r + 1$ rv's whose joint pdf can be expressed as

$$f(\boldsymbol{x}, y; \mu, \theta) = \theta^{-(r+1)} f((x_1 - \mu)/\theta, \ldots, (y - \mu)/\theta) \ . \tag{9.3.1}$$

Let $a_i = E[(X_i - \mu)/\theta]$ and $v_{i,j} = \mathrm{Cov}((X_i - \mu)/\theta,\ (X_j - \mu)/\theta)$, for $i, j = 1, \ldots, r$. Then $E[\boldsymbol{X}] = \mu\mathbf{1} + \theta\boldsymbol{a}$, and $\mathrm{Cov}(X) = \theta^2\boldsymbol{V}$, where $\mathbf{1} = (1, \ldots, 1)'$, $\boldsymbol{a} = (a_1, \ldots, a_r)'$ and $\boldsymbol{V} = (v_{i,j})$, $i, j = 1, \ldots, r$. Further, the BLUE's are given by

$$\hat{\mu} = -\boldsymbol{a}'\boldsymbol{G}\boldsymbol{X} \qquad \text{and} \qquad \hat{\theta} = \mathbf{1}'\boldsymbol{G}\boldsymbol{X} \ , \tag{9.3.2}$$

where $d = (\mathbf{1}'\boldsymbol{V}^{-1}\mathbf{1})(\boldsymbol{a}'\boldsymbol{V}^{-1}\boldsymbol{a}) - (\mathbf{1}'\boldsymbol{V}^{-1}\boldsymbol{a})^2$ and $\boldsymbol{G} = \boldsymbol{V}^{-1}(\mathbf{1}\boldsymbol{a}' - \boldsymbol{a}\mathbf{1}')\boldsymbol{V}^{-1}/d$. From the least-squares theory it is also known that $\mathrm{Var}(\hat{\mu}) = \theta^2\boldsymbol{a}'\boldsymbol{V}^{-1}\boldsymbol{a}/d$, $\mathrm{Var}(\hat{\theta}) = \theta^2\mathbf{1}'\boldsymbol{V}^{-1}\mathbf{1}/d$ and $\mathrm{Cov}(\hat{\mu}, \hat{\theta}) = -\theta^2\mathbf{1}'\boldsymbol{V}^{-1}\boldsymbol{a}/d$.

Now, suppose $a_0 = E[(Y - \mu)/\theta]$ and $w_i = \mathrm{Cov}((X_i - \mu)/\theta,\ (Y - \mu)/\theta)$. The BLUE of $E[Y] = \mu + a_0\theta$, is $\hat{\mu} + a_0\hat{\theta}$. To obtain the BLUP of Y we will have to minimize the variance of $(Y - \boldsymbol{c}'\boldsymbol{X})$ subject to the condition of unbiasedness. This can be easily accomplished by the method of Lagrange multipliers. It turns out that the BLUP of Y is (Goldberger, 1962, p. 371)

$$\hat{Y} = \hat{\mu} + a_0\hat{\theta} + \boldsymbol{w}'\boldsymbol{V}^{-1}(\boldsymbol{X} - \hat{\mu}\mathbf{1} - \hat{\theta}\boldsymbol{a}) \ , \tag{9.3.3}$$

where $\hat{\mu}$, $\hat{\theta}$ are the BLUE's of μ and θ given by (9.3.2) and $\boldsymbol{w}' = (w_1, \ldots, w_r)$.

Another criterion used in connection with estimation is invariance. In the prediction context, while dealing with a location-scale family given by (9.3.1), one would restrict attention to predictors that are invariant with respect to location and scale transformations. For such a family, Kaminsky, Mann, and Nelson (1975, p. 146) obtain an expression for $\tilde{Y}$, the best linear invariant predictor (BLIP) of Y. They have shown that

$$\tilde{Y} = \hat{Y} - \{c_{12}/(1 + c_{22})\}\hat{\theta} \ , \tag{9.3.4}$$

where $c_{12}\theta^2 = \text{Cov}\{\hat{\theta}, (1 - \boldsymbol{w}'\boldsymbol{V}^{-1}\mathbf{1})\hat{\mu} + (a_0 - \boldsymbol{w}'\boldsymbol{V}^{-1}\boldsymbol{a})\hat{\theta}\}$, and $c_{22}\theta^2 = \text{Var}(\hat{\theta})$.

When Y and components of $\boldsymbol{X}$ are uncorrelated, it is clear from the above discussion that

$$\hat{Y} = \hat{\mu} + a_0\hat{\theta} \qquad \text{and} \qquad \tilde{Y} = \hat{Y} - \{c_{12}/(1 + c_{22})\}\hat{\theta} , \tag{9.3.5}$$

with $c_{12}\theta^2 = \text{Cov}\{\hat{\theta}, \hat{\mu} + a_0\hat{\theta}\}$, and $c_{22}\theta^2 = \text{Var}(\hat{\theta})$.

Takada (1981) has developed the general theory of unbiased and invariant prediction for location-scale families. From his work it follows that $\hat{Y}$ is in fact the BUP and $\tilde{Y}$ is the BIP of Y.

9.3.2 Best Unbiased and Invariant Predictors of Exponential Samples

For the $e(\mu, \theta)$ distribution, when $X_{1:n}, \ldots, X_{r:n}$ are observed, the minimum variance unbiased estimators (MVUE's, also the BLUE's) of μ and θ are given by

$$\hat{\mu} = X_{1:n} - \hat{\theta}/n$$

and

$$\begin{aligned} \hat{\theta} &= \left\{\sum_{i=1}^{r}(X_{i:n} - X_{1:n}) + (n-r)(X_{r:n} - X_{1:n})\right\}\Big/(r-1) \\ &= T/(r-1) , \end{aligned} \tag{9.3.6}$$

where $T = \{\sum_{i=2}^{r}(n - i + 1)(X_{i:n} - X_{i-1:n})\}$. Since $\text{Cov}(X_{i:n}, X_{j:n}) = \text{Var}(X_{i:n}) = \theta^2\alpha_2(i, n:n)$ for $i \le j$, $\boldsymbol{V}^{-1}$ in (9.3.2) has a simple form [Arnold, Balakrishnan and Nagaraja (1992, p. 176)] and thus substantial simplifications take place.

For the problem P1, $w_i = \text{Var}(X_{i:n})$, and further simplifications are possible in (9.3.3). Thus the BLUP of $X_{k:n}$ reduces to

$$\hat{X}_{k:n} = X_{r:n} + \{\alpha_1(0, k; n) - \alpha_1(0, r:n)\}\hat{\theta} = X_{r:n} + \alpha_1(r, k; n)\hat{\theta} , \tag{9.3.7}$$

where $\alpha_1(i, n; n)$ is given by (9.1.1).

Now, the mean squared error (MSE) of prediction of the BLUP is $\text{MSE}(\hat{X}_{k:n}) = E(\hat{X}_{k:n} - X_{k:n})^2$. To compute it, we observe from (9.3.6) that, $(r-1)\hat{\theta} \stackrel{d}{=} \theta\sum_{i=2}^{r} Z_i$, and hence, $(r-1)\hat{\theta}/\theta$ has Gamma$(r-1, 1)$ distribution. Further $X_{k:n} - X_{r:n} \stackrel{d}{=} \theta Z_{k-r:n-r}$, and $(X_{k:n} - X_{r:n})$ and $\hat{\theta}$ are independent. On combining all these facts, we observe that

$$\begin{aligned} \text{MSE}(\hat{X}_{k:n}) &= \text{Var}(X_{k-r:n-r}) + \{\alpha_1(r, k; n)\}^2\text{Var}(\hat{\theta}) \\ &= \theta^2\{\alpha_2(r, k; n) + [\alpha_1(r, k; n)]^2/(r-1)\} . \end{aligned}$$

For the BLIP of $X_{k:n}$ it turns out that in (9.3.4), $c_{12} = \alpha_1(r, k; n)/(r-1)$ and $c_{22} = 1/(r-1)$ and thus

$$\tilde{X}_{k:n} = X_{r:n} + (1 - r^{-1})\alpha_1(r, k; n)\hat{\theta} . \tag{9.3.8}$$

The magnitude of decrease in MSE of BLIP over the BLUP is $(r-1)r^{-2}\{\alpha_1(r, k; n)\}^2\theta^2$. As noted earlier, $\hat{X}_{k:n}$ and $\tilde{X}_{k:n}$ are in fact the BUP and BIP of $X_{k:n}$, respectively. Similar simplifications are possible if we want to predict the average of the $n - r$ censored observations. When some of the order statistics among $X_{1:n}, \ldots, X_{r:n}$ are missing, expressions for the predictors are slightly messier. In fact, the general approach described in Section 9.3.1 can handle the problem of predicting any linear function of future order statistics. However, we do not plan to pursue the details here.

For the $e(0,\theta)$ distribution, the forms of the predictors are very similar and in fact are simpler. For that distribution, Kaminsky (1977a) has obtained the BLUP and BLIP of the sth failure time based on the first r failure times when the failed items are immediately replaced. Even though technically this is different than predicting $X_{k:n}$, the resulting expressions are simple and similar.

For the problem P2, Y and $\boldsymbol{X}$ are uncorrelated, and we can use (9.3.5) to determine the BUP and BIPs. It is easily seen (Takada, 1991) that

$$\hat{Y}_{k:m} = X_{1:n} + (1 - n^{-1})\hat{\theta} \text{ and } \tilde{Y}_{k:m} = X_{1:n} + (1 - n^{-1})(1 - r^{-1})\hat{\theta} \ , \tag{9.3.9}$$

where $\hat{\theta}$ is given by (9.3.6).

9.4 Maximum Likelihood Prediction

Kaminsky and Rhodin (1985) apply the principle of maximum likelihood to the joint prediction and estimation of a future order statistic and an unknown parameter in the one-sample problem. The procedure follows the classical maximum likelihood method of estimation. Let $\boldsymbol{X}$ be the vector of observed values and $\boldsymbol{Y}$ be a vector of rv's to be predicted. Suppose $f(\boldsymbol{x}, \boldsymbol{y}; \boldsymbol{\theta})$ is their joint pdf where $\boldsymbol{\theta}$ is the unknown, possibly vector valued, parameter. Let $\boldsymbol{Y}^* = t(\boldsymbol{X})$ and $\boldsymbol{\theta}^* = u(\boldsymbol{X})$ be statistics such that $f(\boldsymbol{x}, \boldsymbol{y}; \boldsymbol{\theta})$ attains its maximum when $\boldsymbol{y} = \boldsymbol{y}^*$ and $\boldsymbol{\theta} = \boldsymbol{\theta}^*$. Then $\boldsymbol{Y}^*$ is called the MLP of $\boldsymbol{Y}$ and $\boldsymbol{\theta}^*$ is called the predictive maximum likelihood estimator (PMLE) of $\boldsymbol{\theta}$. Note that the PMLE of $\boldsymbol{\theta}$ is different than the MLE of $\boldsymbol{\theta}$ based on either $\boldsymbol{X}$ or $(\boldsymbol{X}, \boldsymbol{Y})$. Further, if $\boldsymbol{\theta}$ is known, the MLP of $\boldsymbol{Y}$ is a mode of the conditional distribution of $\boldsymbol{Y}$ given $\boldsymbol{X} = \boldsymbol{x}$.

When the parameters of the exponential distribution are known, for the one-sample problem P1, one has to find the mode of the conditional distribution of $X_{k:n}$ given $X_{r:n}$. The MLP of $X_{k:n}$ is given by

$$\begin{aligned} X^*_{k:n} &= X_{r:n} + \theta \, \text{Mode}(Z_{k-r:n-r}) \\ &= X_{r:n} + \theta \log\{(n-r)/(n-k+1)\} \ . \end{aligned} \tag{9.4.1}$$

The above form assumes $r + 1 < k \le n$. If $k = r + 1$, the MLP of $X_{k:n}$ does not exist.

When the parameters are unknown, the likelihood function $f(\boldsymbol{x}, y; \mu, \theta)$ can be factored as $f(\boldsymbol{x}; \mu, \theta) f(y|\boldsymbol{x}; \mu, \theta)$ where the second factor represents the conditional pdf of $X_{k:n}$ given $X_{r:n}$. It is free of μ and, for a given θ, it is maximized when $y = y_0 = X_{r:n} + \theta \log\{(n-r)/(n-k+1)\}$. For a given θ the first factor is maximized when $\mu = X_{1:n}$. Thus, we have to determine θ for which $f(\boldsymbol{x}, y_0; x_{1:n}, \theta)$ attains its maximum. On differentiation, it is easy to see that this occurs when $\theta = \theta^*$, given by

$$\theta^* = T/(r+1) \ , \tag{9.4.2}$$

where T is the statistic appearing in (9.3.6). Thus, the PMLP of θ is θ^* given above while the MLE of θ based on $\boldsymbol{X}$ is $(r+1)\theta^*/r$. Further, the MLE of θ based on

$(\boldsymbol{X}, X_{k:n})$ coincides with θ^* only when $k = r + 1$. From (9.3.6) and (9.4.2) it is evident that $\theta^* = (r-1)\hat{\theta}/(r+1)$ where $\hat{\theta}$ is the MVUE of θ. Now we can conclude that, when the parameters are unknown, the MLP of $X_{k:n}$ is given by

$$X^*_{k:n} = X_{r:n} + \theta^* \log\{(n-r)/(n-k+1)\} , \tag{9.4.3}$$

where $(r+1)\theta^*/\theta$ has a Gamma$(r-1, 1)$ distribution.

On using (9.3.7), (9.3.8), (9.4.3), and the fact that $\alpha_1(r, k; n) > \log\{(n-r)/(n-k+1)\}$ it is easily seen that MLP < BLIP < BLUP. Consequently, the MLP and BLIP under predict $X_{k:n}$. Further, since the MLP is also an invariant predictor, its MSE is more than that of the BLIP. However, the MSE of BLUP (BUP) can exceed that of the MLP.

For the $e(0, \theta)$ distribution, Kaminsky and Rhodin (1985) show that (9.4.3) holds with $\theta^* = \{\sum_{i=1}^{r-1} X_{i:n} + (n-r+1)X_{r:n}\}/(r+1)$. They have numerically compared the MSE of MLP with that of BUP. The conclusion is that MLP tends to be more efficient when $k < 2r$ or if k is near n, and is likely to be less efficient otherwise. Note that for the $e(0, \theta)$ parent, $(r+1)\theta^*/\theta$ has a Gamma$(r, 1)$ distribution.

9.5 Prediction Intervals and Regions

9.5.1 Definitions and Methods of Determination

We say that an interval $(L(\boldsymbol{X}), U(\boldsymbol{X}))$ is a $100\gamma\%$ prediction interval (PI) for Y if $P_\theta[L(\boldsymbol{X}) < Y < U(\boldsymbol{X})] = \gamma$ for all θ. The limits $L(\boldsymbol{X})$ and $U(\boldsymbol{X})$ are called the lower and the upper prediction limits. By choosing one of them to be $(+/-)$ infinity we obtain a one-sided PI for Y. In the reliability context, quite often the one-sided intervals are relevant. It appears Hewett (1968) was the first to obtain a PI for an exponential population when he considered the problem of predicting a future sample sum based on a random sample from a χ^2 distribution.

Hahn and Meeker (1991, Chapter 2) present an overview of different kinds of statistical intervals, and discuss various types of PI's and their applications. When $\boldsymbol{X}$ and Y are independent, $(L(\boldsymbol{X}), U(\boldsymbol{X}))$, the $100\gamma\%$ PI for Y, can also be interpreted as a γ-expectation tolerance interval.

Patel (1989), in an extensive review of frequentist approach to the PI's, provides a comprehensive catalog of these intervals for common families of distributions. The most popular approach to determine a PI is to use a pivotal quantity that is a function of $\boldsymbol{X}$ and Y such that its probability distribution does not depend on θ. Let $Q(\boldsymbol{X}, Y)$ be such a pivotal quantity and suppose intervals of the type $\{q_1 < Q < q_2\}$ can be inverted into an interval of the form $\{L(\boldsymbol{X}) < Y < U(\boldsymbol{X})\}$. If $P[q_1 < Q < q_2] = \gamma$, then $(L(\boldsymbol{X}), U(\boldsymbol{X}))$ provides a $100\gamma\%$ PI for Y. While it is desirable that Q contain as much information as possible from the observed data, work involved in the computation of its percentiles and the determination of the PI should be reasonable. Basically all the PI's suggested for the exponential samples are based on these considerations. In some of our applications Q is a function of the MLE of θ based on $\boldsymbol{X}$.

Another useful general procedure, due to Faulkenberry (1973), involves the idea of conditioning with respect to a sufficient statistic. His approach assumes the setting where $T(\boldsymbol{X}, Y)$ is sufficient for θ, $P[Y \in R(t) \mid T = t] = \gamma$ and the region $Y \in R(t)$

can be transformed into a region of the form $\{L(\boldsymbol{X}) < Y < U(\boldsymbol{X})\}$. Other approaches include those employing normal approximations and certain test procedures. Brief descriptions of these may be found in Patel (1989).

9.5.2 General Distribution Theory

Several researchers have computed exact and approximate PI's for statistics of interest in both one-sample and two-sample problems. All these intervals invariably depend on some ratio of linear combinations of standard exponential variables. Unaware of the general results derived by Box (1954) in his study of effect of inequality of variance in the one-way classification, these authors have independently derived the distributions of the special ratios of their interest. We will unify all of those derivations with the help of the following three results adapted from the valuable work of Box.

Result 9.1 Let $U_1, \ldots, U_t$ be independent χ^2 random variables with even degrees of freedom, where U_i is $\chi^2_{2\lambda_i}$, $i = 1, \ldots, t$. Let $U = \sum_{i=1}^t a_i U_i$, where the a_i's are nonzero constants. For $i = 1$ to t, define $g_i(y) = \prod_{j \neq i}^t \{1 + [(1-y)a_j/a_i]\}^{-\lambda_j}$ and let

$$\alpha_{ij} = g_i^{(\lambda_i - j)}(0)/((\lambda_i - j)!), \qquad j = 1, \ldots, \lambda_i \ , \tag{9.5.1}$$

where $g_i^{(k)}(0)$ is the kth derivative of $g_i(y)$ evaluated at 0. Then, for all real u,

$$P[U > u] = \sum_{i=1}^{t} \sum_{j=1}^{\lambda_i} \alpha_{ij} \ P[a_i \chi^2_{2\lambda_j} > u] \ . \tag{9.5.2}$$

The basic idea behind the proof of this result is the possibility of expressing the characteristic function of U as a linear function of the characteristic functions of χ^2 variables through a partial function expansion. Details may be found in Box (1954, pp. 291-292). It may be noted that his equation (2.11), which is a modification of (9.5.2) above, tacitly assumes that the a_i's are positive.

Result 9.2 Let the U_i's be as defined in Result 9.1 above. Take $t \geq 2$ and let m be a positive integer less than t. Let $V = (\sum_{i=1}^m c_i U_i)/(\sum_{i=m+1}^t d_i U_i)$, where the constants d_i's are positive. For a real v, let $U = \sum_{i=1}^t a_i U_i$, where $a_i = c_i$ if $1 \leq i \leq m$, and $a_i = -v d_i$, if $m+1 \leq i \leq t$. Then

$$P[V > v] = \begin{cases} \sum_{i=1}^m \sum_{j=1}^{\lambda_i} \alpha_{ij} I_{[c_i > 0]} & \text{if} \quad v \geq 0 \ , \\ \sum_{i=1}^m \sum_{j=1}^{\lambda_i} \alpha_{ij} I_{[c_i > 0]} + \sum_{i=m+1}^t \sum_{j=1}^{\lambda_i} \alpha_{ij} & \text{if} \quad v < 0 \ , \end{cases} \tag{9.5.3}$$

where the α_{ij}'s are as defined in Result 9.1, and I represents the indicator function.

Proof: Note that $P[V > v] = P\left[\sum_{i=1}^m c_i U_i > v \left(\sum_{i=m+1}^t d_i U_i\right)\right]$, since the d_j's are positive. Thus $P[V > v] = P[U > 0]$ where U is as defined in Result 9.2. On using Result 9.1, it then follows that

$$P[V > v] = \sum_{i=1}^{m} \sum_{j=1}^{\lambda_i} \alpha_{ij} P[c_i \chi^2_{2\lambda_j} > 0] + \sum_{i=m+1}^{t} \sum_{j=1}^{\lambda_i} \alpha_{ij} P[-v d_i \chi^2_{2\lambda_j} > 0] \ .$$

This is equivalent to (9.5.3) and thus the proof is complete. □

In quite a few applications, it is time consuming to obtain the percentiles of V because of the iterative process involved, where, at each step, α_{ij}'s are to be evaluated. To avoid this, one can use a χ^2 approximation to a nonnegative linear function of independent χ^2 variables, due to Satterthwaite (1941). Patnaik (1949) has extended it to noncentral χ^2 distributions. The following result, collected from Box (1954), describes the necessary approximations.

Result 9.3 If the a_i's in Result 9.1 are positive, $U = \sum_{i=1}^{t} a_i U_i$ is distributed approximately as $a\chi^2_\nu$ where $a\nu = E[U]$ and $2a^2\nu = \mathrm{Var}(U)$. This means

$$a = \left\{\sum_{j=1}^{t} a_j^2 \lambda_j\right\} \Big/ \left\{\sum_{j=1}^{t} a_j \lambda_j\right\}$$

and

$$\nu = \left\{\sum_{j=1}^{t} a_j \lambda_j\right\}^2 \Big/ \left\{\sum_{j=1}^{t} a_j^2 \lambda_j\right\} .$$

Further, if the c_i's and d_i's in Result 9.2 are all positive, V is approximately distributed as bF_{ν_1,ν_2} where

$$b = \left\{\sum_{j=1}^{m} c_j \lambda_j\right\} \Big/ \left\{\sum_{j=m+1}^{t} d_j \lambda_j\right\} ,$$

$$\nu_1 = \left\{\sum_{j=1}^{m} c_j \lambda_j\right\}^2 \Big/ \left\{\sum_{j=1}^{m} c_j^2 \lambda_j\right\} ,$$

and

$$\nu_2 = \left\{\sum_{j=m+1}^{t} d_j \lambda_j\right\}^2 \Big/ \left\{\sum_{j=m+1}^{t} d_j^2 \lambda_j\right\} .$$

Discussion

In most of our applications we have $t = m+1$, $\lambda_j = 1$ for $j = 1$ to m and $\lambda_{m+1} \equiv \lambda_t = \lambda$, and c_j's are all distinct. Without loss of generality we may take d_{m+1} to be 1. Then substantial simplifications are possible in (9.5.1) and (9.5.3). In fact, in that special case,

$$P[V > v] = \begin{cases} \sum_{i=1}^{m} \alpha_{i1} I_{[c_i>0]} & \text{if } \; v \geq 0 , \\ \sum_{i=1}^{m} \alpha_{i1} I_{[c_i>0]} + \sum_{j=1}^{\lambda} \alpha_{m+1,j} & \text{if } \; v < 0 , \end{cases} \tag{9.5.4}$$

where, from (9.5.1) we have,

$$\alpha_{i1} = \left(\frac{c_i}{c_i + v}\right)^{\lambda} \prod_{j \neq i=1}^{m} \left(\frac{c_i}{c_i - c_j}\right) , \qquad \text{for } i = 1, \ldots, m , \tag{9.5.5}$$

and

$$\alpha_{m+1,j} = g_{m+1}^{(\lambda-j)}(0)/((\lambda-j)!), \qquad j = 1, \ldots, \lambda \ ,$$

with

$$g_{m+1}(y) = \prod_{j=1}^{m} \{1 - [(1-y)c_j/v]\}^{-1} \ .$$

If the c_i's are all nonnegative, V will be a nonnegative variable and thus there is no need to evaluate $\alpha_{m+1,j}$ at all!

Box (1954) has carried out a small numerical study to evaluate the accuracy of the approximations suggested in Result 9.3. His conclusion is that while the χ^2 approximation is fairly good, the F approximation is not of great accuracy. In Result 9.3, if ν is not an integer, one can either take the closest integer as the approximate degrees of freedom or use incomplete gamma function to determine the approximate percentiles. Mann, Schafer, and Singpurwalla (1974, pp. 173-174) note that, when ν_1 and ν_2 are not integers, one can use the relationship between an F variable and a beta variable to approximate the needed percentiles. They also give necessary details.

9.5.3 One-parameter Exponential Distribution

Patel (1989), has enumerated details about various PI's connected with the one-parameter exponential parent population. In view of his work and the limitation of space here, we will summarize the suggested procedures through two tables. Table 9.1 considers one-sample problems (mostly P1) and Table 9.2 deals with essentially two types of two-sample problems. In these tables Q is the pivotal quantity employed and q or the q_i's represent appropriately chosen percentiles of Q. In the third column, along with Q, we indicate how one can obtain its distribution using a relevant result from Section 9.5.2. In the last column we note whether the exact distribution of Q has been derived in the references cited. We also briefly comment on the computational aspects related to the q_i's.

For the purpose of illustration, we will now apply Result 9.2 to obtain the distribution of Q for the first situation considered in Table 9.1. The problem there is of predicting $X_{k:n}$ based on $X_{1:n}, \ldots, X_{r:n}$. Note that $Q = (X_{k:n} - X_{r:n})/T_1 \stackrel{d}{=} (\sum_{i=1}^{k-r} c_i U_i)/U_{k-r+1}$ where $c_i = (n-k+i)^{-1}$ and U_i is χ_2^2 for $i = 1$ to $k-r$, and U_{k-r+1} is χ_{2r}^2. We can use the discussion following Result 9.3 to obtain the distribution of Q. The α_{i1} in (9.5.5) simplifies to $\{1 + (n-k+i)v\}^{-r} \prod_{j \neq i=1}^{k-r} \binom{\frac{n-k+j}{j-i}}{}$, and consequently (9.5.4) yields

$$\begin{aligned} P[Q > v] &= \frac{1}{B(n-k+1, k-r)} \sum_{i=0}^{k-r-1} (-1)^i \binom{k-r-1}{i} \frac{1}{n-k+i+1} \\ &\quad \cdot \frac{1}{\{1 + (n-k+i+1)v\}^r} \\ &= G(v; r, k, n) \ . \end{aligned} \qquad (9.5.6)$$

Lawless (1971) has derived (9.5.6) through direct manipulation. He also discusses modifications necessary for Type I censored samples and for the situation when the failed items are immediately replaced.

Table 9.1: One-sample problems in $e(0,\theta)$ population

$\boldsymbol{X}$	Y	Q	$(L(\boldsymbol{X}), R(\boldsymbol{X}))$	References	Comments
$X_{1:n}, \ldots, X_{r:n}$	$X_{k:n}$	$(X_{k:n} - X_{r:n})/T_1$ (Result 9.2)	$(X_{r:n}, X_{r:n} + qT_1)$	Lawless (1971)	Exact; q tabulated by Bain and Patel (1991)
"	"	$(X_{k:n} - X_{r:n})/T_1$ (Result 9.3)	$(X_{r:n} + q_1T_1, X_{r:n} + q_2T_1)$	Mann and Grubbs (1974)	Approximate; uses F tables
"	"	$(X_{k:n} - X_{r:n})/X_{k:n}$ (Result 9.2)	$(X_{r:n}, X_{r:n} + qX_{r:n})$	Lingappaiah (1973)	Exact; q not tabulated
"	$X_{n:n}$	$(X_{n:n} - X_{r:n})/T_1$ (Result 9.2)	$(X_{r:n} + q_1T_1, X_{r:n} + q_2T_1)$	Takada (1979)	Exact; q_i's tabulated
$X_{i_1:n}, \ldots, X_{i_r:n}$	$X_{k:n}$	$(X_{k:n} - X_{i_r:n})/T_1^*$ (Result 9.2)	$(X_{r:n} + q_1T_1^*, X_{r:n} + q_2T_1^*)$	Kaminsky and Nelson (1974)	Exact; q not tabulated. F approximation

Note: $T_1 = \sum_{i=1}^{r} X_{i:n} + (n-r)X_{r:n}$ is the total time on test and T_1^* is the BLUE of θ based on $\boldsymbol{X}$. The two sided PI's are generally equitailed.

Takada (1985) has introduced the idea of (scale) invariant prediction limit. Let $X_1, \ldots, X_r$, Y have the joint pdf $\theta^{-(r+1)} f(x_1/\theta, \ldots, x_r/\theta, y/\theta)$, $x_i > 0$, $y > 0$. If a prediction limit $L(\boldsymbol{X})$ satisfies $P_\theta[Y > L(\boldsymbol{X})] = \gamma$ and $L(c\boldsymbol{X}) = cL(\boldsymbol{X})$, $c > 0$, he calls $L(\boldsymbol{X})$ to be invariant. We may call $L_0(\boldsymbol{X})$ to be the best invariant lower prediction limit if in the class of invariant prediction limits $E_\theta[Y - L(\boldsymbol{X}) \,|\, Y > L(\boldsymbol{X})]$ is the smallest when $L(\boldsymbol{X}) = L_0(\boldsymbol{X})$. For the problem of predicting $X_{k:n}$ given $X_{1:n}, \ldots, X_{r:n}$, Takada has shown that $L_0(\boldsymbol{X}) = X_{r:n} + qT_1$; that is, Lawless' limits are such limits. For the case of $k = n$, Takada (1979) has considered the problem of minimizing $E_\theta[U(\boldsymbol{X}) - L(\boldsymbol{X})]$ subject to the invariant condition mentioned above. He shows that the shortest interval for $X_{n:n}$ for $e(0, \theta)$ parent is given by $(X_{r:n} + q_1 T_1,\ X_{r:n} + q_2 T_1)$ where the q_i's are chosen to satisfy certain requirements for G given in (9.5.6).

Note that in Table 9.1 above, Lawless (1971) and Lingappaiah (1973) have used different pivotal quantities for the same problem. The natural question is which one is to be preferred. Kaminsky (1977b) has compared the resulting prediction limits and concludes that Lawless' intervals are shorter with probability tending to 1 as the sample size increases. It is to be anticipated since, based on the observed data, T_1 is sufficient for θ.

Table 9.2 summarizes the several approaches to two-sample problems. Shayib, Awad, and Dawagreh (1986) have carried out a simulation study to compare four two-sided PI's for $\bar{Y}_m$ based on the sample mean $\bar{X}_n$. One of them uses the fact that $\bar{Y}_m/\bar{X}_n$ has an F distribution while another one uses the fact that if n is large, this ratio behaves like a χ^2_{2m} variable approximately. Based on three criteria they have formulated, the conclusion is that the latter interval provides the best results even for n as low as 10.

In addition to the above common problems, some special situations involving more than two samples have been investigated in the literature. Based on a Type II censored sample, Hahn (1975) has obtained lower prediction limit for the minimum of s future independent sample means. His pivotal quantity, $\min\{\bar{Y}(1), \ldots, \bar{Y}(s)\}/T_1$, behaves like the minimum of s correlated F rv's, where $\bar{Y}(j)$ is the mean of the j-th future sample. Chou (1989), extending the work of Lawless (1972) to s future independent samples, has obtained lower prediction limit for $\min(Y_{j_1:m_1}, \ldots, Y_{j_s:m_s})$. The pivotal quantity used is the ratio of this statistic and T_1. Meeker and Hahn (1980) have obtained a two-sided PI for the ratio of mean times to failures of two future samples based on mean times to failure of two previous (independent) samples. It involves the distribution of ratio of two independent F rv's. Another problem, relevant to the reliability engineers, is that of predicting future availability of a repairable unit over future n_2 cycles based on the observed availability over n_1 previous cycles. This has been considered by Nelson (1970).

9.5.4 Two-parameter Exponential Distribution

Researchers have investigated the prediction problems discussed in Tables 9.1 and 9.2 for the situation when both the location and scale parameters are unknown. Table 9.3 summarizes these results. For the first PI in Table 9.3, the percentiles of Q are related to those needed for the first PI in Table 9.1. In fact, with $Q = (X_{k:n} - X_{r:n})/T$, $P[Q > v] = G(v; r-1, s-1, n-1)$, where G is given by (9.5.6).

Table 9.2: Two-sample problems in $e(0, \theta)$ population

$\boldsymbol{X}$	Y	Q	$(L(\boldsymbol{X}), R(\boldsymbol{X}))$	References	Comments
$X_{1:n}$, ..., $X_{r:n}$	$Y_{k:m}$	$Y_{k:m}/T_1$ (Result 9.2)	$(qT_1/r, \infty)$	Lawless (1972) Hall and Prairie (1973)	Exact; q_i's tabulated. Graphs to determine q_i
"	$\bar{Y}_m$	$\bar{Y}_m/T_1$	$(qT_1/r, \infty)$,	Lawless (1972)	Exact F distribution
$X_{1:n}$ ..., $X_{n:n}$	$\bar{Y}_m$	Functions of $\bar{Y}_m/\bar{X}_n$	Two-sided	Shayib et al. (1986)	Large sample approximations
$X_{i_1:n}$ ..., $X_{i_r:n}$	$Y_{k:m}$ $(1 \leq k \leq m)$	$Y_{k:m}/T_1^*$ (Result 9.2)	$(q_1T_1^*, q_2T_1^*)$,	Kaminsky and Nelson (1974)	Exact; F approximation for q_i
$X_{i_1:n}$, ..., $X_{i_r:n}$	$\bar{Y}_m$	$\bar{Y}_m/T_1^*$ (Result 9.2)	$(q_1T_1^*, q_2T_1^*)$,	Kaminsky and Nelson (1974)	Exact; F approximation for q_i

Table 9.3: One- and two-sample problems in $e(\mu, \theta)$ population

$\boldsymbol{X}$	Y	Q	$(L(\boldsymbol{X}), R(\boldsymbol{X}))$	References	Comments
$X_{1:n}, \ldots, X_{r:n}$	$X_{k:n}$	$(X_{k:n} - X_{r:n})/T$ (Result 9.2)	$(X_{r:n} + q_1 T, X_{r:n} + q_2 T)$	Likeš (1974)	Exact; Connects to studentized range. q_i's tabulated in Likeš and Nedělka (1973)
"	"	$(X_{k:n} - X_{r:n})/T$ (Result 9.3)	$(X_{r:n} + q_1 T, X_{r:n} + q_2 T)$	Mann and Grubbs (1974)	Approx-imate; uses F tables
"	$Y_{k:m}$	$(Y_{k:m} - X_{1:n})/T$ (Result 9.2)	$(X_{1:n} + q_1 T, X_{1:n} + q_2 T)$	Lawless (1977) Shah and Rathod (1978)	Exact; q_i's not tabulated
"	$\bar{Y}_m$	$(\bar{Y}_m - X_{1:n})/T$ (Result 9.2)	$(X_{1:n} + q_1 T, X_{1:n} + q_2 T)$	Lawless (1977)	Exact; q_i's not tabulated
$X_{q:n} \ldots, X_{r:n}$	$Y_{k:m}$	$(Y_{k:m} - X_{q:n})/T^*$ (Result 9.2)	$(X_{q:n} + aT^*, \infty)$	Ng (1984)	Exact; q not tabulated
$X_{q:n}, \ldots, X_{r:n}$	$\bar{Y}_m$	$(\bar{Y}_m - X_{q:n})/T^*$ (Result 9.2)	$(X_{q:n} + aT^*, \infty)$	Ng (1984)	Exact; q not tabulated
$X_{1:n}, \ldots, X_{r:n}$	minimum of $Y_{j_1:m_1}, \ldots, Y_{j_s:m_s}$	$(Y - X_{1:n})/T$	$(X_{1:n} + qT, \infty)$	Chou and Owen (1989)	Exact

Note: T is given by (9.3.6) and T^* is a linear function of spacings among $X_{q:n}, \ldots, X_{r:n}$.

9.6 Bayes Predictors and Prediction Regions

There is quite a bit of work on prediction problems arising in exponential populations by Bayesian statisticians. In view of a separate chapter devoted to Bayesian inference, our coverage of this topic will be brief. An excellent reference on Bayesian prediction analysis is Aitchison and Dunsmore (1975). They discuss various methods of prediction and use one- and two-parameter exponential parents as examples to illustrate these procedures.

The basic premise for the Bayes analysis is that, given the parameter $\theta \in \Theta$, the observed data, $\boldsymbol{X}$, and the data to be predicted $\boldsymbol{Y}$, are independent, where $\boldsymbol{Y}$ is possibly vector valued. If $f(\theta|\boldsymbol{x})$ is the posterior pdf of θ, the predictive density of $\boldsymbol{Y}$ given the data is

$$f(\boldsymbol{y}|\boldsymbol{x}) = \int_{\Theta} f(\boldsymbol{y}|\theta) f(\theta|\boldsymbol{x})\, d\theta \,. \tag{9.6.1}$$

Bayes predictors and prediction regions are determined from the predictive density given by (9.6.1). The optimal point predictor depends on the choice of the loss function. For example, when the loss function is squared error, the mean of the predictive density is the optimal predictor. A Bayesian prediction region of "cover" γ for $\boldsymbol{Y}$ is A if $\int_A f(\boldsymbol{y}|\boldsymbol{x}) d\boldsymbol{x} = \gamma$, where typically A is chosen such that $A = \{\boldsymbol{y} : f(\boldsymbol{y}|\boldsymbol{x}) > \lambda\}$ [Aitchison and Dunsmore (1975, pp. 69-70)].

Dunsmore (1974) has obtained PI's for P1 and P2 for the exponential distributions. We will now obtain a PI for P1 in the one-parameter case, where apriori $1/\theta$ is assumed to have Gamma(α, β) distribution. This provides the conjugate prior distribution for the problem.

Even though $X_{k:n}$ is not independent of the observed data $\boldsymbol{X}$, $X_{k:n} - X_{r:n}$ and $\boldsymbol{X}$ are independent. Besides, T_1, the total time on test, is sufficient for θ. So we can consider the problem P1 to be equivalent to predicting $Y = X_{k:n} - X_{r:n} \stackrel{d}{=} X_{k-r:n-r}$ having observed $X = T_1$. Given θ, X is Gamma$(r, 1/\theta)$, Y is distributed like $(k-r)$-th order statistic from a random sample of size $(n-r)$ from Gamma$(1, 1/\theta)$, and they are independent. The posterior distribution of $1/\theta$ given $X = x$ is Gamma$(\eta + r, \beta + x)$, and it can be shown that (9.6.1) yields [Dunsmore (1974)]

$$\begin{aligned} f(y|x) \quad &= \quad \frac{(\beta + x)^{\eta + r}}{B(n-k+1, k-r)} \sum_{i=0}^{k-r-1} (-1)^i \binom{k-r-1}{i} \\ &\quad \cdot \frac{1}{\{\beta + x + (n-k+i+1)y\}^{\eta + r + 1}} \,. \end{aligned}$$

We thus obtain

$$\begin{aligned} P[Y > y\,|\,x] \quad &= \quad \frac{(\beta + x)^{\eta + r - 1}}{B(n-k+1, k-r)} \sum_{i=0}^{k-r-1} (-1)^i \binom{k-r-1}{i} \frac{1}{n-k+i+1} \\ &\quad \cdot \frac{1}{\{\beta + x + (n-k+i+1)y\}^{\eta + r}} \end{aligned} \tag{9.6.2}$$

which closely resembles (9.5.6). This can be used to obtain a PI or a point predictor for $X_{k:n}$. If we use squared error loss, the best predictor of Y will be $E[Y|x]$ which can be computed easily from (9.6.2) since $E[Y|x] = \int_0^\infty P[Y > y\,|\,x]\,dy$. In other words,

$$E[Y|x] = \frac{(\beta+x)}{(\eta+r-1)B(n-k+1,k-r)} \sum_{i=0}^{k-r-1} (-1)^i \binom{k-r-1}{i} \cdot \frac{1}{(n-k+i+1)^2} \,. \tag{9.6.3}$$

To evaluate the sum in (9.6.3), we note that, in view of (9.5.6),

$$E[Y|x] = (\beta+x)(r-1)(\eta+r-1)^{-1}E[Q]\,, \tag{9.6.4}$$

where $Q = (X_{k:n} - X_{r:n})/T_1$. Note that Q can be written as the ratio of $Z_{k-r:n-r}$ and an independent Gamma$(r,1)$ variable and consequently $E[Q] = \alpha_1(r,k;n)(r-1)^{-1}$. Hence from (9.6.4) we can conclude that $E[Y|x] = (\beta+x)\alpha_1(r,k;n)(\alpha+r-1)^{-1}$. So the Bayes predictor of $X_{k:n}$ is $X_{k:n}^B = X_{r:n} + (\beta+T_1)\alpha_1(r,k;n)(\eta+r-1)^{-1}$. In contrast, the BLUP will be $X_{r:n} + \alpha_1(r,k;n)T_1 r^{-1}$.

Quite often $f(\boldsymbol{y}|\boldsymbol{x})$ does not have a nice form. Dunsmore (1976) has developed a series expansion for this density around the MLE of θ and has used one-parameter exponential distribution to illustrate his approximation. Bancroft and Dunsmore (1976) derive the predictive distributions in life tests under competing causes of failure and in another paper (1978), they consider the problem of predictive sequential life testing.

During the 1970's, in addition to the work of Dunsmore, Ling and his co-workers published several papers on Bayesian approach to prediction applied to one- and two-parameter exponential populations [Ling (1975), Ling and Siaw (1975), Leong and Ling (1976), Ling and Lee (1977, 1978), Ling and Leong (1977), Ling and Tan (1979)]. These rather hard to access papers derive the predictive distributions for one-sample and two-sample problems with various types of restrictions on the available data. They include situations where some middle order statistics are missing and when the sample values are partially grouped. Some of these papers consider the predictive density of the number of observations falling in certain intervals. Geisser (1985) has also considered such problems.

Another major contributor to this area is Lingappaiah who in a long series of papers [Lingappaiah (1979a,b,c, 1980, 1984, 1986, 1989, 1990, 1991)] has explored one-, two-, and multisample problems with several types of assumptions about the available data. He has considered situations where either outliers are present, or the data is multiply censored, or the average lifelengths are successively increasing, or the sample size is a random variable. He has also considered the prediction of linear combination of selected spacings. In general, the details are quite involved. Geisser (1984, 1990), and Evans and Nigm (1980) have also considered some Bayesian prediction problems associated with the exponential distributions. Other references on this topic are Upadhyay and Pandey (1989), and Shayib and Awad (1990).

9.7 Other Prediction Methods

Just as in the case of estimation, criteria other than unbiasedness and MSE have been used to compare point predictors. We will introduce and apply them for comparing predictors obtained in earlier sections. Bibby and Toutenberg (1977, Chapter 7) discuss the general concepts related to PN and PC criteria introduced below.

9.7.1 PN criterion

Let $\hat{Y}_1$ and $\hat{Y}_2$ be two predictors of the rv Y. We say $\hat{Y}_1$ has higher probability of nearness (PN) to Y than $\hat{Y}_2$ if

$$P_\theta[|\hat{Y}_1 - Y| < |\hat{Y}_2 - Y|] > 0.5 \ . \tag{9.7.1}$$

If (9.7.1) holds for all θ, we say that $\hat{Y}_1$ is preferable to $\hat{Y}_2$ in PN sense. Pitman (1937) had introduced this idea in the context of estimation. Keating et al. (1991) contains several papers related to this measure.

Nagaraja (1986) has compared in terms of the above criterion, the BLUP and BLIP of $X_{k:n}$ for the problem P1 for the $e(\mu, \theta)$ parent. The BLUP, $\hat{X}_{k:n}$, and the BLIP, $\tilde{X}_{k:n}$, are given by (9.3.7) and (9.3.8) respectively. Since $\hat{X}_{k:n} > \tilde{X}_{k:n}$,

$$\begin{aligned} & P[|\hat{X}_{k:n} - X_{k:n}| < |\tilde{X}_{k:n} - X_{k:n}|] \\ = \ & P[X_{k:n} - \hat{X}_{k:n} > 0] \\ & + P[X_{k:n} - \hat{X}_{k:n} < 0 \text{ and } X_{k:n} - \hat{X}_{k:n} + X_{k:n} - \tilde{X}_{k:n} > 0] \\ = \ & P\left[X_{k:n} - X_{r:n} > \frac{\alpha_1(k,r;n)(2r-1)}{2r(r-1)}\,T\right] \end{aligned} \tag{9.7.2}$$

on simplification. This probability can be evaluated by using Result 9.2. He shows that when $k = r + 1$, $P[|\hat{X}_{k:n} - X_{k:n}| < |\tilde{X}_{k:n} - X_{k:n}|] < 0.5$ if and only if $r \geq 3$. Thus, for $r \geq 3$, $\tilde{X}_{r+1:n}$ is preferred to $\hat{X}_{r+1:n}$ in the PN sense. For $k > r + 1$, the probability in (9.7.2) depends on n and for a given n, it increases with k for a fixed r and decreases with r for a fixed k. Numerical computations show that the BLUP is preferred to BLIP if $(k - r)$ is large.

9.7.2 PC criterion

A predictor $\hat{Y}_1$ has higher probability of concentration (PC) around Y that $\hat{Y}_2$ if

$$P_\theta[|\hat{Y}_1 - Y| < c] \geq P_\theta[|\hat{Y}_2 - Y| < c] \ , \tag{9.7.3}$$

for all $c > 0$ with strict inequality for at least one c. If (9.7.3) holds for all θ, $\hat{Y}_1$ is preferable to $\hat{Y}_2$ in the PC sense.

Nagaraja (1986) has also compared the BLUP and BLIP for problem P1. One can find $P[|\hat{X}_{k:n} - X_{k:n}| < c]$ and $P[|\tilde{X}_{k:n} - X_{k:n}| < c]$ using Result 9.1. Resulting expressions are complicated and are not analytically comparable if $k > r+1$. Moreover, numerical computations show that they are not PC-comparable. For $k = r+1$ the BLIP is preferred when $r = 2$; otherwise the predictors are not comparable.

9.7.3 Median Unbiasedness

A predictor $\hat{Y}$ of Y is a median unbiased predictor (MUP) if $P_\theta[\hat{Y} \leq Y] = P_\theta[\hat{Y} \geq Y]$. Takada (1991) has considered the property of median unbiasedness in an invariant prediction problem. Let $X_1, \ldots, X_r$, Y have the joint pdf given by (9.3.1). Suppose $\hat{Y}$ is the BUP of Y and $\hat{\theta}$ is the MVUE of θ. Then $Q_0 = (Y - \hat{Y})/\hat{\theta}$ is a pivotal quantity. Define the predictor $\hat{Y}_0 = \hat{Y} + \text{Med}(Q_0)\hat{\theta}$. Note that $\hat{Y}_0$ is a MUP of Y. Takada has shown that for any predictor $\hat{Y}_1$ given by $\hat{Y}_1 = \hat{Y} + c\hat{\theta}$, $\hat{Y}_0$ is better than $\hat{Y}_1$ in the PN sense.

For the $e(\mu, \theta)$ parent, let us now consider the problem P1. From (9.3.6) and (9.3.7), we have $Q_0 = (X_{k:n} - \hat{X}_{k:n})/\hat{\theta} = (r-1)\{(X_{k:n} - X_{r:n})/T\} - \alpha_1(r, k; n) = (r-1)Q - \alpha_1(r, k; n)$, where $Q = (X_{k:n} - X_{r:n})/T$. Thus, $\text{Med}(Q_0) = (r-1)\text{Med}(Q) - \alpha_1(r, k; n)$, and consequently $X_{r:n} + (r-1)\text{Med}(Q)\hat{\theta}$ is an MUP of $X_{k:n}$. As we have noted in Section 9.5.4, $P[Q > v] = G(v; r-1, s-1, n-1)$ where G is given by (9.5.6). This fact can be used to determine $\text{Med}(Q)$. This MUP is better than either $\hat{X}_{k:n}$ or $\tilde{X}_{k:n}$ in the PN-sense.

For predicting $Y_{k:m}$, one can use (9.3.9) to produce an MUP that is better than both the BUP and BIP. Again, by following Takada (1991), we conclude that $\hat{Y}_{k:m} + \text{Med}\{(Y_{k:m} - \hat{Y}_{k:m})/\hat{\theta}\}T/(r-1)$ is an MUP, and beats both the other two predictors in the PN-sense. The median involved in this MUP can be obtained from $\text{Med}\{(Y_{k:m} - X_{1:n})/T\}$. Thus, Result 9.2 will be handy here as well.

9.7.4 Conditional Median Predictor

Recently Raqab (1992) has introduced the idea of conditional median predictor (CMP). It is known that the BUP of Y having observed $\boldsymbol{X}$ is $E_\theta[Y|\boldsymbol{X}]$ which, of course, possibly depends on θ. Instead of the conditional mean, one can use the conditional median and replace θ by its MVUE to produce a predictor of Y. Raqab calls this a CMP. Let us examine its form for predicting $X_{k:m}$. Note that $\{X_{k:n}|x_{1:n}, \ldots, x_{r:n}\} \stackrel{d}{=} x_{r:n} + \theta Z_{k-r:n-r}$. Thus one would use $X_{r:n} + \hat{\theta}\text{Med}(Z_{k-r:n-r})$ as the CMP of $X_{k:n}$. For a sample size of 10, for various choices of r and k, he has compared the MSE's of the BLUP, MUP and CMP to observe that CMP is better than BLUP in all the cases, and is better than the MUP in most of the cases.

9.8 Miscellaneous Results

When all the parameters of the parent distribution are known, the best predictor of $X_{r+1:n}$ given $X_{r:n}$ and some previous order statistics is simply $E[X_{r+1:n}|X_{r:n}]$. It can be shown that this predictor uniquely determines f. In fact, from Ferguson (1967) it follows that this is the sum of $X_{r:n}$ and a constant if and only if f is exponential. But it is still unknown, without additional assumptions on f, whether for $k > r+1$, $E[X_{k:n}|X_{r:n}] = X_{r:n} + c$ for some constant c, characterizes the exponential parent.

Kaminsky and his co-workers examined issues related to the point prediction of order statistics from several angles. Let now f be a pdf belonging to a location-scale family and the parameters, μ and θ are unknown. Let $r = [np_1]$ and $k = [np_2]$, $0 < p_1 < p_2 < 1$. Under some regularity conditions on the convergence of the first two moments of order statistics, Kaminsky and Nelson (1975b) show that if the BLUP of $X_{k:n}$ does not depend functionally on μ or its BLUE for all p_1, p_2, $0 < p_1 < p_2 < 1$, then f must be exponential.

Kaminsky and Rhodin (1978) introduce the idea of predictive information in $X_{r:n}$ about $X_{k:n}$ relative to that contained in $\boldsymbol{X} = (X_{1:n}, \ldots, X_{r:n})$. It is defined as the ratio of the asymptotic MSE's of the asymptotically BUP of $X_{k:n}$ given just $X_{r:n}$, and given $\boldsymbol{X}$. For a two-parameter exponential distribution with known θ, Kaminsky (1977b) showed that this ratio has a minimum value of 0.9658 for variation of r/n and k/n.

Balasooriya (1989) has used $\hat{X}_{r+1:n}$, defined in (9.3.7), to test whether $X_{r+1:n}$ is an outlier. The test statistic $Q = \{X_{r+1:n} - X_{r:n}\}/\{\hat{X}_{r+1:n} - X_{r:n}\} = (n-r)\{X_{r+1:n} - X_{r:n}\}/\hat{\theta}$, and hence is an F random variable under the null hypothesis of no outliers.

Prediction of upper record values (see Chapter 17 for the definition) from the two-parameter exponential distribution has also been considered. Let Y_i be the i-th upper record value for $i \geq 1$. Note that Y_i is Gamma$(i, 1)$ and these Y_i's have a nice dependence structure. We can use the general theory developed in Section 9.3.1 to obtain the BLUP and BLIP of Y_k when the first r record values are observed. It easily follows that [Ahsanullah (1980)] $\hat{Y}_k = Y_r + \{(k-r)/(r-1)\}(Y_r - Y_1)$ and $\tilde{Y}_k = Y_r + \{(k-r)/r\}(Y_r - Y_1)$. Nagaraja (1986) has carried out PN and PC comparison of these predictors. The results parallel those for order statistics with the sample size n approaching infinity. Bayesian approach for the prediction of a future record value has been pursued by Dunsmore (1983). It is also known that the exponential distribution is the only continuous distribution for which both $E[Y_{r+1}|Y_r]$ and $E[Y_r|Y_{r+1}]$ are linear in the conditioning variables [Nagaraja (1988)].

9.9 Concluding Remarks

In this chapter we have attempted to synthesize and summarize the vast literature on the area of prediction problems as related to the one- and two-parameter exponential distributions. Motivated by applications to reliability and life testing problems, most of these problems involve some functions of order statistics from the current or a future sample. We have not touched upon the prediction problems related to mixtures of exponential populations. Both frequentist and Bayesian approaches have been tried out in that context. Interested readers may refer to Robbins (1980), Ashour and Shoukry (1981), Lingappaiah (1981), Abu-Salih, Ali Khan, and Husein (1987), and Sloan and Sinha (1991) for further information on these results.

As we conclude this chapter it is time for the prediction of the future direction of research on prediction related to the exponential distribution. Obviously it is a prime candidate for the application of any new general prediction method to be suggested in the future. In the context of reliability some interesting problems remain. Recall that the k-th order statistic represents the life length of $(n-k+1)$-out-of-n system. This is a simple coherent system and the challenge now is to find good predictors of lifelengths of coherent systems whose structure functions are known. Hopefully such problems will be tackled in the near future. Another area for future research is that of finding optimal regions for the simultaneous prediction of two or more rvs of interest such as the joint prediction of $X_{r+1:n}$ and $\sum_{i=r+1}^{n} X_{i:n}/(n-r)$ based on $X_{1:n}, \ldots, X_{r:n}$.

References

Abu-Salih, M.S., Ali Khan, M.S., and Husein, A. (1987). Prediction intervals of order statistics for the mixture of two exponential distributions, *Aligarh Journal of Statistics*, **7**, 11-22.

Ahsanullah, M. (1980). Linear prediction of record values for the two parameter exponential distribution, *Annals of the Institute of Statistical Mathematics*, **32**, 363-368.

Aitchison, J. and Dunsmore, I.R. (1975). *Statistical Prediction Analysis*, Cambridge University Press, Cambridge, England.

Arnold, B.C., Balakrishnan, N., and Nagaraja, H.N. (1992). *A First Course in Order Statistics*, John Wiley & Sons, New York.

Ashour, S.K. and Shoukry, S.E. (1981). The Bayesian predictive distribution of order statistics of a future sample from a mixed exponential distribution, *Egyptian Statistical Journal*, **25**, 94-102.

Bain, P.T. and Patel, J.K. (1991). Factors for calculating prediction intervals for samples from a one-parameter exponential distribution, *Journal of Quality Technology*, **23**, 48-52.

Balasooriya, U. (1989). Detection of outliers in the exponential distribution based on prediction, *Communications in Statistics - Theory and Methods*, **18**, 711-720.

Bancroft, G.A. and Dunsmore, I.R. (1976). Predictive distributions in life tests under competing causes of failure, *Biometrika*, **63**, 195-198.

Bancroft, G.A. and Dunsmore, I.R. (1978). Predictive sequential life testing, *Biometrika*, **65**, 609-614.

Bibby, J. and Toutenburg, H. (1977). *Prediction and Improved Estimation in Linear Models*, John Wiley & Sons, Chichester, England.

Box, G.E.P. (1954). Some theorems on quadratic forms applied in the study of analysis of variance problems, I. Effect of inequality of variance in the one-way classification, *Annals of Mathematical Statistics*, **25**, 290-302.

Chou, Y.-M. (1989). One-sided simultaneous prediction intervals for the order statistics of l future samples from an exponential distribution, *Communications in Statistics - Theory and Methods*, **17**, 3995-4003.

Chou, Y.-M. and Owen, D.B. (1989). Simultaneous one-sided prediction intervals for a two-parameter exponential distribution using complete or Type II censored data, *Metrika*, **36**, 279-290.

David, H.A. (1981). *Order Statistics*, Second edition, John Wiley & Sons, New York.

Dunsmore, I.R. (1974). The Bayesian predictive distribution in life testing models, *Technometrics*, **16**, 455-460.

Dunsmore, I.R. (1976). Asymptotic prediction analysis, *Biometrika*, **63**, 627-630.

Dunsmore, I.R. (1983). The future occurrence of records, *Annals of the Institute of Statistical Mathematics*, **35**, 267-277.

Evans, I.G. and Nigm, A.H.M. (1980). Bayesian prediction for the left truncated exponential distribution, *Technometrics*, **22**, 201-204.

Faulkenberry, G.D. (1973). A method of obtaining prediction intervals, *Journal of the American Statistical Association*, **68**, 433-435.

Ferguson, T.S. (1967). On characterizing distributions by properties of order statistics, *Sankhyā, Series A*, **29**, 265-278.

Geisser, S. (1984). Predicting Pareto and exponential observables, *Canadian Journal of Statistics*, **12**, 143-152.

Geisser, S. (1985). Interval prediction for Pareto and exponential observables, *Journal of Econometrics*, **29**, 173-185.

Geisser, S. (1986). Predictive analysis, In *Encyclopedia of Statistical Sciences*, Vol. 7 (Eds., S. Kotz, N.L. Johnson and C.B. Read), pp. 158-170, John Wiley & Sons, New York.

Geisser, S. (1990). On hierarchical Bayes procedures for predicting simple exponential survival, *Biometrics*, **46**, 225-230.

Goldberger, A.S. (1962). Best linear unbiased prediction in the generalized linear regression model, *Journal of the American Statistical Association*, **57**, 369-375.

Hahn, G.J. (1975). A simultaneous prediction limit on the means of future samples from an exponential distribution, *Technometrics*, **17**, 341-346.

Hahn, G.J. and Meeker, W.Q., Jr. (1991). *Statistical Intervals - A Guide to Practitioners*, John Wiley & Sons, New York.

Hahn, G.J. and Nelson, W. (1973). A survey of prediction intervals and their applications, *Journal of Quality Technology*, **5**, 178-188.

Hall, I.J. and Prairie, R. (1973). One-sided prediction intervals to contain at least m out of k future observations, *Technometrics*, **15**, 897-914.

Hewett, J.E. (1968). A note on prediction intervals based on partial observations in certain life test experiments, *Technometrics*, **10**, 850-853.

Hinkley, D.V. (1979). Predictive likelihood, *Annals of Statistics*, **7**, 718-728.

Kalbfleisch, J.D. (1971). Likelihood methods of prediction, In *Foundations of Statistical Inference - A Symposium* (Eds., V.P. Godambe and D.A. Sprott), Holt, Rinehart, and Winston, Toronto, Canada.

Kaminsky, K.S. (1977a). Best prediction of exponential failure times when items may be replaced, *Australian Journal of Statistics*, **19**, 61-62.

Kaminsky, K.S. (1977b). Comparison of prediction intervals for failure times when life is exponential, *Technometrics*, **19**, 83-86.

Kaminsky, K.S., Mann, N.R., and Nelson, P.I. (1975). Best and simplified linear invariant prediction of order statistics in location and scale families, *Biometrika*, **62**, 525-526.

Kaminsky, K.S. and Nelson, P.I. (1974). Prediction intervals for the exponential distribution using subsets of the data, *Technometrics*, **16**, 57-59.

Kaminsky, K.S. and Nelson, P.I. (1975a). Best linear unbiased prediction of order statistic in location and scale families, *Journal of the American Statistical Association*, **70**, 145-150.

Kaminsky, K.S. and Nelson, P.I. (1975b). Characterization of distributions by the form of predictors of order statistics, In *Statistical Distributions in Scientific Work* (Eds., G.P. Patil, S. Kotz, and J.K. Ord), Vol. 3, pp. 113-116, D. Reidel, Dordrecht.

Kaminsky, K.S. and Rhodin, L.S. (1978). The prediction information in the latest failure, *Journal of the American Statistical Association*, **73**, 863-866.

Kaminsky, K.S. and Rhodin, L.S. (1985). Maximum likelihood prediction, *Annals of the Institute of Statistical Mathematics*, **37**, 507-517.

Keating, J.P., Mason, R.L., Rao, C.R., and Sen, P.K. (1991). Pitman's measure of closeness, *Communications in Statistics - Theory and Methods, Special Issue*, **20**(11).

Lawless, J.F. (1971). A prediction problem concerning samples from the exponential distribution, with application to life testing, *Technometrics*, **13**, 725-730.

Lawless, J.F. (1972). On prediction intervals for samples from the exponential distribution and prediction limits for system survival, *Sankhyā, Series B*, **34**, 1-14.

Lawless, J.F. (1977). Prediction intervals for the two parameter exponential distribution, *Technometrics*, **19**, 469-472.

Leong, C.Y. and Ling, K.D. (1976). Bayesian predictive distributions for future observations from exponential variate in a life testing model, *Nanta Mathematica*, **9**, 171-177.

Likeš, J. (1974). Prediction of s-th ordered observation for the two-parameter exponential distribution, *Technometrics*, **16**, 241-244.

Likeš, J. and Nedělka, S. (1973). Note on studentized range in samples from an exponential distribution, *Biometrical Journal*, **15**, 545-555.

Ling, K.D. (1975). On structural prediction distribution for samples from exponential distribution, *Nanta Mathematica*, **8**, 47-52.

Ling, K.D. and Lee, G.C. (1977). Bayesian predictive distributions for samples from exponential distribution based on partially grouped samples: I, *Nanta Mathematica*, **10**, 166-173.

Ling, K.D. and Lee, G.C. (1978). Bayesian predictive distributions for samples from exponential distribution based on partially grouped samples: II, *Nanta Mathematica*, **11**, 55-62.

Ling, K.D. and Leong, C.Y. (1977). Bayesian predictive distributions for samples from exponential distribution: I, *Tamkang Journal of Mathematics*, **8**, 11-16.

Ling, K.D. and Siaw, V.C. (1975). On prediction intervals for samples from exponential distribution, *Nanyang University Journal, Part III*, **9**, 41-46.

Ling, K.D. and Tan, C.K. (1979). On predictive distributions based on samples with missing observations – Exponential distribution, *Nanta Mathematica*, **12**, 173-181.

Lingappaiah, G.S. (1973). Prediction in exponential life testing, *Canadian Journal of Statistics*, **11**, 113-117.

Lingappaiah, G.S. (1979a). Bayesian approach to prediction and the spacings in the exponential distribution, *Annals of the Institute of Statistical Mathematics*, **31**, 391-401.

Lingappaiah, G.S. (1979b). Bayesian approach to the prediction of the restricted range in the censored samples from the exponential population, *Biometrical Journal*, **21**, 361-366.

Lingappaiah, G.S. (1979c). Bayesian approach to the prediction problem in complete and censored samples from the gamma and exponential populations, *Communications in Statistics - Theory and Methods*, **8**, 1403-1424.

Lingappaiah, G.S. (1980). Intermittant life testing and Bayesian approach to prediction with spacings in the exponential model, *Statistica*, **40**, 477-490.

Lingappaiah, G.S. (1981). Mixture of exponential populations and prediction by Bayesian approach, *Revue Roumaine de Mathematiques Pures et Appliquees*, **26**, 753-760.

Lingappaiah, G.S. (1984). Bayesian prediction regions for the extreme order statistics, *Biometrical Journal*, **26**, 49-56.

Lingappaiah, G.S. (1986). Bayes prediction in exponential life-testing when sample size is a random variable, *IEEE Transactions on Reliability*, **35**, 106-110.

Lingappaiah, G.S. (1989). Bayes prediction of maxima and minima in exponential life tests in the presence of outliers, *Industrial Mathematics*, **39**, 169-182.

Lingappaiah, G.S. (1990). Inference with samples from an exponential population in the presence of an outlier, *Publications of the Institute of Statistics, University of Paris*, **35**, 43-54.

Lingappaiah, G.S. (1991). Prediction in exponential life tests where average lives are successively increasing, *Pakistan Journal of Statistics*, **7**, 33-39.

Lloyd, E.H. (1952). Least-squares estimation of location and scale parameters using order statistics, *Biometrika*, **39**, 88-95.

Mann, N.R. and Grubbs, F.E. (1974). Chi-square approximations for exponential parameters, prediction intervals and beta percentiles, *Journal of the American Statistical Association*, **69**, 654-661.

Mann, N.R., Schafer, R.E., and Singpurwalla, N.D. (1974), *Methods for Statistical Analysis of Reliability and Life Data*, John Wiley & Sons, New York.

Meeker, W.Q. and Hahn, G.J. (1980). Prediction intervals for the ratio of normal distribution sample variances and exponential distribution sample means, *Technometrics*, **22**, 357-366.

Nagaraja, H.N. (1986). Comparison of estimators and predictors from two-parameter exponential distribution, *Sankhyā, Series B*, **48**, 11-18.

Nagaraja, H.N. (1988). Some characterizations of continuous distributions based on regressions of adjacent order statistics and record values, *Sankhyā, Series A*, **50**, 70-73.

Nelson, W. (1970). A statistical prediction interval for availability, *IEEE Transactions on Reliability*, **19**, 179-182.

Nelson, W. (1982). *Applied Life Data Analysis*, John Wiley & Sons, New York.

Ng, V.-M. (1984). Prediction intervals for the 2-parameter exponential distribution using incomplete data, *IEEE Transactions on Reliability*, **33**, 188-191.

Patel, J.K. (1989). Prediction intervals - A review, *Communications in Statistics - Theory and Methods*, **18**, 2393-2465.

Patnaik, P.B. (1949). The non central χ^2 and F distributions and their applications, *Biometrika*, **36**, 202-232.

Pitman, E.J.G. (1937). The closest estimate of statistical parameters, *Proceedings of the Cambridge Philosophical Society*, **33**, 212-222.

Raqab, M.Z. (1992). *Predictors of Future Order Statistics from Type II Censored Samples*, Ph.D. Dissertation, The Ohio State University, Department of Statistics.

Robbins, H. (1980). Estimation and prediction for mixtures of the exponential distribution, *Proceedings of the National Academy of Science*, **77**, 2382-2383.

Robinson, G.K. (1991). That BLUP is a good thing: The estimation of random effect, *Statistical Science*, **6**, 15-51.

Satterthwaite, F.E. (1941). Synthesis of variance, *Psychometrika*, **6**, 309-316.

Shah, S.M. and Rathod, V.R. (1978). Prediction intervals for future order statistics in two-parameter exponential distributions, *Journal of the Indian Statistical Association*, **16**, 113-122.

Shayib, M.A. and Awad, A.M. (1990). Prediction interval for the difference between two sample means from exponential population: A Bayesian treatment, *Pakistan Journal of Statistics*, **6**, 1-23.

Shayib, M.A., Awad, A.M., and Dawagreh, A.M. (1986). Large sample prediction intervals for a future sample mean: A comparative study, *Journal of Statistical Computation and Simulation*, **24**, 255-270.

Sloan, J.A. and Sinha, S.K. (1991). Bayesian predictive intervals for a mixture of exponential failure-time distributions with censored samples, *Statistics & Probability Letters*, **11**, 537-545.

Takada, Y. (1979). The shortest invariant prediction interval for the largest observation from the exponential distribution, *Journal of the Japan Statistical Society*, **9**, 87-91.

Takada, Y. (1981). Relation of the best invariant predictor and the best unbiased predictor in location and scale families, *Annals of Statistics*, **9**, 917-921.

Takada, Y. (1985). Prediction limit for observation from the exponential distribution, *Canadian Journal of Statistics*, **13**, 325-330.

Takada, Y. (1991). Median unbiasedness in an invariant prediction problem, *Statistics & Probability Letters*, **12**, 281-283.

Upadhyay, S.K. and Pandey, M. (1989). Prediction limits for an exponential distribution: A Bayes predictive distribution approach, *IEEE Transactions on Reliability*, **38**, 599-602.

CHAPTER 10

Bayesian Inference and Applications

Asit P. Basu[1]

10.1 Introduction

In this chapter, we consider Bayesian inference for the exponential distribution. Here, unlike the classical or non-Bayesian method, the population parameter is considered a random variable. The density function of the exponential distribution, conditional on the parameter θ ($\theta > 0$), is given by

$$f(x \mid \theta) = \frac{1}{\theta}\, e^{-x/\theta} \qquad (x > 0)\ . \tag{10.1.1}$$

Denote this by $X \mid \theta \sim e(\theta)$.

The Bayesian method typically will consist of the following four steps. Given a random sample $\boldsymbol{X} = (X_1, X_2, \ldots, X_n)$ from $f(x \mid \theta)$, obtain the likelihood function $f(\boldsymbol{x} \mid \theta) \equiv L(\theta)$ as a function of θ. State the prior density $g(\theta)$. The prior distribution reflects the prior belief or knowledge about θ before observing the data. The parameters of $g(\theta)$ can be appropriately chosen to reflect prior knowledge about θ. Obtain the posterior density $g(\theta \mid \boldsymbol{x})$ using Bayes' theorem, which is given by

$$g(\theta \mid \boldsymbol{x}) = f(\boldsymbol{x} \mid \theta) g(\theta) \Big/ \int f(\boldsymbol{x} \mid \theta) g(\theta)\, d\theta\ , \tag{10.1.2}$$

where the integral in (10.1.2) is over the parameter space. The posterior density expresses what is known about θ after observing the data. Given the posterior distribution of θ, we then can make appropriate inference. For example, under squared error loss function, $L(\theta, a) = (\theta - a)^2$, the Bayes estimator is the mean of the posterior distribution. For an excellent reference for Bayesian methods, see Berger (1985).

In the next few sections, we will discuss a number of applications. In Section 10.2 we consider the Bayesian inference procedure when the underlying distribution is the one-parameter exponential distribution. In Section 10.3 we obtain Bayesian estimates of reliability, and in Section 10.4 a Bayesian classification problem is discussed.

[1]University of Missouri-Columbia, Columbia, Missouri

10.2 Bayesian Concepts

Let $\boldsymbol{x} = (x_1, x_2, \ldots, x_n)$ be the observed data corresponding to a random sample from (10.1.1). Then, the likelihood function is

$$L(\theta) = f(\boldsymbol{x} \mid \theta) = \frac{1}{\theta^n}\, e^{-n\overline{x}/\theta}\,, \tag{10.2.1}$$

where $\overline{\boldsymbol{X}} = \frac{1}{n}\sum_1^n X_i$ is a sufficient statistic for θ. To obtain the posterior density, we consider the following prior distributions. First, we consider the noninformative prior $g(\theta)$, when no information is available about θ. Let $I(\theta)$ denote the expected Fisher information given by

$$I(\theta) = -E_\theta\left[\frac{d^2 \log f(x \mid \theta)}{d\theta^2}\right] = \frac{1}{\theta^2}\,. \tag{10.2.2}$$

Then, the noninformative prior $g_1(\theta)$ is given by

$$g_1(\theta) = [I(\theta)]^{1/2} = \frac{1}{\theta}\,. \tag{10.2.3}$$

A noninformative prior is used when we do not have any prior knowledge of θ.

To reflect prior opinion, we consider the conjugate prior, namely the inverted gamma distribution with density

$$g_2(\theta) = \frac{\alpha^\nu}{\Gamma(\nu)}\,\theta^{-(\nu+1)}\, e^{-\alpha/\theta}\,, \qquad \theta > 0\,. \tag{10.2.4}$$

Here, the parameters α and ν are chosen to reflect prior knowledge. We denote this by $\theta \sim IG(\alpha, \nu)$. Note $g_1(\theta)$ is a special case of $g_2(\theta)$ with $\alpha = \nu = 0$. The posterior density $g_2(\theta \mid \boldsymbol{x})$, when $g_2(\theta)$ is the prior density, by Bayes' theorem, is given by

$$g_2(\theta \mid \boldsymbol{x}) = \frac{(\alpha + n\overline{x})^{n+\nu}}{\Gamma(n+\nu)}\;\frac{1}{\theta^{n+\nu+1}}\; e^{-(\alpha+n\overline{x})/\theta}\,, \tag{10.2.5}$$

which is $IG(\alpha + n\overline{x},\, n+\nu)$. Similarly, the posterior density $g_1(\theta \mid \overline{x})$ is $IG(n\overline{x}, n)$ and is obtained directly or taking $\alpha = \nu = 0$ in (10.2.5). From now on, we will only consider the conjugate prior $g_2(\theta)$ as the prior density for illustration.

For any other arbitrary prior $g(\theta)$, the posterior distribution may not be readily available in explicit form, as the relevant integrals may not be obtained in closed form. One needs to consider appropriate approximation methods due to Lindley (1980), Tierney and Kadane (1986), and others. See O'Hagan (1994) and Press (1989) for some of these details. However, we will only consider the case of the conjugate prior density $g_2(\theta)$ from now on, and for convenience, we will drop the subscript and denote this by $g(\theta)$. In this case, the Bayes estimate of θ, assuming squared error loss, is given by the posterior mean

$$\widehat{\theta} = E(\theta \mid \boldsymbol{x}) = \frac{\alpha + n\overline{x}}{n+\nu-1}\,. \tag{10.2.6}$$

The posterior variance of the estimate is

$$E\left((\theta - \widehat{\theta})^2 \mid \boldsymbol{x}\right) = \frac{(\alpha + n\overline{x})^2}{(n+\nu-1)^2(n+\nu-2)}\,. \tag{10.2.7}$$

We next consider $100(1-\alpha)\%$ Bayesian intervals for θ. Like the classical method, we may consider an equal-tail $(1-\alpha)$ credible interval (L, U) for θ. This is given by computing L and U satisfying

$$\int_0^L g(\theta \mid \boldsymbol{x}) d\theta = \int_U^\infty g(\theta \mid \boldsymbol{x})\, dx = \frac{\alpha}{2} \,. \tag{10.2.8}$$

Usually, we consider the highest posterior density (HPD) interval. For the exponential distribution the $100(1-\alpha)\%$ HPD interval satisfies the following:

$$\int_L^U g(\theta \mid \boldsymbol{x})\, d\theta = 1 - \alpha \tag{10.2.9}$$

and

$$g(U \mid \boldsymbol{x}) = g(L \mid \boldsymbol{x}) \tag{10.2.10}$$

We can solve for L and U numerically using the equations (10.2.9) and (10.2.10) to obtain the $100(1-\alpha)\%$ HPD interval. To test the null hypothesis $H_0 : \theta \leq \theta_0$ against the alternative hypothesis $H_1 : \ \theta > \theta_0$, we compute the Bayes factor B in favor of H_0, where

$$\begin{aligned} B &= \text{(posterior odds ratio)/(prior odds ratio)} \\ &= (\alpha_0/\alpha_1)/(\Pi_0/\Pi_1) = \frac{\alpha_0 \Pi_1}{\alpha_1 \Pi_0} \,. \end{aligned} \tag{10.2.11}$$

Here, α_0, α_1 are the posterior probabilities with $\alpha_0 = P(\theta \leq \theta_0 \mid \boldsymbol{x})$, $\alpha_1 = P(\theta > \theta_0 \mid \boldsymbol{x})$, $\Pi_0 = P(\theta \leq \theta_0)$, and $\Pi_1 = P(\theta > \theta_0)$. If $B > 1$, we accept H_0, otherwise we reject H_0.

Example 10.1 We consider an illustrative example. Let a random sample of size 10 be obtained from (10.1.1). Let $\overline{x} = 2.9$. We consider the prior distribution $IG(\alpha, \nu)$, where $\alpha = 1$ and $\nu = 1$. Then, the posterior distribution $IG(n\overline{x} + \alpha, n + \nu)$ reduces to $IG(30, 11)$. Note $\frac{2(n\overline{x}+1)}{\theta} = \frac{60}{\theta} \sim \chi^2_{22}$. Then, the maximum likelihood estimate of θ is $\overline{x} = 2.9$, and the Bayesian estimate using (10.2.6) is $\widehat{\theta} = 3$. Using (10.2.7), the posterior variance of the estimate is one.

10.3 Life Testing and Reliability Estimation

In this section, we consider some estimation problems in reliability. As in Section 10.1, let the density of X conditional on the random parameter θ be given by (10.1.1), and let the prior density $g(\theta)$ of θ be $IG(\alpha, \nu)$.

Let x denote the lifetime of a physical system. Then, the reliability function R_1 is the probability that the system will be in operating condition and function at mission time t and is given by

$$R_1 = R_1(t) = P(X > t) = \exp(-t/\theta) \,. \tag{10.3.1}$$

Let us call this mission time reliability model as Model I.

Given a random sample $\boldsymbol{X} = (X_1, X_2, \ldots, X_n)$ from $f(x \mid \theta)$, the Bayes estimate of θ, assuming squared error loss function, is given by (10.2.6). The Bayes estimate of R_1 is given by

$$\widehat{R}_1(t) = \int_0^\infty e^{-t/\theta} g(\theta \mid \boldsymbol{x})\, d\theta = (1 + t/(\alpha + n\overline{x}))^{-(n+\nu)} . \tag{10.3.2}$$

A second definition of reliability is given by following Model II. Suppose the random variable Y denotes the strength of a component subject to a random stress X. Then, the component fails if, at any moment, the applied stress is greater than its strength or resistance. The reliability of the component in this case is given by

$$R_2 = P(X < Y) . \tag{10.3.3}$$

Let X and Y be independent with $X \mid \theta_1 \sim e(\theta_1)$ and $Y \mid \theta_2 \sim e(\theta_2)$. Let the prior distribution of θ_i be $IG(\alpha_i, \nu_i)$, $i = 1, 2$. Then,

$$R_2 = \theta_2/(\theta_1 + \theta_2) = 1/(1 + \lambda) , \tag{10.3.4}$$

where

$$\lambda = \theta_1/\theta_2 . \tag{10.3.5}$$

Let $(X_1, X_2, \ldots, X_{n_1})$ and $(Y_1, Y_2, \ldots, Y_{n_2})$ be two independent random samples from $e(\theta_1)$ and $e(\theta_2)$, respectively. Then, the posterior density $g(\lambda \mid \overline{x}, \overline{y})$ of λ can be derived from the posterior densities of θ_1 and θ_2 and is given by

$$g(\lambda \mid \overline{x}, \overline{y}) = \frac{U^{n_1+\nu_1}}{B(n_1 + \nu_1, n_2 + \nu_2)} \frac{\lambda^{n_2+\nu_2-1}}{(\lambda + U)^{n_1+n_2+\nu_1+\nu_2}} , \ \lambda > 0 . \tag{10.3.6}$$

Here, $U = (\alpha_1 + n_1\overline{x})/(\alpha_2 + n_2\overline{y})$.

We also can obtain the posterior density of R_2 from (10.3.6).

The Bayes estimate of R_2, assuming squared error loss function, is then given by

$$\begin{aligned} \widehat{R}_2 &= E(R_2 \mid \overline{x}, \overline{y}) \\ &= \int_0^\infty (1/(1 + \lambda)) g(\lambda \mid \overline{x}, \overline{y})\, d\lambda, \end{aligned} \tag{10.3.7}$$

which can be computed numerically.

For a survey of Bayesian estimates of more complex systems, see Basu and Tarmast (1987), Basu and Ebrahimi (1991, 1992), and Basu and Thompson (1992, 1993).

10.4 Bayesian Classification Rules

In this section, we consider a classification rule for the exponential population. Consider three exponential populations Π_0, Π_1, and Π_2 and let the density function corresponding to Π_i conditional on the parameters θ_i, be

$$f_i(x \mid \theta_i) = f(x \mid \theta_i) = \frac{1}{\theta_i} e^{-x/\theta_i} , \ (x > 0, \ \theta_i > 0), \ i = 0, 1, 2 . \tag{10.4.1}$$

It is known that $\theta_0 = \theta_i$ for exactly one i ($i = 1$ or 2). Let $\boldsymbol{Z} = (Z_1, Z_2, \ldots, Z_{n_0})$ be a random sample from Π_0. We want to classify Π_0 as Π_1 or Π_2 based on the random sample $\boldsymbol{Z}$.

The problem of classification has been considered widely. See, for example, Anderson (1958) and Press (1972). Basu and Gupta (1974) have considered classification rules for exponential populations. We summarize some of their results here.

In the Bayesian approach, we compute the posterior density $g(\theta_i \mid \overline{x}_i)$ based on the random sample $(X_{i1}, X_{i2}, \ldots, X_{in_i})$ from Π_i, $(i = 1, 2)$. The *predictive probability density* for classifying a future sample $\boldsymbol{Z} = (Z_1, Z_2, \ldots, Z_{n_0})$ into Π_i is calculated using $g(\theta_i \mid \overline{x}_i)$. Finally, the *predictive odds ratio* R_{ij} for classifying $\boldsymbol{Z}$ into Π_i, as compared with Π_j, is calculated. Bayesian classification rule is based on the predictive odds ratio. If $R_{12} > 1$, classify Π_0 as Π_1, and if $R_{12} < 1$, classify Π_0 as Π_2.

For illustrative purposes, let us consider the noninformative or diffused prior density function. In this case, the posterior density is given by

$$g(\theta_i \mid \boldsymbol{x}) = \frac{(n_i\overline{x}_i)^{n_i}}{\theta_i^{n_i+1}\Gamma(n_i)} \exp(-n_i\overline{x}_i/\theta_i), \qquad \theta_i > 0 \,. \tag{10.4.2}$$

The predictive probability density $f(\overline{z} \mid \overline{x}_i, \Pi_i)$ is given by

$$\begin{aligned} f(\overline{z} \mid \overline{x}_i, \Pi_i) &= \int_0^\infty f(\overline{z} \mid \theta_i) g(\theta_i \mid \overline{x}_i)\, d\theta_i \\ &= \frac{(n_0/n_i)^{n_0}}{B(n_0, n_i)} \frac{\overline{z}^{n_0-1}}{\overline{x}_i^{n_0}} \left(1 + \frac{n_0\overline{z}}{n_i\overline{x}_i}\right)^{-(n_0+n_i)} . \end{aligned} \tag{10.4.3}$$

Let q_i denote the prior probability that the sample $\boldsymbol{z}$ is from Π_i, $(i = 1, 2)$. Then, the predictive odds ratio R_{12} for classifying $\boldsymbol{z}$ into Π_1, as compared with Π_2, based on the sample mean $\boldsymbol{z}$ is given by

$$R_{12} = q_1 f(\overline{z} \mid \overline{x}_1, \Pi_1)/(q_2 f(\overline{z} \mid \overline{x}_2, \Pi_2) \,. \tag{10.4.4}$$

For simplicity, assume $q_1 = q_2 = 1/2$. Then, from (10.4.3)

$$\begin{aligned} R_{12} &= f(\overline{z} \mid \overline{x}_1, \Pi_1)/f(\overline{z} \mid \overline{x}_2, \Pi_2) \\ &= \frac{B(n_0, n_2)}{B(n_0, n_1)} (n_2\overline{x}_2 \mid n_1\overline{x}_1)^{n_0} \frac{(1 + n_0\overline{z}/n_2\overline{x}_2)^{n_0+n_2}}{(1 + n_0\overline{z}/n_1\overline{x}_1)^{n_0+n_1}} \,. \end{aligned} \tag{10.4.5}$$

We consider an illustrative example.

Example 10.2 Two independent random samples from two exponential populations, Π_1 and Π_2, are given below.

Sample from Π_1: 413, 14, 58, 37, 100, 65, 9, 169, 447, 184, 36, 201, 118, 34, 31, 18, 18, 67, 57, 62, 7, 22, 34 ($n_1 = 23$).

Sample from Π_2: 23, 261, 87, 7, 120, 14, 62, 47, 225, 71, 246, 21, 42, 20, 5, 12, 120, 11, 3, 14, 71, 11, 14, 11, 16, 90, 1, 16, 52, 95 ($n_2 = 30$).

We want to classify the following sample of size 6 into Π_1 or Π_2.

Sample from Π_0: 50, 254, 5, 283, 35, 12 ($n_0 = 6$).

Here, $n_1 = 23$, $n_2 = 30$, $n_0 = 6$, $\overline{x}_1 = 95.70$, $\overline{x}_2 = 59.60$, $\overline{z} = 106.50$. Substituting these values in (10.4.5), we obtain

$$R_{12} = \frac{f(\overline{z} \mid \overline{x}_1, \Pi_1)}{f(\overline{z} \mid \overline{x}_2, \Pi_2)} = 2.44 .$$

Thus, we classify Π_0 as Π_1, since the predictive odds are slightly more in favor of Π_1 than the prior odds ratio of 1 : 1.

References

Anderson, T.W. (1958). *An Introduction to Multivariate Statistical Analysis*, John Wiley & Sons, New York.

Basu, A.P. and Ebrahimi, N. (1991). Bayesian approach to life testing and reliability estimation using asymmetric loss function, *Journal of Statistical Planning and Inference*, **29**, 21-31.

Basu, A.P. and Ebrahimi, N. (1992). Bayesian approach to some problems in life testing and reliability estimation, In *Bayesian Analysis in Statistics and Econometrics* (Eds., P.K. Goel and N.S. Iyenger), pp. 257-266, Springer-Verlag, New York.

Basu, A.P. and Gupta, A.K. (1974). Classification rules for exponential populations, In *Reliability and Biometry* (Eds., F. Proschan and R.J. Serfling), pp. 637-650, SIAM, Philadelphia.

Basu, A.P. and Gupta, A.K. (1977). Classification rules for exponential populations: Two parameter case, In *Theory and Applications of Reliability* (Eds., C.P. Tsokos and I.N. Shimi), Vol. 1, pp. 507-525, Academic Press, New York.

Basu, A.P. and Tarmast, G. (1987). Reliability of a complex system from Bayesian viewpoint, In *Probability and Bayesian Statistics* (Ed., R. Viertt), pp. 31-38, Plenum, New York.

Basu, A.P. and Thompson, R.D. (1992). Life testing and reliability estimation under asymmetric loss, In *Survival Analysis* (Eds., P.K. Goel and J.P. Klein), pp. 3-10, Kluwer Academic, Hingham, Massachusetts.

Berger, J.O. (1985). *Statistical Decision Theory and Bayesian Analysis*, Vol. 2, Springer-Verlag, New York.

Lindley, D.V. (1980). Approximate Bayesian methods, In *Bayesian Statistics* (Eds., J.M. Bernardo, M. De Groot, D.V. Lindley, and A.F.M. Smith), pp. 223-245, Valencia Press, Spain.

O'Hagan, A. (1994). *Bayesian Inference, Kendall's Advanced Theory of Statistics*, Vol. 2B, Edward Arnold, London.

Press, S.J. (1972). *Applied Multivariate Analysis*, Holt, Rinehart, and Winston, New York.

Press, S.J. (1989). *Bayesian Statistics: Principles, Models, and Applications*, John Wiley & Sons, New York.

Sinha, S.K. (1986). *Reliability and Life Testing*, John Wiley & Sons, New York.

Thompson, R.D. and Basu, A.P. (1993). Bayesian reliability of stress-strength systems, In *Advances in Reliability* (Ed., A.P. Basu), pp. 411-421, North-Holland, Amsterdam.

Tierney, L. and Kadane, J. (1986). Accurate approximations for posterior moments and marginals, *Journal of the American Statistical Association*, **81**, 82-86.

CHAPTER 11

Conditional Inference and Applications

Román Viveros[1]

11.1 Ancillarity and Conditionality: A Brief Overview

Let $\boldsymbol{T}$ be an observable vector of random variables with probability density function (PDF) $f_{\boldsymbol{T}}(\boldsymbol{t};\boldsymbol{\theta})$, where $\boldsymbol{\theta}$ is an unknown possibly vector parameter taking values over a space $\boldsymbol{\Omega}$. Following Kalbfleisch (1982), a statistic, or vector of statistics, $\boldsymbol{A} = \boldsymbol{A}(\boldsymbol{T})$ that has a PDF free of $\boldsymbol{\theta}$ is called an *ancillary* statistic for the estimation of $\boldsymbol{\theta}$. If an information-preserving transformation from $\boldsymbol{T}$ to $(\boldsymbol{S}, \boldsymbol{A})$ exists where $\boldsymbol{A}$ is ancillary, then the joint PDF of $(\boldsymbol{S}, \boldsymbol{A})$ can be written

$$f_{\boldsymbol{S}|\boldsymbol{A}}(\boldsymbol{s};\boldsymbol{\theta} \mid \boldsymbol{a})f_{\boldsymbol{A}}(\boldsymbol{a}), \tag{11.1.1}$$

where the second factor is free of $\boldsymbol{\theta}$. The conditionality principle advocates basing on $\boldsymbol{S}$ any inference (estimation and hypothesis-testing) procedure about $\boldsymbol{\theta}$, and judging the corresponding frequency properties according to $f_{\boldsymbol{S}|\boldsymbol{A}}(\boldsymbol{s};\boldsymbol{\theta} \mid \boldsymbol{a})$ within the reference set of samples for which $\boldsymbol{A}$ is held fixed at its observed value $\boldsymbol{a}$.

In order to avoid certain difficulties raised by Basu (1964), effective and unambiguous use of conditional inference can be made by restricting attention to situations in which $(\boldsymbol{S}, \boldsymbol{A})$ is the minimal sufficient statistic for $\boldsymbol{\theta}$, $\dim(\boldsymbol{S}, \boldsymbol{A}) > \dim(\boldsymbol{\theta})$ and $\boldsymbol{A}$ has maximum dimension. See, e.g. Cox and Hinkley (1974, pp. 31-32). It can thus be seen, as emphasized from its inception by Fisher (1934, 1936), ancillary statistics yield a dimensionality reduction in the process of deriving inferences about $\boldsymbol{\theta}$. In all of the applications considered in this chapter, the conditional reference set equals $\boldsymbol{\Omega}$ in dimension.

An ancillary statistic $\boldsymbol{A}$ is interpreted as an index of precision of a statistical experiment. Thus, certain values of $\boldsymbol{A}$ indicate that the experiment has been relatively informative about $\boldsymbol{\theta}$ while others suggest a less informative outcome. This interpretation is well illustrated in the next section in the context of the exponential model. A vivid

[1]McMaster University, Hamilton, Ontario, Canada

illustration of this role can be seen when estimating the mean of a normal distribution by realizing that the sample size is indeed a random variable on which we unconsciously condition. Note that the foregoing conditional method respects the likelihood principle since $f_{\boldsymbol{S}|\boldsymbol{A}}(\boldsymbol{s};\boldsymbol{\theta} \mid \boldsymbol{a})$, as a function of $\boldsymbol{\theta}$, is proportional to $f_{\boldsymbol{T}}(\boldsymbol{t};\boldsymbol{\theta})$. For more details on ancillary statistics and conditional inference, see Kalbfleisch (1982) and Kiefer (1982), and the references therein.

A general class of problems where structure (11.1.1) finds application is in deriving inferences about location and scale parameters when the observable variable follows a distribution belonging to the family of transformation models. The basic results were first discovered by Fisher (1934) and are explained in detail by Lawless (1982, pp. 533-539). It is in this connection that conditional inference for the exponential model is possible.

11.2 A Simple Example

In the simplest context, the issue of conditional inference arises in relation to the exponential model when three items are put on test at the same time but the time at which the first failure occurs goes unrecorded. Under the assumption that the failure time of each item follows $e(\theta)$, the observed failure times (T_1, T_2) are seen to be the two largest order statistics from an i.i.d. sample of size $n = 3$ from $e(\theta)$. It is readily seen that $(T_1, T_2) \leftrightarrow (S = \bar{T}, A = 2T_1/(T_1 + T_2))$, where $\bar{T} = (T_1 + T_2)/2$ is the sample mean, is an information-preserving transformation for which structure (11.1.1) applies with

$$f_{S|A}(s;\theta \mid a) = 4s(1 - e^{-as/\theta})e^{-2s/\theta}/\theta^2[1 - (1 + a/2)^{-2}], \quad 0 < u < \infty, \tag{11.2.1}$$

and

$$f_A(a) = 3[1 - (1 + a/2)^{-2}], \quad 0 < a < 1. \tag{11.2.2}$$

If the purpose is to set-up confidence intervals for θ, one may conveniently work with $U = S/\theta$ which is a pivotal quantity for θ and whose conditional density can be obtained directly from (11.2.1). Numerical calculations are usually involved in obtaining the required percentage points.

Example 11.1 As a numerical illustration, if one observes $(t_1, t_2) = (19.2, 27.6)$ then $a = 0.82$ yielding

$$f_{U|A}(u \mid a) = 8.05u(1 - e^{-0.82u})e^{-2u}, \quad 0 < u < \infty.$$

From the conditional distribution function of U, by trial and error one obtains $\Pr[U \leq 0.263 \mid a = 0.82] = 0.025$ and $\Pr[U \leq 3.163 \mid a = 0.82] = 0.975$. Thus, solving for θ in $0.263 \leq \bar{t}/\theta \leq 3.163$ gives $7.40 \leq \theta \leq 88.97$ as a conditional 95% confidence interval for θ.

The role of the ancillary statistic A as an index of precision is put in evidence in Figure 11.1 where $f_{U|A}(u \mid a)$ is depicted for several values of A. Large values of A indicate a relatively informative experiment concerning the estimation of θ while small values convey less precision. Compare the various plots of $f_{U|A}(u \mid a)$ with the marginal density of U,

$$f_U(u) = 12e^{-2u}(u - 1 + e^{-u}), \quad 0 < u < \infty,$$

which is also depicted in Figure 11.1. From (11.2.2) it is seen that A has a monotonic increasing density with mean value at $5.5 \ln 1.5 \doteq 0.6344$ and median at $2/3$.

11.3 One-Parameter Exponential Model

In the most general context, the issue of conditional inference for the one-parameter exponential model appears to arise when data are gathered following a Type II progressively censored scheme that contains at least one observation censored on the left. More specifically, suppose n units are randomly selected from a population of units, and each unit is subjected to a life or fatigue test under identical environmental conditions and levels of stress. The first R_0 ($R_0 \geq 1$) failure times are unrecorded and, at every subsequent failure, a pre-specified number of randomly selected units among the surviving ones is withdrawn from the experiment. Let $T_1 \leq T_2 \leq \cdots \leq T_m$ be the observed failure times and $R_1, R_2, ..., R_m$ the corresponding numbers of units removed ($R_i \geq 0,\ \ 1 \leq i \leq m$). Note that $R_0 + m + \sum_{i=1}^{m} R_i = n$.

Assuming that each unit's failure time follows $e(\theta)$, the joint density of $\boldsymbol{T} = (T_1, T_2, ..., T_m)$ can be shown to be

$$f_{\boldsymbol{T}}(\boldsymbol{t};\theta) = C\,[1 - \exp(-t_1/\theta)]^{R_0} \exp\left(-\sum_{i=1}^{m}(R_i+1)t_i/\theta\right)/\theta^m, \tag{11.3.1}$$

for $0 < t_1 \leq t_2 \leq \cdots \leq t_m < \infty$, $0 < \theta < \infty$, where $C = n!(n - R_0 - R_1 - 1)(n - R_0 - R_1 - R_2 - 2)\cdots(n - R_0 - \sum_{i=1}^{m-1}(R_i+1))/R_0!(n - R_0 - 1)!$. It is readily seen from (11.3.1) that

$$(T_1, \sum_{i=2}^{m}(R_i+1)T_i) \tag{11.3.2}$$

is minimal sufficient for the estimation of θ.

In order to derive the required distributional results, it is convenient to consider the spacings

$$D_i = \left[n - R_0 - \sum_{j=1}^{i-1}(R_j+1)\right](T_i - T_{i-1}),$$

for $i = 2, 3, ..., m$. Using standard arguments on transformations of random variables in conjunction with (11.3.1), the following result is established.

Result 11.1 Under the above assumptions:

(i) $T_1, D_2, ..., D_m$ are statistically independent.

(ii) D_i follows $e(\theta)$, $2 \leq i \leq m$.

(iii) T_1 is distributed as the (R_0+1)th order statistic from a random sample of size n from $e(\theta)$.

Viveros and Balakrishnan (1994) already noted the case $R_0 = 0$ of this result. Also, the result generalizes a well-known property of the spacings of the order statistics from a complete exponential sample derived by Sukhatme (1937); see Chapter 3 for details.

Consider the one-to-one transformation from the minimal sufficient statistic (11.3.2) to (S, A) where

$$S = \frac{1}{n - R_0} \sum_{i=1}^{m} (R_i + 1) T_i, \quad A = (n - R_0) T_1 / \sum_{i=1}^{m} (R_i + 1) T_i. \qquad (11.3.3)$$

Note from Result 11.1 that $D = \sum_{i=2}^{m} D_i \sim \text{Gamma}(m-1, \theta)$, and that T_1 and D are independent. Using these facts in conjunction with the relationships $S = [(n - R_0)T_1 + D]/(n - R_0)$ and $A = (n - R_0)T_1/[(n - R_0)T_1 + D]$, one can establish again that (S, A) of (11.3.3) possesses structure (11.1.1) with

$$f_{S|A}(s; \theta \mid a) = K(a) s^{m-1} e^{-(n-R_0)s/\theta} [1 - e^{-as/\theta}]^{R_0} / \theta^m, \quad 0 < s < \infty; \qquad (11.3.4)$$

and

$$f_A(a) = (m-1) \binom{n}{R_0} (1-a)^{m-2} \sum_{j=0}^{R_0} \left\{ (-1)^j \binom{R_0}{j} \left[1 + \frac{j}{n - R_0} a \right]^{-m} \right\}, \quad 0 < a < 1; \qquad (11.3.5)$$

where $K(a)$ is a normalizing constant. Defining

$$w_j = (-1)^j \binom{R_0}{j} / (n - R_0 + ja)^m, \quad 0 \le j \le R_0, \qquad (11.3.6)$$

one can immediately show that

$$K(a) = \left[(m-1)! \sum_{j=1}^{R_0} w_j \right]^{-1}.$$

Conditional confidence intervals about θ can be obtained from the pivotal quantity $U = S/\theta$ and its conditional density; the latter can be derived directly from (11.3.4). The required percentage points of U can be obtained numerically from $f_{U|A}(u \mid a)$. Alternatively, one may resort to the distribution function of U which, after some algebraic manipulations, can be written as

$$F_{U|A}(u \mid a) = \left(\sum_{j=0}^{R_0} w_j \Pr\left[\chi^2_{2m} \le 2(n - R_0 + ja)u \right] \right) \Big/ \left(\sum_{j=0}^{R_0} w_j \right), \qquad (11.3.7)$$

where the w_j's are as given in (11.3.6).

In testing statistical hypotheses about θ such as

$$H_0 : \theta = \theta_0 \quad \text{vs.} \quad H_1 : \theta > \theta_0,$$

one may consider the test statistic $U_0 = S/\theta_0$. The null conditional distribution function of U_0 is as given by (11.3.7).

Example 11.2 In spite of the considerable number of articles dealing with inference procedures for Type II left-censored samples, few real-life data sets have been reported. The closest experiment to the foregoing scheme in which the exponential model has been considered is the experiment on insulating fluid breakdowns reported by Nelson (1982, Table 2.1). Among the $n = 12$ specimens tested at 45 kV, 3 failed before 1 second and the times to breakdown (in seconds) of the remaining $m = 9$ specimens were 2, 2, 3, 9, 13, 47, 50, 55 and 71. To conform these data to the present sampling structure, one should take $R_i = 0$ $(1 \le i \le 9)$ and $R_0 = 3$.

The observed values of S and A, as calculated from (11.3.3), are $s = 28$ and $a = 0.0714$. One finds from (11.3.7) that $\Pr[U \le 0.761 \mid a = 0.0714] = 0.05$ and $\Pr[U \le 2.000 \mid a = 0.0714] = 0.95$. Hence, a 90% conditional confidence interval for the expected life of the specimens is obtained from $0.761 \le s/\theta \le 2.000$ as 14.00 seconds $\le \theta \le$ 36.79 seconds.

Consider the testing of

$$H_0 : \theta = 10 \text{ seconds} \quad \text{vs.} \quad H_1 : \theta > 10 \text{ seconds}.$$

The observed value of the test statistic $U_0 = S/10$ is 2.8. Large values of U_0 will signal departure from H_0 in the direction of H_1, thus a conditional P-value for the test is $\Pr[U_0 \ge 2.8 \mid a = 0.0714] = 0.0011$. Therefore, the data provide significant evidence against a mean life of 10 seconds for the insulating fluid specimens, the evidence indicating a longer average life.

Some general comments:

(a) Although the removal scheme $(R_1, R_2, ..., R_m)$ enters in the calculation of (S, A) of (11.3.3), its effect on the distributions (11.3.4) and (11.3.5) is only through their total in $n - R_0 = m + \sum_{i=1}^{m} R_i$.

(b) If no left censoring takes place, i.e. $R_0 = 0$, it is easily seen from (11.3.4) that S is independent of the ancillary statistic A, thus conditioning has no effect in this case. This fact can also be obtained as a consequence of Basu's Theorem [see e.g., Hogg and Craig (1978, p. 390)] since S is a complete sufficient statistic for θ when $R_0 = 0$.

(c) From the general theory of transformation models, S can be replaced with any statistic satisfying the property of equivariance: $S(aT_1 + b, aT_2 + b, ..., aT_m + b) = aS(T_1, T_2, ..., T_m) + b$ for any real numbers $a > 0$ and $b > 0$. These statistics include the maximum likelihood estimator of θ, the smallest observation T_1 and any weighted average of the observations such as S of (11.3.3), among others. The ancillary statistic A will take different forms depending on the choice of S. Unlike marginal (unconditional) methods, all of these choices will yield equivalent conditional inferences about θ. For details, see Lehmann (1983, Chapter 3) and Lawless (1982, pp. 533-535).

(d) As noted by several authors, the maximum likelihood estimator $\widehat{\theta}$ of θ does not admit an explicit expression unless $R_0 = 0$, i.e. when no left-censored observations are present. In fact, the maximum likelihood equation is obtained from (11.3.4) as

$$\{m\widehat{\theta} + [R_0(1-a) - n]s\}(1 - e^{-as/\widehat{\theta}}) + R_0 as = 0,$$

and yields $\widehat{\theta} = ns/m$ when $R_0 = 0$.

11.4 Exponential Regression

Consider a medical or industrial life experiment in which, associated with every experimental unit there is a number of measured concomitant variables available. If T_i denotes the lifetime of the ith unit, an exponential regression model arises when the error term $T_i/\theta_i(\boldsymbol{x}_i)$ has a standard exponential distribution $e(1)$, where $\theta_i(\boldsymbol{x}_i)$ is the expected lifetime of the unit and $\boldsymbol{x}_i$ is the corresponding $1 \times p$ vector of covariates. In the present situation, a linear regression model is more conveniently developed in terms of the log-lifetimes. In standard notation the model can be written as

$$Y_i = \boldsymbol{x}_i\boldsymbol{\beta} + E_i, \quad f_{E_i}(e) = \exp(e - \exp(e)), \quad 1 \le i \le n, \tag{11.4.1}$$

where $Y_i = \ln T_i$, $\boldsymbol{\beta}$ is a $p \times 1$ vector of regression coefficients and the experimental errors $E_1, ..., E_n$ are assumed independent. Note that $f_{E_i}(e)$ is the standard extreme-value density which is the density of the logarithm of an $e(1)$ variate. Here $\boldsymbol{x}_i\boldsymbol{\beta}$ is the 0.6321-quantile of Y_i.

Assuming no censored observations are present in the experiment, from (11.4.1) the joint density of the log-lifetimes $\boldsymbol{Y} = (Y_1, ..., Y_n)$, given $\boldsymbol{x}_1, ..., \boldsymbol{x}_n$, is

$$\begin{aligned} f(\boldsymbol{y};\boldsymbol{\beta} \mid \boldsymbol{X}) &= \prod_{i=1}^{n} \exp[(y_i - \boldsymbol{x}_i\boldsymbol{\beta}) - \exp(y_i - \boldsymbol{x}_i\boldsymbol{\beta})] \\ &= \exp\left[\sum_{i=1}^{n}(y_i - \boldsymbol{x}_i\boldsymbol{\beta}) - \sum_{i=1}^{n}\exp(y_i - \boldsymbol{x}_i\boldsymbol{\beta})\right], \end{aligned} \tag{11.4.2}$$

where $\boldsymbol{X}$ is the design matrix whose rows are the concomitant vectors.

The conditional approach for inferences about $\boldsymbol{\beta}$ follows from general results on linear regression in the context of transformation models. These results can be found in Verhagen (1961), Fraser (1979), and Lawless (1982, pp. 538-539). The results for the exponential model reported below were worked out and illustrated in a variety of applications by Lawless (1976; 1982, pp. 290-295).

Estimators $\widetilde{\boldsymbol{\beta}} = \widetilde{\boldsymbol{\beta}}(\boldsymbol{Y})$ of $\boldsymbol{\beta}$ are said to possess the property of equivariance if $\widetilde{\boldsymbol{\beta}}(c\boldsymbol{Y} + \boldsymbol{X}\boldsymbol{d}) = c\widetilde{\boldsymbol{\beta}}(\boldsymbol{Y}) + \boldsymbol{d}$ for all scalar c and $p \times 1$ vector of constants $\boldsymbol{d}$. The maximum likelihood, least squares and best linear invariant estimators of $\boldsymbol{\beta}$ all satisfy the property of equivariance. Note that the joint likelihood of $\boldsymbol{\beta}$ is proportional to (11.4.2).

Result 11.2 If $\widetilde{\boldsymbol{\beta}} = \widetilde{\boldsymbol{\beta}}(\boldsymbol{Y})$ are equivariant estimators of $\boldsymbol{\beta}$, then:

(a) The statistics

$$\boldsymbol{A} = (A_1, ..., A_n), \quad A_i = y_i - \boldsymbol{x}_i\widetilde{\boldsymbol{\beta}}, \quad 1 \le i \le n, \tag{11.4.3}$$

are ancillary for $\boldsymbol{\beta}$ and exactly $n - p$ of them are functionally independent.

(b) The quantities

$$\boldsymbol{U} = (U_1, ..., U_p), \quad U_j = \widetilde{\boldsymbol{\beta}}_j - \boldsymbol{\beta}_j, \quad 1 \le j \le p, \tag{11.4.4}$$

are pivotal quantities and their joint conditional density given $\boldsymbol{A} = \boldsymbol{a}$ is

$$f(\boldsymbol{u} \mid \boldsymbol{a}) = K(\boldsymbol{a}, \boldsymbol{X}, n) \exp\left[\sum_{i=1}^{n} \boldsymbol{x}_i\boldsymbol{u} - \sum_{i=1}^{n} \exp(a_i + \boldsymbol{x}_i\boldsymbol{u})\right], \tag{11.4.5}$$

where $K(\boldsymbol{a}, \boldsymbol{X}, n)$ is a normalizing constant.

Proof: See Lawless (1982, pp. 534-536). □

Individual inferences about any regression coefficient β_j are obtained from the pivotal quantity U_j and its marginal conditional density $f(u_j \mid \boldsymbol{a})$ arising from (11.4.5). In general, the required integrations have to be carried out numerically and thus are feasible only when p is small. The simplest of such cases is the simple linear exponential regression model that arises when only one concomitant variable is available ($p = 2$). This case has been treated in detail by Lawless (1982, pp. 291-292) who, using basic calculus, has derived expressions for $f(u_1 \mid \boldsymbol{a})$ and $f(u_2 \mid \boldsymbol{a})$ which are convenient for numerical calculations. A summary of these results follows.

Assume $\boldsymbol{x}_i = (1, x_i)$, $1 \le i \le n$, and designate by β_0 and β_1 the y-intercept and the slope of the linear relationship, respectively. In the sequel, $(\widehat{\beta}_0, \widehat{\beta}_1)$ denotes the maximum likelihood estimator of (β_0, β_1). Thus, the pivotal quantities of interest are

$$U_1 = \widehat{\beta}_0 - \beta_0, \qquad U_2 = \widehat{\beta}_1 - \beta_1.$$

The maximum likelihood equations are obtained from (11.4.2). One equation yields

$$\widehat{\beta}_0 = \ln\left[\frac{1}{n}\sum_{i=1}^{n} t_i \exp(-x_i\widehat{\beta}_1)\right], \tag{11.4.6}$$

while $\widehat{\beta}_1$ has to be computed numerically from

$$\bar{x} - \left[\sum_{i=1}^{n} x_i t_i \exp(-x_i\widehat{\beta}_1)\right] \Big/ \left[\sum_{i=1}^{n} t_i \exp(-x_i\widehat{\beta}_1)\right] = 0, \tag{11.4.7}$$

where $\bar{x}$ is the average of $x_1, ..., x_n$.

Define

$$h(u; \boldsymbol{a}) = \exp(n\bar{x}u) \Big/ \left[\sum_{i=1}^{n} \exp(a_i + x_i u)\right]^n, \quad -\infty < u < \infty. \tag{11.4.8}$$

Integrating u_1 out in (11.4.5) gives

$$f_2(u_2 \mid \boldsymbol{a}) = (n-1)!K(\boldsymbol{a}, \boldsymbol{X}, n)h(u_2; \boldsymbol{a}), \quad -\infty < u_2 < \infty, \tag{11.4.9}$$

from which the expression

$$K(\boldsymbol{a}, \boldsymbol{X}, n) = \left[(n-1)! \int_{-\infty}^{\infty} h(u; \boldsymbol{a}) du\right]^{-1} \tag{11.4.10}$$

is obtained. Note that probability calculations about U_2 involve numerical integration of (11.4.8) on variable u.

After some algebraic simplification, the conditional distribution function of U_1 can be written as

$$\begin{aligned} F_1(u_1 \mid \boldsymbol{a}) &= \Pr[U_1 \le u_1 \mid \boldsymbol{a}] \\ &= \int_{-\infty}^{\infty} (n-1)!K(\boldsymbol{a}, \boldsymbol{X}, n)h(u; \boldsymbol{a}) \Pr\left[\chi^2_{(2n)} \le 2\sum_{i=1}^{n} \exp(a_i + u_1 + x_i u)\right] du. \end{aligned} \tag{11.4.11}$$

Table 11.1: Times to breakdown of $n = 76$ insulating fluid specimens tested at various constant elevated voltage stresses reported by Nelson (1972, Table I)

26 Kv	28 Kv	30 Kv	32 Kv	34 Kv	36 Kv	38 Kv
5.79	68.85	7.74	0.27	0.19	0.35	0.09
1579.52	108.29	17.05	0.40	0.78	0.59	0.39
2323.70	110.29	20.46	0.69	0.96	0.96	0.47
	426.07	21.02	0.79	1.31	0.99	0.73
	1067.60	22.66	2.75	2.78	1.69	0.74
		43.40	3.91	3.16	1.97	1.13
		47.30	9.88	4.15	2.07	1.40
		139.07	13.95	4.67	2.58	2.38
		144.12	15.93	4.85	2.71	
		175.88	27.80	6.50	2.90	
		194.90	53.24	7.35	3.67	
			82.85	8.01	3.99	
			89.29	8.27	5.35	
			100.58	12.06	13.77	
			215.10	31.75	25.50	
				32.52		
				33.91		
				36.71		
				72.89		

A suggestion that results in some simplification in the above calculations is to center the x-values so that $\bar{x} = 0$.

Example 11.3 As an illustration of the inverse power law model, Nelson (1972) reports data consisting of times to breakdown of an insulating fluid subjected to various constant elevated voltage stresses. At each stress level, a number of times to breakdown was observed. The experiment was in fact an accelerated life test, the elevated voltage stresses being employed in order to save time in collecting the breakdown data. The resulting data are reproduced in Table 11.1. The purpose of the experiment was to estimate the relationship between the voltage stress and the failure time distribution of the fluid.

Although these data are more properly analyzed under a Weibull regression model [see Nelson (1972)], Lawless (1976) performs an exponential regression analysis since departure from this model does not appear strong. Denoting by t_i and v_i the observed failure time and voltage stress for the ith specimen ($1 \leq i \leq n$, $n = 76$), the inverse power law model suggests a linear relationship between the logarithm of the mean life time and the log-voltage. We thus consider a linear exponential regression (11.4.1) with $\boldsymbol{x}_i = (1, x_i) = (1, \ln v_i - (\sum_{k=1}^{n} \ln v_k)/n)$, $1 \leq i \leq n$.

The maximum likelihood estimates of β_0 and β_1, which are obtained from (11.4.6) and (11.4.7), are found to be $\widehat{\beta}_0 = 3.020$ and $\widehat{\beta}_1 = -17.704$. Then the observed values $a_i = y_i - \widehat{\beta}_0 - x_i\widehat{\beta}_1$, $1 \leq i \leq n$, are calculated. A short computer program, based on Simpson's Rule, that integrates (11.4.9) numerically and calculates (11.4.11) gives $F_2(-2.031 \mid \boldsymbol{a}) = 0.05$ and $F_2(2.212 \mid \boldsymbol{a}) = 0.95$. Hence, a conditional 90% confidence interval for the slope β_1 is obtained from $-2.031 \leq \widehat{\beta}_1 - \beta_1 \leq 2.212$ as $-19.916 \leq \beta_1 \leq$

-15.673.

Similarly, from the program one obtains $F_1(-0.207 \mid \boldsymbol{a}) = 0.05$ and $F_1(0.173 \mid \boldsymbol{a}) = 0.95$. Thus, a conditional 90% confidence interval for the y-intercept β_0 is obtained from $-0.207 \leq \widehat{\beta}_0 - \beta_0 \leq 0.173$ as $2.847 \leq \beta_0 \leq 3.227$.

11.5 Discussion

The conditional method based on ancillary statistics allows a dimensionality reduction in the process of deriving inferences about unknown parameters. The method is applied here to the one-parameter exponential model when the data are gathered following a Type II progressively censored scheme that includes at least one observation censored on the left. These schemes have been used in life-testing. The conditional method is extended to the estimation of regression coefficients in the linear exponential regression model.

When the underlying model is the two-parameter exponential distribution, one can show that statistics (11.3.2) are jointly sufficient for the threshold and scale parameters. As a result, the issue of conditional inference does not arise for this model.

Some authors have expressed concern about the potential inefficiency of conditional confidence intervals, the arguments being based on the results of Welch (1939). Barnard (1976) has shown that such claims arise from a misunderstanding, and has demonstrated that conditional confidence intervals are not inefficient.

A problem of practical importance is the computational difficulty in the derivation of inferences about regression coefficients in the linear exponential regression model when a moderate-to-large number of covariates is available. This problem remains whether one seeks conditional or unconditional inferences. The actual difficulty is in the calculation of the marginal (conditional or unconditional) density of each pivotal $U_j = \widehat{\beta}_j - \beta_j$. A reliable method for calculating $f_j(u_j \mid \boldsymbol{a})$ would be very helpful here.

References

Barnard, G.A. (1976). Conditional inference is not inefficient, *Scandinavian Journal of Statistics*, **3**, 132-134.

Basu, D. (1964). Recovery of ancillary information, *Sankhyā, Series A*, **26**, 3-16.

Cox, D.R. and Hinkley, D.V. (1974). *Theoretical Statistics*, Chapman and Hall, London.

Fisher, R.A. (1934). Two new properties of mathematical likelihood, *Proceedings of the Royal Society, Series A*, **144**, 285-305.

Fisher, R.A. (1936). Uncertain inference, *Proceedings of the American Academy of Arts and Sciences*, **71**, 245-258.

Fraser, D.A.S. (1979). *Inference and Linear Models*, McGraw-Hill, New York.

Hogg, R.V. and Craig, A.T. (1978). *Introduction to Mathematical Statistics*, Macmillan, New York.

Kalbfleisch, J.D. (1982). Ancillary statistics, In *Encyclopedia of Statistical Sciences, Volume 1* (Eds., S. Kotz and N.L. Johnson), pp. 77-81, John Wiley & Sons, New York.

Kiefer, J. (1982). Conditional inference, In *Encyclopedia of Statistical Sciences, Volume 2* (Eds., S. Kotz and N.L. Johnson), pp. 103-109, John Wiley & Sons, New York.

Lawless, J.F. (1976). Confidence interval estimation in the inverse power law model, *Applied Statistics*, **25**, 128-138.

Lawless, J.F. (1982). *Statistical Models & Methods for Lifetime Data,* John Wiley & Sons, New York.

Lehmann, E.L. (1983). *Theory of Point Estimation*, John Wiley & Sons, New York.

Nelson, W. (1972). Graphical analysis of accelerated life test data with the inverse power law model, *IEEE Transactions on Reliability,* **21**, 2-11.

Nelson, W. (1982). *Applied Life Data Analysis*, John Wiley & Sons, New York.

Sukhatme, P.V. (1937). Tests of significance for samples of the chi-squared population with 2 degrees of freedom, *Annals of Eugenics,* **8**, 52-56.

Verhagen, A.M.W. (1961). The estimation of regression and error-scale parameters when the joint distribution of the errors is of any continuous form and known apart from a scale parameter, *Biometrika*, **48**, 125-132.

Viveros, R. and Balakrishnan, N. (1994). Interval estimation of parameters of life from progressively censored data, *Technometrics*, **36**, 84-91.

Welch, B.L. (1939). On confidence limits and sufficiency, with special reference to parameters of location, *Annals of Mathematical Statistics*, **10**, 58-69.

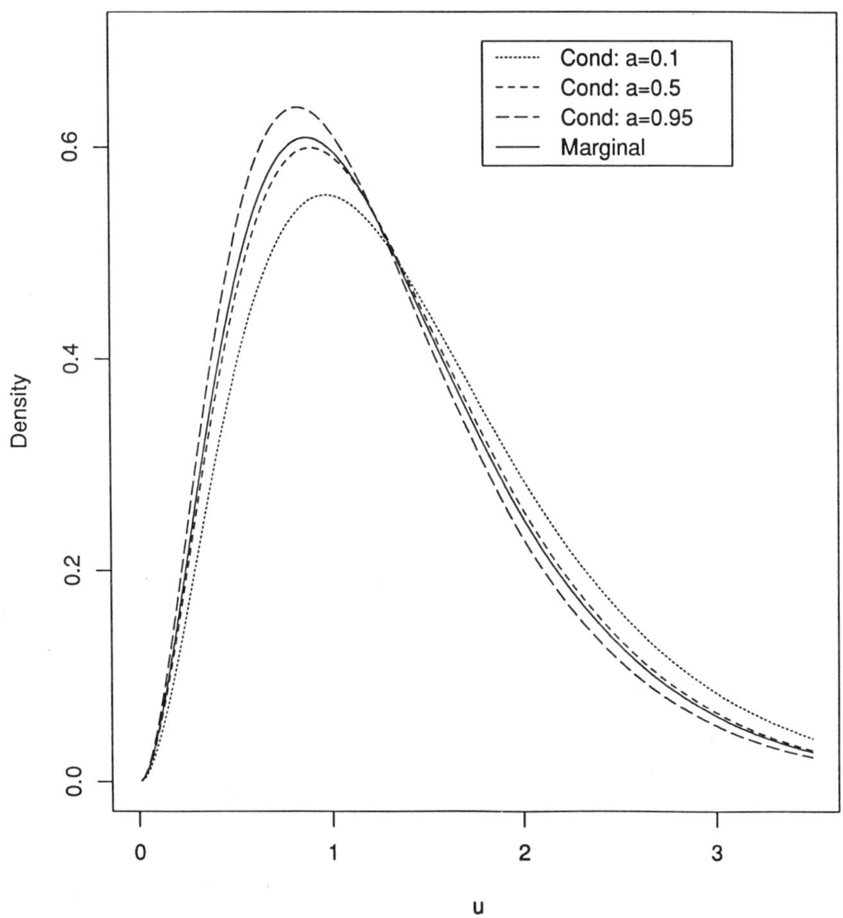

Figure 11.1: Conditional density of U for several values of the ancillary statistic and marginal density of U

CHAPTER 12

Characterizations

Barry C. Arnold[1] and J.S. Huang[2]

12.1 Let Me Count the Ways

The exponential distribution is remarkably friendly. Closed form expressions exist for its density, distribution, moment generating function, mean residual life function, failure rate function, moments of order statistics, record values, etc. So everyone introducing a new concept or functional to classify or organize distributions inevitably uses the exponential distribution as one of their examples. The exponential distribution is often the simplest example or the most analytically tractable. A consequence of its tendency to represent the extreme case is a plethora of characterizations. A chapter is inadequate to catalog the enormous number of characterizations. Azlarov and Volodin (1986) wrote a monograph on the topic with a bibliography involving 143 items. A more up to date bibliography available from the present authors has 275 citations. We can not and should not try to document all of this material in this chapter. Many citations involve refinement of results, eliminating or reducing regularity conditions. For modeling purposes, such refinements are of marginal interest. For example, suppose that an engineer is studying the strength of bundles of fibres. He finds that the shape of the distribution of time to failure does not depend on how many fibres are in the bundle. He probably tried bundles of say 5, 10, 15 fibres, but his experience suggests that similar results would hold for bundles of 7, 13, 27 or whatever number of fibres. A well known characterization result will suggest a Weibull model for time to failure of a bundle (assuming one fibre failure will cause the bundle to fail). The basic result is concerned with the fact that the shape of the distribution of $X_{1:n}$ (the smallest of n i.i.d. X's) doesn't depend on n. The refined version of the result says that you better consider at least two relatively prime choices for n to avoid logarithmically periodic non-Weibull solutions. From a mathematical viewpoint, this is important and interesting. From a modeling view even though the engineer didn't test bundles with relatively prime numbers of fibres, he probably won't be too interested in those esoteric non-

[1] University of California, Riverside, California
[2] University of Guelph, Ontario, Canada

Weibull alternatives. It seems appropriate to mention technical refinements of regularity conditions for various characterizations but to supply proofs which make convenient simplifying assumptions. For example, if assuming a differentiable density allows a 3 line proof to be given and omitting this assumption entails a 5 page argument, we will present the shorter argument.

Even after winnowing out the refinement and technical variation papers, there still remain an enormous variety of exponential characterizations. We have organized our survey, borrowing to some extent the structure suggested by earlier chroniclers of the exponential characterization literature [Galambos and Kotz (1978), Azlarov and Volodin (1986)] in the following broad categories: (i) lack of memory results, (ii) distributional relations among order statistics, (iii) independence of functions of order statistics, (iv) moments of order statistics, (v) geometric compounding, (vi) record value concepts, (vii) miscellaneous characterizations via functionals of distributions.

12.2 Lack of Memory

Focus on non-negative random variables envisioned to represent lifetimes of devices or individuals in stressful environments. Wear and tear suggests that as time goes on, the unit will weaken. The boundary case involves a distribution unaffected by aging. A survival function $\bar{F}$ will be said to exhibit the lack of memory property if for every $x, y \geq 0$ we have

$$\bar{F}(x+y) = \bar{F}(x)\bar{F}(y) \ . \tag{12.2.1}$$

The interpretation is that the conditional probability that a device will survive x more units of time given that it has survived y units of time is exactly the same as the probability that a brand new device will survive x units of time. The device does not remember how old it is!

If we denote the logarithm of the survival function by $\phi(x)$, i.e.

$$\phi(x) = \log \bar{F}(x), \quad x \geq 0 \ ,$$

then the assumed lack of memory implicates that the function ϕ satisfies the celebrated Cauchy functional equation on $\mathbf{R}^+$, i.e.

$$\phi(x+y) = \phi(x) + \phi(y); \quad x, y \geq 0 \ . \tag{12.2.2}$$

Modest assumptions on the behavior of ϕ (right continuity is more than enough) guarantee that the only solutions to this equation are of the form

$$\phi(x) = cx$$

where $c \in \mathbf{R}$. Since $\bar{F}(x)$ is a survival function, the corresponding ϕ is right continuous and satisfies $\phi(0) = 0$ (since $\bar{F}(0) = 1$). We then conclude that

$$\bar{F}(x) = e^{-\lambda x}, \quad x \geq 0 \tag{12.2.3}$$

for some $\lambda > 0$. In other words, the only distribution functions which exhibit the lack of memory property are exponential distributions.

Technical modifications of this observation are possible. For example it suffices to have (12.2.1) hold for all rational x's and y's. Alternatively it is enough to have (12.2.1) hold for all non-negative x's and for some sequence of y's which converges to 0.

In fact it is enough [Marsaglia and Tubilla (1975)] to have (12.2.1) hold for every $x \geq 0$ and for two incommensurable values of y (i.e. values y_1, y_2 such that y_1/y_2 is irrational).

If we envision a random variable X having the distribution F we may write (12.2.1) in the form

$$P[X > x + y | X > y] = P[X > x], \quad x > 0,\ y > 0\ . \tag{12.2.4}$$

A natural analogue of this in a multivariate set up is the following. Let X and Y be two independent, nonnegative random variables satisfying

$$P[X > x + Y | X > Y] = P[X > x], \quad x > 0\ . \tag{12.2.5}$$

Thus a component with lifetime X is still as good as new given that it has outlived a second component whose lifetime is Y. Expressing (12.2.5) by

$$\int_0^\infty \bar{F}(x+y)dG(y) = \int_0^\infty \bar{F}(x)\bar{F}(y)dG(y), \quad x > 0 \tag{12.2.6}$$

where $G(y) = P[Y \leq y]$, we recognize this as the Cauchy functional equation (12.2.1) only in the expectation sense. In the literature this is referred to as the integrated Cauchy functional equation on $\mathbf{R}^+$. It is clear that if (12.2.1) holds, so will (12.2.6), but not vice versa. It is too much to ask for exponentiality of F if G is, for example, degenerate or geometric.

Some special cases were resolved quite some time ago. Krishnaji (1970), showed that if G is exponential and if F is absolutely continuous then (12.2.6) implies exponentiality of F. Krishnaji (1971) showed that if (i) G dominates Lebesgue measure on the positive reals and (ii) F is continuous and is either "New Better than Used", or "New Worse than Used", then F is exponential.

Using different techniques, three papers [Ramachandran (1979), Shimizu (1979), Huang (1981)] arrived at the same result.

If in (12.2.5) Y is nonlattice and if

$$P[Y = 0] < P[X \geq Y]\ , \tag{12.2.7}$$

then X is either degenerate at zero or is exponential. Condition (12.2.7) is unpleasant but is necessary, as was shown in Huang (1981).

A series of further extensions, relating this result to the Choquet-Deny theorem, are given by Rao, Lau, and Ramachandran; see for example, Lau and Rao (1982).

12.3 Distributional Relations Among Order Statistics

Let $X_{k:n}$ be the kth order statistic from an arbitrary population with distribution function F, and let the distribution function of $X_{k:n}$ be denoted by $F_{k:n}$. Thus

$$F_{k:n}(x) = \sum_{j=k}^{n} \binom{n}{j} F^j(x)[1 - F(x)]^{n-j} = \int_0^{F(x)} k\binom{n}{k} t^{k-1}(1-t)^{n-k}dt\ , \tag{12.3.1}$$

and it is clear that F determines $F_{k:n}$ and *vice versa*. If it is given, for instance, that $F_{k:n}$ has the density $f_{k:n}(x) = k\binom{n}{k}x^{k-1}(1-x)^{n-k}$ then F is the uniform (0,1) distribution. Likewise, if $nX_{1:n}$ has the unit exponential density

$$f(x) = e^{-x}, \quad x > 0 ,$$

then X must be unit exponential as well.

Let $Z_{k:n}$ be the normalized spacing

$$Z_{k:n} = (n-k+1)(X_{k:n} - X_{k-1:n}), \quad k = 1, 2, \ldots, n ,$$

where $X_{0:n} \equiv 0$. We can write

$$X_{k:n} = \sum_{i=1}^{k} Z_{i:n}/(n-i+1), \quad k = 1, 2, \ldots, n . \tag{12.3.2}$$

The following remarkable property of the exponential distribution

$$F(x) = 1 - e^{-\lambda x}, \quad x > 0, \lambda > 0 , \tag{12.3.3}$$

was first reported by Sukhatme (1937) [see also Chapter 3].

Theorem 12.1 *If F is exponential (12.3.3), then the normalized spacings $Z_{k:n}$, $k = 1, 2, \ldots, n$, for any $n = 1, 2, \ldots$ are independently and identically distributed with common distribution function F. Thus each order statistic $X_{k:n}$ is the sum (12.3.2) of* independent *exponential random variables.*

It is no great surprise to find many of the exponential characterizations being based on one form or another of this exceedingly simple and neat representation. For the exponential distribution we have, for instance,

$$nX_{1:n} \stackrel{d}{=} X_{1:1}, \quad n = 2, 3, \ldots . \tag{12.3.4}$$

That the converse is also true is the classic exponential characterization of Desu (1971).

Theorem 12.2 *The exponential distribution (12.3.3) is the only nondegenerate distribution satisfying the property (12.3.4).*

A careful look at the proof suggests that it is not necessary to assume identical distribution for *all* n. It is enough that it holds for two integers n_1 and n_2 such that $\ln n_1/\ln n_2$ is irrational [Sethuraman (1965)]. Indeed it is enough that it holds for just one $n > 1$ provided that F has a positive right-hand derivative at the origin [Arnold (1971), Gupta (1973)]. Without this additional differentiability condition the characterization fails. It is even possible to construct counterexamples where $nX_{1:n}$ and $X_{1:1}$ are identically distributed for infinitely many n, $n = 2^i, i = 1, 2, 3, \ldots$ [Huang (1974a)].

So far we seem to be concerned with identical distributions of two order statistics. We can, however, view (12.3.4) as a special case of 'spacings', albeit a rather artificial one, $Z_{1:n} \stackrel{d}{=} Z_{1:1}$. Characterizations based on identical distribution of more general spacings first appeared in Puri and Rubin (1970).

Theorem 12.3 *If X_1, X_2 are i.i.d. with common distribution function F and if*

$$X_{1:1} \stackrel{d}{=} X_{2:2} - X_{1:2} , \tag{12.3.5}$$

then F is either discrete, absolutely continuous, or singular, and no mixture is possible.

(a) If F has a density, then it is exponential.

(b) If F has bounded support, then it either degenerates at zero or it has mass $1/2$ at $X = 0$ and $X = \alpha$ for some $\alpha > 0$.

(c) If F is discrete and if its support is unbounded and is not everywhere dense in $(0, \infty)$, then F has mass $p_0, 0 < p_0 < 1/2$, at the origin and the rest of the mass distributed geometrically on a set of positive lattice points $\alpha < 2\alpha < 3\alpha < \ldots$

However, the question of whether there are other discrete distributions (with dense support) or singular distributions enjoying property (12.3.5) was left open. This question was answered recently by Stadje (1994) who showed that the list of distributions given above in (a) - (c) is exhaustive.

More is available in the $n = 2$ case. Arnold and Isaacson (1976) consider non-degenerate, nonnegative and independent (but not necessarily identically distributed) random variables X and Y, and show that if the distribution function F of X has a right derivative at zero, and if

$$\min(X, Y) \stackrel{d}{=} \lambda X \stackrel{d}{=} (1 - \lambda) Y$$

for some $\lambda \in (0, 1)$, then X is exponential. Somewhat similar, but not having to do with order statistics, is the characterization by Alzaid and Al-Osh (1992) using the relation $X \stackrel{d}{=} U(X_1 + X_2)$ where U is uniform $(0, 1)$.

The most remarkable result is due to Rossberg (1972a). It works for arbitrary spacings and sample size, and does not even require that F be continuous.

Theorem 12.4 *If F is non-lattice and $Z_{1:n} \stackrel{d}{=} Z_{k:n}$ for some n and $k > 1$ then F is exponential.*

The proofs of these results are rather complicated. With additional regularity conditions on F, such as monotone hazard rate or new-better-than-used, the proofs can be greatly simplified. Within the restricted family, say the class of distributions with monotone hazard rate, the exponential distribution can even be characterized by the identical distribution of higher order spacings, $X_{k+j:n} - X_{k:n}, j > 1$. This will be discussed further in Section 12.8.

Without the hazard rate assumption, a characterization based on higher order spacings is known to be achievable but with a different tradeoff, as observed by Gather (1989).

Theorem 12.5 *Let F be a continuous distribution function that is strictly increasing for all $x > 0$. Then F is exponential if and only if*

$$X_{j-i:n-i} \stackrel{d}{=} X_{j:n} - X_{i:n} \tag{12.3.6}$$

for two distinct values j_1 and j_2 of j and some $i, n, 1 \leq i < j_1 < j_2 \leq n, n \geq 3$.

This last characterization was originally due to Ahsanullah (1975), but the proof given there was defective. It is not known whether (12.3.6) holding for just *one* arbitrary j is sufficient to characterize the exponential.

Seshadri, Csörgö, and Stephens (1969) give a different type of characterization, originating from a goodness-of-fit test, as follows.

Theorem 12.6 *Let $X_1, X_2, \ldots, X_n, n \geq 3$, be i.i.d. positive random variables and let $Y_k = (\sum_{i=1}^{k} X_i)/(\sum_{i=1}^{n} X_i), k = 1, 2, \ldots, n-1$. If $Y_1, Y_2, \ldots, Y_{n-1}$ are jointly distributed like $n-1$ order statistics in a sample of size $n-1$ from the uniform $(0,1)$ distribution, then the X_i's are exponentially distributed.*

Again it is possible to trim the requirement a bit. It is necessary to know only that Y_1 and Y_2 are jointly distributed like the first *two* order statistics in a sample of size $n-1$ from *some* distribution [Huang, Arnold, and Ghosh (1979)].

12.4 Independence of Functions of Order Statistics

If X and Y are independent and identically distributed random variables and if $X+Y$ is independent of $X-Y$, then both X and Y must be normal random variables. If instead $X+Y$ is nonnegative and is independent of X/Y, then X and Y are Gamma distributed. These are just some of the well known characterizations of distributions based on properties of independence. The exponential distribution, thanks to the Sukhatme representation, lends itself naturally to such characterizations. Characterizations based on properties of independence seem to have been known much longer than those based on identical distributions.

A parallel to the $n = 2$ case of Theorem 12.1 was first proved by Fisz (1958).

Theorem 12.7 *If F is absolutely continuous and if $X_{1:2}$ and $X_{2:2} - X_{1:2}$ are independent then F is the exponential,*

$$F(x) = 1 - e^{-\lambda(x-\mu)}, \quad x > \mu, \lambda > 0, \quad -\infty < \mu < \infty . \qquad (12.4.1)$$

Notice that unlike Section 12.3, here the solution admits a location parameter since independence is not affected by a location shift. Tanis (1964) then extends Fisz's result to arbitrary n.

Theorem 12.8 *If F is absolutely continuous and if $X_{1:n}$ and $\sum_{i=2}^{n}(X_{i:n} - X_{1:n})$ are independent, then F is exponential.*

Rogers (1963) gives a different extension of Theorem 12.7.

Theorem 12.9 *If F is absolutely continuous and if for some $m = 2, 3, \ldots, n$, $X_{m:n}$ and $X_{m+1:n} - X_{m:n}$ are independent, then F is exponential.*

It is extended by Govindarajulu (1966, 1978).

Theorem 12.10 *If F is (absolutely) continuous and if $X_{m:n}$ and $X_{t:n} - X_{s:n}$ are independent for some $m, s, t, 1 \leq m \leq s < t \leq n$, then F is exponential.*

A word of comment is in order. All four results above were originally given in terms of the power function distribution $F(x) = (x/b)^c$, where b is finite and $c > 0$. In his proof Rogers used only that the regression of $X_{m:n}/X_{m+1:n}$ on $X_{m+1:n}$ is constant. Since $F(0) = 0$, the ratio is bounded, and the expectation exists. Thus constant regression is a weaker requirement than independence. When the result is translated to the exponential, however, the difference $X_{m+1:n} - X_{m:n}$ is no longer bounded, and the existence of its expectation is no longer guaranteed. In his characterization of the Pareto distribution Malik (1970) made the mistake of assuming constancy of the expectation which was not known to exist. It is thus no longer weaker than the assumption of independence. See Crawford (1966) and Wang and Srivastava (1980) on this point. Crawford's observation was to lead to many characterizations based on regression properties.

We now go back to the $n = 2$ case. Let X and Y be independent random variables and suppose that $\min\{X, Y\}$ is independent of $X - Y$. Here the assumption of identical distribution of X and Y is missing, but the independence assumption is now stronger. Ferguson (1964) shows that if X and Y are absolutely continuous then they are both exponential. He (1965) also shows that if either X or Y has a discrete part then X and Y are both geometric.

Crawford (1966) shows that if $\min\{X, Y\}$ is independent of $X - Y$ then either X or Y has a discrete part or they both are absolutely continuous. In other words the two cases covered by Ferguson (1964, 1965) actually exhaust all possibilities.

Assuming $X_1, X_2, \ldots, X_n$ are i.i.d. and continuous (not necessarily absolutely), Ferguson (1967) shows that linear regression of $X_{m+1:n} - X_{m:n}$ on $X_{m:n}$ characterizes a family which includes the exponential and the Pareto.

The case of general spacings was covered by Rossberg (1972b).

Theorem 12.11 *If F is continuous, and if $X_{l:n}$ and $\sum_{i=k}^{m} c_i X_{i:n}$ are independent, where $1 \le l \le k < m \le n$, and $\sum_{i=k}^{m} c_i = 0, c_k \ne 0, c_m \ne 0$, then F is exponential.*

The idea of regression of a spacing on an order statistic has many variants. With the exponential distribution, the intrinsic Markovian nature of the order statistics can take on many guises in the form of conditional expectation, truncated distribution or the lack of memory property. The literature in this area is much richer than that of the last section. For a heuristic treatment of the underlying principles, see Arnold, Balakrishnan, and Nagaraja (1992).

12.5 Moments of Order Statistics

In the theory of characterizations one always strives for weaker and weaker assumptions. The less one has to assume, the more elegant is the result. Returning to Desu's Theorem 12.2, one can not help ask: What if the requirement of identical *distribution* is weakened to identical *expectation*? First we review some basic properties of the expected values of order statistics.

It is clear from (12.3.1) that $F_{k:n}$ is a binomial probability, and as such it satisfies the familiar recurrence relationship of the binomial tail:

$$\left(1 - \frac{k}{n}\right) F_{k:n} + \frac{k}{n} F_{k+1:n} = F_{k:n-1} .$$

It follows that, whenever the expectations exist, the same recurrence applies:

$$\left(1-\frac{k}{n}\right) E[X_{k:n}] + \frac{k}{n} E[X_{k+1:n}] = E[X_{k:n-1}] , \tag{12.5.1}$$

where

$$E[X_{k:n}] = \int_{-\infty}^{\infty} x dF_{k:n}(x) = \int_0^1 F^{-1}(t) k \binom{n}{k} t^{k-1}(1-t)^{n-k} dt ,$$

in which F^{-1} is the inverse function of F,

$$F^{-1}(t) = inf\{x|F(x) \geq t\} .$$

Hereafter we abbreviate $E[X_{k:n}]$ by $\alpha_{k:n}$. Hoeffding (1953) points out that the population is completely determined by the set of expected values of its order statistics.

Theorem 12.12 *Let F possess a finite mean. Then all the $F_{k:n}$'s possess finite means as well, and F is completely determined by*

$$\{\alpha_{k:n}|k = 1, 2, \ldots, n;\; n = 1, 2, \ldots\} . \tag{12.5.2}$$

It is possible to require less. Chan (1967) and Konheim (1971) noticed that the sequence

$$\{\alpha_{1:n}|n = 1, 2, 3, \ldots\} \tag{12.5.3}$$

suffices to characterize F. Pollak (1973) extends it to a more general sequence,

$$\{\alpha_{k_n:n}|n = 1, 2, 3, \ldots\} , \tag{12.5.4}$$

where k_n is any integer between 1 and n. As a special case of Chan's result, the following result follows.

Theorem 12.13 *If for some $c > 0$,*

$$\mu_{1:n} = c/n, \quad n = 1, 2, \ldots ,$$

then F is the exponential (12.3.3) with $\theta = 1/c$.

Superficially this new characterization is no stronger than Desu's. In order to apply Chan's result one needs to assume finiteness of $E|X|$, a condition not needed by Desu. It can be shown, however, that identical distribution of $nX_{1:n}$ already implies a finite expectation for F. It follows therefore that specializing Chan's theorem to the exponential is already enough as an improvement over Desu's.

The recurrence relationship (12.5.1) means that it is sufficient to know any two of the three values $\alpha_{k:n}, \alpha_{k+1:n}$ and $\alpha_{k:n-1}$. Thus knowing the sequence (12.5.3) amounts to knowing the entire triangular array (12.5.2). The same can be said for (12.5.4). A better way to understand Chan's characterization theorem is to realize that it is equivalent to the L_1-completeness of the sequence of polynomials

$$1, x, x^2, x^3, \ldots . \tag{12.5.5}$$

In Pollak's case, instead of (12.5.5) we have

$$P_0, P_1, P_2, P_3, \ldots , \tag{12.5.6}$$

where P_n is the polynomial of degree n defined by

$$P_{n-1}(x) = x^{k_n-1}(1-x)^{n-k_n}, \qquad n = 1, 2, 3, \ldots .$$

As long as the degree of the polynomials starts with zero, the sequence (12.5.6) will generate, much like (12.5.5), the entire class of polynomials. In other words, both sequences are L_1-complete since they generate the same linear space. A celebrated theorem in analysis [Muntz (1914)] says that a subsequence of (12.5.5) is already L_1-complete. It yields the following improvement.

Theorem 12.14 *If $\{n_i\}$ is a sequence of distinct positive integers such that $\sum n_i^{-1} = \infty$ and $E|X_{k:n_1}| < \infty$ for some k, then F is uniquely determined by*

$$\{E[X_{k:n}] \mid n = n_1, n_2, \ldots\} . \qquad (12.5.7)$$

This leads to a twofold improvement over Desu's result [Huang (1974)].

Theorem 12.15 *$E[X_{1:n}] = c/n$ for all prime numbers n if and only if F is exponential (12.3.3), with $\theta = 1/c$.*

It is tempting to consider extension of (12.5.7) to a more general sequence such as

$$\{E[X_{k:n}^m] \mid n = n_1, n_2, n_3, \ldots\} ,$$

where m is some positive odd integer. Since $(X_{k:n})^m = (X^m)_{k:m}$, it amounts to characterizing the new variable X^m by the same method. Indeed for any monotone function ϕ, characterization of X is equivalent to that of $\phi(X)$, and so the possibilities are endless.

Expected values of the spacings

$$\{E[X_{k:n} - X_{k-1:n}] \mid n = k, k+1, k+2, \ldots\} \qquad (12.5.8)$$

can be used to characterize F as well [see Govindarajulu, Huang, and Saleh (1975)]. Since $X_{k:n} - X_{k-1:n}$ is invariant with respect to a shift of origin, except for $k = 1$, it is clear that F can be characterized only up to a location shift.

12.6 Characterizations Involving Geometric Compounding

Suppose that $X_1, X_2, \ldots$ are i.i.d. non-negative random variables and that N, independent of the X_i's, has a geometric(p) distribution. Thus

$$P[N = n] = pq^{n-1}, \quad n = 1, 2, \ldots . \qquad (12.6.1)$$

We now define a geometrically compounded random variable Y_p by

$$Y_p = \sum_{i=1}^{N} X_i . \qquad (12.6.2)$$

If the X_i's are i.i.d. $e(\lambda)$ random variables, it is readily verified that the Laplace transform of the distribution of Y assumes the form

$$\begin{aligned}\varphi_{Y_p}(t) &= E[E[e^{-tY_p} \mid N]] \\ &= \sum_{n=1}^{\infty}(1+\frac{t}{\lambda})^{-n}pq^{n-1} \\ &= \left(1+\frac{t}{p\lambda}\right)^{-1} . \end{aligned} \tag{12.6.3}$$

It follows that

$$pY_p \stackrel{d}{=} X_1, \qquad \forall\, p \in (0,1) . \tag{12.6.4}$$

To what extent does this property characterize the exponential distribution?

Suppose that $X_1, X_2, \ldots$ are i.i.d. non-negative random variables with Laplace transform φ and suppose that (12.6.4) holds for every p. By conditioning on N (as in the development of (12.6.3)) we conclude that

$$\varphi_{Y_p}(t) = p\varphi(t)/[1 - q\varphi(t)]$$

and so, if (12.6.4) holds, we have,

$$\varphi(t) = p\varphi(pt)/[1 - q\varphi(pt)] \qquad \forall\, p . \tag{12.6.5}$$

Now $\varphi(t)$ is continuous and $\varphi(0) = 1$. Thus there is an interval $I = [0, t_0]$ in which $\varphi(t) \neq 0$. In I define a new function $\phi(t)$ as follows

$$\phi(t) = 1 - [\varphi(t)]^{-1} . \tag{12.6.6}$$

From (12.6.5) we find that ϕ satisfies the following equation

$$p\phi(t) = \phi(pt), \qquad p \in (0,1), \qquad t \in I . \tag{12.6.7}$$

Since ϕ is continuous it follows that

$$\phi(t) = \gamma t \tag{12.6.8}$$

for some real γ. Since $\varphi(t) \leq 1$ it follows that γ in (12.6.8) must be negative, so we write $\varphi(t) = -t/\lambda$ for some positive λ and conclude that

$$\varphi(t) = (1 + t/\lambda)^{-1} .$$

Thus if, for every p, p times a geometric minimum of i.i.d. non-negative random variables $X_1, X_2, \ldots$ has the same distribution as X_1, then the common distribution of the X_i's is $e(\lambda)$. It is in fact enough to assume that (12.6.4) holds for two distinct values of p, say p_1 and p_2, chosen such that $\{p_1^j/p_2^k : j = 1, 2, \ldots, k = 1, 2, \ldots\}$ is dense in $(0,1)$. If (12.6.4) holds for just one value of p then provided we assume that $E[X]$ exists (i.e., that $\varphi(t)$ is right differentiable at 0) then again we may conclude that the common distribution of the X's is exponential [details may be found in Arnold (1973)].

An alternative interpretation of this result is possible in terms of thinned renewal processes. We may say that a renewal process undergoes thinning if each event is independently recorded with probability p. We say that a renewal process is preserved under thinning if the thinned process is, except for a change in time scale, identically distributed as the original process. Exponential interarrival times identify a renewal process as a Poisson process. Our characterization result is equivalent to the statement that Poisson processes are the only renewal processes preserved under thinning.

The Poisson process scenario may be used to introduce another geometric compounding characterization. A renewal process may be randomly "split" into k processes, if each event in the original process is assigned to one of the k new processes. A particular event is assigned to process 1 with probability p_1, to process 2 with probability p_2 and, eventually, to process k with probability p_k where $p_i > 0 \ \forall\ i$ and $\sum_{i=1}^{k} p_i = 1$. The k thinned processes so derived will again be renewal processes. They will, in general, not be independent processes. It is however well known that if the original renewal process were a Poisson process with intensity λ, then the k thinned processes *are* independent. In fact they are Poisson processes with intensities $\lambda p_1, \lambda p_2, \ldots, \lambda p_k$ respectively. This is characteristic property of the Poisson process. It may be restated as a result dealing with multivariate geometric compounding as follows.

First we envision an experiment with k possible outcomes $1, 2, \ldots, k$ with associated probabilities $p_1, p_2, \ldots, p_k$ (with $\sum_{i=1}^{k} p_i = 1$). The experiment is repeated until all k of the outcomes have appeared at least once. For each i, let N_i denote the number of experiments up to and including the one which yielded the first outcome of type i. The vector $\boldsymbol{N} = (N_1, N_2, \ldots, N_k)$ has a multivariate geometric distribution and we write $\boldsymbol{N} \sim \mathcal{G}(\boldsymbol{p})$. Now consider a sequence of i.i.d. non-negative random variables $X_1, X_2, \ldots$ assumed to be independent of $\boldsymbol{N}$. Define a k-dimensional random vector $\boldsymbol{Y}$ by

$$Y_i = \sum_{j=1}^{N_i} X_j, \quad i = 1, 2, \ldots, k \,. \tag{12.6.9}$$

Such a vector $\boldsymbol{Y}$ has a compound multivariate geometric distribution. The characterization theorem takes the following form.

Theorem 12.16 *$Y_1, Y_2, \ldots, Y_k$ in (12.6.9) are independent random variables iff X_1, $X_2, \ldots$ are exponential random variables.*

Proof: If the X_i's are exponential then the result follows by envisioning a randomly split Poisson process and using the fact that a Poisson(λ) sum of multinomial($1, \boldsymbol{p}$) random variables has independent Poisson(λp_i) coordinates. The converse is proved by observing that, in fact, independence of just Y_1 and Y_2 is enough to guarantee an exponential distribution of the X_j's. Write down the joint Laplace transform of Y_1, Y_2 in terms of the Laplace transform $\varphi(t)$ of the X_j's by conditioning on the outcome of the first experiment: outcome 1 with probability p_1, and outcome 2 with probability p_2 and something else with probability $p_0 = 1 - p_1 - p_2$. The independence of Y_1, Y_2 can be shown to imply that $\phi(t) = 1 - [\varphi(t)]^{-1}$ satisfies the Cauchy functional equation in an interval $[0, t_0]$. It follows that $\phi(t) = -t/\lambda$ for some $\lambda > 0$ and consequently that the X_j's are exponential (λ). Further details and discussion may be found in Arnold (1975). □

A limit theorem may be associated with the characterization based on (12.6.4). The idea involved is repeated geometric compounding. Begin with a sequence of i.i.d. non-negative random variables with common Laplace transform $\varphi_0(t)$. Define $\varphi_1(t)$ to be the Laplace transform of a geometric (p_1) sum of i.i.d. random variables with common Laplace transform $\varphi_0(t)$ and, more generally, $\varphi_k(t)$ to be the Laplace transform of a geometric (p_k) sum of i.i.d. random variables with common Laplace transform $\varphi_{k-1}(t)$. Now let Y_k denote a random variable with Laplace transform $\varphi_k(t)$, $k = 0, 1, 2, \ldots$. Obviously

$$\varphi_k(t) = p_k \varphi_{k-1}(t)/[1 - q_k \varphi_{k-1}(t)] \ . \tag{12.6.10}$$

Theorem 12.17 *Suppose that $\varphi_0(t)$ has a finite right derivative at zero and that $\varphi_1, \varphi_2, \ldots$ are as defined in (12.6.10). If X_k has Laplace transform $\varphi_k(t)$, $k = 1, 2, \ldots$ then as $k \to \infty$,*

$$\left(\prod_{j=1}^{k} p_j \right) X_k \xrightarrow{d} e(\lambda)$$

for some $\lambda > 0$ provided that $\prod_{j=1}^{\infty} p_j$ diverges to 0.

12.7 Record Value Concepts

Suppose that $X_1, X_2, \ldots$ is a sequence of i.i.d. continuous random variables. An observation X_j will be called a record if it is bigger than all preceding X_i's, i.e. if $X_j > X_i \ \forall \ i < j$. By convention the first observation is the zero'th record R_0 and the successive records are denoted by $R_1, R_2, \ldots$ (see Chapter 17 for details). If the X_i's are $e(\lambda)$ random variables, then the lack of memory property ensures that the record increments $R_i - R_{i-1}, i = 1, 2, \ldots$ are themselves i.i.d. $e(\lambda)$ random variables and consequently for each n,

$$R_n \sim Gamma\left(n + 1, \frac{1}{\lambda}\right) \ . \tag{12.7.1}$$

If the common distribution of the X_i's is $F(x)$, then a straightforward change of variables yields the following expression for the distribution of R_n

$$P[R_n \le r] = 1 - [1 - F(r)] \sum_{k=0}^{n} \{-\log(1 - F(r))\}^k / k! \ . \tag{12.7.2}$$

Two curious properties of record value sequences for exponential variables have potential for characterizing the exponential distribution. For exponential variables, the interrecord times are identically distributed and they are independent random variables. The first of these is indeed a characteristic property of the exponential distribution.

Theorem 12.18 *Let $\{R_n\}_{n=0}^{\infty}$ denote the record value sequence corresponding to i.i.d. random variables $\{X_n\}_{n=1}^{\infty}$ with common continuous distribution F. Suppose that for some $j \ge 1$, R_0 and $R_j - R_{j-1}$ are identically distributed. Then $F(x) = 1 - e^{-\lambda x}, x > 0$ for some $\lambda > 0$, i.e. the X_n's are exponentially distributed.*

Proof: By conditioning on R_{j-1} we may write for $x > 0$,

$$\begin{aligned} \bar{F}(x) &= P[R_0 > x] = P[R_j - R_{j-1} > x] \\ &= \int_0^{\infty} [\bar{F}(x + y)/\bar{F}(y)] dF_{R_{j-1}}(y) \ . \end{aligned}$$

This can be rearranged to assume the form

$$\int_0^\infty \bar{F}(x+y)dH(y) = \int_0^\infty \bar{F}(x)\bar{F}(y)dH(y) \tag{12.7.3}$$

where

$$dH(y) = \left(\frac{1}{\bar{F}(y)}\right) dF_{R_{j-1}}(y) \ .$$

But we recognize this as an integrated Cauchy functional equation. The assumed continuity of F implies that H is continuous and consequently the only solution to (12.7.3) are of the form $\bar{F}(x) = e^{-\lambda x}, \ x > 0$. □

Slightly more complicated arguments are required if we address the question of whether the identical distribution of $R_k - R_j$ and R_{k-j} implies that the common distribution of the X_i's is exponential. Ahsanullah (1987) has shown that this is the case under the additional assumption that F is either New Better than Used or New Worse than Used (F is new better (worse) than used if $\bar{F}(x+y) \leq (\geq)\bar{F}(x)\bar{F}(y) \ \forall \ x, y > 0$).

What if we merely postulate that for some $j < k, R_j - R_{j-1}$ and $R_k - R_{k-1}$ are identically distributed (assuming the common distribution of X_i's, F, is continuous as usual). Since the distributions of $R_j - R_{j-1}$ and $R_k - R_{k-1}$ are unchanged by adding a constant to all X_i's, we can only hope to characterize the exponential distribution, up to translation, so we will assume for convenience that the support of F is $(0, \infty)$.

In this case we may immediately write,

$$\int_0^\infty \frac{\bar{F}(y+x)}{\bar{F}(y)} dF_{R_{j-1}}(y) = \int_0^\infty \frac{\bar{F}(y+x)}{\bar{F}(y)} dF_{R_{k-1}}(y) \qquad \forall \ x \ , \tag{12.7.4}$$

and consequently we have

$$\int_0^\infty [\bar{F}(y+x) - \bar{F}(x)\bar{F}(y)]d\tilde{H}(y) = 0 \qquad \forall \ x$$

where [using (12.7.2)]

$$d\tilde{H}(y) = \sum_{i=j}^{k-1} \frac{(-\log \bar{F}(y))^i}{i!}$$

is a signed measure. Again we recognize the integrated Cauchy functional equation and conclude that $\bar{F}(x) = e^{-\lambda x}, \quad x > 0$.

Next we turn to independence of spacings. If the X_i's are exponentially distributed then we observed that the sequence $R_0, R_1 - R_0, R_2 - R_1, \ldots$ consists of independent identically distributed random variables. As we have just seen, the identical distribution of any two of $R_0, R_1 - R_0, \ldots$ is enough to guarantee that $\bar{F}_X(x) = e^{-\lambda x}, \ x > 0$. Houchens (1984) has made the remarkable observation that if the X_i's have a common extreme value distribution, i.e. if

$$F_X(x) = 1 - \exp[-e^{(x-\mu)/\sigma}], \qquad x \in \mathbf{R} \ ,$$

then the sequence $R_1 - R_0, R_2 - R_1, \ldots$ consists of independent exponential random variables. They are *not* identically distributed since $E[R_i - R_{i-1}] = \sigma/i$, but they are independent. In this extreme value case, R_0 is of course not an exponential random variable and nor is it independent of any $R_i - R_{i-1}$. The example does rule out some otherwise plausible putative exponential characterizations. If $R_i - R_{i-1}$ and $R_j - R_{j-1}$ are of the same type (i.e. if these distributions differ at most by location and scale) we *cannot* conclude that the X_i's are exponential. If $R_i - R_{i-1}$ and $R_j - R_{j-1}$ are independent, we cannot conclude that the X_i's are exponential. Independence of R_0 and $R_i - R_{i-1}$ is a different matter however. We have the following general result.

Theorem 12.19 *Let $\{R_n\}_{n=0}^{\infty}$ denote the record value sequence corresponding to i.i.d. random variables $\{X_n\}_{n=1}^{\infty}$ with common absolutely continuous distribution function $F(x)$, with support $(0, \infty)$. Suppose that for some $0 \le j \le k < \ell$ the random variables R_j and $R_\ell - R_k$ are independent. Then, $F(x) = 1 - e^{-\lambda x}$, $x > 0$ for some $\lambda > 0$.*

The proof of this theorem in its full generality is a complicated calculus exercise culminating in the inevitable Cauchy functional equation. Nayak (1981) supplies the details. The special case $j = k = 0, \ell = 1$ is easily resolved. In this case the joint density of (R_0, R_1) is

$$f_{R_0,R_1}(r_0, r_1) = \frac{f(r_0)}{\bar{F}(r_0)} f(r_1), \quad 0 < r_0 < r_1 < \infty .$$

So the joint density of $R_0, R_1 - R_0$ is

$$f_{R_0,R_1-R_0}(r_0, t) = \frac{f(r_0)}{\bar{F}(r_0)} f(r_0 + t), \quad 0 < r_0 < \infty, 0 < t < \infty ,$$

and since these are independent random variables we can write

$$\frac{f(r_0)}{\bar{F}(r_0)} f(r_0 + t) = f(r_0)g(t)$$

where $g(t)$ is the marginal density of $R_1 - R_0$. It follows that

$$f(r_0 + t) = \bar{F}(r_0)g(t) \quad \forall\, r_0, t > 0 .$$

We recognize this as Pexider's equation [Pexider (1903)] which only admits exponential solutions.

It is possible to characterize the exponential distribution using moments or conditional moments of records. If we are given that, for some $c > 0$, $E[R_n] = nc$ for every n, then we may conclude that $\bar{F}(x) = e^{-\lambda x}, x > 0$ provided that we assume $E|X_1|^{1+\epsilon} < \infty$ for some $\epsilon > 0$. This is a sufficient condition to ensure that the expected record sequence $\{E[R_n]\}_{n=1}^{\infty}$ determines F uniquely [Kirmani and Beg (1984)].

Closely related results are of the following form. If $E[R_{n+1} - R_n | R_n]$ does not depend on R_n, then $\bar{F}(x) = e^{-\lambda x}$ [Nagaraja (1977)] and if $\mathrm{Var}(R_{n+1} - R_n | R_n)$ does not depend on R_n, then $\bar{F}(x) = e^{-\lambda x}$ [Ahsanullah (1981)].

12.8 Miscellaneous Characterizations

The lack of memory property of the exponential distribution can be rephrased in terms of *residual life*, as discussed in reliability theory. Viewing the left hand side of (12.2.4) as the distribution of the residual life X_y of X at y, we can characterize the exponential by

$X_y \stackrel{d}{=} X$ for all $y > 0$. More generally, characterizations can be based on the moments of X_y [Hall and Wellner (1981)], or on the quantiles of X_y [Gupta and Langford (1984), Joe (1985)]. There is a vast amount of literature on the *mean residual life function*,

$$\mu(x) = E[X - x|X > x] = E[X|X > x] - x ,$$

or equivalently, the conditional expectation $E[X|X > x]$. For details, see Meilijson (1972), Hall and Wellner (1981), Mukherjee and Roy (1985), and Oakes and Desu (1990). Since the mean residual life function determines the distribution, it is easy to confirm that a constant mean residual life function characterizes the exponential distribution.

Parallel to the truncated distributions mentioned above, there is an equally rich field of literature on conditional distributions involving order statistics. Contributions include improvements over the results based on independence presented in Section 12.4. Ferguson (1967) noted that in the proof of Theorem 12.9, the full strength of independence is not needed. All that is needed is the constancy of the regression of $X_{m+1:n} - X_{m:n}$ on $X_{m:n}$.

Another reliability concept useful for the characterization of distributions is the hazard (or failure) rate function,

$$\rho_F(x) = \rho(x) = \frac{f(x)}{\bar{F}(x)} .$$

The well known inversion formuli

$$\bar{F}(x) = \exp\{-\int_0^x \rho(t)dt\} ,$$

$$\bar{F}(x) = \frac{\mu(0)}{\mu(x)} \exp\{-\int_0^x \frac{1}{\mu(t)}\, dt\}$$

show that knowing any one of the three, $\bar{F}(x), \rho_F(x)$ and $\mu(x)$, amounts to knowing the other two. The exponential distribution is, of course, uniquely characterized by its constant hazard rate function (see Chapter 2).

Several of the characterizations dealt with in earlier sections (especially those dealing with order statistics) will admit easier proofs or sometimes more general results provided that we are willing to narrow the class of distributions under scrutiny to some reliability class such as the class of distributions with monotone hazard rate or the new-better-than-used class. The exponential is frequently extremal in such classes and this has been exploited in several recent papers by Ahsanullah.

Another fruitful source of exponential characterizations involves the study of renewal processes. The Poisson process with exponential inter-arrival times can be characterized among renewal processes by properties of its forward and backward recurrence time distributions, see Cinlar and Jagers (1973) and Isham, Shanbhag, and Westcott (1975) and the references therein.

References

Ahsanullah, M. (1975). A characterization of the exponential distribution, In *Statistical Distributions in Scientific Work* (Eds., G.P. Patil, S. Kotz, and J.K. Ord), Vol. **3**, pp. 131-135, D. Reidel, Dordrecht.

Ahsanullah, M. (1981). On a characterization of the exponential distribution by weak homoscedasticity of record values, *Biometrical Journal*, **23**, 715-717.

Ahsanullah, M. (1987). Record statistics and the exponential distribution, *Pakistan Journal of Statistics, Series A*, **3**, 17-40.

Alzaid, A.A. and Al-Osh, M.A. (1992). Characterization of probability distributions based on the relation $X \stackrel{d}{=} U(X_1 + X_2)$, *Sankhyā, Series B*, **53**, 188-190.

Arnold, B.C. (1971). Two characterizations of the exponential distribution, *Unpublished manuscript*, Iowa State University, Ames, Iowa.

Arnold, B.C. (1973). Some characterizations of the exponential distribution by geometric compounding, *SIAM Journal of Applied Mathematics*, **24**, 242-244.

Arnold, B.C. (1975). A characterization of the exponential distribution by multivariate geometric compounding, *Sankhyā, Series A*, **37**, 164-173.

Arnold, B.C., Balakrishnan, N., and Nagaraja, H.N. (1992). *A First Course in Order Statistics*, John Wiley & Sons, New York.

Arnold, B.C. and Isaacson, D. (1976). On solutions to $\min(X,Y) \stackrel{d}{=} aX$ and $\min(X,Y) \stackrel{d}{=} aX \stackrel{d}{=} bY$, *Zeitschrift Wahrscheinlichkeitstheorie verwandte Gebiete*, **35**, 115-119.

Azlarov, T.A. and Volodin, N.A. (1986). *Characterization Problems Associated With the Exponential Distribution*, Springer-Verlag, New York.

Chan, L.K. (1967). On a characterization of distributions by expected values of extreme order statistics, *American Mathematical Monthly*, **74**, 950-951.

Cinlar, E. and Jagers, P. (1973). Two mean values which characterize the Poisson process, *Journal of Applied Probability*, **10**, 678-681.

Crawford, G.B. (1966). Characterization of geometric and exponential distributions, *Annals of Mathematical Statistics*, **37**, 1790-1795.

Desu, M.M. (1971). A characterization of the exponential distribution by order statistics, *Annals of Mathematical Statistics*, **42**, 837-838.

Ferguson, T.S. (1964). A characterization of the exponential distribution, *Annals of Mathematical Statistics*, **35**, 1199-1207.

Ferguson, T.S. (1965). A characterization of the geometric distribution, *American Mathematical Monthly*, **72**, 256-260.

Ferguson, T.S. (1967). On characterizing distributions by properties of order statistics, *Sankhyā, Series A*, **29**, 265-278.

Fisz, M. (1958). Characterization of some probability distributions, *Skandinavisk Aktuarietidskrift*, **41**, 65-67.

Galambos, J. and Kotz, S. (1978). *Characterizations of Probability Distributions*, Lecture Notes in Mathematics No. 675, Springer-Verlag, Berlin.

Gather, U. (1989). On a characterization of the exponential distribution by properties of order statistics, *Statistics & Probability Letters*, **7**, 93-96.

Govindarajulu, Z. (1966). Characterization of the exponential and power distributions, *Skandinavisk Aktuarietidskrift*, **49**, 132-136.

Govindarajulu, Z. (1978). Correction to "Characterization of the exponential and power distributions," *Scandinavian Actuarial Journal*, 175-176.

Govindarajulu, Z., Huang, J.S., and Saleh, A.K.Md.E. (1975). Expected value of the spacings between order statistics, In *A Modern Course on Statistical Distributions in Scientific Work* (Eds., G.P. Patil, S. Kotz, and J.K. Ord), pp. 143-147, D. Reidel, Dordrecht.

Gupta, R.C. (1973). A characteristic property of the exponential distribution, *Sankhyā, Series B*, **35**, 365-366.

Gupta, R.C. and Langford, E.S. (1984). On the determination of a distribution by its median residual life function: A functional equation, *Journal of Applied Probability*, **21**, 120-128.

Hall, W.J. and Wellner, J.A. (1981). Mean Residual Life, In *Statistics and Related Topics* (Eds., M. Csórgö, D. Dawson, J.N.K. Rao, and A.K.Md.E. Saleh), pp. 169-184, North-Holland, Amsterdam.

Hoeffding, W. (1953). On the distribution of the expected values of the order statistics, *Annals of Mathematical Statistics*, **24**, 93-100.

Houchens, R.L. (1984). Record value theory and inference, *Ph.D. Thesis*, University of California, Riverside.

Huang, J.S. (1974a). On a theorem of Ahsanullah and Rahman, *Journal of Applied Probability*, **11**, 216-218.

Huang, J.S. (1974b). Characterization of the exponential distribution by order statistics, *Journal of Applied Probability*, **11**, 605-608.

Huang, J.S. (1981). On a 'Lack of memory' property, *Annals of the Institute of Statistical Mathematics*, **33**, 131-134.

Huang, J.S., Arnold, B.C., and Ghosh, M. (1979). On characterizations of the uniform distribution based on identically distributed spacings, *Sankhyā, Series B*, **41**, 109-115.

Isham, V., Shanbhag, D.N., and Westcott, M. (1975). A characterization of the Poisson process using forward recurrence times, *Mathematical Proceedings of the Cambridge Philosophical Society*, **78**, 513-516.

Joe, H. (1985). Characterizations of life distributions from percentile residual lifetimes, *Annals of the Institute of Statistical Mathematics*, **37**, 165-172.

Kirmani, S.N.U.A. and Beg, M.I. (1984). On characterization of distributions by expected records, *Sankhyā, Series A*, **46**, 463-465.

Konheim, A.G. (1971). A note on order statistics, *American Mathematical Monthly*, **78**, 524.

Krishnaji, N. (1970). Characterization of the Pareto distribution through a model of under-reported income, *Econometrica*, **38**, 251-255.

Krishnaji, N. (1971). Note on a characterizing property of the exponential distribution, *Annals of Mathematical Statistics*, **42**, 361-362.

Lau, K.S. and Rao, C.R. (1982). Integrated Cauchy functional equation and characterizations of the exponential law, *Sankhyā, Series A*, **44**, 72-90.

Malik, H.J. (1970). A characterization of the Pareto distribution, *Skandinavisk Aktuarietidskrift*, 115-117.

Marsaglia, G. and Tubilla, A. (1975). A note on the 'Lack of memory' property of the exponential distribution, *Annals of Probability*, **3**, 353-354.

Meilijson, I. (1972). Limiting properties of the mean residual life function, *Annals of Mathematical Statistics*, **43**, 354-357.

Mukherjee, S.P. and Roy, D. (1985). Some characterizations of the exponential and related life distributions, *Calcutta Statistical Association Bulletin*, 189-197.

Muntz, C.H. (1914). Uber den approximationssatz von Weierstrass, Schwarz-Festschrift.

Nagaraja, H.N. (1977). On a characterization based on record values, *Australian Journal of Statistics*, **19**, 70-73.

Nayak, S.S. (1981). Characterization based on record values, *Journal of the Indian Statistical Association*, **19**, 123-127.

Oakes, D. and Desu, M.M. (1990). A note on residual life, *Biometrika*, **77**, 409-410.

Pexider, J.V. (1903). Notiz uber funktional theoreme, *Monatsh. Math. Phys.*, **14**, 293-301.

Pollak, M. (1973). On equal distributions, *Annals of Statistics*, **1**, 180-182.

Puri, P.S. and Rubin, H. (1970). A characterization based on the absolute difference of two I.I.D. random variables, *Annals of Mathematical Statistics*, **41**, 2113-2122.

Ramachandran, B. (1979). On the 'Strong memorylessness property' of the exponential and geometric probability laws, *Sankhyā, Series A*, **41**, 244-251.

Rogers, G.S. (1963). An alternative proof of the characterization of the density Ax^B, *American Mathematical Monthly*, **70**, 857-858.

Rossberg, H.-J. (1972a). Characterization of the exponential and the Pareto distributions by means of some properties of the distributions which the differences and quotients of order statistics are subject to, *Mathematisch Operationsforschung und Statistik, Series Statistics*, **3**, 207-215.

Rossberg, H.-J. (1972b). Characterization of distribution functions by the independence of certain functions of order statistics, *Sankhyā, Series A*, **34**, 111-120.

Seshadri, V., Csörgö, M., and Stephens, M.A. (1969). Tests for the exponential distribution using Kolmogorov-type statistics, *Journal of the Royal Statistical Society, Series B*, **31**, 499-509.

Sethuraman, J. (1965). On a characterization of the three limiting types of the extreme, *Sankhyā, Series A*, 357-364.

Shimizu, R. (1979). On a lack of memory property of the exponential distribution, *Annals of the Institute of Statistical Mathematics*, **31**, 309-313.

Stadje, W. (1994). A characterization of the exponential distribution involving absolute differences of i.i.d. random variables, *Proceedings of the American Mathematical Society* (To appear).

Sukhatme, P.V. (1937). Tests of significance for samples of the χ^2 population with two degrees of freedom, *Annals of Eugenics*, **8**, 52-56.

Tanis, E.A. (1964). Linear forms in the order statistics from an exponential distribution, *Annals of Mathematical Statistics*, **35**, 270-276.

Wang, Y.H. and Srivastava, R.C. (1980). A characterization of the exponential and related distributions by linear regression, *Annals of Statistics*, **8**, 217-220.

CHAPTER 13

Goodness-of-Fit Tests

Samuel Shapiro[1]

13.1 Introduction

The subject of assessing whether a particular distributional model can be used to analyze a data set is one which has received a good deal of attention in the past several years. This topic is especially important for models such as the exponential that are used to predict lifetimes of products or assess the reliability of systems where the arbitrary use of an improper model can yield answers which are seriously in error. The question of concern is whether or not it is reasonable to use a particular model for formulating answers to a given problem rather than the issue of whether or not the data has a specific distribution; the answer to this latter question is "no" for any real set of data.

Since we shall assume that a testing procedure is to be used for "real" data sets any procedures which require information that is not readily available, such as values of unknown parameters, will be omitted from the chapter. This eliminates the very early simple hypothesis procedures such as the Kolmogorov-Smirnov test. Thus only composite tests for the exponential distribution will be described. The choice of which test is "best" is difficult or impossible to state since the relative performance of a test such as its power will depend on the true unknown underlying distribution, the probability of a Type I error, and the sample size. Choices of which test to select are generally made depending on the availability of a procedure on a computer, the ease of use or the prejudice of the user. This chapter contains descriptions of several procedures which have been chosen for all of the above reasons. Since the number of available tests is large and increases with each passing year, a selection has been made which is far from complete but represents, from a power or ease of use standpoint, some of the best omnibus procedures available. The concluding section contains a summary of some of the power studies which have appeared in the literature and reference to some procedures which are powerful against specific alternatives.

This chapter pertains to the following problem. Given a sample from some unknown

---

[1]Florida International University, Miami, Florida

distribution, is it reasonable to represent the data by the model

$$f(x;\mu,\lambda) = \lambda\, e^{-\lambda(x-\mu)}, \qquad x \geq \mu,\ \lambda > 0,\ -\infty < \mu < \infty\ , \tag{13.1.1}$$

where μ and λ are unknown constants. The formal procedures test the null hypothesis that the distribution is model (13.1.1) with arbitrary parameters versus the alternative hypothesis that it is equal to some other model. These formal test procedures allow one to set the probability of a Type I error. In the standard test of hypothesis rationale one chooses a small value for this probability; however since the decision to reject the null hypothesis only means looking for another model which is usually a minor concern, the probability of a Type I error is often set higher than that used with other types of statistical tests. A probability as high as 0.20 may not be unreasonable for some applications.

In developing a test for a distributional model one chooses some characteristic of the null case and determines if that characteristic can be used to distinguish the null case from a wide range of alternatives. The only practical consideration is how good the discrimination is; i.e. the power under a wide range of alternatives. The particular characteristic chosen is irrelevant as long as it "works". Obviously the characterizations described in Chapter 12 are candidates on which to base such a test. In many of the cases the null distribution for finite samples is intractable and the test percentiles for these procedures have been generated by Monte Carlo simulation. In some cases the simulation results have been given as tables of percentiles and in others, regression models were fitted to selected percentiles as a function of sample size and the resulting regression equations have been given. References to these results are given for each of the selected procedures; however actual results have been included in this chapter only for those tests where the percentiles can be summarized as a simple function of sample size. The calculations involved in some of the test procedures are illustrated in Section 13.8.

The history of general goodness-of-fit tests starts with the chi-squared test of Karl Pearson and includes such early omnibus procedures as the Kolmogorov-Smirnov and Cramer-von Mises tests. The original forms of these procedures are presently only of theoretical interest and are generally not used today with the exponential distribution since their power is lower than modern day tests and they are limited to use with simple hypotheses. Since these procedures are described in many references, a discussion of them will be omitted from this chapter. A discussion of testing specifically for the exponential distribution dates back to Epstein (1960) and since then there have been a plethora of procedures developed. An excellent reference for a detailed description of many of these tests is D'Agostino and Stephens (1986). A handbook description of some of the procedures can be found in Shapiro (1990) and a brief overview of the many tests which existed in 1984 was given by Spurrier (1984).

13.2 Graphical Procedures

Graphical procedures have the advantage that they allow the analyst to "see" the data and to make a subjective evaluation as to the appropriateness of the hypothesized model. Any peculiarities in the data can be readily detected and one is not faced simply with a decision to accept or reject a given model based on a single number. Furthermore if the model is deemed inappropriate then the plot can supply information as to what might be a more appropriate model. There are several types of plots commonly used, the two most popular are probability and hazard plots. The ideas behind both are similar. For a detailed description see Nelson (1982).

A probability plot can be considered as a plot of the ordered sample observations against the expected values of the order statistics of the standardized (scale parameter of one and location parameter of zero) null distribution. The paper is scaled so that the ordered observations can be plotted against a quantity such as $p_i = (i - 1/2)/n$ or other similar value and the scaling converts this quantity to an approximation of the expected value of the ith order statistic. If the sample came from the null distribution the plot will be approximately linear with the slope of the line yielding an estimate of the scaling parameter and the intercept giving an estimate of the location parameter. Thus, for model (13.1.1) the plot can be represented by

$$X_{(i)} = \mu + \lambda\,\alpha_i \tag{13.2.1}$$

where $X_{(i)}$, $i = 1, 2, \ldots, n$, are the ordered sample observations and α_i are the expected values from an exponential distribution with μ, the origin parameter, equal to zero and λ, the scale parameter, equal to 1. The analyst then has to determine if the plot is reasonably linear; there is no specified probability of a Type I error and two individuals looking at the same plot can come to different decisions. However with practice, reasonably good decisions can be made. There will be deviations from linearity due to random sampling fluctuations, but the larger the sample size the greater the tendency toward a straight line. However, if the assumed distribution is incorrect, the plotted points will fluctuate around a nonlinear curve and non-exponentiality is indicated by systematic departures from linearity. Outliers are easily detected. Some examples of the variability in plots when sampling from a null distribution and a comparison with samples from alternative distributions can be found in Chapter 8 of Hahn and Shapiro (1967). It is important to remember when evaluating the plot that the observations are not independent once they have been ordered and that runs of points on one side of the regression line do not necessarily mean that the null distribution should be rejected. Such runs can not be evaluated by the usual run tests for randomness. Furthermore, the variance of each point is not the same; the points in the upper tail of the exponential distribution have a higher variance than those in the restricted lower tail. This fact should be remembered when assessing the deviation from a straight line and a greater leeway must be given for points in the upper tail. Detailed instructions for making an exponential probability plot using semi-logarithmic paper can be found in Shapiro (1990). Special graph papers for probability plots for the exponential distribution are available commercially and are usually called *chi-squared 2 df paper*.

The theory behind hazard plotting is similar. The empirical cumulative hazard rate function, $H^*(t)$, for an exponential distribution will plot as a straight line versus time. For the exponential distribution the hazard rate function

$$\rho(t) = f(t)/\overline{F}(t) = \lambda$$

is constant and the cumulative hazard rate function is

$$H(t) = \int_0^t \rho(u)du = \lambda t\ .$$

A plot of the cumulative hazard rate function versus t will yield a straight line which passes through the origin. Therefore a plot of the empirical hazard rate function versus time will yield an approximate straight line if the sample came from the exponential distribution. Hazard paper is scaled so that a plot of the ordered sample observations (accounting for the non-failed items) against p_i, the same plotting positions as would be used if making a probability plot, would result in an approximate straight line if the null model is correct. The advantage of a hazard plot over a probability plot is that it can handle the situation where the data set contains some cases whose order numbers are unknown (i.e., non-failed units whose exposure times are smaller than some of the failed units). For details, see Nelson (1982) or Nelson (1986).

Another alternative to the probability plot, denoted the P-P probability plot, was suggested by Gan and Koehler (1990). This plot is constructed by plotting

$$Z_i = F\{(X_{(i)} - \hat{\mu})/\hat{\lambda}\} = 1 - \exp\{-(X_{(i)} - \hat{\mu})/\hat{\lambda}\}$$

versus p_i, $i = 1, 2, 3, \ldots, n$, where the p_i's are the plotting positions mentioned previously, $X_{(i)}$'s are the ordered observations, and $\hat{\mu}$ and $\hat{\lambda}$ are the estimates of the origin and scaling parameters,

$$\hat{\lambda} = n(\overline{X} - X_{(1)})/(n-1) \qquad \text{and} \qquad \hat{\mu} = X_{(1)} - \hat{\lambda}/n\ .$$

The advantage of the P-P plot is that samples from the null distribution will fall close to a forty five degree straight line while those from alternative distributions will depart systematically from this line. Thus the assessment is made easier since only one straight line represents the null hypothesis unlike the case for a probability plot where any straight line is acceptable.

13.3 Regression Tests

One method of making an objective assessment in conjunction with a probability plot is to consider (13.2.1) as a regression model and construct procedures to assess the linearity of the plot. One of the earliest attempts at this for the exponential distribution was proposed by Shapiro and Wilk (1972) where they compared a scaled estimate of the variance obtained from the slope of the line with an estimate of the variability which did not depend on the plot. In the rationale of the analysis of variance, these two estimates are compared by means of a ratio. The derivation of the test statistic proceeds as follows.

Let $Y_{(1)} \leq Y_{(2)} \leq \cdots \leq Y_{(n)}$ denote the order statistics from a sample if size n from the standard exponential. Denote the expected value of $Y_{(i)}$ by α_i, $i = 1, 2, \ldots, n$, and the covariance of $Y_{(i)}$ and $Y_{(j)}$ by β_{ij} and let

$$\boldsymbol{\alpha}' = (\alpha_1, \alpha_2, \ldots, \alpha_n) \text{ and } \boldsymbol{B} = (\beta_{ij}) \text{ an } n \times n \text{ matrix.}$$

Let $\boldsymbol{X}' = (X_{(1)}, X_{(2)}, \ldots, X_{(n)})$ denote a vector of sample order statistics. Then (13.2.1) represents the relationship between $Y_{(i)}$ and $X_{(i)}$. Using this model the best unbiased estimate of λ, linear in the order statistics, is the generalized least-squares estimator (see Chapter 5)

$$\lambda^* = \frac{\mathbf{1}'\boldsymbol{B}^{-1}(\mathbf{1}\boldsymbol{\alpha}' - \boldsymbol{\alpha}\mathbf{1}')\boldsymbol{B}^{-1}\boldsymbol{X}}{(\mathbf{1}'\boldsymbol{B}^{-1}\mathbf{1})(\boldsymbol{\alpha}'\boldsymbol{B}^{-1}\boldsymbol{\alpha}) - (\mathbf{1}'\boldsymbol{B}^{-1}\boldsymbol{\alpha})^2}$$

where $\mathbf{1}' = (1, 1, \ldots, 1)$. For the exponential distribution, this reduces to

$$\lambda^* = n(\overline{X} - X_{(1)})/(n-1) .$$

The test statistic, known as the W-test, is

$$W = [(n-1)/n]^{1/2} \lambda^{*2} / S^2$$

where $S^2 = \sum(X_{(i)} - \overline{X})^2$ and $X_{(1)}$ is the smallest sample value. This numerator is the generalized least-squares estimator of the slope of the regression line which has been multiplied by a constant and the denominator is the total sum of squares about the mean. The numerator is an appropriate estimate of the scale parameter only if the regression line is linear while the quantity in the denominator does not depend on the regression line. Tables of percentiles of the test statistic can be found in Shapiro and Wilk (1972) and Shapiro (1990) for samples up to size 100. This procedure yields a two-tail test and is appropriate when both μ and λ of (13.1.1) are unspecified. When μ is known, the test is modified as suggested by Stephens (1978). The known value of μ is treated as a data point and the sample size becomes $n+1$. The test is carried out as above.

A shortcoming of this test is that it is not consistent for some alternatives. D'Agostino and Stephens (1986) point out that nW converges in probability to 1 for the exponential distribution. There are other distributions where the statistic also converges to 1 and hence this test will have no power against these alternatives. Angus (1989) also pointed out this problem and suggested using the test on the normalized spacings (13.4.1) defined below. Sarkadi (1975) also noted the inconsistency of the procedure and suggested using the analog of the W-test for normality given by Shapiro and Francia (1972). Thus he recommends using

$$W^* = \frac{\left\{\sum_{i=1}^{n} \sum_{j=1}^{i} X_{(i)}/(n-j+1) - \sum_{i=1}^{n} X_{(i)}\right\}^2}{S^2}$$

where the $X_{(i)}$'s and S^2 are defined above. No table of percentiles was given.

The Shapiro-Wilk W-procedure can be improved if one knows whether the possible alternatives to the exponential have only increasing (decreasing) failure rate (IFR or DFR) functions since the test can then be used as one tail procedure. Alternative distributions with increasing failure rate functions yield values of the test statistic which tend to lie in the upper tail of the null distribution while the opposite is true for those with decreasing failure rate functions.

A modification of the above procedure is to compute the correlation coefficient for model (13.2.1) and use it as a measure of goodness-of-fit. (The Shapiro-Wilk procedure above is in fact a weighted correlation coefficient.) The idea of using the correlation coefficient from a probability plot was suggested by Filliben (1975) and an approximation for the percentiles for the exponential distribution was published by Gan (1985). The rationale of this procedure is that when sampling from the null population, the sample correlation coefficient between the ordered observations and the expected values of the order statistics will be close to one while samples from an alternative distribution will yield a smaller value of the correlation coefficient. Thus the procedure results in a lower tail test. The test statistic is defined as

$$r^2 = \left[\sum(X_{(i)} - \overline{X})(\alpha_i - \overline{\alpha})\right]^2 \Big/ \left[\sum(X_{(i)} - \overline{X})^2 \sum(\alpha_i - \overline{\alpha})^2\right] , \qquad (13.3.1)$$

Table 13.1: Values of A_p and B_p for the statistics k^2 and k_0^2

p	Statistic	A	B
0.001	1	0.5574	0.3207
	2	0.6578	0.1912
0.005	1	0.6464	0.4145
	2	0.6382	0.2565
0.01	1	0.4366	0.4784
	2	0.6794	0.2959
0.05	1	0.8781	0.6837
	2	0.8797	0.4540
0.10	1	0.8334	0.8380
	2	1.0049	0.5780

Statistic 1 in the table is k^2 and 2 is k_0^2.

where $\alpha_i = F^{-1}[i/(n+1)] = -\ln[1 - i/(n+1)]$. One advantage of this procedure is that r^2 can be computed from a censored sample.

Two modifications of (13.3.1) based on the P-P plot were suggested by Gan and Koehler (1990). The two statistics are both correlation coefficients based on the P-P plot. The first suggestion is the statistic k^2 obtained by substituting respectively Z_i defined earlier, and $\overline{Z}$, the mean of the Z_i's, for $X_{(i)}$ and $\overline{X}$; and p_i and $\overline{p}$ for α_i and $\overline{\alpha}$ in (13.3.1). The second version of the statistic, k_0^2, is obtained by substituting 1/2 for $\overline{X}$ and $\overline{\alpha}$ in k^2. Percentiles for these statistics can be obtained from a simple regression approximation obtained by the authors from a simulation study. The formula

$$1 - k_p^2 = (A_p + B_p n)^{-1}$$

gives an approximation of the lower pth percentile of the null distribution of the test statistic. Values of A_p and B_p for the two procedures are presented in Table 13.1. Values of the coefficients are also given for the case when μ is known. This test statistic can also be used with censored samples. Another correlation type test was proposed by Smith and Bain (1976). They suggested computing the product moment correlation coefficient, r, between the order statistics and $-\ln[1 - \{i/(n+1)\}]$ and using an upper tail test with the statistic $1 - r^2$. Another regression procedure for use with censored samples was suggested by Smith and Bain (1976).

13.4 Spacing Tests

A property of the order statistics from an exponential distribution is that the weighted spacings

$$Y_i = (n - i + 1)(X_{(i)} - X_{(i-1)}), \qquad i = 1, 2, \ldots, n \;, \tag{13.4.1}$$

where $X_{(1)} \leq X_{(2)} \leq \cdots \leq X_{(n)}$ and $X_{(0)} = \mu$, are independently and identically distributed exponential variates with origin parameter zero and scale parameter λ (see Chapter 3).

Gnedenko, Belyayev, and Solovyev (1969) used the spacings to construct an F-test for the exponential null hypothesis. In effect, the recommended test assesses whether the hazard rate function is constant over time. The spacings are divided into two groups, the first r and the remaining s. The test statistic is

$$g(r,s) = s\sum_{i=1}^{r} Y_i/\{r\sum_{i=r+1}^{n} Y_i\} \tag{13.4.2}$$

where the Y_i's are defined in (13.4.1). Under the null hypothesis the numerator and denominator are sums of independent exponential random variables with the same scale parameter and hence each $2\lambda Y_i$ has a chi-squared distribution with two degrees of freedom. Thus each of the sums in (13.4.2) when multiplied by 2λ are chi-squared variates with $2r$ and $2s$ degrees of freedom, respectively, and the ratio has an F-distribution with degrees of freedom $2r$ and $2s$. The test is equivalent to assessing the null hypothesis that the hazard rate function (λ) is a constant over the two intervals versus the alternative that it is not. If μ is unknown, then the first spacing is not available; however, the test can be run with the spacings that are available. Obviously, this procedure can be used with censored samples. The value of r must be selected and the authors give some guidelines for making that choice.

Similar tests were suggested by Fercho and Ringer (1972), Harris (1976), and Lin and Mudholkar (1980). Power comparisons by Lin and Mudholkar (1980) showed that their bi-variate F-procedure and (13.4.2) were the best of this group; the former procedure offered protection against alternatives with non-monotone hazard rate functions such as the log-normal. When using the bi-variate F-test, the spacings defined in (13.4.1) are divided into three groups. Two F ratios are obtained; the first, F_1, is the ratio of the sum of the first group of r spacings to the middle group of $n-2r$ spacings and the second F_2, is a similar ratio comparing the upper group of r spacings to the middle. If the hazard rate function is constant over the three intervals, the pair of statistics obtained has a bi-variate F-distribution (this follows from the same argument as for (13.4.2)). The authors give suggestions as to the best choice of r. The percentiles of the test statistic can be obtained from a uni-variate F-distribution by using a theorem due to Hewett and Bulgren (1971) which showed that

$$P[a \le F_1 \le b,\ a \le F_2 \le b] \le \{P[a \le F \le b]\}^2 ,$$

where F has an F-distribution with $2r$ and $2(n-2r)$ degrees of freedom and n is the total number of spacings. This procedure does not require knowledge of the origin parameter μ and can be used with censored samples.

Two other procedures based on spacings were given by Brain and Shapiro (1983). In these procedures, the weighted spacings in (13.4.1) are regressed against their order number. Under the null hypothesis, the regression line will have a slope of zero and two test statistics were constructed to assess this hypothesis. (The plot resembles the mirror image of the hazard rate function which is a constant for the exponential distribution.) One version of the test statistic uses a form of the Laplace test suggested by Cox and Lewis (1966, p. 53). In this form, the test statistic evaluates the null hypothesis of no slope by testing whether the linear contrast due to slope is significantly different from zero. The second procedure adds to the first a quadratic contrast which will protect against models with hazard rate functions which are parabolic or reverse parabolic and hence whose linear component may be close to zero. The first procedure, when μ is unknown, is

$$z = \frac{[12/(m-2)]^{1/2}\sum a_i Y_i}{\sum Y_i} \tag{13.4.3}$$

where the Y_i's are defined in (13.4.1), m is the number of observations (there are $m-1$ spacings) and $a_i = i - m/2$. The distribution of z is approximately standard normal even for relatively small samples and the test is two-tailed. The quadratic component is obtained from

$$z_q = [5/\{4(m+1)(m-2)(m-3)\}]^{1/2}\,\frac{12\sum a_i^2 Y_i - m(m-2)\sum Y_i}{\sum Y_i} \tag{13.4.4}$$

where all terms are as defined in (13.4.3). The second test statistic is obtained by combining (13.4.3) and (13.4.4) yielding

$$Z^* = z^2 + z_q^2 \ .$$

Since both components of Z^* are obtained from orthogonal contrasts and are squares of independent approximately standard normal variates, it has a distribution which converges with increasing sample sizes to chi-squared with 2 df. The following equations are used for adjusting the 90th, 95th, and 97.5th percentiles for small samples sizes but require that n be greater than 14.

$$\begin{aligned} 90th: \quad & Z^*_{0.90} &= \ & 4.605 - 2.5/m \\ 95th: \quad & Z^*_{0.95} &= \ & 5.991 - 1.25/m \\ 97.5th: \quad & Z^*_{0.975} &= \ & 7.378 + 3.333/m \end{aligned}$$

Since this procedure is based on the available spacings, it can be used when μ is either known or not or with censored or complete samples. When μ is known, it is treated as if it is an observation; see Brain and Shapiro (1983) for the details. Shapiro and Balakrishnan (1994) have recently proposed a similar test for testing exponentiality based on progressively Type-II censored samples.

Another procedure suggested by Tiku (1980) is based on using the spacings and the analysis-of-variance rationale suggested by Shapiro and Wilk cited above. Two estimates of the standard deviation based on the spacings are computed and the proposed statistic for complete samples is

$$Z = 2\sum(n-1-i)\,D_i/(n-2)\sum D_i \ , \tag{13.4.5}$$

where $D_i = (n-i)(X_{(i+1)} - X_{(i)})$ and the $X_{(i)}$'s are the ordered observations. The distribution of $Z/2$ is the same as the mean of $(n-2)$ i.i.d. uniform random variates which was given by Hall (1927) and is cited in Kendall and Stuart (1958, p. 257). This test is easily modified to handle censored samples.

13.5 EDF Tests

Several tests for the exponential distribution make use of the fact that the probability integral transformation

$$F(x) = \int_0^x f(y)dy$$

can be used to convert a set of observations to the uniform distribution given that $f(y)$ is known. The early tests such as the Kolmogorov-Smirnov, Cramer-von-Mises and Anderson-Darling made use of this transformation for testing simple null hypotheses. Since the procedure is basically a test for the uniform distribution, a single set of percentiles can be used for all $f(y)$. However, if the parameters are unknown and are replaced by their estimators, then the null distribution of the statistic is dependent on the form of the density and separate percentiles are needed for each null case. Tests based on the EDF when the parameters are unknown were suggested by Lilliefors (1969) (Kolmogorov-Smirnov), van Soest (1969) (Cramer-von Mises) and Finkelstein and Schafer (1971). A summary of the adaptations of these tests for the exponential distribution when the parameter values are unknown was given by Spinelli and Stephens (1987) and their results are presented below.

The first step in calculating the test statistic when both μ and λ are unknown is to determine the estimators

$$\mu^* = X_{(1)} - \lambda^*/n$$

and

$$\lambda^* = n(\overline{X} - X_{(1)})/(n-1) \; .$$

Then, let $W_i = (X_{(i)} - \mu^*)/\lambda^*$ and $Z_i = 1 - \exp(-W_i)$ for $i = 1, 2, \ldots, n$. Note that $X_{(1)}$ is the smallest sample value.

The Cramer-von Mises statistic is computed from

$$W^2 = \sum [Z_i - (2i-1)/2n]^2 + 1/(12n) \; , \qquad (13.5.1)$$

the Watson statistic from

$$U^2 = W^2 - n(\overline{Z} - 1/2)^2 \; , \qquad (13.5.2)$$

and the Anderson-Darling statistic from

$$A^2 = -n^{-1} \sum (2i-1)\{\ln Z_i + \ln(1 - Z_{n+1-i})\} - n \; . \qquad (13.5.3)$$

In order to simplify the use of these procedures, modifications were suggested so that the percentiles are independent of n. The modifications are to multiply each of the statistics above by the corresponding function of n shown below.

$$\begin{aligned} W^2 &: \quad (1 + 2.8/n - 3/n^2) \\ U^2 &: \quad (1 + 2.3/n - 3/n^2) \\ A^2 &: \quad (1 + 5.4/n - 11/n^2) \; . \end{aligned}$$

These procedures are upper-tail tests and the percentiles are presented in Table 13.2.

The functions were derived from Monte Carlo studies. The asymptotic distributions of these procedures were given in Stephens (1974, 1976). These procedures can be modified for censored samples; see D'Agostino and Stephens (1986, p. 141) and Pettitt (1977). Durbin (1975) suggested a procedure that eliminated the need to estimate the origin parameter μ. If the smallest order statistic is subtracted from each of the observations, then the test can be carried out based on a sample of $n - 1$ from an exponential distribution with only the scale parameter unknown and the appropriate EDF procedure can then be used.

Table 13.2: Upper tail percentiles for the EDF tests

Statistic	0.25	0.15	0.10	0.05	0.025	0.01
W^2	0.116	0.148	0.175	0.222	0.271	0.338
U^2	0.090	0.112	0.129	0.159	0.189	0.230
A^2	0.736	0.916	1.062	1.321	1.591	1.959

13.6 Tests Based on a Characterization

A test for the exponential hypothesis which is based on a characterization was suggested by Angus (1982). The test is based on the property that if $1 - F(2X) = [1 - F(X)]^2$ for every X greater than zero, then $F(X)$ is exponential. He suggested three test statistics using this concept and evaluated them by means of a simulation power study. Of the three he suggested, the use of

$$T = \frac{1}{n} \sum [F_n(2X_i) - \{(n-i)/n\}^2]^2 \qquad (13.6.1)$$

where the $X_{(i)}$, $i = 1, 2, \ldots, n$, are the ordered observations, and $F_n(u)$ is the number of observations greater than u divided by n, the sample size. When the origin parameter μ is unknown, it is necessary to subtract the smallest observation from each of the other ordered observations and run the test as a sample of size $n - 1$. The 90th and 95th test percentiles were given for nT.

13.7 Power Comparisons

Most of the articles referenced above contained some power comparisons which attempted to show that the proposed procedure was at least as good as some of the competitors that were available at that time. However it is difficult to make an overall comparison of these studies since authors did not use the same alternative distributions, sample sizes, test levels or competing procedures. Also several authors have run power comparisons assuming that μ equals zero but have not used this information when evaluating the procedures where this fact is not assumed and this information should be added to the assessment; hence the results are not comparable.

In an extensive power study reported in D'Agostino and Stephens (1986), they concluded that the statistics (13.5.1) and (13.5.3) were the best of the group of EDF procedures and gave power results similar to the Shapiro-Wilk W test. Tiku (1980) concluded from his power studies that Z defined in (13.4.5) had power similar to these three. In a further study of Tiku's Z-test, Balakrishnan (1983) compared its power to a procedure suggested by Durbin (1975) and the Shapiro-Wilk W-test. He concluded that the Tiku procedure was more powerful than that proposed by Durbin and in many cases more powerful than the W-test. Angus (1982) showed that (13.5.3) gave power results similar to the bi-variate F-test of Lin and Mudholkar (1980). Brain and Shapiro (1983) showed that the power of Z^* was superior to the Shapiro-Wilk W-test, the bi-variate F-test of Lin and Mudholkar (1980), and the F-test of Gnedenko, Belyayev, and

Table 13.3: Times between arrivals (in minutes)

1.80	3.43	3.98	4.23	4.65
2.89	3.48	4.06	4.34	4.84
2.93	3.57	4.11	4.37	4.91
3.03	3.85	4.13	4.53	4.99
3.15	3.92	4.16	4.62	5.17

Solovyev (1969). It is also clear from these power studies that the relative rankings of the procedures vary depending on the form of the alternative distribution: whether it is symmetric or skewed or heavy- or light-tailed. The procedures referenced in the above paragraphs are probably the best choices for an omnibus test when nothing is known about the possible alternative distribution.

If the possible alternatives can be specified or if one just wishes to assure that the data did not come from an IFR (DFR) alternative, then a procedure can possibly be selected that will have maximum power against such specified alternatives. As noted previously in such instances the Shapiro-Wilk W-test can be used as a one-tail test which will increase its power against some specified alternatives. Statistics for testing against IFR and IFRA alternatives were suggested by Bickel and Doksum (1969) and Proschan and Pike (1967). A test which can be used against DFRA alternatives was proposed by Barlow (1968) and a procedure which is good against the "new better than used" alternative was given by Hollander and Proschan (1972).

Tests of exponentiality can be designed to have optimum power against one specific alternative. Uthoff (1970) showed that thc Shapiro-Wilk W-test is the most powerful scale and location invariant test of the exponential versus the normal distribution. Shorack (1972) presented a most powerful test for the gamma alternative and Thoman, Bain, and Antle (1969) described a test with high power for Weibull alternatives.

The above procedures are appropriate for the case where both μ and λ were unknown. If one or more of these parameters are known, then such information should be used as part of the test. When μ, the origin is known, it should be included as if it were a data point in the spacing tests and in the Shapiro-Wilk procedure. This in effect increases the number of observations by one and improves the power of the procedure. Modifications for the EDF tests for this case can be found in D'Agostino and Stephens (1986). The case where λ is known would occur rarely; procedures for this situation can be found in D'Agostino and Stephens (1986).

13.8 Examples of Calculations

The following set of data will be used to illustrate several of the test procedures which are referenced above. The data are the times between arrivals of 25 customers at a facility and are taken from Wadsworth (1990, p. 611). The data have been listed in ascending order in Table 13.3.

A quantile plot of the above data is shown in Figure 13.1 where the ordered observations are plotted against $F^{-1}(p_i)$ where $p_i = (i - 0.5)/25$. Examination of the plot clearly indicates that there is a systematic departure from a straight line and the exponential assumption is not tenable.

The Shapiro-Wilk W-test requires the quantities $X_{(1)} = 1.8$, $\overline{X} = 3.966$ and $S^2 = 14.9784$. Thus

$$W = [25/24](3.966 - 1.80)^2/14.9784 = 0.3263 \ .$$

The values of $W_{0.025}$ and $W_{0.975}$ are 0.0218 and 0.0836. Since 0.3263 is far above the 0.975 percentile one would conclude that the use of an exponential model is unwarranted for this data set.

The Filliben procedure (13.3.1) requires computing the square of the correlation coefficient between the ordered observations and the quantities $\alpha_i = -\ln[1 - i/(n+1)]$. The result is $r^2 = 0.7283$. The Gan and Koehler modification replaces the observations by

$$Z_i = 1 - \exp\{-(X_{(i)} - \hat{\mu})/\hat{\lambda}\}$$

and the α_i's by p_i's $= (i - 0.5)/n$. For example, for the above set of data, $\hat{\lambda} = (25/24)(3.966 - 1.80) = 2.256$, $\hat{\mu} = 1.8 - 2.256/25 = 1.7098$, $Z_1 = 1 - \exp\{-(1.8 - 1.7098)/2.256\} = 0.0392$ and $p_1 = 0.02$. Computing the other Z_i's, one obtains the square of the correlation coefficient to be 0.7494. Using Table 13.1, the 0.001 percentile is 0.8834 and since the computed value is less than this the same conclusion reached previously holds.

The Brain and Shapiro test requires obtaining the weighted spacings $Y_j = (n - j)(X_{(j+1)} - X_{(j)})$; for example, $Y_1 = 24(2.89 - 1.80) = 26.16$. Then the following sums are obtained;

$$\sum Y_i = 53.448 \qquad \sum a_i Y_i = -413.52 \qquad \sum a_i^2 Y_i = 4646.4 \ .$$

Using $m = 25$ in (13.4.3) and (13.4.4) yields $z = -5.588$, $z_q = 4.564$ and $Z^* = 52.1$. The upper 0.975 percentile of the distribution of the test statistic, Z^*, is 7.51 and hence the hypothesis of an exponential model is untenable. Using the same weighted spacing for the Gnedenko, Belyayev, and Solovyev procedure using $r = 12$ yields the two sums 48.52 and 4.93. Thus using (13.4.2) yields

$$F = [48.52/12]/[4.93/12] = 9.85$$

which exceeds the upper 0.01 percentile of an F-distribution with 24 and 24 degrees of freedom.

The distance tests use the same Z_i values computed for the Gan-Koehler procedure. Using these values in (13.5.1) yields the Cramer-von Mises test statistic $W^2 = 1.017$. This value is multiplied by the given factor to yield 1.107 which can be checked against the tabulated 0.01 percentile of 0.338. The result is the same as for the previous procedures.

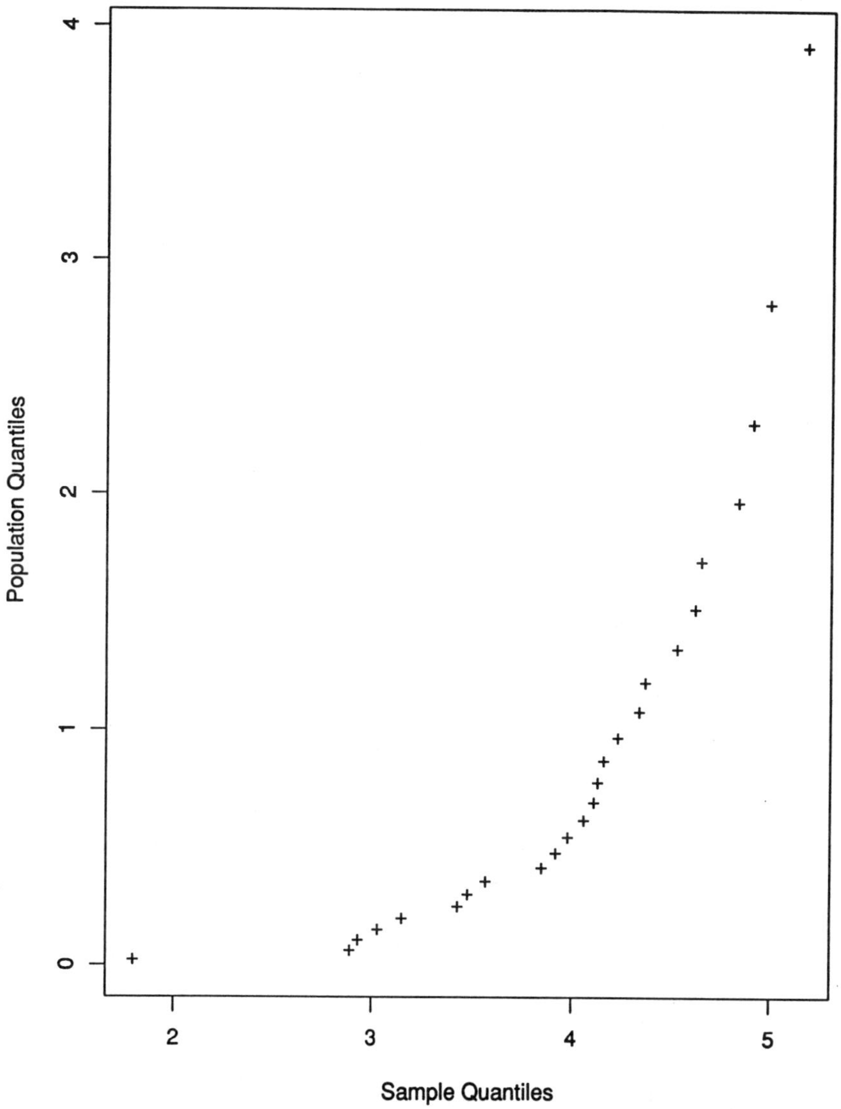

Figure 13.1: Quantile plot of time between arrivals of customers time (minutes)

13.9 Summary

In summary, as D'Agostino and Stephens (1986) state, "test statistics should be regarded as giving information about the data and their parent population, rather than as tools for formal testing procedures". In order to maximize the amount of information gained, it is always a good rule to use one of the plotting techniques described above with a formal test to extract the maximum information from the data. If one is not satisfied with the fit, then alternative models should be selected and assessed.

References

Angus, J.E. (1982). Goodness-of-fit tests for exponentiality based on a loss-of memory type functional equation, *Journal of Statistical Planning and Inference*, **6**, 241-251.

Angus, J.E. (1989). A note on the use of the Shapiro-Wilk test of exponentiality for complete samples, *Communications in Statistics - Theory and Methods*, **18**, 1819-1830.

Balakrishnan, N. (1983). Empirical power study of a multi-sample test of exponentiality based on spacings, *Journal of Statistical Computation and Simulation*, **18**, 265-271.

Barlow, R.E. (1968). Likelihood ratio tests for restricted families of probability distributions, *Annals of Mathematical Statistics*, **39**, 547-560.

Bickel, P.J. and Doksum, K.A. (1969). Tests for monotone failure rate based on normalized spacings, *Annals of Mathematical Statistics*, **40**, 1216-1235.

Brain, C.W. and Shapiro, S.S. (1983). A regression test for exponentiality: Censored and complete samples, *Technometrics*, **25**, 69-76.

Cox, D.R. and Lewis, P.A. (1966). *The Statistical Analysis of a Series of Events*, Methuen, London.

D'Agostino, R.B. and Stephens, M.A. (Eds.) (1986). *Goodness-of-Fit Techniques*, Marcel Dekker, New York.

Durbin, J. (1975). Kolmogorov-Smirnov tests when parameters are estimated with application to tests of exponentiality and tests on spacings, *Biometrika*, **62**, 5-22.

Epstein, B. (1960). Tests for the validity of the assumption that the underlying distribution of life is exponential, *Technometrics*, **2**, 83-101 and 167-183.

Fercho, W.W. and Ringer, L.J. (1972). Small sample power of some tests of constant failure rate, *Technometrics*, **14**, 713-724.

Filliben, J.J. (1975). The probability plot correlation coefficient test for normality, *Technometrics*, **17**, 111-117.

Finkelstein, J.M. and Schafer, R.E. (1971). Improved goodness of fit tests, *Biometrika*, **58**, 641-645.

Gan, F.F. (1985). Goodness-of-fit statistics for location-scale distributions, *Unpublished Ph.D. dissertation*, Iowa State University, Department of Statistics.

Gan, F.F. and Koehler, K.J. (1990). Goodness-of-fit tests based on P-P probability plots, *Technometrics*, **32**, 289-303.

Gnedenko, B.V., Belyayev, Y.K., and Solovyev, A.D. (1969). *Mathematical Methods of Reliability*, Academic Press, New York.

Hahn, G.J. and Shapiro, S.S. (1967). *Statistical Models in Engineering*, John Wiley & Sons, New York.

Hall, P. (1927). Distribution of the mean of samples from a rectangular population, *Biometrika*, **19**, 240.

Harris, C.M. (1976). A note on testing for exponentiality, *Naval Research Logistic Quarterly*, **23**, 169-175.

Hewett, J.E. and Bulgren, W.G. (1971). Inequalities for some multivariate F-distributions with applications, *Technometrics*, **13**, 397-402.

Hollander, M. and Proschan, F. (1972). Testing whether new is better than used, *Annals of Mathematical Statistics*, **43**, 1136-1146.

Kendall, M.G. and Stuart, A. (1958). *The Advanced Theory of Statistics*, Hafner, New York.

Lilliefors, H.W. (1969). On the Kolmogorov-Smirnov test for the exponential distribution with mean unknown, *Journal of the American Statistical Association*, **64**, 387-389.

Lin, C.C. and Mudholkar, G.S. (1980). A test for exponentiality based on the bivariate F distribution, *Technometrics*, **22**, 79-82.

Nelson, W. (1982). *Applied Life Data Analysis*, John Wiley & Sons, New York.

Nelson, W. (1986). *How to Analyse Data with Simple Plots*, Volume 1, ASQC Basic References in Quality Control: Statistical Techniques, American Society for Quality Control, Milwaukee.

Pettitt, A.N. (1977). Tests for the exponential distribution with censored data using Cramer-von Mises statistics, *Biometrika*, **64**, 629-632.

Proschan, F. and Pike, K. (1967). Tests for monotone failure rate, In *Proceedings of the 5th Berkeley Symposium in Mathematics, Statistics and Probability*, Vol. 3, 293-312, University of California Press, Berkeley.

Sarkadi, K. (1975). The consistency of the Shapiro-Francia test, *Biometrika*, **62**, 445-450.

Shapiro, S.S. (1990). *How To Test Normality and Other Distributional Assumptions*, Volume 3, ASQC Basic References in Quality Control: Statistical Techniques, American Society for Quality Control, Milwaukee.

Shapiro, S.S. and Balakrishnan, N. (1994). Testing for exponentiality with progressively censored data, *Submitted for publication.*

Shapiro, S.S. and Francia, R.S. (1972). Approximate analysis of variance test for normality, *Journal of the American Statistical Association*, **67**, 215-225.

Shapiro, S.S. and Wilk, M.B. (1972). An analysis of variance test for the exponential distribution (complete samples), *Technometrics*, **14**, 355-370.

Shorack, G.R. (1972). The best test of exponentiality against gamma alternatives, *Journal of the American Statistical Association*, **67**, 213-214.

Smith, R.M. and Bain, L.J. (1976). Correlation type goodness of fit statistics with censored data, *Communications in Statistics - Theory and Methods*, **5**, 119-132.

Spinelli, J.J. and Stephens, M.A. (1987). Tests for exponentiality when origin and scale parameters are unknown, *Technometrics*, **29**, 471-476.

Spurrier, J.D. (1984). An overview of tests for exponentiality, *Communications in Statistics - Theory and Methods*, **13**, 1635-1654.

Stephens, M.A. (1974). EDF statistics for goodness of fit and some comparisons, *Journal of the American Statistical Association*, **69**, 730-737.

Stephens, M.A. (1976). Asymptotic results for goodness-of-fit statistics with unknown parameters, *Annals of Statistics*, **4**, 357-368.

Stephens, M.A. (1978). On the W test for exponentiality with origin known, *Technometrics*, **20**, 33-35.

Thoman, D.R., Bain, L.J., and Antle, C.E. (1969). Inference on the parameters of the Weibull distribution, *Technometrics*, **11**, 445-460.

Tiku, M.L. (1980). Goodness of fit statistics based on the spacings of complete or censored samples, *Australian Journal of Statistics*, **22**, 260-275.

Uthoff, V.A. (1970). An optimum test property of two well known statistics, *Journal of the American Statistical Association*, **65**, 1597-1600.

van Soest, J. (1969). Some goodness of fit tests for the exponential distribution, *Statistica Neerlandica*, **23**, 41-51.

Wadsworth, H.M. (1990). *Handbook of Statistical Methods for Engineers and Scientists*, McGraw-Hill, New York.

CHAPTER 14

Outlier Models and Some Related Inferential Issues

U. Gather[1]

14.1 Introduction

The problem of outliers in random data sets is a very interesting, important and common one. Nevertheless there is no formal and generally accepted definition of what is meant by an outlier. Terms like *outlier, spurious observation, contaminant, gross error* and others are used with different and overlapping meanings. On an intuitive basis, however, most of us agree when calling an outlier a data point which is surprisingly far away from the main group of the data, or which deviates from the pattern, set by the majority of the data [see Barnett and Lewis (1984, 1994), Hampel et al. (1986)]. When aiming at a more formal definition or framework, it is the *purpose* of the analysis which should lead us.

Two main purposes are in the focus when dealing with outliers in statistical data. One concerns the estimation of unknown parameters or other *statistical inference* which should be performed in an optimal way *despite* the possible *occurrence of some outliers* in the data. The other is the detection or *identification of outliers* in some sample *as a main issue.* It is obvious that both tasks can only be dealt with properly after having specified some data generating statistical model and after having formalized the goal in well defined statistical terms. For that reason, in addition to our usual working hypothesis H_0 which claims that we are looking at a random sample $x_1, \ldots, x_n$ from some (exponential) distribution, we will first introduce some so-called *outlier generating models* which should explain better than H_0 the occurrence of some outliers in the sample $x_1, \ldots, x_n$. Such models are used quite often in simulations to investigate different statistical procedures or the performance of outlier identification procedures in the presence of outliers.

Of course, usually nobody knows the mechanism which has created the outliers, and assuming an even parametric model for the occurrence of outliers is much more dubious

[1]University of Dortmund, D-44221 Dortmund, Germany

than doing so for the *good* observations. Nevertheless, such outlier generating models are sometimes seen as the first and only feasible step when developing and comparing procedures which should either detect or accommodate outliers.

It is the main purpose of this chapter to present a brief survey on outlier generating models and to show that different procedures can turn out depending on the choice of the outlier generating model. Of course this is all done for the special case of exponential r.v.'s under the null hypothesis H_0. For a more general treatment, see Gather (1990) and Gather and Kale (1992).

14.2 Outlier Generating Models

The general assumption, common to all outlier generating models, is that no longer all data points come independently and identically distributed from some target distribution $F \in \mathcal{F}$, but a few, say k, observations have foreign sources, say $G_1, \ldots, G_k$, where $G_1, \ldots, G_k \in \mathcal{G}_k(F)$, which more likely lets those observations from $G_1, \ldots, G_k$ stand out as outliers. The observations from $G_1, \ldots, G_k \neq F$ are called *contaminants* whereas we would like to call the observations from F *regular observations*.

$$X_1, X_2, \ldots, X_j, \ldots, X_{n-1}, X_n$$

$n-k$	k
i.i.d. $F \in \mathcal{F}$	corresponding to $G_1, \ldots, G_k$
regular observations	contaminants

¿From this general assumption we get specific models when making precise what kind of conditions are put on k, $\mathcal{F}$, $G_1, \ldots, G_k$, the indices of the *contaminants*, and on the structure of dependence of the underlying random variables $X_1, \ldots, X_n$. We will state all such models as supermodels containing H_0. The following is the simplest one.

Identified outliers model

For the exponential case, this means that $\mathcal{F} = \{1 - e^{-x/\theta};\ \theta > 0\}$, k is known and fixed, say for simplicity $k = 1$, $X_1, \ldots, X_n$ are independent and the index of the contaminant is also known. If we assume further that the distribution function of the *contaminant* is

$$G(x) = F(b^{-1}x) = 1 - e^{-(b\theta)^{-1}x}, \qquad x \in \mathbf{R}$$

for some $b \geq 1$, then, without loss of any generality, the joint distribution function of $X_1, \ldots, X_n$ is given by

$$F_{(X_1,\ldots,X_n)}(x_1, \ldots, x_n) = \left[\prod_{i=1}^{n-1}(1 - e^{-x_i/\theta})\right](1 - e^{-x_n/b\theta}),$$
$$x > 0,\ \theta > 0,\ b \geq 1\ . \qquad (14.2.1)$$

Of course, such a model is quite unrealistic in practice, since the number k of contaminants and the indices of the contaminants are usually not known. This *identified outliers model* is nevertheless useful for some simulation studies as long as permutation invariant procedures are involved. For an explicit use of this model, one may refer to Veale (1975), Anscombe (1960) and Gather (1986b).

The following models give up the assumption that the indices of the contaminants are known.

Unidentified outliers model

In this model, the underlying random variables are still independent, but there is an unknown function $s: \{1,\ldots,n\} \rightarrow \{0,1,\ldots,k\}$ with $\#\{s^{-1}(0)\} = n-k$, which for $s(i) = 0$ has the interpretation $X_i \sim F$ and for $s(i) = j$ means that $X_i \sim G_j$, $j = 1,\ldots,k$, $i = 1,\ldots,n$. The joint distribution of $X_1,\ldots,X_n$ is then given by

$$F_{(X_1,\ldots,X_n)}(x_1,\ldots,x_n) = \prod_{i=1}^{n} G_{s(i)}(x_i) \ , \tag{14.2.2}$$

where $G_0 = F \in \mathcal{F}$, $(G_1,\ldots,G_k) \in \mathcal{G}_k(F)$. This model was introduced for the normal distribution by Ferguson (1961) who assumed that $X_i \sim N(\mu+\Delta a_{s(i)}\cdot\sigma, \sigma^2)$, $i = 1,\ldots,n$ with $a_0 = 0$ and $a_1,\ldots,a_k$ known, $\mu \in \mathbf{R}$, $\sigma^2 > 0$, $\Delta \geq 0$, i.e., $F = N(\mu,\sigma^2)$ and $G_j = N(\mu+\Delta a_j\sigma, \sigma^2)$, $j = 1,\ldots,k$. For the exponential distribution, this Ferguson-type model is given by

$$X_i \sim 1 - \exp\left[-(\theta\cdot(\exp(\Delta a_{s(i)})))^{-1}x\right], \qquad i = 1,\ldots,n \ , \tag{14.2.3}$$

where $a_0 = 0$, $a_1,\ldots,a_k > 0$ and $X_1,\ldots,X_n$ are independent random variables.

In the special case of identical

$$G_1 = G_2 = \cdots = G_k = G(F) = 1 - \exp(-x/\theta\, b), \qquad x \geq 0,\ b > 1,\ \theta > 0 \ ,$$

this is also called a k-outlier model [David (1979)] for an exponential distribution. Some more inferential procedures under such a k-outlier model are discussed in the next chapter.

Exchangeable outliers model

If we give up the assumption of independence of the $X_1,\ldots,X_n$, but claim that any set $\{i_1,\ldots,i_k\}$ of k indices chosen from $\{1,\ldots,n\}$ is equally likely to belong to the contaminants $X_{i_1},\ldots,X_{i_k}$ and that, once given those indices $i_1,\ldots,i_k$, it holds that

$$X_{i_j} \sim F,\ j \neq 1,\ldots,k, \qquad X_{i_1},\ldots,X_{i_k} \sim G \ ,$$

then we are talking of an **exchangeable outliers model**, where the joint distribution function of $X_1,\ldots,X_n$ is given by

$$\begin{aligned} F_{(X_1,\ldots,X_n)}(x_1,\ldots,x_n) &= \frac{1}{\binom{n}{k}} \sum \prod_{i\neq i_1,\ldots,i_k} (1-e^{-x_i/\theta}) \\ &\quad \cdot \prod_{j\in\{i_1,\ldots,i_k\}} (1-e^{-x_j/\theta\, b}) \end{aligned} \tag{14.2.4}$$

with the sum taken over all $\binom{n}{k}$ possible choices of $\{i_1,\ldots,i_k\}$ from $\{1,\ldots,n\}$. This model was introduced by Joshi (1972); see also Kale and Sinha (1971).

In both, in the Ferguson-type and in the exchangeable outliers model with $b > 1$, one can prove, roughly spoken, that the k largest order statistics $X_{n-k+1:n}, \ldots, X_{n:n}$ possess the highest probability of being the contaminants [Kale (1975); Mount and Kale (1973); Gather (1979)]. It is easy to see that any permutation invariant statistic, especially the vector of order statistics $(X_{1:n}, \ldots, X_{n:n})$ has a distribution which is independent of assuming an unidentified type-, identified- or exchangeable outliers model as long as the involved distributions F, $G_1, \ldots, G_k$ are the same.

This fact and the idea that usually one only suspects the smallest or largest order statistics of being the contaminants are the reasons to introduce the following model.

Labelled outliers model

Suppose for simplicity that $G_1 = \cdots = G_k = G$. Assuming that only the k largest extremes can be contaminants, one gives the following distribution function for the vector of order statistics $(X_{1:n}, \ldots, X_{n:n})$ of $X_1, \ldots, X_n$:

$$F_{(X_{1:n},\ldots,X_{n:n})}(x_1, \ldots, x_n) = \{C_n(F,G)\}^{-1} \left[\prod_{i=1}^{n-k} F(x_i)\right] \left[\prod_{i=n-k+1}^{n} G(x_i)\right] \cdot \mathbf{I}_{\mathbf{R}^n_{\leq}}(x_1, \ldots, x_n)\ , \quad (14.2.5)$$

where $\mathbf{R}^n_{\leq}$ denotes the order cone in $\mathbf{R}^n$ and $\mathbf{I}_A$ denotes the indicator function of a set A, $C_n(F,G)$ is just the probability that an independent sample of $(n-k)$ observations from F and k observations from G is observed in $\mathbf{R}^n_{\leq}$. For $F = G$, we have

$$C_n(F,F) = \frac{1}{n!}\ .$$

In the exponential case where as assume $F(x) = 1 - e^{-x/\theta}$, $G(x) = 1 - e^{-x/b\,\theta}$, we get

$$C_n(F,G) = \prod_{i=1}^{n-k} \left[\frac{1}{i + b^{-1}}\right]\ .$$

In general, $C_n(F,G)$ cannot be represented in closed form. This labelled outliers model goes back to Barnett and Lewis (1978); see also Fieller (1976) for the exponential case, and Gather (1979) where it is treated formally.

Roughly spoken, the labelled model is often described as: $X_{1:n}, \ldots, X_{n-k:n}$ come from F and $X_{n-k+1:n}, \ldots, X_{n:n}$ come from G. This has led to a confusion with an order statistics slippage outliers model introduced by Tiku (1975), which is quite often used in simulations [see Jain and Pingel (1981), Tiku, Tan and Balakrishnan (1986)].

Order statistics slippage outliers model

One assumes in a first step that random variables $X_1, \ldots, X_n$ are observed, which are independent and identically distributed according to F. Then the k largest of them are shifted by an amount $a > 0$ or multiplied by some number $b > 1$, such that we get, for the scale inflation case, the following distribution function for the order statistics:

$$F_{(X_{1:n},\ldots,X_{n:n})}(x_1,\ldots,x_n) = n!\left[\prod_{i=1}^{n-k} F(x_i)\right]\left[\prod_{i=n-k+1}^{n} F(x_i/b)\right]\cdot \mathbf{I}_{\mathbf{R}^n_{\leq\cdot b;n-k+1}}(x_1,\ldots,x_n)\,, \quad (14.2.6)$$

where

$$\mathbf{R}^n_{\leq\cdot b;n-k+1} = \{(x_1,\ldots,x_n)\in\mathbf{R}^n;\ x_n \geq x_{n-1} \geq \cdots \geq x_{n-k+1} \geq x_{n-k}\cdot b,\ x_{n-k} \geq x_{n-k-1} \geq \cdots \geq x_1\}\,.$$

For the exponential case, we have

$$F_{(X_{1:n},\ldots,X_{n:n})}(x_1,\ldots,x_n) = n!\left[\prod_{i=1}^{n-k}(1-e^{-x_i/\theta})\right]\left[\prod_{i=n-k+1}^{n}(1-e^{-x_i/b\,\theta})\right]\cdot \mathbf{I}_{\mathbf{R}^n_{\leq\cdot b;n-k+1}}(x_1,\ldots,x_n)\,.$$

From this, it is immediately seen that the Tiku model is not identical with the labelled outliers model which, in the exponential case, has distribution function

$$F_{(X_{1:n},\ldots,X_{n:n})}(x_1,\ldots,x_n) = \left[\prod_{i=1}^{n-k}\frac{1}{i+b^{-1}}\right]^{-1}\left[\prod_{i=1}^{n-k}(1-e^{-x_i/\theta})\right]\left[\prod_{i=n-k+1}^{n}(1-e^{-x_i/b\,\theta})\right]\cdot \mathbf{I}_{\mathbf{R}^n_{\leq}}(x_1,\ldots,x_n)\,.$$

This shows, that for observations from a Tiku model, say with $k = 1$, we always have $x_{n:n} \geq x_{n-1:n}\cdot b$ for some $b > 1$, wheras for the labelled model we only claim $x_{n:n} \geq x_{n-1:n}$.

In the context of robust statistical inference, another deviation from the working hypothesis H_0 is also quite often used.

Mixture model

Take the number k of contaminants not as fixed but as an observation of a binomial $B(n,p)$ random variable for some small $p \in (0,1)$. After having observed k, one observes $X_1,\ldots,X_n$ from an exchangeable model with k contaminants.

In the exponential case, this yields for the joint distribution function of $X_1,\ldots,X_n$

$$F_{(X_1,\ldots,X_n)}(x_1,\ldots,x_n) = \sum_{k=0}^{n}\binom{n}{k}p^k(1-p)^{n-k}\sum\binom{n}{k}^{-1}\prod_{i\neq i_1,\ldots,i_k}(1-e^{-x_i/\theta})\cdot\prod_{j\in\{i_1,\ldots,i_k\}}(1-e^{-x_j/\theta\,b}),\quad x_i > 0,$$

with the second sum taken over all $\binom{n}{k}$ possible choices of $\{i_1,\ldots,i_k\}$ from $\{1,\ldots,n\}$.

Changing the order of summation, we get

$$F_{(X_1,\ldots,X_n)}(x_1,\ldots,x_n) = \prod_{i=1}^{n}\left[(1-p)(1-e^{-x_i/\theta}) + p(1-e^{-x_i/b\,\theta})\right] \quad (14.2.7)$$

which is a product of p-mixtures of $F(x) = (1 - e^{-x/\theta})$ with $G(x) = (1 - e^{-x/b\theta})$.

It should be noted that (14.2.7) is of the same structure as H_0, i.e., (14.2.7) is also a model of independent and identically distributed random variables. In this respect, a mixture model is not a genuine outlier model. It does also not produce *enough* surprisingly large observations. This is illustrated by the following example which has been given by Tukey (1960) for the normal distribution.

Consider a $N(0,1)$ random variable X, a random variable $Y \sim N(0,9)$ and the mixture $Z \sim 0.1 \cdot N(0,9) + 0.9 \cdot N(0,1)$; then for $I = \{x; -2.53 \leq x \leq 2.53\}$ we have

$$\begin{aligned} P(X \notin I) &= 0.0114 \\ P(Y \notin I) &= 0.401 \\ P(Z \notin I) &= 0.0491 \;. \end{aligned}$$

This shows that although we expect in a sample of 100 observations from Z about 10 observations from $G = N(0,9)$, there are only four observations to be expected outside the 0.01-fractiles of the target standard normal distribution. More details on the mixture exponential model and their applications may be found in Chapter 19.

In fact, none of the above models describe outliers via their position in the sample with respect to the target distribution. This has been the reason to consider the following model which is quite different in spirit [Davies and Gather (1993)].

δ-outliers model

Let the target distribution F have a density f. Then for $\delta \in (0,1)$, we call

$$\text{out}(\delta, f) = \{x \in \mathbf{R};\; f(x) < y(\delta)\} \;,$$

where $y(\delta) = \sup\{y;\; P(f(X) < y) \leq \delta\}$, with $X \sim F$, the *δ-outlier region for f*.

For a $N(\mu, \sigma^2)$ distribution, we simply get

$$\text{out}\left(\delta,\; \frac{1}{\sigma}\,\varphi\left(\frac{x-\mu}{\sigma}\right)\right) = \left\{x;\; |x-\mu| \geq z_{1-\frac{\delta}{2}}\sigma\right\} \;,$$

where $\Phi(z_q) = q$.

For an exponential distribution $F(x) = 1 - e^{-x/\theta}$, $x > 0$, we have

$$\text{out}\left(\delta, \frac{1}{\theta}\; e^{-x/\theta}\right) = \{x;\; x \geq -\theta \log \delta\} \;.$$

Now by definition, a *δ-outliers model* describes a sample with $n-k$ regular observations, which come independently and identically distributed from some target distribution F with density f, say $F(x) = 1 - e^{-x/\theta}$, $x > 0$, and with k observations which are put arbitrarily into the δ-outlier region for f. No other restrictions are put on these nonregular observations. They need not be independent nor do we know which elements of our sample are nonregular. Of course we may also find some regular observations in the δ-outlier region for f.

If one wants to take the sample size n into account, one can choose $\delta = \delta_n = 1 - (1 - \widetilde{\delta})^{1/n}$ for some $\widetilde{\delta} \in (0,1)$. This implies that for independent and identically distributed random variables $X_1, \ldots, X_n$ with density f, one gets $P(X_i \notin \text{out}(\delta_n, f)$ for all $i = 1, \ldots, n) \geq 1 - \tilde{\delta}$, that is, for H_0 we find an observation x_i in the δ_n-outlier region for f only with probability $\tilde{\delta}$.

It should be noted that if the target distribution were completely known, then the corresponding δ-outlier region will also be completely known. However, in the case of an exponential distribution with the scale parameter θ unknown, it is obvious that the determination of $\text{out}(\delta, f)$ is related to the estimation of θ. The δ_n-outlier model meets best the situation which we have in practice. No distributional assumption is anticipated for the outliers only that they are in positions outside the main body of the data and considered as surprising w.r.t. the target distribution. The surprisingness of a position is naturally depending on the sample size and is captured by specifying δ_n (an observation of size 10 in an exponential sample with $\theta = 1$ is less surprising when the sample size is 1000 than for a sample of size 5). The δ_n-outlier model was motivated mainly by the aim of finding a formal approach to the detection of an unknown number of outliers in some sample. Only assuming that the main body of the data (i.e. at least $\frac{n}{2} + 1$ points) comes i.i.d. from F with density f, this task can be seen as wanting to identify all observations in $\text{out}(\delta_n, f)$ for given δ_n, and that independently of the mechanism which has located these observations in that region. We will come back to this approach in 14.4.

As we have seen now, there are quite a few different models with the purpose of describing the occurrence of outliers in random samples. They all have been employed to derive outlier identification procedures as well as statistical procedures to be used in outlier situations. They are also used in simulation studies where we want to compare statistical procedures in outlier situations. It is obviously a matter of great importance to always specify exactly which model is being used, since different models will yield quite different results in general. As pointed out above, we want to show this effect for some inferential issucs and we start with deriving tests for outliers on the basis of the different models just defined.

14.3 Outlier Tests - Testing for Contamination

Under the assumption of the above outlier generating models, all formulated as super-models including the working hypothesis H_0, one can derive tests for outliers. We will treat some examples of such derivations here. To be precise, these tests should rather be called tests for contamination or tests for (k) contaminants than outlier tests, which however is their more common and most often used name.

14.3.1 Outlier testing in an unidentified outliers model

For the normal distribution, we find this topic extensively treated in Ferguson (1961). The general case of a location/scale family is considered in Gather (1989). We first cite the general result here in order to apply this to the exponential case.

Assume a Ferguson-type model where the joint density $f(x_1, \ldots, x_n)$ of $(X_1, \ldots, X_n)$ is given by

$$f(x_1, \ldots, x_n) = \prod_{j=1}^{n} f_{i_j}(x_j), \qquad (x_1, \ldots, x_n) \in \mathbf{R}^n \;, \tag{14.3.1}$$

where, for $i = 1, 2, \ldots, n$, we have either

$$f_i(x) = \frac{1}{\sigma}\, f_0\left(\frac{x - (\mu + \Delta\, \alpha_i \sigma)}{\sigma}\right) \qquad \text{(model A)}$$

or

$$f_i(x) = \frac{1}{\sigma \exp \Delta\, a_i}\, f_0\left(\frac{x - \mu}{\sigma \exp \Delta\, a_i}\right) \qquad \text{(model B)}\ ,$$

with unknown parameters $\mu \in \mathbf{R}$, $\sigma^2 > 0$, $\Delta \in \mathbf{R}$, $(i_1, \ldots, i_n) \in \Pi_n$ (Π_n is the set of all permutations on $\{1, \ldots, n\}$) and $a_1, \ldots, a_k > 0$ given numbers, $a_{k+1} = \cdots = a_n = 0$ and f_0 a given density on $\mathbf{R}$. Note, that for $\Delta = 0$, (14.3.1) represents the joint density of an independent and identically distributed sample from a distribution with density $\frac{1}{\sigma}\, f_0\left(\frac{x-\mu}{\sigma}\right)$.

We consider the testing problem

$$H_0 :\ \Delta = 0 \text{ vs } H_1 :\ \Delta > 0\ , \tag{14.3.2}$$

where one wants to find out if we are dealing with an independent and identically distributed sample or if there exist k contaminants from differently far shifted or differently strongly inflated distributions. A generalization of Ferguson's result of 1961, proven in Gather (1989), states that for the general testing problem (14.3.2), under some regularity conditions on f_0, there exists a locally optimal invariant unbiased test of a given size α. Invariance is meant here with respect to the group of transformations generated by all permutations of $X_1, \ldots, X_n$, all additions of some constant $a \in \mathbf{R}$ to each X_i, $i = 1, \ldots, n$, and all multiplications of each X_i by some positive constant b.

This locally optimal test is given by

$$\begin{aligned} &\varphi(x_1, \ldots, x_n; a_1, \ldots, a_k) \\ &= \begin{cases} 1, \\ 0, \end{cases} \quad T_n = \left.\frac{d^m \log g(\boldsymbol{y}(x_1, \ldots, x_n); \Delta, \boldsymbol{a})}{d\Delta^m}\right|_{\Delta=0} \begin{matrix} > \\ \le \end{matrix}\ c \end{aligned} \tag{14.3.3}$$

with c determined by $E_{H_0}\varphi = \alpha$, where

$$\begin{aligned} \boldsymbol{y}(x_1, \ldots, x_n) &= (y_2, \ldots, y_{n-1}) = I(x_1, \ldots, x_n) \\ &:= \left(\frac{x_{2:n} - x_{1:n}}{x_{n:n} - x_{1:n}}, \ldots, \frac{x_{n-1:n} - x_{1:n}}{x_{n:n} - x_{1:n}}\right) \\ &\in [0,1]^{n-2}_{\le} = \left\{(y_2, \ldots, y_{n-1}) \in [0,1]^{n-2},\ y_2 \le \cdots \le y_{n-1}\right\}\ , \end{aligned}$$

and where

$$g(y_2, \ldots, y_{n-1}; \Delta, \boldsymbol{a}) = \int_0^\infty \int_0^\infty v^{n-2} \sum_{v=(v_1, \ldots, v_n) \in \Pi_n} f_i(y_{v_i} v + t)\, dt\, dv$$

denotes the joint density of $I(X_1, \ldots, X_n)$, $y_1 = 0$, $y_n = 1$, which does not depend on μ and σ^2 but on Δ and $a_1, \ldots, a_n$, and for which we assume that there exists a positive integer m such that

$$\left.\frac{d^j\, g(\boldsymbol{y}; \Delta, \boldsymbol{a})}{d\Delta^j}\right|_{\Delta=0} = 0 \qquad \text{for all } j = 1, \ldots, m-1$$

and $\frac{d^m g(\boldsymbol{y};\Delta,\boldsymbol{a})}{d\Delta^m}$ is continuous, finite and nonzero in some neighborhood of $\Delta = 0$.

Under some regularity conditions, the test statistic T_n depends only on $\sum a_i, \sum a_i^2, \ldots, \sum a_i^m$, and for many distributions it does not depend on $a_1, \ldots, a_n$ at all. This is the case for a normal parent for instance, where for model A the statistic T_n is equal to the sample skewness, and for model B to the sample kurtosis. In the case of an exponential parent, the Ferguson model B claims that each of the independent and identically distributed X_i has density

$$f_i(x) = \frac{1}{\theta \exp(\Delta a_i)} \exp - \left(\frac{x-\mu}{\theta \exp(\Delta a_{v_i})} \right), \qquad x > \mu,\ i = 1, \ldots, n$$

with $\mu \in \mathbf{R}$, $\theta > 0$, $\Delta \geq 0$, $(v_1, \ldots, v_n) \in \Pi_n$ unknown and $a_1, \ldots, a_n > 0$, $1 \leq k \leq n-1$ given, $a_{k+1} = \cdots = a_n = 0$ (cf. 14.2.3).

Under these assumptions, we get for the exponential distribution

$$g(\boldsymbol{y}; 0, \boldsymbol{a}) = c(a_1, \ldots, a_n) \left(\sum_{i=2}^{n-1} y_i \right)^{-(n-1)}, \qquad \left. \frac{dg(\boldsymbol{y}; \Delta, \boldsymbol{a})}{d\Delta} \right|_{\Delta=0} = 0$$

and

$$\left. \frac{d^2 \log g(\boldsymbol{y}; \Delta, \boldsymbol{a})}{d\Delta^2} \right|_{\Delta=0} = \sum_{i=1}^{n} (a_i - \overline{a})^2 \cdot \frac{n \sum y_i^2}{\left(\sum y_i\right)^2} .$$

Hence for an exponential parent, there is a test statistic $V_n = \frac{n \sum y_i^2}{\left(\sum y_i\right)^2}$ which is independent of k and $a_1, \ldots, a_k$ for testing the hypothesis H_0 (independent and identically distributed) against the alternative that there is some number k of contaminants from scale inflated distributions in the sample and this yields a locally most powerful invariant unbiased test for this testing problem. It should be noted that $\frac{n^{3/2} V_n - 2n^{1/2}}{2\sqrt{5}}$ is asymptotically standard normally distributed [Gather and Helmers (1983)].

The independence of V_n from k and $a_1, \ldots, a_k$ implies also that a rejection of H_0 does not give us any information about the number nor about the size of contaminations. We only know after rejecting H_0 that there are some contaminants in the sample at hand without knowing which ones. If in a Ferguson type model all contaminants have the same parent G, i.e. $a_1 = \cdots = a_k$, that is if we consider a k-outlier model, then other kinds of outlier tests can also be derived. Suppose a k-outlier model with k contaminants from $G(x) = 1 - e^{-x/b\theta}$, $x > 0$, $\theta > 0$, $b \geq 1$ and the remaining regular observations from $F(x) = 1 - e^{-x/\theta}$, $x > 0$, $\theta > 0$. Then a Maximum Likelihood Ratio Test for $H_0 :\ b = 1$ vs $H_1 :\ b > 1$ exists and it rejects H_0 if and only if

$$T_{n,k} := \sum_{i=n-k+1}^{n} x_{i:n} \,/\, \overline{x}$$

exceeds some critical value [Gather and Kale (1988)]. For $k = 1$, $T_{n,k}$ was already suggested by Cochran (1941) and it is probably one of the best known outlier test statistic which has been used and investigated for the exponential distribution by many authors in the field. It turns out also under other performance criteria and other outlier generating models as 'the' test statistic as we will see below. The computation of critical values for $T_{n,k}$ is well solved since the distribution of $T_{n,k}$ has been calculated under H_0

and even under the alternative model H_1 in the k-outlier model [see Chikkagoudar and Kunchur (1983)].

The rejection of H_0 here gives us the information that in the sample $x_1, \dots, x_n$ which we are considering there are k contaminants from a distribution with larger scale, but without specifying which observations are the contaminants. It is of course quite natural to suspect the k largest extremes and we have mentioned above that indeed these ones have the largest probability of coming from G and not from F. However, if for a k-outlier model, one tries to find out which of the observations are contaminants, if any, one should rather solve a multidecision problem with hypothesis

$$H_0: \ X_1, \dots, X_n \text{ i.i.d. with } F(x) = 1 - e^{-x/\theta}$$

against the $\binom{n}{k}$ alternatives for all choices $\{i_1, \dots, i_k\}$ from $\{1, \dots, n\}$

$$H_{1(i_1,\dots,i_k)}: \ X_j \sim F, \ j \neq i_1, \dots, i_k \text{ and } X_{i_v} \sim F\left(\frac{\cdot}{b}\right), \quad v = 1, \dots, k.$$

For $k = 1$, among all multidecision procedures which are invariant with respect to permutations and affine linear transformations and which decide for H_0 if H_0 is true with probability $1 - \alpha$, the following rule maximizes the probability of deciding for H_{1i} if H_{1i} is true, $i = 1, \dots, n$:

> Take a decision for H_{1i} if $x_i = x_{n:n}$, and $T_{n,1} = x_{n:n}/\bar{x}$ is larger than c_α; decide for H_0 if $T_{n,1} \leq c_\alpha$, where c_α is such that $P(T_{n,1} > c_\alpha \mid H_0) = \alpha$.

For a proof, see Paulson (1952), David (1981), Kudo (1956), Truax (1953), and Doornbos (1976). Hence, the test statistic $T_{n,1}$ of the MLR test has also an optimality property in the multidecision framework.

14.3.2 Outlier testing in a labelled slippage model

Since the labelled model is often implicitly assumed, it is a very important one. Also, the tests to be derived here are tests which have been suggested much earlier on an intuitive background or in other framework also.

Remember that under a labelled exponential model, where we have k contaminants from a scale inflated exponential, the joint density of $(X_{1:n}, \dots, X_{n:n})$ is

$$\begin{aligned} &f_{\text{labelled},k}(x_1, \dots, x_n) \\ &= \left(\prod_{i=1}^{n-k} \frac{1}{i + b^{-1}}\right)^{-1} \left(\prod_{i=1}^{n-k} \frac{1}{\theta}\, e^{-x_i/\theta}\right) \left(\prod_{i=n-k+1}^{n} \frac{1}{b\,\theta}\, e^{-x_i/b\theta}\right) \mathbf{I}_{[0,\infty]^n_{\leq}}(x_1, \dots, x_n)\,. \end{aligned}$$

This can also be written as

$$\begin{aligned} &f_{\text{labelled},k}(x_1, \dots, x_n) \\ &= C_n(\theta, b) \exp\left[-\frac{1-b}{b\,\theta} \sum_{i=n-k+1}^{n} x_i - \frac{1}{\theta} \sum_{i=1}^{n} x_i\right] \mathbf{I}_{[0,\infty]^n_{\leq}}(x_1, \dots, x_n)\,, \end{aligned}$$

yielding a density of a two-parametric exponential family with corresponding convex natural parameter space. For $\eta_1 = \frac{1-b}{b\theta} = 0$, i.e., for $b = 1$, the statistic $V = \sum_{i=1}^n x_{i:n}$ is complete and sufficient for $\eta_2 = -\frac{1}{\theta}$. Hence the test

$$\widetilde{\varphi}(x_{1:n}, \ldots, x_{n:n}) = \begin{cases} 1, \\ 0, \end{cases} T = \sum_{i=n-k+1}^{n} x_{i:n} \begin{matrix} \geq \\ < \end{matrix} c_{n,\alpha}^k(v) ,$$

where $c_{n,\alpha}^k(v)$ for each $V = v$ is determined to make the test of size α, is a UMPU level-α test for $H_0 : b = 1$ versus $H_1 : b > 1$. After transforming the test statistic T into $T_{n,k} = \sum_{i=n-k+1}^n x_{i:n} / \sum_{i=1}^n x_{i:n}$ which for any fixed $V = v$ is a strictly isotonic transformation of T, we see that

$$\varphi(x_{1:n}, \ldots, x_{n:n}) = \begin{cases} 1, \\ 0, \end{cases} T_{n,k} \begin{matrix} > \\ \leq \end{matrix} c_{\alpha,n}^k ,$$

is equivalent to $\widetilde{\varphi}$ with the already well known test statistic $T_{n,k}$ which has a distribution independent of θ. Thus $T_{n,k}$ is test statistic of a UMPU test for k upper scale inflated outliers in the labelled exponential model (cf. Gather (1979) for $k = 1$, and Gather and Kale (1981) for arbitrary k where the proofs are given for general exponential families including the normal). The distribution of $T_{n,k}$, as mentioned above, has been derived by some authors in the field. It should also be noted that $T_{n,k}$ is not only test statistic of an MLR-test in the unidentified outlier model but also of an MLR-test in the labelled slippage model [Fieller (1976) and Kimber (1979)], stating again that there are many optimality properties of $T_{n,k}$.

The derivation of outlier tests in other models cannot be considered as satisfactory. First, because of the expression $\mathbf{I}_{\mathbf{R}^n_{\leq \cdot b; n-k+1}}(x_1, \ldots, x_n)$ depending on the unknown parameter b, present in the density of $(X_{1:n}, \ldots, X_{n:n})$ under the Tiku model [see Eq. (14.2.6)], there is no chance to derive MLR- or UMPU tests in this model. However, some procedures have been suggested on an intuitive background for this model [Tiku, Tan and Balakrishnan (1986)]. We have a similar situation for the mixture model, where only in the case of a known scale parameter θ, Durairajan and Kale (1982) have derived an LMP similar level-α test. But even in the labelled model where we are in the best position of having derived a UMPU test for k upper outliers, when using $T_{n,k}$ as a test statistic, we need a specification of the number k of contaminants. Therefore, when applying the test based on $T_{n,k}$, one has to envisage two possible dangers:

- the possibility of not rejecting $H_0 : b = 1$ in favour of a labelled model with k contaminants although a model with even $m > k$ contaminants is true [which is related to the so-called *masking effect*, first observed by Pearson and Chandra Sekar (1936) and called masking effect by Murphy (1951)].

- the possibility of deciding for a labelled model with k outliers when instead a labelled model with $m < k$ outliers is true [related to the so-called *swamping effect*, first discussed by Fieller (1976)].

Both are indeed severe dangers.

Suppose, for example, we have a sample $x_1, \ldots, x_{10}$ assumed under ideal conditions to result from an exponential distribution with scale parameter $\theta = 1$. Let $x_{1:10}, \ldots, x_{8:10}$ be grouped close together near 1 and $x_{9:10} = 7.5$ such that

$x_{9:10} / \sum_{i=1}^{9} x_{i:10}$ exceeds the critical value $c^1_{0.05,9} = 0.4775$. If now $x_{10:10} < 1.69\, x_{9:10}$, then $x_{10:10} / \sum_{i=1}^{10} x_{i:10} < c^1_{0.05,10} = 0.445$, then masking occurs. Similarly, if $x_{1:10}, \ldots, x_{8:10}$ are close to one, $x_{9:10} = 7$ such that $x_{9:10} / \sum_{i=1}^{9} x_{i:10}$ does not exceed the critical value $c^1_{0.05,9} = 0.4775$. But if $x_{10:10} > 8$, then $(x_{8:10} + x_{9:10}) / \sum_{i=1}^{10} x_{i:10}$ exceeds the critical value $c^2_{0.05,10} = 0.643$ and the swamping effect occurs here.

14.4 Identification of Outliers

From the above examples of masking and swamping, it is clear that rather then *testing* if a fixed number k of contaminants is present in a sample $x_1, \ldots, x_n$ (which under ideal conditions is supposed to come independently and identically from $1 - e^{-x/\theta}$, $\theta > 0$) one may want to identify either all contaminants or the α_n-outliers with respect to $F(x) = 1 - e^{-x/\theta}$ for some $\alpha_n = 1 - (1 - \tilde{\alpha})^{1/n}$ (c.f. 14.2). There is not much to this topic for the exponential case in current literature. The most comprehensive references are Hawkins (1980) and again Barnett and Lewis (1994). Indeed to detect an unknown number k of contaminants or all α_n-outliers, respectively, in an exponential sample one can think of some different approaches which are only briefly sketched here:

- Methods consisting of the estimation of k followed by some block test, where the estimation of k can be based e.g. on MLE's [Gather and Kale (1988)], or on gaps [Tietjen and Moore (1972), Tiku (1975), Jain and Pingel (1981)].

- Consecutive inwards- or outwards testing, which should formally be treated as solution (step up or step down) of an appropriate multiple testing problem [see Gather and Pigeot (1994)]. Inwards testing, however, to avoid masking, must be based on test statistics using robust scale estimates [Davies and Gather (1993)], or performed as in Marasinghe (1985). Outwards testing, as suggested by Kimber (1982), Sweeting (1983, 1986) and others is less prone to masking, but as pointed out in Davies and Gather (1993) can also miss very large outliers. But actually it is rather the problem of determining the critical values for each outside step in order to keep global and multiple level, which makes outwards testing not very attractive for practioners. Especially at this point, one should note again, that different outlier generating models result in quite different procedures.

- Estimation of the indices of the contaminats, either by MLE [Gather and Kale (1988)] or Akaike's Information Criterion [Kitagawa (1979)].

- A Prediction-type approach like in Balasooriya (1989).

- One-step rules based on robust estimates like the following: if $x_i/\hat{\theta} > c(\alpha)$, then x_i is identified as an 'outlier', $i = 1, \ldots, n$. (especially as an α_n-outlier, where $\alpha_n = 1 - (1 - \alpha)^{1/n}$). Here $\hat{\theta}$ should be some robust high break down point - estimator like in Davies and Gather (1993) for the normal case.

The development and study of rules to identify contaminants or α_n-outliers for the exponential distribution which themselves involve robust estimates is under current investigation of the author.

14.5 Estimation of the Scale Parameter of an Exponential Distribution in Outlier Situations

There are contributions in the literature to the problem of finding estimators of the scale parameter θ of an exponential distribution in outlier situations. Among those are the papers by Thall (1979) and by Kimber (1983) where in the classical spirit of robustness, influence functions, gross error sensitivities and asymptotic relative efficiencies of some trimmed means and M-estimators are determined. To a much larger extent we find contributions to the 'accomodation of outliers' which assume some outlier generating model and then derive and compare suitable estimators, mainly L-estimators. This topic was firstly treated by Kale and Sinha (1971) and Joshi (1972) who found out that under H_0

$$T_n^* = \frac{1}{n+1}\sum_{i=1}^{n} x_{i:n}$$

has smallest MSE among all L-estimators and that the same is true for

$$\frac{1}{m+1}\left\{\sum_{j=1}^{m-1} x_{j:n} + (n-m+1)x_{m:n}\right\} = W_{n,m}$$

among all L-estimators of type $\sum_{j=1}^{m} c_j x_{j:n}$, with fixed $m < n$.

These results have initiated an extensive investigation of all kinds of trimmed and Winsorized means, various L-estimators and Anscombe-type pretest estimators under almost all outlier generating models from above. Most of the work relates to finding good choices of the weights c_j or comparing estimators with respect to MSE's under different outlier models. We cite some of the proposals before giving a closer look to Anscombe-type estimators.

For some special values of b and n in an exchangeable model with k small, MSE's have been compared by Kale and Sinha (1971) and Joshi (1972) resulting in recommending W_{n,m^*} with 'optimally' determined m^*. Kale (1975) on the other hand proposes a trimmed mean

$$Tr_{n,m} = \frac{1}{m}\sum_{j=1}^{m} x_{j:m}, \qquad m < n.$$

Chikkagoudar and Kunchur (1980) suggest using

$$CK_n = \frac{1}{n}\sum_{j=1}^{n}\left(1 - \frac{2j}{n(n+1)}\right) x_{j:m}$$

as long as $b \leq 2.2$ and $Tr_{n,n-1}$ for 'large' values of b.

In Kimber (1983), some selected estimators have also been compared through a Monte Carlo study assuming an unidentified outliers model with $k = 1, 2$ contaminants from a scale inflated exponential where the scale inflation factor b ranges from 1, 4, 8 to ∞. He recommends the 10% or 20% - trimmed mean to estimate θ.

Moment estimators have been studied by Joshi (1988), again under an exchangeable outlier model with $k = 1$; roughly spoken he ends up by proposing CK_n.

From (1991) tries to find weights c_j which minimize $MSE(\sum_{j=1}^{n} c_j x_{j:n})$ under the same model, and suggests

$$F_n = \sum_{j=1}^{M} c_1^* x_{j:n} + \sum_{j=M+1} c_2^* x_{j:n}$$

with 'optimal' choices of M and c_1^*, c_2^*. From shows also that F_n has smaller MSE than CK_n and $W_{n,m}$ (with optimal m).

Assuming an unidentified outliers model with $k \geq 1$ and b unknown, Gather and Kale (1988) derive maximum likelihood estimators for θ.

As a general remark we point out that in the exponential case, it is possible, though tedious, to compute explicitly mean square errors (and other moments) of many estimators, especially L-estimators, under some outlier generating models. These calculations have been carried out in Gather (1984, 1986a, 1986b) [see also Balakrishnan (1994)], where explicit expressions are given for trimmed and Winsorized means under the identified, unidentified and labelled outlier model for $k = 1$. We have, for example,

$$\begin{aligned} MSE(\sum_{i=1}^{n} c_i x_{i:n} | \text{ labelled }) &= \sum_{i=1}^{n} \frac{1}{(n+b-i)^2} \left(\sum_{k=i}^{n} c_k \right)^2 \\ &+ \left(\sum_{i=1}^{n} \frac{1}{n+b-i} \left(\sum_{k=i}^{n} c_k \right) - 1 \right)^2 . \end{aligned}$$

Under the unidentified outliers models the formulas are very lengthy and hence are not presented here.

However, it is even possible to calculate the MSE's of some Anscombe-type pretest estimators of the form

$$T_n(c) = T_n = \begin{cases} \frac{1}{m} \left(\sum_{i=1}^{n-1} x_{i:n} + y \right) & \text{if } V(X_1, \ldots, X_n) > c \\ T_n^* = \frac{1}{n+1} \sum_{i=1}^{n} x_{i:n} & \text{otherwise,} \end{cases}$$

where m is a weight (like n, $n-1$, $n+1$), y is a replacement of $x_{n:n}$ (like $y = 0$, $y = x_{n-1:n}$, etc.), and V is an outlier test statistic like $X_{n:n} / \overline{X}$ or $\frac{X_{n:n} - X_{n-1:n}}{X_{n-1:n} - X_{1:n}}$. Such estimators are defined piecewise, choosing either T_n^* or some trimmed (Winsorized) estimator, depending on how large the test statistic V of some outlier test turns out.

One can then be interested in applying a premium-protection approach as initiated by Anscombe (1960) to compare different estimators of the type T_n. Anscombe suggested to determine the critical value c for $V(X_1, \ldots, X_n)$ such that the so-called

$$\begin{aligned} \text{Premium}(T_n(c)) &= \frac{MSE(T_n(c)) - MSE(T_n^*)}{MSE(T_n^*)} \\ &= \frac{MSE(T_n(c))}{MSE(T_n^*)} - 1 \end{aligned}$$

under the assumption of independent and identically distributed variables becomes equal to some given value $p \in (0, 1)$. This means that under the null model one is willing to accept a decrease of efficiency for $T_n(c)$ compared with $T_n^* = \frac{1}{n+1}\sum_{i=1}^n X_{i:n}$, which among all linear estimators $\sum_{i=1}^n a_i X_{i:n}$ has the smallest MSE under the null model. Thus one gets $c = c(T_n, p)$. In a second step then, one compares for all the above types of estimators with $c = c(T_n, p)$ the values for the so-called

$$\begin{aligned}\text{Protection}(T_n(c)) &= \frac{MSE(T_n^* \mid H_1) - MSE(T_n(c) \mid H_1)}{MSE(T_n^* \mid H_1)} \\ &= 1 - \frac{MSE(T_n(c) \mid H_1)}{MSE(T_n^* \mid H_1)},\end{aligned}$$

where H_1 denotes some outlier generating model. That is, the Protection$(T_n(c))$ is a measure of the gain or loss of the MSE under H_1 when $T_n(c)$ is used instead of T_n^*.

As a typical result [see Gather (1986b), Gather and Benda (1989)] which can be seen from elaborate numerical calculations of Protection$(T_n(c))$, for $p = 0.01$ and 0.05, $n = 3, \ldots, 50$, $b = 1, \ldots, 10$ in a labelled or Ferguson-type model with $k = 1$, one gets that

$$\widetilde{T}_n = \begin{cases} \frac{1}{n+1}\left(\sum_{i=1}^{n-1} x_{i:n} + \frac{c}{n-c}\sum_{i=1}^{n-1} x_{i:n}\right) & \text{if } x_{n:n}/\overline{x} > c \\ T_n^* & \text{otherwise} \end{cases}$$

where $c = c(T_n, p)$ gives the largest protection values in all of the above situations among a large class of estimators T_n of the above type.

Of course, when b gets close to one, T_n^* must be optimal. There are similar results for $k > 1$. Here, in case that $k > 1$ is known, one can recommend

$$T_n^k(c) = \begin{cases} \frac{n}{(n+1)(n-c)}\sum_{i=1}^{n-k} x_{i:n} & \text{if } T_{n,k} > c \\ T_n^* & \text{otherwise} \end{cases}$$

where $T_{n,k}$ is as in Section 14.3.

Some more results on estimators under a k-outliers labelled model are given in the next chapter, wherein some tables of the values of MSE's and the gain or loss in efficiency for various linear estimators are presented.

References

Anscombe, F.J. (1960). Rejection of outliers, *Technometrics*, **2**, 123-147.

Balakrishnan, N. (1994). Order statistics from non-identical exponential random variables and some applications (with discussions), *Computational Statistics & Data Analysis*, **18**, 203-253.

Balasooriya, U. (1989). Detection of outliers in the exponential distribution based on prediction, *Communications in Statistics - Theory and Methods*, **18**, 711-720.

Barnett, V. and Lewis, T. (1978). *Outliers in Statistical Data*, First edition, John Wiley & Sons, Chichester, England.

Barnett, V. and Lewis, T. (1984). *Outliers in Statistical Data*, Second edition, John Wiley & Sons, Chichester, England.

Barnett, V. and Lewis, T. (1994). *Outliers in Statistical Data*, Third edition, John Wiley & Sons, Chichester, England.

Chikkagoudar, M.S. and Kunchur, S.M. (1980). Estimation of the mean of an exponential distribution in the presence of an outlier, *Canadian Journal of Statistics*, **8**, 59-63.

Chikkagoudar, M.S. and Kunchur, S.M. (1983). Distributions of test statistics for multiple outliers in exponential samples, *Communications in Statistics - Theory and Methods*, **12**, 2127-2142.

Cochran, W.G. (1941). The distribution of the largest of a set of estimated variances as a fraction of their total, *Annals of Eugenics*, **11**, 47-52.

David, H.A. (1979). Robust estimation in the presence of outliers, In *Robustness in Statistics* (Eds., R.L. Launer and G.N. Wilkinson), pp. 61-74, Academic Press, New York.

David, H.A. (1981). *Order Statistics*, Second edition, John Wiley & Sons, New York.

Davies, P.L. and Gather, U. (1993). The identification of multiple outliers (with discussion), *Journal of the American Statistical Association*, **88**, 782-792.

Doornbos, R. (1976). *Slippage Tests*, Second edition, Mathematical Centre Tracts No. 15, Mathematisch Centrum, Amsterdam.

Durairajan, T.M. and Kale, B.K. (1982). Locally most powerful similar test for mixing proportion, *Sankhyā, Series A*, **44**, 153-161.

Ferguson, T.S. (1961). On the rejection of outliers, *Proceedings of the Fourth Berkeley Symposium on Mathematical Statistics and Probability*, Vol. 1, pp. 253-287, University of California Press, Berkeley.

Fieller, N.R.J. (1976). Some problems related to the rejection of outlying observations, *Ph.D. Thesis*, University of Hull, U.K.

From, S.G. (1991). Mean square error efficient estimation of an exponential mean under an exchangeable single outlier model, *Communications in Statistics - Simulation and Computation*, **20**, 1073-1084.

Gather, U. (1979). Über Ausreissertests und Ausreisseranfälligkeit von Wahrscheinlichkeitsverteilungen, *Doctoral Dissertation*, Aachen Technical University, Germany.

Gather, U. (1984). Tests and estimators in outlier models (in german), *Habilitation Thesis*, Aachen Technical University, Germany

Gather, U. (1986a). Robust estimation of the mean of the exponential distribution in outlier situations, *Communications in Statistics - Theory and Methods*, **15**, 2323-2345.

Gather, U. (1986b). Estimation of the mean of the exponential distribution under the labelled outlier-model, *Methods in Operations Research*, **53**, 535-546.

Gather, U. (1989). Testing for multisource contamination in location/scale families, *Communications in Statistics - Theory and Methods*, **18**, 1-34.

Gather, U. (1990). Modelling the occurence of multiple outliers, *Allg. Statist. Archiv*, **74**, 413-428.

Gather, U. and Benda, N. (1989). Adaptive estimation of expected lifetimes when outliers are present, *Manuscript*, FB Statistik, University of Dortmund, Germany.

Gather, U. and Helmers, M. (1983). A locally most powerful test for outliers in samples from the exponential distribution, *Methods in Operations Research*, **47**, 39-47.

Gather, U. and Kale, B.K. (1981). UMP Tests for r upper outliers in samples from exponential families, *Proceedings of the Golden Jubilee Conference, Indian Statistical Institute*, 270-278.

Gather, U. and Kale, B.K. (1988). Maximum likelihood estimation in the presence of outliers, *Communications in Statistics - Theory and Methods*, **17**, 3767-3784.

Gather, U. and Kale, B.K. (1992). Outlier generating models – a review, In *Contributions to Stochastics* (Ed., N. Venugopal), pp. 57-85, Wiley (Eastern), New Delhi, India.

Gather, U. and Pigeot, I. (1994). Identifikation von Ausreissern als multiples Testproblem, In *Medizinische Informatik: Ein integrierender Teil arztunterstützender Technologien (38. Jahrestagung der GMDS, Lübeck, September 1993)*, (Eds., Pöppl, S.J., Lipinski, H.-G., Mansky, T.), MMV Medizin Verlag, München, Germany, 474-477.

Gather, U. and Rauhut, B.O. (1990). The outlier behaviour of probability distributions, *Journal of Statistical Planning and Inference*, **26**, 237-252.

Hampel, F.R., Ronchetti, E.M., Rousseeuw, P.J., and Stahel, W.A. (1986). *Robust Statistics: The Approach Based on Influence Functions*, John Wiley & Sons, New York.

Hawkins, D.M. (1980). *Identification of outliers*, Chapman and Hall, London.

Jain, R.B. and Pingel, L.A. (1981). A procedure for estimating the number of outliers, *Communications in Statistics - Theory and Methods*, **10**, 1029-1041.

Joshi, P.C. (1972). Efficient estimation of a mean of an exponential distribution when an outlier is present, *Technometrics*, **14**, 137-144.

Joshi, P.C. (1988). Estimation and testing under an exchangeable exponential model with a single outlier, *Communications in Statistics - Theory and Methods*, **17**, 2315-2326.

Kale, B.K. (1975). Trimmed means and the method of maximum likelihood when spurious observations are present, In *Applied Statistics* (Ed., R.P. Gupta), North-Holland, Amsterdam.

Kale, B.K. and Sinha, S.K. (1971). Estimation of expected life in the presence of an outlier observation, *Technometrics*, **13**, 755-759.

Kimber, A.C. (1979). Tests for a single outlier in a gamma sample with unknown shape and scale parameters, *Applied Statistics*, **28**, 243-250.

Kimber, A.C. (1982). Tests for many outliers in an exponential sample, *Applied Statistics*, **32**, 304-310.

Kimber, A.C. (1983). Comparison of some robust estimators of scale in gamma samples with known shape, *Journal of Statistical Computation and Simulation*, **18**, 273-286.

Kitagawa, G. (1979). On the use of AIC for the detection of outliers, *Technometrics*, **21**, 193-199.

Kudo, A. (1956). On the invariant multiple decision procedures, *Bulletin of Mathematical Statistics*, **6**, 57-68.

Marasinghe, M.G. (1985). A multistage procedure for detecting several outliers in linear regression, *Technometrics*, **27**, 395-399.

Mount, K.S. and Kale, B.K. (1973). On selecting a spurious observation, *Canadian Mathematical Bulletin*, **16**, 75-78.

Murphy, R.B. (1951). On Tests for outlying observations, *Ph.D. Thesis*, Princeton University.

Paulson, E. (1952). A optimum solution to the k-sample slippage problem for the normal distribution, *Annals of Mathematical Statistics*, **23**, 610-616.

Pearson, E.S. and Chandra Sekar, C. (1936). The efficiency of statistical tools and a criterion for the rejection of outlying observations, *Biometrika*, **28**, 308-320.

Sweeting, T.J. (1983). Independent scale-free spacings for the exponential and uniform distribution, *Statistics & Probability Letters*, **1**, 115-119.

Sweeting, T.J. (1986). Asymptotically independent scale–free spacings with applications to discordancy testing, *Annals of Statistics*, **14**, 1485-1496.

Thall, P.F. (1979). Huber-sense robust M-estimation of a scale parameter, with application to the exponential distribution, *Journal of the American Statistical Association*, **74**, 147-152.

Tietjen, G.L. and Moore, R.H. (1972). Some Grubbs-type statistics for the detection of several outliers, *Technometrics*, **14**, 583- 597.

Tiku, M.L. (1975). A new statistic for testing suspected outliers, *Communications in Statistics*, **A4**, 737-752.

Tiku, M.L., Tan, W.Y., and Balakrishnan, N. (1986). *Robust Inference*, Marcel Dekker, New York.

Truax, D.R. (1953). An optimum slippage test for the variances of k, normal distributions, *Annals of Mathematical Statistics*, **24**, 669-674.

Tukey, J.W. (1960). A survey of sampling from contaminated distributions, In *Contributions to Probability and Statistics* (Ed., I. Olkin), Stanford University Press, Stanford, California.

Veale, J.R. (1975). Improved estimation of expected life when one identified spurious observation may be present, *Journal of the American Statistical Association*, **70**, 398-401.

CHAPTER 15

Extensions to Estimation Under Multiple-Outlier Models

N. Balakrishnan and A. Childs[1]

15.1 Introduction

During the past twenty years or so, many papers have appeared on the efficient estimation of the mean of an exponential distribution when outliers are possibly present in the sample. Due to the algebraic and computational difficulties faced when dealing with order statistics from multiple-outlier models, most of these papers addressed the case of a single outlier. For example, Kale and Sinha (1971) discussed the Winsorized mean

$$W_{m,n} = \frac{1}{m+1}\left[\sum_{i=1}^{m-1} X_{i:n} + (n-m+1)X_{m:n}\right], \qquad (15.1.1)$$

under the exchangeable single-outlier model. Joshi (1972) computed the optimal value m^* of m and relative efficiency for the Winsorized mean in the presence of a single outlier. Kale (1975) discussed the trimmed mean

$$T_{m,n} = \frac{1}{m}\sum_{i=1}^{m} X_{i:n}, \qquad (15.1.2)$$

and compared the efficiency of $T_{n-1,n}$ and $W_{n-1,n}$ when a single outlier is present in the sample. And Chikkagoudar and Kunchur (1980) proposed the following estimator of θ,

$$\widetilde{\theta} = \frac{1}{n}\sum_{i=1}^{n}\left[1 - \frac{2i}{n(n+1)}\right] X_{i:n}, \qquad (15.1.3)$$

and compared it with the Winsorized mean under the single-outlier model.

[1]McMaster University, Hamilton, Ontario, Canada

Most of these developments have been reviewed by Professor Ursula Gather in Chapter 14; for an elaborate treatment of this topic, interested readers may also refer to the recent book by Barnett and Lewis (1994). A great emphasis has been placed in literature on the single-outlier model. Even when the multiple-outlier situation was addressed, the discussion was inevitably on the behaviour of the proposed estimation and testing methods when one, two or no outliers are present. Only recently, the work of Balakrishnan (1994) made it possible to discuss the efficient estimation of the exponential mean in the presence of multiple outliers.

In this chapter, we discuss the results of Balakrishnan (1994) and some other recent papers which use these results to study the efficiency of various estimators of the exponential mean in the presence of multiple outliers and to determine optimal estimators under multiple-outlier models.

In Section 2 we present the recurrence relations, derived by Balakrishnan (1994), satisfied by the single and product moments of order statistics from a sample of I.NI.D. (independent and non-identically distributed) exponential random variables, and how these can be used to address the problem of efficient estimation of the mean under the multiple-outlier model. In Section 3 we discuss various statistics that have been proposed as estimators of the exponential mean, and in Section 4 we will see, from tables of the MSE (mean square error), that the optimal estimator depends on the sample size, the number of outliers, and how pronounced these outliers are. In Section 5 we explain how to efficiently estimate the exponential mean based on a sample of independent exponential random variables which possibly contains one or more outliers. Two illustrative examples are presented in Section 6. Finally, in Section 7 extensions to the estimation of location and scale parameters of the double exponential distribution (when possibly multiple outliers are present in the sample) are briefly discussed.

15.2 The Recurrence Relations

Suppose that $X_1, X_2, \ldots, X_n$ are independent random variables, with each X_i having an exponential distribution with mean θ_i,

$$f_i(x;\theta_i) = \frac{1}{\theta_i}\, e^{-x/\theta_i}, \qquad x>0,\ \theta_i>0,\ i=1,2,\ldots,n\ . \tag{15.2.1}$$

Let $X_{1:n} \le X_{2:n} \le \cdots \le X_{n:n}$ denote the order statistics obtained by arranging the n X_i's in increasing order of magnitude.

Let us denote the single moments $E(X_{r:n}^k)$ by $\mu_{r:n}^{(k)}$, $1 \le r \le n$ and $k = 1, 2, \ldots$, and the product moments $E(X_{r:n}X_{s:n})$ by $\mu_{r,s:n}$ for $1 \le r < s \le n$. Let us also use $\mu_{r:n-1}^{[i](k)}$ and $\mu_{r,s:n-1}^{[i]}$ to denote the single and product moments of order statistics arising from $n-1$ variables obtained by deleting X_i from the original n variables $X_1, X_2, \ldots, X_n$.

Recently, by making use of the differential equations

$$f_i(x) = \frac{1}{\theta_i}\,\{1 - F_i(x)\}, \qquad i = 1, 2, \ldots, n\ , \tag{15.2.2}$$

and permanent expressions for the density and joint density of order statistics from I.NI.D. exponential random variables, Balakrishnan (1994) derived the following recurrence relations for single moments:

(a) For $n = 1, 2, \ldots$ and $k = 0, 1, 2, \ldots$,

$$\mu_{1:n}^{(k+1)} = \frac{k+1}{\left(\sum_{i=1}^{n} \frac{1}{\theta_i}\right)} \mu_{1:n}^{(k)} , \tag{15.2.3}$$

(b) For $2 \leq r \leq n$ and $k = 0, 1, 2, \ldots$,

$$\mu_{r:n}^{(k+1)} = \frac{1}{\left(\sum_{i=1}^{n} \frac{1}{\theta_i}\right)} \left\{ (k+1)\,\mu_{r:n}^{(k)} + \sum_{i=1}^{n} \frac{1}{\theta_i}\, \mu_{r-1:n-1}^{[i](k+1)} \right\} ; \tag{15.2.4}$$

and the following four recurrence relations for product moments:

(c) For $n = 2, 3, \ldots$,

$$\mu_{1,2:n} = \frac{1}{\left(\sum_{i=1}^{n} \frac{1}{\theta_i}\right)} \left\{ \mu_{1:n} + \mu_{2:n} \right\} , \tag{15.2.5}$$

(d) For $2 \leq r \leq n-1$,

$$\mu_{r,r+1:n} = \frac{1}{\left(\sum_{i=1}^{n} \frac{1}{\theta_i}\right)} \left\{ (\mu_{r:n} + \mu_{r+1:n}) + \sum_{i=1}^{n} \frac{1}{\theta_i}\, \mu_{r-1,r:n-1}^{[i]} \right\} , \tag{15.2.6}$$

(e) For $3 \leq s \leq n$,

$$\mu_{1,s:n} = \frac{1}{\left(\sum_{i=1}^{n} \frac{1}{\theta_i}\right)} \left\{ \mu_{1:n} + \mu_{s:n} \right\} , \tag{15.2.7}$$

and

(f) For $2 \leq r < s \leq n$ and $s - r \geq 2$,

$$\mu_{r,s:n} = \frac{1}{\left(\sum_{i=1}^{n} \frac{1}{\theta_i}\right)} \left\{ (\mu_{r:n} + \mu_{s:n}) + \sum_{i=1}^{n} \frac{1}{\theta_i}\, \mu_{r-1,s-1:n-1}^{[i]} \right\} . \tag{15.2.8}$$

These recurrence relations will enable one to compute all the single and product moments of order statistics from I.NI.D. exponential random variables in a simple recursive manner.

The recurrence relations for the multiple-outlier model are obtained as a special case of the I.NI.D. relations by taking

$$\theta_1 = \theta_2 = \cdots = \theta_{n-p} = \theta$$

and

$$\theta_{n-p+1} = \theta_{n-p+2} = \cdots = \theta_n = \tau > \theta .$$

Here, we have that $X_1, X_2, \ldots, X_{n-p}$ are independent observations from an exponential distribution with mean θ,

$$f(x;\theta) = \frac{1}{\theta}\, e^{-x/\theta}\,, \qquad x > 0,\ \theta > 0\,, \tag{15.2.9}$$

and $X_{n-p+1}, \ldots, X_n$ arise from the same distribution but with mean $\tau > \theta$,

$$g(x;\tau) = \frac{1}{\tau}\, e^{-x/\tau}\,, \qquad x > 0;\ \tau > 0\,. \tag{15.2.10}$$

This special case of the I.NI.D. exponential model is known as a multiple-outlier exponential model with a slippage of p observations, as pointed out in Chapter 14. In this case, the p outlying observations have a mean τ that has slipped to the right of the remaining observations whose mean is θ; see Barnett and Lewis (1994). This specific multiple-outlier model was introduced by David (1979).

Let $\mu_{r:n}^{(k)}[p]$ $(1 \le r \le n)$, $\mu_{r,s:n}[p]$ $(1 \le r < s \le n)$ denote the single and product moments of order statistics from the above p-outlier model. Then recurrence relations (a) to (f) reduce to the following:

(a) For $n \ge 1$ and $k = 0, 1, 2, \ldots,$

$$\mu_{1:n}^{(k+1)}[p] = \frac{k+1}{\left[\frac{n-p}{\theta} + \frac{p}{\tau}\right]}\, \mu_{1:n}^{(k)}[p]\,; \tag{15.2.11}$$

(b) For $2 \le r \le n$ and $k = 0, 1, 2, \ldots,$

$$\mu_{r:n}^{(k+1)}[p] = \frac{1}{\left[\frac{n-p}{\theta} + \frac{p}{\tau}\right]} \left\{ (k+1)\mu_{r:n}^{(k)}[p] + \frac{n-p}{\theta}\, \mu_{r-1:n-1}^{(k+1)}[p] + \frac{p}{\tau}\, \mu_{r-1:n-1}^{(k+1)}[p-1] \right\}; \tag{15.2.12}$$

(c) For $n \ge 2$,

$$\mu_{1,2:n}[p] = \frac{1}{\left[\frac{n-p}{\theta} + \frac{p}{\tau}\right]} \left\{ \mu_{1:n}[p] + \mu_{2:n}[p] \right\}; \tag{15.2.13}$$

(d) For $2 \le r \le n-1$,

$$\mu_{r,r+1:n}[p] = \frac{1}{\left[\frac{n-p}{\theta} + \frac{p}{\tau}\right]} \left\{ \mu_{r:n}[p] + \mu_{r+1:n}[p] + \frac{n-p}{\theta}\, \mu_{r-1,r:n-1}[p] + \frac{p}{\tau}\, \mu_{r-1,r:n-1}[p-1] \right\}; \tag{15.2.14}$$

(e) For $3 \le s \le n$,

$$\mu_{1,s:n}[p] = \frac{1}{\left[\frac{n-p}{\theta} + \frac{p}{\tau}\right]} \left\{ \mu_{1:n}[p] + \mu_{s:n}[p] \right\}; \tag{15.2.15}$$

(f) For $2 \le r < s \le n$ and $s - r \ge 2$,

$$\mu_{r,s:n}[p] = \frac{1}{\left[\frac{n-p}{\theta} + \frac{p}{\tau}\right]} \left\{ \mu_{r:n}[p] + \mu_{s:n}[p] + \frac{n-p}{\theta}\, \mu_{r-1,s-1:n-1}[p] + \frac{p}{\tau}\, \mu_{r-1,s-1:n-1}[p-1] \right\}. \quad (15.2.16)$$

It is these recurrence relations that Balakrishnan (1994) derived that will allow one to compute all the single moments and product moments of order statistics from the multiple-outlier exponential model in a simple recursive manner. Hence, one can finally deal with the elusive problem of efficient estimation of the mean of the exponential distribution in the presence of multiple outliers.

It should be noted that the above recurrence relations which were derived as a special case of the I.NI.D. recursion relations can be proved directly by starting with the density and joint density functions of order statistics from the multiple-outlier model. These density functions can be obtained by using standard multinomial arguments, and they avoid the use of permanents. This alternative more direct approach is discussed in Childs and Balakrishnan (1995). They note that their direct approach might be applicable in different situations where the indirect approach used by Balakrishnan (1994) cannot be used. This would be the case if the more general I.NI.D. recursion relations do not hold, but the multiple-outlier ones do.

15.3 The Estimators

Balakrishnan (1994) used his recursive numerical algorithm to discuss, as estimators of θ, the Winsorized mean,

$$W_{m,n} = \frac{1}{m+1} \left[\sum_{i=1}^{m-1} X_{i:n} + (n-m+1) X_{m:n} \right], \quad (15.3.1)$$

the Trimmed mean,

$$T_{m,n} = \frac{1}{m} \sum_{i=1}^{m} X_{i:n}, \quad (15.3.2)$$

and the estimator of Chikkagoudar and Kunchur (1980),

$$\widetilde{\theta} = \frac{1}{n} \sum_{i=1}^{n} \left[1 - \frac{2i}{n(n+1)} \right] X_{i:n}, \quad (15.3.3)$$

and provided tables of relative efficiencies, bias, and mean square error.

He found that the Chikkagoudar-Kunchur estimator performed well for larger values of $\xi = \frac{\theta}{\tau}$ which is when the outliers are not very pronounced, but performed poorly for small values of ξ, which is when the outlying observations have a much larger mean than the regular observations. This poor performance for small values of ξ is due to the fact that the Chikkagoudar-Kunchur estimator does not give a small enough weight to the larger observations, which for small ξ are likely to be the outliers. On the other hand, the trimmed mean performed very well for small values of ξ due to the fact that the trimmed mean for values of $m < n$ does not include the observations that are most likely to be the outliers; however, this higher protection (for small values of ξ) is achieved by the trimmed mean at a higher premium (for large values of ξ).

A modification of the Chikkagoudar-Kunchur estimator,

$$\widetilde{\theta}_{(k)} = \frac{1}{n} \sum_{i=1}^{n} \left[1 - \frac{(k+1)(i+k-1)^{(k)}}{(n+k)^{(k+1)}} \right] X_{i:n} \ , \ k = 1, 2, \ldots, \tag{15.3.4}$$

where $a^{(k)} = a(a-1)\cdots(a-k+1)$, was introduced by Balakrishnan and Barnett (1994). This estimator, which reduces to the Chikkagoudar-Kunchur estimator when $k = 1$, gives less weight for increasing k to the larger observations as compared with the smaller observations. Thus, this estimator can be expected to be an improvement of the Chikkagoudar-Kunchur estimator for small values of ξ. In fact, Balakrishnan and Barnett (1994) found, by examining tables of bias and mean square error that the estimators $\widetilde{\theta}_{(k)}, k = 2, 3, \ldots, 7$ are indeed an improvement over the Chikkagoudar-Kunchur estimator when p is large and/or ξ is small.

Another extension of the Chikkagoudar-Kunchur estimator is discussed by Childs and Balakrishnan (1995). The Chikkagoudar-Kunchur estimator is a linear combination

$$\widetilde{\theta} = \sum_{i=1}^{n} \beta_i \widetilde{\theta}_i \tag{15.3.5}$$

of separate estimators

$$\widetilde{\theta}_i = \overline{X} - \frac{X_{i:n}}{n} \tag{15.3.6}$$

obtained by removing one observation at a time from $\overline{X}$, with weights

$$\beta_i = \frac{2i}{n(n+1)} \ . \tag{15.3.7}$$

Each estimator $\widetilde{\theta}_i$ is assigned a weight i, and $\left(\sum_{i=1}^{n} i\right)^{-1} = \frac{2}{n(n+1)}$ is the normalization factor. As a generalization of this estimator, Childs and Balakrishnan (1995) have considered a linear combination

$$\widetilde{\theta}^{(k)} = \sum_{i_1=1}^{n-k+1} \sum_{i_2=i_1+1}^{n-k+2} \cdots \sum_{i_k=i_{k-1}+1}^{n} \beta_{i_1,i_2,\ldots,i_k} \widetilde{\theta}_{i_1,i_2,\ldots,i_k} \tag{15.3.8}$$

of separate estimators

$$\widetilde{\theta}_{i_1,i_2,\ldots,i_k} = \overline{X} - \frac{X_{i_1:n}}{n} - \frac{X_{i_2:n}}{n} - \cdots - \frac{X_{i_k:n}}{n} \tag{15.3.9}$$

obtained by removing k observations at a time from $\overline{X}$, with weights

$$\begin{aligned} \beta_{i_1,i_2,\ldots,i_k} &= \frac{2(k-1)!(i_1+i_2+\cdots+i_k)}{n(n^2-1)(n-2)(n-3)\cdots(n-k+1)} \\ &= \frac{2(i_1+i_2+\cdots+i_k)}{k(n+1)\binom{n}{k}} \ . \end{aligned} \tag{15.3.10}$$

In this case, each estimator $\widetilde{\theta}_{i_1,i_2,\ldots,i_k}$ is assigned a weight $i_1 + i_2 + \cdots + i_k$ and

$$\left(\sum_{i_1=1}^{n-k+1} \sum_{i_2=i_1+1}^{n-k+2} \cdots \sum_{i_k=i_{k-1}+1}^{n} (i_1 + i_2 + \cdots + i_k) \right)^{-1} = \frac{2}{k(n+1)\binom{n}{k}}$$

is the normalization factor. The resulting estimator,

$$\begin{aligned} \widetilde{\theta}^{(k)} &= \frac{1}{n} \sum_{i=1}^{n} \left[1 - \frac{2i(n-k) + (k-1)n(n+1)}{n(n^2-1)} \right] X_{i:n} \\ &= \frac{1}{n} \sum_{i=1}^{n} \left[1 - \frac{k-1}{n-1} - \frac{2i(n-k)}{n(n^2-1)} \right] X_{i:n} \qquad (15.3.11) \end{aligned}$$

which gives less weight to the larger observations for $k > 1$ as compared with the Chikkagoudar-Kunchur estimator and which reduces to the Chikkagoudar-Kunchur estimator when $k = 1$, can be expected to be more efficient for increasing values of p and/or decreasing values of ξ.

15.4 Optimal Estimators of θ Under the Multiple-Outlier Model

For each estimator discussed in the previous section, the question arises as to which form of that estimator is optimal. For example, for a given value of n, p and ξ, which trimmed mean $T_{m,n}$, $m = 1, 2, \ldots, n$ is the most efficient? Also, of the four estimators discussed, which optimal estimator is the most efficient for a given value of n, p and ξ? Table 15.1 is designed to answer these questions. It presents, for $n = 5(5)20$, $p = 1(1)4$, and $\xi = 0.20(0.05)0.50$, 0.75, the optimal form of each estimator and its bias and mean square error. Also included for comparison is the full sample Winsorized mean $W_{n,n} = T_n$. More extensive tables can be found in Childs and Balakrishnan (1995) who have included multiple forms of each estimator, more outliers (up to 6), and higher sample sizes (up to 50).

From Table 15.1 we see that for larger values of ξ, starting at around $\xi = 0.35$ or 0.40, the estimator $\widetilde{\theta}^{(k^*)}$ is usually the most efficient for every value of n and p, although occasionally $\widetilde{\theta}_{(k^*)}$ has a slightly smaller MSE. For smaller values of ξ, it is usually the trimmed mean that is the most efficient for each value of n and p.

Similar observations hold for higher sample sizes and more outliers as discussed in Childs and Balakrishnan (1995).

15.5 Efficient Estimation of θ Under the Multiple-Outlier Model

We have seen in Section 15.4 that the optimal estimator of θ depends on the number of outliers and how pronounced these outliers are, i.e., the value of ξ. But when one has a sample of independent exponential random variables which possibly contains outliers, it is generally not known beforehand how many outliers there are or, even if there are outliers, what the value of ξ is. There are various methods available for determining the number of outliers p, perhaps the simplest being the $Q-Q$ plot. Once p has been determined, how does one estimate ξ? One way of estimating ξ for the multiple-outlier model, which is an extension of the method given by Joshi (1972) for the single-outlier model, is discussed in Balakrishnan (1994). It is as follows.

If the number of outliers p is known, then the trimmed mean based on the first $n-p$ observations $T_{n-p,n}$ can be used as an initial estimate of θ, i.e., $T_{n-p,n} \approx \theta$. Further, with p outliers in the sample, the expected value of $n\overline{X} = \sum_{i=1}^{n} X_i$ is $(n-p)\theta + p\tau$, i.e., $n\overline{X} \approx (n-p)\theta + p\tau$. Hence the ratio

$$\frac{pT_{n-p,n}}{n\overline{X} - (n-p)T_{n-p,n}} \left(\approx \frac{p\theta}{(n-p)\theta + p\tau - (n-p)\theta} = \frac{\theta}{\tau} = \xi \right) \tag{15.5.1}$$

can be used as an initial estimate of ξ. Once ξ is estimated, Table 15.1 can be used to determine the optimal value m^* of m for the trimmed mean $T_{m,n}$. With this optimal value m^*, $T_{m^*,n}$ can be used in place of $T_{n-p,n}$ in (15.5.1) above to obtain a revised estimate of ξ. This iterative process can be repeated until the value of ξ becomes stable. One could also use any of the other three estimators discussed in Section 15.3 in place of the trimmed mean in this process. Once the value of ξ is estimated, Table 15.1 can then be used to determine which estimator is the most efficient, and that estimator can be used to estimate θ and its mean square error can be determined from Table 15.1.

15.6 Examples

1. Consider the following fifteen observations,

1.89, 1.97, 2.01, 2.21, 2.32, 2.43, 3.29, 4.32, 4.89, 5.64, 6.89, 7.30, 12.34, 12.87, 15.64.

A $Q-Q$ plot (see Figure 15.1) reveals that the three largest observations may be outliers. Thus, according to the method described above, with $p = 3$, we use $T_{12,15}$ as an initial estimator of θ. From (15.5.1) we get $\xi \approx 0.276$. According to Table 15.1, $T_{13,15}$ is the optimal trimmed mean for this value of ξ. Now we use $T_{13,15}$ in (15.5.1) to get $\xi \approx 0.403$. For this value of ξ, Table 15.1 shows that again $T_{13,15}$ is the optimal trimmed mean; hence the iterations stop here and we use 0.403 as our estimate of ξ. For this value of ξ we see from Table 15.1 that $\widetilde{\theta}^{(4)}$ is the most efficient estimator with MSE approximately $0.0741\,\theta^2$. The relative efficiency is 2.27, 1.62, 1.07, and 1.28 compared with the full sample Winsorized mean, the optimal winsorized mean, the optimal trimmed mean and the optimal $\widetilde{\theta}_{(k)}$, respectively. The estimator of θ based on $\widetilde{\theta}^{(4)}$ is $\theta \approx \widetilde{\theta}^{(4)} = 4.09$ and its root mean square error is $4.09\sqrt{0.0741} = 1.113$.

2. Consider the following data,

0.27, 0.30, 0.52, 0.53, 0.64, 1.04, 1.25, 1.28, 1.35, 2.07, 2.16, 3.43, 4.71, 5.63, 5.9, 6.05, 6.60, 6.70, 12.89, 15.39.

A $Q-Q$ plot (see Figure 15.2) reveals that the two largest observations may be outliers. Using the method described above we get, using $T_{18,20}$ as an initial estimate of θ, $\xi \approx 0.198$. For this value of ξ we see from Table 15.1 that in fact $T_{18,20}$ is the optimal trimmed mean and so the iterative process stops at the first step. When the outliers are this pronounced we would naturally expect the trimmed mean to be the optimal estimator since it does not include the largest observations which are quite likely to be the outliers. In fact, Table 15.1 confirms that the optimal trimmed mean is the most efficient estimator for this small value of ξ, with MSE approximately $0.0576\,\theta^2$. The relative efficiency of the trimmed mean as compared with the other optimal estimators can also be seen from Table 15.1. The relative efficiency is 4.6, 1.44, 2.49, and 1.38 as compared with T_n, W_{m^*}, $\widetilde{\theta}_{(k^*)}$, and $\widetilde{\theta}^{(k^*)}$, respectively. The estimate of θ based on the trimmed mean is $\theta \approx T_{18,20} = 2.80$, and its root mean square error is $2.80\sqrt{0.0576} = 0.672$.

15.7 Extensions to the Double Exponential Model

The recurrence relations for the exponential model discussed in Section 15.2 can also be used to discuss the efficient estimation of the location and scale parameters of the double exponential model in the presence of multiple outliers.

Balakrishnan (1989) derived equations relating the moments and product moments of two sets of I.NI.D. random variables, whose distribution functions and density functions are related by the equations,

$$F_i^*(x) = 2F_i(x) - 1 \text{ and } f_i^*(x) = 2f_i(x), \; x \geq 0, \; i = 1, 2, \ldots, n\,. \tag{15.7.1}$$

For the special case of the multiple-outlier model, Balakrishnan's (1989) results reduce to the following equations which can be used to relate the moments of order statistics from the multiple-outlier exponential model ($\nu_{r:n}^{(k)}[p]$ and $\nu_{r,s:n}[p]$), to those of the multiple-outlier double exponential model ($\mu_{r:n}^{(k)}[p]$ and $\mu_{r,s:n}[p]$). Bear in mind that the outliers in this case are the scale-outliers (location parameter being the same for the entire sample).

For $1 \leq r \leq n$ and $k = 1, 2, \ldots$,

$$\begin{aligned} \mu_{r:n}^{(k)}[p] &= 2^{-n}\Bigg\{\sum_{t=0}^{p}\Bigg(\sum_{i=p-t}^{\min(r-1,n-t)} \binom{p}{t}\binom{n-p}{n-i-t}\nu_{r-i:n-i}^{(k)}[t] \\ &\quad + (-1)^k \sum_{i=\max(r,t)}^{n-p+t} \binom{p}{t}\binom{n-p}{i-t}\nu_{i-r+1:i}^{(k)}[t]\Bigg)\Bigg\}\,, \end{aligned} \tag{15.7.2}$$

and for $1 \leq r < s \leq n$,

$$\begin{aligned} \mu_{r,s:n}[p] &= 2^{-n}\Bigg\{\sum_{t=0}^{p}\Bigg(\sum_{i=p-t}^{\min(r-1,n-t)} \binom{p}{t}\binom{n-p}{n-i-t}\nu_{r-i,s-i:n-i}[t] \\ &\quad - \sum_{i=\max(r,p-t)}^{\min(s-1,n-t)} \binom{p}{t}\binom{n-p}{n-i-t}\nu_{s-i:n-i}[t]\;\nu_{i-r+1:i}[p-t] \\ &\quad + \sum_{i=\max(s,t)}^{n-p+t} \binom{p}{t}\binom{n-p}{i-t}\nu_{i-s+1,i-r+1:i}[t]\Bigg)\Bigg\}\,. \end{aligned} \tag{15.7.3}$$

These equations, for $p = 1$, reduce to those used by Balakrishnan and Ambagaspitiya (1988) to study the efficiency of various estimators of the location and scale parameters of the double exponential distribution when a single scale-outlier is present in samples of size up to 20. More recently, Childs and Balakrishnan (1995) extended these results using the above equations to include sample sizes up to 30 and to accommodate up to 5 outliers.

References

Balakrishnan, N. (1989). Recurrence relations among moments of order statistics from two related sets of independent and non-identically distributed random variables, *Annals of the Institute of Statistical Mathematics*, **41**, 323-329.

Balakrishnan, N. (1994). Order statistics from non-identical exponential random variables and some applications (with discussions), *Computational Statistics & Data Analysis*, **18**, 203-253.

Balakrishnan, N. and Ambagaspitiya, R.S. (1988). Relationships among moments of order statistics from two related outlier models and some applications, *Communications in Statistics - Theory and Methods*, **17**, 2327-2341.

Balakrishnan, N. and Barnett, V. (1994). Outlier-robust estimation of the mean of an exponential distribution, *Submitted for publication.*

Barnett, V. and Lewis, T. (1994). *Outliers in Statistical Data*, Third edition, John Wiley & Sons, Chichester.

Chikkagoudar, M.S. and Kunchur, S.H. (1980). Estimation of the mean of an exponential distribution in the presence of an outlier, *Canadian Journal of Statistics*, **8**, 59-63.

Childs, A. and Balakrishnan, N. (1995). Some extensions in the robust estimation of parameters of exponential and double exponential distributions in the presence of multiple outliers, In *Handbook of Statistics - 15: Robust Methods* (Eds., C.R. Rao and G.S. Maddala), North-Holland, Amsterdam (To appear).

David, H.A. (1979). Robust estimation in the presence of outliers, In *Robustness in Statistics* (Eds., R.L. Launer and G.N. Wilkinson), pp. 61-74, Academic Press, New York.

Joshi, P.C. (1972). Efficient estimation of the mean of an exponential distribution when an outlier is present, *Technometrics*, **14**, 137-143.

Kale, B.K. (1975). Trimmed means and the method of maximum likelihood when spurious observations are present, In *Applied Statistics* (Ed., R.P. Gupta), pp. 177-185, North-Holland, Amsterdam.

Kale, B.K. and Sinha, S.K. (1971). Estimation of expected life in the presence of an outlier observation, *Technometrics*, **13**, 755-759.

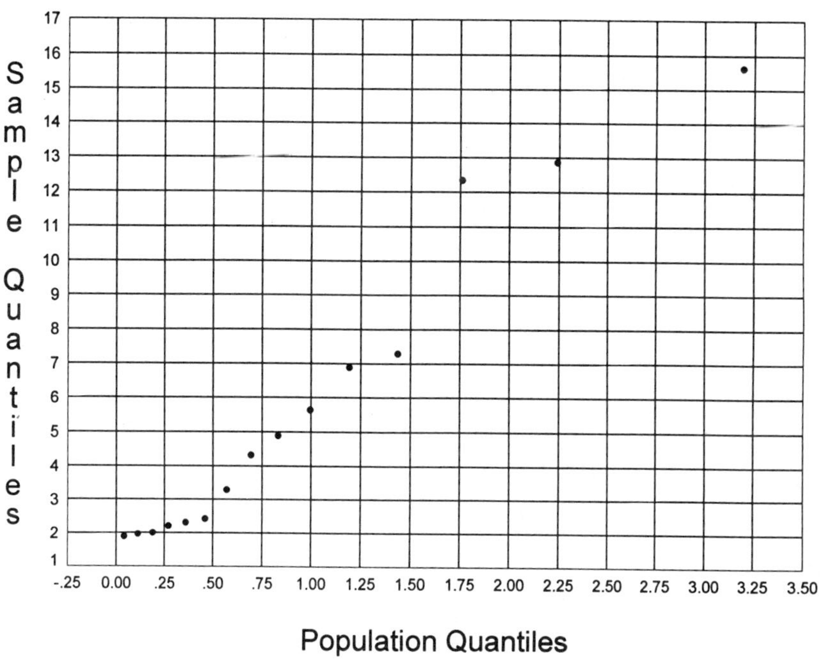

Figure 15.1: $Q - Q$ plot for Example 1

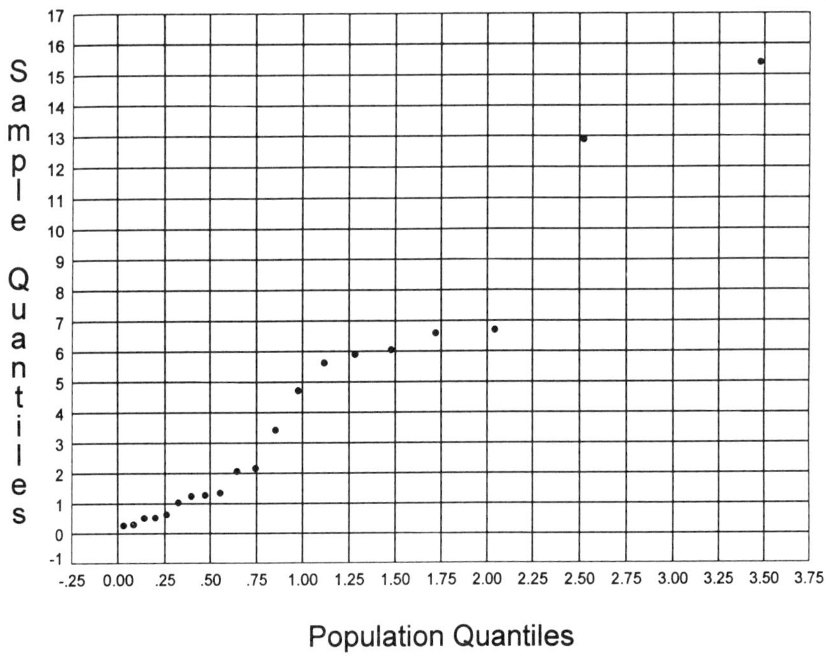

Figure 15.2: $Q - Q$ plot for Example 2

Table 15.1: Bias and mean square error of optimal estimators of θ for $n = 5(5)20$, $p = 1(1)4$, $\xi = 0.20(0.05)0.50$ and 0.75

n=5								
ξ		p=1				p=2		
0.20			Bias	MSE			Bias	MSE
	T_n		0.5000	1.0556			1.1667	2.8333
	W_{m^*}	4	0.0589	0.2958		2	0.0294	0.5359
	T_{m^*}	4	-0.1163	0.2144		3	-0.2214	0.2595
	$\theta_{(k^*)}$	7	-0.0303	0.2346		7	0.4143	0.7010
	$\theta^{(k^*)}$	3	-0.3578	0.2650		3	-0.0733	0.2618
0.25								
	T_n		0.3333	0.6667			0.8333	1.6667
	W_{m^*}	4	0.0293	0.2764		2	-0.0081	0.4965
	T_{m^*}	4	-0.1386	0.2106		3	-0.2579	0.2584
	$\theta_{(k^*)}$	7	-0.1057	0.1993		7	0.2303	0.4295
	$\theta^{(k^*)}$	2	-0.1380	0.2364		3	-0.2113	0.2164
0.30								
	T_n		0.2222	0.4691			0.6111	1.0741
	W_{m^*}	4	0.0027	0.2606		2	-0.0423	0.4641
	T_{m^*}	4	-0.1589	0.2080		3	-0.2898	0.2602
	$\theta_{(k^*)}$	7	-0.1581	0.1866		7	0.1054	0.3021
	$\theta^{(k^*)}$	2	-0.2059	0.2101		3	-0.3037	0.2177
0.35								
	T_n		0.1429	0.3583			0.4524	0.7415
	W_{m^*}	4	-0.0214	0.2477		3	0.0857	0.4132
	T_{m^*}	4	-0.1774	0.2064		3	-0.3181	0.2639
	$\theta_{(k^*)}$	7	-0.1972	0.1830		7	0.0142	0.2380
	$\theta^{(k^*)}$	2	-0.2547	0.2023		2	-0.0548	0.2222
0.40								
	T_n		0.0833	0.2917			0.3333	0.5417
	W_{m^*}	4	-0.0432	0.2373		3	0.0425	0.3739
	T_{m^*}	4	-0.1942	0.2056		4	-0.0169	0.2606
	$\theta_{(k^*)}$	5	-0.1980	0.1816		7	-0.0556	0.2050
	$\theta^{(k^*)}$	2	-0.2914	0.2026		2	-0.1296	0.1948
0.45								
	T_n		0.0370	0.2497			0.2407	0.4156
	W_{m^*}	4	-0.0631	0.2288		3	0.0041	0.3434
	T_{m^*}	4	-0.2097	0.2055		4	-0.0636	0.2370
	$\theta_{(k^*)}$	3	-0.1810	0.1797		7	-0.1111	0.1886
	$\theta^{(k^*)}$	1	-0.0936	0.1936		2	-0.1881	0.1848

Table 15.1 (continued)

n=5 (Continued)												
ξ			p=1				p=2				p=3	
0.50			Bias	MSE			Bias	MSE			Bias	MSE
	T_n		0.0000	0.2222			0.1667	0.3333				
	W_{m^*}	4	-0.0813	0.2219		4	0.0671	0.3059				
	T_{m^*}	4	-0.2240	0.2058		4	-0.1030	0.2223				
	$\theta_{(k^*)}$	2	-0.1737	0.1776		7	-0.1564	0.1815				
	$\theta^{(k^*)}$	1	-0.1244	0.1827		2	-0.2350	0.1841				
0.75												
	T_n		-0.1111	0.1728			-0.0556	0.1852				
	W_{m^*}	5	-0.1111	0.1728		5	-0.0556	0.1852				
	T_{m^*}	4	-0.2808	0.2120		4	-0.2379	0.2064				
	$\theta_{(k^*)}$	1	-0.2184	0.1719		1	-0.1698	0.1695				
	$\theta^{(k^*)}$	1	-0.2184	0.1719		1	-0.1698	0.1695				
n=10												
0.20												
	T_n		0.2727	0.3554			0.6364	0.8843			1.0000	1.6777
	W_{m^*}	8	0.0219	0.1336		6	0.0871	0.2077		5	0.1805	0.3164
	T_{m^*}	9	-0.0832	0.1035		8	-0.1533	0.1181		7	-0.2170	0.1409
	$\theta_{(k^*)}$	7	0.0286	0.1438		7	0.2990	0.3351		7	0.5868	0.7118
	$\theta^{(k^*)}$	4	-0.2055	0.1475		5	-0.1517	0.1474		6	-0.1713	0.1424
0.25												
	T_n		0.1818	0.2397			0.4545	0.5372			0.7273	0.9835
	W_{m^*}	9	0.0456	0.1294		7	0.1094	0.1931		5	0.1434	0.2864
	T_{m^*}	9	-0.0986	0.1033		8	-0.1781	0.1211		8	-0.0433	0.1285
	$\theta_{(k^*)}$	7	-0.0284	0.1168		7	0.1773	0.2135		7	0.3944	0.4146
	$\theta^{(k^*)}$	3	-0.1371	0.1255		4	-0.0924	0.1333		5	-0.1027	0.1300
0.30												
	T_n		0.1212	0.1809			0.3333	0.3609			0.5455	0.6309
	W_{m^*}	9	0.0273	0.1230		7	0.0826	0.1788		6	0.1655	0.2604
	T_{m^*}	9	-0.1123	0.1034		8	-0.1997	0.1247		8	-0.0904	0.1208
	$\theta_{(k^*)}$	7	-0.0674	0.1052		7	0.0948	0.1556		7	0.2647	0.2692
	$\theta^{(k^*)}$	2	-0.0628	0.1173		4	-0.1661	0.1228		5	-0.1953	0.1264

Table 15.1 (continued)

n=10 (Continued)										
ξ		p=1				p=2			p=3	
0.35			Bias	MSE		Bias	MSE		Bias	MSE
	T_n		0.0779	0.1479		0.2468	0.2619		0.4156	0.4330
	W_{m^*}	9	0.0109	0.1179	8	0.0971	0.1671	6	0.1285	0.2340
	T_{m^*}	9	-0.1247	0.1038	9	-0.0124	0.1174	8	-0.1285	0.1189
	$\theta_{(k^*)}$	7	-0.0960	0.1003	7	0.0347	0.1258	7	0.1707	0.1915
	$\theta^{(k^*)}$	2	-0.0978	0.1071	3	-0.0887	0.1130	4	-0.1139	0.1129
0.40										
	T_n		0.0455	0.1281		0.1818	0.2025		0.3182	0.3140
	W_{m^*}	9	-0.0036	0.1138	8	0.0696	0.1534	7	0.1372	0.2105
	T_{m^*}	9	-0.1359	0.1044	9	-0.0425	0.1104	8	-0.1603	0.1202
	$\theta_{(k^*)}$	7	-0.1180	0.0984	7	-0.0112	0.1100	7	0.0993	0.1477
	$\theta^{(k^*)}$	2	-0.1242	0.1027	3	-0.1350	0.1075	4	-0.1736	0.1124
0.45										
	T_n		0.0202	0.1156		0.1313	0.1650		0.2424	0.2391
	W_{m^*}	9	-0.0167	0.1105	8	0.0453	0.1429	7	0.1018	0.1889
	T_{m^*}	9	-0.1460	0.1051	9	-0.0677	0.1065	8	-0.1877	0.1233
	$\theta_{(k^*)}$	4	-0.1046	0.0971	7	-0.0476	0.1016	7	0.0430	0.1225
	$\theta^{(k^*)}$	2	-0.1447	0.1011	2	-0.0527	0.1050	3	-0.0903	0.1035
0.50										
	T_n		0.0000	0.1074		0.0909	0.1405		0.1818	0.1901
	W_{m^*}	10	0.0000	0.1074	9	0.0512	0.1316	8	0.1016	0.1682
	T_{m^*}	9	-0.1552	0.1060	9	-0.0894	0.1045	9	-0.0183	0.1139
	$\theta_{(k^*)}$	3	-0.1054	0.0957	7	-0.0773	0.0974	7	-0.0027	0.1080
	$\theta^{(k^*)}$	1	-0.0564	0.0982	2	-0.0858	0.0993	3	-0.1338	0.1014
0.75										
	T_n		-0.0606	0.0927		-0.0303	0.0964		0.0000	0.1019
	W_{m^*}	10	-0.0606	0.0927	10	-0.0303	0.0964	10	0.0000	0.1019
	T_{m^*}	10	0.0333	0.1089	9	-0.1662	0.1071	9	-0.1413	0.1044
	$\theta_{(k^*)}$	1	-0.1125	0.0922	1	-0.0840	0.0922	2	-0.0834	0.0925
	$\theta^{(k^*)}$	1	-0.1125	0.0922	1	-0.0840	0.0922	1	-0.0555	0.0939

Table 15.1 (continued)

n=15														
ξ		p=1			p=2			p=3			p=4			
0.20			Bias	MSE		Bias	MSE		Bias	MSE		Bias	MSE	
	T_n		0.1875	0.1875		0.4375	0.4375		0.6875	0.8125		0.9375	1.3125	
	W_{m^*}	13	0.0301	0.0844	11	0.0952	0.1206	9	0.1463	0.1701	7	0.1819	0.2351	
	T_{m^*}	14	-0.0669	0.0687	13	-0.1220	0.0774	13	-0.0054	0.0852	12	-0.0692	0.0847	
	$\theta_{(k^*)}$	7	0.0345	0.0982	7	0.2333	0.2068	7	0.4390	0.4062	7	0.6508	0.7031	
	$\theta^{(k^*)}$	4	-0.1042	0.0954	6	-0.1149	0.1042	7	-0.0776	0.1053	8	-0.0738	0.1026	
0.25														
	T_n		0.1250	0.1328		0.3125	0.2734		0.5000	0.4844		0.6875	0.7656	
	W_{m^*}	13	0.0187	0.0819	11	0.0760	0.1133	9	0.1206	0.1564	8	0.1845	0.2126	
	T_{m^*}	14	-0.0790	0.0689	13	-0.1414	0.0799	13	-0.0484	0.0785	12	-0.1140	0.0845	
	$\theta_{(k^*)}$	7	-0.0102	0.0801	7	0.1409	0.1354	7	0.2965	0.2422	7	0.4560	0.4039	
	$\theta^{(k^*)}$	3	-0.0730	0.0834	5	-0.1005	0.0909	6	-0.0758	0.0929	7	-0.0763	0.0920	
0.30														
	T_n		0.0833	0.1050		0.2292	0.1901		0.3750	0.3177		0.5208	0.4878	
	W_{m^*}	14	0.0265	0.0790	12	0.0781	0.1060	10	0.1205	0.1430	8	0.1522	0.1918	
	T_{m^*}	14	-0.0897	0.0693	14	0.0045	0.0785	13	-0.0819	0.0767	13	0.0058	0.0865	
	$\theta_{(k^*)}$	7	-0.0406	0.0722	7	0.0784	0.1013	7	0.2004	0.1619	7	0.3250	0.2556	
	$\theta^{(k^*)}$	3	-0.1065	0.0771	4	-0.0721	0.0823	5	-0.0573	0.0851	6	-0.0620	0.0849	
0.35														
	T_n		0.0536	0.0894		0.1696	0.1433		0.2857	0.2242		0.4018	0.3320	
	W_{m^*}	14	0.0141	0.0762	12	0.0594	0.0996	11	0.1173	0.1308	9	0.1483	0.1716	
	T_{m^*}	14	-0.0992	0.0697	14	-0.0213	0.0736	13	-0.1093	0.0773	13	-0.0379	0.0791	
	$\theta_{(k^*)}$	7	-0.0628	0.0686	7	0.0330	0.0837	7	0.1309	0.1189	7	0.2305	0.1751	
	$\theta^{(k^*)}$	2	-0.0581	0.0719	4	-0.1162	0.0780	5	-0.1175	0.0800	5	-0.0382	0.0806	
0.40														
	T_n		0.0313	0.0801		0.1250	0.1152		0.2188	0.1680		0.3125	0.2383	
	W_{m^*}	14	0.0031	0.0739	13	0.0592	0.0931	12	0.1119	0.1197	10	0.1404	0.1532	
	T_{m^*}	14	-0.1077	0.0703	14	-0.0425	0.0709	13	-0.1326	0.0792	13	-0.0736	0.0766	
	$\theta_{(k^*)}$	7	-0.0797	0.0670	7	-0.0015	0.0742	7	0.0781	0.0946	7	0.1590	0.1287	
	$\theta^{(k^*)}$	2	-0.0776	0.0689	3	-0.0720	0.0723	4	-0.0790	0.0741	5	-0.0986	0.0753	
0.45														
	T_n		0.0139	0.0742		0.0903	0.0975		0.1667	0.1325		0.2431	0.1792	
	W_{m^*}	14	-0.0067	0.0722	13	0.0412	0.0876	12	0.0862	0.1089	11	0.1296	0.1363	
	T_{m^*}	14	-0.1154	0.0709	14	-0.0605	0.0697	14	-0.0017	0.0758	13	-0.1036	0.0769	
	$\theta_{(k^*)}$	4	-0.0701	0.0661	7	-0.0288	0.0691	7	0.0365	0.0805	7	0.1027	0.1010	
	$\theta^{(k^*)}$	2	-0.0928	0.0675	3	-0.1001	0.0704	3	-0.0375	0.0721	4	-0.0603	0.0717	

Table 15.1 (continued)

n=15 (Continued)													
ξ		p=1			p=2			p=3			p=4		
0.50			Bias	MSE		Bias	MSE		Bias	MSE		Bias	MSE
	T_n		0.0000	0.0703		0.0625	0.0859		0.1250	0.1094		0.1875	0.1406
	W_{m*}	15	0.0000	0.0703	14	0.0396	0.0820	13	0.0786	0.0991	12	0.1165	0.1211
	T_{m*}	14	-0.1224	0.0716	14	-0.0760	0.0692	14	-0.0271	0.0719	13	-0.1295	0.0788
	$\theta_{(k*)}$	3	-0.0708	0.0652	7	-0.0509	0.0665	7	0.0029	0.0723	7	0.0572	0.0843
	$\theta^{(k*)}$	1	-0.0362	0.0664	2	-0.0496	0.0677	3	-0.0714	0.0686	4	-0.1017	0.0703
0.75													
	T_n		-0.0417	0.0634		-0.0208	0.0651		0.0000	0.0677		0.0208	0.0712
	W_{m*}	15	-0.0417	0.0634	15	-0.0208	0.0651	15	0.0000	0.0677	15	0.0208	0.0712
	T_{m*}	15	0.0222	0.0706	14	-0.1313	0.0724	14	-0.1137	0.0705	14	-0.0959	0.0692
	$\theta_{(k*)}$	1	-0.0758	0.0631	1	-0.0558	0.0632	2	-0.0555	0.0634	4	-0.0613	0.0636
	$\theta^{(k*)}$	1	-0.0758	0.0631	1	-0.0558	0.0632	1	-0.0358	0.0642	2	-0.0861	0.0641
n=20													
0.20													
	T_n		0.1429	0.1202		0.3333	0.2653		0.5238	0.4830		0.7143	0.7732
	W_{m*}	18	0.0300	0.0612	15	0.0723	0.0832	13	0.1185	0.1118	11	0.1577	0.1484
	T_{m*}	19	-0.0568	0.0515	18	-0.1031	0.0576	18	-0.0135	0.0599	17	-0.0651	0.0606
	$\theta_{(k*)}$	7	0.0331	0.0727	7	0.1912	0.1434	7	0.3528	0.2688	7	0.5174	0.4516
	$\theta^{(k*)}$	4	-0.0654	0.0699	6	-0.0475	0.0795	8	-0.0680	0.0813	9	-0.0395	0.0828
0.25													
	T_n		0.0952	0.0884		0.2381	0.1701		0.3810	0.2925		0.5238	0.4558
	W_{m*}	18	0.0203	0.0595	16	0.0708	0.0784	14	0.1139	0.1035	12	0.1511	0.1352
	T_{m*}	19	-0.0669	0.0518	18	-0.1194	0.0596	18	-0.0471	0.0568	17	-0.1007	0.0618
	$\theta_{(k*)}$	7	-0.0036	0.0601	7	0.1164	0.0962	7	0.2386	0.1634	7	0.3629	0.2630
	$\theta^{(k*)}$	3	-0.0476	0.0615	5	-0.0513	0.0682	7	-0.0837	0.0709	8	-0.0672	0.0715
0.30													
	T_n		0.0635	0.0723		0.1746	0.1217		0.2857	0.1958		0.3968	0.2945
	W_{m*}	19	0.0238	0.0578	17	0.0687	0.0741	15	0.1079	0.0957	13	0.1424	0.1228
	T_{m*}	19	-0.0757	0.0521	19	-0.0031	0.0563	18	-0.0736	0.0565	18	-0.0068	0.0603
	$\theta_{(k*)}$	7	-0.0284	0.0544	7	0.0660	0.0736	7	0.1618	0.1117	7	0.2589	0.1696
	$\theta^{(k*)}$	3	-0.0747	0.0568	4	-0.0391	0.0618	6	-0.0803	0.0641	7	-0.0726	0.0649

Table 15.1 (continued)

n=20 (Continued)													
ξ			p=1			p=2			p=3			p=4	
0.35			Bias	MSE		Bias	MSE		Bias	MSE		Bias	MSE
	T_n		0.0408	0.0633		0.1293	0.0945		0.2177	0.1415		0.3061	0.2040
	W_{m^*}	19	0.0137	0.0559	17	0.0532	0.0699	16	0.1007	0.0884	14	0.1320	0.1113
	T_{m^*}	19	-0.0836	0.0525	19	-0.0233	0.0537	18	-0.0956	0.0575	18	-0.0407	0.0570
	$\theta_{(k^*)}$	7	-0.0464	0.0518	7	0.0294	0.0619	7	0.1063	0.0840	7	0.1840	0.1188
	$\theta^{(k^*)}$	2	-0.0408	0.0536	4	-0.0756	0.0574	5	-0.0660	0.0591	6	-0.0654	0.0602
0.40													
	T_n		0.0238	0.0578		0.0952	0.0782		0.1667	0.1088		0.2381	0.1497
	W_{m^*}	19	0.0048	0.0545	18	0.0501	0.0659	16	0.0804	0.0814	15	0.1204	0.1006
	T_{m^*}	19	-0.0906	0.0530	19	-0.0400	0.0525	19	0.0143	0.0582	18	-0.0686	0.0563
	$\theta_{(k^*)}$	7	-0.0601	0.0506	7	0.0017	0.0555	7	0.0642	0.0684	7	0.1273	0.0895
	$\theta^{(k^*)}$	2	-0.0562	0.0516	3	-0.0470	0.0542	4	-0.0450	0.0559	5	-0.0501	0.0568
0.45													
	T_n		0.0106	0.0544		0.0688	0.0679		0.1270	0.0883		0.1852	0.1154
	W_{m^*}	19	-0.0030	0.0534	18	0.0358	0.0626	17	0.0723	0.0750	16	0.1077	0.0907
	T_{m^*}	19	-0.0968	0.0535	19	-0.0541	0.0519	19	-0.0090	0.0547	18	-0.0923	0.0572
	$\theta_{(k^*)}$	4	-0.0527	0.0500	7	-0.0201	0.0521	7	0.0311	0.0593	7	0.0828	0.0721
	$\theta^{(k^*)}$	2	-0.0682	0.0507	3	-0.0697	0.0525	4	-0.0771	0.0540	4	-0.0296	0.0549
0.50													
	T_n		0.0000	0.0522		0.0476	0.0612		0.0952	0.0748		0.1429	0.0930
	W_{m^*}	20	0.0000	0.0522	19	0.0322	0.0591	18	0.0638	0.0691	17	0.0945	0.0818
	T_{m^*}	19	-0.1025	0.0540	19	-0.0664	0.0519	19	-0.0288	0.0528	19	0.0102	0.0569
	$\theta_{(k^*)}$	3	-0.0534	0.0493	7	-0.0378	0.0503	7	0.0044	0.0541	7	0.0468	0.0615
	$\theta^{(k^*)}$	1	-0.0266	0.0500	2	-0.0342	0.0511	3	-0.0467	0.0518	4	-0.0639	0.0525
0.75													
	T_n		-0.0317	0.0481		-0.0159	0.0491		0.0000	0.0506		0.0159	0.0527
	W_{m^*}	20	-0.0317	0.0481	20	-0.0159	0.0491	20	0.0000	0.0506	20	0.0159	0.0527
	T_{m^*}	20	0.0167	0.0522	19	-0.1099	0.0546	19	-0.0963	0.0532	19	-0.0826	0.0522
	$\theta_{(k^*)}$	1	-0.0572	0.0479	1	-0.0418	0.0481	3	-0.0526	0.0482	4	-0.0463	0.0483
	$\theta^{(k^*)}$	1	-0.0572	0.0479	1	-0.0418	0.0481	1	-0.0264	0.0487	2	-0.0630	0.0485

CHAPTER 16

Selection and Ranking Procedures

S. Panchapakesan[1]

16.1 Introduction

Selection and ranking problems first came under systematic investigation by statistical researchers in the early 1950s. The classical tests for homogeneity hypotheses were not formulated to address, in many situations, the experimenter's real goal, which is often to rank several competing populations (treatments, systems, etc.) or to select the best among them. The development of the selection and ranking theory was a result of attempts to formulate the decision problem appropriately to answer such realistic goals.

Over the past 40 years, the selection and ranking literature has steadily grown with many ramifications. An important part of these developments is the study of selection and ranking problems for specific families of distributions. In this chapter, we discuss procedures for selection from one-parameter and two-parameter exponential populations, and also some related inference problems. As an aid to the general reader, we provide (Section 16.2) a brief introduction to selection and ranking methodology explaining the salient features.

In Section 16.3, we discuss selection from one-parameter (scale) exponential populations. Single-stage procedures are discussed based on complete as well as Type I, Type II, and randomly censored samples. Section 16.4 deals with selection from two-parameter exponential populations in terms of their guaranteed lifetimes. Single-stage, two-stage, three-stage, and purely sequential procedures are discussed. Comparison of exponential guaranteed lifetimes with a standard or a control is discussed in Section 16.5. The final two sections are concerned with estimation concerning selected populations (Section 16.6) and some concluding remarks (Section 16.7).

16.2 Selection and Ranking Formulations

Selection and ranking problems have generally been studied by using either the *indifference zone* (IZ) formulation of Bechhofer (1954) or the *subset selection* (SS) formulation

[1]Southern Illinois University, Carbondale, Illinois

due mainly to Gupta (1956). In the former approach the number of populations to be selected is specified in advance, while in the latter it is random. Suppose that we have k (≥ 2) populations $\Pi_1, \ldots, \Pi_k$, where Π_i is characterized by the distribution function F_{θ_i} and θ_i is a real-valued parameter taking a value in the set Θ, $i = 1, \ldots, k$. The θ_i are assumed to be unknown. Let $\theta_{[1]} \leq \ldots \leq \theta_{[k]}$ denote the ordered θ_i and $\Pi_{(i)}$ be the population associated with $\theta_{[i]}$, $i = 1, \ldots, k$. We define $\Pi_{(j)}$ to be better than $\Pi_{(i)}$ if $i < j$, that is, $\theta_{[i]} \leq \theta_{[j]}$. We assume that there is no prior knowledge about the true pairing of the ordered and unordered θ_i.

Consider the basic problem of selecting the *best* population which is the one associated with the largest θ_i. In the IZ approach, *one* of the k populations is selected as the best. Let $\Omega = \{\boldsymbol{\theta} | \boldsymbol{\theta} = (\theta_1, \ldots, \theta_k),\ \theta_i \in \Theta,\ i = 1, \ldots, k\}$ be the parameter space and $\Omega_{\delta^*} = \{\boldsymbol{\theta} | \delta(\theta_{[k]}, \theta_{[k-1]}) \geq \delta^* > 0\}$, where $\delta(\theta_{[k]}, \theta_{[k-1]})$ is an appropriate measure of the separation between the best population $\Pi_{(k)}$ and the second best $\Pi_{(k-1)}$. A *correct selection* (CS) occurs if the selected population is the best population. Let $P(CS|R)$ denote the probability of a correct selection (PCS) using the rule R. It is required that a valid rule R satisfies:

$$P(CS|R) \geq P^* \text{ whenever } \boldsymbol{\theta} \in \Omega_{\delta^*} \ . \tag{16.2.1}$$

The constants δ^* and $P^*(k^{-1} < P^* < 1)$ are specified in advance by the experimenter. For a rule R based on a single sample of fixed size n from each population, one has to determine the minimum sample size n for which (16.2.1) holds. The region Ω_{δ^*} is called the *preference zone* (PZ), and its complement w.r.t. Ω is the *indifference zone* (IZ).

In the SS approach for selecting the best population, the goal is to select a non-empty subset of the k populations so that the selected subset includes the best population (which results in CS) with a minimum guaranteed probability $P^*(k^{-1} < P^* < 1)$. In other words, any valid rule R should meet the requirement:

$$P(CS|R) \geq P^* \text{ for all } \boldsymbol{\theta} \in \Omega \ . \tag{16.2.2}$$

It is assumed that, in the case of a tie for the best population, one of the contenders is tagged as the best. The size S of the selected subset is not specified in advance but is determined by the data. The usual measures of the performance of a valid rule are the expected subset size $E(S)$ and the expected number of non-best populations, which is equal to $E(S)$–PCS.

The probability requirements (16.2.1) and (16.2.2) are known as the P^*-conditions. In order to obtain the constant(s) associated with a proposed rule R so that the P^*-condition is met, one needs to evaluate the infinum of the PCS over Ω_{δ^*} or Ω depending on the formulation. Any configuration of $\boldsymbol{\theta}$ for which this infinum is attained is called a *least favorable configuration* (LFC).

There are several variations and generalizations of the basic goal in both IZ and SS formulations. For detailed discussions of various aspects of the theory and related problems, the reader is referred to Bechhofer, Kiefer and Sobel (1968), Büringer, Martin and Schriever (1980), Gibbons, Olkin and Sobel (1977), Gupta and Huang (1981), and Gupta and Panchapakesan (1979). The last authors have given a comprehensive survey of developments in theory with an extensive bibliography. A categorical bibliography is provided by Dudewicz and Koo (1982).

In the succeeding sections, we discuss specific selection procedures relating to exponential distributions. When variations and modifications occur with regard to goals and model assumptions, these will be explained at relevant places.

16.3 Selection From One-Parameter (Scale) Exponential Populations

Let $\Pi_1, \ldots, \Pi_k$ be $k(\geq 2)$ independent exponential populations $E(\theta_i)$, $i = 1, \ldots, k$, where the density associated with Π_i is

$$f(x;\theta_i) = \frac{1}{\theta_i} \exp(-x/\theta_i), x \geq 0, \theta_i > 0; i = 1, \ldots, k \; . \tag{16.3.1}$$

Let $\theta_{[1]} \leq \ldots \leq \theta_{[k]}$ denote the ordered values of the (unknown) parameters θ_i. We describe procedures for selecting the population associated with $\theta_{[k]}$ and $\theta_{[1]}$, which are denoted by $\Pi_{(k)}$ and $\Pi_{(1)}$, respectively.

16.3.1 Single-Stage Procedures Based on Complete Samples

Let $\overline{X}_1, \ldots, \overline{X}_k$ denote the means of random samples of size n from $\Pi_1, \ldots, \Pi_k$, respectively. For selecting a subset containing $\Pi_{(k)}$, we have the rule

$$R_1 : \text{ Select } \Pi_i \text{ if and only if } \overline{X}_i \geq c \max_{1\leq j\leq k} \overline{X}_j \; , \tag{16.3.2}$$

where c is the largest number in $(0, 1)$ for which the P^*-condition is met. This procedure was proposed by Gupta (1963) for selecting from gamma populations with scale parameters θ_i and a common known shape parameter α. His results apply to our case with $\alpha = 1$. The *LFC* is given by $\theta_1 = \cdots = \theta_k$ and the constant c is determined by

$$\int_0^\infty G_\nu^{k-1}(x/c)dG_\nu(x) = P^* \; , \tag{16.3.3}$$

where G_n denotes the *cdf* of a *chi*-square random variable with $\nu = 2n$ degrees of freedom. The value of c satisfying (16.3.3) can be obtained from the tables of Gupta (1963) for $k = 2(1)11$, $n = 1(1)20$ and $P^* = 0.75$, 0.90, 0.95, 0.99.

Gupta's analogous procedure for selecting a subset containing $\Pi_{(1)}$ is

$$R_2 : \text{ Select } \Pi_i \text{ if and only if } \overline{X}_i \leq \frac{1}{c'} \min_{1\leq j\leq k} \overline{X}_j \; , \tag{16.3.4}$$

where $0 < c' < 1$ is the largest number for which the P^*-condition is met. The constant c' is given by

$$\int_0^\infty [1 - G_\nu(c'x)]^{k-1} dG_\nu(x) = P^* \tag{16.3.5}$$

with $\nu = 2n$, and the value of c' satisfying (16.3.5) can be obtained for $k = 2(1)11$, $n = 1(1)25$, and $P^* = 0.75$, 0.90, 0.95, 0.99 from the tables of Gupta and Sobel (1962b) provided in the context of selecting the normal population with the smallest variance which was studied by them in a companion paper (1962a).

The exponential family $\{F(x, \theta)\}$ is stochastically increasing in θ, i.e., $F(x, \theta_i)$ and $F(x, \theta_j)$ are distinct for $\theta_i \neq \theta_j$, and $F(x, \theta_i) \geq F(x, \theta_j)$ for all x when $\theta_i < \theta_j$. This implies that ranking in terms of θ (which is the mean, the standard deviation, and the reciprocal of the hazard rate) is equivalent to ranking in terms of α-quantile for any $0 < \alpha < 1$.

Under the indifference zone formulation, we can take the PZ to be $\{\boldsymbol{\theta} : \theta_{[k]} \geq \delta^* \theta_{[k-1]}, \delta^* > 1\}$. In this case, the natural procedure for selecting $\Pi_{(k)}$ is

$$R_3 : \text{ Select the population that yields the largest } \overline{X}_i \ . \tag{16.3.6}$$

It can be shown that the LFC is given by $\theta_{[1]} = \ldots = \theta_{[k-1]} = \theta_{[k]}/\delta^*$. The sample size required to meet the P^*-condition is the smallest positive integer n for which

$$\int_0^\infty G_\nu^{k-1}(\delta^* x)\, dG_\nu(x) \geq P^* \tag{16.3.7}$$

with $\nu = 2n$. By noting that the left-hand sides of (16.3.3) and (16.3.7) are same for $\delta^* = c^{-1}$, the tables of Gupta (1963) can be used to determine n in (16.3.7) for appropriate values of δ^*. Bristol and Desu (1990) have discussed the procedure R_3 in the special case of $k = 2$ populations. They have also considered a slightly modified goal.

The problem of selecting $\Pi_{(1)}$ under the IZ approach is handled in an analogous manner.

16.3.2 Single-Stage Procedures Based on Censored Samples

Consider a life testing situation where a sample of n items from each population is put on test and the sample is censored at a specified, say, rth failure, $1 \leq r < n$. This is known as Type II censoring. Type I censored data arise by terminating the test a time T specified in advance.

Let us first consider Type II censored data. Let $X_{i[1]} \leq X_{i[2]} \leq \ldots \leq X_{i[r]}$ denote the observed failure times from Π_i, $i = 1, \ldots, k$. Let

$$T_i = X_{i[1]} + X_{i[2]} + \ldots X_{i[r]} + (n-r)X_{i[r]}, i = 1, \ldots, k \ .$$

The T_i are the so-called *total life statistics*. It is well-known that $2T_i/\theta_i$ has a *chi*-square distribution with $2r$ degrees of freedom.

For selecting $\Pi_{(k)}$, Huang and Huang (1980) proposed the subset selection procedure

$$R_4 : \text{ Select } \Pi_i \text{ if and only if } T_i \geq c \max_{1 \leq j \leq k} T_j \ , \tag{16.3.8}$$

where the constant $0 < c < 1$ satisfying the P^*-condition is given by (3.3) with $\nu = 2r$. Thus it can be obtained from the tables of Gupta (1963) as detailed in the case of R_1.

Berger and Kim (1985) considered selecting t best populations, namely, $\Pi_{(k-t+1)}, \ldots, \Pi_{(k)}$, with the preference zone $\{\boldsymbol{\theta} : \theta_{[k-t+1]} \geq \delta^* \theta_{[k-t]}, \delta^* > 1\}$. They proposed the natural procedure

$$R_5 : \text{ Select the populations which yield the } t \text{ largest } T_j\text{'s} \ . \tag{16.3.9}$$

Here P^* is specified such that $1/\binom{k}{t} < P^* < 1$. The LFC is $\delta^*\theta_{[1]} = \cdots = \delta^*\theta_{[k-t]} = \theta_{[k-t+1]} = \ldots = \theta_{[k]}$. The number of failures r required to satisfy the P^*-condition is the minimum r for which

$$(k-t)\int_0^\infty G_\nu^{k-t-1}(u)[1 - G_\nu(u)]^t\, dG_\nu(u) \geq P^* \ , \tag{16.3.10}$$

where $\nu = 2r$ and G_ν is the *cdf* of χ^2 with ν degrees of freedom. Tables in Bechhofer and Sobel (1954) can be used to determine r.

Now, we consider selection procedures under Type I censoring. Let X_{ij}, $j = 1, \ldots, n$, denote the failure times of n items taken from Π_i, $i = 1, \ldots, k$. Let $C_{ij} = 1$ if $X_{ij} < T$ and $C_{ij} = 0$ otherwise. Then, $N_i = \sum_{j=1}^{n} C_{ij}$ is the number of items that fail before the specified time T at which the test is terminated.

For the items from Π_i, we observe N_i and the ordered failure times $Y_{i1} \leq \cdots \leq Y_{iN_i}$. Let $Y_i = \sum_{j=1}^{N_i} Y_{ij} + (n - N_i)T$, which is the total life time of the n items from Π_i up to time T. For selecting the population $\Pi_{(k)}$ associated with the largest θ_i, Gupta and Liang (1993) derived a Bayes rule under a prior density $g(\boldsymbol{\theta}) = \prod_{i=1}^{k} g_i(\theta_i)$. They used a loss function $L^*(\boldsymbol{\theta}, i)$, corresponding to selection of Π_i when the true configuration is $\boldsymbol{\theta}$, given by $L^*(\boldsymbol{\theta}, i) = L(\theta_{[k]} - \theta_i)$, where $L(x)$ is nondecreasing in x with $L(0) = 0$. In the special case of $L(x) = x$, i.e. $L^*(\boldsymbol{\theta}, i) = \theta_{[k]} - \theta_i$, they derived a modified selection rule designed to make a selection earlier than the termination time T of the experiment.

Kim (1988) has studied a subset selection procedure for selecting $\Pi_{(k)}$ based on randomly censored data. As before, let X_{ij}, $j = 1, \ldots, n$, denote the failure times of the n items from Π_i, $i = 1, \ldots, k$. Let Z_{ij} be the censoring random variables which are independent of the X_{ij}, and are *iid* with exponential *cdf* $G(z) = 1 - \exp(-z/\xi)$, $z \geq 0$, $\xi > 0$. The observable random variables are then $Z_{ij} = \min(X_{ij}, Y_{ij})$. Let $\delta_{ij} = 1$ if $X_{ij} \leq Y_{ij}$, and $\delta_{ij} = 0$ otherwise. Finally, let $D_i = \sum_j \delta_{ij}$. The maximum likelihood estimator of θ_i is given by

$$T_i = \begin{cases} \sum_{j=1}^{n} Z_{ij}/D & \text{if } D > 0 \\ \infty & \text{if } D = 0 \,. \end{cases}$$

The distribution of T_i has an absolutely continuous part with density given by

$$f(t; \theta_i) = \sum_{\alpha=1}^{n} \binom{n}{\alpha} p_i^{\alpha} q_i^{n-\alpha} g_\alpha(t; \lambda_i), \quad 0 < t < \infty$$

and a discrete part with probability mass q_i^n at $t = \infty$, where

$$\begin{aligned} p_i &= 1 - q_i = \xi/(\xi + \theta_i), \\ \lambda_i &= (\xi + \theta_i)/\xi\theta_i, \text{ and} \\ g_\alpha(t; \lambda_i) &= \frac{(\alpha\lambda_i)^n t^{n-1}}{\Gamma(n)} \exp\{-\alpha\lambda_i t\} \,. \end{aligned}$$

For selecting a subset containing $\Pi_{(k)}$, Kim (1988) studied the rule

$$R_6 : \text{ Select } \Pi_i \text{ if and only if } T_i \geq c \max_{1 \leq j \leq k} T_j \,, \tag{16.3.11}$$

where $0 < c < 1$ is to be determined subject to the P^*-condition. Since the distribution of T_i is stochastically increasing in θ_i, the LFC for the rule R_6 is given by $\theta_1 = \cdots = \theta_k = \theta$ (say). Denoting the PCS at the LFC by $P_\theta(CS|R_6)$, its infinum over θ does not depend on the censoring parameter ξ. Kim (1988) has shown that for $\inf_\theta P_\theta(CS|R_6) \geq P^* > 1/k$, P^* should be specified such that

$$P^* < 1 - k^{-1/(k-1)} + k^{-k/(k-1)} .$$

He has tabulated the minimum PCS for $k = 2, 3$, $c = .02(.04).98$ and $n = 1(1)10$. Kim (1988) has also established some desirable properties of R_6.

Since, in a life testing experiment, the failure times occur in increasing order, the Type II censoring at the rth failure is called censoring on the right. Huang and Huang (1980) have also considered subset selection procedures for selecting $\Pi_{(k)}$ defined in terms of the ranges of doubly-censored samples and in terms of linear combinations of order statistics of the samples.

16.4 Selection From Two-Parameter (Location and Scale) Exponential Populations

Let $\Pi_1, \ldots, \Pi_k$ be k (≥ 2) independent exponential populations $E(\mu_i, \theta_i)$, $i = 1, \ldots, k$, where the density associated with Π_i is

$$f(x; \mu_i, \theta_i) = \frac{1}{\theta_i} \exp\{-\frac{(x - \mu_i)}{\theta_i}\}, \quad x \geq \mu_i,\ -\infty < \mu_i < \infty,\ \theta_i > 0 . \qquad (16.4.1)$$

When the location parameters μ_i are known, the problem reduces to the one-parameter case by taking $Y = X - \mu_i$. So, we now assume that the μ_i are unknown and the θ_i may or may not be known. Although technically the μ_i could be any real numbers, in the context of life-length distributions they are positive and are referred to as the *guaranteed lifetimes*.

16.4.1 Single-Stage Procedures for Selecting the Populations With the Largest (Smallest) Guaranteed Lifetimes

It is assumed that all the populations have a *common* scale parameter θ. Let $\mu_{[1]} \leq \cdots \leq \mu_{[k]}$ denote the ordered μ_i. We will discuss subset selection as well as indifference zone procedures with various goals.

Raghavachari and Starr (1970) considered the goal of selecting the t best populations, namely, those associated with $\mu_{[k-t+1]}, \ldots, \mu_{[k]}$, using the IZ formulation. The preference zone is $\boldsymbol{\mu} = \{(\mu_1, \ldots, \mu_k) : \mu_{[k-t+1]} - \mu_{[k-t]} \geq \delta^* > 0\}$. Of course, the guaranteed minimum PCS P^* is specified such that $1/\binom{k}{t} < P^* < 1$. It is assumed that the common scale parameter θ is *known* and so is taken to be 1 without loss of generality. Let X_{ij}, $j = 1, \ldots, n$, denote a random sample of n observations from Π_i, $i = 1, \ldots, k$. Define $Y_i = \min_{1 \leq i \leq n} X_{ij}$, $i = 1, \ldots, k$. Let $Y_{[1]} \leq \ldots \leq Y_{[k]}$ denote the ordered Y_i. Raghavachari and Starr (1970) studied the rule

$$R_7 : \text{ Select the } t \text{ populations associated with the } t \text{ largest } Y_i. \qquad (16.4.2)$$

The LFC for this rule is given by

$$\begin{cases} \mu_{[1]} = \ldots = \mu_{[k-t]}; \\ \mu_{[k-t+1]} = \ldots = \mu_{[k]}; \\ \mu_{[k-t+1]} - \mu_{[k-t]} = \delta^* . \end{cases} \qquad (16.4.3)$$

The minimum sample size n for which the infinum of the PCS over the preference zone is $\geq P^*$ is the smallest integer n for which $n\delta^* \geq -\log\nu$, where ν $(0 < \nu < 1)$ is the solution to the equation

$$(1-\nu)^{k-t} + (k-t)\nu^{-t} I(\nu; t+1, k-t) = P^* \ , \tag{16.4.4}$$

where

$$I(z; a, b) = \int_0^z u^{a-1}(1-u)^{b-1}\, du, \quad a, b > 0; \quad 0 \leq z \leq 1.$$

Raghavachari and Starr (1970) have tabulated the ν-values for $k = 2(1)15, t = 1(1)k-1$, and $P^* = 0.90, 0.95, 0.975, 0.99$. In particular, for selecting the best population ($t = 1$), the equation (16.4.4) reduces to

$$(\nu k)^{-1}[1-(1-\nu)^k] = P^* \ . \tag{16.4.5}$$

When the common scale parameter θ is not known, it is not possible to determine in advance the sample size needed for a single stage procedure in order to guarantee the P^*-condition. This necessitates at least a two-stage procedure. Such procedures are discussed in Section 16.4.2.

Bofinger (1991) has considered selecting the population associated with the largest μ_i using a *preliminary test approach* in which a decision of not selecting any population is allowed. First, assume that the common θ is *known*. As in the case of rule R_7 in (16.4.2), Y_i is the minimum of the n independent observations X_{ij}, $j = 1, \ldots, n$, from Π_i, $i = 1, \ldots, k$. Bofinger (1991) proposed the rule

$$R_8 \ : \ \text{Select } \Pi_i \text{ that yields the largest } Y_i \text{ provided } Y_{[k]} - Y_{[k-1]} > d\theta; \text{ make no selection otherwise.} \tag{16.4.6}$$

The constant d is determined so that $P(NWS|R_8) \equiv Pr\{\text{not making a wrong selection}\}$ is at least P^*. The LFC is given by $\mu_1 = \ldots = \mu_k = \mu$ and, for this configuration, $P(NWS|R_8)$ is independent of μ and is equal to $(1-k^{-1})e^{-d}$. Thus $d = \log[(k-1)/kP^*]$.

When the common θ is unknown, the rule R_8 is modified by replacing θ with its usual unbiased estimator

$$\hat{\theta} = \sum_{i=1}^{k}\sum_{j=1}^{n}(X_{ij} - Y_i)/k(n-1) \ . \tag{16.4.7}$$

In this case, d is given by

$$1 - (1-k^{-1})(1+d/\nu)^{-\nu} = P^* \ , \tag{16.4.8}$$

where $\nu = k(n-1)$.

Bofinger (1991) has also studied analogous procedures for selecting the population having the smallest μ_i. These are not discussed here. Further, Chen and Mithongtae (1986) have discussed selecting the population having the largest μ_i using a preliminary test in the case of common known θ. Their approach is different from that of Bofinger (1991). However, a more efficient (in terms of the sample size determination) procedure was proposed by Leu and Liang (1990) who used a slightly modified definition of the preference zone and nonselection zone in order to cover both known and unknown θ cases. We describe their procedure below.

The parameter space

$$\Omega = \{(\boldsymbol{\mu}, \theta) \mid \boldsymbol{\mu} = (\mu_1, \dots, \mu_k), -\infty < \mu_i < \infty, i = 1, \dots, k; \theta > 0\}$$

is partitioned into the preference zone $\Omega(PZ)$, the nonselection zone $\Omega(NZ)$, and the indifference zone $\Omega(IZ)$, where

$$\Omega(PZ) = \{(\boldsymbol{\mu}, \theta) \mid \mu_{[k]} - \mu_{[k-1]} \geq \delta^* \theta\} ,$$

and

$$\Omega(NZ) = \{(\boldsymbol{\mu}, \theta) \mid \mu_{[k]} = \mu_{[k-1]}\} .$$

The possible decisions are to select *one* population as the best or not select any. Let AS denote the event of making a selection and CS denote the event of selecting the best population, namely, the one having the largest μ_i. For any procedure R, it is required that

$$P(AS|R) \leq \alpha \text{ for all } (\boldsymbol{\mu}, \theta) \in \Omega(NZ) \tag{16.4.9}$$

and

$$P(CS|R) \geq P^* \text{ for all } (\boldsymbol{\mu}, \theta) \in \Omega(PZ) , \tag{16.4.10}$$

where $\alpha \in (0,1)$ and $P^* \in (k^{-1}, 1)$ are specified in advance. Leu and Liang (1990) studied, in the case of known θ, the rule

$$R_9 : \begin{array}{l} \text{Select the } \Pi_i \text{ yielding } Y_{[k]} \text{ if } Y_{[k]} - Y_{[k-1]} > \lambda\theta/n; \\ \text{otherwise, make no selection,} \end{array} \tag{16.4.11}$$

where λ and n are determined so that the probability requirements (16.4.9) and (16.4.10) are met. The requirement (16.4.9) is met by taking $\lambda = -\log\gamma$ and LFC for $P(CS|R_9)$ is given by $\mu_{[1]} = \dots = \mu_{[k-1]} = \mu_{[k]} - \delta^*\theta$. It is shown that the smallest sample size n needed is given by $n = \langle -\log(\gamma v)/\delta^* \rangle$, where $\langle x \rangle$ denotes the smallest integer $\geq x$, and the values of v are available for $P^* = 0.90$, 0.95, 0.975, 0.99 and $k = 2(1)15$ in Raghavachari and Starr (1970).

When θ is unknown, Leu and Liang (1990) studied a procedure which is R_9 with $\hat{\theta}$ given by (16.4.7) in the place of θ. In this case, the requirement (16.4.9) is met by taking λ such that $\gamma = (\nu/(\nu + \lambda))^\nu$, where $\nu = k(n-1)$. For obtaining the final solution, one can use the numerical methods of Chen and Mithongtae (1986).

Bofinger (1992) used confidence intervals for multiple comparisons with the best population to define subset selection procedures for selecting the best population which is either Π_i having the largest μ_i or Π_i having the smallest μ_i. Consider the case of the largest μ_i and assume that θ is unknown. By considering simultaneous lower confidence bounds for the differences $\phi_i = \mu_i - \max\limits_{j \neq i} \mu_j$, Bofinger (1992) proposed the rule

$$R_{10} : \text{Select } \Pi_i \text{ if and only if } Y_i \geq \max_j (Y_j - e_i\hat{\theta}/n_j) , \tag{16.4.12}$$

where n_i is the size of the sample from Π_i, $\hat{\theta}$ is now given by

$$\hat{\theta} = \sum_{i=1}^{k} \sum_{j=1}^{n_i} (X_{ij} - Y_i)/(N - k - 1) \text{ with } N = \sum n_i ,$$

and the e_i are to be determined so that the $PCS \geq P^*$ (see Bofinger (1992) for details). For the case of equal sample sizes ($n_1 = \ldots = n_k = n$), Chen and Vanichbuncha (1991) suggest a similar subset selection procedure, but their method generally results in more populations being selected (since they, in effect, use less constrained intervals). Bofinger (1992) has also proposed a similar rule for selecting Π_i having the smallest μ_i which selects Π_i if and only if $Y_i \leq Y_{[1]} + g_i\hat{\theta}/n_i$.

A modified goal in subset selection is to select a subset containing all good populations, where a good population is one that is close enough (to be defined) to the best population. For the problem of selecting the population with $\mu_{[k]}$, let any Π_i be defined *good* if $\mu_i \geq \mu_{[k]} - \delta$, where $\delta > 0$ is a specified constant. Gill and Sharma (1993) considered selecting a subset containing all the good populations (which results in a correct selection) with a minimum guaranteed PCS equal to P^*. When the common scale parameter θ is known, their rule is

$$R_{11} : \text{ Select } \Pi_i \text{ if and only if } Y_i \geq Y_{[k]} - \delta_1 - q\theta/n , \tag{16.4.13}$$

where $q = q(k, P^*)$ required to satisfy the P^*-condition is given by $q = -\log[1 - (P^*)^{1/(k-1)}]$.

When θ is unknown, Gill and Sharma (1993) proposed a rule which is R_{11} with q and θ replaced by $q' = q'(k, n, P^*)$ and $\hat{\theta}$ (given by (16.4.7), respectively. The values of q' can be obtained from the tables of Chen and Dudewicz (1984) for selected values of k, n, and P^*.

Berger and Kim (1985) studied subset selection procedures for selecting the population associated with the largest μ_i based on Type I and Type II censored data. It is assumed that the common scale parameter θ is known. Let $Y_i = \min(X_{i1}, \ldots, X_{in})$, $i = 1, \ldots, k$, as before. We first consider Type I censoring where the censoring is done at time T. Define $Z_i = \min(Y_i, T)$, $i = 1, \ldots, k$. Berger and Kim (1985) proposed the rule

$$R_{12} : \text{ Select } \Pi_i \text{ if and only if } Z_i \geq Z_{[k]} - d\theta/n , \tag{16.4.14}$$

where $Z_{[1]} \leq \cdots \leq Z_{[k]}$ are the ordered Z_i and the constant of $d \geq 0$ is determined subject to the P^*-condition. The infimum of the PCS is shown to attain at $\boldsymbol{\mu} = (0, \ldots, 0) = \boldsymbol{\mu}_0$ (say). This infimum is

$$P_{\boldsymbol{\mu}_0}(CS|R_{12}) = e^d[k^{-1}(1 - e^{-nT/\theta})^k - k^{-1}(1 - e^{-d})^k + e^{-nT/\theta}] . \tag{16.4.15}$$

Let P_0 denote the value of $P_{\boldsymbol{\mu}_0}(CS|R_{12})$ in (16.4.15) for $d = 0$. Then, for $k^{-1} < P^* < P_0$, the P^*-condition is strictly satisfied by taking $d = 0$. For $P_0 \leq P^* < 1$, the d-value is obtained by equating the right-hand side of (16.4.15) to P^*. These d-values are tabulated by Berger and Kim (1985) for $k = 2(1)5$, $nT/\theta = 0.25(0.25)4.00(1.00)10.00$ and $P^* = 0.75(0.05)0.95$, 0.99.

Let us now consider Type II censored data. In terms of life testing, let n items be put on test from each population. The censoring occurs at the r_ith failure for the sample from Π_i, $i = 1, \ldots, k$. For selecting the population having the largest μ_i, Ofosu (1974) proposed the subset selection rule

$$R_{13}: \text{ Select } \Pi_i \text{ if and only if } Y_i \geq Y_{[k]} - d\theta/n \ , \tag{16.4.16}$$

where Y_i is the first failure time observed in the sample from Π_i, $i = 1, \ldots, k$, and the constant $d > 0$ satisfying the P^*-condition is given by

$$e^d[1 - (1 - e^{-d})^k] = kP^* \ . \tag{16.4.17}$$

16.4.2 Two-Stage Procedures for Selecting the Population With the Largest (Smallest) Guaranteed Lifetime

In Section 16.4.1, we discussed a procedure R_7 defined in (16.4.2) for selecting the populations having the t largest μ_i, assuming that the common scale parameter θ was known. If θ is unknown, it is not possible (under the indifference zone formulation) to determine the sample size n for a single-stage procedure. We need at least a two-stage procedure, where the first stage samples are utilized for estimating θ. Desu, Narula and Villarreal (1977) proposed such a procedure for selecting the best population (i.e. $t = 1$) analogous to that of Bechhofer, Dunnett and Sobel (1954) for selecting the normal population with the largest mean from several normal populations having a common unknown variance. These are non-elimination type procedures. The first-stage samples are purely used to estimate θ without eliminating any population from further consideration. On the contrary, Kim and Lee (1985) have studied an elimination type two-stage procedure analogous to that of Gupta and Kim (1984) for the normal means problem.

The procedure R_{14} of Kim and Lee (1985) consists of two stages as follows:

Stage 1: Take n_0 independent observations from each Π_i $(1 \leq i \leq k)$, and compute $Y_i^{(1)} = \min_{1 \leq j \leq n_0} X_{ij}$, and a pooled estimate $\hat{\theta}_0$ of θ, namely,

$$\hat{\theta}_0 = \sum_{i=1}^{k} \sum_{j=1}^{n_0} (X_{ij} - Y_i^{(1)})/k(n_0 - 1) \ . \tag{16.4.18}$$

Obtain the subset I of $\{1, \ldots, k\}$ defined by

$$I = \{i | Y_i^{(1)} \geq \max_{1 \leq j \leq k} Y_j^{(1)} - (2k(n_0 - 1)\hat{\theta}_0 h/\delta^*)^+\} \ ,$$

where $a^+ = a$ for $a \geq 0$ and $a^+ = 0$ for $a < 0$, and the constant $h(> 0)$ is a design constant to be determined.

(a) If I has only one element, then stop sampling and decide that the population with $Y_{[k]}^{(1)}$ is the best.

(b) If I has more than one element, then go to the second stage.

Stage 2: Take $N - n_0$ additional observations X_{ij} from each Π_i for $i \in I$, where

$$N = \max\{n_0, < 2k(n_0 - 1)\hat{\theta}_0 h/\delta^* >\},$$

and $\langle y \rangle$ denotes the smallest integer greater than or equal to y. Now compute, for the overall samples, $Y_i = \min_{1 \le j \le N} X_{ij}$ and select the population Π_i associated with the largest Y_i, $i \in I$.

The constant h used in the procedure R_{14} is given by

$$\int_0^\infty \{1 - (1 - \alpha(x))^k\}^2 / \{k^2 \alpha^2(x)\} f_\nu(x)\, dx = P^* \ ,$$

where $\alpha(x) = \exp\{-hx\}$ and $f_\nu(x)$ is the χ^2 density with $\nu = 2k(n_0 - 1)$ degrees of freedom. The h-values have been tabulated by Kim and Lee (1985) for $P^* = 0.95$, $k = 2(1)5(5)20$, and $n_0 = 2(1)30$.

In Section 16.4.1, we discussed the procedure R_9, defined in (16.4.11), of Leu and Liang (1990) for selecting the population associated with $\mu_{[k]}$ using a preliminary test when θ is known, and mentioned their modification of the rule for the case of unknown θ. A single-stage procedure in this case was possible because the preference zone $\Omega(PZ)$ was defined in units of θ, which is the standard deviation. Suppose that $\Omega(PZ) = \{(\boldsymbol{\mu}, \theta) : \mu_{[k]} - \mu_{[k-1]} \ge \delta^*\}$ as in Chen and Mithongtae (1986) who considered the problem in the known θ case; their procedure, however, is more conservative (in terms of sample size determination) than the procedure R_9 of Leu and Liang (1990) and cannot be modified, as is done in the case of R_9, to suit the case of unknown θ. With $\Omega(PZ)$ defined as above, and $\Omega(NZ)$ and $\Omega(IZ)$ defined as in the case of R_9, Leu and Liang (1990) have proposed a two-stage procedure satisfying the probability requirements (16.4.9) and (16.4.10), which is not described here.

Liang and Panchapakesan (1992) proposed a procedure for selecting good populations. Any population Π_i is called good if $\mu_i \ge \mu_{[k]} - \delta^*$, where $\delta^* > 0$ is a specified constant. A correct selection occurs if any selected subset contains *only* good populations. It is required of a rule that $PCS \ge P^*$, $k^{-1} < P^* < 1$. The procedure of Liang and Panchapakesan (1992), which is analogous to that of Santner (1976) for the case of normal means, is as follows:

R_{15} :

(1) Take n_0 (arbitrary) independent observations $X_{i1}, \ldots, X_{in_0}$ from Π_i, $i = 1, \ldots, k$, and define $\hat{\theta}_0$ as in (16.4.18).

(2) Find $d = d(P^*, \nu, k)$ satisfying

$$Pr[U_j \le U_k + dW/\nu, j = 1, \ldots, k-1] = P^* \ , \tag{16.4.19}$$

where the U_j are i.i.d. each having a standard exponential distribution $E(1)$, and W, independent of the U_j, has a χ^2 distribution with $\nu = 2k(n-1)$ degrees of freedom.

(3) Fix N by *any* stopping rule of the form

$$N = \max\{n, [\hat{\theta}_0 \max\{h, d/\delta^*\}] + 1\} \ , \tag{16.4.20}$$

where $[y]$ denotes the greatest integer $\le y$. Then take an additional $N - n_0$ observations from each Π_i. A particular choice of h which achieves a second requirement is described later.

(4) Select all Π_i's for which

$$Y_i \geq Y_{[k]} - \delta^* + d\hat{\theta}_0/N,$$

where Y_i is the smallest of the N observations from Π_i, $i = 1, \ldots, k$.

Liang and Panchapakesan (1992) have shown that the P^*-condition is satisfied by the rule R_{15}. For given k, n_0 and P^*, the constant d satisfying (16.4.19) can be obtained from Desu, Narula, and Villareal (1977) who have tabulated the values of d/ν for $k = 2(1)6$; $n = 2(1)30$; $P^* = 0.95$, and from Bristol, Chen, and Mithongtae (1992, Table 3) who have tabulated the values of d/ν for $k = 3(1)16, 21$; $n = 2(1)10$, 15, 20, 30, 60; $P^* = 0.90, 0.95, 0.99$.

Let $0 < \Delta < \delta^*$ and $0 < p < 1$, and let $P(N)$ denote the proportion selected from the populations having $\mu_i \geq \mu_{[k]} - \Delta$ (which are *sufficiently good* populations). Liang and Panchapakesan (1992) have considered and provided formulas for the determination of h in (16.4.20) by requiring that $E[P(N)] \geq p$ for all $(\boldsymbol{\mu}, \theta) \in \Omega$, where $0 < p < 1$ is specified in advance.

Two-stage procedures have also been discussed by Mukhopadhyay (1984, 1986) and Mukhopadhay and Hamdy (1984), who have considered the case of unknown and unequal θ_i. These results are not discussed here.

16.4.3 Three-Stage and Sequential Procedures

We first consider sequential procedures for selecting the population associated with the largest μ_i under the indifference zone formulation assuming that the populations have a common scale parameter θ. The preference zone is $\Omega(\delta^*) = \{(\boldsymbol{\mu}, \theta) : \mu_{[k]} \geq \mu_{[k]} - \delta^*, \delta^* > 0\}$.

When θ is *known*, Miller and Mukhopadhyay (1990a) proposed a likelihood based sequential procedure. They treated the problem by assuming that the LFC $\mu_{[1]} = \ldots = \mu_{[k-1]} = \mu_{[k]} - \delta^*$ holds and requiring to choose one of the hypotheses $H_i : \mu_i = \mu_{[1]}, i = 1, \ldots, k$. In the special case of $k = 2$ populations, they have examined a few asymptotic properties and moderate sample performances by simulation. They have also studied by simulation moderate sample performances in the case of $k = 3$ populations.

When θ is *unknown*, Miller and Mukhopadhyay (1990a) developed an analogue of the sequential probability ratio test for selecting one of the k hypothesis $H_i : \mu_i = \mu_{[k]}$, $i = 1, \ldots, k$, assuming, as in the known θ case, that the LFC $\mu_{[1]} = \ldots = \mu_{[k-1]} = \mu_{[k]} - \delta^*$ held. Moderate sample performances was studied by simulation when $k = 2$.

Mukhopadhyay (1986) studied a sequential procedure for selecting the population associated with $\mu_{[k]}$, when θ is unknown. To describe his procedure, let $\{X_{ij} : j = 1, 2, \ldots\}$ be a sequence of i.i.d. observable random variables from Π_i, $i = 1, \ldots, k$. Having recorded $\{X_{i1}, \ldots, X_{in}\}$ from Π_i, we let $Y_{in} = \min(X_{i1}, \ldots, X_{in})$ and

$$W_n = \sum_{i=1}^{k} \sum_{j=1}^{n} (X_{ij} - Y_{in})/k(n-1), \quad i = 1, \ldots, k\ .$$

The sequential procedure of Mukhopadhyay (1986) is

R_{16}: Take an initial sample of m_0 observations from each Π_i. Let the stopping time N be defined by

$$N = \inf\{n \geq m_0 : n \geq aW_n/\delta^*\}\ ,$$

where a is the solution of

$$\int_0^\infty \{1-\exp(-t-a)\}^{k-1}\exp(-t)\,dt = P^* \;. \tag{16.4.21}$$

When the sequential procedure stops, select the population that yielded the largest Y_{iN}.

The constant a satisfying (16.4.21) can be obtained from Desu and Sobel (1968), and Raghavachari and Starr (1970). The sequential procedure terminates with probability 1. Mukhopadhyay (1986) has studied some second order asymptotic ($\delta^* \to 0$) properties of R_{16} and derived the rate of convergence of the asymptotic distribution of the normalized stopping variable.

A purely sequential procedure such as R_{16} is operationally inconvenient for implementation. In order to cut down the sampling operations considerably, Mukhopadhyay and Solanky (1992) proposed an *accelerated sequential procedure*. The idea, originally proposed by Hall (1983) for a different problem, is to implement the sequential procedure R_{16} but terminate it 'early' in the process. At this stage, we take all the 'remaining' samples in one single batch from each population. Mukhopadhyay and Solanky (1992) have made moderate and small sample comparisons between R_{16} and certain three-stage procedures of Mukhopadhyay (1986, 1987). They have also compared these two methodologies with the accelerated version of R_{16}. These comparisons show that the accelerated sequential selection procedure can save considerable amount of sampling operations, and yet can be competitive with R_{16} while at the same time outperforming the three-stage scheme.

Mukhopadhyay (1987) proposed a three-stage procedure for selecting the population associated with the largest μ_i when the common scale parameter θ is unknown under the indifference zone formulation. He studied some second-order characteristics of his procedure and suggested a modification that would yield $PCS = P^* + o(\delta^*)$ as $\delta^* \to 0$.

Finally, Miller and Mukhopadhyay (1990b) studied likelihood-based sequential rules for selecting the population associated with the smallest θ_i under the indifference zone formulation assuming that the location parameters μ_i are either all known, or all unknown, or some known and some unknown. Their specific results relate to $k = 2$ populations.

For a detailed treatment of multistage selection and ranking procedures with emphasis on second-order asymptotics, reference should be made to relevant chapters in the recent monograph by Mukhopadhyay and Solanky (1994).

16.5 Selection With Respect to a Standard or Control

Although the experimenter is generally interested in selecting the best of k (≥ 2) competing categories or treatments, it is possible that in some situations even the best one among them may not be good enough to warrant its selection. Such a situation arises when the goodness of a treatment is defined in comparison with a known standard or a control population. Here we may just refer to either one as the control.

Let $\Pi_1, \ldots, \Pi_k$ be the k (experimental) populations with associated distribution functions $F(x, \theta_i)$, $i = 1, \ldots, k$, respectively, where the θ_i are unknown. Let θ_0 be the parameter of the control population Π_0 with distribution function $F(x, \theta_0)$. Several different goals have been considered in the literature. For example, we may want to select the best experimental population (i.e. the Π_i associated with the largest θ_i) provided that it is better than the control (i.e. $\theta_{[k]} > \theta_0$), and not select any of them otherwise. Alternatively, one may select a subset of random size of the k populations that includes all Π_i's better than the control.

One can also use different criteria to define a good population. For example, Π_i may be called good if $\theta_i > \theta_0 + \Delta$ or if $|\theta_i - \theta_0| \leq \Delta$ for some $\Delta > 0$.

16.5.1 Comparison of Exponential Guaranteed Lifetimes With a Control

Let $\Pi_1, \ldots, \Pi_k$ be k experimental populations and Π_0 be the control population, where Π_i has the two-parameter exponential distribution $E(\mu_i, \theta_i)$ whose density is given in (16.4.1), $i = 0, 1, \ldots, k$. It is assumed that $\mu_1, \ldots, \mu_k$ are unknown. The control guaranteed life time μ_0 may or may not be known. Also, the scale parameters θ_i, $i = 0, 1, \ldots, k$, may or may not be known and can possibly be unequal.

Bristol and Desu (1985) considered selecting $\Pi_{(k)}$, the Π_i associated with $\mu_{[k]}$ when it is better than the control. Let $\Omega_C = \{\boldsymbol{\mu} : \mu_{[k]} \leq \mu_0 + d_0^*\}$ and $\Omega_E = \{\boldsymbol{\mu} : \mu_{[k]} \geq \mu_0 + d_1^*, \mu_{[k]} - \mu_{[k-1]} \geq d_2^*\}$, where $d_0^* < d_1^*$ and $d_2^* > 0$. Let $(k+1)^{-1} < P_0^*$, $P_1^* < 1$. When no experimental population is selected, we will say Π_0 is selected. Any selection rule is required to satisfy:

$$P\{\Pi_0 \text{ is selected } \} \geq P_0^* \text{ whenever } \boldsymbol{\mu} \in \Omega_C \tag{16.5.1}$$

and

$$P\{\Pi_{(k)} \text{ is selected } \} \geq P_1^* \text{ whenever } \boldsymbol{\mu} \in \Omega_E \ . \tag{16.5.2}$$

Bristol and Desu (1985) assumed that μ_0 is *known* and proposed the rule

$$\begin{aligned} R_{17} \quad : \quad & \text{Select the } \Pi_i \text{ associated with } Y_{[k]} \text{ if } Y_{[k]} > \mu_0 + d \\ & \text{and select } \Pi_0 \text{ otherwise,} \end{aligned} \tag{16.5.3}$$

where Y_i is the smallest of n independent observations from Π_i, $i = 1, \ldots, k$.

When all θ_i's are equal to a known θ, Bristol and Desu (1985) have tabulated the values of d and $M = -\log\{1 - (P_0^*)^{1/k}\}/(d - d_0^*)$ for $P_1^* = P_2^* = 0.90$, $d_0^* = 0$, $d_1^* = 0.1$, 0.2, $d_2^* = 0.1$ and $k = 2(1)10$. The sample size n required is $M\theta$ either rounded up or down, whichever satisfies (16.5.3).

When θ is unknown, Bristol and Desu (1985) proposed a two-stage procedure R_{18} which has the following steps:

1. Take an initial sample of n_0 independent observations, $X_{i1}, \ldots, X_{in_0}$, from Π_i, $i = 1, \ldots, k$. Let $Y_i^{(1)} = \min\{X_{i1}, \ldots, X_{in_0}\}$ and $\hat{\theta}_0$ be the pooled estimator of θ given in (16.4.18). Define

$$N = \max\{n_0, \langle 2hk(n_0 - 1)\hat{\theta}_0 \rangle\} \tag{16.5.4}$$

where $\langle y \rangle$ denotes the smallest integer $\geq y$.

2. Take $N - n_0$ additional observations from each Π_i and let Y_i be smallest of all N observations from Π_i, $i = 1, \ldots, k$. Now use the selection rule R_{17} in (16.5.3).

The constants d and h of the rule R_{18} are chosen so that the probability requirements (16.5.1) and (16.5.2) are satisfied. See Bristol and Desu (1985) for a discussion of the determination of h and d.

Bristol, Chen, and Mithongtae (1992) have discussed the above problem when μ_0 is unknown. The treatment of the problem is similar to that of Bristol and Desu (1985). When θ is known, we replace μ_0 in the rule R_{17} in (16.5.3) with its estimator Y_0 which is the smallest of n observations from Π_0. When θ is unknown, the initial sample of size n_0 is taken also from Π_0 and the pooled estimator $\hat{\theta}_0$ now includes the sample from Π_0 and k in (16.5.4) will now be $k + 1$. Bristol, Chen, and Mithongtae (1992) have tabulated d and $m = n/\theta$ in the θ known case for $d_0^* = 0$, $d_2^* = 1$, $d_1^* = 0.5(0.5)3$, $P_0^* = P_1^* = 0.90, 0.95, 0.99$, and $k = 2(1)10, 12, 15, 20$. In the case of unknown θ, they have tabulated the required values of h and d for $d_0^* = 0$, $d_2^* = 1$, $d_1^* = 0.5, 0.75, 1(0.5)3$, $P_0^* = P_1^* = 0.90, 0.95, 0.99$, $k = 2(1)15, 20$ and $n_0 = 2(1)10, 15, 20, 30, 60$.

Lam and Ng (1990) proposed an improvement of a two-stage procedure for selecting the best two-parameter exponential population originally proposed by Mukhopadhyay and Hamdy (1984). This improvement permits having fewer samples in the second stage. In this context, Lam and Ng (1990) also introduced a two-stage multiple comparisons procedure for comparing k experimental populations with a control when the scale parameters θ_i are unknown and are possibly unequal. This procedure will not be detailed here.

Bofinger (1992) considered the goals of selecting a subset of the experimental population that contains (a) *all* the populations better than Π_0, and (b) *only* populations that are better than Π_0. She proposed procedures based on joint confidence intervals for $\mu_i - \mu_0$. These are single-stage procedures.

Finally, we consider a situation where we may know that, for the experimental populations, $\mu_1 \leq \mu_2 \leq \ldots \leq \mu_k$ although the values of the μ_i are unknown. This is typical, for example, of experiments involving different dose levels of a drug so that the treatment effect, will have a known ordering. Let Π_i be called good if $\mu_i \geq \mu_0$ and bad otherwise. Gupta and Leu (1986) considered selecting a subset of the k experimental populations so that all good populations are included the selected subset with a guaranteed probability P^*. As we can see, any reasonable procedure R should have the property: If R selects Π_i, then it selects all populations Π_j with $j > i$. This is called the *isotonic* behavior of R. One should then consider procedures based on the isotonic estimators of the μ_i, which are also the maximum likelihood estimators of the μ_i, $i = 1, \ldots, k$, with restriction that $\mu_1 \leq \mu_2 \leq \ldots \leq \mu_k$. Gupta and Leu (1986) considered all the cases arising from each of μ_0 and θ being known or unknown. They have also compared these isotonic procedures with some other procedures in terms of the expected number of bad populations included in the selected subset.

16.6 Estimation Concerning Selected Populations

Consider selecting the exponential population associated with the largest (smallest) θ_i from k populations, $E(\theta_i)$, $i = 1, \ldots, k$. Let $X_1, \ldots, X_k$ be the observations from these. The natural rule selects the Π_i that yielded the largest (smallest) X_i. The problem of interest is to estimate $\theta_M(\theta_J)$, the θ_i of the selected population. This is important in practice because we are interested in not only selecting the population with the largest (smallest) guaranteed lifetime but also in estimating the guaranteed lifetime of the selected population. Sackrowitz and Samuel-Cahn (1984) studied this problem. They used the loss function $(a - \theta_M)^2/\theta_M^2$ for θ_M and a corresponding loss for θ_J. They have obtained $UMVU$ and minimax estimators and have also discussed certain other reasonable estimators. If we take a random sample of size n from each Π_i, then the sum of the observations Y_i has a gamma distribution with scale parameter θ_i and shape parameter n. In this case, the results of Vellaisamy and Sharma (1988) are applicable. They considered estimation after selection of the gamma population with the largest mean based on a sample of specified size from each. However, their results were obtained only for the case of $k = 2$ populations and later extended by them (1989) for $k \geq 2$. They showed that a minimax estimator dominates the natural estimator and the $UMVUE$ for the squared error loss.

Jeyaratnam and Panchapakesan (1986) considered estimation after a subset selection from exponential populations, $E(\theta_i)$, $i = 1, \ldots, k$. Let $M = \sum_{i=1}^{k} I_i\theta_i \Big/ \sum_{i=1}^{k} I_i$, where $I_i = 1$ if Π_i is selected and $I_i = 0$ otherwise. They called it the *average worth* of the selected subset. They considered the rule which selects Π_i if and only if $Y_i \geq c \max_{1\leq j\leq k} Y_j$, where the Y_i are the sample means based on n observations from each Π_i and $0 < c < 1$ is the constant which guarantees the usual P^*-condition. They considered estimating M in the case of $k = 2$. They showed the 'natural' estimator to be positively biased and compared its performance with that of another estimator obtained by making some adjustments for the bias.

In Sections 16.4 and 16.5, we discussed selection procedures of Bofinger (1992) based on confidence intervals. Reversing this approach, Mukhopadhyay, Hamdy, and Darmanto (1988) and Mukhopadhyay and Solanky (1993) obtained simultaneous confidence regions for the guaranteed lifetimes μ_i (besides the normal means case) after the stopping time N has been determined for the purely sequential and three-stage procedures for selecting the best population. They have studied asymptotic second-order characteristics of the proposed confidence regions. This problem is referred to by these authors as *estimation after selection and ranking*, which is different from estimation of the parameter(s) of the selected population(s).

16.7 Concluding Remarks

In the preceding sections, we have described several selection procedures. However, we have not dealt with desirable properties and performance evaluations of these procedures. For several single-stage procedures, properties such as monotonicity and minimaxity have been established by appealing to earlier results for location and scale parameters. However, there is scope for further investigations concerning new procedures and performance comparisons.

References

Bechhofer, R.E. (1954). A single-sample multiple decision procedure for ranking means of normal populations with known variances, *Annals of Mathematical Statistics*, **25**, 16-39.

Bechhofer, R.E., Dunnett, C.W., and Sobel, M. (1954). A two sample multiple decision procedure for ranking means of normal populations with a common unknown variance, *Biometrika*, **41**, 170-176.

Bechhofer, R.E., Kiefer, J., and Sobel, M. (1968). *Sequential Identification and Ranking Procedures (with Special Reference to Koopman-Darmois Populations)*, University of Chicago Press, Chicago.

Bechhofer, R.E. and Sobel, M. (1954). A single-sample multiple-decision procedure for ranking variances of normal populations, *Annals of Mathematical Statistics*, **25**, 273-289.

Berger, R.L. and Kim, J.S. (1985). Ranking and subset selection procedures for exponential populations with type-I and type-II censored data, In *Frontiers of Modern Statistical Inference Procedures* (Ed., E.J. Dudewicz), pp. 425-455 (with discussion), American Sciences Press, Columbus, Ohio.

Bofinger, E. (1991). Selecting "demonstrably best" or "demonstrably worst" exponential populations, *Australian Journal of Statistics*, **33**, 183-190.

Bofinger, E. (1992). Comparisons and selection of two-parameter exponential populations, *Australian Journal of Statistics*, **34**, 65-75.

Bristol, D.R., Chen, H.J., and Mithongtae, J.S. (1992). Comparing exponential guarantee times with a control, In *Frontiers of Modern Statistical Inference Procedures - II* (Eds., E.J. Dudewicz et al.), pp. 95-151 (with discussion), American Sciences Press, Syracuse, New York.

Bristol, D.R. and Desu, M.M. (1985). Selection procedures for comparing exponential guarantee times with a standard, In *Frontiers of Modern Statistical Inference Procedures* (Ed., E.J. Dudewicz), pp. 307-324 (with discussion), American Sciences Press, Columbus, Ohio.

Bristol, D.R. and Desu, M.M. (1990). Comparison of two exponential distribution, *Biometrical Journal*, **32**, 267-276.

Büringer, H., Martin, H., and Schriever, K.-H. (1980). *Nonparametric Sequential Procedures*, Birkhauser, Boston.

Chen, H.J. and Dudewicz, E.J. (1984). A new subset selection theory for all best populations and its applications, In *Developments in Statistics and Its Applications* (Eds., A.M. Abouammah et al.), pp. 63-87, King Saud University Press, Riyadh, Saudi Arabia.

Chen, H.J. and Mithongtae, J. (1986). Selection of the best exponential distribution with a preliminary test, *American Journal of Mathematics and Management Sciences*, **6**, 219-249.

Chen, H.J. and Vanichbuncha, K. (1991). Multiple comparisons with the best exponential distribution, In *The Frontiers of Statistical Scientific Theory & Industrial Applications - Vol. II* (Eds., A. Öztürk et al.), pp. 53-76, American Sciences Press, Columbus, Ohio.

Desu, M.M., Narula, S.C., and Villareal, B. (1977). A two-stage procedure for selecting the best of k exponential distributions, *Communications in Statistics - Theory and Methods*, **6**, 1233-1230.

Desu, M.M. and Sobel, M. (1968). A fixed-subset size approach to a selection problem, *Biometrika*, **55**, 401-410. Corrections and amendments: **63** (1976), 685.

Dudewicz, E.J. and Koo, J.O. (1982). The complete categorized guide to statistical selection and ranking procedures, *Series in Mathematical and Management Sciences*, Vol. 6, American Sciences Press, Columbus, Ohio.

Gibbons, J.D., Olkin, I., and Sobel, M. (1977). *Selecting and Ordering Populations: A New Statistical Methodology*, John Wiley & Sons, New York.

Gill, A.N. and Sharma, S.K. (1993). A selection procedure for selecting good exponential populations, *Biometrical Journal*, **35**, 361-369.

Gupta, S.S. (1956). On a decision rule for a problem in ranking means, *Mimeograph Series No. 150*, Institute of Statistics, University of North Carolina, Chapel Hill, North Carolina.

Gupta, S.S. (1963). On a selection and ranking procedure for gamma populations, *Annals of the Institute of Statistical Mathematics*, **14**, 199-216.

Gupta, S.S. and Huang, D.-Y. (1981). Multiple decision theory: Recent developments. *Lecture Notes in Statistics*, Vol. 6, Springer-Verlag, New York.

Gupta, S.S. and Kim, W.-C. (1984). A two-stage elimination type procedure for selecting the largest of several normal means with a common unknown variance, In *Design of Experiments: Ranking and Selection* (Eds., T.J. Santner and A.C. Tamhane), pp. 77-93, Marcel Dekker, New York.

Gupta, S.S. and Leu, L.-Y. (1986). Isotonic procedures for selecting populations better than a standard: two-parameter exponential distributions, In *Reliability and Quality Control* (Ed., A.P. Basu), pp. 167-183, Elsevier, Amsterdam.

Gupta, S.S. and Liang, T. (1993). Selecting the best exponential population based on type-I censored data: A Bayesian approach, In *Advances in Reliability* (Ed., A.P. Basu), pp. 171-180, Elsevier, Amsterdam.

Gupta, S.S. and Panchapakesan, S. (1979). *Multiple Decision Procedures: Theory and Methodology of Selecting and Ranking Populations.* John Wiley & Sons, New York.

Gupta, S.S. and Sobel, M. (1962a). On selecting a subset containing the population with the smallest variance, *Biometrika*, **49**, 495-507.

Gupta, S.S. and Sobel, M. (1962b). On the smallest of several correlated F-statistics, *Biometrika*, **49**, 509-523.

Huang, W.-T. and Huang, K.-C. (1980). Subset selections of exponential populations based on censored data, *Proceedings of the Conference on Recent Developments in Statistical Methods and Applications*, pp. 237-254, Directorate-General of Budget, Accounting and Statistics, Executive Yuan, Taipei, Taiwan, Republic of China.

Jeyaratnam, S. and Panchapakesan, S. (1986). Estimation after subset selection from exponential populations, *Communications in Statistics - Theory and Methods*, **10**, 1869-1878.

Kim, J.S. (1988). A subset selection procedure for exponential populations under random censoring, *Communications in Statistics - Theory and Methods*, **17**, 183-206.

Kim, W.-C. and Lee, S.-H. (1985). An elimination type two-stage selection procedure for exponential distributions, *Communications in Statistics - Theory and Methods*, **14**, 2563-2571.

Lam, K. and Ng, C.K. (1990). Two-stage procedures for comparing several exponential populations with a control when the scale parameters are unknown and unequal, *Sequential Analysis*, **9**, 151-164.

Leu, L.-Y. and Liang, T. (1990). Selection of the best with a preliminary test for two-parameter exponential distributions, *Communications in Statistics - Theory and Methods*, **19**, 1443-1455.

Liang, T. and Panchapakesan, S. (1992). A two-stage procedure for selecting the δ^*-optimal guaranteed lifetimes in the two-parameter exponential model, In *Multiple Comparisons, Selection, and Applications in Biometry, A Festschrift in Honor of Charles W. Dunnett* (Ed., F.M. Hoppe), pp. 353-365, Marcel Dekker, New York.

Miller, D.S. and Mukhopadhyay, N. (1990a). Likelihood based sequential procedures for selecting the best exponential population, *Metron*, **48**, 255-282.

Miller, D.S. and Mukhopadhyay, N. (1990b). Sequential likelihood rules for selecting the negative exponential population having the smallest scale parameter, *Calcutta Statistical Association Bulletin*, **39**, 31-44.

Mukhopadhyay, N. (1984). Sequential and two-stage procedures for selecting the better exponential population covering the case of scale parameters being unknown and unequal, *Journal of Statistical Planning and Inference*, **9**, 33-44.

Mukhopadhyay, N. (1986). On selecting the best exponential population, *Journal of Indian Statistical Association*, **24**, 31-41.

Mukhopadhyay, N. (1987). Three-stage procedures for selecting the best exponential population, *Journal of Statistical Planning and Inference*, **16**, 345-352.

Mukhopadhyay, N. and Hamdy, H.I. (1984). Two-stage procedures for selecting the best exponential population when the scale parameters are unknown and unequal, *Sequential Analysis*, **3**, 51-74.

Mukhopadhyay, N., Hamdy, H.I., and Darmanto, S. (1988). Simultaneous estimation after selection and ranking and other procedures, *Metrika*, **35**, 275-286.

Mukhopadhyay, N. and Solanky, T.K.S. (1992). Accelerated sequential procedure for selecting the best exponential population, *Journal of Statistical Planning and Inference*, **32**, 347-361.

Mukhopadhyay, N. and Solanky, T.K.S. (1993). Estimation after selection and ranking, *Technical Report No. 93-32*, The University of Connecticut, Storrs, CT 06269.

Mukhopadhyay, N. and Solanky, T.K.S. (1994). *Multistage Selection and Ranking Procedures: Second-Order Asymptotics*, Marcel Dekker, New York.

Ofosu, J.B. (1974). On selection procedures for exponential distributions, *Bulletin of Mathematical Statistics*, **16**, 1-9.

Raghavachari, M. and Starr, N. (1970). Selection problems for some terminal distributions, *Metron*, **28**, 185-197.

Sackrowitz, H.B. and Samuel-Cahn, E. (1984). Estimation of the mean of a selected negative exponential population, *Journal of the Royal Statistical Society, Series B*, **46**, 242-249.

Santner, T.J. (1976). A two-stage procedure for selecting δ^*-optimal means in the normal model, *Communications in Statistics - Theory and Methods*, **5**, 283-292.

Vellaisamy, P. and Sharma, D. (1988). Estimation of the mean of the selected gamma population, *Communications in Statistics - Theory and Methods*, **17**, 2797-2817.

Vellaisamy, P. and Sharma, D. (1989). A note on the estimation of the mean of the selected gamma population, *Communications in Statistics - Theory and Methods*, **18**, 555-560.

CHAPTER 17

Record Values

M. Ahsanullah[1]

17.1 Introduction

Let $\{X_n,\ n \geq 1\}$ be a sequence of independent and identically distributed random variables with cumulative distribution function $F(x)$ and the corresponding probability density function $f(x)$. Set $Y_n = \max(X_1, \ldots, X_n),\ n \geq 1$. We say Y_j is an upper record value of $\{X_n,\ n \geq 1\}$ if $Y_j > Y_{j-1},\ j > 1$. By definition, $X_1 \equiv Y_1$ is an upper record value. Thus the upper record values in the sequence $\{X_n,\ n \geq 1\}$ are the successive maxima. For example, consider the weighing of objects on a scale missing it's spring. An object is placed on this scale and its weight measured. The 'needle' indicates the correct value but does not return to zero when the object is removed. If various objects are placed on the scale, only the weights greater than the previous ones can be recorded. These recorded weights are the upper record value sequence. Suppose we consider a sequence of products which may fail under stress. We are interested to determine the minimum failure stress of the products sequentially. We test the first product until it fails with stress less than X_1 then we record its failure stress, otherwise we consider the next product. In general we will record stress X_m of the mth product if $X_m < \min(X_1, \ldots, X_{m-1}),\ m > 1$. The recorded failure stresses are the lower record values. One can go from lower records to upper records by replacing the original sequence of random variables by $\{-X_i,\ i \geq 1\}$ or if $P(X_i > 0) = 1$ by $\{1/X_i,\ i \geq 1)$. Unless mentioned otherwise, we will call the upper record values as record values. The indices at which the record values occur are given by the record times $\{U(n)\},\ n > 0$, where $U(n) = \min\{j \mid j > U(n-1),\ X_j > X_{U(n-1)},\ n > 1\}$ and $U(1) = 1$. The record times of the sequence $\{X_n,\ n \geq n\}$ are the same as those for the sequence $\{F(X_n),\ n \geq 1\}$. Since $F(X)$ has uniform distribution, it follows that the distribution of $U(n),\ n \geq 1$, does not depend on F. Chandler (1952) introduced record values and record times. Feller (1966) gave some examples of record values with respect to gambling problems. Properties of record values of i.i.d. random variables have been extensively studied in the literature. See Ahsanullah (1988), Arnold and

---

[1]Rider University, Lawrenceville, New Jersey

Balakrishnan (1989), Arnold, Balakrishnan, and Nagaraja (1992), Nagaraja (1988) and Nevzorov (1988) for recent reviews.

17.2 Distribution of Record Values

Many properties of the record value sequence can be expressed in terms of the function $R(x)$, where $R(x) = -\log \overline{F}(x)$, $0 < \overline{F}(x) < 1$ and $\overline{F}(x) = 1 - F(x)$. All the logarithms used in this chapter are natural logarithms.

If we define $F_n(x)$ as the distribution function of $X_{U(n)}$ for $n \geq 1$, then we have

$$F_1(x) = P\left[X_{U(1)} \leq x\right] = F(x) \,, \tag{17.2.1}$$

and

$$\begin{aligned} F_2(x) &= P\left[X_{U(2)} \leq x\right] \\ &= \int_{-\infty}^{x} \int_{-\infty}^{y} \sum_{i=1}^{\infty} (F(u))^{i-1} dF(u)\, dF(y) \\ &= \int_{-\infty}^{x} \int_{-\infty}^{y} \frac{dF(u)}{1 - F(u)}\, dF(y) = \int_{-\infty}^{x} R(y)\, dF(y) \,. \end{aligned} \tag{17.2.2}$$

The distribution function $F_3(x)$ of $X_{U(3)}$ is given by

$$\begin{aligned} F_3(x) &= P\left[X_{U(3)} \leq x\right] \\ &= \int_{-\infty}^{x} \int_{-\infty}^{y} \sum_{i=0}^{\infty} (F(u))^{i} R(u)\, dF(u)\, dF(y) \\ &= \int_{-\infty}^{x} \int_{-\infty}^{y} \frac{R(u)\, dF(u)}{1 - F(u)}\, dF(y) \\ &= \int_{-\infty}^{x} \frac{(R(y))^2}{2!}\, dF(y) \,. \end{aligned} \tag{17.2.3}$$

It can similarly be shown that the distribution function $F_n(x)$ of $X_{U(n)}$ is

$$\begin{aligned} F_n(x) &= P\left[X_{U(n)} \leq x\right] \\ &= \int_{-\infty}^{x} \int_{-\infty}^{y} \sum_{i=0}^{\infty} (F(u))^{i} \frac{(R(u))^{n-2}}{(n-2)!}\, dF(u)\, dF(y) \\ &= \int_{-\infty}^{x} \frac{(R(y))^{n-1}}{(n-1)!}\, dF(y) \,. \end{aligned} \tag{17.2.4}$$

If $F(x)$ has the density $f(x)$, then $f_n(x)$, the density function of $X_{U(n)}$, $n \geq 1$ is

$$f_n(x) = \frac{(R(x))^{n-1}}{(n-1)!} f(x), \; n = 1, 2, \ldots \,. \tag{17.2.5}$$

The joint density $f(x_1, \ldots, x_n)$ of the n records $X_{U(1)}, \ldots, X_{U(n)}$ is given by

$$\begin{aligned} f(x_1, \ldots, x_n) &= \rho(x_1) \ldots \rho(x_{n-1})\, f(x_n) \,, \\ & \qquad -\infty < x_1 < \cdots < x_{n-1} < x_n < \infty \\ &= 0 \,, \qquad \text{otherwise} \,, \end{aligned} \tag{17.2.6}$$

where $\rho(x) = \frac{d}{dx} R(x) = f(x)/(1-F(x))$ is the hazard rate. The joint density of $X_{U(i)}$ and $X_{U(j)}$ is

$$f_{i,j}(x_i, x_j) = \frac{(R(x_i))^{i-1}}{(i-1)!} \rho(x_i) \frac{(R(x_j)-R(x_i))^{j-i-1}}{(j-i-1)!} f(x_j) , \quad -\infty < x_i < x_j \le \infty . \tag{17.2.7}$$

The conditional density of $X_{U(j)} \mid X_{U(i)} = x_i$, is

$$f(x_j \mid X_{U(i)} = x_i) = \frac{(R(x_j)-R(x_i))^{j-i-1}}{(j-i-1)!} \frac{f(x_j)}{1-F(x_i)} , \quad -\infty < x_i < x_j < \infty . \tag{17.2.8}$$

Now, let us consider the two-parameter exponential population with density function

$$\begin{aligned} f(x) &= \tfrac{1}{\theta} e^{-(x-\mu)/\theta} \quad , x \ge \mu \\ &= 0 \quad , \text{otherwise} , \end{aligned} \tag{17.2.9}$$

$$F(x) = 1 - e^{-(x-\mu)/\theta}, \qquad x \ge \mu ,$$

and

$$\rho(x) = f(x)/(1-F(x)) = \frac{1}{\theta} .$$

Theorem 17.1 *Let X_i, $i = 1, 2, \ldots$ be independently and exponentially distributed with $\mu = 0$ and $\theta = 1$. Suppose $\xi_i = \frac{X_{U(i)}}{X_{U(i+1)}}$, $i = 1, 2, m-1$; then ξ_i's are independent.*

Proof: The joint density of $X_{U(1)}, X_{U(2)}, \ldots, X_{U(m)}$ is

$$f(x_1, x_2, \ldots, x_m) = e^{-x_m}, \ 0 < x_1 < x_2 < \cdots < x_m < \infty .$$

Let us use the transformation

$$\xi_0 = X_{U(1)} , \qquad \text{and} \qquad \xi_i = \frac{X_{U(i)}}{X_{U(i+1)}} , i = 1, 2, \ldots, m-1 .$$

The Jacobian of the transformation is

$$J = \left| \frac{\partial(X_{U(1)}, X_{U(2)}, \ldots, X_{U(m)})}{\partial(\xi_0, \xi_1, \ldots, \xi_{m-1})} \right| = \frac{\xi_0^{m-1}}{\xi_1^m \, \xi_2^{m-1} \cdots \xi_{m-1}^2} .$$

We can write the joint density of ξ_i, $i = 0, 1, \ldots, m-1$, as

$$f(e_0, e_1, \ldots, e_{m-1}) = \frac{e_0^{m-1}}{e_1^m \, e_2^{m-1} \cdots e_{m-1}^2} \, e^{-\left(\frac{e_0}{e_1 \cdots e_{m-1}}\right)} .$$

Now integrating the above expression with respect to e_0, we obtain the joint density of ξ_i, $i = 1, \ldots, m-1$, as

$$f(e_1, \ldots, e_{m-1}) = \Gamma(m) \, e_2 \ldots e_{m-1}^{m-2} .$$

Thus ξ_i, $i = 1, 2, \ldots, m-1$, are independent and

$$P[\xi_k \le x] = x^k , \qquad 1 \le k \le m .$$

□

Corollary 17.1 *Let* $W_k = (\xi_k)^k$, $k = 1, 2, \ldots, m-1$; *then* $W_1, W_2, \ldots, W_{m-1}$ *are i.i.d. random variables.*

Using Eq. (17.2.5) and noting $R(x) = (x-\mu)/\theta$, we have

$$\begin{aligned} f_n(x) &= \tfrac{\theta^{-n}}{\Gamma(n)}(x-\mu)^{n-1}e^{-(x-\mu)/\theta} \quad , \ x \geq \mu \ , \\ &= 0 \quad , \ \text{otherwise} \ . \end{aligned} \tag{17.2.10}$$

The joint density of $X_{U(m)}$ and $X_{U(n)}$, $n > m$, is

$$\begin{aligned} f_{m,n}(x,y) &= \frac{\theta^{-n}}{\Gamma(m)} \frac{(x-\mu)^{m-1}}{\Gamma(n-m)} (y-x)^{n-m-1} e^{-(x-\mu)/\theta} \ , \\ & \qquad \mu \leq x < y < \infty \ , \\ &= 0, \qquad \text{otherwise} \ . \end{aligned} \tag{17.2.11}$$

It can be shown that $X_{U(n)} - X_{U(n-1)}$ and $X_{U(m)} - X_{U(m-1)}$ are identically distributed for $1 < m < n < \infty$. Ahsanullah (1991) showed that $X_{U(m)} \stackrel{d}{=} X_{U(m-1)} + U$, $(m > 1)$, where U is independent of $X_{U(m)}$ and $X_{U(m-1)}$ and is identically distributed as X_i's, iff $F(x) = 1 - \exp(x/\theta)$. For $\mu = 0$, it is easy to verify $P[X_{U(n)} > wX_{U(n-1)}] = w^{-n}$, $w > 1$. The conditional density of $X_{U(n)} \mid X_{U(m)} = x$ is

$$\begin{aligned} f(y \mid X_{U(m)} = x) &= \frac{\theta^{m-n}}{\Gamma(n-m)} (y-x)^{n-m-1} e^{-(y-x)/\theta} \\ & \qquad \mu \leq x < y < \infty \ , \\ &= 0 \ , \qquad \text{otherwise} \ . \end{aligned} \tag{17.2.12}$$

Thus $P[X_{U(n)} - X_{U(m)} = y \mid X_{U(m)} = x]$ does not depend on x. It can be shown that if $\mu = 0$, then $X_{U(n)} - X_{U(m)}$ is identically distributed as $X_{U(n-m)}$, $n > m$. It follows from (17.2.12) that $X_{U(n)} - X_{U(m)}$ and $X_{U(m)}$ $(n > m)$ are independent. Ahsanullah and Kirmani (1991) showed that, if $N = \min\{j \mid j > 1, \ X_j < X_1\}$, then NX_N and X_1 are identically distributed if $F(x) = 1 - \exp(-x/\theta)$, $x > 0$, $\theta > 0$. These properties are characterizing properties of the exponential distribution. For various characterizations of the exponential distribution, see Chapter 12.

17.3 Moments of Record Values

Without any loss of generality, we will consider in this section the standard exponential population, $e(0,1)$, with density $f(x) = \exp(-x)$, $0 \leq x < \infty$, in which case we have $f(x) = 1 - F(x)$.

From (17.2.10) it is obvious that $X_{U(n)}$ can be written as the sum of n i.i.d. random variables $V_1, V_2, \ldots, V_n$ each of which is distributed as $e(0,1)$. It is easy to verify that

$$\begin{aligned} E[X_{U(n)}] &= n \ , \\ \text{Var}(X_{U(n)}) &= n \ , \qquad \text{and} \\ \text{Cov}(X_{U(m)}, X_{U(n)}) &= m \ , \qquad m < n \ . \end{aligned} \tag{17.3.1}$$

Let $P_{m,n}$ be the correlation between $X_{U(m)}$ and $X_{U(n)}$, then $P_{m,n} = (m/n)^{1/2}$, $m < n$.

Some simple recurrence relations satisfied by single and product moments of record values are given by the following theorems. For details see Balakrishnan and Ahsanullah (1994).

Theorem 17.2 *For $n \geq 1$ and $r = 0, 1, 2, \ldots,$*

$$E[X_{U(n)}^{r+1}] = E[X_{U(n-1)}^{r+1}] + (r+1)E[X_{U(n)}^{r}] \tag{17.3.2}$$

and consequently, for $0 \leq m \leq n-1$, we can write

$$E[X_{U(n)}^{r+1}] = E[X_{U(m)}^{r+1}] + (r+1) \sum_{p=m+1}^{n} E[X_{U(p)}^{r}] \tag{17.3.3}$$

with $E[X_{U(0)}^{r+1}] = 0$ and $E[X_{U(n)}^{0}] = 1$.

Proof: For $n \geq 1$ and $r = 0, 1, \ldots,$ we have from (17.2.5)

$$\begin{aligned} E[X_{U(n)}^{r}] &= \frac{1}{\Gamma(n)} \int_0^\infty x^r \{R(x)\}^{n-1} f(x)\, dx \\ &= \frac{1}{\Gamma(n)} \int_0^\infty x^r \{R(x)\}^{n-1} \{1 - F(x)\} dx\ , \end{aligned}$$

since $f(x) = 1 - F(x)$. Upon integrating by parts treating x^r for integration and the rest of the integrand for differentiation, we obtain

$$\begin{aligned} E[X_{U(n)}^{r}] &= \frac{1}{\Gamma(n)(r+1)} \left[\int_0^\infty x^{r+1} \{R(x)\}^{n-1} f(x)\, dx \right. \\ &\qquad \left. -(n-1) \int_0^\infty x^{r+1} \{R(x)\}^{n-2} f(x)\, dx \right] \\ &= \frac{1}{r+1} \left[\int_0^\infty x^{r+1} \frac{1}{\Gamma(n)} \{R(x)\}^{n-1} f(x)\, dx \right. \\ &\qquad \left. - \int_0^\infty x^{r+1} \frac{1}{\Gamma(n-1)} \{R(x)\}^{n-2} f(x)\, dx \right] \\ &= \frac{1}{r+1} \left\{ E[X_{U(n)}^{r+1}] - E[X_{U(n-1)}^{r+1}] \right\}\ , \end{aligned}$$

which, when rewritten, gives the recurrence relation in (17.3.2). Then, by repeatedly applying the recurrence relation (17.3.2), we simply derive the recurrence relation in (17.3.3).

Remark 17.1 The recurrence relation in (17.3.2) can be used in a simple way to compute all the single moments of all record values. Once again, using the property that $f(y) = 1 - F(y)$, we can derive some simple recurrence relations for the product moments of record values, as follows.

Theorem 17.3 *For $m \geq 1$ and $r,\ s = 0, 1, 2, \ldots,$*

$$E[X_{U(m)}^{r} X_{U(m+1)}^{s+1}] = E[X_{U(m)}^{r+s+1}] + (s+1)E[X_{U(m)}^{r} X_{U(m+1)}^{s}]\ , \tag{17.3.4}$$

and for $1 \leq m \leq n-2,\ r, s = 0, 1, 2, \ldots,$

$$E[X_{U(m)}^{r} X_{U(n)}^{s+1}] = E[X_{U(m)}^{r} X_{U(n-1)}^{s+1}] + (s+1)E[X_{U(m)}^{r} X_{U(n)}^{s}]\ . \tag{17.3.5}$$

Proof: Let us consider for $1 \leq m < n$ and $r, s = 0, 1, 2, \ldots$,

$$E[X^r_{U(m)} X^s_{U(n)}] = \frac{1}{\Gamma(m)\Gamma(n-m)} \int_0^\infty x^r \{R(x)\}^{m-1} \frac{f(x)}{1-F(x)} I(x)\, dx \,, \quad (17.3.6)$$

where

$$\begin{aligned} I(x) &= \int_x^\infty y^s \{R(y) - R(x)\}^{n-m-1} f(y)\, dy \\ &= \int_x^\infty y^s \{R(y) - R(x)\}^{n-m-1} \{1 - F(y)\}\, dy \end{aligned}$$

since $f(y) = 1 - F(y)$. Upon integrating by parts treating y^s for integration and the rest of the integrand for differentiation, we obtain when $n = m+1$ that

$$I(x) = \frac{1}{s+1} \left[\int_x^\infty y^{s+1} f(y)\, dy - x^{s+1}\{1 - F(x)\} \right] ,$$

and when $n \geq m+2$ that

$$\begin{aligned} I(x) &= \frac{1}{s+1} \left[\int_x^\infty y^{s+1} \{R(y) - R(x)\}^{n-m-1} f(y)\, dy \right. \\ &\quad \left. -(n-m-1) \int_x^\infty y^{s+1} \{R(y) - R(x)\}^{n-m-2} f(y)\, dy \right] . \quad (17.3.7) \end{aligned}$$

Upon substituting the above expressions of $I(x)$ in Eq. (17.3.6) and simplifying, we obtain when $n = m+1$ that

$$E[X^r_{U(m)} X^s_{U(m+1)}] = \frac{1}{s+1} \left\{ E[X^r_{U(m)} X^{s+1}_{U(m+1)}] - E[X^{r+s+1}_{U(m)}] \right\} ,$$

and when $n \geq m+2$ that

$$E[X^r_{U(m)} X^s_{U(n)}] = \frac{1}{s+1} \left\{ E[X^r_{U(m)} X^{s+1}_{U(n)}] - E[X^r_{U(m)} X^{s+1}_{U(n-1)}] \right\} .$$

The recurrence relations in (17.3.4) and (17.3.5) follow readily when the above two equations are rewritten.

Remark 17.2 By repeated application of the recurrence relation in (17.3.5), with the help of the relation in (17.3.4), we obtain for $n \geq m+1$ that

$$E[X^r_{U(m)} X^{s+1}_{U(n)}] = E[X^{r+s+1}_{U(m)}] + (s+1) \sum_{p=m+1}^{n} E[X^r_{U(m)} X^s_{U(p)}] \,. \quad (17.3.8)$$

Corollary 17.2 *For $n \geq m+1$,*

$$Cov(X_{U(m)}, X_{U(n)}) = Var(X_{U(m)}) \,.$$

Corollary 17.3 *By repeated application of the recurrence relations in (17.3.4) and (17.3.5), we also obtain for $m \geq 1$*

$$E[X^r_{U(m)} X^{s+1}_{U(m+1)}] = \sum_{p=0}^{s+1} (s+1)^{(p)} E[X^{r+s+1-p}_{U(m)}] ,$$

and for $1 \le m \le n-2$

$$E[X^r_{U(m)} X^{s+1}_{U(n)}] = \sum_{p=0}^{s+1} (s+1)^{(p)} E[X^r_{U(m)} X^{s+1-p}_{U(n-1)}] \,,$$

where $(s+1)^{(0)} = 1$ *and* $(s+1)^{(i)} = (s+1)s \ldots (s+1-i+1)$, *for* $i \ge 1$.

Remark 17.3 The recurrence relations in Eqs. (17.3.4) and (17.3.5) can be used in a simple way to compute all the product moments of all record values.

Theorem 17.4 *For* $m \ge 2$ *and* $r, s = 0, 1, 2, \ldots,$

$$E[X^{r+1}_{U(m-1)} X^s_{U(m)}] = E[X^{r+s+1}_{U(m)}] - (r+1) E[X^r_{U(m)} X^s_{U(m+1)}] \,; \tag{17.3.9}$$

and for $2 \le m \le n-2$ *and* $r, s = 0, 1, 2, \ldots,$

$$E[X^{r+1}_{U(m-1)} X^s_{U(n-1)}] = E[X^{r+1}_{U(m)} U^s_{U(n-1)}] - (r+1) E[X^r_{U(m)} X^s_{U(n)}] \,. \tag{17.3.10}$$

Corollary 17.4 *By repeated application of the recurrence relation in (17.3.10), with the help of the relation in (17.3.9), we obtain for* $2 \le m \le n-1$ *and* $r, s = 0, 1, 2, \ldots$

$$E[X^{r+1}_{U(m-1)} X^s_{U(n-1)}] = E[X^{r+s+1}_{U(n-1)}] - (r+1) \sum_{p=m}^{n-1} E[X^r_{U(p)} X^s_{U(n)}] \,.$$

Corollary 17.5 *By repeated application of the recurrence relations in (17.3.9) and (17.3.10), we also obtain for* $m \ge 2$

$$E[X^{r+1}_{U(m-1)} X^s_{U(m)}] = \sum_{p=0}^{r+1} (-1)^p (r+1)^p E[X^{r+s+1-p}_{U(m+p)}] \,,$$

and for $2 \le m \le n-2$

$$E[X^{r+1}_{U(m-1)} X^s_{U(n-1)}] = \sum_{p=0}^{r+1} (-1)^p (r+1)^p E[X^{r+1-p}_{U(m+p)} X^s_{U(n-1+p)}] \,,$$

where $(r+1)^{(i)}$ *is as defined earlier.*

It is also important to mention here that this approach can easily be adopted to derive recurrence relations for product moments involving more than two record values. For example, proceeding along the lines of Theorem 17.3, we can easily show that for $1 \le m_1 < m_2 < \cdots < m_k$ and $r_1, r_2, \ldots, r_{k+1} = 0, 1, 2, \ldots$

$$\begin{aligned} E\left[\prod_{i=1}^{k} X^{r_i}_{U(m_i)} X^{r_{k+1}+1}_{U(m_k+1)}\right] &= E\left[\prod_{i=1}^{k-1} X^{r_i}_{U(m_i)} X^{r_k + r_{k+1}+1}_{U(m_k)}\right] \\ &\quad + (r_{k+1}+1) E\left[\prod_{i=1}^{k} X^{r_i}_{U(m_i)} X^{r_{k+1}}_{U(m_k+1)}\right] \,, \end{aligned}$$

and that for $1 \leq m_1 < m_2 < \cdots < m_{k+1} - 2$ and $r_1, r_2, \ldots, r_{k+1} = 0, 1, 2, \ldots$

$$E\left[\prod_{i=1}^{k} X_{U(m_i)}^{r_i} X_{U(m_{k+1})}^{r_{k+1}+1}\right] = E\left[\prod_{i=1}^{k} X_{U(m_i)}^{r_i} X_{U(m_{k+1}-1)}^{r_{k+1}+1}\right] + (r_{k+1}+1)E\left[\prod_{i=1}^{k+1} X_{U(m_i)}^{r_i}\right] .$$

17.4 Inter-Record Times

Let $\Delta_r = U(r+1) - U(r)$, $r = 1, 2, \ldots$, then Δ_r is called the rth inter-record time. Since $U(1) = 1$, we have $U(r) = 1 + \Delta_1 + \cdots + \Delta_{r-1}$, $r = 1, 2, \ldots$

Lemma 17.1 *For any* $n \geq 1$, $P[\Delta_n < \infty] = 1$.

Proof:

$$\begin{aligned} P[\Delta_1 = \infty] &= P\left[\bigcap_{i=2}^{\infty} X_i \leq X_1\right] \\ &= \lim_{m\to\infty} P\left[\bigcap_{i=2}^{m} X_i \leq X_1\right] \\ &= \lim_{m\to\infty} \frac{1}{m} = 0 . \end{aligned}$$

Similarly it can be shown that for $n \geq 2$, $P(\Delta_n = \infty) = 0$. □

Lemma 17.2

$$P[\Delta_n = k] = \sum_{i=0}^{k-1} \binom{k-1}{i} (-1)^i \frac{1}{(2+i)^n} , \qquad k = 1, 2, \ldots$$

Proof:

$$\begin{aligned} P[\Delta_1 = k \mid X_1 = x] &= P[U(2) = k+1 \mid X_1 = x] \\ &= (1 - e^{-x})^{k-1} e^{-x} \qquad \text{with } F(x) = 1 - e^{-x} \end{aligned}$$

and

$$\begin{aligned} P[\Delta_1 = k] &= \int_0^{\infty} (1 - e^{-x})^{k-1} e^{-x} e^{-x}\, dx \\ &= \int_0^1 t^{k-1}(1-t)\, dt \\ &= \frac{1}{k} - \frac{1}{k+1} = \frac{1}{k(k+1)} ; \end{aligned}$$

$$P[\Delta_1 \mid X_1 = x] = \frac{1}{e^{-x}} = e^x ,$$

$$\begin{aligned}\mathrm{Var}(\Delta_1 \mid X_1) &= \frac{1-e^{-x}}{e^{-2x}} = e^{2x} - e^{x} ,\\ E[\Delta_1] &= \infty .\end{aligned}$$

Similarly, it can be shown that

$$\begin{aligned}P[\Delta_n = k \mid X_{U(n)} = x_n] &= (1-e^{-x_n})^{k-1} e^{-x_n} ,\\ E[\Delta_n \mid X_{U(n)} = x_n] &= \frac{1}{e^{-x_n}} = e^{x_n} ,\\ \mathrm{Var}(\Delta_n \mid X_{U(n)} = x_n) &= \frac{1-e^{-x_n}}{e^{-2x_n}} = e^{2x_n} - e^{x_n} .\end{aligned}$$

$$\begin{aligned}P[\Delta_n = k] &= \int_0^\infty (1-e^{-x_n})^{k-1} e^{-x_n} \frac{x_n^{n-1}}{\Gamma(n)} e^{-x_n}\, dx_n\\ &= \sum_{i=0}^{k-1} \int_0^\infty \binom{k-1}{i} (-1)^i e^{-(2+i)x_n} \frac{x_n^{n-1}}{\Gamma(n)}\, dx_n\\ &= \sum_{i=0}^{k-1} \binom{k-1}{i} (-1)^i \frac{1}{(2+i)^n} ,\end{aligned}$$

$$\begin{aligned}P[\Delta_n > s] &= \int_0^\infty \sum_{k=s+1}^{\infty} (1-e^{-x_n})^{k-1} e^{-x_n} \frac{x_n^{n-1}}{\Gamma(n)} e^{-x_n}\, dx_n\\ &= \int_0^\infty \sum_{s=0}^{\infty} (1-e^{-x_n})^{s} \frac{x_n^{n-1}}{\Gamma(n)} e^{-x_n}\, dx_n\\ &= \sum_{i=0}^{s} \binom{s}{i} (-1)^i \frac{1}{(1+i)^n} . \qquad (17.4.1)\end{aligned}$$

□

Theorem 17.5 *For* $n = 1, 2, \ldots$

$$E[\Delta_n^\alpha] = \infty, \ \alpha \geq 1$$

and

$$E[\Delta_n^\alpha] < \infty, \ \alpha < 1 .$$

Proof:

$$\begin{aligned}E[\Delta_n] &= \sum_{s=0}^{\infty} P[\Delta_n > s]\\ &= \int_0^\infty \sum_{s=0}^{\infty} (1-e^{-x_n})^{s} \frac{x_n^{n-1}}{\Gamma(n)} e^{-x_n}\, dx_n\\ &= \int_0^\infty \frac{x_n^{n-1}}{\Gamma(n)}\, dx_n = \infty .\end{aligned}$$

For $\alpha > 1$,

$$E[\Delta_n^\alpha] \geq E[\Delta_n] = \infty \ .$$

For $0 < \alpha < 1$,

$$\begin{aligned} E[\Delta_n^\alpha] &= \int_0^\infty \sum_{k=1}^\infty k^\alpha (1-e^{-x_n})^{k-1} e^{-x_n} \, \frac{x_n^{n-1}}{\Gamma(n)} \, e^{-x_n} \, dx_n \\ &= \int_0^\infty \frac{e^{-2t} t^{n-1}}{\Gamma(n)} \sum_{k=1}^\infty k^\alpha (1-e^{-t})^{k-1} dt \ . \end{aligned}$$

Now

$$\begin{aligned} \sum_{k=1}^\infty k^\alpha (1-e^{-t})^{k-1} &\leq \int_0^\infty (s+1)^\alpha e^{-s\log(1-e^{-t})^{-1}} \, ds \\ &= \int_1^\infty u^\alpha e^{-(u-1)\log(1-e^{-t})^{-1}} \, du \\ &\leq (1-e^{-1}) \int_1^\infty u^\alpha e^{-u\log(1-e^{-t})^{-1}} \, du \\ &= (1-e^{-1})^{-1} \, \frac{1+\alpha}{(-\log(1-e^{-t}))^{\alpha+1}} \\ &\leq (1-e^{-1})^{-1}(1+\alpha) \, e^{(1+\alpha)t} \ . \end{aligned}$$

Thus

$$E[\Delta_n^\alpha] \leq (\Gamma(n))^{-1}(1+\alpha) \int_0^\infty (1-e^{-t})^{-1} \, e^{-(1-\alpha)t} \, t^{n-1} \, dt \ .$$

Since $x \leq x(1-e^{-x})^{-1} \leq x+1$, we have

$$\begin{aligned} E[\Delta_n^\alpha] &\leq (\Gamma(n))^{-1}(1+\alpha) \int_0^\infty e^{-(1-\alpha)x}(x+1)x^{n-2} \, dt \\ &= (1-\alpha)^{-n}(1+\alpha)(1+(1-\alpha)n^{-1}) < \infty \ . \end{aligned}$$

□

Let $P_g(n,t)$ be the probability generating function of Δ_n ($|t| < 1$), then

$$\begin{aligned} P_g(1,t) &= \sum_{x=1}^\infty P[\Delta_1 = x] t^x = \sum_{x=1}^\infty \frac{1}{x(x+1)} \, t^x \\ &= 1 + \frac{1-t}{t} \, \ln(1-t) \ , \end{aligned}$$

$$\begin{aligned} E[t^{\Delta_n} \mid X_{U(n)} = x_n] &= \sum_{k=1}^\infty t^k (F(x_n))^{k-1}(1-F(x_n)) \\ &= t \, \frac{1-F(x_n)}{1-t\,F(x_n)} \end{aligned}$$

Table 17.1: Values of $P[\Delta_n < s]$

	n					
s	1	2	3	4	5	6
2	0.5000	0.2500	0.1250	0.0625	0.0313	0.0156
5	0.8000	0.5433	0.3323	0.1890	0.1039	0.0552
10	0.9000	0.7071	0.4936	0.3143	0.1870	0.1058
15	0.9333	0.7788	0.5803	0.3926	0.2459	0.1451
20	0.9500	0.8201	0.6365	0.4484	0.2912	0.1773
30	0.9667	0.8668	0.7071	0.5251	0.3586	0.2283
50	0.9800	0.9100	0.7813	0.6153	0.4460	0.3002
100	0.9900	0.9481	0.8573	0.7209	0.5615	0.4062
200	0.9950	0.9706	0.9095	0.8047	0.6657	0.5137
500	0.9980	0.9864	0.9522	0.8836	0.7780	0.6453
1000	0.9950	0.9925	0.9712	0.9235	0.8425	0.6985
5000	0.9998	0.9982	0.9916	0.9734	0.9353	0.7798

and

$$\begin{aligned} P_g(n,t) &= \int_{-\infty}^{\infty} t\,\frac{1-F(x)}{1-t\,F(x)}\,\frac{(-\ln(1-F(x))^{n-1}}{\Gamma(n)}\,f(x)\,dx \\ &= \int_0^{\infty} t\,\frac{1-u}{1-tu}\,\frac{(-\ln(1-u))^{n-1}}{\Gamma(n)}\,du \\ &= 1-(1-t)\int_0^{\infty}\frac{1}{1-tu}\,\frac{(-\ln(1-u))^{n-1}}{\Gamma(n)}\,du\ . \end{aligned}$$

The following table gives the values of $P[\Delta_n < s]$ for some selected values of n. From (17.4.1), we have $P[\Delta_n < s] = 1 - \sum_{i=0}^{s-1}\binom{s}{i}(-1)^i\,\frac{1}{(1+i)^n}$.

17.5 Linear Estimation of Location and Scale Parameters

17.5.1 Best Linear Unbiased Estimates

Suppose $X_{U(1)}, X_{U(2)}, \ldots, X_{U(m)}$ are the m record values from an i.i.d. sequence $e(\mu,\theta)$. Let $Y_i = (X_{U(i)} - \mu)/\theta,\ i = 1, 2, \ldots, m$, then

$$E[Y_i] = i = \mathrm{Var}(Y_i), \qquad i = 1, 2, \ldots, m,$$

and

$$\mathrm{Cov}(Y_i, Y_j) = \min(i,j)\ .$$

Let $\boldsymbol{X} = (X_{U(1)}, X_{U(2)}, \ldots, X_{U(m)})$, then

$$E[\boldsymbol{X}] = \mu\boldsymbol{L} + \theta\boldsymbol{\delta}\ ,$$

$$\mathrm{Var}(\boldsymbol{X}) = \theta^2\boldsymbol{V}\ ,$$

where

$$L = (1,1,\ldots,1)' \ ,\delta = (1,2,\ldots,m)' \ ,$$

$$V = (V_{ij}) \ , V_{ij} = \min(i,j) \ , \qquad i,j = 1,2,\ldots,m \ .$$

The inverse V^{-1} $(= V^{ij})$ can be expressed as

$$\begin{aligned} V^{ij} \quad = \quad & 2 \quad \text{if } i = j = 1,2,\ldots,m-1 \\ & 1 \quad \text{if } i = j = m \\ & -1 \quad \text{if } |i-j| = 1, \ i,j = 1,2,\ldots,m \\ & 0 \quad \text{otherwise} \ . \end{aligned}$$

The minimum variance linear unbiased estimates (MVLUEs) μ^*, θ^* of μ and θ, respectively, [see David (1981); Balakrishnan and Cohen (1991)] are

$$\begin{aligned} \mu^* &= -\delta' V^{-1}(L\delta' - \delta L')V^{-1}X/\Delta \ , \\ \theta^* &= L'V^{-1}(L\delta' - \delta L')V^{-1}X/\Delta \ , \end{aligned}$$

where $\Delta = (L'V^{-1}L)(\delta'V^{-1}\delta) - (L'V^{-1}\delta)^2$ and

$$\begin{aligned} \text{Var}(\mu^*) &= \theta^2 \delta' V^{-1}\delta/\Delta \ , \\ \text{Var}(\theta^*) &= \theta^2 L'V^{-1}L/\Delta \ , \\ \text{Cov}(\mu^*,\theta^*) &= -\theta^2 L'V^{-1}\delta/\Delta \ . \end{aligned}$$

It can be shown that

$$\begin{aligned} L'V^{-1} &= (1,0,0,\ldots,0) \ , \\ \delta'V^{-1} &= (0,0,0,\ldots,1) \ , \\ \delta'V^{-1}\delta &= m \qquad \text{and} \qquad \Delta = m-1 \ . \end{aligned}$$

On simplification we get

$$\begin{aligned} \mu^* &= (mX_{U(1)} - X_{U(m)}) \,/\, (m-1) \ , \\ \theta^* &= (X_{U(m)} - X_{U(1)}) \,/\, (m-1) \ , \end{aligned} \tag{17.5.1}$$

and

$$\begin{aligned} \text{Var}(\mu^*) &= m\theta^2/(m-1) \ , \\ \text{Var}(\theta^*) &= \theta^2/(m-1) \ , \\ \text{Cov}(\mu^*,\theta^*) &= -\theta^2/(m-1) \ . \end{aligned} \tag{17.5.2}$$

17.5.2 Best Linear Invariant Estimators

The best linear invariant (in the sense of minimum mean squared error and invariance) estimators (BLIEs) μ^{**} and θ^{**} are

$$\mu^{**} = \mu^* - \theta^* \left(\frac{E_{12}}{1 + E_{22}} \right)$$

and

$$\theta^{**} = \theta^*/(1+E_{22}) \ ,$$

where μ^* and θ^* are the MVLUEs of μ and θ and

$$\begin{pmatrix} \mathrm{Var}(\mu^*) & \mathrm{Cov}(\mu^*,\theta^*) \\ \mathrm{Cov}(\mu^*,\theta^*) & \mathrm{Var}(\theta^*) \end{pmatrix} = \theta^2 \begin{pmatrix} E_{11} & E_{12} \\ E_{12} & E_{22} \end{pmatrix} .$$

The mean-squared-errors of these estimators are

$$\mathrm{MSE}(\mu^{**}) = \theta^2 \left(E_{11} - E_{12}^2(1+E_{22})^{-1}\right)$$

and

$$\mathrm{MSE}(\theta^{**}) = \theta^2 E_{22}(1+E_{22})^{-1} \ .$$

We have

$$E(\mu^{**}-\mu)(\theta^{**}-\theta) = \theta^2 E_{12}(1+E_{22})^{-1} \ .$$

Using the value of E_{11}, E_{12} and E_{22} from (17.5.2), we obtain

$$\begin{aligned} \mu^{**} &= \left\{(m+1)X_{U(1)} - X_{U(m)}\right\}/m \ , \\ \theta^{**} &= \left\{X_{U(m)} - X_{U(1)}\right\}/m \ , \end{aligned}$$

$$\mathrm{Var}(\mu^{**}) = \frac{m+1}{m}\,\theta^2$$

and

$$\mathrm{Var}(\theta^{**}) = \frac{m-1}{m^2}\,\theta^2 \ .$$

17.6 Prediction of Record Values

We will predict the sth record value based on the first m record values for $s > m$. Let $\boldsymbol{W}' = (W_1, W_2, \ldots, W_m)$, where $\theta^2 W_i = \mathrm{Cov}(X_{U(i)}, X_{U(s)})$, $i = 1,\ldots,m$ and $\alpha^* = E[(X_{U(s)} - \mu)/\theta]$. The best linear unbiased predictor of $X_{U(s)}$ [see Ahsanullah (1980)] is $X^*_{U(s)}$, where

$$X^*_{U(s)} = \mu^* + \theta^*\alpha^* + \boldsymbol{W}'\boldsymbol{V}^{-1}(\boldsymbol{X} - \mu^*\boldsymbol{L} - \theta^*\boldsymbol{\delta}) \ .$$

It can be shown that $\boldsymbol{W}'\boldsymbol{V}^{-1}(\boldsymbol{X} - \mu^*\boldsymbol{L} - \theta^*\boldsymbol{\delta}) = 0$. Thus

$$\begin{aligned} X^*_{U(s)} &= \mu^* + s\theta^* \\ &= \left\{(s-1)X_{U(m)} + (m-s)X_{U(1)}\right\}/(m-1) \ . \end{aligned}$$

Also,

$$\begin{aligned} E[X^*_{U(s)}] &= \mu + s\theta \ , \\ \mathrm{Var}(X^*_{U(s)}) &= \theta^2(m+s^2-2s)/(m-1) \ , \end{aligned}$$

and

$$\begin{aligned} \mathrm{MSE}(X^*_{U(s)}) &= E[X^*_{U(s)} - X_{U(s)}]^2 \\ &= \theta^2(s+m)(s-1)/(m-1) \ . \end{aligned}$$

Let $X^{**}_{U(s)}$ be the best linear invariant predictor of $X_{U(s)}$. Then it can be shown [see Ahsanullah (1980)] that

$$X^{**}_{U(s)} = X^{*}_{U(s)} - C_{12}(1+E_{22})^{-1}\theta^* ,$$

where

$$C_{12}\theta^2 = \operatorname{Cov}\left(\theta^*, (\boldsymbol{L} - \boldsymbol{W}'\boldsymbol{V}^{-1}\boldsymbol{L})\mu^* + (\alpha^* - \boldsymbol{W}'\boldsymbol{V}^{-1}\boldsymbol{\delta})\theta^*\right)$$

and $\theta^2 E_{22} = \operatorname{Var}(\theta^*)$.

On simplification, we get

$$X^{**}_{U(s)} = \frac{m-s}{m} X_{U(1)} + \frac{s}{m} X_{U(m)} ,$$

and

$$E[X^{**}_{U(s)}] = \mu + \left(\frac{ms+m-s}{m}\right)\theta ,$$

and

$$\operatorname{Var}(X^{**}_{U(s)}) = \theta^2(m^2+ms^2-s^2)/m^2 ,$$

and

$$\operatorname{MSE}(X^{**}_{U(s)}) = \operatorname{MSE}(X^{*}_{U(s)}) - \frac{(s-m)^2}{m(m-1)}\,\theta^2 .$$

It is well known that the best (unrestricted) least squares predictor $\overline{X}^*_{U(s)}$ of $X_{U(s)}$ is

$$\begin{aligned}\overline{X}^*_{U(s)} &= E[X_{U(s)} \mid X_{U(1)}, \ldots, X_{U(m)}] \\ &= X_{U(m)} + (s-m)\theta .\end{aligned}$$

But $\overline{X}^*_{U(s)}$ depends on the unknown parameter θ. If we substitute the minimum variance unbiased estimate θ^* for θ, then $\overline{X}^*_{U(s)}$ becomes equal to $X^*_{U(s)}$. Now

$$\begin{aligned}E[\overline{X}^*_{U(s)}] &= \mu + s\theta = E[X_{U(s)}] ,\\ \operatorname{Var}(\overline{X}^*_{U(s)}) &= m\theta^2 ,\end{aligned}$$

and

$$\operatorname{MSE}(\overline{X}^*_{U(s)}) = E[\overline{X}^*_{U(s)} - X_{U(s)}]^2 = (s-m)\theta^2 .$$

By considering the mean square errors of $X^*_{U(s)}$, $X^{**}_{U(s)}$ and $\overline{X}^*_{U(s)}$, it can be shown that

$$\operatorname{MSE}(\overline{X}^*_{U(s)}) < \operatorname{MSE}(X^{**}_{U(s)}) < \operatorname{MSE}(X^{*}_{U(s)}) .$$

17.7 δ-Exceedance Records

Balakrishnan, Balasubramanian, and Panchapakesan (1994) introduced the δ-exceedance record model. An observation X_j will be called a δ-exceedance upper (lower) record if it is larger (smaller) than the previous record by the quantity δ. The indices at which the (upper) record values occur are given by the record times $\{U(n,\delta)\}$, $n>0$, where $U(n,\delta) = \min\{j \mid j > U(n-1,\delta),\ X_j > X_{U(n-1,\delta)} + \delta\}$ and $U(1,\delta)=1$. If $\delta = 0$, then

$U(n,\delta) = U(n)$ and the record times are independent of the distribution function F. If $\delta > 0$, then the probability distribution function of $U(n,\delta)$ will depend on n, δ and F.

Assuming the standard exponential distribution with pdf $f(x) = e^{-x}$, $0 \le x < \infty$, we can write the joint density of $X_{U(1,\delta)}, X_{U(2,\delta)}, \ldots, X_{U(n,\delta)}$ as

$$f_{X_{U(1,\delta)},\ldots,X_{U(n,\delta)}}(x_1, x_2, \ldots, x_n) = e^{n\delta}\, e^{-x_n}\,,$$
$$x_1 > 0,\ x_{i+1} > x_i + \delta,\ i = 1, \ldots, n-1,\ \delta \ge 0\,. \tag{17.7.1}$$

Upon integrating out the variables $x_1, \ldots, x_{n-1}$, we get the density of $X_{U(n,\delta)}$ as

$$\begin{aligned} f_{X_{U(n,\delta)}}(x_n) &= \frac{1}{(n-1)!}\, e^{-x_n+(n-1)\delta}\, (x_n - (n-1)\delta)^{n-1}\,, \\ & \qquad x_n > (n-1)\delta,\ \delta \ge 0\,. \end{aligned} \tag{17.7.2}$$

(17.7.2) equals to (17.2.10) if we take $\mu = 0$, $\theta = 1$ and $\delta = 0$. (17.7.2) implies that $X_{U(n,\delta)} - (n-1)\delta$ is distributed as gamma with shape parameter n and scale parameter 1. Hence,

$$E[X_{U(n,\delta)}] = n + (n-1)\delta\,,$$
$$\mathrm{Var}(X_{U(n,\delta)}) = n\,.$$

Considering the joint density of $X_{U(m,\delta)}$ and $X_{U(n,\delta)}$, $n > m$, it can be shown that

$$\mathrm{Cov}(X_{U(m,\delta)}, X_{U(n,\delta)}) = \mathrm{Var}(X_{U(m,\delta)}) = m\,.$$

Let $\Delta_r(\delta) = U(r+1,\delta) - U(r,\delta)$, $r = 1, 2, \ldots,$ $\delta \ge 0$, with $U(1,\delta) = 1$. Now

$$\begin{aligned} P[\Delta_1(\delta) = k \mid X_1 = x] &= P[U(2,\delta) = k+1 \mid X_1 = x] \\ &= (1 - e^{-(x+\delta)})^{k-1}\, e^{-(x+\delta)}\,. \end{aligned}$$

Thus,

$$\begin{aligned} E[\Delta_1(\delta) \mid X_1 = x] &= e^{x+\delta}\,, \\ \mathrm{Var}[\Delta_1(\delta) \mid X_1 = x] &= e^{2(x+\delta)} - e^{x+\delta}\,. \end{aligned}$$

Similarly it can be shown that

$$\begin{aligned} P[\Delta_n(\delta) = k \mid X_{U(n,\delta)} = x_n] &= (1 - e^{-(x_n+\delta)})^{k-1}\, e^{-(x_n+\delta)}\,, \\ E[\Delta_n(\delta) \mid X_n = x_n] &= e^{x_n+\delta}\,, \\ \mathrm{Var}(\Delta_n(\delta) \mid X_n = x_n) &= e^{2(x_n+\delta)} - e^{x_n+\delta}\,, \end{aligned}$$

and

$$\begin{aligned} P[\Delta_n(\delta) = k] &= \int_0^\infty (1 - e^{-(x_n+\delta)})^{k-1}\, e^{-(x_n+\delta)} f_{U(n,\delta)}(x_n)\, dx_n \\ &= \sum_{i=0}^{k-1} \binom{k-1}{i} (-1)^i \int_{(n-1)\delta}^{\infty} e^{-(i+1)(x_n+\delta)} \end{aligned}$$

$$
\begin{aligned}
&\cdot \frac{1}{\Gamma(n)} e^{-(x_n-(n-1)\delta)}(x_n-(n-1)\delta)^{n-1}\,dx_n \\
&= \sum_{i=0}^{k-1}\binom{k-1}{i}(-1)^i e^{n\delta}\int_{(n-1)\delta}^{\infty}\frac{e^{-(2+i)(x_n+\delta)}}{\Gamma(n)} \\
&\quad \cdot (x_n-(n-1)\delta)^{n-1}\,dx_n \\
&= \sum_{i=0}^{k-1}\binom{k-1}{i}(-1)^i e^{n\delta}\int_0^{\infty}\frac{t^{n-1}}{\Gamma(n)} e^{-(2+i)(t+n\delta)}\,dt \\
&= \sum_{i=0}^{k-1}\binom{k-1}{i}(-1)^i e^{-(1+i)n\delta}\frac{1}{(2+i)^n}\,. \qquad (17.7.3)
\end{aligned}
$$

If $\delta = 0$, then (17.7.3) equals the expression of $P[\Delta = k]$ as given in Lemma 17.2.

References

Ahsanullah, M. (1977). A characteristic property of the exponential distribution, *Annals of Statistics*, **5**, 580-582.

Ahsanullah, M. (1979). Characterization of the exponential distribution by record values, *Sankhyā, Series B*, **41**, 116-121.

Ahsanullah, M. (1980). Linear prediction of record values for the two parameter exponential distribution, *Annals of the Institute of Statistical Mathematics*, **32**, 363-368.

Ahsanullah, M. (1981a). Record values of the exponentially distributed random variables, *Statistische Hefte*, **2**, 121-127.

Ahsanullah, M. (1981b). On a characterization of the exponential distribution by weak homoscedasticity of record values, *Biometrical Journal*, **23**, 715-717.

Ahsanullah, M. (1982). Characterization of the exponential distribution by some properties of record values, *Statistische Hefte*, **23**, 326-332.

Ahsanullah, M. (1987a). Two characterizations of the exponential distribution, *Communications Statistics - Theory and Methods*, **16**, 375-381.

Ahsanullah, M. (1987b). Record statistics and the exponential distribution, *Pakistan Journal of Statistics, Series A*, **3**, 17-40.

Ahsanullah, M. (1988). *Introduction to Record Statistics*, Ginn Press, Needham Heights, Massachusetts.

Ahsanullah, M. (1991). Some characteristic properties of the record values from the exponential distribution, *Sankhyā, Series B*, **53**, 403-408.

Ahsanullah, M. and Kirmani, S.N.U.A. (1991). Characterizations of the exponential distribution by lower record values, *Communications in Statistics - Theory and Methods*, **20**, 1293-1299.

Arnold, B.C. and Balakrishnan, N. (1989). *Relations, Bounds and Approximations for Order Statistics*, Lecture Notes in Statistics No. 53, Springer-Verlag, New York.

Arnold, B.C., Balakrishnan, N., and Nagaraja, H.N. (1992). *A First Course in Order Statistics*, John Wiley & Sons, New York.

Azlarov, T.A. and Volodin, N.A. (1986). *Characterization Problems Associated with the Exponential Distribution*, Springer-Verlag, New York.

Balakrishnan, N. and Ahsanullah, M. (1994). Relations for single and product moments of record values from exponential distribution, *Journal of Applied Statistical Science* (To appear).

Balakrishnan, N., Balasubramanian, K., and Panchapakesan, S. (1994). δ-exceedance records, *Submitted for publication.*

Barton, D.E. and Mallows, C.L. (1965). Some aspects of the random sequence, *Annals of Mathematical Statistics*, **36**, 236-260.

Chandler, K.M. (1952). The distribution and frequency of record values, *Journal of the Royal Statistical Society, Series B*, **14**, 220-228.

Cinlar, E. (1975). *Introduction to Stochastic Processes*, Prentice-Hall, New Jersey.

Feller, W. (1966). *An Introduction to Probability Theory and Its Applications*, Vol. II, John Wiley & Sons, New York.

Galambos, J. (1987). *The Asymptotic Theory of Extreme Order Statistics*, Second edition, Krieger, Malabar, Florida.

Galambos, J. and Seneta, E. (1975). Record times, *Proceedings of the American Mathematical Society*, **50**, 383-387.

Glick, N. (1978). Breaking records and breaking boards, *American Mathematical Monthly*, **85**, 2-26.

Gnedenko, B. (1943). Sur la distribution limite du terme maximum d'une serie aletoise, *Annals of Mathematics*, **44**, 423-453.

Goldberger, A.S. (1962). Best linear unbiased predictors in the generalized linear regression model, *Journal of the American Statistical Association*, **57**, 369-375.

Grosswald, E. and Kotz, S. (1981). An integrated lack of memory property of the exponential distribution, *Annals of the Institute of Statistical Mathematics*, **33**, 205-214.

Gupta, R.C. (1984). Relationships between order statistics and record values and some characterization results, *Journal of Applied Probability*, **21**, 425-430.

Kakosyan, A.V., Klebanov, L.B., and Melamed, J.A. (1984). *Characterization of Distribution by the Method of Intensively Monotone Operators*, Lecture Notes in Mathematics No. 1088, Springer-Verlag, New York.

Kirmani, S.N.U.A. and Beg, M.I. (1983). On characterization of distributions by expected records, *Sankhyā, Series A*, **48**, 463-465.

Nagaraja, H.N. (1988). Record values and related statistics - a review, *Communications in Statistics - Theory and Methods*, **17**, 2223-2238.

Nevzorov, V.B. (1988). Records, *Theory of Probability and Its Applications*, **32**, 201-228.

CHAPTER 18

Related Distributions and Some Generalizations

Norman L. Johnson,[1] Samuel Kotz,[2] and N. Balakrishnan[3]

18.1 Introduction

We have already seen, in the preceding chapters, reference being made to several univariate distributions. In this chapter, we describe some specific univariate distributions that are directly related to the exponential distribution. Since any continuous distribution may be related to the exponential distribution by means of the inverse probability integral transformation, we elicit here only those that serve as interesting models and are not trivially related to the exponential. The related multivariate distributions are discussed in Chapters 20, 21 and 22.

18.2 Transformed Distributions

Let X be a standard exponential random variable with probability density function $f_X(x) = e^{-x}$, $x > 0$. Then, by reflecting this density about the origin ($x = 0$), we obtain the standard *double exponential distribution* with density function

$$f_X(x) = \frac{1}{2}\, e^{-|x|}, \qquad -\infty < x < \infty \,. \tag{18.2.1}$$

This is also called the standard *First Law of Laplace* (the *Second Law* is the unit normal distribution), or simply *Laplace distribution.*

Several other important continuous distributions are related to exponential distribution by simple transformations. Some notable examples are presented below in Table 18.1.

---

[1]University of North Carolina, Chapel Hill, North Carolina
[2]University of Maryland, College Park, Maryland
[3]McMaster University, Hamilton, Ontario, Canada

Table 18.1: Transformations of exponential distributions

Transformation $X =$	Limits for y	Sdf of Y, $\overline{F}_Y(y) = 1 - F_Y(y)$	Pdf of Y, $f_Y(y) = dF_Y(y)/dy$	Name of distribution
Y^c $(c > 0)$	$(0, \infty)$	e^{-y^c}	$cy^{c-1}e^{-y^c}$	Weibull
$c \log Y$ $(c > 0)$	$(1, \infty)$	y^{-c}	cy^{-c-1}	Pareto
e^Y	$(-\infty, \infty)$	e^{-e^y}	$e^y\, e^{-e^y}$	Extreme Value*
$-\log(1 - Y)$	$(0, 1)$	$1 - y$	1	Uniform

* Extreme value distributions are also, sometimes, called 'double exponential' distributions, causing on occasion confusion with Laplace distributions.

Of course, as it has already been mentioned, *any* univariate continuous distribution can be transformed into *any* other univariate continuous distribution, by the transformation $y = F_Y^{-1}(F_X(x))$. It is the *simplicity* of the transformations in Table 18.1 that is noteworthy. Judicious use of such transformations makes it possible to apply well-established techniques, developed for exponentially distributed variables, in the analysis of data for which one of the related distributions provides a suitable model. In the cases of Weibull and Pareto, it is necessary to know the value of the shape parameter c, but even if this value is not accurately known, exploratory calculations using wisely chosen assumed values for c can often provide useful information. For an elaborate treatment of these distributions, interested readers may refer to Johnson, Kotz, and Balakrishnan (1994, 1995).

The linear function

$$T = \sum_{i=1}^{k} \theta_i X_i \ , \tag{18.2.2}$$

where $\theta_i > 0$, no two θ_i's are equal, and the X_i's are mutually independent with a common standard exponential distribution, has the probability density function

$$f_T(t) = \sum_{i=1}^{k} \left\{ \prod_{j \neq i} (\theta_i - \theta_j)^{-1} \right\} \theta_i^{k-2}\, e^{-t/\theta_i}, \qquad t > 0 \ . \tag{18.2.3}$$

The distribution (18.2.3) has found applications in psychology, queueing theory, and reliability. It is, in fact, a special case of *mixture* of exponential distributions, discussed in Section 18.4. It has also been called a *general gamma* or a *general Erlang* distribution.

18.3 Quadratic Forms

The quadratic form

$$Q = \boldsymbol{Z}'\boldsymbol{A}\boldsymbol{Z} = \sum_{i=1}^{2k} \sum_{j=1}^{2k} a_{ij} Z_i Z_j \qquad (a_{ij} = a_{ji}) \tag{18.3.1}$$

in $2k$ mutually independent standard normal variables $(Z_1, Z_2, \ldots, Z_{2k}) = \boldsymbol{Z}'$, where A is a positive definite symmetric $(2k \times 2k)$ matrix, is distributed as

$$\sum_{i=1}^{2k} \phi_i U_i^2 \qquad (U_i\text{'s} \rightarrow N(0,1) \text{ and independent}) \tag{18.3.2}$$

where $\phi_1, \phi_2, \ldots, \phi_{2k}$ are the roots of the equation $|\boldsymbol{A} - \phi \boldsymbol{I}| = 0$ (the *latent roots* or *eigenvalues* of $\boldsymbol{A}$). If these roots are equal in pairs, i.e.,

$$\phi_{2h-1} = \phi_{2h} = \frac{1}{2}\,\theta_h \qquad (h = 1, 2, \ldots, k)\,, \tag{18.3.3}$$

and $\theta_h \neq \theta_g$ for $h \neq g$, then Q is distributed as T in (18.2.2) since $U_{2h-1}^2 + U_{2h}^2$ is distributed as $2X_h$, where X_h has the standard exponential distribution.

This result underlies the genesis of exponential distributions as the resultant distribution of squared aiming error from independent perpendicular errors in two dimensions with common normal distribution having zero expected value (*unbiased*). If the errors are still independent unbiased and normal but possess different variances, the squared resultant error is distributed as a linear function

$$\sigma_1^2 U_1^2 + \sigma_2^2 U_2^2 \tag{18.3.4}$$

where U_1 and U_2 are independent standard normal variables. If, further the expected values of the components are ξ_1, ξ_2 (say), then the distribution of squared aiming error is that of

$$(\xi_1 + \sigma_1 U_1)^2 + (\xi_2 + \sigma_2 U_2)^2 = \sigma_1^2 \left(U_1 + \frac{\xi_1}{\sigma_1}\right)^2 + \sigma_2^2 \left(U_2 + \frac{\xi_2}{\sigma_2}\right)^2 . \tag{18.3.5}$$

If $\sigma_1 = \sigma_2 = \sigma$ (but $(\xi_1, \xi_2) \neq (0,0)$), we have the distribution of

$$\sigma^2 \left\{ \left(U_1 + \frac{\xi_1}{\sigma}\right)^2 + \left(U_2 + \frac{\xi_2}{\sigma}\right)^2 \right\} . \tag{18.3.6}$$

This distribution depends only on σ^2 and $(\xi_1^2 + \xi_2^2)/\sigma^2$. The distribution of $\left\{ \left(U_1 + \frac{\xi_1}{\sigma}\right)^2 + \left(U_2 + \frac{\xi_2}{\sigma}\right)^2 \right\}$ depends only on $(\xi_1^2 + \xi_2^2)/\sigma^2$; it is the noncentral chi-squared distribution with two degrees of freedom and noncentrality parameter $(\xi_1^2 + \xi_2^2)/\sigma^2$.

18.4 Mixtures of Exponential Distributions

Mixtures of exponential distributions are often used to represent variation from a pure exponential distribution in models of lifetime distributions. They correspond to the concept of a population composed of several subpopulations, within each of which there is a constant hazard rate. If there are k such populations $\Pi_1, \Pi_2, \ldots, \Pi_k$ in proportions $p_1, p_2, \ldots, p_k$ $(p_1 + p_2 + \cdots + p_k = 1)$ and in Π_j there is a constant hazard rate λ_j, then the cumulative distribution function for the lifetime distribution in the whole population is

$$F_T(t) = \sum_{j=1}^{k} p_j(1 - e^{-\lambda_j t}), \qquad (p_j, \lambda_j > 0;\ t > 0)\ , \tag{18.4.1}$$

and the probability density function is

$$f_T(t) = \sum_{j=1}^{k} p_j \lambda_j\, e^{-\lambda_j t}, \qquad t > 0\ . \tag{18.4.2}$$

The mixture distribution (18.4.1) does not have a constant hazard rate. The hazard rate function is

$$\rho_T(t) = \left\{ \sum_{j=1}^{k} p_j \lambda_j\, e^{-\lambda_j t} \right\} \Big/ \left\{ \sum_{j=1}^{k} p_j e^{-\lambda_j t} \right\}\ . \tag{18.4.3}$$

It is of interest to observe that the hazard rate $\rho_T(t)$ in (18.4.3) is a weighted average of the individual hazard rates $\{\lambda_i\}$ with weights $p_i e^{-\lambda_i t}/\{\sum_{j=1}^{k} p_j e^{-\lambda_j t}\}$. More details on the mixture-exponential distributions and their applications may be found in Chapter 19.

It is not, of course, essential that all p_j's in (18.4.1) be positive (though if they are not, they cannot reasonably be regarded as 'proportions'). This more general class is sometimes called *generalized hyperexponential*, and sometimes *generalized mixed exponential* distributions. For these distributions, the p_j's and λ_j's must satisfy certain conditions to ensure that $F_T(t)$ in (18.4.1) can, indeed, be a cumulative distribution function. The conditions

$$\lim_{t\to 0} F_T(t) = 0 \qquad \text{and} \qquad \lim_{t\to\infty} F_T(t) = 1 \tag{18.4.4}$$

are easily seen to be satisfied. We also need

$$f_T(t) = \frac{d}{dt}\, F_T(t) = \sum_{j=1}^{k} p_j \lambda_j e^{-\lambda_j t} \ge 0 \text{ for all } t > 0\ . \tag{18.4.5}$$

Without loss of any generality, we arrange to have $\lambda_1 < \lambda_2 < \cdots < \lambda_k$. Steutel (1967) obtained the necessary set of conditions

$$\sum_{j=1}^{k} p_j \lambda_j \ge 0 \qquad \text{and} \qquad p_1 > 0\ . \tag{18.4.6}$$

Bartholomew (1969) showed that the set of conditions

$$\sum_{j=1}^{r} p_j \lambda_j \ge 0 \qquad (r = 1, 2, \ldots, k) \tag{18.4.7}$$

is sufficient to ensure that $F_T(t)$ can be a cumulative distribution function. He had also revised this condition to

$$\sum_{j=1}^{k} p_j\lambda_j \geq 0 \qquad \text{and} \qquad \sum_{j=2}^{k-2} p_j\lambda_j \times \prod_{i=k-j+1}^{k} (\lambda_i - \lambda_j) \geq 0 . \tag{18.4.8}$$

Harris, Marchal, and Botta (1992) arrived at the following sufficient condition. Writing

$$f_T(t) = f_T^+(t) + f_T^-(t) = \left(\sum_{p_j>0} + \sum_{p_j<0} \right) p_j\lambda_j e^{-\lambda_j t} , \tag{18.4.9}$$

the condition is

$$\{\min(\lambda_j; p_j < 0) - \lambda_1\} \log\left(\frac{|f_T^-(0)|}{p_1\lambda_1} \right) < 0 . \tag{18.4.10}$$

18.5 Gamma Distributions

Exponential distributions are a special case of gamma distributions - they *belong to the gamma family*. In fact, the general standard gamma probability density function

$$f_X(x) = \frac{1}{\Gamma(\alpha)}\, e^{-x}\, x^{\alpha-1} , \qquad x > 0,\ \alpha > 0 , \tag{18.5.1}$$

becomes the standard exponential density function when the shape parameter $\alpha = 1$.

Exponential distributions, therefore, belong to the Pearson Type III class (*gamma family*) of distributions, and also *a fortiori* to the broader *natural exponential family*. (The adjective *exponential* is used in a somewhat different sense in the two cases, though in both cases the mathematical form of the probability density function is referred to.)

For elaborate treatments of the gamma distribution, we refer the readers to Bowman and Shenton (1988) and to Chapter 17 of Johnson, Kotz, and Balakrishnan (1994).

Thorin (1977) introduced the concept of a *generalized gamma convolution* which was subsequently refined by Bondesson (1979). This family of distributions, arising from a convolution of k independent gamma $(\alpha_i,\ \beta_i)$ variables, naturally includes the convolution of k independent $e(\theta_i)$ variables as a special case. The definition of this family can be formalized in such a way that it includes the limits of convolutions as $k \to \infty$ with the scale parameters θ_i's following a mixed nonnegative distribution. Interested readers may refer to the recent volume on this topic by Bondesson (1992).

18.6 Linear-Exponential Distributions

The usage of a distribution with its hazard rate function being a low-order polynomial has been suggested in the fields of life-testing and reliability; see, for example, Bain (1974), Gross and Clark (1975), and Lawless (1982). The linear-exponential distributions with their hazard rates varying as a linear function form one such family of distributions. Broadbent (1958) and Carbone, Kellerhouse, and Gehan (1967) have illustrated the usefulness of the linear-exponential distributions as survival models.

The linear-exponential distribution, with an increasing hazard rate, has its probability density function as

$$f(x) = (\lambda + vx)e^{-(\lambda x + vx^2/2)}, \qquad 0 < x < \infty,\ \lambda, v > 0\ , \tag{18.6.1}$$

and its cumulative distribution function as

$$F(x) = 1 - e^{-(\lambda x + vx^2/2)}, \qquad 0 < x < \infty,\ \lambda, v > 0\ . \tag{18.6.2}$$

For $v < 0$, we have the linear-exponential distribution with a decreasing hazard rate; but, in this case, the support of the distribution also changes in order to guarantee the non-negativity of the density $f(x)$ in (18.6.1). Observe that the density (18.6.1) becomes an exponential density function when $v = 0$.

Balakrishnan and Malik (1986) studied order statistics from the linear-exponential distributions. Sen and Bhattacharyya (1995) have addressed inferential issues relating to these distributions.

18.7 Wear-Out Life Distributions

Consider a system having two phases, with Phase I being the phase in which the components have a constant failure rate λ and Phase II being the phase in which the failure rate is greater than λ and increasing. Phase I may be considered as a mature stable phase of the system, while Phase II may be considered as a wear-out phase. Zacks (1984) considered a wear-out life distribution that possesses this feature, and is based on the following nondecreasing hazard rate function

$$\rho(t; \lambda, \alpha, \tau) = \begin{cases} \lambda & \text{if } \ 0 \le t \le \tau_+, \\ \lambda + \lambda^\alpha \alpha (t - \tau_+)^{\alpha - 1} & \text{if } \ \tau_+ < t, \end{cases} \tag{18.7.1}$$

where λ is the constant failure rate in Phase I $(0 < \lambda < \infty)$, α is a shape parameter $(1 \le \alpha < \infty)$, and τ is the change point parameter, $\tau_+ = \max(0, \tau)$. Realize that if $\tau \le 0$, the system is already in Phase II. The distribution represents a superposition of a Weibull hazard rate function on a constant hazard function for $t \ge \tau_+$. For this reason, this distribution is also sometimes called the *Weibull-exponential distribution.* Note that for the case $\alpha = 1$, the distribution shifts at τ from an exponential distribution with mean $1/\lambda$ to an exponential distribution with mean $1/(2\lambda)$.

The cumulative distribution function of the wear-out life distribution corresponding to the hazard rate in (18.7.1) is given by

$$F(t; \lambda, \alpha, \tau) = 1 - e^{-\lambda t - (\lambda (t - \tau_+)_+)^\alpha}, \qquad t \ge 0\ , \tag{18.7.2}$$

and the probability density function is

$$f(t; \lambda, \alpha, \tau) = \lambda\, e^{-\lambda t} \left\{1 + \alpha(\lambda(t - \tau_+)_+)^{\alpha - 1}\right\} e^{-\lambda^\alpha (t - \tau_+)_+^\alpha}\ , \ t \ge 0\ . \tag{18.7.3}$$

As τ gets very large (as compared to $1/\lambda$), the wear-out life distribution in (18.7.2) becomes essentially an exponential distribution with mean $1/\lambda$. This corresponds to the case when the system is far from its wear-out phase.

For inferential issues relating to this distribution, we refer the readers to Zacks (1984); also see Chapter 24.

18.8 Ryu's Generalized Exponential Distributions

Ryu (1993) recently extended the bivariate exponential distribution of Marshall and Olkin (see Chapter 20) to an absolutely continuous model, not necessarily possessing the memoryless-property. The marginal distribution of this generalized bivariate model has been presented by Ryu (1993) as a generalized exponential distribution. The probability density function of this generalized distribution is

$$f(x) = \{\lambda_1 + \lambda_{12}(1 - e^{-sx})\}\, e^{-\lambda_1 x - \lambda_{12} x + \frac{\lambda_{12}}{s}(1-e^{-sx})}, \quad x > 0\ , \tag{18.8.1}$$

and the survival function is

$$\overline{F}(x) = e^{-\lambda_1 x - \lambda_{12} x + \frac{\lambda_{12}}{s}(1-e^{-sx})}, \quad x > 0\ . \tag{18.8.2}$$

The hazard rate function of this distribution,

$$\rho(x) = \frac{f(x)}{\overline{F}(x)} = \lambda_1 + \lambda_{12}(1 - e^{-sx}) \tag{18.8.3}$$

is increasing, in general. As $s \to \infty$, the generalized exponential density function in (18.8.1) tends to the exponential density function with constant hazard rate $\lambda_1 + \lambda_{12}$; also, as $s \to 0$, the generalized density in (18.8.1) becomes the exponential density function with constant hazard rate λ_1.

18.9 Brittle Fracture Distributions

Based on the statistical theory of extreme values and the Griffith theory of brittle fracture, Black et al. (1989) derived the brittle fracture distributions which include exponential and Weibull distributions as special cases. The cumulative distribution function of this distribution is given by

$$F(x;\alpha,\beta,r) = 1 - e^{-\alpha x^{2r}} e^{-\beta/x^2}, \qquad x > 0,\ \alpha > 0,\ \beta \geq 0,\ r > 0\ , \tag{18.9.1}$$

and the probability density function is given by

$$f(x;\alpha,\beta,r) = 2\alpha\, x^{2r-1}\left(\frac{\beta}{x^2} + r\right) e^{-(\beta/x^2) - \alpha x^{2r} e^{-\beta/x^2}}, \qquad x > 0\ . \tag{18.9.2}$$

Here, the underlying variable X represents the breaking stress or strength.

Clearly, for the case $\beta = 0$ and $c = 2r$, the brittle fracture distribution (18.9.1) becomes the Weibull distribution presented in Table 18.1. For the case $\beta = 0$ and $r = 1/2$, (18.8.1) becomes an exponential distribution with mean $\theta = 1/\alpha$.

For further information on the brittle fracture distributions, we refer the readers to Black et al. (1989) and Black, Durham, and Padgett (1990).

18.10 Geometric and Poisson Distributions

Suppose the random variable X is distributed as $e(\theta)$, and $Y = [X]$, the integer part of X. Then, the random variable Y is distributed as Geometric (p), where $p = 1 - e^{-1/\theta}$.

This may be easily seen by considering, for $y = 0, 1, 2, \ldots$,

$$\begin{aligned} P[Y = y] &= P[y \leq X < y + 1] \\ &= F_X(y + 1) - F_X(y) \\ &= e^{-y/\theta} - e^{-(y+1)/\theta} \\ &= (1 - e^{-1/\theta})(e^{-1/\theta})^y \ . \end{aligned}$$

Furthermore, it may be shown in this case that the random variables $[X]$ and $\langle X \rangle = X - [X] = X - Y$, the fractional part of X, are statistically independent. These properties have been utilized by Steutel and Thiemann (1989) in establishing some distributional results for geometric order statistics similar to those presented in Chapter 3 for the exponential order statistics.

Exponential distributions are also related to Poisson distributions, in the following way. Consider a replacement process on a single component, in which each item has the same lifetime exponential distribution with hazard rate λ. As soon as the item fails, it is replaced by another. The number of replacements needed (that is, the number of failures) in time τ has a Poisson distribution with expected value $\lambda\tau$. For this reason, such a process is usually termed a *Poisson process*, though it would be equally reasonable to call it an *exponential process*.

In addition to all the related distributions listed above, it is important to bear in mind that the exponential distribution also forms a boundary for many classes of distributions defined on the basis of reliability properties. For example, the *IFR (Increasing Failure Rate)* and the *DFR (Decreasing Failure Rate)* families of distributions naturally have the exponential distribution (with constant failure rate) at their boundaries. Reference may be made to Johnson, Kotz, and Balakrishnan (1995, Chapter 33) and the books by Barlow and Proschan (1981), Pečarić, Proschan, and Tong (1992), and Shaked and Shanthikumar (1994) for an elaborate treatment of this subject.

References

Bain, L.J. (1974). Analysis for the linear failure rate distribution, *Technometrics*, **16**, 551-559.

Balakrishnan, N. and Malik, H.J. (1986). Order statistics from the linear-exponential distribution, Part I: Increasing hazard rate case, *Communications in Statistics - Theory and Methods*, **15**, 179-203.

Barlow, R.E. and Proschan, F. (1981). *Statistical Theory of Reliability and Life Testing: Probability Models*, Second edition, To Begin With, Silver Spring, Maryland.

Bartholomew, D.J. (1969). Sufficient conditions for a mixture of exponentials to be a probability density function, *Annals of Mathematical Statistics*, **40**, 2183-2188.

Black, C.M., Durham, S.D., Lynch, J.D., and Padgett, W.J. (1989). A new probability distribution for the strength of brittle fibers, *Fiber-Tex 1989*, The Third Conference on Advanced Engineering Fibers and Textile Structures for Composites, *NASA Conference Publication 3082*, 363-374.

Black, C.M., Durham, S.D., and Padgett, W.J. (1990). Parameter estimation for a new distribution for the strength of brittle fibers: A simulation study, *Communications in Statistics - Simulation and Computation*, **19**, 809-825.

Bondesson, L. (1979). A general result on infinite divisibility, *Annals of Probability*, **7**, 965-979.

Bondesson, L. (1992). *Generalized Gamma Convolutions and Related Classes of Distributions and Densities*, Springer-Verlag, New York.

Bowman, K.O. and Shenton, L.R. (1988). *Properties of Estimators for the Gamma Distribution*, Marcel Dekker, New York.

Broadbent, S. (1958). Simple mortality rates, *Applied Statistics*, **7**, 86-95.

Carbone, P.O., Kellerhouse, L.E., and Gehan, E.A. (1967). Plasmacytic myeloma: A study of the relationship of survival to various clinical manifestations and anomalous protein type in 112 patients, *American Journal of Medicine*, **42**, 937-948.

Gross, A.J. and Clark, V.A. (1975). *Survival Distributions: Reliability Applications in the Biomedical Sciences*, John Wiley & Sons, New York.

Harris, C.M., Marchal, W.G., and Botta, R.F. (1992). A note on generalized hyperexponential distributions, *Communications in Statistics - Stochastic Models*, **8**, 179-191.

Johnson, N.L., Kotz, S., and Balakrishnan, N. (1994). *Continuous Univariate Distributions - Volume 1*, Second edition, John Wiley & Sons, New York.

Johnson, N.L., Kotz, S., and Balakrishnan, N. (1995). *Continuous Univariate Distributions - Volume 2*, Second edition, John Wiley & Sons, New York.

Lawless, J.F. (1982). *Statistical Models & Methods for Lifetime Data*, John Wiley & Sons, New York.

Pečarić, J.E., Proschan, F., and Tong, Y.L. (1992). *Convex Functions, Partial Orderings, and Statistical Applications*, Academic Press, San Diego.

Ryu, K. (1993). An extension of Marshall and Olkin's bivariate exponential distribution, *Journal of the American Statistical Association*, **88**, 1458-1465.

Sen, A. and Bhattacharyya, G.K. (1995). Inference procedures for the linear failure rate model, *Journal of Statistical Planning and Inference* (to appear).

Shaked, M. and Shanthikumar, J.G. (1994). *Stochastic Orders and Their Applications*, Academic Press, San Diego.

Steutel, F.W. (1967). Note on the infinite divisibility of exponential mixtures, *Annals of Mathematical Statistics*, **38**, 1303-1305.

Steutel, F.W. and Thiemann, J.G.F. (1989). On the independence of integer and fractional parts, *Statistica Neerlandica*, **43**, 53-59.

Thorin, O. (1977). On the infinite divisibility of the ratio distribution, *Scandinavian Actuarial Journal*, 31-40.

Zacks, S. (1984). Estimating the shift to wear-out systems having exponential-Weibull life distribution, *Operations Research*, **32**, 741-749.

CHAPTER 19

Mixtures - Models and Applications

G.J. McLachlan[1]

19.1 Introduction

19.1.1 Exponential Mixture Model

Normal mixture distributions have been applied widely in practice to model heterogeneous data, as surveyed in McLachlan and Basford (1988). In some circumstances however, the adoption of component densities that are normal is inappropriate, such as in the modelling of failure time data. For data of this type, mixtures of exponential distributions can play a useful role [Mendenhall and Hader (1958)]. The components in the mixture correspond to the distinct causes of failure, which are taken to act in a mutually exclusive manner. For example, Choi (1979) used a two-component mixture model to study the toxicity of chemical agents used in chemotherapy. The two causes for failure, which was death, was toxicity of the chemical agent and regrowth of the tumour. Other references on the use of mixture models in this spirit in survival analysis include Farewell (1977, 1982), Farewell and Sprott (1988), Gordon (1990), Kuk (1992), Kuk and Chen (1992), and McLachlan et al. (1993). For examples on the use of exponential mixtures in a discriminant analysis context, the reader is referred to McLachlan (1992, Chapter 7).

The assumption that the underlying distribution is a mixture of exponential distributions is widely invoked in applied science and the social sciences. It is frequently adopted to model the distribution of time to failure in those situations where the hazard function is observed empirically to decline with time. Recently, Heckman, Robb, and Walker (1990) have presented nonparametric tests for to distinguish mixtures of exponentials models from more general models with declining hazards.

The random variable T is said to have a finite mixture distribution if its probability density function (p.d.f.) $f(t)$ is of the form

$$f(t) = \sum_{i=1}^{g} \pi_i f_i(t) ,$$

[1] The University of Queensland, Queensland 4072, Australia

where each $f_i(t)$ is a p.d.f. and $\pi_1, \ldots, \pi_g$ denote the mixing proportions, which are nonnegative and sum to one. These component density functions $f_i(t)$ are with respect to arbitrary measure on $\mathbf{R}^p$, so that $f_i(t)$ can be a mass function by the adoption of counting measure.

In the particular case where the ith component density $f_i(t)$ is exponential with mean θ_i, we have

$$f_i(t) = \theta_i^{-1}\, e^{-t/\theta_i}\, I_{(0,\infty)}(t) \qquad (\theta_i > 0)$$

for $i = 1, \ldots, g$, where the indicator function $I_{(0,\infty)}(t) = 1$ for $t > 0$ and is zero elsewhere. We let

$$\boldsymbol{\Psi} = (\pi_1, \ldots, \pi_{g-1}, \theta_1, \ldots, \theta_g)'$$

be the vector containing all the parameters, apart from π_g, which is given by

$$\pi_g = 1 - \sum_{i=1}^{g-1} \pi_i \ .$$

We shall henceforth write the exponential mixture density as

$$f(t; \boldsymbol{\Psi}) = \sum_{i=1}^{g} \pi_i f_i(t; \boldsymbol{\Psi}) \ , \tag{19.1.1}$$

where

$$f_i(t; \boldsymbol{\Psi}) = \theta_i^{-1} e^{-t/\theta_i}\, I_{(0,\infty)}(t)$$

for $i = 1, \ldots, g$. The distribution function corresponding to $f(t; \boldsymbol{\Psi})$ and to $f_i(t; \boldsymbol{\Psi})$ is denoted by $F(t; \boldsymbol{\Psi})$ and $F_i(t; \boldsymbol{\Psi})$, respectively.

In a later section, we are to generalize this definition of a finite exponential mixture model to include the case where the mixing proportions may be negative. However, we shall focus exclusively on finite mixtures of exponentials. Results and examples on infinite mixtures of exponentials are discussed, for example, by Jewell (1982).

19.2 Maximum Likelihood Estimation

In this section we consider estimation of $\boldsymbol{\Psi}$ by maximum likelihood. In order to be able to estimate all the parameters in $\boldsymbol{\Psi}$ from some observed data distributed according to (19.1.1), it is necessary that they should be identifiable. Teicher (1961) has shown that finite mixtures of exponentials are identifiable; see Titterington, Smith, and Makov (1985) for a lucid account of the concept of identifiability for mixtures.

Let

$$\overline{F}(t; \boldsymbol{\Psi}) = 1 - F(t; \boldsymbol{\Psi})$$

and

$$\overline{F}_i(t; \boldsymbol{\Psi}) = 1 - F_i(t; \boldsymbol{\Psi}) \qquad (i = 1, \ldots, g) \ .$$

In a survival analysis context, $\overline{F}(t; \boldsymbol{\Psi})$ is the survival function. In this context, a study undertaken to observe a random sample $T_1, \ldots, T_n$ from the exponential mixture model (19.1.1) will generally be terminated in practice before all of these random variables are able to be observed. We let

$$\boldsymbol{y} = (\boldsymbol{u}_1', \ldots, \boldsymbol{u}_n')' \tag{19.2.1}$$

denote the observed data, where $\boldsymbol{u}_j = (w_j, \delta_j)'$ and $\delta_j = 0$ or 1 according as the observation T_j is censored or uncensored at w_j ($j = 1, \ldots, n$). That is, if the observation T_j is uncensored, its observed value t_j is equal to w_j, while if it is censored at w_j, then t_j is some value greater than w_j ($j = 1, \ldots, n$).

The log-likelihood for $\boldsymbol{\Psi}$ that can be formed under the exponential mixture model (19.1.1) on the basis of the observed data $\boldsymbol{y}$ is given by

$$\log L(\boldsymbol{\Psi}) = \sum_{j=1}^{n} \{\delta_j \log f(w_j; \boldsymbol{\Psi}) + (1 - \delta_j) \log \overline{F}(w_j; \boldsymbol{\Psi})\} . \tag{19.2.2}$$

An estimated $\widehat{\boldsymbol{\Psi}}$ of $\boldsymbol{\Psi}$ is provided by an appropriate root of the likelihood equation

$$\partial \log L(\boldsymbol{\Psi}) / \partial \boldsymbol{\Psi} = \mathbf{0} .$$

19.2.1 *EM* Algorithm

The computation of a solution or solutions of the likelihood equation can be undertaken within the *EM* framework [Dempster, Laird, and Rubin (1977)] by declaring the complete-data vector $\boldsymbol{x}$ to be

$$\boldsymbol{x} = (t_1, \ldots, t_n, \boldsymbol{z}_1', \ldots, \boldsymbol{z}_n')' ,$$

where $\boldsymbol{z}_1, \ldots, \boldsymbol{z}_n$ denote the vectors of zero-one indicator variables defining the mixture component-membership of $t_1, \ldots, t_n$, respectively. In this framework, each observation T_j is viewed as having arisen from the ith component of the exponential mixture (19.1.1) with a prior probability π_i ($i = 1, \ldots, g$). That is, given $\boldsymbol{z}_j$, the probability density function of T_j is

$$\sum_{i=1}^{g} z_{ij} f_i(t_j; \boldsymbol{\Psi}) ,$$

where $z_{ij} = (\boldsymbol{z}_j)_i$ is 1 or 0, according as T_j comes or does not come from the ith component of the exponential mixture model to be fitted ($i = 1, \ldots, g;\ j = 1, \ldots, n$).

We suppose now that the observations have been relabelled, so that $T_1, \ldots, T_r$ denote the uncensored observations and $T_{r+1}, \ldots, T_n$ the $n - r$ censored observations. Then the complete-data vector $\boldsymbol{x}$ is declared to be

$$\boldsymbol{x} = (\boldsymbol{y}', t_{r+1}, \ldots, t_n, \boldsymbol{z}_1', \ldots, \boldsymbol{z}_n')' ,$$

where $\boldsymbol{y}$ is the incomplete-data vector (the observed data).

The complete-data log-likelihood $\log L_C(\boldsymbol{\Psi})$ that can be formed from $\boldsymbol{x}$ is

$$\begin{aligned} \log L_C(\boldsymbol{\Psi}) &= \sum_{i=1}^{g} \sum_{j=1}^{n} z_{ij} \log\{\pi_i f_i(t_j; \boldsymbol{\Psi})\} \\ &= \sum_{i=1}^{g} \left(n_i \log \pi_i - n_i \log \theta_i - \theta_i^{-1} \sum_{j=1}^{n} z_{ij} t_j \right) , \end{aligned}$$

where

$$n_i = \sum_{j=1}^{n} z_{ij} \qquad (i = 1, \ldots, g) .$$

The EM algorithm is applied to this problem by treating the t_j $(j = r+1, \ldots, n)$ and the $\boldsymbol{z}_j$ $(j = 1, \ldots, n)$ as missing data. It is easy to program and proceeds iteratively in two steps, E (for expectation) and M (for maximization).

19.2.2 E-step

Using some initial value for $\boldsymbol{\Psi}$, say $\boldsymbol{\Psi}^{(0)}$, the E-step requires the calculation of

$$H(\boldsymbol{\Psi}; \boldsymbol{\Psi}^{(0)}) = E[\log L_C(\boldsymbol{\Psi}) \mid \boldsymbol{y}, \boldsymbol{\Psi}^{(0)}] , \tag{19.2.3}$$

the expectation of the complete-data log-likelihood $\log L_C(\boldsymbol{\Psi})$, conditional on the observed data and the initial fit $\boldsymbol{\Psi}^{(0)}$ for $\boldsymbol{\Psi}$.

This step effectively involved taking the conditional expectations

$$E[Z_{ij} \mid \boldsymbol{u}_j, \boldsymbol{\Psi}^{(0)}] \qquad (i = 1, \ldots, g) \tag{19.2.4}$$

for $j = 1, \ldots, n$, and also

$$E[Z_{ij} T_j \mid \boldsymbol{u}_j, \boldsymbol{\Psi}^{(0)}] \qquad (i = 1, \ldots, g) \tag{19.2.5}$$

for $j = r+1, \ldots, n$; that is, for each censored observation.

The expectation in (19.2.4) is easily computed to give

$$E[Z_{ij} \mid \boldsymbol{u}_j, \boldsymbol{\Psi}^{(0)}] = \tau_i(w_j; \boldsymbol{\Psi}^{(0)}) \tag{19.2.6}$$

for $j = 1, \ldots, r$ (an uncensored observation), and

$$E[Z_{ij} \mid \boldsymbol{u}_j, \boldsymbol{\Psi}^{(0)}] = \phi_i(w_j; \boldsymbol{\Psi}^{(0)}) \tag{19.2.7}$$

for $j = r+1, \ldots, n$ (a censored observation). Here for a given i $(i = 1, \ldots, g)$,

$$\tau_i(w_j; \boldsymbol{\Psi}) = \pi_i f_i(w_j; \boldsymbol{\Psi}) / f(w_j; \boldsymbol{\Psi})$$

is the posterior probability that the jth observation belongs to the ith component, given $T_j = w_j$ $(j = 1, \ldots, r)$, while

$$\phi_i(w_j; \boldsymbol{\Psi}) = \pi_i \overline{F}_i(w_j; \boldsymbol{\Psi}) / \overline{F}(w_j; \boldsymbol{\Psi})$$

is the posterior probability that the jth observation belongs to the ith component of the mixture, given $T_j > w_j$ $(j = r+1, \ldots, n)$.

For a censored observation T_j, the expectation in (19.2.5) for a given i $(i = 1, \ldots, g)$ can be calculated to give

$$E[Z_{ij} T_j \mid \boldsymbol{u}_j, \boldsymbol{\Psi}^{(0)}] = \phi_i(w_j; \boldsymbol{\Psi}^{(0)}) \nu_i(w_j; \boldsymbol{\Psi}^{(0)}) , \tag{19.2.8}$$

where

$$\nu_i(w_j; \boldsymbol{\Psi}^{(0)}) = E[T_j \mid T_j > w_j, T_j \in G_i, \boldsymbol{\Psi}^{(0)}]$$

is the conditional expectation of T_j, given that it is greater than w_j and comes from G_i (the ith component of the exponential mixture), using the fit $\boldsymbol{\Psi}^{(0)}$ for $\boldsymbol{\Psi}$.

By the lack of memory property for the exponential distribution (see, for example, Chapters 1 and 2)

$$\begin{aligned} \nu_i(w_j;\boldsymbol{\Psi}^{(0)}) &= w_j + E[T_j \mid T_j \in G_i, \boldsymbol{\Psi}^{(0)}] \\ &= w_j + \theta_i^{(0)} \qquad (i = 1,\ldots,g) \end{aligned} \tag{19.2.9}$$

for $j = r+1,\ldots,n$.

On using the results (19.2.4) - (19.2.9) in (19.2.3), we have that

$$\begin{aligned} H(\boldsymbol{\Psi};\boldsymbol{\Psi}^{(0)}) &= \sum_{i=1}^{g}\Bigg\{ n_i \log \pi_i - n_i \log \theta_i - \theta_i^{-1}\sum_{j=1}^{r} \tau_i(w_j;\boldsymbol{\Psi}^{(0)})w_j \\ &\quad -\theta_i^{-1}\sum_{j=r+1}^{n} \phi_i(w_j;\boldsymbol{\Psi}^{(0)})(w_j + \theta_i^{(0)})\Bigg\} . \end{aligned} \tag{19.2.10}$$

This completes the E-step.

19.2.3 M-step

On the M-step first time through, the intent is to choose $\widehat{\boldsymbol{\Psi}}^{(1)}$ to maximize $H(\boldsymbol{\Psi};\boldsymbol{\Psi}^{(0)})$, as given by (19.2.10). One nice feature of the EM algorithm is that the solution to the M-step often exists in closed form, which is the case here.

The estimate $\widehat{\boldsymbol{\Psi}}^{(1)}$ is given by

$$\widehat{\boldsymbol{\Psi}}^{(1)} = (\widehat{\pi}_1^{(1)},\ldots,\widehat{\pi}_{(g-1)}^{(1)},\widehat{\theta}_1^{(1)},\ldots,\widehat{\theta}_g^{(1)})' ,$$

where

$$\widehat{\pi}_i^{(1)} = \left\{\sum_{j=1}^{r} \tau_i(w_j;\boldsymbol{\Psi}^{(0)}) + \sum_{j=r+1}^{n} \phi_i(w_j;\boldsymbol{\Psi}^{(0)})\right\}\Big/ n \tag{19.2.11}$$

and

$$\widehat{\theta}_i^{(1)} = \left\{\sum_{j=1}^{r} \tau_i(w_j;\boldsymbol{\Psi}^{(0)})w_j + \sum_{j=r+1}^{n} \phi_i(w_j;\boldsymbol{\Psi}^{(0)})(w_j + \theta_i^{(0)})\right\}\Big/ n\widehat{\pi}_i^{(1)} \tag{19.2.12}$$

for $i = 1,\ldots,g$.

The E- and M-steps are alternated repeatedly, where in their subsequent executions, the initial fit $\boldsymbol{\Psi}^{(0)}$ is replaced by the current fit for $\boldsymbol{\Psi}$, say $\widehat{\boldsymbol{\Psi}}^{(k)}$, on the $(k+1)$th cycle. Another nice feature of the EM algorithm is that the likelihood $L(\boldsymbol{\Psi})$ for the incomplete-data specification can never be decreased after an EM sequence. Hence,

$$L(\boldsymbol{\Psi}^{(k+1)}) \geq L(\boldsymbol{\Psi}^{(k)}) ,$$

which implies that $L(\boldsymbol{\Psi}^{(k)})$ converges to some L^*, since the sequence of likelihood values is bounded above for exponential mixtures.

19.3 Partially Classified Data

19.3.1 Introduction

Up to now, we have considered the fitting of an exponential mixture model on observed data that are unclassified with respect to their component membership of the mixture. In some situations in practice, there are also some data that are classified with respect to the components of the mixture. Situations of this type are common in survival analysis. For example, suppose that the random variable T denotes the time to failure and that there are g distinct causes of this failure. Further, suppose that when failure occurs, the cause of failure is known, so that the cause of failure is only unknown for those failure times that are censored. We shall give an example of this type in Section 19.9.

19.3.2 *EM* Algorithm

We now extend the application of the *EM* algorithm, as described in the previous section, to solve the likelihood equation where there are also available m survival times of known origin with respect to the components of the mixture. These classified data are denoted by $\boldsymbol{c}_{n+1}, \ldots, \boldsymbol{c}_{n+m}$, where $\boldsymbol{c}_j = (t_j, \boldsymbol{z}_j')'$ and where t_j is the jth classified observation on T and $\boldsymbol{z}_j$ is its known indicator vector, defining its component-membership of the mixture. As before, $z_{ij} = (\boldsymbol{z}_j)_i$, which is 1 or 0, according as T_j comes or does not come from the ith component of the mixture ($i = 1, \ldots, g$; $j = n+1, \ldots, n+m$).

Then the solutions for π_i and θ_i on the first cycle of the *EM* algorithm, corresponding to (19.2.11) and (19.2.12), are given by

$$\widehat{\pi}_i^{(1)} = \left\{ \sum_{j=1}^{r} \tau_i(w_j; \boldsymbol{\Psi}^{(0)}) + \sum_{j=r+1}^{n} \nu_i(w_j; \boldsymbol{\Psi}^{(0)}) + m_i \right\} \Big/ (n+m)$$

and

$$\begin{aligned} \widehat{\theta}_i^{(1)} &= \left\{ \sum_{j=1}^{r} \tau_i(w_j; \boldsymbol{\Psi}^{(0)}) w_j + \sum_{j=r+1}^{n} \phi_i(w_j; \boldsymbol{\Psi}^{(0)})(w_j + \theta_i^{(0)}) \right. \\ &\qquad \left. + \sum_{j=n+1}^{n+m} z_{ij} t_j \right\} \Big/ (n+m)\widehat{\pi}_i^{(1)}, \end{aligned}$$

where

$$m_i = \sum_{j=n+1}^{n+m} z_{ij}$$

for $i = 1, \ldots, g$.

19.4 Testing for the Number of Components

With some applications the *a priori* information on the group structure extends to knowing the number g of underlying groups in the population, and hence the number of components in the mixture model. For example, in the screening for a disease, there is an obvious dichotomy of the population into disease-free and diseased groups. However, in some situations, where the possible groups represent the various causes of a disease, the assumed number of groups g may not be the actual number, as there may be further, as of yet undetected, causes.

The problem of assessing the true value of g is an important but very difficult problem. In the present framework of finite mixture models, an obvious approach is to use the likelihood ratio statistic Λ to test for the smallest value of the number g of components in the mixture compatible with the data. Unfortunately with mixture models, regularity conditions do not hold for $-2 \log \Lambda$ to have its usual asymptotic null distribution of chi-squared with degrees of freedom equal to the difference between the number of parameters under the null and alternative hypotheses; see McLachlan and Basford (1988, Chapter 1) and the references therein, including McLachlan (1987a) and Quinn, McLachlan, and Hjort (1987).

One way of assessing the null distribution is to use a resampling method, which can be viewed as a particular application of the general bootstrap approach of Efron (1979, 1982); see also Efron and Tibshirani (1993). More specifically, for the test of the null hypothesis of $g = g_0$ versus the alternative of $g = g_1$, the log-likelihood ratio statistic can be bootstrapped as follows. Proceeding under the null hypothesis, a so-called bootstrap sample of size n is generated by sampling with replacement from the n observations $\boldsymbol{u}_1, \ldots, \boldsymbol{u}_n$ in the observed sample $\boldsymbol{y}$. The value of $-2 \log \Lambda$ is calculated for the bootstrap sample after fitting mixture models for $g = g_0$ and $g = g_1$ in turn to it. This process is repeated independently a number of times K, and the replicated values of $-2 \log \Lambda$ evaluated from the successive bootstrap samples can be used to assess the bootstrap, and hence the true, null distribution of $-2 \log \Lambda$. In particular, it enables an approximation to be made to the achieved level of significance P corresponding to the value of $-2 \log \Lambda$ evaluated from the original sample. The value of the jth order statistic of the K replications can be taken as an estimate of the quantile of order $j/(K+1)$, and the P-value can be assessed by reference with respect to the ordered bootstrap replications of $-2 \log \Lambda$. Actually, the value of the jth order statistic of the K replications is a better approximation to the quantile of order $(3j-1)/(3K+1)$ [Hoaglin (1985)].

In the present situation of exponential components, the null distribution of Λ will not depend on any unknown parameters for $g_0 = 1$. In this case there will be no difference between the bootstrap and true null distributions of $-2 \log \Lambda$. In such cases where it is the actual statistic $-2 \log \Lambda$ and not its bootstrap analogue that is being resampled, the resampling may be viewed as an application of the Monte Carlo approach to the construction of a hypothesis test having an exact level of desired significance. This approach was proposed originally by Barnard (1963); see Hope (1968) and Hall and Titterington (1989). Aitkin, Anderson, and Hinde (1981) appear to have been the first to apply the resampling approach in the context of finite mixture models.

19.5 Provision of Standard Errors

19.5.1 Observed Information Matrix

We shall let $\boldsymbol{I}(\boldsymbol{\Psi})$ denote the matrix of the negative of the second order partial derivatives of the log-likelihood function $\log L(\boldsymbol{\Psi})$ with respect to the elements of the parameter vector $\boldsymbol{\Psi}$; that is,

$$\boldsymbol{I}(\boldsymbol{\Psi}) = -\partial^2 \log L(\boldsymbol{\Psi})/\partial\boldsymbol{\Psi}\partial\boldsymbol{\Psi}' .$$

The matrix $\boldsymbol{I}(\widehat{\boldsymbol{\Psi}})$ is referred to as the observed information matrix, and the matrix $\mathcal{I}(\boldsymbol{\Psi}) = E[\boldsymbol{I}(\boldsymbol{\Psi})]$ as the expected information matrix. Under regularity conditions, the inverse of the asymptotic covariance matrix of the maximum likelihood estimate $\widehat{\boldsymbol{\Psi}}$ of $\boldsymbol{\Psi}$ is given by $\mathcal{I}(\boldsymbol{\Psi})$. It is common in practice to estimate the inverse of the covariance matrix of a maximum likelihood solution by the observed information matrix, rather than the expected information matrix evaluated at the solution; see Efron and Hinkley (1978).

For applications of the EM algorithm to data viewed as being incomplete, Louis (1982) has shown how $\boldsymbol{I}(\widehat{\boldsymbol{\Psi}})$ can be expressed in terms of the gradient and curvature of the complete-data log-likelihood function evaluated at the solution. Specifically,

$$\boldsymbol{I}(\widehat{\boldsymbol{\Psi}}) = \boldsymbol{I}_c(\widehat{\boldsymbol{\Psi}}; \boldsymbol{y}) - \boldsymbol{K}(\widehat{\boldsymbol{\Psi}}; \boldsymbol{y}) , \tag{19.5.1}$$

where

$$\begin{aligned} \boldsymbol{I}_c(\boldsymbol{\Psi}; \boldsymbol{y}) &= E[\boldsymbol{I}_c(\boldsymbol{\Psi}) \mid \boldsymbol{y}; \boldsymbol{\Psi}] \\ &= E[-\partial^2 \log L_C(\boldsymbol{\Psi})/\partial\boldsymbol{\Psi}\partial\boldsymbol{\Psi}' \mid \boldsymbol{y}; \boldsymbol{\Psi}] , \end{aligned}$$

and where

$$\boldsymbol{K}(\boldsymbol{\Psi}; \boldsymbol{y}) = E[\boldsymbol{h}(\boldsymbol{\Psi})\boldsymbol{h}(\boldsymbol{\Psi})' \mid \boldsymbol{y}; \boldsymbol{\Psi}]$$

and

$$\boldsymbol{h}(\boldsymbol{\Psi}) = \partial \log L_C(\boldsymbol{\Psi})/\partial\boldsymbol{\Psi}$$

is the gradient vector of the complete-data log-likelihood function $L_C(\boldsymbol{\Psi})$.

More recently, Meilijson (1989) and Meng and Rubin (1991) have considered the problem of obtaining the observed information matrix within the EM framework.

19.5.2 Approximating the Observed Information Matrix

The use of the observed information matrix $\boldsymbol{I}(\widehat{\boldsymbol{\Psi}})$, whether calculated directly from the incomplete-data log-likelihood, or indirectly in terms of the complete-data log-likelihood as given by (19.5.1), involves the calculation of second-order partial derivatives. Although the calculation of the second-order partial derivatives and the conditional expectations in (19.5.1) is feasible for mixtures of exponentials, it is computationally more convenient to compute the observed information matrix $\boldsymbol{I}(\widehat{\boldsymbol{\Psi}})$, using an approximation that requires only the gradient vector of the complete-data log-likelihood. This approximation, which has been used by various workers [Berndt et al. (1974), Redner and Walker (1984), McLachlan and Basford (1988, Chapter 1), Meilijson (1989), and Jones and McLachlan (1992), among others], is

$$\begin{aligned} \boldsymbol{I}(\widehat{\boldsymbol{\Psi}}) \approx & \sum_{j=1}^{n} \{\partial \log L_C(\widehat{\boldsymbol{\Psi}}; \tilde{t}_j, \tilde{\boldsymbol{z}})/\partial\boldsymbol{\Psi}\}\{\partial \log L_C(\widehat{\boldsymbol{\Psi}}; \tilde{t}_j, \tilde{\boldsymbol{z}}) / \partial\boldsymbol{\Psi}\}' \\ & + \sum_{j=n+1}^{n+m} \{\partial \log L_C(\widehat{\boldsymbol{\Psi}}; t_j, \boldsymbol{z}_j)/\partial\boldsymbol{\Psi}\}\{\partial \log L_C(\widehat{\boldsymbol{\Psi}}; t_j, \boldsymbol{z}_j)/\partial\boldsymbol{\Psi}\}' , \end{aligned}$$

where $\partial \log L_C(\boldsymbol{\Psi}; t_j, \boldsymbol{z}_j)/\partial \boldsymbol{\Psi}$ denotes the complete-data log-likelihood formed from just the jth observation $\boldsymbol{x}_j = (t_j, \boldsymbol{z}_j')'$, $j = 1, \ldots, n+m$. Also, $\partial \log L_C(\widehat{\boldsymbol{\Psi}}; \tilde{t}_j, \tilde{\boldsymbol{z}}_j)/\partial \boldsymbol{\Psi}$ is equal to $\partial \log L_C(\boldsymbol{\Psi}; t_j, \boldsymbol{z}_j)/\partial \boldsymbol{\Psi}$ evaluated at $\boldsymbol{\Psi} = \widehat{\boldsymbol{\Psi}}$, $t_j = \tilde{t}_j$, and $\boldsymbol{z}_j = \tilde{\boldsymbol{z}}_j$, where

$$\tilde{t}_j = \delta_j w_j + (1-\delta_j)\phi_i(w_j; \widehat{\boldsymbol{\Psi}})(w_j + \widehat{\theta}_i)$$

and

$$\tilde{\boldsymbol{z}}_j = \delta_j \boldsymbol{\tau}(w_j; \widehat{\boldsymbol{\Psi}}) + (1-\delta_j)\boldsymbol{\phi}(w_j; \widehat{\boldsymbol{\Psi}})$$

for $j = 1, \ldots, n$. The vector

$$\boldsymbol{\tau}(w_j; \boldsymbol{\Psi}) = ((\tau_1(w_j; \boldsymbol{\Psi}), \ldots, \tau_g(w_j; \boldsymbol{\Psi}))'$$

contains the posterior probabilities of component membership of the mixture for the r uncensored observations with $\delta_j = 1$ $(j = 1, \ldots, r)$, while

$$\boldsymbol{\phi}(w_j; \boldsymbol{\Psi}) = ((\phi_1(w_j; \boldsymbol{\Psi}), \ldots, \phi_g(w_j; \boldsymbol{\Psi}))'$$

is the corresponding vector for the $n - r$ censored observations with $\delta_j = 0$ $(j = r + 1, \ldots, n)$.

For a g-component exponential mixture, the elements of $\partial \log L_C(\boldsymbol{\Psi}; t_j, \boldsymbol{z}_j)/\partial \boldsymbol{\Psi}$ are defined by

$$\partial \log L_C(\boldsymbol{\Psi}; t_j, \boldsymbol{z}_j)/\partial \pi_i = z_{ij}\pi_i^{-1} - z_{gj}\pi_g^{-1} \qquad (i = 1, \ldots, g-1)\ ,$$

and

$$\partial \log L_C(\boldsymbol{\Psi}; t_j, \boldsymbol{z}_j)/\partial \theta_i = -z_{ij}\theta_i^{-1} + z_{ij}\theta_i^{-2}t_j \qquad (i = 1, \ldots, g)\ .$$

19.5.3 Bootstrap Approach

As noted by Day (1969), among others, for mixture models the sample size n has to be very large for the asymptotic theory of maximum likelihood to be applicable. Hence the information-based methods above may be of limited value in practice in their provision of standard errors. We therefore consider another approach to this problem, using the bootstrap.

As before, we let $\widehat{\boldsymbol{\Psi}}$ be the estimate of $\boldsymbol{\Psi}$ obtained by fitting a g-component exponential mixture model $f(t; \boldsymbol{\Psi})$ to the original observed data $\boldsymbol{y}$. We proceed with the generation of the bootstrap samples as in Section 19.4, except that now we allow for the possible presence of some classified observations in the observed data $\boldsymbol{y}$. Generate K bootstrap samples $\boldsymbol{y}_1^*, \ldots, \boldsymbol{y}_K^*$ by resampling with replacement from the $n+m$ observations $\boldsymbol{u}_1, \ldots, \boldsymbol{u}_n, \boldsymbol{c}_{n+1}, \ldots, \boldsymbol{c}_{n+m}$ in $\boldsymbol{y}$. Let $\widehat{\boldsymbol{\Psi}}_1^*, \ldots, \widehat{\boldsymbol{\Psi}}_K^*$ be the estimate of $\boldsymbol{\Psi}$ calculated from $\boldsymbol{y}_1^*, \ldots, \boldsymbol{y}_K^*$, respectively. Then the covariance matrix of $\widehat{\boldsymbol{\Psi}}$ can be assessed by the sample covariance matrix of $\widehat{\boldsymbol{\Psi}}_1^*, \ldots, \widehat{\boldsymbol{\Psi}}_K^*$, given by

$$\boldsymbol{V} = \sum_{k=1}^{K}(\widehat{\boldsymbol{\Psi}}_k^* - \overline{\boldsymbol{\Psi}}^*)(\widehat{\boldsymbol{\Psi}}_k^* - \overline{\boldsymbol{\Psi}}^*)'/(K-1)\ ,$$

where

$$\overline{\boldsymbol{\Psi}}^* = \sum_{k=1}^{K}\widehat{\boldsymbol{\Psi}}_k^*/K\ .$$

The standard error of the ith element of $\hat{\boldsymbol{\Psi}}$, $\hat{\boldsymbol{\Psi}}_i = (\hat{\boldsymbol{\Psi}})_i$, is therefore assessed by $(\boldsymbol{V})_{ii}^{1/2}$. A nominal $100(1-\alpha)\%$ confidence interval for $\boldsymbol{\Psi}_i = (\boldsymbol{\Psi})_i$ is given by

$$\left[\widehat{CDF}_i^{-1}\left(\frac{1}{2}\,\alpha\right), \widehat{CDF}_i^{-1}\left(1-\frac{1}{2}\,\alpha\right)\right] ,$$

where

$$\widehat{CDF}_i^{-1}(\alpha) = \sup\{u : \widehat{CDF}_i(u) \leq \alpha\}$$

and where

$$\widehat{CDF}_i(u) = \#\{\hat{\boldsymbol{\Psi}}_{ik}^* \leq u\}/K$$

is the empirical distribution function of the K bootstrap replications of $\hat{\boldsymbol{\Psi}}_i$, denoted by $\hat{\boldsymbol{\Psi}}_{ik}^* = (\hat{\boldsymbol{\Psi}}_k^*)_i$ for $k = 1, \ldots, K$.

19.6 Homogeneity of Mixing Proportions

With some applications of finite mixture models to life-testing problems, the data are taken from each of several strata (say, B), and a mixture model with common exponential components is fitted to each. A consequent consideration is whether the B mixtures have the same mixing proportions; that is, whether these mixtures are identical. Examples of this type may be found in Choi (1979), McLachlan, Lawoko, and Ganesalingam (1982), and McLachlan and Basford (1988, Chapter 4).

Let

$$f^{(b)}(t; \boldsymbol{\Psi}_b) = \sum_{i=1}^{g} \pi_{ib} f_i(t; \boldsymbol{\Psi}_b) ,$$

denote the exponential mixture model to be fitted to the data $\boldsymbol{y}_b$ from the bth strata, where

$$\boldsymbol{\Psi}_b = (\boldsymbol{\xi}_b', \theta_1, \ldots, \theta_g)'$$

and where

$$\boldsymbol{\xi}_b = (\pi_{1b}, \ldots, \pi_{g-1,b})'$$

for $b = 1, \ldots, B$. The null hypothesis H_0 that these B exponential mixtures have the same mixing proportions can be expressed as

$$H_0 : \ \boldsymbol{\xi}_1 = \boldsymbol{\xi}_2 = \cdots = \boldsymbol{\xi}_B \ .$$

We let $\hat{\boldsymbol{\Psi}}_b = (\hat{\boldsymbol{\xi}}_b', \hat{\theta}_1, \ldots, \hat{\theta}_g)'$ denote the maximum likelihood estimate of $\boldsymbol{\Psi}$ based on $\boldsymbol{y}_1, \ldots, \boldsymbol{y}_B$. That is, it is the maximizer of

$$\sum_{b=1}^{B} \log L(\boldsymbol{\Psi}_b; \boldsymbol{y}_b) ,$$

where $\log L(\boldsymbol{\Psi}_b; \boldsymbol{y}_b)$ is the log-likelihood formed from $\boldsymbol{y}_b$, and so corresponds to (19.2.2). Also, we let $\hat{\boldsymbol{\Psi}}_0 = (\hat{\boldsymbol{\xi}}_0', \tilde{\theta}_1, \ldots, \tilde{\theta}_g)'$, where now under H_0, $\hat{\boldsymbol{\xi}}_0$ is the maximum likelihood estimate of the common value ξ_0 of $\xi_1, \ldots, \xi_B$, and $\tilde{\theta}_i$ is the maximum likelihood estimate of θ_i $(i = 1, \ldots, g)$.

The likelihood ratio test rejects H_0 at a nominal level α if

$$2\sum_{b=1}^{B}\{\log L(\widehat{\boldsymbol{\Psi}}_b;\boldsymbol{y}_b) - \log L(\widehat{\boldsymbol{\Psi}}_0;\boldsymbol{y}_b)\} > c\ ,$$

where c denotes the $(1-\alpha)$th quantile of the chi-squared distribution with $(g-1)(B-1)$ degrees of freedom.

19.7 Generalized Mixtures of Exponentials

We now generalize the definition of a finite mixture of exponential distributions to allow for negative mixing proportions π_i. Many authors have used this formulation as introduced by Erlang; see Brockmeyer, Halstrom, and Jensen (1948). If $T_1,\ldots,T_g$ denote g independent exponentially distributed random variables with respective means $\theta_1,\ldots,\theta_g$, then

$$T = \sum_{i=1}^{g} T_i$$

has the general Erlang density. It is a special case of the exponential mixture model (19.1.1), in which the mixing proportions are given by

$$\pi_i = \prod_{\substack{k=1\\k\neq i}}^{g} \theta_i/(\theta_i - \theta_k) \qquad (i = 1,\ldots,g)\ .$$

It is assumed that $\theta_i \neq \theta_k$ for all $i \neq k$.

As generalized mixtures of exponentials have negative mixing proportions, they will not always be a proper density function. Steutel (1967) and Bartholomew (1969), and more recently, Harris, Marchal, and Botta (1992), have investigated conditions for which a generalized exponential mixture is a density function; also see Chapter 18.

19.8 Some Further Extensions

Exponential mixture models can be extended in a variety of other ways. They concern extensions of the model to allow for the effect of different characteristics of the entities on which the data have been sampled. Suppose that $\boldsymbol{a}_j$ is a d-dimensional covariate vector containing measurements made on these characteristics for the jth entity. Then the mixing proportions can be formulated to depend on $\boldsymbol{a}_j$ by using the logistic model, under which the ith mixing proportion is taken to be of the form

$$\pi_i(\boldsymbol{a}_j;\boldsymbol{\beta}) = \exp(\beta_{0i} + \boldsymbol{\beta}_i'\boldsymbol{a}_j)\bigg/\left\{1 + \sum_{h=1}^{g-1}\exp(\beta_{0h} + \boldsymbol{\beta}_h'\boldsymbol{a}_j)\right\} \tag{19.8.1}$$

for $i = 1,\ldots,g-1$, and

$$\pi_g(\boldsymbol{a}_j;\boldsymbol{\beta}) = 1 - \sum_{i=1}^{g-1}\pi_i(\boldsymbol{a}_j;\boldsymbol{\beta})\ .$$

Here

$$\boldsymbol{\beta} = (\beta_{01}, \boldsymbol{\beta}_1', \ldots, \beta_{0,g-1}, \boldsymbol{\beta}_{g-1}')'$$

is a vector of unknown parameters to be inferred from the data. It can be estimated by maximum likelihood within the EM framework as described in Section 19.2. But now the solution to the M-step will not exist in closed form, as the estimate of $\boldsymbol{\beta}$ has to be computed iteratively.

A further extension of the exponential mixture model is to allow the expected failure time θ_i for the ith component to depend also on the covariate vector $\boldsymbol{a}_j$ by setting

$$\theta_i(\boldsymbol{a}_j) = \exp(\gamma_{0i} + \boldsymbol{\gamma}_i'\boldsymbol{a}_j) \qquad (i = 1, \ldots, g)$$

for the jth entity that is sampled, where the γ_{0i} and the $\boldsymbol{\gamma}_i$ are to be inferred from the data, along with the other unknown parameters in the model.

19.9 Example

We present here an example from McGiffin et al. (1994b) on the fitting of exponential mixture models to some actual survival data. The data are drawn from the study of McGiffin et al. (1994a) on 2,100 patients who underwent 2,366 aortic valve replacements with a variety of allograft, xenograft, and mechanical valves. The study of McGiffin et al. (1994b) focussed on the identification of risk factors for reoperation following an initial operation for replacement of the aortic valve. All patients who underwent a reoperation were reentered as new cases into the study. In this way the 2,100 patients gave rise to 2,366 survival times.

The example reported here concerns the time in months T to reoperation for any reason following an operation for the replacement of the aortic valve with a valve of a particular variety, namely a viable allograft valve. One of the reasons for reoperation is degeneration of the prosthesis (the replacement valve); another is due to infection of the prosthesis.

There were 320 valve replacements with a viable allograft valve, of which 16 resulted in reoperations. There were 44 patients who subsequently died without needing a reoperation. The remaining 260 survival times were all censored in that at the end of the study they were recorded on patients still living, without having undergone a reoperation.

The distribution of the time T to reoperation was modelled by a modified version of the exponential mixture model (19.1.1) with $g = 2$ components, where the second component represents the density of time to reoperation. The work of McGiffin et al. (1994b) suggests that the distribution of time to reoperation (for those patients who undergo a reoperation before death) can be modelled adequately by the exponential distribution in the case of a replacement viable allograft valve. This is because the hazard function is essentially constant and does not depend significantly on characteristics of the patient, such as the age at operation. The first component density was taken to have a degenerate distribution that places mass one at infinity. This is equivalent to setting the failure time at infinity for those patients who die without undergoing a reoperation. It avoids having to model the distribution of time to death. This class of mixture models, which allows for long-term survival, has been applied by Anscombe (1961) and Goldman (1984), among others, for exponential component distributions.

Table 19.1: Results of fit

Logistic parameters β_0	β_1	Estimated time to reoperation θ_2
-0.861	0.005	104.62
(0.789)	(0.001)	(24.382)

Further references may be found in Meeker (1987) and, more recently, Ghitany (1993), who have summarized the literature on this class, starting with the initial work by Boag (1949) and Berkson and Gage (1952).

In McGiffin et al. (1994b), the mixture exponential model (19.1.1) was modified further by using the extension (19.8.1) to model the mixing proportions as a logistic function of the age a of the patient at the time of the initial operation to replace his or her aortic valve. Accordingly, π_1 was modelled as

$$\pi_1(a_j;\boldsymbol{\beta}) = \exp(\beta_0 + \beta_1 a_j)/\{1 + \exp(\beta_0 + \beta_1 a_j)\} \ ,$$

where a_j denotes the age at operation (in months) of the jth patient.

With these modifications to the exponential mixture model, the probability $\overline{F}(t;\theta_2, \boldsymbol{\beta}, a)$ that a patient with a replacement viable allograft valve will not have to undergo a reoperation for any reason within time t of the initial operation is given by

$$\overline{F}(t;\theta_2,\boldsymbol{\beta},a) = \pi_1(a;\boldsymbol{\beta}) + \pi_2(a;\boldsymbol{\beta})\overline{F}_2(t;\theta_2) \ ,$$

where

$$\overline{F}_2(t;\theta_2) = e^{-t/\theta_2}$$

is the survival function for time to reoperation for those patients who will undergo a reoperation before death. As t tends to infinity, $\overline{F}(t;\theta_2,\boldsymbol{\beta},a)$ thus tends to $\pi_1(a;\boldsymbol{\beta})$, which is the probability that the patient will die without having to undergo a reoperation for any reason. Equations (19.2.11) and (19.2.12) were modified appropriately to allow for the fact that $\overline{F}_1(t;\theta_1)$ has been set to be identically equal to one. The actual fitting of the model was carried out using the computer program given in McLachlan et al. (1993).

The results of the fit are displayed in Table 19.1, where the standard errors for the estimates of the parameters are displayed in parentheses. The latter were computed by applying the bootstrap approach of Efron (1979, 1982), as described in Section 19.5.3. In this application, $K = 100$ bootstrap samples were generated. This approach was used also to provide the 70% confidence limits for the estimate of the survival function $\overline{F}_2(t;\widehat{\theta}_2)$, as plotted in Figure 19.1. For example, it can been seen that a patient having an aortic valve replaced with a viable allograft has a conditional probability of 0.58 of going 5 years and beyond the initial operation without a reoperation, given that a reoperation is necessary before death. In Figure 19.2, we have plotted $\pi_2(a;\widehat{\boldsymbol{\beta}})$, the estimated probability that a patient will undergo a reoperation, versus the age a of the patient at time of operation.

References

Aitkin, M., Anderson, D., and Hinde, J. (1981). Statistical modelling of data on teaching styles (with discussion), *Journal of the Royal Statistical Society, Series A*, **144**, 414-461.

Anscombe, F.J. (1961). Estimating a mixed-exponential response law, *Journal of the American Statistical Association*, **56**, 493-502.

Barnard, G.A. (1963). Contribution to the discussion of paper by M.S. Bartlett, *Journal of the Royal Statistical Society, Series B*, **25**, 294.

Bartholomew, D.J. (1969). Sufficient conditions for a mixture of exponentials to be a probability density function, *Annals of Mathematical Statistics*, **40**, 2183-2188.

Berkson, J. and Gage, R.P. (1952). Survival curve for cancer patients following treatment, *Journal of the American Statistical Association*, **47**, 501-515.

Berndt, E.R., Hall, B.H., Hall, R.E., and Hausman, J.A. (1974). Estimation and inference in nonlinear structural models, *Annals of Economic and Social Measurement*, **3**, 653-665.

Boag, J.Q. (1949). Maximum likelihood estimates of the proportion of patients cured by cancer therapy, *Journal of the Royal Statistical Society, Series B*, **11**, 15-53.

Brockmeyer, F., Halstrom, H.L., and Jensen, A. (1948). The life and works of A.K. Erlang, *Transactions of the Danish Academy of Science*, No. 2.

Choi, S.C. (1979). Two-sample tests for compound distributions for homogeneity of mixing proportions, *Technometrics*, **21**, 361-365.

Cox, D.R. and Oakes, D. (1984). *Analysis of Survival Data*, Chapman & Hall, London.

Day, N.E. (1969). Estimating the components of a mixture of normal distributions, *Biometrika*, **56**, 463-474.

Dempster, A.P., Laird, N.M., and Rubin, D.B. (1977). Maximum likelihood from incomplete data via the EM algorithm (with discussion), *Journal of the Royal Statistical Society, Series B*, **39**, 1-38.

Efron, B. (1979). Bootstrap methods: Another look at the jackknife, *Annals of Statistics*, **7**, 1-26.

Efron, B. (1982). *The Jackknife, the Bootstrap and Other Resampling Plans*, SIAM, Philadelphia.

Efron, B. and Tibshirani, R.J. (1993). *An Introduction to the Bootstrap*, Chapman & Hall, London.

Efron, B. and Hinkley, D.V. (1978). Assessing the accuracy of the maximum likelihood estimator: Observed versus expected Fisher information (with discussion), *Biometrika*, **65**, 457-487.

Everitt, B.S. and Hand, D.J. (1981). *Finite Mixture Distributions*, Chapman & Hall, London.

Farewell, V.T. (1977). A model for a binary variable with time-censored observations, *Biometrika*, **64**, 43-46.

Farewell, V.T. (1982). The use of mixture models for the analysis of survival data with long-term survivors, *Biometrics*, **38**, 1041-1046.

Farewell, V.T. and Sprott, D. (1988). The use of a mixture model in the analysis of count data, *Biometrics*, **44**, 1191-1194.

Ghitany, M.E. (1993). On the information matrix of exponential mixture models with long-term survivors, *Biometrical Journal*, **35**, 15-27.

Goldman, A.I. (1984). Survivorship analysis when cure is a possibility: a Monte Carlo study, *Statistics in Medicine*, **3**, 153-163.

Gordon, N.H. (1990). Application of the theory of finite mixtures for the estimation of 'cure' rates of treated cancer patients, *Statistics in Medicine*, **9**, 397-407.

Hall, P. and Titterington, D.M. (1989). The effect of simulation order on level accuracy and power of Monte Carlo tests, *Journal of the Royal Statistical Society, Series B*, **51**, 459-467.

Harris, C.M., Marchal, W.G., and Botta, R.F. (1992). A note on generalized hyper-exponential distributions, *Communications in Statistics - Stochastic Models*, **8**, 179-191.

Heckman, J.J., Robb, R., and Walker, J.R. (1990). Testing the mixture of exponentials hypothesis and estimating the mixing distribution by the method of moments, *Journal of the American Statistical Association*, **85**, 582-589.

Hoaglin, D.C. (1985). Using quantiles to study shape, In *Exploring Data Tables, Trends and Shapes* (Eds., D.C. Hoaglin, F. Mosteller, and J.W. Tukey), pp. 417-460, John Wiley & Sons, New York.

Hope, A.C.A. (1968). A simplified Monte Carlo significance test procedure, *Journal of the Royal Statistical Society, Series B*, **30**, 582-598.

Jewell, N.P. (1982). Mixtures of exponential distributions, *Annals of Statistics*, **10**, 479-484.

Jones, P.N. and McLachlan, G.J. (1992). Improving the convergence rate of the EM algorithm for a mixture model fitted to grouped truncated data, *Journal of Statistical Computation and Simulation*, **43**, 31-44.

Kuk, A.Y.C. (1992). A semiparametric mixture model for the analysis of competing risks data, *Australian Journal of Statistics*, **34**, 169-180.

Kuk, A.Y.C. and Chen, C.-H. (1992). A mixture model combining logistic regression with proportional hazards regression, *Biometrika*, **79**, 531-541.

Louis, T.A. (1982). Finding the observed information matrix when using the EM algorithm, *Journal of the Royal Statistical Society, Series B*, **44**, 226-233.

McGiffin, D.C., O'Brien, M.F., Galbraith, A.J., McLachlan, G.J., Stafford, E.G., Gardiner, M.A.H., Pohlner, P.G., Early, L., and Kear, L. (1994a). An analysis of risk factors for death and mode-specific death following aortic valve replacement using allograft, xenograft and mechanical valves, *Journal of Cardiac and Thoracic Surgery*, **106**, 895-911.

McGiffin, D.C., O'Brien, M.F., Galbraith, A.J., McLachlan, G.J., Stafford, E.G., Gardiner, M.A.H., Pohlner, P.G., Early, L., and Kear, L. (1994b). An analysis of risk factors for reoperation following aortic valve replacement using allograft, xenograft and mechanical valves, *Journal of Cardiac and Thoracic Surgery* (To appear).

McLachlan, G.J. (1987). On bootstrapping the likelihood ratio test statistic for the number of components in a normal mixture, *Applied Statistics*, **36**, 318-324.

McLachlan, G.J. (1992). *Discriminant Analysis and Statistical Pattern Recognition*, John Wiley & Sons, New York.

McLachlan, G.J., Adams, P., McGiffin, D.C., and Galbraith, A.J. (1993). Fitting mixtures of Gompertz distributions to censored survival data, *Submitted for publication.*

McLachlan, G.J. and Basford, K.E. (1988). *Mixture Models: Inference and Applications to Clustering*, Marcel Dekker, New York.

McLachlan, G.J., Lawoko, C.R.O., and Ganesalingam, S. (1982). On the likelihood ratio test for compound distributions for homogeneity of mixing proportions, *Technometrics*, **24**, 331-334.

Meeker, W.Q. (1977). Limited failure population life tests: Application to integrated circuit reliability, *Technometrics*, **29**, 51-65.

Meilijson, I. (1989). A fast improvement to the EM algorithm on its own terms, *Journal of the Royal Statistical Society, Series B*, **51**, 127-138.

Mendenhall, W. and Hader, R.J. (1958). Estimation of parameters of mixed exponentially distributed failure time distributions from censored life-test data, *Biometrika*, **45**, 504-519.

Meng, X.L. and Rubin, D.B. (1991). Using EM to obtain asymptotic variance-covariance matrices: The SEM algorithm, *Journal of the American Statistical Association*, **86**, 899-909.

Quinn, B.G., McLachlan, G.J., and Hjort, N.L. (1987). A note on the Aitkin-Rubin approach to hypothesis testing in mixture models, *Journal of the Royal Statistical Society, Series B*, **49**, 311-314.

Redner, R.A. and Walker, H.F. (1984). Mixture densities, maximum likelihood and the EM algorithm, *SIAM Review*, **26**, 195-239.

Steutel, F.W. (1967). Note on the infinite divisibility of exponential mixtures, *Annals of Mathematical Statistics*, **38**, 1303-1305.

Teicher, H. (1961). Identifiability of mixtures, *Annals of Mathematical Statistics*, **32**, 244-248.

Titterington, D.M., Smith, A.F.M., and Makov, U.E. (1985). *Statistical Analysis of Finite Mixture Distributions*, John Wiley & Sons, New York.

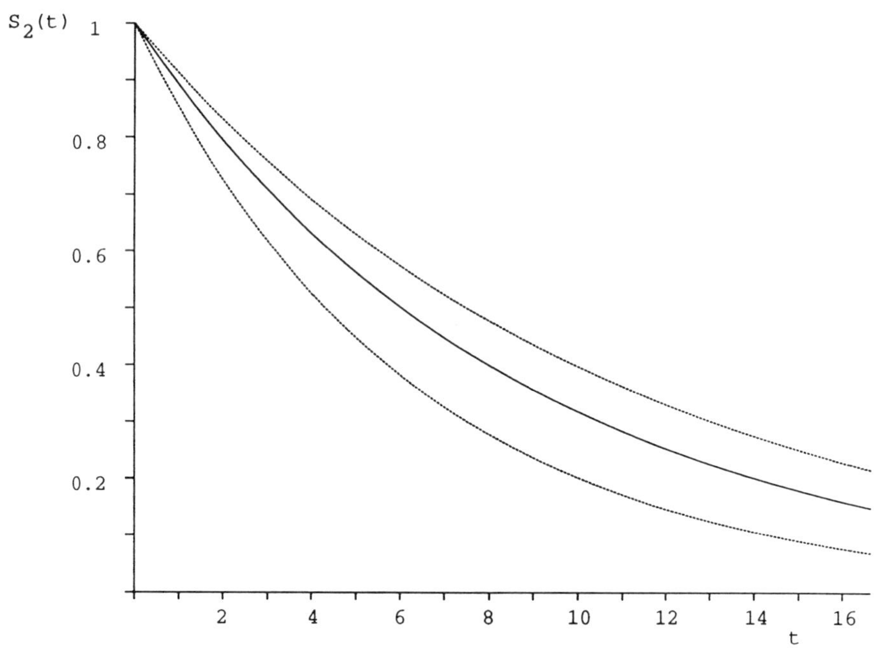

Figure 19.1: Plot of survival function $\overline{F}_2(t)$ for time to reoperation following aortic valve replacement with a viable allograft valve. Dotted curves define the 70% confidence limits

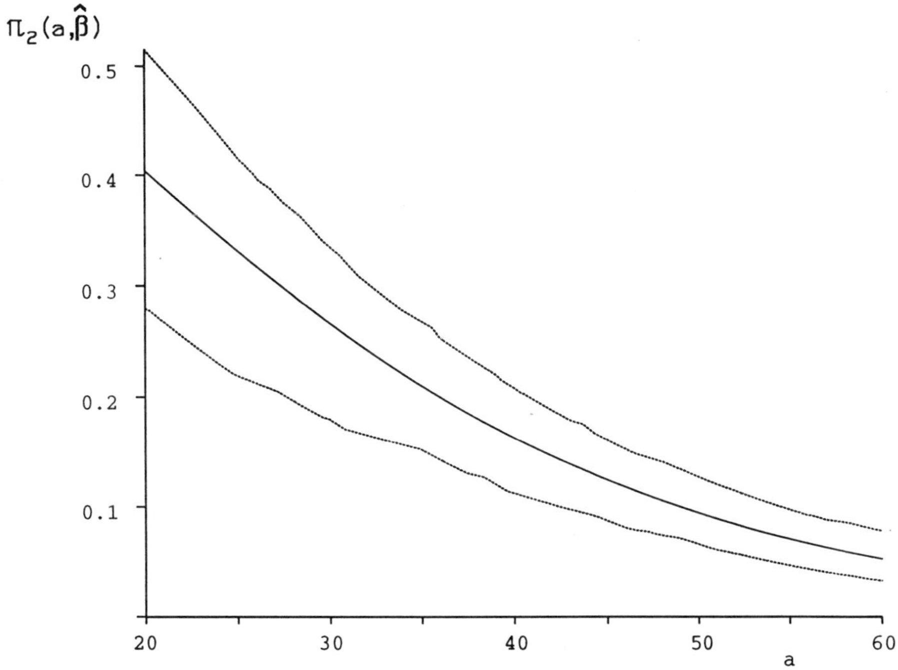

Figure 19.2: Plot of the estimated probability $\pi_2(a; \widehat{\beta})$ that a patient will undergo a reoperation following aortic valve replacement with a viable allograft valve for various levels of the age a (in years) at time of initial operation. Dotted curves define the 70% confidence limits

CHAPTER 20

Bivariate Exponential Distributions

Asit P. Basu[1]

20.1 Introduction

Let X follow the univariate exponential distribution with density function

$$f(x) = \lambda \exp(-\lambda x), \quad x \geq 0, \lambda > 0, \tag{20.1.1}$$

and distribution function

$$F(x) = 1 - \exp(-\lambda x), \quad x \geq 0. \tag{20.1.2}$$

Let us denote this by $X \sim e(\lambda)$. The survival function (or the reliability function) is $\bar{F}(x) = \exp(-\lambda x)$.

The exponential distribution (20.1.1) has a number of properties as seen in earlier chapters. Three of these are given below:

P1) $F(x)$ is absolutely continuous.

P2) $F(x)$ has a constant failure rate function $\rho(x)$ given by $\rho(x) = f(x)/\bar{F}(x) = \lambda$.

P3) F satisfies the loss of memory or the lack of memory (LMP) property. That is,

$$\bar{F}(x+t) = \bar{F}(x)\bar{F}(t) \text{ for all } x, t > 0. \tag{20.1.3}$$

Because of the importance of the univariate distribution as a physical model, it is natural to consider multivariate extensions. However, unlike the multivariate normal distribution, there is no single natural multivariate extension. In this chapter, for simplicity, we consider only a number of bivariate exponential distributions. Many of the models have multivariate extensions. However, the notations become quite complex. We would like a bivariate distribution $F(x, y)$ to possess properties similar to (or properties that are multivariate extensions of) properties of the univariate exponential distribution. Thus, we would want a bivariate exponential distribution to be physically motivated and be continuous with bivariate loss of memory property (BLMP) and constant bivariate failure rate. We define BLMP and bivariate failure rate as follows.

[1]University of Missouri-Columbia, Columbia, Missouri

A bivariate distribution $F(x, y)$ satisfies the BLMP if

$$\bar{F}(xt, yt) = \bar{F}(x, y)\bar{F}(t, t) \qquad (20.1.4)$$

for all $x, y, t > 0$, where $\bar{F}(x, y) = P(X > x, Y > y)$ is the bivariate survival function.

The bivariate failure rate of an absolutely continuous bivariate distribution function with density $f(x, y)$ is given by

$$\rho(x, y) = f(x, y)/\bar{F}(x, y), \qquad (20.1.5)$$

where $\bar{F}(x, y) > 0$.

The marginal distributions should be univariate exponential distributions. We list below a number of distributions that have been proposed as models.

Since both x and y are non-negative, the boundary values of the bivariate exponential distribution should be given by

$$F(x, 0) = F(0, y) = F(0, 0) = 0 \text{ and } F(\infty, \infty) = 1 \qquad (20.1.6)$$

Further discussions on the bivariate and multivariate exponential distributions are provided in Chapters 21, 22 and 23.

20.2 Gumbel's Distributions

Gumbel (1960) is one of the first to consider bivariate exponential distributions. He proposed three, which are absolutely continuous models. None of these are physically motivated. We consider them below.

Model 1. The joint distribution of (X, Y) is given by

$$F(x, y) = 1 - e^{-x} - e^{-y} + e^{-x-y-\delta xy}, \quad x \geq 0, y \geq 0, 0 \leq \delta \leq 1. \qquad (20.2.1)$$

Here, X and Y are independent when $s = 0$. Also, the marginal distributions are exponential distributions. Here, the joint density function is

$$f(x, y) = [\exp -(x + y + \delta xy)][(1 + \delta x)(1 + \delta y) - \delta] \qquad (20.2.2)$$

Model 2. Gumbel's second model is given by

$$F(x, y) = (1 - e^{-x})(1 - e^{-y})[1 + \alpha e^{-x-y}], \quad x \geq 0,\ y \geq 0,\ -1 \leq \alpha \leq 1. \qquad (20.2.3)$$

Here, the joint density is given by

$$f(x, y) = e^{-x-y}[1 + \alpha(2e^{-x} - 1)(2e^{-y} - 1)]\ . \qquad (20.2.4)$$

X and Y are independent when $\alpha = 0$.

Model 3. Gumbel's third model is given by distribution function

$$F(x, y) = 1 - e^{-x} - e^{-y} + \exp[-(x^m + y^m)^{1/m}] \qquad (20.2.5)$$

and density function

$$f(x, y) = \exp[-(x^m + y^m)^{1/m}](x^m + y^m)^{(1/m)-2} x^{m-1} y^{m-1} [(x^m + y^m)^{1/m} + m - 1], \quad x \geq 0,\ y \geq 0,\ m \geq 1\ . \qquad (20.2.6)$$

Here, $m = 1$ corresponds to the case of independence.

20.3 Freund's Model

Freund (1961) is one of the first to consider a model that is physically motivated. Let X and Y denote the lifetimes of components A and B of a two-component system. To start with, X and Y are independent with $X \sim e(\alpha)$ and $Y \sim e(\beta)$. If A fails first, the lifetime of B is changed to $e(\beta')$ because of extra stress. If B fails first, the distribution of A changes to $e(\alpha')$. Assume $\alpha + \beta - \beta' \neq 0$. The joint density of (X, Y), under the above assumption, is given by

$$f(x,y) = \begin{array}{ll} \alpha\beta' \exp[-\beta' y - (\alpha + \beta - \beta')x] & \text{if } 0 < x < y \\ \alpha'\beta \exp[-\alpha' x - (\alpha + \beta - \alpha')y] & \text{if } 0 < y < x. \end{array} \tag{20.3.1}$$

However, the marginal distributions of X and Y are not univariate exponential distributions in general. These are mixtures or weighted averages of exponential distributions and are given by

$$f(x) = \frac{(\alpha - \alpha')(\alpha + \beta)e^{-(\alpha+\beta)x}}{\alpha + \beta - \alpha'} + \frac{\alpha'\beta e^{-\alpha' x}}{\alpha + \beta - \alpha'}, \tag{20.3.2}$$

provided $\alpha + \beta - \alpha' \neq 0$, and

$$g(y) = \frac{(\beta - \beta')(\alpha + \beta)\exp\{-(\alpha + \beta)y\}}{\alpha + \beta - \beta'} + \frac{\alpha\beta' \exp(-\beta' y)}{\alpha + \beta - \beta'}. \tag{20.3.3}$$

Here, X and Y are independent if and only if $\alpha = \alpha'$ and $\beta = \beta'$. Note if $\beta > \beta'$ (or if $\alpha > \alpha'$), the expected life of component B(A) improves when component A(B) fails. In this case, $\alpha + \beta - \beta' > 0$ and the marginal densities are mixtures of exponential densities. On the other hand, if $\beta < \beta'$ ($\alpha < \alpha'$), the expected lifetime of B(A) decreases after the failure of component A(B). In this case, $\alpha + \beta - \beta' < 0$ ($\alpha + \beta - \alpha' < 0$) and the marginal densities are weighted averages of exponential densities.

20.4 Marshall-Olkin Model

The joint survival function for the Marshall-Olkin model (1967) is given by

$$\begin{aligned} \bar{F}(x,y) &= P(X > x, Y > y) = \exp(-\lambda_1 x - \lambda_2 y - \lambda_3 \max(x,y)), \\ & \qquad x > 0,\ y > 0,\ \lambda_1 > 0,\ \lambda_2 > 0,\ \lambda_3 \geq 0\ . \end{aligned} \tag{20.4.1}$$

We shall call this the BVE distribution. Here, the marginal distributions are univariate exponential distributions, and λ_3=0 corresponds to the case where X and Y are independent.

The BVE can be derived in a number of different ways. We consider a system consisting of two components, A and B. Let there be three independent Poisson processes, P_1, P_2, and P_3, with parameters λ_1, λ_2, and λ_3, respectively. If an event occurs according to the first process, A fails. If an event occurs in the second process, B fails; and if an event occurs in the third process, both A and B fail simultaneously. Let U, V, and W be the occurrence time of the first event in P_1, P_2, and P_3, respectively, and let X be failure time of component A and Y be failure time of component B. Then, $X = \min(U, W)$,

$Y = \min(V, W)$, and $\bar{F}(x, y) = P(X > x, Y > y) = P(U > x, V > y, W > \max(x, y))$, which is given by (20.4.1).

The BVE has a number of important properties. It satisfies the bivariate loss of memory property as given in (20.1.4). In fact, BVE is the only solution of the functional equation (20.1.4) subject to the requirement of exponential marginal distributions.

However, BVE has both an absolutely continuous and a singular part, which can be expressed as

$$\bar{F}(x, y) = \frac{\lambda_1 + \lambda_2}{\lambda} \bar{F}_a(x, y) + \frac{\lambda_{12}}{\lambda} \bar{F}_s(x, y), \tag{20.4.2}$$

where $\lambda = \lambda_1 + \lambda_2 + \lambda_3$,

$$\bar{F}_s(x, y) = \exp[-\lambda \max(x, y)] \tag{20.4.3}$$

is a singular distribution, and

$$\bar{F}_a(x, y) = \frac{\lambda}{\lambda_1 + \lambda_2} \exp[-\lambda_1 x - \lambda_2 y - \lambda_3 \max(x, y)] - \frac{\lambda_3}{\lambda_1 + \lambda_2} \exp[-\lambda \max(x, y)] \tag{20.4.4}$$

is absolutely continuous.

Note that $\min(X, Y) \sim e(\lambda)$, a result similar to the case when X and Y have independent exponential distributions.

20.5 Block-Basu Model

The model proposed by Block and Basu (1974) is closely related to the BVE and the model proposed by Freund. Let us call this the ACBVE. The BVE has univariate exponential marginals, and it satisfies the BLMP. However, it is not absolutely continuous. Block and Basu (1974) showed that it is not possible to have an absolutely continuous bivariate distribution that satisfies BLMP and has exponential marginal distributions. Since absolute continuity is a desirable criterion, Block and Basu proposed the ACBVE where the marginals are mixtures or weighted averages of exponential distributions. Here, the joint survival function is given by

$$\begin{aligned} \bar{F}(x, y) &= \frac{\lambda}{\lambda_1 + \lambda_2} \exp[-\lambda_1 x - \lambda_2 y - \lambda_3 \max(x, y)] \\ &\quad -(\lambda_3/(\lambda_1 + \lambda_2)) \exp[-\lambda \max(x, y)], \\ &\quad x,\ y > 0,\ \lambda_1,\ \lambda_2 > 0, \text{ and } \lambda_3 \geq 0\,. \end{aligned} \tag{20.5.1}$$

The joint density is given by

$$f(x, y) = \begin{cases} \frac{\lambda_1 \lambda(\lambda_2 + \lambda_3)}{\lambda_1 + \lambda_2} \exp[-\lambda_1 x - (\lambda_2 + \lambda_3) y] & \text{if } x < y \\ \frac{\lambda_2 \lambda(\lambda_1 + \lambda_3)}{\lambda_1 + \lambda_2} \exp[-(\lambda_1 + \lambda_3) x - \lambda_2 y] & \text{if } x > y \end{cases} \tag{20.5.2}$$

Here, the marginal survival functions are given by

$$\begin{aligned} \bar{F}_1(x) &= \tfrac{\lambda}{\lambda_1 + \lambda_2} \exp[-(\lambda_1 + \lambda_3) x] - \tfrac{\lambda_3}{\lambda_1 + \lambda_2} \exp(-\lambda), \quad x > 0 \\ \bar{F}_2(y) &= \tfrac{\lambda}{\lambda_1 + \lambda_2} \exp[-(\lambda_2 + \lambda_3) y] - \tfrac{\lambda_3}{\lambda_1 + \lambda_2} \exp(-\lambda), \quad y > 0. \end{aligned} \tag{20.5.3}$$

Note the ACBVE is the absolutely continuous component of the BVE. However, it is not a special case, since the BVE has exponential marginals and the ACBVE does not. The ACBVE is a special case of the Freund model as it can be derived in the same way as in Section 20.3 with special choice of the parameters. Assume $\alpha < \alpha'$ and $\beta < \beta'$ in Freund's model and let

$$\alpha = \lambda_1 + \lambda_3(\lambda_1/(\lambda_1+\lambda_2)), \quad \alpha' = \lambda_1+\lambda_3, \quad \beta = \lambda_2 + \lambda_3(\lambda_2/(\lambda_1+\lambda_2)), \quad \text{and } \beta' = \lambda_2+\lambda_3. \tag{20.5.4}$$

Substituting these values of α, α', β, and β' in (20.3.1), we obtain (20.5.2).

The ACBVE, like BVE, has the BLMP and $\min(x,y) \sim e(\lambda)$. Also, λ_3=0 corresponds to the case when X and Y are independent. However, unlike the BVE, the ACBVE is absolutely continuous. Other properties of the ACBVE are given in Block and Basu (1974).

20.6 Other Distributions

Friday and Patil (1977) obtain a general distribution, which they call the BEE distribution. The BEE distribution includes as special cases the BVE, ACBVE, Freund's model, and a model due to Proschan and Sullo (1974). Here, the joint survival distribution is given by

$$\bar{F}(x,y) = \alpha_0 \bar{F}_A(x,y) + (1-\alpha_0)\bar{F}_S(x,y), \tag{20.6.1}$$

where $0 < \alpha_0 \leq 1$ and $\bar{F}_A(x,y)$ is the joint survival function corresponding to the density function (20.3.1) of Freund's model. Here, $\bar{F}_S(x,y)$ is given by

$$\bar{F}_S(x,y) = \exp[-(\alpha+\beta)\max(x,y)], \quad x,y > 0 \tag{20.6.2}$$

When α_0=1, (20.6.1) reduces to the Freund's distribution. BVE is obtained by defining α, α', β, and β' as in (20.5.4) and letting $\alpha_0 = (\lambda_1+\lambda_2)/\lambda$.

For a survey of other bivariate exponential distributions, see Basu (1988) and the references therein. These models are due to Arnold (1975), Downton (1970), Hawkes (1972), Paulson (1973), Raftery (1984), Sarkar (1987), and Tosch and Holmes (1980).

Klein has described in Chapter 21 the exponential frailty models. For a related class of distributions, see Lindley and Singpurwalla (1986) and Bandyopadhyay and Basu (1990).

References

Arnold, B.C. (1975). Multivariate exponential distributions based on hierarchical successive damage, *Journal of Applied Probability,* **12,** 142-147.

Bandyopadhyay, D. and Basu, A.P. (1990). On a generalization of a model by Lindley and Singpurwalla, *Advances in Applied Probability,* **22**, 498-500.

Basu, A.P. (1988). Multivariate exponential distributions and their applications in reliability, In *Handbook of Statistics* (Eds., P.R. Krishnaiah and C.R. Rao), **7,** Elsevier Science Publishers, New York.

Block, H.W. and Basu, A.P. (1974). A continuous bivariate exponential extension, *Journal of of the American Statistical Association,* **69,** 1031-1037.

Downton, F. (1970). Bivariate exponential distributions in reliability theory, *Journal of the Royal Statistical Society, Series B,* **32,** 408-417.

Freund, J. (1961). A bivariate extension of the exponential distribution, *Journal of the American Statistical Association,* **56,** 971-977.

Friday, D.S. and Patil, G.P. (1977). A bivariate exponential model with applications to reliability and computer generation of random variables, In *Theory and Applications of Reliability, Vol. I.* (Eds., C.P. Tsokos and I. Shimi), pp. 527-549, Academic Press, New York.

Gumbel, E.J. (1960). Bivariate exponential distribution, *Journal of the American Statistical Association,* **55,** 698-707.

Hawkes, A.G. (1972). A bivariate exponential distribution with applications to reliability, *Journal of the Royal Statistical Society, Series B,* **34,** 129-131.

Johnson, N.L. and Kotz, S. (1977). *Distributions in Statistics: Continuous Multivariate Distributions,* John Wiley & Sons, New York.

Marshall, A.W. and Olkin, I. (1967). A multivariate exponential distribution, *Journal of the American Statistical Association,* **62,** 30-44.

Paulson, A.S. (1974). A characterization of the exponential distribution and a bivariate exponential distribution, *Sankhyā, Series A,* **35,** 69-78.

Proschan, F. and Sullo, P. (1974). Estimating the parameters of a bivariate exponential distribution in several sampling situations, In *Reliability and Biometry* (Eds., F. Proschan and R. J. Serfling), pp. 423-440, SIAM, Philadelphia.

Proschan, F. and Sullo, P. (1976). Estimating the parameters of a multivariate exponential distribution, *Journal of the American Statistical Association,* **71,** 465-472.

Raftery, A.E. (1984). A continuous multivariate exponential distribution, *Communications in Statistics - Theory and Methods,* **13,** 947-965.

Sarkar, S.K. (1987). A continuous bivariate exponential distribution, *Journal of the American Statistical Association,* **82,** 667-675.

Tosch, T.J. and Holmes, P.T. (1980). A bivariate failure model, *Journal of the American Statistical Association,* **75,** 415-417.

CHAPTER 21

Inference for Multivariate Exponential Distributions

John P. Klein[1]

21.1 Introduction

In this chapter we provide a survey of techniques for inference about parameter values for multivariate exponential distributions. We shall focus, in most cases, on the bivariate model for ease of exposition and indicate when multivariate extensions can be found.

Section 21.2 describes inference for the Marshall-Olkin (1967) model. This distribution has exponential marginals and allows for simultaneous failures. Section 21.3 discusses the bivariate model of Freund (1961). This model arises quite naturally in many contexts where failure of one component of a system increases the strain on the surviving component. Section 21.4 deals with the absolutely continuous bivariate exponential distribution of Block and Basu (1974). This model is a special case of Freund's model and is the absolutely continuous part of the Marshall-Olkin model. Section 21.5 looks at inference for a class of models induced by individuals in groups (lots, litters, families, etc.) sharing unobserved common random effects. An estimation scheme based on the EM-algorithm is described here. Finally, Section 21.6 surveys results for one of Gumbel's (1960) bivariate exponential model and for the bivariate model proposed by Downton (1960).

21.2 Inference for the Marshall-Olkin Model

The Marshall and Olkin (1967) bivariate exponential distribution (MOBVE) has a joint survival function

$$S(x,y) = \exp[-\lambda_1 x - \lambda_2 y - \lambda_3 \max\{x,y\}], \qquad x,y>0,\ \lambda_i \geq 0\ , \qquad (21.2.1)$$

where at least one λ_i is positive. For this model the marginal distributions of (X,Y) are exponential with hazard rates $(\lambda_1+\lambda_3)$ and $(\lambda_2+\lambda_3)$, respectively. The correlation

[1]The Medical College of Wisconsin, Milwaukee, Wisconsin

is $\rho = \lambda_3/\lambda$, where $\lambda = \lambda_1 + \lambda_2 + \lambda_3$. The distribution is not absolutely continuous with respect to the usual Lebesgue measure since $P[X = Y] = \rho$. The density function is

$$f(x,y) = \begin{cases} \lambda_1(\lambda_2 + \lambda_3)S(x,y) & \text{if } y > x > 0 \\ \lambda_2(\lambda_1 + \lambda_3)S(x,y) & \text{if } x > y > 0 \\ \lambda_3 S(x,x) & \text{if } x = y > 0 \end{cases} \tag{21.2.2}$$

21.2.1 Estimation of Model Parameters

To estimate the parameters of the MOBVE suppose we have a sample of size n from (21.2.1). Let n_1, n_2, and n_3 be the number of observations with $\{X < Y\}$, $\{X > Y\}$, and $\{X = Y\}$, respectively. Let $S_X = \sum X_j$, $S_Y = \sum Y_j$, $S_1 = \sum \min\{X_j, Y_j\}$, and $S_2 = \sum \max\{X_j, Y_j\}$. Note that (n_1, n_2, n_3) has a multinomial distribution with parameters n, λ_1/λ, λ_2/λ, λ_3/λ, and S_X, S_Y, and S_1 are gamma random variables with shape parameter n and scale parameter $(\lambda_1 + \lambda_3)$, $(\lambda_2 + \lambda_3)$, and λ, respectively. The statistics $(n_1, n_2, S_X, S_1, S_2)$ are a set of minimal sufficient statistics, but they are not complete.

Arnold (1968) considers estimation when only the minimum of X and Y is observed. He shows that in this case n_1, n_2 and S_1 are complete sufficient statistics and the minimum variance unbiased estimators of $(\lambda_1, \lambda_2, \lambda_3)$ are

$$\widehat{\lambda}_{A_i} = \frac{n-1}{n}\frac{n_i}{S_1}, \qquad i = 1, 2, 3 . \tag{21.2.3}$$

These are consistent estimators with variances $[\lambda\lambda_i(n-1) + \lambda_i^2]/(n(n-2))$, $i = 1, 2, 3$ if $n > 2$.

In the complete data case, parameter estimation has been considered by several authors including Bemis, Bain, and Higgins (1972), Proschan and Sullo (1974, 1976), and Bhattacharyya and Johnson (1971, 1973). In this case the log-likelihood function is

$$\begin{aligned} L \;=\; & -\lambda_1 S_X - \lambda_2 S_Y - \lambda_3 S_2 + n_1 \ln[\lambda_1(\lambda_2 + \lambda_3)] \\ & + n_2 \ln[\lambda_2(\lambda_1 + \lambda_3)] + n_3 \ln[\lambda_3] . \end{aligned} \tag{21.2.4}$$

Maximum likelihood estimates (MLEs) are found by solving the score equations, namely

$$\begin{aligned} n_1/\lambda_1 + n_2/(\lambda_1 + \lambda_3) - S_X &= 0 , \\ n_1/(\lambda_2 + \lambda_3) + n_2/\lambda_2 - S_Y &= 0 , \end{aligned}$$

and

$$n_1/(\lambda_2 + \lambda_3) + n_2/(\lambda_1 + \lambda_3) + n_3/\lambda_3 - S_2 = 0 .$$

If all the $n_i > 0$, then the maximum likelihood estimators of the λ's are the unique solution to these equations which is found numerically. If $n_3 = 0$, then the MLE's of $(\lambda_1, \lambda_2, \lambda_3)$ is $(n/S_X, n/S_Y, 0)$. If $n_3 = 0$ and either $n_1 = 0$ or $n_2 = 0$, the MLE exists but is not unique, while if $n_3 > 0$ and either $n_1 = 0$ or $n_2 = 0$, the MLE does not exist since the supremum of (21.2.4) is not obtained in the parameter space.

Bemis, Bain, and Higgins (1972) give the method of moments estimators, namely,

$$\begin{aligned}\widehat{\lambda}_{B_1} &= [n/S_X - n_3/S_Y]/[1+n_3/n]\ ,\\ \widehat{\lambda}_{B_2} &= [n/S_Y - n_3/S_X]/[1+n_3/n]\ ,\\ \widehat{\lambda}_{B_3} &= n_3[1/S_X + 1/S_Y]/[1+n_3/n]\ .\end{aligned}$$

Proschan and Sullo (1976) propose an estimator based on the first iteration in a maximization of the likelihood equation. Their estimators are

$$\begin{aligned}\widehat{\lambda}_{P_1} &= n\{n_1/(n_1+n_3)\}/S_X\ ,\\ \widehat{\lambda}_{P_2} &= n\{n_2/(n_2+n_3)\}/S_Y\ ,\end{aligned}$$

and

$$\widehat{\lambda}_{P_3} = n_3[1 + n_2/(n_1+n_3) + n_1/(n_2+n_3)]/S_2\ .$$

All of the above estimators are trivariate normal with asymptotic covariance matrices found in the respective papers when the n_i's are non-zero. Proschan and Sullo (1976) compare the asymptotic relative efficiency (ARE) of the estimators $\widehat{\lambda}_A$, $\widehat{\lambda}_B$, $\widehat{\lambda}_P$ to the MLE of λ based on either the ratio of the traces or determinants of the limiting covariance matrix. All of the estimators have been extended to the multivariate case.

A special case of the MOBVE is the symmetric MOBVE with $\lambda_1 = \lambda_2 = \theta$. Bhattacharyya and Johnson (1971, 1973) show that for this restricted model the MLEs of θ and λ_3 are uniquely determined, if $n_3 < n$, by $\widehat{\theta} = 2n/(S_X + S_Y)$ and $\widehat{\lambda}_3 = 0$ if $n_3 = 0$ and if $n_3 > 0$ by $\widehat{\lambda}_3 = [\{n^2(S_2-S_1)^2 + 4n_3(2n-n_3)S_XS_Y\}^{1/2} - n(S_2-S_1)]/(2S_XS_Y)$ and $\widehat{\theta} = (n-n_3)\widehat{\lambda}_3/[n_3 + \widehat{\lambda}_3 S_1]$. If $n_3 = n$, the MLE does not exist.

Klein and Basu (1985) have considered the problem of estimating the reliability function (21.2.1) for the MOBVE distribution. The natural estimator of $S(x,y)$ is found by substitution of one of the four estimators into (21.2.1), namely, $\widehat{S}_{SUB}(x,y) = \exp[-\widehat{\lambda}_1 x - \widehat{\lambda}_2 y - \widehat{\lambda}_3 \max\{x,y\}]$. This estimator is biased. Several bias correction methods have been suggested. Let Σ denote the estimated covariance of $(\widehat{\lambda}_1, \widehat{\lambda}_2, \widehat{\lambda}_3)$. The first bias corrected estimator is based on a first order Taylor series approximation to S. That is, $\widehat{S}_{TS}(x,y) = \widehat{S}_{SUB}(x,y)/[1+\sigma^2/2]$, where $\sigma^2 = (x,y,\max(x,y))\Sigma(x,y,\max(x,y))'$. The second estimator is based on the asymptotic normality of $(\widehat{\lambda}_1, \widehat{\lambda}_2, \widehat{\lambda}_3)$ which implies that $\widehat{S}_{SUB}(x,y)$ should be approximately log-normal. The estimator is $\widehat{S}_{LN}(x,y) = \widehat{S}_{SUB}(x,y)\exp\{\sigma^2/2\}$. A final bias reduced estimator was found by using the jackknife. That is, $\widehat{S}_{JACK}(x,y) = n\,\widehat{S}_{SUB}(x,y) - \frac{n-1}{n}\sum_{j=1}^n \widehat{S}^j_{SUB}(x,y)$, where $\widehat{S}^j_{SUB}(x,y)$ is the substitution estimator based on the sample of size $n-1$ with the jth observation deleted. Based on a Monte Carlo study, they recommend using the jackknife estimator based on either the method of moments estimators or the Proschan and Sullo estimator.

Ebrahimi (1987) has considered the problem of accelerated life testing for the MOBVE model. In accelerated life tests, products are tested at various high stresses, $V_1, \ldots, V_J$, where it is expected that the products will have relatively short lifetimes. The results of these tests are extrapolated to a usage stress, V_0. This is done to reduce test time and costs. He assumes that the failure distribution for a parallel system at each stress follows a MOBVE with $\lambda_{ij} = C_i V_j^P$, $i = 1,2,3$; $j = 1,\ldots,J$. Maximum likelihood, method of moment, and a first iteration of the likelihood estimator in the spirit of Proschan and Sullo estimators of $C_i, i = 1, 2, 3$ are developed in the cases where P is known and unknown. Klein and Basu (1982) have considered this problem for series systems where only the minimum of X and Y is observed for each system.

21.2.2 Tests of Hypothesis

A variety of authors have considered the inference problems based on the Marshall-Olkin Bivariate exponential model. In this section we shall consider tests for the hypothesis of independence, $\lambda_3 = 0$; for symmetry, $\lambda_1 = \lambda_2$, and goodness-of-fit tests.

Consider first the problem of testing the hypothesis $H_0 : \lambda_3 = 0$ or equivalently that $\rho = \lambda_3/\lambda = 0$ where $\lambda = \lambda_1 + \lambda_2 + \lambda_3$, against $H_1 : \lambda_3 > 0$. As noted by Weier and Basu (1981), if we observe any observations along the diagonal where $X = Y$ then we must have $\lambda_3 > 0$, and the null hypothesis should be rejected. The probability of seeing a pair with $X_i = Y_i$ is ρ. This test has a Type I error of zero and power of $1 - (1 - \rho)^n$.

Bemis, Bain, and Higgins (1972) show that the most powerful test of H_0 against $H_1 : \rho = \rho_0 > 0$, when λ_1 and λ_2 are known, rejects if either $n_3 > 0$ or if $T = \rho_0(1 + \rho_0)^{-1}(\delta_1 + \delta_2)S_1 + n_1 \ln\{(1 - \rho_0\delta_2/\delta_1)/(1 - \rho_0\delta_1/\delta_2)] > C(\alpha)$, where $\delta_i = \lambda_i + \lambda_3$, $i = 1, 2$. The critical value $C(\alpha)$ is determined from the power function of the test, namely,

$$
\begin{aligned}
Q(\rho) &= 1 - (1 - \rho)^n + (1 - \rho)^n \sum_{i=1}^{n} P[n_1 = j \mid n_3 = 0] \\
&\quad \cdot P[T > C(\alpha) \mid n_1 = j,\, n_3 = 0] \,.
\end{aligned}
$$

Here $P[n_1 = j \mid n_3 = 0] = \binom{n}{j}(\delta_1 - \rho\delta_2)^j(\delta_2 - \rho\delta_1)^{(n-j)}[(\delta_1 + \delta_2)(1 - \rho)]^{-n}$, and $P[T > C(\alpha) \mid n_1 = j, n_3 = 0] = P[\chi^2_{2n} > 2(1 + \rho^{-1})(1 + \rho)\rho^{-1}(C(\alpha) - j \ln\{(1 - \rho_0\delta_2/\delta_1)/(1 - \rho_0\delta_1/\delta_2))]$, where χ^2_{2n} is a chi-squared random variable with $2n$ degrees of freedom. The power function is monotone decreasing in $C(\alpha)$ so that critical values can be found for any level of significance.

Bhattacharyya and Johnson (1973) show that the α level uniformly most powerful test (UMP) of $H_0 : \lambda_3 = 0$ versus $H_1 : \lambda_3 > 0$ in the symmetric MOBVE model with $\lambda_1 = \lambda_2$ is given by: reject H_0 if $n_3 > 0$ or if $2S_1/(S_2 - S_1) > F_{2n,2n,\alpha}$, the upper α percentage point of a central F distribution with $(2n, 2n)$ degrees of freedom.

Tests of the hypothesis that $\lambda_3 = \lambda_3^0$ can be based on the large sample trivariate normality of all four estimators $(\widehat{\lambda}_1, \widehat{\lambda}_2, \widehat{\lambda}_3)$ discussed in Section 21.2.1. The test statistic is $[\widehat{\lambda}_3 - \lambda_3^0]/\sigma_3$, where σ_3 is an estimate of the standard deviation of $\widehat{\lambda}_3$ obtained from the appropriate limiting covariance matrix. When testing $\lambda_3^0 = 0$ this approach fails since all estimators yield estimates of λ_3 of zero except when $n_3 > 0$, when, as noted earlier, we should reject the null hypothesis. The suggestion of Weier and Basu (1981) is to use as a test of independence when $n_3 = 0$, the large sample test based on the Block-Basu model discussed in Section 21.3.2, since this model corresponds to the Marshall-Olkin model conditional on $X \neq Y$. When $\lambda_3^0 > 0$, Hanagal (1992) and Hanagal and Kale (1991b) have developed the appropriate tests based on the maximum likelihood, Arnold, and Bemis, Bain, and Higgins estimators in Section 21.2.1.

A second hypothesis of interest is the test of symmetry of the marginals of the MOBVE, namely $H_0 : \lambda_1 = \lambda_2$. Here the tests are based on the large sample distribution of $(\widehat{\lambda}_1, \widehat{\lambda}_2, \widehat{\lambda}_3)$. Details of these tests are found in Hanagal (1992).

Several papers have dealt with the problems of testing for goodness-of-fit of the Marshall-Olkin bivariate exponential model. Csörgö and Welsh (1989) propose a test of the hypothesis that the data is from a MOBVE with arbitrary values of $(\lambda_1, \lambda_2, \lambda_3)$ but fixed values of $\delta_i = \lambda_i/\lambda$, $i = 1, 2, 3$ against a global alternative. Their proposed test compares the moment generating function of the MOBVE to the empirical moment generating function. Critical values of the test statistics, based on a Monte Carlo study are found in their paper.

Several authors have considered tests of the hypothesis of MOBVE against the alternative that (X, Y) has a distribution which falls in some particular aging class. Basu and Ebrahimi (1984) consider testing H_0 : (X, Y) MOBVE against H_1 : (X, Y) are Bivariate New Better than Used (BNBU) (but not MOBVE). A survival function is said to be BNBU-I if $S(x+t, y+t) \leq S(x, t)S(y, t)$, and BNBU-II if $S(x+t, x+t) \leq S(x, x)S(t, t)$ for all $x \geq 0$, $y \geq 0$, $t \geq 0$. Tests against either BNBU formulation are based on comparing weighted empirical estimates of $S(x+t, y+t) - S(x, t)S(y, t)$.

Basu and Habibullah (1987) have developed a test of H_0 : (X, Y) MOBVE against H_1 : (X, Y) have a Bivariate Increasing Failure Average (BIFRA) distribution. (X, Y) have a BIFRA distribution if and only if $S^\alpha(x, y) \leq S(\alpha x, \alpha y)$ for all $x, y > 0$, $0 < \alpha < 1$. Sen and Jain (1990) provide a test of MOBVE versus the alternative that (X, Y) have a bivariate harmonic new better than used in expectation (BHNBUE) distribution. (X, Y) have a BHNBUE if $\int_0^\infty S(x+t, y+t)\,dt \leq \int_0^\infty \exp[-\lambda_1(x+t) - \lambda_2(y+t) - \lambda_3 \max\{x+t, y+t\}]dt$, for $x, y > 0$.

21.3 Inference for the Freund Model

The bivariate exponential extension of Freund (1961) has a joint probability density function given by

$$f(x, y) = \begin{cases} \alpha\beta' \exp\{-\beta' y - (\alpha + \beta - \beta')x\} & \text{for} \quad 0 < x < y \\ \beta\alpha' \exp\{-\alpha' x - (\alpha + \beta - \alpha')y\} & \text{for} \quad 0 < y < x\ . \end{cases} \tag{21.3.1}$$

This model arises if one considers a two component parallel system with X and Y the times to failure of the two components A and B, respectively. When both components are functioning the times to component failure are independent exponential random variables with hazard rates α and β. Failure of the A component changes the failure rate of the surviving component to β'. Similarly, if the B component fails first the failure rate of A is changed to α'. The component life lengths are independent if and only if $\alpha = \alpha'$ and $\beta = \beta'$. The marginal times to failure of the components are mixtures of exponentials.

21.3.1 Estimation of Model Parameters

Estimates of model parameters are given in Freund (1961) for the uncensored case and in Leurgans et al. (1980) for the censored data case. Censoring for a bivariate system can arise in two ways. Let (C_X, C_Y) be a bivariate censoring time. We observe $T_1 = \min(X, C_X)$, $\delta_1 = \{1 \text{ if } X \leq C_X,\ 0 \text{ otherwise}\}$, and $T_2 = \min(Y, C_Y)$, $\delta_2 = \{1 \text{ if } Y \leq C_Y,\ 0 \text{ otherwise }\}$. If $C_X = C_Y$ then there is univariate censoring with a single clock for the complete bivariate system. If $C_X \neq C_Y$, then each component has its own internal censoring clock and we have bivariate censoring. Suppose we have a sample $(T_{1i}, \delta_{1i}, T_{2i}, \delta_{2i}),\ i = 1, \ldots, n$.

Under univariate censoring, Leurgans et al. (1980) show that the maximum likelihood estimators of $(\alpha, \alpha', \beta, \beta')$ are given by

$$\begin{aligned} \widehat{\alpha} &= \frac{\sum_{i=1}^n \delta_{1i}(1-\delta_{2i}[T_{1i} > T_{2i}])}{\sum_{i=1}^n \min(T_{1i}, T_{2i})}\,,\ \widehat{\beta} = \frac{\sum_{i=1}^n \delta_{2i}(1-\delta_{1i}[T_{1i} < T_{2i}])}{\sum_{i=1}^n \min(T_{1i}, T_{2i})}\,, \\ \widehat{\alpha}' &= \frac{\sum_{i=1}^n \delta_{1i}\delta_{2i}[T_{1i} > T_{2i}]}{\sum_{i=1}^n [T_{1i} > T_{2i}](T_{1i} - T_{2i})}\,,\ \widehat{\beta}' = \frac{\sum_{i=1}^n \delta_{1i}\delta_{2i}[T_{1i} < T_{2i}]}{\sum_{i=1}^n [T_{1i} < T_{2i}](T_{2i} - T_{1i})}\,, \end{aligned} \tag{21.3.2}$$

where $[\cdot]$ denotes the indicator function. When there is bivariate censoring, the log-likelihood is not linear in the parameters and parameter estimates must be found by numerical methods [see Leurgans et al. (1980) for details].

When there is no censoring Freund (1961) shows that $E[\widehat{\alpha}] = (n/(n-1))\alpha$ and $\mathrm{Var}(\widehat{\alpha}) = n\alpha[\alpha n + \beta(n-1)]/\{(n-1)^2(n-2)]$, with a similar result for $\widehat{\beta}$. For $\widehat{\beta}'$, in the uncensored case the estimator is of the form $r/\sum(Y_i - X_i)$, where r is the number of systems in which the A component fails first and the sum is over those cases. Conditional on r we have $E[\widehat{\beta}' \mid r] = (r/(r-1))\beta'$ if $r > 1$, and $\mathrm{Var}(\widehat{\beta}' \mid r) = r^2\beta'/\{(r-1)^2(r-2)\}$ if $r > 2$. While the estimates of β' are clearly biased, Freund shows that $1/\widehat{\beta}'$ is an unbiased estimator of $1/\beta'$. Again similar properties hold for $\widehat{\alpha}'$.

Hanagal and Kale (1992) show, in the complete data case, that the maximum likelihood estimators $(\widehat{\alpha}, \widehat{\beta}, \widehat{\alpha}', \widehat{\beta}')$ are complete sufficient statistics and that they have asymptotically a multivariate normal distribution with a variance matrix which is diagonal with entries $n^{-1}[\alpha(\alpha+\beta), \beta(\alpha+\beta), \alpha'^2(\alpha+\beta)/\beta, \beta'^2(\alpha+\beta)/\alpha]$.

For the symmetric Freund model with $\alpha = \beta = \theta_1$ and $\alpha' = \beta' = \theta_2$ the complete sufficient statistics are $W_1 = \sum \min\{X_i, Y_i\}$ and $W_2 = \sum \mid X_i - Y_i \mid$. The maximum likelihood estimates are $\widehat{\theta}_1 = n/[2W_1]$ and $\widehat{\theta}_2 = n/W_2$. These estimators have asymptotically bivariate exponential distributions with variance diagonal $[\theta_1^2/n, \theta_2^2/n]$.

Weier (1981) has analyzed a reparametrization of the symmetric Freund model by a Bayesian approach. He models the post first failure hazard rate θ_2 by $\rho\theta_1,\ \rho > 0$. Here $\rho = 1$ implies independence, $\rho > 1$ corresponds to an additional work load on the surviving component, while $p < 1$ corresponds to a decrease work load. Two prior distributions are used for (θ_1, ρ). The first is the conjugate prior with θ_1 having a gamma (ω_1, π_1) distribution and $\rho \mid \theta_1$ a gamma $(\omega_2, \theta_1\pi_2)$ distribution. The posterior distribution for this prior is given by

$$\begin{aligned} f(\theta_1, \rho \mid W_1, W_2) &= \frac{(2W_1 + \pi_1)^{n+\omega_1}}{\Gamma(n+\omega_1)}\, \theta_1^{n+\omega_1-1} \exp\{-\theta_1(2W_1 + \pi_1)\} \\ &\quad \cdot \frac{(W_2 + \pi_2)^{n+\omega_2}}{\Gamma(n+\omega_2)}\, \theta_1(\theta_1\lambda)^{n+\omega_2-1} \exp\{-\theta_1\rho(W_2 + \pi_2)\}\,. \end{aligned}$$

The Bayes estimates under squared error loss are $\widetilde{\theta}_1 = \frac{n+\omega_1}{2W_1+\pi_1}$ and $\widetilde{\rho} = \frac{n+\omega_2}{n+\omega_1-1}\,\frac{2W_1+\pi_1}{W_2+\omega_2}$.

The second prior distribution suggested by Weier is the Jeffrey's vague prior, $f(\theta_1, \rho) = n/(\theta_1\rho)$, which yields a posterior distribution for θ_1 and ρ to be

$$f(\theta_1, \rho \mid W_1, W_2) = \frac{[2W_1W_2]^n}{\Gamma(n)^2}\, \theta_1^{2n-1}\rho^{n-1} \exp\{-2W_1\theta_1 - W_2\theta_1\rho\}\,.$$

The Bayes estimates for this prior under squared error loss reduce to $\tilde{\theta}_1 = n/[2W_1]$ and $\tilde{\rho} = \frac{2nW_1}{(n-1)W_2}$. Note that $\tilde{\theta}_1$ here is the maximum likelihood estimator and $\tilde{\rho} = [(n-1)/n]$ times the maximum likelihood estimator.

Weier also provides estimators of $p(u,v) = P[\min(X,Y) > u, \max(X,Y) - \min(X,Y) > v]$, for his model. This is the probability neither component will fail prior to u, and given a failure the remaining component will survive at least v. Here $p(x,y) = \exp[-2u\theta_1 - v\theta_1\rho]$. The Bayes estimate of this quantity, under squared error loss is found to be $\tilde{p}(u,v) = \frac{W_1 W_2}{(W_1+u)((W_2+v)}$ under the vague prior, and $\tilde{p}(u,v) = \left\{\frac{W_1+\pi_1/2}{W_1+u+\pi_1/2}\right\}^{n+\omega_1} \cdot \left\{\frac{W_2+\pi_2}{W_2+v+\pi_2}\right\}^{n+\omega_2}$ under the informative prior.

21.3.2 Tests of Hypothesis

O'Neill (1985) has considered the problem of testing for symmetry or independence in the Freund bivariate exponential model when there is no censoring. The test for symmetry is a test of the null hypothesis H_0 : $\alpha = \beta$ and $\alpha' = \beta'$ versus the alternative hypothesis H_A : either $\alpha \neq \beta$ or $\alpha' \neq \beta'$. A likelihood ratio test is based on the statistic

$$\lambda = \frac{n^{2n} v^r u^{n-r}}{2^n r^{2r} (n-r)^{2(n-r)} (u+v)^n}, \tag{21.3.3}$$

where $u = \sum(X_i - \min\{X_i, Y_i\})$ and $v = \sum(Y_i - \min\{X_i, Y_i\})$. For large samples $-2\log(\lambda)$ is approximately chi-squared with two degrees of freedom. O'Neill finds the small sample size distribution of λ and provides exact critical points for samples of size two to forty.

A second test of symmetry, proposed by Hanagal and Kale (1992) is based on the Wald statistic. Here the test statistic is

$$\chi^2 = \frac{n[\hat{\alpha} - \hat{\beta}]^2}{[\hat{\alpha} + \hat{\beta}]^2} + \frac{n[\hat{\alpha}' - \hat{\beta}']^2}{[\hat{\alpha} + \hat{\beta}]\left[\frac{\hat{\alpha}'^2}{\hat{\beta}} + \frac{\hat{\beta}'^2}{\hat{\alpha}}\right]}$$

which has an approximate chi-squared with two degrees of freedom distribution under H_0.

An alternative test of symmetry for this model is based on the test for symmetry in the Block-Basu model discussed in Section 21.4.

Tests for independence requires testing of the hypothesis H_0 : $\alpha = \alpha'$ and $\beta = \beta'$. This test can be performed in a straight forward manner by noting that $Z = \min(X,Y)$ and $W = \max(X,Y) - \min(X,Y)$ are two independent exponential random variables with failure rates $(\alpha + \beta)$ and $(\alpha' + \beta')$, respectively. Any test of the equality of two independent exponential random variables based on Z_i, W_i, will provide a test of independence in the Freund model.

21.4 Inference for the Block-Basu Model

Block and Basu (1974) proposed the so called Absolutely Continuous Bivariate Exponential (ACBVE) distribution. This model has joint probability density function

$$f(x,y)=\begin{cases}\frac{\lambda_1\lambda(\lambda_2+\lambda_3)}{(\lambda_1+\lambda_2)}\exp\{-\lambda_1 x-(\lambda_2+\lambda_3)y\} & \text{if } y>x>0\\ \frac{\lambda_2\lambda(\lambda_1+\lambda_3)}{(\lambda_1+\lambda_2)}\exp\{-(\lambda_1+\lambda_3)x-\lambda_2 y\} & \text{if } x>y>0\end{cases} \qquad (21.4.1)$$

for $\lambda_1,\ \lambda_2>0,\ \lambda_3\geq 0$, and $\lambda=\lambda_1+\lambda_2+\lambda_3$. This model is related to the Marshall-Olkin model discussed in Section 21.2 since it is the absolutely continuous part of that model. It is a reparametrization of Freund's model discussed in Section 21.3. Here, we have $\alpha=\lambda_1+\lambda_3[\lambda_1/(\lambda_1+\lambda_2)]$, $\beta=\lambda_2+\lambda_3[\lambda_2/(\lambda_1+\lambda_2)]$, $\alpha'=\lambda_1+\lambda_3$, and $\beta'=\lambda_2+\lambda_3$. Note that in this model we have the restriction that $\alpha\leq\alpha'$, and $\beta\leq\beta'$. As in the Freund model the marginals of this model are mixtures of exponentials. For the ACBVE X and Y are independent if and only if λ_3 is zero.

21.4.1 Estimation of Model Parameters

Based on a random sample (X_i,Y_i) from the ACBVE distribution, the likelihood function is

$$\begin{aligned}L(\lambda_1,\lambda_2,\lambda_3) &= \left\{\frac{\lambda}{(\lambda_1+\lambda_2)}\right\}^n\{\lambda_1(\lambda_2+\lambda_3)\}^{n_1}\{\lambda_2(\lambda_1+\lambda_3)\}^{n_2}\\ &\quad\cdot\exp(-\lambda_1 S_X-\lambda_2 S_Y-\lambda_3 U_1\}\ ,\end{aligned}$$

where $S_X=\sum X_i$, $S_Y=\sum Y_i$ and $U_1=\sum\max(X_i,Y_i)$. Maximum likelihood estimates are found by numerically solving the score equations given by

$$\begin{aligned}\frac{n}{\lambda}-\frac{n}{\lambda_1+\lambda_2}+\frac{n_1}{\lambda_1}+\frac{n_2}{\lambda_1+\lambda_3}-S_X &= 0\ ,\\ \frac{n}{\lambda}-\frac{n}{\lambda_1+\lambda_2}+\frac{n_2}{\lambda_2}+\frac{n_3}{\lambda_2+\lambda_3}-S_Y &= 0\ ,\\ \frac{n}{\lambda}+\frac{n_1}{\lambda_1+\lambda_3}+\frac{n_2}{\lambda_1+\lambda_3}-U_1 &= 0\ .\end{aligned}$$

Hanagal and Kale (1991a) provide the Fisher information matrix for the maximum likelihood estimators. They also proposed consistent method of moment like estimators based on $E[X]$, $E[Y]$ and $E[\min\{X,Y\}]$. These estimators are

$$\begin{aligned}\widehat{\lambda}_1 &= \frac{n}{\sum\min(X_i,Y_i)}+\frac{n_1}{\sum\min(X_i,Y_i)-S_Y}\ ,\\ \widehat{\lambda}_2 &= \frac{n}{\sum\min(X_i,Y_i)}+\frac{n_2}{\sum\min(X_i,Y_i)-S_X}\ ,\end{aligned}$$

and

$$\widehat{\lambda}_3=\frac{n_1}{\sum\min(X_i,Y_i)-S_Y}+\frac{n_2}{\sum\min(X_i,Y_i)-S_X}-\frac{n}{\sum\min(X_i,Y_i)}\ .$$

For the symmetric ACBVE with $\lambda_1=\lambda_2=\theta$, Mehrotra and Michalek (1976) show that the complete sufficient statistics for θ and λ_3 are $U_1=\sum\max(X_i,Y_i)$ and $U_2=\sum(X_i+Y_i)$. The maximum likelihood estimators are given by

$$\widehat{\theta}=n\left\{\frac{1}{U_2-U_1}-\frac{1}{2U_1-U_2}\right\}\ \text{and}\ \widehat{\lambda}_3=n\left\{\frac{2}{2U_1-U_2}-\frac{1}{U_2-U_1}\right\}\ .$$

These estimators are biased by a factor of $n/(n-1)$, so that the unique minimum variance unbiased estimators of θ and λ_3 are $[(n-1)/n]\widehat{\theta}$ and $[(n-1)/n]\widehat{\lambda}_3$, respectively. For $n \geq 3$, we have

$$\begin{aligned}
\mathrm{Var}(\widehat{\theta}) &= \frac{n^2(5\theta^2 + 6\theta\lambda_3 + \lambda_2^3)}{(n-1)^2(n-2)} , \\
\mathrm{Var}(\widehat{\lambda}_3) &= \frac{n^2(8\theta^2 + 12\theta\lambda_3 + 5\lambda_2^3)}{(n-1)^2(n-2)} , \\
\mathrm{Cov}(\widehat{\theta}, \widehat{\lambda}_3) &= \frac{-n^2(6\theta^2 + 8\theta\lambda_3 + 3\lambda_2^3)}{(n-1)^2(n-2)} .
\end{aligned}$$

Estimation of the system reliability, $S(x,y) = P[X > x, Y > y]$, for the ACBVE was considered Klein and Basu (1985). For the symmetric ACBVE model, a unique minimum variance unbiased estimator of $S(x,y)$ can be found by using the Rao-Blackwell and Lehmann-Scheffe theorems. The estimator is $\{1 - \frac{x}{T}\}^{n-1}$, for $0 \leq x = y \leq T$. When $x < y < T$ the estimator is given by

$$\begin{aligned}
S(x,y) &= \left\{1 - \frac{y}{T}\right\}^{n-1} + \frac{n-1}{2V^{n-1}T^{n-1}} \sum_{k=1}^{n-1} (-1)^k \frac{\binom{k}{n-1}(V - y + T)^{n-1-k}}{n+k-1} \\
&\quad \cdot \left\{(T-x)^{n+k-1} - (T-y)^{n+k-1}\right\} ,
\end{aligned}$$

with the $x > y$ case defined by symmetry. Here $T = U_2 - U_1 = \sum \min(X_i, Y_i)$ and $V = 2U_1 - U_2 = \sum \mid X_i - Y_i \mid$.

For the asymmetric ACBVE, there are no complete sufficient statistics and so the above construction fails. The maximum likelihood estimator of $S(x,y)$ can be found by substitution of the MLE into the functional form for $S(x,y)$. Klein and Basu show that the jackknife version of this statistic is approximately unbiased for sample sizes as small as 10.

Basu and Ebrahimi (1987) have considered accelerated life tests based on the ACBVE distribution. They assume that for a stress level V_i, $i = 0, \ldots, J$, the rate coefficients in the ACBVE follow the same power rule translation function with $\lambda_{ik} = c_k V_i^p$, $k = 1, 2, 3$. Maximum likelihood estimators of c_1, c_2, c_3 and p are found by numerical methods.

21.4.2 Tests of Hypothesis

The ACBVE has been suggested by Gross and Lam (1981) as a model for paired data in biomedical applications. In these applications a patient has measurements of the time to some event, say relapse, under two treatments. The times for the ith patient are (X_i, Y_i). Gross and Lam assume that these bivariate times on the ith patient follow an ACBVE distribution with parameters $(\lambda_{1i}, \lambda_{2i}, \lambda_{3i})$, where the λ's may vary from patient to patient. Of primary interest is a test of the hypothesis of no treatment effect, namely, $H_0 : \lambda_{1i} = \lambda_{2i}$, $i = 1, \ldots, n$. Their test is based on $T_i = Y_i / X_i$, $X_i > 0$, which has marginal density

$$f_T(t) = \begin{cases} \frac{[1+\xi+\xi_3][\xi+\xi_3]}{[1+\xi][1+(\xi+\xi_3)t]^2} & \text{if} \quad t > 1 , \\ \frac{\xi[1+\xi+\xi_3][1+\xi_3]}{[1+\xi][1+\xi_3+\xi t]^2} & \text{if} \quad 0 < t < 1 , \end{cases} \tag{21.4.2}$$

where $\xi = \lambda_{2i}/\lambda_{1i}$ and $\xi_3 = \lambda_{3i}/\lambda_{1i}$. A likelihood ratio test of $H_0 : \xi = 1$ is suggested by Gross and Lam. Under H_0, the likelihood is

$$\begin{aligned} L_0(\xi_3) &= \{.5(2+\xi_3)(1+\xi_3)\}^n \prod_{i\le r}\{(1+\xi_3)+t_i\}^{-2} \\ &\quad \cdot \prod_{i>r+1} \{(1+(\xi_3+1)t_i\}^{-2} , \end{aligned} \tag{21.4.3}$$

where the observations $(t_1, \ldots, t_r)$ are between 0 and 1 and $(t_{r+1}, \ldots, t_n)$ are greater than 1. The unrestricted likelihood is

$$\begin{aligned} L(\xi,\xi_3) &= [\xi(1+\xi_3)]^r(\xi+\xi_3)^{n-r}\left[\frac{1+\xi+\xi_3}{1+\xi}\right]^n \prod_{i\le r}\{(1+\xi_3)+\xi t_i\}^{-2} \\ &\quad \cdot \prod_{i>r+1} \{(1+(\xi_3+\xi)t_i\}^{-2} . \end{aligned} \tag{21.4.4}$$

Maximum likelihood estimates for ξ_3 under H_0 and for ξ and ξ_3 under H_1 are found by numerically maximizing (21.4.3) and (21.4.4), respectively. The likelihood ratio test statistic is then $-2\ln[L_0(\xi_3)/L(\xi,\xi_3)]$ which has an approximate chi-squared with one degree of freedom distribution under H_0.

Another test of symmetry for identically distributed ACBVE data is proposed in Hanagal and Kale (1991a, 1992). This statistic is based on the Wald statistic for testing $H_0 : \lambda_1 = \lambda_2$. For this test, the required Fisher information matrix is found in Hanagal and Kale (1991a).

Tests for independence in the ACBVE model have been considered by Gupta, Mehrotra, and Michalek (1984) and Mehrotra and Michalek (1976) in the symmetric case with $\lambda_1 = \lambda_2 = \theta$. In this case, as noted earlier, the complete sufficient statistics are $T = \sum \min(X_i, Y_i)$ and $V = \sum \mid X_i - Y_i \mid$. Gupta, Mehrotra, and Michalek (1984) show that in this restricted model the likelihood ratio test of $H_0 : \lambda_3 = 0$ reduces to $X = R(1-R)$ where $R = V/(2T+V)$ and that under the null hypothesis R has a beta distribution with parameters n and n. This fact allows for computation of an exact critical value, c, for X by solving the equation $P[R(1-R) \le c \mid H_0] = \alpha$. That is, one must solve the equation

$$1-\alpha = \int_{B(c)}^{A(c)} \frac{(2n-1)!}{[(n-1)!]^2}\,[x(1-x)]^{n-1}\,dx$$

for c, where $A(c) = [1+(1-4c)^{1/2}]/2$ and $B(c) = [1-(1-4c)^{1/2}]/2$.

When the marginals are not identical there are no complete sufficient statistics nor closed form expressions for the parameter estimates. Tests of independence are based on the standard large sample theory. Wald tests of $H_0 : \lambda_3 = 0$ are discussed in Hanagal and Kale (1991a). Weier and Basu (1981) provide a power comparison of this test as compared to the nonparametric independence tests based on either Kendall's τ or Spearman's ρ.

Gross and Lam (1981) briefly discuss a procedure for checking if the ACBVE provides a reasonable fit to data based on the properties of the model. They suggest first testing the hypothesis that the minimum of X and Y has an exponential distribution. Next a nonparametric test of independence is performed on $\mathrm{Min}(X_i, Y_i)$ and $X_i - Y_i$. Finally a goodness-of-fit test is performed on $V_i = X_i - Y_i$, which should, under the ACBVE model, have a distribution function given by

$$F_V(v) = \begin{cases} 1 - \lambda_2(\lambda_1 + \lambda_2)^{-1} \exp\{-(\lambda_1 + \lambda_3)v\} & \text{if} \quad v \geq 0 \\ \lambda_1(\lambda_1 + \lambda_2)^{-1} \exp\{(\lambda_1 + \lambda_3)v\} & \text{if} \quad v < 0 . \end{cases}$$

Rejection of any of these three tests suggest that the ACBVE is not appropriate.

21.5 Inference for Exponential Frailty Models

The use of multivariate models derived from a model which incorporates a shared unobserved random effect, called a frailty, which induces association between individuals, has found increasing use in a variety of applications [see Klein et al. (1992) or Costigan and Klein (1993) for a detailed review of these applications]. These unobservable random effects can be interpreted as environmental, genetic, or batch effects. For the multivariate frailty model we suppose we have K components in a system with life lengths, $\boldsymbol{X} = (X_1, \ldots, X_K)$. Suppose that given an unobserved frailty, W, that these K life lengths are independent with hazard rates $W\lambda_j$, $j = 1, \ldots, K$. The joint survival function of $\boldsymbol{X}$ at time $\boldsymbol{x} = (x_1, \ldots, x_K)$ is given by the Laplace transform of the $\sum x_j \lambda_j$.

Three models for W are popular in the literature. These are the gamma distribution [Clayton and Cuzick (1985), Oakes (1989)], the positive stable distribution [Hougaard (1986), Wang, Klein, and Moeschberger (1993)], and the inverse Gaussian distribution [Hougaard, Harvard, and Holm (1992), Whitmore and Lee (1991), Li, Klein, and Moeschberger (1992)]. For the gamma frailty model with mean 1 and variance α, the joint survival function is

$$S(\boldsymbol{x}) = \left[1 + \alpha \sum_{j=1}^{K} x_j \lambda_j\right]^{-1/\alpha}, \qquad \alpha \geq 0 . \tag{21.5.1}$$

Note that $\alpha = 0$ corresponds to independent exponentials in this model.

For the positive stable distribution with Laplace transform $LP[u] = \exp\{-u^{\phi}\}$, $0 > \phi \leq 1$, the joint survival function is

$$S(\boldsymbol{x}) = \exp\left\{-\left[\sum_{j=1}^{K} x_j \lambda_j\right]^{\phi}\right\} . \tag{21.5.2}$$

Here, $\phi = 1$ corresponds to independence and the marginals have an exponential distribution with hazard rate $\phi\lambda_j$. This model was originally proposed by Gumbel (1960) in a different context.

For the inverse Gaussian distribution with density function $f(w) = (\pi\eta)^{-1/2} \exp[2/\eta]$ $\cdot w^{-3/2} \exp[-\eta^{-1}w - \eta^{-1}w^{-1}]$, $\eta \geq 0$, the joint survival function is

$$S(\boldsymbol{x}) = \exp\left\{2\eta^{-1} - 2\left[\eta^{-2} + \eta^{-1}\sum_{j=1}^{K} x_j \lambda_j\right]^{1/2}\right\} . \tag{21.5.3}$$

Here, $\eta = 0$ corresponds to independence. Note that for both the gamma frailty model and the inverse Gaussian model the marginal distributions are not exponential.

21.5.1 Estimation

To find maximum likelihood estimators of model parameters for the frailty models described above the EM algorithm of Dempster, Laird, and Rubin (1977) can be applied. Sample data consists of the time on study for the jth individual in the ith group, T_{ij}, and an indicator δ_{ij} of whether this time is an event time ($\delta_{ij} = 0$) or a censoring time ($\delta_{ij} = 1$), $j = 1, \dots, K$, $i = 1, \dots, n$. To apply the EM algorithm to joint estimation of $\lambda_1, \dots, \lambda_K$ and a generic frailty which has probability density function $g(w \mid \nu)$ where ν is a generic dependence parameter, note that if we could observe the frailty W_i, common to the K individuals in the ith group, the augmented log-likelihood (up to an additive constant) can be written as

$$\begin{aligned} L(\nu, \lambda_1, \dots, \lambda_K \mid ((T_{ij}, \delta_{ij}),\ j = 1, \dots, K, W_i),\ i = 1, \dots, n) \\ = L_1(\nu \mid W_i) + \sum_{j=1}^{K} L_{2j}(\lambda_j \mid ((T_{ij}, \delta_{ij}), W_i))\ , \end{aligned} \tag{21.5.4}$$

where $L_1(\nu \mid W_i) = \sum_{i=1}^n \ln[g(\nu \mid W_i)]$, $L_{2j}(\lambda_j \mid ((T_{ij}, \delta_{ij}), W_i)) = D_j \ln[\lambda_j] - \lambda_j \sum_{i=1}^n W_i T_{ij}$, and $D_j = \sum_{i=1}^n \delta_{ij}$.

To apply the algorithm one first computes $E[(21.5.4)]$ with respect to the observable data and the current parameter values. For L_{2j} this means we need to find $\widehat{W_i} = E[W_i \mid DATA]$. For the gamma model one can show that, conditional on the data, the W_i's are independent gamma random variables with shape parameter $A_i^G = [\alpha^{-1} + d_i]$ and scale parameter $C_i^G = [\alpha^{-1} + \sum_{j=1}^K \lambda_j T_{ij}]$, where $d_i = \sum_{j=1}^K \delta_{ij}$, so $\widehat{W_i} = A_i / C_i$. For the inverse Gaussian model one can show that the W_i's are conditionally independent generalized inverse Gaussian random variables with parameters (A_i^I, C_i^I, η), where $A_i^I = d_i - 1/2$ and $C_i^I = \eta^{-1} + \sum_{j=1}^K \lambda_j T_{ij}$. It follows that $\widehat{W_i} = [\eta C_i^I]^{-1/2} R[2\eta^{-1/2}(C_i^I)^{1/2}; A_i^I]$, where $R[x; \varphi] = \Omega_{\varphi+1}(x)/\Omega_\varphi(x)$ where $\Omega_\varphi(x)$ is the modified Bessel function of the third kind with index φ. For the positive stable model one can show that $\widehat{W_i} = E[W_i^{d_i+1} \exp\{-W_i \sum_{j=1}^K \lambda_j T_{ij}\}]/E[W_i^{d_i} \exp\{-W_i \sum_{j=1}^k \lambda_j T_{ij}\}]$. Wang, Klein, and Moeschberger (1993) show that $E[W_i^q \exp\{-W_i s\}] = (\phi s^{\phi-1}) \exp\{-s^\phi\} J[q, s]$, $q = 0, 1, \dots$; $s > 0$, where $J[q, s] = \sum_{j=0}^{q-1} b_{q,k}$ and $b_{q,k}$ is a polynomial of degree k in $1/\phi$ defined recursively by $b_{q,0} = 1$; $b_{q,k} = b_{q-1,k} + b_{q-1,k-1}\{(q-1)\phi^{-1} - (q-k)\}$, $k = 1, \dots, q-2$; and $b_{q,q-1} = \phi^{1-q}\Gamma[q - \phi]/\Gamma[1 - \phi]$.

The M-Step of the algorithm involves maximization of $E[(21.5.4)]$ with respect to the λ's and ν. The updated estimate of λ_j is given by $D_j \left[\sum_{i=1}^n \widehat{W_i} T_{ij}\right]^{-1}$. For the gamma model the updated estimate of α is found by maximizing

$$\begin{aligned} E[L_1(\alpha) \mid DATA] &= -n[\alpha^{-1} \ln[\alpha] + \ln(\Gamma[\alpha^{-1}]) \\ &\quad + \sum_{i=1}^{n} \{[\alpha^{-1} + d_i - 1] E[\ln(W_i) \mid DATA] - \widehat{W_i} \alpha^{-1}\}\ , \end{aligned} \tag{21.5.5}$$

where $E[\ln(W_i) \mid DATA] = \Psi[A_i^G] - \ln[C_i^G]$, with $\Psi[\cdot]$ as the digamma function. For the inverse Gaussian model, we have

$$E[L_1(\eta) \mid DATA] = n\{-(1/2)\ln[\eta] - 1/\eta\} - \sum_{i=1}^{n} \eta^{-1}\{\widehat{W_i} + E[W_i^{-1} \mid DATA]\} \,, \tag{21.5.6}$$

where $E[W_i^{-1} \mid DATA] = [\eta C_i^I]^{-1/2} R[2\eta^{-1/2}(C_i^I)^{1/2}; -A_i^I]$. For these two models, the estimation proceeds as follows: Initial estimates of ν and $(\lambda_1, \ldots, \lambda_K)$ are given. Based on these estimates one computes $\widehat{W_i}$, and either $E[\ln(W_i) \mid DATA]$ or $E[W_i^{-1} \mid DATA]$. The new estimate of λ_j of $D_j \left[\sum_{i=1}^{n} \widehat{W_i} T_{ij}\right]^{-1}$ is computed and either (21.5.5) or (21.5.6) is maximized to obtain the updated estimate of ν. This process is repeated until convergence.

For the positive stable model a modified EM algorithm is used. Fix a value of ϕ. An initial guess at $\lambda_j(\phi)$ is made, we estimate W_i and compute an updated estimate of the λ's as before. A new estimate of W_i is made based on the updated estimate of λ and this process continues until convergence to the vector $\lambda_1(\phi), \ldots, \lambda_K(\phi)$ which maximizes $\sum L_{2j}(\lambda_j)$ for this value of ϕ. This maximization gives a value of the profile likelihood based on the unaugmented likelihood. A one dimensional search of this profile likelihood is carried out to find the maximum likelihood estimates.

For any of the three models, standard errors of the estimates are found by usual information calculations based on the unaugmented likelihood. The relevant likelihoods are:

$$\begin{aligned} L[\alpha, \lambda_1, \ldots, \lambda_K] &= \sum_{j=1}^{K} D_j \ln[\lambda_j] + \sum_{i=1}^{n} \left\{ \sum_{h=1}^{d_i} \ln[1 + h\alpha] \right. \\ &\quad \left. -[\alpha^{-1} + d_i] \ln \left[1 + \alpha \sum_{j=1}^{K} \lambda_j T_{ij} \right] \right\} \end{aligned}$$

for the Gamma model;

$$\begin{aligned} L[\phi, \lambda_1, \ldots, \lambda_K] &= \sum_{j=1}^{K} D_j \ln[\lambda_j] + \sum_{i=1}^{n} \left\{ d_i \left(\ln[\phi] + (\phi - 1) \ln \left[\sum_{j=1}^{K} \lambda_j T_{ij} \right] \right) \right. \\ &\quad \left. - \left[\sum_{j=1}^{K} \lambda_j T_{ij} \right]^{\phi} + \ln \left[J \left[d_i, \sum_{j=1}^{k} \lambda_j T_{ij} \right] \right] \right\} \end{aligned}$$

for the positive stable frailty; and

$$\begin{aligned} L[\eta, \lambda_1, \ldots, \lambda_K] &= \sum_{j=1}^{k} D_j \ln[\lambda_j] + \sum_{i=1}^{n} \left\{ 2/\eta - \ln[\eta]/4 - d_i \ln[\eta]/2 \right. \\ &\quad \left. - A_i^I \ln[C_i^I]/2 + \ln[\Omega_{d_i - 1/2}[2(C_i^I/\eta)^{1/2}] \right\} \end{aligned}$$

for the inverse Gaussian.

An alternate approach to finding maximum likelihood estimators for these three models is to directly maximize the above unaugmented likelihoods. For details of this approach and an extension of these results to allow for assessment of covariates effects, see the papers by Costigan and Klein (1993) for the gamma model, Li, Klein, and Moeschberger (1993) for the inverse Gaussian model, and Wang, Klein, and Moeschberger (1993) for the positive stable model.

Whitmore and Lee (1991) have considered the multivariate inverse Gaussian model in a slightly different context. They assume that for each of n batches of K components the hazard rate of all components share a common random hazard rate S_i and that this hazard rate is drawn independently from a two parameter inverse Gaussian distribution with density $g(s \mid \theta, \nu) = \frac{1}{\sqrt{2\pi \nu s^3}} \exp\left[-\frac{(\theta s - 1)^2}{2\nu s}\right]$, for $\theta, \nu > 0$. They show, in the uncensored data case that the posterior density of S_i given the event times T_{ij} ($j = 1, \ldots, K$), θ and ν follows a generalized inverse gamma distribution and they provide an estimate of S_i based on the posterior mean of this distribution. Maximum likelihood estimation of the frailty parameters is performed by numerical maximization of the likelihood which is of a form similar to the likelihood for the one parameter inverse Gaussian model described above.

21.6 Inference for Other Multivariate Exponential Models

Barnett (1985) has considered maximum likelihood and method of moments estimators for Gumbel's (1960) bivariate exponential model with density function

$$f(x, y) = [(1 + \theta x)(1 + \theta y) - \theta] \exp\{-x - y - \theta xy\} .$$

Based on a sample (X_i, Y_i) he shows that the maximum likelihood estimator of θ is the solution of the equation

$$\sum_{i=1}^{n} \frac{u_i + 2\theta\, t_i}{1 + \theta\, u_i + \theta^2 t_i} = \sum_{i=1}^{n} t_i \ , \quad \text{where } t_i = X_i Y_i \text{ and } u_i = X_i + Y_i - 1 \ .$$

The method of moments estimator is found by solving the equation $\theta^{-1} \exp[\theta^{-1}] \int_1^{\infty} z^{-1} \cdot \exp\{-\theta^{-1} z\} dz = r + 1$, where r is the sample correlation coefficient. Barnett derives the asymptotic variance of these two estimates and compares their small sample properties.

Several papers have appeared on estimation of parameters for Downton's (1970) exponential distribution. This model in the bivariate case has joint density function $f(x, y) = \frac{\mu_1 \mu_2}{1-\rho} \exp\{-\frac{\mu_1 x + \mu_2 y}{1-\rho}\} I_0\{\frac{2[\rho \mu_1 \mu_2 x y]^{1/2}}{1-\rho}\}$, where $I_M[\cdot]$ is the modified Bessel function of the first kind of Mth order. Nagao and Kadoya (1971) show that the maximum likelihood estimators of μ_1 and μ_2 are the reciprocals of the sample means, $\overline{X}^{-1}$ and $\overline{Y}^{-1}$, respectively. They show that the maximum likelihood estimator of ρ is the solution to the equation

$$\sqrt{\rho} = \frac{1}{n} \sum_{i=1}^{n} \frac{I_1\{\frac{2[\rho \mu_1 \mu_2 X_i Y_i]^{1/2}}{1-\rho}\}}{I_0\{\frac{2[\rho \mu_1 \mu_2 X_i Y_i]^{1/2}}{1-\rho}\}} [\rho \mu_1 \mu_2 X_i Y_i]^{1/2} \ , \qquad (21.6.1)$$

and they suggest an iterative scheme for solving (21.6.1). Nagao and Kadoya (1971) and Al-Saadi and Young (1980) show that a method of moments estimator of ρ is $[\sum X_i Y_i/(n\overline{X}\overline{Y})]^{-1}$. Al-Saadi and Young (1980) compare the performance of the maximum likelihood, method of moments, a reduced bias version of the method of moments estimator, and the product-moment correlation coefficient.

Tests for independence in the Downton model are considered in Al-Saadi, Scrimshae, and Young (1979). They consider tests based on the statistic $S = \sum[(X_i - \overline{X})(Y_i - \overline{Y})]/(n\overline{X}\,\overline{Y})$, the product-moment correlation coefficient, R, the correlation coefficient based on exponential scores of the joint ranks of (X_i, Y_i), R_e, and Spearman's rank correlation. Critical values for S, R and R_e are found by a Monte Carlo method. Al-Saadi and Young (1982) consider the multivariate versions of the statistics S and R_e for the equicorrelated multivariate version of Downton's model.

References

Al-Saadi, S.D. and Young, D.H. (1980). Estimators for the correlation coefficient in a bivariate exponential distribution, *Journal of Statistical Computation and Simulation*, **11**, 13-20.

Al-Saadi, S.D. and Young, D.H. (1982). A test for independence in a multivariate exponential distribution with equal correlation coefficients, *Journal of Statistical Computation and Simulation*, **14**, 219-227.

Al-Saadi, S.D., Scrimshae, D.F., and Young, D.H. (1979). Tests for independence of exponential variables, *Journal of Statistical Computation and Simulation*, **9**, 217-233.

Arnold, B.C. (1968). Parameter estimation for a multivariate exponential distribution, *Journal of the American Statistical Association*, **63**, 848-852.

Barnett, V. (1985). The bivariate exponential distribution: A review and some new results, *Statistica Neerlandica*, **39**, 343-356.

Basu, A.P. and Ebrahimi, N. (1984). Testing whether survival function is bivariate new better than used, *Communications in Statistics - Theory and Methods*, **13**, 1839-1849.

Basu, A.P. and Ebrahimi, N. (1987). On a bivariate accelerated life test, *Journal of Statistical Planning and Inference*, **16**, 297-304.

Basu, A.P. and Habibullah, A. (1987). A test for bivariate exponentiality against BIFRA alternative, *Calcutta Statistical Association Bulletin*, **36**, 79-84.

Bemis, B.M., Bain, L.J., and Higgins, J.J. (1972). Estimation and hypothesis testing for the bivariate exponential distribution, *Journal of the American Statistical Association*, **67**, 927-929.

Bhattacharyya, G.K. and Johnson, R.A. (1971). Maximum likelihood estimation and hypothesis testing in the bivariate exponential model of Marshall and Olkin, *Technical Report 276*, Department of Statistics, University of Wisconsin, Madison.

Bhattacharyya, G.K. and Johnson, R.A. (1973). On a test of independence in a bivariate exponential distribution, *Journal of the American Statistical Association*, **68**, 704-706.

Block, H.W. and Basu, A.P. (1974). A continuous bivariate exponential extension, *Journal of the American Statistical Association*, **69**, 1031-1037.

Clayton, D. and Cuzick, J. (1985). Multivariate generalizations of the proportional hazards model (with discussion), *Journal of the Royal Statistical Society, Series A*, **148**, 82-117.

Costigan, T. and Klein, J.P. (1993). Multivariate survival analysis based on frailty models, In *Advances in Reliability* (Ed., A.P. Basu), North-Holland, Amsterdam.

Csörgö, S. and Welsh, A.H. (1989). Testing for exponential and Marshall-Olkin distributions, *Journal of Statistical Planning and Inference*, **23**, 287-300.

Dempster, A.P., Laird, N.M., and Rubin, D.B. (1977). Maximum likelihood from incomplete data via the EM algorithm (with discussion), *Journal of the Royal Statistical Society, Series B*, **39**, 1-38.

Downton, F. (1960). Bivariate exponential distributions in reliability theory, *Journal of the Royal Statistical Society, Series B*, **32**, 408-417.

Ebrahimi, N. (1987). Analysis of bivariate accelerated life test data for the bivariate exponential distribution, *American Journal of the Mathematical and Management Sciences*, **6**, 175-190.

Freund, J. (1961). A bivariate extension of the exponential distribution, *Journal of the American Statistical Association*, **56**, 971-977.

Gross, A.J. and Lam, C.F. (1981). Paired observations from a survival distribution, *Biometrics*, **37**, 505-511.

Gumbel, E.J. (1960). Bivariate exponential distributions, *Journal of the American Statistical Association*, **55**, 698-707.

Gupta, R.C., Mehrotra, K.G., and Michalek, J.E. (1984). A small sample test for an absolutely continuous bivariate exponential model, *Communications in Statistics - Theory and Methods*, **13**, 1735-1740.

Hanagal, D.D. (1992). Some inference results in bivariate exponential distributions based on censored samples, *Communications in Statistics - Theory and Methods*, **21**, 1273-1295.

Hanagal, D.D. and Kale, B.K. (1991a). Large sample tests of independence for an absolutely continuous bivariate exponential model, *Communications in Statistics - Theory and Methods*, **20**, 1301-1313.

Hanagal, D.D. and Kale, B.K. (1991b). Large sample tests of l_3 in the bivariate exponential distribution, *Statistics & Probability Letters*, **12**, 311-313.

Hanagal, D.D. and Kale, B.K. (1992). Large sample tests for testing symmetry and independence in some bivariate exponential models, *Communications in Statistics - Theory and Methods*, **21**, 2625-2643.

Hougaard, P. (1986). A class of multivariate failure time distributions, *Biometrika*, **73**, 671-678.

Hougaard, P., Harvard, B., and Holm, N. (1992). Assessment of dependence in the life times of twins, In *Survival Analysis: State of the Art* (Eds., J.P. Klein and P.K. Goel), pp. 77-98, Kluwer Academic Publishers, Boston.

Klein, J.P. and Basu, A.P. (1982). Accelerated life tests under competing Weibull causes of failure, *Communications in Statistics - Theory and Methods*, **11**, 2271-2287.

Klein, J.P. and Basu, A.P. (1985). On estimating reliability for bivariate exponential distributions, *Sankhyā, Series B*, **47**, 346-353.

Klein, J.P., Moeschberger, M.L., Li, Y.I., and Wang, S.T. (1992). Estimating random effects in the Framingham heart study, In *Survival Analysis: State of the Art* (Eds., J.P. Klein and P.K. Goel), pp. 99-120, Kluwer Academic Publishers, Boston.

Leurgans, S., Tsai, T., Wei, Y., and Crowley, J. (1982). Freund's bivariate exponential distribution and censoring, In *Survival Analysis* (Eds., R.A. Johnson and J. Crowley), pp. 230-242, IMS Lecture Notes.

Li, Y.I., Klein, J.P., and Moeschberger, M.L. (1993). Semi-parametric estimation of covariate effects using the inverse Gaussian frailty model, *Technical Report*, The Ohio State University, Columbus.

Marshall, A.W. and Olkin, I. (1967). A multivariate exponential distribution, *Journal of the American Statistical Association*, **62**, 30-44.

Mehrotra, K.G. and Michalek, J.E. (1976). Estimation of parameters and tests of independence in a continuous bivariate exponential distribution, *Technical Report*, Syracuse University, Syracuse.

Nagao, M. and Kadoya, M. (1971). Two-variate exponential distribution and its numerical table for engineering applications, *Bulletin of Disaster Prevention Research Institute*, **20**, 183-215.

Oakes, D. (1989). Bivariate survival models induced by frailties, *Journal of the American Statistical Association*, **84**, 487-493.

O'Neill, T.J. (1985). Testing for symmetry and independence in a bivariate exponential distribution, *Statistics & Probability Letters*, **3**, 269-274.

Proschan, F. and Sullo, P. (1974). Estimating the parameters of a bivariate exponential distribution in several sampling situations, In *Reliability and Biometry* (Eds., F. Proschan and R.J. Serfling), pp. 423-440, SIAM, Philadelphia.

Proschan, F. and Sullo, P. (1976). Estimating the parameters of a multivariate exponential distribution, *Journal of the American Statistical Association*, **71**, 465-472.

Sen, K. and Jain, M.B. (1990). A test for bivariate exponentiality against BHNBUE alternative, *Communications in Statistics - Theory and Methods*, **19**, 1827-1835.

Wang, S.T., Klein, J.P., and Moeschberger, M.L. (1993). Semi-parametric estimation of covariate effects using the positive stable frailty model, *Technical Report*, The Ohio State University, Columbus.

Weier, D.R. (1981). Bayes estimation for bivariate survival models based on the exponential distribution, *Communications in Statistics - Theory and Methods*, **10**, 1415-1427.

Weier, D.R. and Basu, A.P. (1980). Testing for independence in multivariate exponential distributions, *Australian Journal of Statistics*, **22**, 276-288.

Weier, D.R. and Basu, A.P. (1981). On tests of independence under bivariate exponential models, In *Statistical Distributions in Scientific Work - Vol. 5* (Eds., C.P. Taillie, G.P. Patil, and B. Baldessari), pp. 169-180, D. Reidel, Dordrecht.

Whitmore, G.A. and Lee, M.-L.T. (1991). A multivariate survival distribution generated by an inverse Gaussian mixture of exponentials, *Technometrics*, **33**, 39-50.

CHAPTER 22

Optimal Tests in Multivariate Exponential Distributions

Ashis SenGupta[1]

22.1 Introduction and Summary

Various generalizations [Basu (1988); Block and Savits (1981)] of the univariate exponential distribution to model life times of the components of a multivariate system have been proposed, as already seen in Chapters 20 and 21. The usefulness of any such model depends on its probabilistic enhancement and its amenability to statistical inference. Tests for parameters of the underlying distribution are of prime interest. Often, however such tests have either been ad-hoc in nature based on intuitive grounds or not available at all. This is partly due to the complexity of the multivariate distributions suggested in reliability theory. Thus it is important and useful to develop statistical tests for such multivariate models which should possess desirable optimal properties, e.g. uniformly or locally best power, robustness, etc. While such univariate theory is well known, the multivariate techniques are only recently emerging.

Reliability theory abounds with many multivariate distributions. Here we consider some of them, giving compact representations of their densities and develop optimal tests for their parameters. These include the Marshall-Olkin, Block-Basu, Freund, Arnold-Strauss, Gumbel I, II, and III, models, etc. Non-regular models are also discussed. Also, well known results as well as very recent ones are reviewed. Strikingly, in many cases even UMPU tests exist. For most other cases, even where non-trivial sufficient statistics do not exist, we present here a unified theory based on locally best (LMP or LBI), or asymptotically locally best (C_α-) test for the important testing problems of independence and of symmetry. Note that uniqueness of such a test will imply its admissibility. Most of these tests turn out to be surprisingly extremely simple. It is hoped that this unified approach can be successfully exploited to cover many other models also.

We will frequently use below the standard abbreviations such as MLE (maximum

[1] Applied Statistics Division, Indian Statistical Institute, Calcutta

likelihood estimator), LRT (likelihood ratio test), UMPU (uniformly most powerful unbiased), REF (regular exponential family), and CEF (curved exponential family).

22.2 Independent Exponentials

The trivial generalization to a multivariate exponential model from a univariate one may be that of independent exponential components. Even in this simplest model, however, optimality of tests in simultaneous inference can be quite interesting. Let there be k independent random samples where the ith sample is drawn from the density $f(x;\mu_i,\theta_i)$, $i=1,\dots,k$,

$$f(x;\mu,\theta)=\theta^{-1}\exp[-(x-\mu)/\theta], \qquad x>\mu,\ 0<\theta,\ \mu<\infty\ . \tag{22.2.1}$$

Let observations be taken under Type-II censoring scheme. Samanta (1986) considers tests for equality of μ_i's, under scale symmetry, i.e., $\theta_1=\theta_2=\cdots=\theta_k=\theta$, θ known and unknown, test for scale symmetry, μ_i's unknown and test for identical distributions. Samanta showed that a well known test in the case of θ known and LR test based on the statistic ℓ_n for each of the remaining three problems are all optimal in the sense of Bahadur efficiency, i.e., they attain their corresponding optimal exact slopes. The optimality was established either by invoking the main result in Killeen, Hettmansperger, and Sievers (1972, 1974) or by verifying conditions A and B in Hsieh (1979). Original versions of Bahadur (1971) were simplified in Bahadur and Raghavachari (1972) who gave condition A as

$$\liminf_n n^{-1}\ell_n \ge J(\theta) \text{ w.p. } 1$$

when $\boldsymbol{\theta}$ obtains for each $\theta\in\Omega-\Omega_0$ and where $2J(\theta)$ is the maximal (optimal) slope at θ. Condition B in Hsieh was a special version of the corresponding condition in Bahadur and it requires that under $\theta\in\Omega_0$, ℓ_n^{-1} is distributed as a product of independent beta random variables. Recall [Bahadur (1971)] that Bahadur efficiency in contrast to Pitman efficiency, deals with alternatives away from the null. Kourouklis (1988) unified and generalized the above optimality results to exhibit that they hold if only condition A is satisfied - condition B being superfluous.

22.3 Marshall-Olkin (MO) Model

The density of (X_1,X_2) following the bivariate exponential (BVE) distribution of Marshall and Olkin (1967) with respect to the Lebesgue measure on R_2 is given by

$$\begin{aligned} f(x_1,x_2) &= \lambda_i(\lambda-\lambda_i)\exp[-(\lambda_1x_1+\lambda_2x_2)-\lambda_3x_j]\ , \\ &\qquad x_i<x_j,\ x_i>0,\ i\neq j=1,2\ , \end{aligned}$$

and the density on the diagonal $x_1=x_2=x$ with respect to the Lebesgue measure on R_1 is given by,

$$\begin{aligned} f(x,x) &= \lambda_3\exp(-\lambda x)\ , \qquad x_1=x_2=x\ge 0\ , \\ &\qquad \lambda_1,\lambda_2>0,\ \lambda_3\ge 0,\ \lambda=\sum_{i=1}^{3}\lambda_i\ . \end{aligned} \tag{22.3.1}$$

The above is not a member of the exponential family. $(n_1, n_2, \sum X_1, \sum X_2, \sum M_{12})$ constitute a jointly minimal sufficient but not complete statistic [Bhattacharyya and Johnson (1971)]. Symmetry corresponds to $\lambda_1 = \lambda_2$ while independence corresponds to $\lambda_3 = 0$. Also note that $\text{Corr}(X_1, X_2) = \lambda_3/\lambda$ is a monotone increasing function of λ_3 and hence it is of some interest to test for non-zero values of λ_3, say, $H_0' : \lambda_3 = \lambda_3^0$ (known). Henceforth, unless otherwise stated, the summations will always extend over the entire sample of n pairs of observation vectors. Denote $\max(x_1, x_2)$ by m_{12}, $n_i = \sum I(x_i < x_j)$, $i, j = 1, 2$, $n_3 = \sum I(x_1 = x_2)$ and $n = n_1 + n_2 + n_3$. Then the log-likelihood is

$$\ln L = \sum_{i=1}^{2} n_i \ln \lambda_i(\lambda - \lambda_i) + n_3 \ln \lambda_3 - \sum \left(\sum_{i=1}^{2} \lambda_i x_i \right) - \lambda_3 \sum m_{12} . \qquad (22.3.2)$$

22.3.1 Tests for Independence

We present optimal tests for $H_0 : \lambda_3 = 0$ vs $H_1 : \lambda_3 > 0$ under part or no knowledge regarding the nuisance parameters λ_1, λ_2. Observe that, since $P(X_1 = X_2) = \lambda_3/\lambda$, so certainly $n_3 > 0$ leads to outright rejection of H_0, and thus also does not contribute to the Type I error.

(a) **Symmetry completely known ($\lambda_1 = \lambda_2 = \lambda_0$, known).** Bemis, Bain, and Higgins (1972) have derived the UMP test for this case given by, assuming that $\gamma_0 = \lambda_0 + \lambda_3$ is known,

$$\omega : n_3 > 0 \qquad \text{or} \qquad \sum \min(x_1, x_2) > C/4\gamma_0 .$$

Power of this test is,

$$P(\rho) = 1 - (1 - \rho)^n + (1 - \rho)^n P[\chi^2_{2n} > C/(1 + \rho)], \; \rho = \lambda_3/\lambda .$$

(b) **Symmetry known ($\lambda_1 = \lambda_2 = \lambda_0$, say, unknown).** Using a complete class theorem to incorporate $n_3 > 0$ as outright rejection of H_0, one translates the problem of testing independence to that of testing of equality of variances of two independent normal populations for which it is well known that a UMP test exists. Then a UMP test exists here also, given by [Bhattacharyya and Johnson (1973)]

$$\omega : n_3 > 0 \qquad \text{or} \qquad 2T_1/V > F_\alpha$$

where $T_1 = \sum \max(x_1, x_2)$, $V = \sum[\max(x_1, x_2) - \min(x_1, x_2)]$ and F_α is the upper $100\alpha\%$ point of the central $F(2n, 2n)$ distribution. This test is equivalent to a test proposed earlier by Bemis, Bain, and Higgins. The power of this test can be also easily computed using only the distribution function of the central $F(2n, 2n)$ distribution. Comparison of this UMP test with several other tests in terms of Pitman's asymptotic relative efficiency (ARE) was made by Weier and Basu (1981). They related this comparison also to the Block-Basu model and we will accordingly present it as Result 22.1 and succeeding comments in Section 22.6 later.

(c) λ_1, λ_2 **unequal and unknown.** No test has been derived here based on optimality considerations. However, several estimators of λ_3 have been proposed, e.g., say $\widehat{\lambda}_{3,A}$ by Arnold (1968), $\widehat{\lambda}_{3,B}$ by Bemis, Bain, and Higgins (1972), $\widehat{\lambda}_{3,P}$ by Proschan and Sullo (1976), and $\widehat{\lambda}_3$, the MLE. With these estimators $\widehat{\lambda}$ are all asymptotically trivariate normal with mean $(\lambda_1, \lambda_2, \lambda_3)$ and hence large-sample tests for $H_0 : \lambda_3 = 0$ or $H_0' : \lambda_3 = \lambda_3^0$ (known) may be based on them. A comparison [Hanagal and Kale (1991a)] between such tests based on $\widehat{\lambda}_{3,A}$, $\widehat{\lambda}_{3,B}$ and $\widehat{\lambda}_3$ seems to indicate the superiority of the test based on MLE $\widehat{\lambda}_3$. However, whereas $\widehat{\lambda}_{3,A}$ and $\widehat{\lambda}_{3,B}$ are trivial to compute, $\widehat{\lambda}_3$ is not available even in a closed form and require iterative techniques. This motivates the derivation of the Neyman's (1954) C_α-test which avoids the computation of $\widehat{\lambda}_3$ and also provides an optimal, asymptotically LMP test.

We introduce the following notations. Let $\phi_i = \partial \ln L/\partial\lambda_i|_{\lambda_3=0}$, $i = 1,2,3$ and assume that ϕ_i's satisfy Cramér conditions given in Neyman (1954). Let β_i^0 be the first order partial regression coefficient of ϕ_i on ϕ_3 computed under $H_0 : \lambda_3 = 0$, $i = 1, 2$. Let $V(Y)$, $C(Y, Z)$ and $\text{Disp}_0(\boldsymbol{\phi})$ denote the variance of Y, covariance of Y and Z, and dispersion matrix of $(\phi_1, \phi_2, \phi_3)' = \boldsymbol{\phi}$, computed under H_0. Let (I_{ij}^0) denote the Fisher information matrix for $\boldsymbol{\phi}$ under H_0. Let $\widehat{\lambda}_i$ denote the MLE of $\lambda_i = 1,2,3$. It is very helpful to note that, when (I_{ij}^0) exists, $\text{Disp}_0(\boldsymbol{\phi}) = (I_{ij}^0)$. Let $S = \phi_3 - \sum_{i=1}^2 \beta_i^0\phi_i$ and $\sigma_0(S)$ denote the standard deviation of S computed under H_0. The superfix '$\sim$', e.g., $\widetilde{\phi}_i$, $\widetilde{\beta}_i^0$, $\widetilde{\sigma}_0(S)$, etc., will denote that the quantity is being evaluated at $\lambda_i = \widetilde{\lambda}_i$, a $\sqrt{n}$-consistent estimator of λ_i, $i = 1, 2$, under H_0. We note that β_i^0 simplifies to the usual regression coefficients if $I_{12} = 0$, and further $\beta_i^0 = 0$ if λ_3 is orthogonal to λ_i, i.e., $I_{3i}^0 = 0$, while $\widetilde{\phi}_i = 0$ if $\widetilde{\lambda}_i = \widehat{\lambda}_i^0$, $i = 1, 2$, the MLE under H_0. However, simplification in $\widetilde{\sigma}_0(S)$ is achieved when $\beta_i^0 = 0$, and not for $\widetilde{\phi}_i = 0$, $\phi_i \neq 0$, $i = 1, 2$. Then, Neyman's C_α-test for $H_0 : \lambda_3 = 0$ vs $H_1 : \lambda_3 > 0$ is given by,

$$\omega : \widetilde{T} = \widetilde{S}/\widetilde{\sigma}_0(S) > z_\alpha ,$$

z_α being the upper $100\alpha\%$ point of the standard normal distribution. The asymptotic power of this asymptotically LMP test is easily available in terms of only the normal distribution function.

For the MO model, $I_{12} = I_{12}^0 = 0$, leading to simplification in computing β_i^0, i.e., $\beta_i^0 = C(\phi_3, \phi_i)/V(\phi_i)$, $i = 1, 2$. Further, $\widehat{\lambda}_i^0$ is simply $\bar{x}_i^{-1}$, $i = 1, 2$. Thus, $\widetilde{T}$ simplifies a lot to,

$$\widetilde{T} = \widetilde{\phi}_3/[\widetilde{V}(\phi_3) - \sum_{i=1}^{2} \widetilde{C}^2(\phi_3, \phi_i)/\widetilde{V}(\phi_i)]^{1/2} .$$

However, note that ϕ_3 computed directly from (22.3.2), as also I_{33}, are not defined. This is expected since, as we noted earlier, $n_3 > 0$ leads to outright rejection of H_0. Hence, we modify the usual C_α-test to incorporate this fact and thereby eliminating the difficulties with the original ϕ_3 and I_{33}. Let $\widetilde{T}'_{MO} = \widetilde{T}|_{n_3=0}$ where in computing $\widetilde{T}'_{MO}$, the $'$ denotes that we take the likelihood conditional on $n_3 = 0$. Then, we [SenGupta (1993)] propose a modification of the usual C_α-test as,

$$\omega(n_3, \widetilde{T}'_{MO}) : \; n_3 > 0 \text{ or } \widetilde{T}'_{MO} > z_\alpha ,$$

where, $\phi_3' = n^{-1}\sum(n_1x_2 + (n - n_1)x_1 - nm_{12}) + (\sum x_1)(\sum x_2)/\sum(x_1 + x_2)$. Now $(I_{ij}^{0\prime})$ may be computed directly, e.g. $I_{ii}^{0\prime} = n\lambda_i^{-2}$, etc. Also, λ_1 and λ_2 are still orthogonal.

Thus simplifications result as before. Also, $V'(\cdot)$, $C'(\cdot)$ etc. can be computed from $(I_{ij}^{0\,\prime})$ using the conditional MLEs $\lambda_i' = \bar{x}_i^{-1}$ in $I_{ij}^{0\,\prime}$. The above test is quite simple to implement.

22.3.2 Tests for ρ, Correlation (X_1, X_2)

Note that for known $\lambda_1, \lambda_2, H_0' : \rho_0\ (\neq 0)$ vs $H_1' : \rho > \rho_0$ is equivalent to $H_0 : \lambda_3 = \lambda_3^0\ (\neq 0)$ vs $H_1 : \lambda_3 > \lambda_3^0$. We consider here tests for $H_0 : \lambda_3 = \lambda_3^0$. The general case where λ_1, λ_2 are unknown is tackled by suitably modifying case (c) of Section 22.3.1. Other cases need suitable modifications also.

Large sample tests based on the estimators $\widehat{\lambda}_{3,A}$, $\widehat{\lambda}_{3,B}$, $\widehat{\lambda}_{3,P}$ and $\widehat{\lambda}_3$ may be used with their usual disadvantages. However, a simple optimal C_α-test [SenGupta (1993)] may be constructed as shown below, even though $\lambda_3^0 \neq 0$. Observe that $I_{12} = 0$. Further, under H_0, the likelihood equations are separable, each yielding its corresponding $\widehat{\lambda}_i^0$ as simply the appropriate root to a quadratic equation,

$$\lambda_i(\lambda_i + \lambda_3^0)\sum x_i - \lambda_i(n_1 + n_2) - \lambda_3^0 n_i = 0, \qquad i = 1, 2\ .$$

Then, like in case (c) of Section 22.3.1, β_i^0 reduce to simply the usual regression coefficients, and $\widetilde{S}$ to simply ϕ_3 (using $\widehat{\lambda}_i^0$). And further, unlike there, the situation $n_3 > 0$ needs no special consideration and $I_{33}^0 < \infty$. These result in major simplifications to the test statistic T as compared to even the corresponding one for the simpler hypothesis, $H_0 : \lambda_3 = \lambda_3^0 = 0$. The optimal C_α-test reduces to the very simple form,

$$\omega : \widetilde{T} = \frac{\left[n_3 {\lambda_3^0}^{-1} + \sum_{i=1}^{2} n_i(\widehat{\lambda}_i^0 + \lambda_3^0) - \sum m_{12}\right]}{\left[n(\widehat{I}_{33}^0 - \sum_{i=1}^{2}(\widehat{I}_{3i}^0)^2/\widehat{I}_{ii}^0)\right]^{1/2}} > z_\alpha\ .$$

As in general, the asymptotic power of this test can be expressed in terms of just the normal distribution function.

22.3.3 Tests for Symmetry

We consider tests for $H_0 : \lambda_1 = \lambda_2$ vs $H_1 : \lambda_1 \neq \lambda_2$, under part or no knowledge about λ_3.

(a) **Independence known.** ($\lambda_3 = 0$ given). Here (22.3.1) simplifies to a two-parameter REF as product of two independent $e(\lambda_i)$, with sufficient statistics $(\sum x_1, \sum x_2)$. Hence a UMPU test exists here given by,

$$\omega : R = \sum x_1 / \sum(x_1 + x_2) < C_1 \qquad \text{or} \qquad > C_2$$

where, under H_0, $R \sim \text{Beta}(n, n)$, and the cut-off points C_1, C_2 are then easily obtained.

(b) λ_3 **unknown.** There is no simplification in (22.3.1) here, even under H_0, - a singular component in the density still being retained. Hence, the usual large sample tests based on MLE or the GLR require iterative computations. As in Section

22.3.1, here also we propose a C_α-test which is easily computed and possesses the optimality property stated above. Let $\gamma = \lambda_1 - \lambda_2$. Then, our hypotheses are $H_0 : \gamma = 0$ vs $H_1 : \gamma \neq 0$ and the reparametrization is $\boldsymbol{\theta} = (\gamma, \lambda_0, \lambda_3)'$, say.

Unlike in Section 22.3.1, however, $\widehat{\boldsymbol{\theta}}^0$ are not explicitly available and still requires iterative computation. Nevertheless, $\widetilde{\boldsymbol{\theta}}$ may be taken as the $\sqrt{n}$-consistent estimators of Arnold (1968), modified for the case $\lambda_1 = \lambda_2 = \lambda_0$, say and then, $(\widetilde{I}^0_{ij})$ from $\widetilde{\boldsymbol{\theta}}$ accordingly easily obtained. Unfortunately then, $\widetilde{\phi}_{\theta_t}$ is no longer zero. Further, neither β_i^0 nor ϕ_i^0 is zero, $i = 2, 3$. Hence, the general form of the C_α-test must be used without any further simplifications.

22.4 Marshall-Olkin-Hyakutake (MOH) Model

Recently Hyakutake (1990) proposed incorporating location parameters in the BVE of the MO model. Its joint survival function and the density are, respectively,

$$\overline{F}(x_1, x_2) = \exp\left[-\sum_{i=1}^{2} \lambda_i(x_i - \mu_i) - \lambda_3 \max_i(x_i - \mu_i)\right]$$

and

$$f(x_1, x_2) = \lambda_i(\lambda - \lambda_i)\overline{F}(x_1, x_2), \qquad 0 < x_i - \mu_i < x_j - \mu_j,\ i \neq j,\ i, j = 1, 2$$

with respect to the Lebesgue measure on R_2, while with respect to that on R_1,

$$f(x, x) = \lambda_3 \overline{F}(x, x),\ 0 < x_1 = x_2 = x,\ \lambda_1, \lambda_2 > 0,\ \lambda_3 \geq 0,\ \lambda = \sum_{i=1}^{3} \lambda_i\ . \qquad (22.4.1)$$

As in the MO model, the above is not a member of REF, independence corresponds to $\lambda_3 = 0$, $\text{Corr}(X_1, X_2) = \lambda_3/\lambda$, and symmetry corresponds to $\lambda_1 = \lambda_2$ with $\mu_1 = \mu_2$. Here, the additional case of testing for structure of location parameters is of interest. This is the only such case that has appeared in the literature, but the tests therein are not known to have any optimal property. We note [SenGupta (1993)] that our unified approach of deriving locally best tests here also may be quite useful.

22.4.1 Tests for Independence

We indicate how optimal tests may be constructed for $H_0 : \lambda_3 = 0$ vs $H_1 : \lambda_3 > 0$ for some situations usually encountered in practice. Observe that, here also, $P(X_1 = X_2) = \lambda_3/\lambda$, so that $n_3 > 0$ leads to outright rejection of H_0 without contributing to the Type I error. Even under scale homogeneity the presence of μ_1, μ_2 destroys the REF nature of the density when $n_3 = 0$. Thus the derivation of a UMP test following the approach for the MO model creates problems. Observe that when μ_1, μ_2 are unknown, the usual C_α-test does not apply due to the non-regular nature of the density. And when λ_1, λ_2 are unknown, a LBI test is not applicable since λ_i's are not, in general, scale parameters. Additionally, one now needs to consider the conditional likelihood given $n_3 = 0$. We propose pursuing the approach of deriving a LBI test when scale symmetry is known and a location invariant C_α-test when all parameters are unknown, as in Section 22.4.2 below, but here, based on the conditional likelihood.

22.4.2 Tests for ρ, Correlation (X_1, X_2)

For known λ_1, λ_2, $H_0' : \rho = \rho_0 \ (\neq 0)$ vs $H_1' : \rho > \rho_0$ is equivalent to $H_0 : \lambda_3 = \lambda_3^0$ vs $H_1 : \lambda_3 > \lambda_3^0$. We indicated how to modify the C_α-test using invariance and following Section 22.3.2 when λ_1, λ_2 are unknown, and LBI test using Wijsman's representation when $\lambda_1 = \lambda_2 = \lambda_0$, unknown.

(a) **Scale homogeneity known ($\lambda_1 = \lambda_2 = \lambda_0$, say, unknown)**. Observe that here λ_0 is still a scale parameter while μ_1 and μ_2 are location parameters. A natural group under which invariance should be invoked is $(a, \boldsymbol{b}) = g_{ab} : \ (a, \boldsymbol{b}) \cdot (\boldsymbol{x}) = (ax_1 + b_1, ax_2 + b_2)$, $a > 0$, $b_1, b_2 \in R$. Then, one can proceed to derive the LBI test exploiting the simple structure of the group $(a, \boldsymbol{b})$ resulting in simplifications in the Wijsman's representation of the ratio of the densities of the maximal invariant under H_0 and H_1. This LBI test is our proposed optimal test.

(b) $\lambda_1, \lambda_2, \mu_1, \mu_2$ **all unknown**. As before, $n_3 > 0$ needs no separate consideration and the corresponding conditional distribution does not arise. Further, Hyakutake shows that the MLE of $\boldsymbol{\mu}$, in general and so under H_0 is $\widehat{\boldsymbol{\mu}}^0 = (x_{1,1}, x_{2,1})$, $x_{i,1}$ being $\min(x_{i1}, \ldots, x_{in})$, $i = 1, 2$. Under H_0, the MLEs, λ_i^0 are obtained very easily exactly as in Section 22.3.2 with x_i replaced by $x_i - \mu_i$. Further, orthogonality of λ_1, λ_2 is retained here. But we do not have our usual ϕ's for μ_1 and μ_2. However, μ_1 and μ_2 are still location parameters. Thus, the problem may be first reduced by invariance using the maximal invariants for μ_1, μ_2 with density $f_I(x; \boldsymbol{\lambda})$, say. But, λ_1, λ_2 are still not scale parameters for $f_I(x; \boldsymbol{\lambda})$. Here as an optimal test for independence, we suggest the corresponding C_α-test derived from this location invariant density $f_I(x; \boldsymbol{\lambda})$.

22.4.3 Tests for Scale Homogeneity

C_α-tests may be constructed here as in Section 22.3.3, but after invoking location invariance as in Section 22.4.2 above. However, the original difficulties with λ_0^0 and λ_3^0 as in there remains and as before, here also the invariant C_α-test needs to be worked out in its full generality without any further simplification.

22.4.4 Tests for Location Homogeneity

Consider $H_0 : \ \gamma = \mu_1 - \mu_2 = 0$ vs $H_1 : \ \gamma \neq 0$. Under scale homogeneity, Hyakutake proposes two two-step conditional tests and one asymptotic unconditional test. However, their optimality properties are unknown. The non-regular form of the density here is a deterring factor for the derivation of the usual C_α-test and a generalization of it to subsume such cases needs to be derived.

22.5 Multivariate Marshall-Olkin (M-MO) Model

The survival function for this model with $m_{1k} = \max(x_1, \ldots, x_k)$ is,

$$\overline{F}(\boldsymbol{x}) = \exp\left[-\sum_{i=1}^{k} \lambda_i x_i - \lambda_{k+1} m_{1k}\right], \ x_i, \lambda_i > 0, \lambda_{k+1} \geq 0, \ \sum_{i=1}^{k+1} \lambda_i = \lambda \ .$$

Symmetry corresponds to $\lambda_1 = \cdots = \lambda_k$, i.e., $\gamma_i = \lambda_i - \lambda_k = 0$, $i = 1, \ldots, k-1$ while mutual independence to $\lambda_{k+1} = 0$. Also $P(X_1 = \cdots = X_k) = \lambda_{k+1}/\lambda$.

22.5.1 Tests for Independence and Symmetry

As for the bivariate case, Hanagal (1991) has proposed the usual large sample tests based on the two CAN estimators $\widehat{\lambda}_{k+1}$, the MLEs, and $\widehat{\lambda}_{k+1,A}$ of Arnold and also on $\widehat{\lambda}_{k+1,C}$, the number of observations with simultaneous failure of all k components. These were also used to test $H_0 : \lambda_{k+1} = \lambda^0_{k+1}$ ($\neq 0$) and for testing symmetry. In a simulation study, the test based on MLE performed best.

For testing independence, unlike the bivariate distribution, existence of UMP test for the symmetric case is, however, not known. We propose that optimal C_α-tests should be used for both the symmetric and general case of all parameters unknown. These can be derived exactly as in Section 22.3.1 and the same simplifications (complication) are retained. For testing symmetry, however, a multiparameter generalization of C_α-test is called for, which again admits of no simplification.

22.6 Block-Basu (BB) Model

The density of (X_1, X_2) following the absolutely continuous bivariate exponential (ACBVE) distribution of Block and Basu (1974) is given by

$$\begin{aligned} f(x_1, x_2) &= [(\lambda\lambda_i(\lambda - \lambda_i))/(\lambda_1 + \lambda_2)] \exp[-(\lambda_1 x_1 + \lambda_2 x_2) - \lambda_3 x_j] \,, \\ & \qquad x_i < x_j,\ x_i, \lambda_i > 0,\ i \neq j = 1, 2,\ \lambda_3 \geq 0,\ \lambda = \sum_{i=1}^{3} \lambda_i \,. \end{aligned} \tag{22.6.1}$$

Denote $\max(x_1, x_2)$ by m_{12} and for a sample of size n, let $n_1 = \sum I(x_1 < x_2)$ and $n_2 = n - n_1$. The log-likelihood is then

$$\begin{aligned} \ln L &= n \ln(\lambda/(\lambda_1 + \lambda_2)) + \sum_{i=1}^{2} n_i \ln(\lambda_i(\lambda - \lambda_i)) \\ & \quad - \sum \left(\sum_{i=1}^{2} \lambda_i x_i \right) - \lambda_3 \sum m_{12} \,. \end{aligned} \tag{22.6.2}$$

The ACBVE is the absolutely continuous part of BVE of Marshall and Olkin and can be represented as a special case of Freund's model in Section 22.12 under a specific reparametrization. It is a member of a (3,4) curved exponential family with sufficient statistic $[\sum I(x_1 < x_2), \sum x_1, \sum x_2, \sum (x_1 - x_2)\, I(x_1 < x_2)]$ for $(\lambda_1, \lambda_2, \lambda_3)$. λ_1, λ_2 are not scale parameters except under independence. Symmetry corresponds to $\lambda_1 = \lambda_2$ while independence corresponds to $\lambda_3 = 0$. In general, marginals are not exponentials but rather each is a negative mixture of exponentials. However, under independence they are exponentials as in MO model, which we will find greatly facilitates and simplifies the derivation of the corresponding optimal C_α-test.

22.6.1 Tests for Independence

We present optimal tests for $H_0 : \lambda_3 = 0$ vs $H_1 : \lambda_3 > 0$.

(a) λ_1, λ_2 **known**. Even here, (22.6.2) does not reduce to a REF and UMP test seems to be a remote proposition. However, the LMP test [SenGupta (1993)] reduces to an appealing simple form given by

$$\omega : T_1 \equiv (\lambda_2^{-1} - \lambda_1^{-1})n_1 - \sum m_{12} > C \ .$$

λ_1, λ_2 not being scale parameters necessitates, unfortunately, the computation of the cut-off points C for each different pair (λ_1, λ_2). Further, the exact distribution of T_1 is complicated, though the distribution of n_1 is Binomial $(n, \lambda_1(\lambda_1 + \lambda_2)^{-1})$ and that of M_{12} can also be obtained.

(b) **Symmetry known** ($\lambda_1 = \lambda_2 = \lambda_0$, **say, unknown**). Let $U = \sum |x_1 - x_2|$, $V = 2\sum \min(x_1, x_2)$, and $T = U + V$. Also, let $\theta_1 = \lambda_0 + \lambda_3/2$, $\theta_2 = \lambda_0 + \lambda_3$. Then, (22.6.1) reduces to

$$f_s(x_1, x_2) = \theta_1\theta_2 \exp[-2\theta_1 \min(x_1, x_2) - \theta_2|x_1 - x_2|] \ , \quad x_1, x_2 \geq 0 \ . \tag{22.6.3}$$

Here, Mehrotra and Michalek (1976) obtain the MLE $\widehat{\lambda}_3$ which can be used as a test statistic. Gupta, Mehrotra, and Michalek (1984) obtained GLRT for H_0 based on U/T, recommending a two-sided critical region. But, note that our hypotheses are equivalent to $H_0' : \theta_1 = \theta_2$ vs $H_1' : \theta_2 > \theta_1$. And, it is now immediate from (22.6.3), a member of a two-parameter REF, that the UMPU test for independence is given by, indeed the one-sided test which rejects for small values of U/T, a ratio of two independent gamma variables.

Comparison of various test statistics through Pitman's asymptotic relative efficiency (ARE) leads to the following result.

Result 22.1 [Theorem 1, Weier and Basu (1981); Basu (1988)] For the ACBVE model of Block-Basu when testing for independence with the statistics T = Kendall's τ, R = Spearman's ρ, U = UMP statistic and M = Maximum likelihood statistic, in the symmetric case,

$$\text{ARE(T,U)} = \text{ARE(R,U)} = \text{ARE(T,M)} = \text{ARE(R,M)} = .5 \text{ and } \text{ARE(M,V)} = 1.$$

Weier and Basu also note that when testing for independence under the Marshall-Olkin and Block-Basu models, the power functions, P_{MO} and P_{BB} respectively, are related by

$$P_{MO} = 1 - (1 - \rho)^n + (1 - \rho)^n \, P_{BB}$$

where $\rho = \lambda_3/\lambda$, the Corr(X_1, X_2) under the MO model. It then follows that the efficiency of 1 in Result 22.1 also holds for the MO model.

Further, their study with several alternative distributions, e.g. bivariate Weibull, normal, lognormal, seems to suggest the non-robustness of the parametric tests above.

(c) λ_1, λ_2 **unequal and unknown**. As in the MO model, here also no test has been proposed yet through optimality requirements. A large sample test based on MLE $\widehat{\lambda}_3$ which rejects H_0 when $\sqrt{n}\widehat{\lambda}_3/(\widehat{I}^{33})^{1/2} > z_\alpha$, $(\widehat{I}^{ij})$ being the MLE of the inverse of the Fisher information matrix, (I_{ij}), may be suggested [Hanagal and Kale (1991)]. $\widehat{\lambda}_3$ may be improved by using its jackknifed version, $\widehat{\lambda}_{3J}$ [Klein and Basu (1985)], which has reduced bias. However, it is well known that 'good' estimators need not yield 'good' tests. Further, both $\widehat{\lambda}_3$ and $\widehat{\lambda}_{3J}$ need to be obtained through iterative techniques. Alternatively, observe that under H_0, λ_1, λ_2 are scale parameters whose $\sqrt{n}$-consistent estimators therein, $\widehat{\lambda}_1$, $\widehat{\lambda}_2$ are easily obtained as $\bar{x}_1^{-1}$ and $\bar{x}_2^{-1}$ respectively. As in the MO model, here also these considerations motivate us [SenGupta (1993)] to propose a C_α-test as an optimal test and lends hope to it having a closed-form easily-computable test statistic.

Similar notations as in case (c) of Section 22.3.1 will be used here. We restrict to $0 < \lambda_i \leq b_i < \infty$, so that the Cramér character of the derivatives ϕ_i, $i = 1, 2, 3$ is ensured. Then let,

$$\widetilde{S}_{BB} = \widetilde{\phi}_3 - \sum_{i=1}^{2} \widetilde{\beta}_i^0 \widetilde{\phi}_i \qquad \text{and} \qquad \widetilde{T}_{BB} = \widetilde{S}_{BB}/\widetilde{\sigma}_0(S_{BB}) \,.$$

As in the MO model, ϕ_3 is not orthogonal to ϕ_i, and hence $\beta_i \neq 0$, $i = 1, 2$. Further, unlike the MO model, ϕ_1 and ϕ_2 are not orthogonal, i.e., $I_{12} \neq 0$, in general, which can add to the computations for $\widetilde{S}_{BB}$. Fortunately, however, under H_0, $I_{12} \equiv I_{12}^0 = 0$. So, β_i^0 simplifies as in the MO model. Also,

$$\phi_3 = n/(\lambda_1 + \lambda_2) + n_1/\lambda_2 + n_2/\lambda_1 - \sum m_{12}, \quad \phi_i = n/\lambda_i - \sum x_i, \qquad i = 1, 2 \,.$$

As in the MO model, the regression term in $\widetilde{S}_{BB}$ also may be made zero by taking the $\sqrt{n}$-consistent estimator of λ_i as its MLE $\widehat{\lambda}_i^0 = \bar{x}_i^{-1}$, $i = 1, 2$. Hence, $\widetilde{T}$ here has the same form as for the MO model. Now, here one may compute these $V(\phi_i)$ and $C(\phi_3, \phi_i)$ directly. For example, under H_0, X_1 and X_2 are independent $e(\lambda_i)$, $i = 1, 2$ respectively and M_{12} is the maximum of such two random variables. Then, $V(X_i) = \lambda_i^{-2}$, $i = 1, 2$, $C(X_1, X_2) = 0$. Also $V(n_1) = n\lambda_1\lambda_2(\lambda_1 + \lambda_2)^{-2}$. Further, for $C(X_1, n_1)$ is suffices to obtain $E[X_1 I(X_1 < X_2)] = \lambda_1/(\lambda_1 + \lambda_2)^2$. Similarly, other expressions can be obtained. More easily recall that $\text{Disp}_0(\phi) = (I_{ij}^0)$ and interestingly, though (I_{ij})'s for MO and BB models are in general not the same, I_{ij}^0, $i, j = 1, 2$ in fact coincide. Thus, the C_α-test reduces to

$$\omega : \widetilde{T}_{BB} = \frac{n\bar{x}_1\bar{x}_2(\bar{x}_1 + \bar{x}_2)^{-1} + n_1\bar{x}_2 + n_2\bar{x}_1 - \sum m_{12}}{\left[n\left(\widehat{I}_{33}^0 - \sum_{i=1}^2 (\widehat{I}_{3i}^0)^2/\widehat{I}_{ii}^0\right)\right]^{1/2}} > z_\alpha$$

where,

$$I_{33}^0 = (\lambda_1 + \lambda_2)^{-2} + (\lambda_1^3 + \lambda_2^3)/\lambda_1^2\lambda_2^2(\lambda_1 + \lambda_2), \qquad I_{ii}^0 = \lambda_i^{-2} \,,$$
$$I_{3i}^0 = (\lambda_1 + \lambda_2)^{-2} + (\lambda_1 + \lambda_2 - \lambda_i)/[\lambda_i^2(\lambda_1 + \lambda_2)], \qquad i = 1, 2$$

and '$\widehat{\ }$'s are obtained by replacing λ_i by $\bar{x}_i^{-1}$, $i = 1, 2$. Note that, as in the MO model, the test statistic here also is extremely simple to compute.

22.6.2 Tests for Symmetry

Here, $H_0 : \lambda_1 = \lambda_2 = \lambda_0$ (unknown), say, greatly simplifies (22.6.1). Let, $Y_1 = \min(X_1, X_2)$ and $Y_2 = |X_1 - X_2|$, $\eta_1 = 2\lambda_0 + \lambda_3$, $\eta_2 = \lambda_0 + \lambda_3$, $0 < \eta_2 < \eta_1 < 2\eta_2$. Then it is easily seen [e.g. Hanagal and Kale (1993)] that under H_0, (22.6.1) reduces to the product of the two independent exponentials, $e(y_i; \eta_i)$. Hence, $\widehat{\lambda}_0^0 = \bar{y}_1^{-1} - \bar{y}_2^{-1}$ and $\widehat{\lambda}_3^0 = 2\bar{y}_2^{-1} - \bar{y}_1^{-1}$. Let $\gamma = \lambda_1 - \lambda_2$. We test $H_0 : \gamma = 0$ vs $H_1 : \gamma \neq 0$.

(a) **Independence known** ($\lambda_3 = 0$). Here the situation is identical to case (a) of Section 22.3.3 and hence exactly the same test is UMPU here with this added robustness property.

(b) λ_3 **unknown**. As in the MO model, here also a test based on MLE or GLR requires iterative computations. However, unlike it, here further simplifications in C_α result due to the extremely simple form $\widehat{\lambda}_0^0$ and $\widehat{\lambda}_3^0$ of the required MLEs. Thus, using these, $\widetilde{\phi}_i = 0$, $i = 2, 3$. Let, $S = \partial \ln L / \partial \gamma|_{\gamma=0}$. Then,

$$S = (n_1 - n/2)\lambda_0^{-1} + n_2(\lambda_0 + \lambda_3)^{-1} + n(2\lambda_0 + \lambda_3)^{-1} - \sum x_1 \ .$$

Now, $\widetilde{S}$ and $\widetilde{\sigma}_0(S)$ can be obtained easily using $\widehat{\lambda}_0^0$ and $\widehat{\lambda}_3^0$.

Weier (1981) has proposed a bivariate exponential model related to BB model and Freund's model. No testing problem has been considered.

22.7 A Trivariate Block-Basu (T-BB) Model

A trivariate model (T-BB) was proposed by Weier and Basu (1980) as a special case of the trivariate Marshall-Olkin model. Its density is

$$\begin{aligned} & f(x_1, x_2, x_3) \\ = \ & [(3\lambda_0 + \lambda_4)(2\lambda_0 + \lambda_4)(\lambda_0 + \lambda_4)/6] \exp\left[-\lambda_0 \sum_{i=1}^{3} x_i - \lambda_4 \max(x_1, x_2, x_3) \right], \\ & x_i > 0,\ i = 1, 2, 3,\ \lambda_0 > 0,\ \lambda_4 \geq 0 \ . \end{aligned} \tag{22.7.1}$$

This is also the appropriate special case of the multivariate generalization and hence of BB model, given by Block (1975) in his equation (5.3). Independence corresponds to $\lambda_4 = 0$. It is a distribution under symmetry, i.e., under independence, marginals have the same distribution, $e(\lambda_0)$. Note that as in the bivariate situation, case (b) of Section 22.6.1, (22.7.1) is indeed a member of a two-parameter REF.

22.7.1 Tests for Independence and Symmetry

Consider testing $H_0 : \lambda_4 = 0$ vs $H_1 : \lambda_4 > 0$, i.e., H_0 : Total independence of X_1, X_2, X_3 vs H_1 : Pairwise positive correlations. Weier and Basu suggest two test statistics. The first one is derived by following the derivation of the UMP test in the bivariate symmetric MO model [Bhattacharyya and Johnson (1973)] adapted to the corresponding BB model and then using multiple tests. More specifically, (λ_0, λ_4) is reparametrized to some $(\theta_1, \theta_2, \theta_3)$ constrained by $\theta_3 = 4\theta_2 - \theta_1$, which now expresses H_0 and H_1 as equivalently, $H_0 : \theta_1 = \theta_2 = \theta_3$ vs $H_1 : \theta_1 < \theta_2 < \theta_3$. Then they use UMP tests based on F statistics, F_i, to test $H_{01} : \theta_i = \theta_{i+1}$ vs $H_{1i} : \theta_i < \theta_{i+1}$, $i = 1, 2, 3$, $\theta_4 \equiv \theta_1$. However, they note that this is not a typical multiple comparison type problem due to ordered alternatives, decision of any H_{0i} influences the others and F_i's are not independent. For simplicity, a single F_i may be used and they advocate F_2 then. For the second statistic L_E, they use the LR approach but ignoring the ordering under the alternative. Hence this is not the LR test. The constraint, $\theta_3 = 4\theta_2 - \theta_1$, on the parameter space, however, is incorporated through the use of Lagrangian multiplier. They also perform power comparison through simulations.

Note, however, that neither F_2 nor L_E is known to have any optimality property. But L_E lends itself nicely to a test, which again is not the LR test, incorporating the order restriction. We point out that, without the order restriction, L_E is indeed equivalent to the Lagrange multiplier test of Aitchison and Silvey (1960) and accordingly resembles the optimal generalized multiparameter C_α-test of Buehler and Puri (1966).

Strikingly however, we note that due to the REF nature of (22.7.1) in fact even an unconditional UMP test can be constructed for this problem. It is given by [SenGupta (1993)]

$$\omega : \ m_{13}/t > c, \ m_{13} = \sum \max(x_1, x_2, x_3), \ t = \sum \left(\sum_{i=1}^{3} x_i \right) ,$$

where c, the cut-off point is obtained from the distribution of M_{13}/T, $(T < 3M_{13})$ under H_0, i.e., when X_i's are i.i.d. $e(\lambda_0)$. Thus c can be easily obtained. In another direction, one can invoke invariance here. Let $\lambda_4^* = \lambda_4/\lambda_0$. Then H_0 and H_1 in terms of λ_4 and λ_4^* are identical. Further, from (22.7.1), now λ_0 becomes the common scale parameter for X_i's, and M_{13}/T and λ_4^* a maximal invariant statistic and parameter respectively. The UMPI test is then given by,

$$\omega : \ M_{13}/T > K \ ,$$

where K is obtained from the distribution of M_{13}/T under H_0.

Let $p_d = 6\lambda_0^2/[(3\lambda_0 + \lambda_4)(2\lambda_0 + \lambda_4)]$, the probability of an observed triple in the symmetric trivariate MO model lying off the diagonals $X_1 = X_2$, $X_2 = X_3$, $X_3 = X_1$ and $X_1 = X_2 = X_3$. Then, as in the bivariate case, here also it follows that, the power functions for T-MO and T-BB models are related by $P_{\text{T-MO}} = 1 - p_d^n + p_d^n P_{\text{T-BB}}$. As before computations for power need to be done only for any one of these models.

Optimal tests for a specific value of λ_0, the scale-symmetry parameter, can be constructed exactly as in the corresponding bivariate case, for both λ_4 known or unknown, Section 22.6.2.

22.8 Multivariate Block-Basu (M-BB) Model

Consider now the multivariate case. Block (1975) gave its density in the form

$$f(\boldsymbol{x}; \boldsymbol{\lambda}) = [(\lambda_{i_1} + \lambda_{k+1})/\alpha] \left[\prod_{r=2}^{k} \lambda_{i_r} \right] \left[\exp\left(-\sum_{r=1}^{k} \lambda_{i_r} x_{i_r} - \lambda_{k+1} x^{(k)} \right) \right] ,$$
$$x_{i_1} > x_{i_2} > \cdots > x_{i_k}, \ i_1 \neq i_2 \neq \cdots \neq i_k = 1, 2, \ldots, k ,$$

$$\alpha = \sum \cdots \sum_{i_1 \neq \cdots \neq i_k = 1}^{k} \left[\left(\prod_{r=2}^{k} \lambda_{i_r} \right) \Big/ \prod_{r=2}^{k} \left(\sum_{j=1}^{r} \lambda_{i_j} + \lambda_{k+1} \right) \right] , \qquad (22.8.1)$$

where $x^{(k)}$ is the kth *order statistic*. The striking feature of this elegant representation is that the single parameter λ_{k+1} dictates the dependency of $X_1, \ldots, X_k$ with $\lambda_{k+1} = 0$ corresponding to total independence. Symmetry corresponds to $\lambda_{i_1} = \cdots = \lambda_{i_k}$.

22.8.1 Tests for Independence

We consider tests for $H_0 : \lambda_{k+1} = 0$ vs $H_1 : \lambda_{k+1} > 0$ under part or no knowledge regarding the other parameters.

(a) **Symmetry known ($\lambda_{i_1} = \cdots = \lambda_{i_k} = \lambda_0$, say unknown).** Here (22.8.1) reduces to a two-parameter REF, and an unconditional UMP test exists, as given in Section 22.7.1,

$$\omega : \sum x^{(k)} > C .$$

The cut-off point C is obtained from the distribution of $\sum X^{(k)}/T$, $T = \sum \sum_{r=1}^{k} X_{i_r}$, which is a member of the one-parameter exponential family.

(b) **λ_{i_j}'s unequal and unknown.** Details of the usual testing procedure of using the standardized MLE are available from Hanagal (1993). However, as in the bivariate case, this calls for iterative techniques.

Several tests were suggested by Weier and Basu (1980) by exploiting their alternative representation of the density to that of Block (1975). Let $x^{(\ell)}$ denote the ℓth smallest component of k-variate vector $(x_1, \ldots, x_k)$, $k = 1, \ldots, \ell$. Then they present the density as

$$f(x_1, \ldots, x_k) = \left[\sum_{\ell=1}^{k} C_\ell / k! \right] \exp \left[- \sum_{\ell=1}^{k} (C_\ell - C_{\ell+1}) x^{(\ell)} \right] , \qquad (22.8.2)$$

where C_i's, $i = 1, \ldots, k$ are parameters, $C_{k+1} \equiv 0$. Let $\theta_i = C_i/(k - i + 1)$ and $U_i = (k - i + 1) \sum_{j=1}^{n} \left(x_j^{(i)} - x_j^{(i-1)} \right)$, $x_j^{(0)} \equiv 0$, U_i's being independent Gamma(θ_i), with densities $g(u_i, \theta_i)$, $i = 1, \ldots, k$. Then, as in T-BB, the hypotheses regarding independence become $H_0 : \theta_1 = \cdots = \theta_k$ vs $H_1 : \theta_1 < \cdots < \theta_k$, and the likelihood function in terms of U_i's becomes $\prod_{i=1}^{k} g(u_i, \theta_i)$. The testing problem thus reduces to that of homogeneity of ordered gamma distributions, for which Barlow et al. (1972) present the LR test. This is based on

$$\bar{\chi}^2 = -2 \ln L = 2 \sum n \ln(\bar{U}/n\widehat{\mu}_i^*), \qquad \bar{U} = \sum U_i / k ,$$

where, $\widehat{\mu}_i^*$ are the isotonic regression estimators of μ_i, $i = 1, \ldots, k$. Weier and Basu (1980) suggest using LR statistics obtained from the two limiting results on $\bar{\chi}^2$-distribution or some combination of the estimators $\widehat{\theta}_i = n/U_i$, $i = 1, \ldots, k$ or L_N, the general LR test under normality.

An appealing simple C_α-test results here. Note that using similar notations, all the discussions regarding orthogonality of parameters, simple forms for MLEs, S_{BB} and T_{BB} in case (c) of Section 22.6.1 are valid here also, resulting in a very simple C_α-test. The required (I_{ij}^0) can be obtained easily from (I_{ij}) [Hanagal (1993)] using $\lambda_{k+1} = 0$. The C_α-test reduces to the simple elegant form [SenGupta (1993)]

$$\omega : \widetilde{T}_{\text{M-BB}} = \frac{\left(\sum_{i=1}^k \bar{x}_i^{-1}\right)^{-1} + \sum_{i=1}^k n_i \bar{x}_i - \sum x^{(k)}}{\left[n\left(\widehat{I}_{k+1\ k+1}^0 - \sum_{i=1}^k (\widehat{I}_{k+1\ i}^0)^2 / \widehat{I}_{ii}^0\right)\right]^{1/2}} > z_\alpha \ .$$

22.9 Scaled Exponential Dependency Parameter (BK) Models

The densities of several absolutely continuous distributions with scaled exponential marginals and indexed by a dependency parameter θ can be presented in a unified form as,

$$\begin{aligned} f_\theta(x_1, x_2) &= \lambda_1 \lambda_2 \exp[-(\lambda_1 x_1 + \lambda_2 x_2)] g(\lambda_1 x_1, \lambda_2 x_2; \theta) \ , \\ & \qquad x_i, \lambda_i > 0, \ i = 1, 2, \ \theta \geq 0 \ . \end{aligned} \tag{22.9.1}$$

Independence arises possibly as a limiting case, as $\theta \downarrow 0$. This family includes the parametric families of Gumbel (Type I and II) (1960), Frank (1979), and Cook and Johnson (1981). Optimal LBI test of independence was obtained by Bilodeau and Kariya (1993) for this family in general and applied specifically to the above-mentioned parametric models. They have established the following result.

Result 22.2 Under certain mild regularity conditions the LBI test of independence for (22.9.1) is

$$\omega : T_n(y) = n^{-1} \sum E_b[\dot{g}(b_1 y_{1k}, b_2 y_{2k})] > C, \ y_{jk} = x_{jk} / \sum x_j, \ k = 1, \ldots, n$$

and b_j, $j = 1, 2$ are independent Gamma(n). Under some further regularity conditions, $\sqrt{n}[T_n - \mu(\theta)] \to N(0, V(\theta))$, where $\mu(\theta) = E_\theta[\dot{g}(x_1, x_2)] = E_\theta[\lim_{\theta \downarrow 0} \partial g(x_1, x_2; \theta)/\partial\theta]$, and

$$\begin{aligned} V(\theta) &= \text{Var}_\theta[\dot{g}(x_1, x_2) - x_1 \zeta_1(\theta) - x_2 \zeta_2(\theta)] \ , \\ \zeta_j(\theta) &= E_t[x_j \dot{g}_j(x_1, x_2)], \ j = 1, 2 \ . \end{aligned}$$

For contiguous alternatives $\theta_n = \omega/\sqrt{n}$, $\omega > 0$, the asymptotic nonnull distribution of $\sqrt{n}[T_n - \mu(0)]$ is $N(\omega\mu'(0), v(0))$.

The above result is a direct consequence, without necessitating any expansion and limiting operations, of invoking scale invariance (with respect to λ_1 and λ_2) and applying Wijsman's (1967) representation for the ratio of the densities of the maximal invariant under $H_0 : \theta = 0$ and $H_1 : \theta = \theta$. Now consider the specific distributions mentioned above.

(a) **Gumbel's Type I (GI) model**

$$g(\cdot) = [1 + \theta(2\exp(-\lambda_1 x_1) - 1)(2\exp(-\lambda_2 x_2) - 1)] .$$

Here, $T_n = \sum[2(1+y_{1k})^{-n} - 1][2(1+y_{2k})^{-n} - 1]$ and $\sqrt{n}[T_n - \theta/9] \xrightarrow{D} N(0, (\theta^3 - 84\theta^2 + 648)/5832]$.

(b) **Gumbel's Type II (GII) model**

$$\begin{aligned} g(\cdot) &= \exp(-\theta\lambda_1\lambda_2 x_1 x_2)[(1+\theta\lambda_1 x_1)(1+\theta\lambda_2 x_2) - \theta] \\ T_n &= 1 - n\sum y_{1k}y_{2k}, \text{ and } \mu(\theta) = 1 - J_1(\theta) , \\ v(\theta) &= 2J_1^3(\theta) + 3J_1^2(\theta) + 4[2\theta J_1^2(\theta) + (1-\theta)J_1(\theta) - 1]/\theta^2 , \end{aligned}$$

where $J_1(\theta) = \int_0^\infty [\exp(-x)/(1+\theta x)]dx$.

(c) **Frank's (F) model**

$$\begin{aligned} g(\cdot) &= \frac{\theta \ln(1+\theta)(1+\theta)^{F(\lambda_1 x_1)+F(\lambda_2 x_2)}}{\left[\theta + \left((1+\theta)^{F(\lambda_1 x_1)} - 1\right)\left((1+\theta)^{F(\lambda_2 x_2)} - 1\right)\right]^2} , \\ F(x) &= 1 - \exp(-x) \end{aligned}$$

Here $\partial g(\cdot)/\partial\theta$ is expanded around $\theta = 0$ to $0(\theta)$, which then yields

$$T_n = -(2n)^{-1}\sum \left[2(1+y_{1k})^{-n} - 1\right]\left[2(1+y_{2k})^{-n} - 1\right] .$$

Then, $T_n^* = -2T_n$ is the LBI statistic for GI model and hence, under H_0, $\sqrt{n}\,T_n^* \xrightarrow{D} N(0, 1/9)$.

(d) **Cook-Johnson (CJ) model**

$$\begin{aligned} g(\cdot) &= (1+\theta)[(F(\lambda_1 x_1)F(\lambda_2 x_2))^{1+\theta}(F(\lambda_1 x_1)^{-\theta} + F(\lambda_2 x_2)^{-\theta} - 1)^{2+1/\theta}]^{-1} \\ T_n &= n^{-1}\sum[1 - \omega_n(y_{1k})][1 - \omega_n(y_{2k})] , \\ \omega_n(y) &= \sum_{j=1}^{\infty}[j(1+jy)^n]^{-1}, \ 0 < y < 1 . \end{aligned}$$

However, the asymptotic normality of T_n could not be established since a required regularity condition could not be established.

22.10 Gumbel III (GIII) Model

Gumbel (1960) proposed three models, of which LBI tests for two of them, GI and GII have been already presented under the unified treatment in Section 22.9. The density of his third model, GIII model, is,

$$\begin{aligned} f(x_1,x_2) &= (\lambda_1\lambda_2)(\lambda_1\lambda_2x_1x_2)^{\theta-1}[(\lambda_1x_1)^\theta+(\lambda_2x_2)^\theta]^{1/\theta-2} \\ &\cdot[((\lambda_1x_1)^\theta+(\lambda_2x_2)^\theta)^{1/\theta}+\theta-1]\exp[-((\lambda_1x_1)^\theta+(\lambda_2x_2)^\theta)^{1/\theta}], \\ &\quad 0<x_i,\ \lambda_i<\infty,\ i=1,2,\ 1\le\theta\le\infty\ . \end{aligned} \tag{22.10.1}$$

The above is not a member of the exponential family, nor does it possess any non-trivial sufficient statistic. However, very recently Lu and Bhattacharyya (1991a,b) have enhanced it through physical motivations and proposed some simple inference procedures. Marginals are $e(\lambda_i)$, $i=1,2$. Independence corresponds to $\theta=1$ and symmetry to $\lambda_1=\lambda_2$.

22.10.1 Tests for Independence

Note that (22.10.1) may be subsumed in the treatment of Section 22.9 for LBI tests. However, the associated $g(\cdot)$ is complicated. The MLE requires non-trivial iterative computations. Lu and Bhattacharyya (1991a) proposed testing independence, i.e., H_0 : $\gamma\equiv1/\theta=1$ and also H_0 : $\gamma=\gamma_0$ by exploiting the associated confidence interval, presumably one-sided, based on their suggested pivotal quantity. This is to be iteratively computed and the percentage points are to be obtained through simulation. We show below how a C_α-test reduces to an elegant form, probably the simplest so far encountered, even for this complicated density. This success is partly due to Theorem 2.1 of Lu and Bhattacharyya (1991a) which exhibits (I_{ij}) in a closed form. Let $\boldsymbol{\lambda}=(\lambda_1,\lambda_2,\gamma)'$ and define ϕ_i accordingly. Even though ϕ_i's and I_{ij}'s have somewhat lengthy expressions it turns out fortunately that both admit of tremendous simplifications under H_0, e.g., $(I_{ij}^0)=\text{Diag}(\lambda_1^2,\lambda_2^2,C)$. Thus the regression term vanishes and $\text{Var}(S)=\text{Var}(\phi_3^0)=\text{constant}=C$, free of λ_1,λ_2, for which there is no need for estimation. Then the C_α-test reduces to

$$\omega:\left\{\sum\ln\left[\left(\sum_{i=1}^2 z_i\right)/z_1z_2\right]-\left(\sum_{i=1}^2 z_i\right)^{-1}\left[1+\sum_{i=1}^2 z_i\ln z_i\right]\right\}\Big/\sqrt{nC}>z_\alpha\ ,$$

where $z_i=\widehat{\lambda}_i^0X_i=(\bar{x}_i)^{-1}X_i$, $i=1,2$, and where the MLE $\bar{x}_i$ under H_0 are $\sqrt{n}$-consistent estimators.

22.10.2 Tests for Symmetry

Lu and Bhattacharyya (1991a) suggest two tests, one based on the binomial count $\sum I(X_1<X_2)$, and another on suitable estimators of transformed parameters, to give simple tests. Here also one may attempt to construct a C_α-test as above.

22.11 Gumbel III Model Under a Type II Censoring Scheme

Lu and Bhattacharyya (1991b) have suggested a Type-II censoring scheme under (22.10.1). We may consider this as a generalization of the Type II scheme for the independent exponentials in Section 22.2.

22.11.1 Tests for Independence

(a) **Symmetry known.** Lu and Bhattacharyya (1991b) demonstrate that here the likelihood function, L, constitutes a member of the two-parameter REF. Hence UMPU tests, possibly conditional, are expected for both $H_0 : \gamma = 0$ and $H_0 : \gamma = \gamma_0$ (known). By suitable reparametrizations they establish that for both these situations, in fact unconditional simple UMPU tests whose cut-off points are available from the central F distribution result. The general case with $\lambda_1 \neq \lambda_2$ has not been considered.

22.11.2 Tests for Symmetry

(a) $\lambda_1 = \lambda_2 = \lambda_0$, **say** γ **unknown.** Exploiting the REF nature of L here, Lu and Bhattacharyya (1991b) obtain unconditional UMPU tests for $H_0 : \lambda_0 = \lambda_0^*$ (known) vs $H_1 : \lambda_1 \neq \lambda_0^*$ and also for $H_0' : \lambda_0 \geq \lambda_0^*$ vs $H_1' : \lambda_0 < \lambda_0^*$. The required cut-off points are obtained by referring to a central χ^2-distribution.

(b) $\lambda_1 \neq \lambda_2$ **and** γ **unknown.** Here L constitutes (3,4) CEF with sufficient statistic $(D, \boldsymbol{T})$ say and hence, as before, standard theory for REF does not apply. Based on D and $\boldsymbol{T}$ separately they obtain two UMPU unconditional tests and obtain the pooled test statistic, $(1 - p_0)U - p_0\bar{D}$, say, where p_0 is so determined as to render the test asymptotically best in the sense of maximum Pitman ARE. We note in connection with Section 22.2, that had U and $\bar{D}$ been independent, then the best test in the sense of maximizing Bahadur efficiency, could be easily obtained. This test then with maximum exact Bahadur slope, could be obtained, as is well known, by Fisher's method of optimally pooling the individual tests.

22.12 Freund (F) Model

The density for the Freund model is

$$\begin{aligned} f(x_1, x_2) &= \beta_1\beta_2\beta_1'\beta_2' \exp[-\beta_j'(x_j - x_i) - (\beta_1 + \beta_2)x_i] \,, \\ & \qquad 0 < x_j < x_i, \ \beta_i, \beta_i' > 0, \ i, j = 1, 2 \,. \end{aligned} \tag{22.12.1}$$

This is a member of the four-parameter REF with complete sufficient statistic $(n_1 = \sum I(x_1 < x_2), \sum x_1, \sum x_2, \sum (x_1 - x_2) I(x_1 < x_2))$. The marginals are not exponentials but rather proper mixtures of exponentials, e.g., the marginal density of X_1 is

$$f_1(x_1) = p(\beta_1 + \beta_2)\exp(-(\beta_1 + \beta_2)x_1) + (1 - p)\beta_1' \exp(-\beta_1' x), \ x > 0 \,,$$

where

$$\begin{aligned} p &= (\beta' - \beta_1')(\beta_1 + \beta_2 - \beta_1')^{-1}, \text{ when } \beta_1 + \beta_2 \neq \beta_1' \\ &= (\beta_1 + \beta_1'\beta_2 x)\exp(-\beta_1' x), \text{ when } \beta_1 + \beta_2 = \beta_1' \,. \end{aligned}$$

BB model is a special case of F model as seen by taking

$$\lambda_1 = (\beta_1 + \beta_2) - \beta_2', \ \lambda_2 = (\beta_1 + \beta_2) - \beta_1' \text{ and } \lambda_3 = (\beta_1' + \beta_2') - (\beta_1 + \beta_2) \,.$$

Symmetry corresponds to $\beta_1 = \beta_2$ and $\beta_1' = \beta_2'$ while independence to $\beta_1 = \beta_1'$ and $\beta_2 = \beta_2'$. Thus both the null hypotheses of independence and symmetry correspond to two-parameter composite ones. No optimal test for this model is available in the literature. We exhibit how optimal C_α-tests can be conveniently derived here.

22.12.1 Tests for Independence

To test $H_0 : \boldsymbol{\theta}' = (\theta_1, \theta_2) = (\beta_1 - \beta_1', \beta_2 - \beta_2') = (0,0)$ vs $H_1 : \theta \neq \mathbf{0}$, define $\beta_i = \eta_i + \theta_i$, $i = 1,2$. Then, under H_0, (22.12.1) retains its REF nature, but now a member of two-parameter REF with $(\sum x_1, \sum x_2)$ as a complete sufficient statistic for the nuisance parameter $\boldsymbol{\eta}' = (\eta_1, \eta_2)$. One may try to pose this as a one parameter problem i.e., $H_0 : \delta^2 = \boldsymbol{\theta}'\boldsymbol{\theta} = 0$ vs $H_1' : \delta^2 \neq 0$ and construct tests based on suitable quadratic-form statistics. Alternatively, one can treat this as a genuine multiparameter problem where both the parameter being tested and the nuisance parameter present are vector parameters, and consider corresponding tests [SenGupta (1991)]; e.g. directional derivative tests of John (1971), locally most mean power tests of SenGupta and Vermeire (1986), etc. These can be exploited only after the removal of the nuisance parameters β_{10}, β_{20}. But considerations of neither similarity nor invariance seem to lead to simplifications. Alternatively, we consider multivariate generalizations of Neyman's C_α-test, e.g. see Buehler and Puri (1966), Hall and Mathiason (1990). Let $\boldsymbol{\gamma} = (\boldsymbol{\theta}', \boldsymbol{\eta}')'$, $\boldsymbol{\theta}$ being the parameter of interest. Local alternatives are given by $\gamma_n = (\theta_n, \eta_n)$ where $\theta_n = \theta + h_\theta/\sqrt{n}$, and similarly for η_n. Let, $\sqrt{n}\, S_{\boldsymbol{\gamma}}$ denote the vector of first order derivative. Assume that, under $\boldsymbol{\gamma}$, $S_n \xrightarrow{D} N(0, B)$, for some $B > 0$. Partition S into (S_θ, S_η), and B into blocks $B_{\theta\theta}$, $B_{\theta\eta}$, $B_{\eta\theta}$ and $B_{\eta\eta}$, B^{-1} having blocks $B^{\theta\theta}$, etc. accordingly. Let $S_\theta^* = S_\theta - B_{\theta\eta} B_{\eta\eta}^{-1} S_\eta$ be the effective score for θ and $B_\theta^* = (B^{\theta\theta})^{-1} = B_{\theta\theta} - B_{\theta\theta} B_{\eta\eta}^{-1} B_{\eta\theta}$ the effective information. Then under the assumptions of S_θ to have Cramér function components and some other regularity assumptions, a generalized Neyman C_α-test, i.e., Neyman-Rao or effective score-test has the test statistic $Q_n = S_\theta^{*\prime}(\tilde{\gamma}_n^0)\bar{B}^{\theta\theta} S_\theta^*(\tilde{\gamma}_n^0)$, where $\tilde{\gamma}_n^0$ is a $\sqrt{n}$-consistent estimator under H_0 and $\bar{B}^{\theta\theta}$ is consistent under H_0. Then it follows from LeCam's third lemma that under γ_n, $Q_n \xrightarrow{D} \chi^2_{d_\theta}(\delta_\theta^2)$ with $\delta_\theta^2 = h_\theta' B_\theta^* h_\theta$ and $\gamma = \gamma^0$, $(B_\theta^*)^{-1} = B_{\theta\theta}$. Further, this test, as are LR, Wald, and Rao tests, is efficient within a class of quadratic form tests. Hall and Mathiason include examples from reliability, e.g. involving Weibull models. Orthogonality of parameters, use of MLEs and other remarks made earlier for the one-parameter case extends here also.

The MLEs for the parameters β_i, β_i' both under H_0 and in general, are available [Freund (1961)] in simple closed forms and hence so is true for θ_i, η_i, $i = 1,2$. The parameters β_i, β_i', $i = 1,2$ are mutually orthogonal and hence so are η_1, η_2. It can be shown further that only non-orthogonal pairs are (θ_i, η_i), $i = 1,2$ and all other 4 pairs are orthogonal. Also the MLEs are easily obtained as

$$\begin{aligned}
\widehat{\theta}_1 &= n_1/\sum \min(x_1, x_2) - n_2/\sum_{x_1 > x_2} (x_1 - x_2)\ , \\
\widehat{\theta}_2 &= n_2/\sum \min(x_1, x_2) - n_1/\sum_{x_1 < x_2} (x_2 - x_1)\ , \\
\widehat{\eta}_1 &= n_2/\sum_{x_1 > x_2} (x_1 - x_2),\ \widehat{\eta}_2 = n_1/\sum_{x_1 < x_2} (x_2 - x_1)\ .
\end{aligned}$$

(I_{ij}) can be easily computed now. These give tremendous simplifications for computing both S_θ^* and $\bar{B}^{\theta\theta}$ (from the MLEs, $\widehat{I}_{ij}$). It is now a routine matter to write down Q_n which has admitted of significant simplifications.

22.12.2 Tests for Symmetry

Consider H_0 : $\boldsymbol{\theta}' = (\theta_1, \theta_2) = (\beta_1 - \beta_2, \beta_1' - \beta_2') = (0,0)$ vs H_1 : $\theta \neq \mathbf{0}$. The nuisance parameter vector $\boldsymbol{\eta}$ is now given by $(\eta_1, \eta_2) = (\beta_2, \beta_2')$. We proceed exactly as for the case of independence and hence further details are omitted. The resulting [SenGupta (1993)] generalized C_α-test is based on the corresponding Q_n, which also admits of significant simplifications.

22.13 Arnold-Strauss (AS) Model

Arnold and Strauss (1988) proposed their model in which the conditionals, instead of the marginals as usually suggested, were posited to be exponentials. The density for the bivariate random vector (X_1, X_2) is given by

$$\begin{aligned} f(x_1, x_2) &= C(\lambda_3)\lambda_1\lambda_2 \exp\left[-\left(\sum_{i=1}^{2} \lambda_i x_i + \lambda_1\lambda_2\lambda_3 x_1 x_2\right)\right] , \\ C^{-1}(\lambda_3) &= \int_0^\infty [(1+\lambda_3 u)\exp(u)]^{-1} du, \quad x_i, \lambda_i > 0,\ i = 1,2,\ \lambda_3 \geq 0 . \end{aligned} \tag{22.13.1}$$

This is a member of the three-parameter REF with sufficient statistic $(\sum x_1, \sum x_2, \sum x_1 x_2)$ for $(\lambda_1, \lambda_2, \lambda_3)$. In this form, λ_i's, $i = 1,2$ are scale parameters, and λ_3, the dependency parameter. Independence corresponds to $\lambda_3 = 0$ while symmetry to $\lambda_1 = \lambda_2$. Marginals are not exponentials unless under independence. No testing procedure has been developed for this model yet and we develop some optimal tests here.

22.13.1 Tests for Independence

Due to the REF nature of (22.13.1), elegant results can be expected. For λ_1, λ_2 known, the UMP test is given by, $\omega : T = \sum x_1 x_2 > C$, where C is obtained easily noting that X_1, X_2 are independent $e(\lambda_i)$. Under symmetry, $\lambda_1 = \lambda_2 = \lambda_0$ (unknown). λ_0 is still a scale parameter, as also are λ_1, λ_2 appearing in the general case of both unknown and unequal. It suffices to consider this latter case. Here, we note that both the UMPU which is conditional and, when invariance is to be invoked, a LBI, which is unconditional, test can be derived in elegant forms [SenGupta (1993)].

Let $\gamma = \lambda_1\lambda_2\lambda_3$ and $T_i = \sum x_i$, $i = 1,2$, $T_3 = \sum x_1 x_2$. Then (22.13.1) still retains its REF nature in terms of $(\lambda_1, \lambda_2, \gamma)$. Thus the size-$\alpha$ UMPU test for independence, i.e., for H_0 : $\gamma = 0$ vs H_1 : $\gamma > 0$ reduces simply to

$$\omega : \ T_3 < C(t_1, t_2),\ P_{\gamma=0}[T_3 < C(t_1,t_2) \mid (T_1, T_2) = (t_1, t_2)] = \alpha \ ,$$

i.e., C is determined as usual, from the conditional distribution of T_3 given $(T_1, T_2) = (t_1, t_2)$, which is again a member of the one-parameter REF.

When invariance is to be invoked, the form (22.13.1) using λ_3 is most convenient. Using Wijsman's representation or directly, it follows that LBI test is given by

$$\omega : \; T_n = n \sum y_1 y_2 < C, \; y_i = x_i / \sum x_i, \qquad i = 1, 2 \; .$$

The exact null and non-null distributions of T_n are somewhat complicated. However, since under H_0, $X_i \sim e(\lambda_i)$, $i = 1, 2$, T_n can be simulated easily and C thus obtained. For power computations, one can use the usual rejection method suggested by Arnold and Strauss in general. We note however, that this distribution is *tailor-made* for generating observations through Gibb's sampling. Details are available from SenGupta (1993). The asymptotic null distribution of $\sqrt{n}\, T_n$ under contiguous alternatives $\lambda_{3n} = \eta/\sqrt{n}$ is $N(0, 1)$. Note that this test coincides with that for GII model which has the same null density as for this AS model. Hence it is null and optimality robust against the GII model.

22.13.2 Tests for Symmetry

To test $H_0 : \; \gamma = \lambda_1 - \lambda_2 = 0$ vs $H_1 : \; \gamma \neq 0$, let $\lambda_1 = \lambda_2 = \lambda_0$ (unknown) and $\delta = \lambda_1 \lambda_2 \lambda_3$. Then, under H_0 also, (22.13.1) retains its REF nature. For independence given, the problem is that of testing homogeneity of means of independent exponentials and hence the UMPU test is given as before. When δ is unknown, again a UMPU test exists given by

$$\omega : T_1 < C_1(t_1 + t_2, t_3) \text{ or } > C_2(t_1 + t_2, t_3), \; T_i = \sum X_i, \; i = 1, 2 \; ,$$

where $C_i(\cdot)$ are computed from the conditional distribution of T_1 given, say, $\boldsymbol{T}_{12} = (T_1 + T_2, T_3) = (t_1 + t_2, t_3) = \boldsymbol{t}_{12}$, a member of one-parameter REF, as

$$E_{\gamma=0}[\phi(T_1, \boldsymbol{T}_{12}) \mid \boldsymbol{t}_{12}] = \alpha, \; E_{\gamma=0}[T_1 \phi(T_1, \boldsymbol{T}_{12}) \mid \boldsymbol{t}_{12}] = \alpha E_{\gamma=0}[T_1 \mid \boldsymbol{t}_{12}] \; ,$$

where ϕ is the critical function corresponding to ω above. The conditional distribution involved, though a member of the one-parameter REF, may be non-trivial to compute. Here one may pursue the approach for optimal C_α-test. However, even under H_0, neither are the MLEs available in closed forms and nor is there orthogonality among the parameters. One may use however the $\sqrt{n}$-consistent moment estimators in the general form of the C_α-test without any further reduction.

A multivariate generalization of AS model, e.g. as in Arnold and Strauss (1988, 1991), does not seem to lend itself to simple optimal estimators or tests.

22.14 Mixture Models

Recall that marginals of F model are proper mixture distributions. Bivariate proper mixture models or extensions have also been proposed. Notably, Friday and Patil (1977) proposed a model (FP) that subsumes both MO and F models. It is given by $\overline{F}_{FP} = p\overline{F}_F + (1-p)F^S_{MO}$, where F^S_{MO} is the singular distribution, $\exp[-(\lambda_1 + \lambda_2)\max(x_1, x_2)]$. Special choices of p lead to F and MO models, with p in general being a function of five model parameters. Mixture models may arise when different batches of a specific item are amalgamated or when outliers are suspected. Even when p does not depend on the parameters of the component distributions a general test for H_0 : No mixture vs H_1 : Mixture, is fraught with non-standard situations and non-regularity problems. For example, $H_0 : p = 1$ requires test for parameter on the boundary and not an interior point in the parameter space. Here, neither MLE nor LRT have standard asymptotic properties. A LMP similar/invariant test may be derived here, in certain situations, along the lines of SenGupta and Pal (1991, 1993). The null hypothesis of no mixture often reduces to the union of several null hypotheses involving p and component parameters. Optimal test for each such hypothesis may be constructed and optimally pooled. A treatment for a Weibull and double exponential mixture in the univariate case is given by Rachev and SenGupta (1992). Some of the methods therein are general and readily extend to the bi(multi-)variate models. We refer the reader to the details given there. In any case, construction of optimal statistical tests for mixture models in the context of reliability theory also is a challenging problem where the development is only in its infancy.

22.15 Non-Regular Models

Our unified approach above assumed the existence of the score statistic ϕ_θ and we also often used the information matrix based on the second order derivatives. However, there are models where even the first derivatives need not exist. Moran-Downton (MD) bivariate model constitutes one such example. Regarding problems with the information matrix, the two-parameter exponential model of Section 22.3 is a well-known example. The bivariate Hashino (1985) (H) model is a generalized bivariate Freund's model which incorporates the univariate two-parameter exponential model. Its density is given by,

$$f(x_1, x_2) = \begin{cases} \beta_1\beta_2\beta_1'\beta_2' \exp[-\beta_j'(x_j - x_i) - (\beta_1+\beta_2)x_i], & 0 \le x_j \le x_i,\ x_j \le \alpha, \\ b_1b_2b_1'b_2' \exp[-b_j'(x_j - \mu) - (b_1+b_2)(x_i - \mu)], & \alpha \le x_i \le x_j\ . \end{cases} \tag{22.15.1}$$

Here one may use Chapman-Robbins bounds for the efficiency of the estimators. Note that I_{ij} is not needed for C_α-test, since any locally $\sqrt{n}$-consistent estimators of the variance of S will do. Work on this by the author is under progress.

Moran (1967) proposed a bivariate exponential distribution which was later popularized by Downton (1970) in the context of reliability studies and as such has been usually associated with his name. The density of MD model is

$$f(x_1, x_2) = (1-\theta)^{-1} I_0[2(x_1x_2\theta)^{1/2}/(1-\theta)] \exp[-(x_1+x_2)/(1-\theta)]\ , \tag{22.15.2}$$

where $I_0(\cdot)$ is the modified Bessel function of the first kind of order zero. Independence corresponds to $\theta = 0$, and $\mathrm{Corr}(X_1, X_2) = \theta$. Scale parameters are often introduced by letting $X_i = \lambda_i Y_i$, $i = 1, 2$. (22.15.2) is of interest as quite a general distribution in that it arises from the shocks to failure following various geometric distributions. The above MD model was generalized by Hawkes (1972) and Paulson (1973) and is a special case of the bivariate gamma distribution discussed earlier by Krishnamoorthy and Parthasarathy (1951).

Let us consider testing $H_0 : \theta = 0$ vs $H_1 : \theta > 0$. Here $\partial \ln f/\partial\theta$ creates a problem at $\theta = 0$ as it involves the term $(x_1x_2/\theta)^{1/2}$. A similar problem was encountered by

Crowder (1990) in the context of reliability studies based on a modified Weibull model. One approach is to reparametrize. Failing this, the component of the score function resulting in its singularity is isolated. The LMP test is then modified as

$$T = n^{-1}\phi_\theta^* = n^{-1}\left[\phi_\theta - \sum (x_1 x_2/\theta)^{1/2}\right]\Big|_{\theta=0} = -\sum_{i=1}^{2} \bar{x}_i > C \ .$$

Note that under H_0, X_i's are independent $e(1)$ and hence C is easily obtained. The exact power behaviour may be studied through simulation easily since $X_1 + X_2$ is the sum of two independent $e(1+\sqrt{\theta})^{-1}$ and $(1-\sqrt{\theta})^{-1}$. This representation is useful in studying the consistency of the test T. For unknown λ_1, λ_2, LBI test of Section 22.9 may be modified using a similar approach on the $g(\cdot)$ here.

Thus, even for non-regular distributions, sometimes suitably modified LBI or C_α-tests may be constructed, where optimality properties need to be however, separately established.

22.16 Other Models

There are various other models; for example, see Raftery (1984), Sarkar (1987) [see also Sinha and Sinha (1993)]. These models were not treated above mainly because no optimality result is yet available for them. Our unified approach as above may be attempted therein.

References

Aitchison, J. and Silvey, S.D. (1960). Maximum-likelihood estimation procedures and associated tests of significance, *Journal of the Royal Statistical Society, Series B*, **22**, 154-171.

Arnold, B.C. (1968). Parameter estimation for a multivariate exponential distribution, *Journal of the American Statistical Association*, **63**, 848-852.

Arnold, B.C. and Strauss, D. (1988). Bivariate distributions with exponential conditionals, *Journal of the American Statistical Association*, **83**, 522-527.

Arnold, B.C. and Strauss, D. (1991). Bivariate distributions with conditionals in prescribed exponential families, *Journal of the Royal Statistical Society, Series B*, **53**, 365-375.

Bahadur, R.R. (1971). *Some Limit Theorems in Statistics*, SIAM, Philadelphia.

Bahadur, R.R. and Raghavachari, M. (1972). Some asymptotic properties of likelihood ratios on general sample spaces, In *Proceedings of the Sixth Berkeley Symposium*, Vol. 1, pp. 129-152.

Barlow, R.E., Bartholomew, D.J., Bremner, J.M., and Brunk, H.D. (1972). *Statistical Inference Under Order Restrictions*, John Wiley & Sons, New York.

Basu, A.P. (1988). Multivariate exponential distributions and their applications in reliability, In *Handbook of Statistics*, Vol. 7 (Eds., P.R. Krishnaiah and C.R. Rao), Elsevier, Amsterdam.

Bemis, B.M., Bain, L.J., and Higgins, J.J. (1972). Estimation and hypothesis testing for the parameters of a bivariate exponential distribution, *Journal of the American Statistical Association*, **67**, 927-929.

Bhattacharyya, G.K. and Johnson, R.A. (1971). Maximum likelihood estimation and hypothesis testing in the bivariate exponential model of Marshall and Olkin, *Technical Report No. 276*, Department of Statistics, University of Wisconsin-Madison.

Bhattacharyya, G.K. and Johnson, R.A. (1973). On a test of independence in a bivariate exponential distribution, *Journal of the American Statistical Association*, **68**, 704-706.

Bilodeau, M. and Kariya, T. (1993). LBI tests of independence in bivariate exponential distributions, Presented at the *Seventh International Conference on Multivariate Analysis*, New Delhi.

Block, H.W. (1975). Continuous multivariate exponential extensions, In *Reliability and Fault Tree Analysis* (Eds., R.E. Barlow, J.B. Fussell, and N.D. Singpurwalla), SIAM, Philadelphia.

Block, H.W. and Basu, A.P. (1974). A continuous bivariate exponential extension, *Journal of the American Statistical Association*, **69**, 1031-1037.

Block, H.W. and Savits, T.H. (1981). Multivariate distributions in reliability theory and life testing, In *Statistical Distributions in Scientific Work*, Vol. 5 (Eds., C. Taillie, G.P. Patil, and B.A. Baldessari), pp. 271-288, Reidel, Dordrecht.

Buehler, W.J. and Puri, P.S. (1966). On optimal asymptotic tests of composite hypotheses with several constraints, *Z. Wahrsch. Verw. Gebiete*, **5**, 71-88.

Cook, R.D. and Johnson, M.E. (1981). A family of distributions for modelling nonelliptically symmetric multivariate data, *Journal of the Royal Statistical Society, Series B*, **43**, 210-218.

Crowder, M. (1990). On some nonregular tests for a modified Weibull model, *Biometrika*, **77**, 499-506.

Downton, F. (1970). Bivariate exponential distributions in reliability theory, *Journal of the Royal Statistical Society, Series B*, **32**, 408-417.

Frank, M.J. (1979). On the simultaneous associativity of $F(x,y)$ and $x+y-F(x,y)$, *Aequationes Math.*, **19**, 194-226.

Freund, J.E. (1961). A bivariate extension of the exponential distribution, *Journal of the American Statistical Association*, **56**, 971-977.

Friday, D.S. and Patil, G.P. (1977). A bivariate exponential model with applications to reliability and computer generation of random variables, In *Theory and Applications of Reliability*, Vol. 1 (Eds., C.P. Tsokos and I. Shimi), Academic Press, New York.

Gumbel, E.J. (1960). Bivariate exponential distributions, *Journal of the American Statistical Association*, **55**, 698-707.

Gupta, R.C., Mehrotra, K.G., and Michalek, J.E. (1984). A small sample test for an absolutely continuous bivariate exponential model, *Communications in Statistics - Theory and Methods*, **13**, 1735-1740.

Hall, W.J. and Mathiason, D.J. (1990). On large sample estimation and testing in parametric models, *International Statistical Review*, **58**, 77-97.

Hanagal, D. (1991). Large sample test of independence and symmetry in the multivariate exponential distribution, *Journal of the Indian Statistical Association*, **29**, 89-93.

Hanagal, D. (1993). Some inference results in an absolutely continuous multivariate exponential model of Block, *Statistics & Probability Letters* (To appear).

Hanagal, D. and Kale, B.K. (1991a). Large sample tests of λ_3 in the bivariate exponential distribution *Statistics & Probability Letters*, **12**, 311-313.

Hanagal, D. and Kale, B.K. (1991b). Large sample tests of independence for absolutely continuous bivariate exponential distribution, *Communications in Statistics - Theory and Methods*, **20**, 1301-1313.

Hanagal, D. and Kale, B.K. (1993). Large sample tests for testing symmetry and independence in some bivariate exponential models, *Communications in Statistics - Theory and Methods* (To appear).

Hashino, M. (1985). Formulation of the joint return period of two hydrologic variates associated with a Poisson process, *Journal of Hydroscience and Hydraulic Engineering*, **3**, 73-84.

Hawkes, A.G. (1972). A bivariate exponential distribution with applications to reliability, *Journal of the Royal Statistical Society, Series B*, **34**, 129-131.

Hsieh, H.K. (1979). On asymptotic optimality of likelihood ratio tests for multivariate normal distributions, *Annals of Statistics*, **7**, 592-598.

Hyakutake, H. (1990). Statistical inferences on location parameters of bivariate exponential distributions, *Hiroshima Mathematical Journal*, **20**, 525-547.

John, S. (1971). Some optimal multivariate tests, *Biometrika*, **58**, 123-127.

Killeen, T.J., Hettmansperger, T.P., and Sievers, G.Z. (1972). An elementary theorem on probability of large deviations, *Annals of Mathematical Statistics*, **43**, 181-192. Correction, *Annals of Statistics* (1974), 1357.

Klein, J.P. and Basu, A.P. (1985). Estimating reliability for bivariate exponential distributions, *Sankhyā, Series B*, **47**, 346-353.

Kourouklis, S. (1988). Asymptotic optimality of likelihood ratio tests for exponential distributions under Type II censoring, *Australian Journal of Statistics*, **30**, 111-114.

Krishnamoorthy, A.S. and Parthasarathy, M. (1951). A multivariate gamma type distribution, *Annals of Mathematical Statistics*, **22**, 549-557.

Lu, J. and Bhattacharyya, G.K. (1991a). Inference procedures for bivariate exponential model of Gumbel, *Statistics & Probability Letters*, **12**, 37-50.

Lu, J. and Bhattacharyya, G.K. (1991b). Inference procedures for a bivariate exponential model of Gumbel based on life test of component and system, *Journal of Statistical Planning and Inference*, **27**, 383-396.

Marshall, A.W. and Olkin, I. (1967). A multivariate exponential distribution, *Journal of the American Statistical Association*, **62**, 30-44.

Mehrotra, K.G. and Michalek, J.E. (1976). Estimation of parameters and tests of independence in a continuous bivariate exponential distribution, *Unpublished manuscript*.

Moran, P.A.P. (1967). Testing for correlation between non-negative variates, *Biometrika*, **54**, 385-394.

Neyman, J. (1959). Optimal asymptotic tests of composite statistical hypotheses, In *Probability and Statistics* (Ed., W. Grenander), pp. 213-234, John Wiley & Sons, New York.

Paulson, A.S. (1973). A characterization of the exponential distribution and bivariate exponential distribution, *Sankhyā, Series A*, **35**, 69-78.

Proschan, F. and Sullo, P. (1974). Estimating the parameters of a bivariate exponential distribution in several sampling situations, In *Reliability and Biometry* (Eds., F. Proschan and R.J. Serfling), pp. 423-440, SIAM, Philadelphia.

Proschan, F. and Sullo, P. (1976). Estimating the parameters of a multivariate exponential distribution, *Journal of the American Statistical Association*, **71**, 465-472.

Rachev, S.T. and SenGupta, A. (1992). Geometric stable distributions and Laplace-Weibull mixtures, *Statistics and Decisions*, **10**, 251-271.

Raftery, A.E. (1984). A continuous multivariate exponential distribution, *Communications in Statistics - Theory and Methods*, **13**, 947-965.

Samanta, M. (1986). On asymptotic optimality of some tests for exponential distributions, *Australian Journal of Statistics*, **28**, 164-172.

Sarkar, S.K. (1987). A continuous bivariate exponential distribution, *Journal of the American Statistical Association*, **82**, 667-675.

SenGupta, A. (1991). A review of optimality of multivariate tests, In "Special Issue on Multivariate Optimality and Related Topics" (Eds., S.R. Jammalamadaka and A. SenGupta), *Statistics & Probability Letters*, **12**, 527-535.

SenGupta, A. (1993). On construction of optimal tests in several multivariate exponential distributions, *Submitted for publication*.

SenGupta, A. and Pal, C. (1991). Locally optimal tests for no contamination in standard symmetric multivariate normal mixtures, In "Special Issue on Reliability Theory" (Eds., A.P. Basu), *Journal of Statistical Planning and Inference*, **29**, 145-155.

SenGupta, A. and Pal, C. (1993). Optimal tests for no contamination in symmetric multivariate normal mixtures, *Annals of the Institute of Statistical Mathematics*, **45**, 137-146.

SenGupta, A. and Vermeire, L. (1986). Locally optimal tests for multiparameter hypotheses, *Journal of the American Statistical Association*, **81**, 819-825.

Sinha, B.K. and Sinha, B.K. (1993). An application of bivariate exponential models and related inference, *Journal of Statistical Planning and Inference* (To appear).

Weier, D.R. (1981). Bayes estimation for a bivariate survival model based on exponential distributions, *Communications in Statistics - Theory and Methods*, **10**, 1415-1427.

Weier, D.R. and Basu, A.P. (1980). Testing for independence in multivariate exponential distributions, *Australian Journal of Statistics*, **22**, 276-288.

Weier, D.R. and Basu, A.P. (1981). On tests of independence under bivariate exponential models, In *Statistical Distributions in Scientific Work*, Vol. 5 (Eds., C. Taillie, G.P. Patil, and B.A. Baldessari), pp. 169-180, Reidel, Dordrecht.

Wijsman, R.A. (1967). Cross-sections of orbits and applications to densities of maximal invariants, In *Proceedings of the Fifth Berkeley Symposium*, Vol. 1, pp. 389-400.

CHAPTER 23

Accelerated Life Testing with Applications

Asit P. Basu[1]

23.1 Introduction

Accelerated life testing of a product is often used to reduce test time by subjecting the product to a higher stress. The resulting data is analyzed, and information about the performance of the product at normal usage condition is obtained. Accelerated testing usually involves subjecting test items to conditions more severe than encountered in normal use. This results in shorter test times, reduced costs, and decreased mean lifetimes for test items. In engineering applications, accelerated test conditions are produced by testing items at higher than normal temperature, voltage, pressure, load, etc. In bioassay problems, accelerated life tests arise when large doses of some chemical or radiological agent are given. In both cases, the data collected at higher stresses are used to extrapolate to some lower stress where testing is not feasible. In most engineering applications, a usage or design stress is known. In biological applications, an additional problem is estimating a safe dose level for the agent.

Several authors have considered the problem of accelerated life tests assuming a single mode of failure. For a bibliography, see Mann (1972), Mann, Schafer and Singpurwalla (1974), and Nelson (1982, 1990). The underlying failure distributions considered are the exponential, Weibull, normal, and log-normal.

In this chapter we present some recent results where the distribution assumed is the exponential distribution. First, we consider the case when there is a single cause of failure. Later, we consider problems where multiple causes of failure are present. The problem of competing risks, that is, problems where competing mechanisms of failure are present, occurs quite naturally. As an example, consider a p component series system where the system fails as soon as one of the p components fails. Section 23.2 describes some acceleration models, where a single cause of failure is present. Section 23.3 considers a corresponding estimation problem. The problem of competing mechanisms of failure is described in Section 23.4. Bivariate accelerated testing is considered in Section 23.5.

[1] University of Missouri-Columbia, Columbia, Missouri

23.2 The Models

Let $f(t, \lambda)$ be the probability density function (pdf) of the time to failure of an item at risk in an environment characterized by a vector of stresses $\boldsymbol{V}$. Let

$$\lambda = g(\boldsymbol{V}, \boldsymbol{\alpha})$$

where the functional form g is known except for a vector $\boldsymbol{\alpha}$ of unknown constants. It is assumed that changing $\boldsymbol{V}$ affects the value of λ only and *not* the functional form of f. The following three models have been used extensively in the literature.

(a) The Power rule model is derived using kinetic theory and activation energy. Here,

$$\lambda = V^B \exp(A). \tag{23.2.1}$$

(b) The Arrhenius reaction rate model,

$$\lambda = \exp(A - B/V). \tag{23.2.2}$$

(c) The Eyring model for a single stress is derived from principles of quantum mechanics. In this case,

$$\lambda = V \exp(A - B/V). \tag{23.2.3}$$

Here, A and B are unknown parameters to be estimated in order to make inference about λ at use condition.

Basu and Klein (1985) have considered a more general model, which includes the above models as special cases. The model is given by

$$\lambda(\boldsymbol{V}, \boldsymbol{B}) = \exp\left(\sum_{l=0}^{k} \alpha_l \beta_l(V)\right), \tag{23.2.4}$$

where $\beta_0(V) = 1$ and $\beta_l(V)$, $l = 1, 2, \ldots, k$, are k functions of the stress. The β_l's are assumed to be nondecreasing in engineering contexts. The power rule model is a special case of (23.2.4) with $k = 1$, $\alpha_0 = A$, $\alpha_1 = B$, and $\beta_1(V) = \ln V$. Model (23.2.2) is obtained when $k = 1$, $\beta_1(V) = -1/V$, and (23.2.3) corresponds to the case when $k = 2$, $\beta_1(V) = -1/V$, $\beta_2(V) = \ln V$ and $\alpha_2 = 1$.

The underlying distribution is assumed to be exponential with probability density function

$$f(x) = \lambda e^{-\lambda x}, \qquad x > 0,\ \lambda > 0 \tag{23.2.5}$$

so that λ is the failure rate. We denote this by $X \sim e(\lambda)$.

23.3 Estimation for Power Rule Model

In this section we consider parametric estimates of the parameters using the power rule model. Let n_i items be put on test at constant application of stress V_i $(i = 1, 2, \ldots s)$. The test at stress V_i is terminated after r_i failures have occurred, with ordered failure times $x_{1i}, x_{2i}, \ldots, x_{ri}$. That is, let us assume that we have s independent Type-II censored samples where the ith sample is from $e(\lambda_i), i = 1, 2, \ldots, s$. Let $\theta_i = 1/\lambda_i$. It is well known that the unique minimum variance unbiased estimator of θ_i is

$$\hat{\theta}_i = \left[\sum_{j=1}^{r_i} x_{ji} + (n_i - r_i) x_{r_i} \right] \Big/ r_i \ , \tag{23.3.1}$$

where $E(\hat{\theta}_i) = \theta_i$ and $\text{Var}(\hat{\theta}_i) = \theta_i^2 / r_i$ (see Chapter 4).

The estimator $\hat{\theta}_i$ has a gamma density function $g(\hat{\theta}_i)$ with parameters θ_i and r_i, so that

$$g(\hat{\theta}_i) = \frac{\exp(-r_i \hat{\theta}_i / \theta_i)}{\Gamma(r_i)} \left(\frac{r_i}{\theta_i} \right)^{r_i} \hat{\theta}_i^{r_i - 1} , \tag{23.3.2}$$

$\hat{\theta}_i \geq 0, r_i \geq 1$. Reparametrizing the power rule model (23.2.1) in terms of θ, with $C = e^{-A}$, we have

$$\theta_i = \frac{C}{V_i^B}, \quad C > 0, \quad i = 1, 2, \ldots, p \ . \tag{23.3.3}$$

Using (23.3.2) the likelihood function for estimating C and B assuming that the θ_i's are independent, is given by

$$L = L(B, C) = \prod_{i=1}^{s} \frac{1}{\Gamma(r_i)} \left[\frac{r_i}{C} V_i^B \right]^{r_i} \hat{\theta}_i^{r_i - 1} \exp \left[\frac{-r_i \hat{\theta}_i}{C} V_i^B \right] . \tag{23.3.4}$$

The maximum likelihood estimators $\hat{B}$ and $\hat{C}$ of B and C, respectively, are obtained by maximizing L in (23.3.4). The invariance principle of maximum likelihood estimators is used to estimate θ at use stress V_u. Asymptotic normality of maximum likelihood estimators is then used to make inference about θ at use stress V_u.

The parameters B and C can also be estimated using the least-squares method. Let

$$\hat{\theta}_i = \theta_i \epsilon_i$$

for $i = 1, 2, \ldots, s$ where the ϵ_i's are random errors. From (23.3.3),

$$\ln \hat{\theta}_i = \ln C - B \ln V_i + \ln \epsilon_i .$$

Define $\boldsymbol{\alpha}' = (\ln C, -B)$, $\boldsymbol{\epsilon}' = (\ln \epsilon_1, \ln \epsilon_2, \ldots, \ln \epsilon_s)$, $\boldsymbol{Y}' = (\ln \hat{\theta}_1, \ln \hat{\theta}_2, \ldots, \ln \hat{\theta}_s)$, and

$$\boldsymbol{X} = \begin{pmatrix} 1 & \ln V_1 \\ 1 & \ln V_2 \\ \vdots & \vdots \\ 1 & \ln V_s \end{pmatrix} .$$

The model of interest is

$$\boldsymbol{Y} = \boldsymbol{X\alpha} + \boldsymbol{\epsilon} \ .$$

Assume that conditions of the generalized Gauss-Markoff theorem are satisfied. That is, assume

(1) $E(\boldsymbol{\epsilon}) = \boldsymbol{0}$

(2) $E(\boldsymbol{\epsilon\epsilon}') = \sigma^2 \boldsymbol{V}$,

where σ^2 is a constant and $\boldsymbol{V}$ is a known positive-definite matrix. If, in addition, the vector $\boldsymbol{\epsilon}$ is normal, then $\log \hat{\theta}$ will be normally distributed.

23.4 Competing Causes of Failure

Most of the works on accelerated life tests assume that a product has only a single mode of failure. However, there are many problems where there are several causes of failure for the same component. For each item, data consist of the time at which the item fails and knowledge of which component failed. This is the competing risks problem. Identifiability of parameters assuming independence is well known. See Basu and Ghosh (1980), Basu and Klein (1982), and Chapter 29 for a survey of identifiability problems. The problem here is to estimate the parameters of the failure distribution of each component at normal use conditions from data collected in an accelerated life test.

Let $V_1, V_2, \ldots, V_s$ be the stress levels at which the accelerated life test is to be conducted. At each stress level we assume that the item survives until one of its p components fails. Let $X_{i1}, X_{i2}, \ldots, X_{ip}$ denote the life lengths of the p components of an item on test at stress V_i. Assume that the X_{ij}'s are independent exponential random variables with hazard rate (23.2.1). That is, let

$$\lambda_{ij} = V_i^{Bj} \exp(A_j), \quad j = 1, \ldots, p, \quad i = 1, \ldots, s. \tag{23.4.1}$$

For an item put on test at stress V_i, we observe Y_i = minimum $(X_{i1}, \ldots, X_{ip})$ and an indicator variable which describes which of the p components failed. The method of maximum likelihood is used to find estimators of A_j and B_j for various censoring schemes.

Type I Censoring

At stress level V_i, n_i items are put on test, $i = 1, \ldots, s$. Testing is continued until some fixed time τ_i. These τ_i's may differ from one stress level to another to allow for increased testing time at low stresses. Such a testing scheme is a Type-I censored accelerated life test.

At stress V_i suppose r_i items have failed prior to time τ_i, $i = 1, \ldots, s$. Let $Y_{i1}, Y_{i2}, \ldots, Y_{ir_i}$ denote the corresponding failure times. That is, $Y_{ik} = \min(X_{i1k}, X_{i2k}, \ldots, X_{ipk})$ where X_{ijk} is the lifetime of the jth component of the kth item which failed prior to time τ_i at stress V_i, $j = 1, \ldots, p$, $k = 1, \ldots, r_i$, $l = 1, \ldots, s$. Let m_{ij} denote the number of items which failed due to failure of component j, $j = 1, \ldots, p$. Note that $r_i = \sum_{j=1}^{p} m_{ij}$. Define the total time on test by

$$T_i = \sum_{k=1}^{r_i} Y_{ik} + (n_i - r_i)\tau_i, \qquad i = 1, \ldots, p\,. \tag{23.4.2}$$

Type II Censoring

For Type-II censoring, testing continues until a preassigned number of failures is observed. At each stress level, V_i, $i = 1, \ldots, s$, n_i items are put on test. Let Y_{ik} denote the failure time of the kth item put on test, as in the case of Type-I censoring. For this testing scheme, testing at stress V_i continues until a preassigned number, r_i, of items has failed. Let $Y_{i(1)}, \ldots, Y_{i(r_i)}$ be the ordered failure times of the r_i failures, and let m_{ij} denote the number of items which fail due to failure of component j, $j = 1, \ldots, p$. For

Type-II censoring, the total time on test is

$$T_i = \sum_{k=1}^{r_i} Y_{i(k)} + (n_i - r_i)Y_{i(r_i)}, \qquad i = 1, 2, \ldots, s \ . \tag{23.4.3}$$

We may also consider other types of censoring.

Details of maximum likelihood method for estimating the parameters A_i and B_i are given in Klein and Basu (1982).

Estimation of Population Parameters at use Conditions

Suppose an accelerated life test has been conducted by one of the censoring schemes discussed. Let $\hat{A}_j$ and $\hat{B}_j$ be the maximum likelihood estimators of A_j and B_j, $j = 1, \ldots, p$. Let $\sum_j$ be the inverse of information matrix obtained. Let $\hat{\sum}_j$ be the corresponding estimator of $\sum_j$. Based on this information we wish to estimate each component's mean survival time and survival function at the use conditions. In addition, we shall obtain estimators of the survival functions for the entire system of components or a subsystem of components under use conditions.

Let V_u denote the stress under use conditions. Let λ_{uj} and μ_{uj} denote the hazard rate and mean survival time at this stress. By (23.4.1)

$$\lambda_{uj} = V_u^{B_j} \exp A_j , \qquad j = 1, \ldots, p \tag{23.4.4}$$

and

$$\mu_{uj} = V_u^{-B_j} \exp(-A_j), \qquad j = 1, \ldots, p \ . \tag{23.4.5}$$

By the invariance principle of maximum likelihood estimators, the maximum likelihood estimators of λ_{uj} and μ_{uj} are

$$\hat{\mu}_{uj} = V_u^{\hat{B}_j} \exp(\hat{A}_j), \qquad j = 1, \ldots, p \tag{23.4.6}$$

and

$$\hat{\mu}_{uj} = V_u^{-\hat{B}_j} \exp(-\hat{A}_j), \quad j = 1, \ldots, p, \tag{23.4.7}$$

respectively. For large sample sizes, at all the stress levels, $(\hat{A}_j, \hat{B}_j)$ are asymptotically bivariate normally distributed with mean vector (A_j, B_j) and covariance matrix $\sum_j$, $j = 1, \ldots, p$. Since $\ln \hat{\lambda}_{uj}$ is linear in $\hat{A}_j$ and $\hat{B}_j$, it has an asymptotic normal distribution with mean $\ln \lambda_{uj}$ and variance

$$\sigma_{uj}^2 = \sigma_{11}^{(j)} + \sigma_{22}^{(j)} (\ln V_u)^2 + 2\sigma_{12}^{(j)} \ln V_u \ . \tag{23.4.8}$$

Similarly, $\ln \hat{\mu}_{uj}$ is asymptotically normal with mean $\ln \mu_{uj}$ and variance σ_{uj}^2.

A simulation study was conducted by Klein and Basu (1982) to compare the effects of censoring on the estimation procedure. The estimates seem to perform reasonably well even with moderate sample sizes.

23.5 Bivariate Accelerated Life Tests

In this section we consider bivariate failure data when there are two failure times for each sampling unit. Let X and Y denote the failure times of a two-component system with component lifetimes X and Y. The s stresses $V_1, V_2, \ldots, V_s$ are selected and life tests are conducted at constant application of the selected stresses. We wish to use this information to make inference about the component lifetimes under normal stress conditions.

We assume (X, Y) follow a specified bivariate exponential distribution like the BVE model of Marshall-Olkin (1967) or the Block-Basu (1974) ACBVE model. See Chapter 20 for a description of these.

Ebrahimi (1987) has considered an accelerated testing problem for the BVE model and Basu and Ebrahimi (1987) have considered the ACBVE model. We summarize here the results of Basu and Ebrahimi.

Let (X, Y) follow the ACBVE distribution. Here, the joint density of (X, Y) at stress level V is given by

$$f(x, y; V_j) = f_j(x, y) = \begin{cases} \frac{\lambda_{1j}\lambda_j(\lambda_{2j}+\lambda_{3j})}{\lambda_{1j}+\lambda_{2j}} \exp\{-\lambda_{1j}x - (\lambda_{2j} + \lambda_{3j})y\} & \text{if } x < y \\ \frac{\lambda_{2j}\lambda_j(\lambda_{1j}+\lambda_{3j})}{\lambda_{1j}+\lambda_{2j}} \exp\{-(\lambda_{1j} + \lambda_{3j})x - \lambda_{2j}y\} & \text{if } x > y, \end{cases} \tag{23.5.1}$$

where $\lambda_j = \lambda_{1j} + \lambda_{2j} + \lambda_{3j}$, $j = 0, 1, 2, \ldots, s$. Here, V_0 is the stress at use condition. We also assume that the correlation coefficient $\rho(X, Y)$ between X and Y is the same at all stress levels and

$$\lambda_{ij} = c_i V_j^p, \qquad i = 1, 2, 3;\ j = 0, 1, \ldots, s$$

where c_1, c_2, c_3 and p are constants to be determined. It can be shown that

$$E(X) = \frac{(c_1 + c_2 + c_3)(c_1 + c_2) + c_2 c_3}{(c_1 + c_2 + c_3)(c_1 + c_2)(c_1 + c_3)} V_j^{-p} \tag{23.5.2}$$

and

$$E(Y) = \frac{(c_1 + c_2 + c_3)(c_1 + c_2) + c_1 c_3}{(c_1 + c_2 + c_3)(c_1 + c_2)(c_2 + c_3)} V_j^{-p}. \tag{23.5.3}$$

Using the above model we can estimate c_1, c_2, c_3 and p by the method of maximum likelihood. The relevant details are given in Basu and Ebrahimi (1987).

References

Basu, A.P. and Ebrahimi, N. (1987). On a bivariate accelerated life test, *Journal of Statistical Planning and Inference*, **16**, 297-304.

Basu, A.P. and Ghosh, J.K. (1980). Identifiability of distributions under competing risks and complementary risks models, *Communications in Statistics - Theory and Methods*, **A9**, 1515-1525.

Basu, A.P. and Klein, J.P. (1982). Some recent results in competing risks theory, *Proceedings on Survival Analysis*, IMS monograph series No. **2**, 216-229.

Basu, A.P. and Klein, J.P. (1985). A model for life testing and for estimating safe dose levels, In *Statistical Theory and Data Analysis* (Ed., Matusita), North-Holland, Amsterdam.

Block, H. and Basu, A.P. (1974). A continuous bivariate exponential extension, *Journal of the American Statistical Association,* **69,** 1031-1037.

Ebrahimi, N. (1987). Analysis of bivariate accelerated life test data from the bivariate exponential of Marshall and Olkin, *American Journal of Mathematical and Management Sciences,* **6,** 175-190.

Klein, J.P. and Basu, A.P. (1982). Accelerated life testing under competing exponential failure distributions, *IAPQR Transactions,* **7,** 1-20.

Mann, N.R. (1972). Design of over-stress life-test experiments when failure times have a two-parameter Weibull distribution, *Technometrics,* **14,** 437-451.

Mann, N.R., Schafer, R.E., and Singpurwalla, N.D. (1974). *Methods for Statistical Analysis of Reliability and Life Data,* John Wiley & Sons, New York.

Marshall, A.W. and Olkin, I. (1967). A multivariate exponential distribution, *Journal of the American Statistical Association,* **62,** 30-44.

Nelson, W. B. (1982). *Applied Life Data Analysis,* John Wiley & Sons, New York.

Nelson, W. (1990). *Accelerated Testing, Statistical Models, Test Plans, and Data Analyses,* John Wiley & Sons, New York.

CHAPTER 24

System Reliability and Associated Inference

S. Zacks[1]

24.1 Introduction

System reliability is the probability that a system will function correctly throughout the mission period. Thus, reliability is a decreasing function of time. We distinguish between the reliability of unrepairable systems and those of repairable ones. Repairable systems are being renewed each time a repair takes place. For this reason, it is better to talk about the *availability* of repairable systems, $A(t)$, which is the probability that the system is up at time t. The reliability function, $R(t)$, is the probability that a system will not fail for at least t units of time, after its up cycle starts. Non-repairable systems have only one up cycle. For such systems the reliability and availability functions coincide. For further reading on this subject see Zacks (1992), Ascher and Feingold (1984), Barlow and Proschan (1965, 1975), Gnedenko, Belyayev, and Solovyev (1969) and others.

The present chapter is devoted to system reliability, and the associated statistical inference. In the present chapter we define a system as an integrated collection of subsystems, modules or components, which function together to attain specific results. Systems fail because some or all of its components do. We distinguish between the *time till failure* (TTF) of the components and that of the system, which will be called global TTF.

In Section 24.2 we discuss the connection between the reliability of components of a system and that of the system as a whole. The system reliability is given as a function of the reliability of its components. It is not always easy or simple to formulate this functional dependence. We will focus attention on the relatively simpler cases of systems with independent components' TTF, and simple coherent structure functions, like those of series, parallel, crosslinked, etc. We further assume that the components' TTF are exponentially distributed random variables. The theory can be easily generalized to more general TTF distributions, such as the Weibull, extreme value and others. Since

[1]Binghamton University, Binghamton, New York

the present volume is devoted to the exponential distributions, we restrict attention to the exponential distributions. When the components have exponential TTF, the system's TTF is generally not exponentially distributed. This makes a big difference in the statistical inference. If random samples are available of the TTF of the components, the inference might be simpler and more efficient than if the information available is only on the global TTF of the whole system.

In Section 24.3 we discuss the problem of estimating the reliability of a system, whose functional form is explicitly given, based on samples of equal size, n, of TTF of the components. We start with maximum likelihood estimators (MLE). The problem of best unbiased estimation is discussed too. It is shown that the most convenient, and asymptotically most efficient method of estimation is that of the MLE. The modification required for censored observation is discussed in Section 24.3.3.

In Section 24.4 we show how the bootstrapping procedure of sampling resampling [Efron (1982)] can be applied to obtain bootstrapping estimates of the standard-error of the MLE. The procedure yields also confidence intervals for the reliability function. This is a computer intensive technique, which replaces the analytical technique required to obtain the asymptotic variance of the MLE. It could be especially useful for complex systems, when large samples of TTF values are available for the components. We demonstrate numerically that, in the case of a double crosslinked system of six components, the bootstrapping with samples of size $n = 20$, on each component, yields biased estimators. If the samples are of size $n = 50$, the bootstrapping estimates are quite good.

Section 24.5 is devoted to the problem of estimating system reliability, when only global observations on the system's TTF are available. The observations might be censored. As mentioned earlier, the distribution of the global TTF, when the components have exponential failure times, is generally not exponential. Non-parametric survival analysis is often performed. The literature on this topic is vast [see Kalbfleisch and Prentice (1980)].

We present the results of Zacks (1986) on the estimation of the system reliability function, when the system is an r-out-of-k identical component, and the observations are global and time censored. We present the MLE of the common components' mean time till failure, θ. The formula for the asymptotic variance of the MLE of θ is quite complicated. This formula is needed to obtain the asymptotic variance of the MLE of the system reliability. Again, if the number of observations is sufficiently large, bootstrapping could yield good approximation. In practice people often fit to the global failure times a convenient distribution, like Weibull, log-normal, etc. The analysis is then performed on the fitted model and not on the complicated one. An example of such a case is the following. Suppose that we have a 27-out-of-30 system. The components are independent and identically distributed as exponential with mean $\theta = 100$. The global TTF are distributed like the 27th order statistic from a sample of size 30. Simulating a random sample of 100 such independent TTF's, we obtained values to which a log-normal distribution fitted excellently. An analysis based on the log-normal distribution is ad-hoc and has no relevance to the system under consideration. Section 24.6 is devoted to the problem of detecting a shift to the wear-out phase of components. In Section 24.7 we discuss a problem of detecting a shift from the phase of "infant mortality" to the mature (stable) phase of a system. Both in Sections 24.6 and 24.7 Bayes sequential procedures are developed. The results of Section 24.7 have applications in determining the duration of a "burn-in" procedure [see Jensen and Petersen (1982)].

24.2 System Reliability Functions

We consider in the present section the problem of expressing the reliability of a system as a function of the reliability of its components (subsystems, modules, etc.). Let $R_1, R_2, \cdots, R_k$ be the reliability values of k components of a system. We wish to express the reliability of the system as a function $R_{\text{sys}} = \psi(R_1, \cdots, R_k)$. This may not be a simple task, especially when the systems are complex, and the operations of their components are not independent. In the present chapter we consider a relatively simple structure function, which will be discussed below. The reader is referred for further details to Barlow and Proschan (1975), Ireson (1982), Gertsbakh (1989) among others. We are especially interested in reliability functions which are *coherent*, i.e., $\psi(R_1, \cdots, R_k)$ is a non-decreasing function in each one of its components (variables). We consider structure functions of systems with *independent* components. We say that a system (subsystem) is *structured in series*, if the failure of any component results in immediate system failure. In such a case, under independence,

$$R_{\text{sys}} = \psi_s(R_1, \cdots, R_n) = \prod_{i=1}^{n} R_i \ . \qquad (24.2.1)$$

A system is *structured in parallel* if the system fails when all its components fail. In such a case

$$R_{\text{sys}} = \psi_p(R_1, \cdots, R_n) = 1 - \prod_{i=1}^{n}(1 - R_i) \ . \qquad (24.2.2)$$

Generally, system structures can be described by structure functions which are compositions of series and parallel structure functions. For example consider a crosslinked system of four components, Zacks (1992, p. 48). The structure function of this system is

$$R_{\text{sys}} = R_2R_3 + R_3R_4 - R_2R_3R_4 + (1 - R_3)R_1R_2 \ . \qquad (24.2.3)$$

It is easy to check that the function (24.2.3) is strictly increasing in each one of its variables, $R_1, \cdots, R_4$, over the domain, $0 < R_1, \cdots, R_4 < 1$. Thus, it represents a coherent system. Assuming that the life length of each component of the given system is a random variable having exponential distribution, with *mean time till failure* (MTTF) θ_i $(i = 1, \cdots, 4)$, the reliability function of the crosslinked system described above is, as a function of time,

$$\begin{aligned} R_{\text{sys}}(t;\boldsymbol{\theta}) &= \exp\left\{-t\left(\frac{1}{\theta_1} + \frac{1}{\theta_2}\right)\right\} + \exp\left\{-t\left(\frac{1}{\theta_2} + \frac{1}{\theta_3}\right)\right\} \\ &\quad + \exp\left\{-t\left(\frac{1}{\theta_3} + \frac{1}{\theta_4}\right)\right\} - \exp\left\{-t\left(\frac{1}{\theta_1} + \frac{1}{\theta_2} + \frac{1}{\theta_3}\right)\right\} \\ &\quad - \exp\left\{-t\left(\frac{1}{\theta_2} + \frac{1}{\theta_3} + \frac{1}{\theta_4}\right)\right\} . \end{aligned} \qquad (24.2.4)$$

In a similar fashion one can express the reliability function of more complicated systems. We emphasize again that reliability functions like (24.2.4) are valid when the TTF of different components are *independent* random variables. In cases of dependent TTF, one can attempt in certain situations to obtain bounds for the reliability function [see Barlow and Proschan (1975)].

24.3 Estimating System Reliability

In the present section we discuss the problem of estimating a reliability function $R_{\text{sys}}(t;\boldsymbol{\theta})$, where the components' MTTF $\theta_1, \cdots, \theta_k$ are unknown. A random sample $\{T_{i1}, \cdots, T_{in}\}$ of times till failure is given for each component of the system. A minimal sufficient statistic is $\overline{\boldsymbol{T}} = (\overline{T}_1, \cdots, \overline{T}_k)$, where $\overline{T}_i = \frac{1}{n}\sum_{j=1}^{n} T_{ij}$ is the mean of the i-th sample. Estimators of $R_{\text{sys}}(t;\boldsymbol{\theta})$ are functions of $\overline{\boldsymbol{T}}$.

24.3.1 Maximum Likelihood Estimators

Maximum likelihood estimator (MLE) of $\boldsymbol{\theta} = (\theta_1, \cdots, \theta_k)'$ is $\widehat{\boldsymbol{\theta}} = \overline{\boldsymbol{T}}$ [see Zacks (1992, p. 135)]. Moreover, the inverse of the Fisher Information Matrix (FIM), $II(\boldsymbol{\theta})$, is the diagonal $k \times k$ matrix $\Sigma(\boldsymbol{\theta}) = \frac{1}{n}\,\text{diag}(\theta_i^2; i = 1, \cdots, k)$. Thus, by the invariance principle for MLE [see Zacks (1992, p. 114)], the MLE of $R_{\text{sys}}(t;\boldsymbol{\theta})$ is

$$\widehat{R}_{\text{sys}}(t;\boldsymbol{\theta}) = R_{\text{sys}}(t;\widehat{\boldsymbol{\theta}}) \ . \tag{24.3.1}$$

Furthermore, whenever $R_{\text{sys}}(t;\boldsymbol{\theta})$ is an analytic function of $\theta_1, \cdots, \theta_k$ (it is sufficient to require continuous second-order partial derivatives) of the form

$$R_{\text{sys}}(t;\boldsymbol{\theta}) = \psi(e^{-t/\theta_1}, \cdots, e^{-t/\theta_k}),$$

as, for example, in Eq. (24.2.4), the MLE, (24.3.1) is strongly consistent, asymptotically normal estimator of $R_{\text{sys}}(t;\boldsymbol{\theta})$, with asymptotic variance

$$AV\{\widehat{R}_{\text{sys}}(t;\boldsymbol{\theta})\} = \frac{1}{n}\sum_{i=1}^{k} \theta_i^2 D_i^2(t;\boldsymbol{\theta}) \ , \tag{24.3.2}$$

where

$$D_i(t;\boldsymbol{\theta}) = \frac{\partial}{\partial \theta_i} R_{\text{sys}}(t;\boldsymbol{\theta}), \quad i = 1, \cdots, k \ . \tag{24.3.3}$$

We illustrate this on the crosslinked system, whose reliability function is given by (24.2.4). The corresponding asymptotic variance of the MLE is

$$\begin{aligned} &AV\{\widehat{R}(t;\theta_1, \cdots, \theta_4)\} \\ &= \frac{t^2}{n}\left\{\frac{1}{\theta_1^2}\left[\exp\left\{-t\frac{\theta_1+\theta_2}{\theta_1\theta_2}\right\} - \exp\left\{-t\frac{\theta_1\theta_2+\theta_1\theta_3+\theta_2\theta_3}{\theta_1\theta_2\theta_3}\right\}\right]^2\right. \\ &\quad + \frac{1}{\theta_2^2}\left[\exp\left\{-t\frac{\theta_1+\theta_2}{\theta_1\theta_2}\right\} + \exp\left\{-t - \frac{\theta_2+\theta_3}{\theta_2\theta_3}\right\}\right. \end{aligned}$$

$$- \exp\left\{ -t\frac{\theta_1\theta_2+\theta_1\theta_3+\theta_2\theta_3}{\theta_1\theta_2\theta_3} \right\} - \exp\left\{ -t\frac{\theta_2\theta_3+\theta_2\theta_4+\theta_3\theta_4}{\theta_2\theta_3\theta_4} \right\}\Bigg]^2$$
$$+ \frac{1}{\theta_3^2}\Bigg[\exp\left\{ -t\frac{\theta_2+\theta_3}{\theta_2\theta_3} \right\} + \exp\left\{ -t\frac{\theta_3+\theta_4}{\theta_3\theta_4} \right\}$$
$$- \exp\left\{ -t\frac{\theta_1\theta_2+\theta_1\theta_3+\theta_2\theta_3}{\theta_1\theta_2\theta_3} \right\} - \exp\left\{ -t\frac{\theta_2\theta_3+\theta_2\theta_4+\theta_3\theta_4}{\theta_2\theta_3\theta_4} \right\}\Bigg]^2$$
$$+ \frac{1}{\theta_4^2}\Bigg[\exp\left\{ -t\frac{\theta_3+\theta_4}{\theta_3\theta_4} \right\} - \exp\left\{ -t\frac{\theta_2\theta_3+\theta_2\theta_4+\theta_3\theta_4}{\theta_2\theta_3\theta_4} \right\}\Bigg]^2 \Bigg\} . \quad (24.3.4)$$

For the special case of $t = 100$, $\theta_1 = 100$, $\theta_2 = \theta_3 = 200$ and $\theta_4 = 250$, $R_{\text{sys}}(t;\boldsymbol{\theta}) = 0.6156$ and the asymptotic variance (24.3.4) is $AV = 0.06125/n$. When the values of $\theta_1, \cdots, \theta_k$ are unknown, their MLE, $\widehat{\theta}_i$, can be substituted in the formula of the asymptotic variance. This yields an MLE of the asymptotic variance.

We see that even in a relatively simple structure, as that of the above crosslinked system, the formula for the asymptotic variance is quite long. In Section 24.4 we will discuss the bootstrapping procedure and its estimates of the variance of the MLE. This is a simple computer intensive technique which can replace the cumbersome analysis.

24.3.2 Uniformly Minimum Variance Unbiased Estimators

It is well known that the uniformly minimum variance unbiased (UMVU) estimator of $e^{-t/\theta}$ is $(1 - \frac{t}{T})^n$ [see Zacks and Even (1966)]. In certain cases, where the system reliability function can be expressed as

$$R_{\text{sys}}(t;\boldsymbol{\theta}) = \sum_{(i_1,\cdots,i_k)} c(i_1,\cdots,i_k) \exp\left\{ -t\sum_{l=1}^{k} \frac{i_l}{\theta_l} \right\} , \quad (24.3.5)$$

where $i_l = 0, 1$, such as Eq. (24.2.4), the UMVU of $R_{\text{sys}}(t;\boldsymbol{\theta})$ is, in the independence case,

$$\widetilde{R}_{\text{sys}}(t;\boldsymbol{\theta}) = \sum_{(i_1,\cdots,i_k)} c(i_1,\cdots,i_k) \prod_{l=1}^{k} \left(1 - \frac{t}{T_l}\right)^{i_l n} . \quad (24.3.6)$$

There are, however, cases in which one cannot obtain the UMVU estimator of $R_{\text{sys}}(t;\boldsymbol{\theta})$ by substituting the corresponding UMVU estimators of its components. Example of such a case is that of an r-out-of-k system of identical components, which has the reliability function

$$R_{\text{sys}}(t;\theta) = \sum_{j=r}^{k} \binom{k}{j} e^{-jt/\theta}(1 - e^{-t/\theta})^{k-j} . \quad (24.3.7)$$

A UMVU estimator of (24.3.7) when n observations are available on the times till failure of a component, is (when $n \geq k$) $P\left\{T_{r:k}^{(k)} > t \mid \sum_{i=1}^{n} T_i\right\}$, where $T_{r:k}^{(k)}$ is the rth order statistic in a subsample of size k. In Section 24.5 we will discuss the problem of estimating (24.3.7) when the observations are *system* failure times. The variance function of the UMVU is generally quite complicated, even in simple cases [see Zacks and Even (1966)].

Approximation to the variance function can be obtained by bootstrapping. Generally, the UMVU estimators of $R_{\text{sys}}(t;\boldsymbol{\theta})$ are asymptotically equivalent to the corresponding maximum-likelihood estimators. It seems advantageous, therefore, from various points of view, to use the MLE rather than the UMVU estimators, for estimating system reliability. Similar comments can be made about Bayesian estimation.

24.3.3 Maximum Likelihood Estimators Under Censoring

There are two types of censoring, time censoring (Type I) and frequency censoring (Type II). Consider here the case of frequency-censoring, in which the observations on the ith component terminate at the rth failure, $1 < r < n$. It is well known [see Zacks (1992, p. 127)] that the MLE of θ_i $(i = 1, \cdots, k)$ is

$$\widehat{\theta}_{r,n}^{(i)} = \frac{T_{r,n}^{(i)}}{r}, \tag{24.3.8}$$

where $T_{r,n}^{(i)}$ is the total time on test for the ith component, i.e.,

$$T_{r,n}^{(i)} = \sum_{j=1}^{r} T_{(j)}^{(i)} + (n-r)T_{(r)}^{(i)}, \tag{24.3.9}$$

and $0 < T_{(1)}^{(i)} < T_{(2)}^{(i)} < \cdots < T_{(r)}^{(i)}$ are the ordered failure times. Moreover, $\widehat{\theta}_{r,n}^{(i)} \sim \frac{\theta_i}{2r}\chi_{2r}^2$. Thus, $\text{Var}(\widehat{\theta}_{r,n}^{(i)}) = \theta_i^2/r$, $i = 1, \cdots, k$. Substituting $\widehat{\theta}_{r,n}^{(i)}$ for θ_i in $R_{\text{sys}}(t;\boldsymbol{\theta})$, one gets the MLE of the system reliability, with an asymptotic variance ($n \to \infty$ and $r \to \infty$) given by substituting r for n in Eq. (24.3.2).

If censoring is time censoring the situation is more complicated. An MLE of θ_i may not exist, if no failure of the ith component has been observed during the allotted time for the experiment.

24.4 Bootstrapping Estimates

In a bootstrapping procedure, a random sample of size n, *with replacement*, is drawn from each one of the original samples independently. Let $\{T_{ij}^*, j = 1, \cdots, n\}$, $i = 1, \cdots, k$ be the values of the bootstrapping samples, and $\overline{T}_i^* = \frac{1}{n}\sum_{j=1}^{n} T_{ij}^*$, their sample means. Let $E^*[\cdot]$ and $\text{Var}^*(\cdot)$ denote the bootstrapping expectation and variance. Obviously,

$$E^*[\overline{T}_i^*] = \overline{T}_i, \quad i = 1, \cdots, k \tag{24.4.1}$$

and

$$\text{Var}^*(\overline{T}_i^*) = \frac{\widehat{\sigma}_i^2}{n}, \quad i = 1, \cdots, k \tag{24.4.2}$$

where $\overline{T}_i$ is the mean of the original ith sample, and

$$\widehat{\sigma}_i^2 = \frac{1}{n}\sum_{j=1}^{n}(T_{ij} - \overline{T}_i)^2 \tag{24.4.3}$$

is the variance of the original ith sample. Given $\overline{T}_1^*, \cdots, \overline{T}_k^*$ we compute the bootstrapping estimator of the system reliability, namely,

$$R^* = R_{\text{sys}}(t; \overline{T}_1^*, \cdots, \overline{T}_k^*) \ . \tag{24.4.4}$$

This resampling procedure is repeated independently M times, yielding a sample $\{R_1^*, \cdots, R_M^*\}$ of bootstrapping estimates. Usually M is taken to be very large (in the thousands).

Let $F_M^*(R)$ denote the empirical distribution based on $R_1^*, \cdots, R_M^*$, namely,

$$F_M^*(R) = \frac{1}{M}\sum_{j=1}^{M} I\{R_j^* \le R\} \ . \tag{24.4.5}$$

The bootstrapping sample mean $\overline{R}_M^*$ and the bootstrapping sample variance

$$V_M^* = \frac{1}{M-1}\sum_{j=1}^{M}(R_j^* - \overline{R}_M^*)^2 \tag{24.4.6}$$

are $*$-unbiased and $*$-consistent estimators of $E^*[R_{\text{sys}}(t; \overline{T}_1^*, \cdots, \overline{T}_k^*)]$ and $\text{Var}^*(R_{\text{sys}}(t; \overline{T}_1^*, \cdots, \overline{T}_k^*))$. As in Section 24.3, let $D_i(t; \boldsymbol{\theta})$ denote the partial derivative of $R_{\text{sys}}(t; \boldsymbol{\theta})$, with respect to θ_i $(i = 1, \cdots, k)$. As before, we assume that $R_{\text{sys}}(t; \boldsymbol{\theta})$ has continuous second order partial derivatives $\frac{\partial^2}{\partial\theta_i\partial\theta_j}(t; \boldsymbol{\theta})$. For large sample size n, the delta method yields that

$$\text{Var}^*(R_{\text{sys}}(t; \overline{T}_1^*, \cdots, \overline{T}_k^*)) = \frac{1}{n}\sum_{j=1}^{n}\widehat{\sigma}_j^2 D_j^2(t; \overline{T}_1, \cdots, \overline{T}_k) + o\left(\frac{1}{n}\right) \ . \tag{24.4.7}$$

Moreover, assuming that the life distributions are exponential, $\overline{\boldsymbol{T}} \to \boldsymbol{\theta}$ a.s. as $n \to \infty$, $\widehat{\sigma}_j^2 \to \theta^2$ a.s. as $n \to \infty$, and

$$\text{Var}^*(R_{\text{sys}}(t; \overline{T}_1^*, \cdots, \overline{T}_k^*) = \frac{1}{n}\sum_{j=1}^{n}\theta_j^2 D_j^2(t; \boldsymbol{\theta}) + o\left(\frac{1}{n}\right) \ . \tag{24.4.8}$$

Comparing Eq. (24.4.7) with Eq. (24.3.2) we see that V_M^* is a consistent estimator of the asymptotic variance $\text{AV}(\widehat{R}_{\text{sys}}(t; \boldsymbol{\theta}))$. Notice that $(F_M^{*-1}(\alpha/2), F_M^{*-1}(1-\alpha/2))$ is a bootstrapping confidence interval, at level $1-\alpha$ for $R_{\text{sys}}(t; \boldsymbol{\theta})$.

We illustrate the bootstrapping procedure on the crosslinked system whose reliability function is given by (24.2.4). Four independent samples of size $n = 50$ each, were generated from exponential distributions with means $\theta_1 = 100$, $\theta_2 = \theta_3 = 200$ and $\theta_4 = 250$. The reliability of the system at $t = 100$ was estimated by bootstrapping the four samples, with $M = 1000$ repetitions. In Section 24.3 we have seen that the asymptotic variance of the MLE, under these conditions is AV$= 0.06125/50 = 0.001225$. The bootstrapping procedure yielded an estimate of AV, $V_M^* = 0.001307$. A 95% bootstrapping confidence interval for R_{sys} is $(0.514, 0.659)$. The true value of system reliability under these conditions is $R_{\text{sys}} = 0.616$.

Another example is that of a double-crosslinked system [see Zacks (1992, p. 53)] having a reliability function

$$\begin{aligned} R_{\text{sys}}(t;\theta_1,\cdots,\theta_6) &= \exp\left\{-\frac{t}{\theta_4}\right\}\left[\exp\left\{-\frac{t}{\theta_5}\right\}\left(\exp\left\{-\frac{t}{\theta_3}\right\}\right.\right. \\ &+ \left.\exp\left\{-\frac{t}{\theta_6}\right\} - \exp\left\{-t\left(\frac{1}{\theta_3}+\frac{1}{\theta_6}\right)\right\}\right) \\ &+ \left.\left(1-\exp\left\{-\frac{t}{\theta_5}\right\}\right)\exp\left\{-t\left(\frac{1}{\theta_2}+\frac{1}{\theta_3}\right)\right\}\right] \\ &+ \left(1-\exp\left\{-\frac{t}{\theta_4}\right\}\right)\exp\left\{-t\left(\frac{1}{\theta_1}+\frac{1}{\theta_2}+\frac{1}{\theta_3}\right)\right\}. \end{aligned} \tag{24.4.9}$$

We simulated six independent samples of size $n = 20$ from exponential distributions with means $\theta_1 = \theta_2 = 150$, $\theta_3 = \theta_4 = 200$, $\theta_5 = \theta_6 = 250$. We are interested in the system reliability at $t = 100$. The true reliability value is $R_{\text{sys}} = 0.479$. Bootstrapping with $M = 1000$ repetitions yields $\overline{R}_M^* = .3843$ with 95% confidence interval $(0.278, 0.482)$.

We see that the bootstrapping, based on samples of size $n = 20$, yields underestimates (negative bias). For $n = 50$ we obtain $R_M^* = 0.4651$ with a 95% confidence interval $(0.395, 0.529)$. This shows that, when the samples are sufficiently large, the bootstrapping estimates are quite accurate.

24.5 Estimating The Reliability Function of An r-out-of-k System: Censored Global Observations

In the present section we discuss the parametric estimation of the reliability function of an r-out-of-k system, whose components operate independently, and have identical exponential life time, with MTTF θ (unknown). The observations, however, are on times till failure of the whole system. Time till failure data on the components are not available. Furthermore, the observations are censored at time t_0. The theory presented here is based on the paper of Zacks (1986).

In an r-out-of-k system, the global failure time is that of the r-th failure among its k components. Thus, if $T_1, \cdots, T_k$ are the failure times of the k components, and $T_{1:k} < T_{2:k} < \cdots < T_{k:k}$ are the corresponding order statistics. The system (global) failure time is $S = T_{r:k}$. The observed random variable is $\widetilde{S} = \min\{t_0, S\}$. The distribution function of $\widetilde{S}$, $G_{r:k}(x;\theta,t_0)$, is absolutely continuous on $(0, t_0)$, with density

$$\begin{aligned} g_{r:k}(x;\theta,t_0) &= r\binom{k}{r}\frac{1}{\theta}\exp\left\{-(k-r+1)\frac{x}{\theta}\right\} \\ &\cdot\left(1-\exp\left\{-\frac{x}{\theta}\right\}\right)^{r-1} 1_{(0,t_0)}(x)\,, \end{aligned} \tag{24.5.1}$$

and has a jump at t_0 of height

$$P_\theta\{T_{r:k} \geq t_0\} = \mathrm{Bin}(r-1; k, 1-e^{-t_0/\theta}) \,, \tag{24.5.2}$$

where $\mathrm{Bin}(j; n, p)$ designates the distribution function of the binomial distribution, at j, with parameters n and p. Furthermore, the expected value and variance of $\widetilde{S}$ [see Zacks (1986)] are

$$M(\theta, t_0) = E[\widetilde{S}] = \theta \sum_{j=0}^{r-1} \frac{1}{k-j}[1 - \mathrm{Bin}(j; k, 1-e^{-t_0/\theta})] \,, \tag{24.5.3}$$

and

$$\begin{aligned} \sigma^2(\theta, t_0) &= \mathrm{Var}(\widetilde{S}) \\ &= \theta^2 \Big\{ 2 \sum_{j=0}^{r-1} \frac{1}{k-j} \sum_{l=0}^{j} \frac{1}{k-l} \Big[1 - \mathrm{Bin}(l; k, 1-e^{-t_0/\theta}) \Big] \\ &\quad -2\frac{t_0}{\theta} \sum_{j=0}^{r-1} \frac{1}{k-j} \mathrm{Bin}(j; k, 1-e^{-t_0/\theta}) \\ &\quad - \Big(\sum_{j=0}^{r-1} \frac{1}{k-1} [1 - \mathrm{Bin}(j; k, 1-e^{-t_0/\theta})] \Big)^2 \Big\} \,. \end{aligned} \tag{24.5.4}$$

Let $\widetilde{S}_1, \cdots, \widetilde{S}_n$ be a sample of n i.i.d. random variables. Let $I_j = I\{\widetilde{S}_j < t_0\}$, $j = 1, \cdots, n$, and $K_n = \sum_{j=1}^{n} I_j$ be the number of uncensored observations. It is shown in Zacks (1986) that if $K_n \geq 1$, there exists a unique finite MLE of θ, given by the root of the equation

$$\begin{aligned} \widehat{\theta} &= \frac{n}{K_n} \sum_{i=1}^{n} I_i \widetilde{S}_i - \frac{r-1}{K_n} \sum_{i=1}^{n} \frac{I_i \widetilde{S}_i}{1 - e^{-\widetilde{S}_i/\widehat{\theta}}} \\ &\quad + \frac{t_0}{K_m}(m - K_m)(k-r+1) \frac{\mathrm{bin}(r-1; k, 1-e^{-t_0/\widehat{\theta}})}{\mathrm{Bin}(r-1; k, 1-e^{-t_0/\widehat{\theta}})} \,, \end{aligned} \tag{24.5.5}$$

where $\mathrm{bin}(j; n, p)$ is the probability function of $\mathrm{Bin}(j; n, p)$. This root can be computed by the Newton-Raphson iterative method. If $K_n = 0$ an MLE does not exist. Notice that the system reliability is, in the present case,

$$R_{\mathrm{sys}}(t; \theta) = \sum_{j=r}^{k} \binom{k}{j} e^{-jt/\theta} (1 - e^{-t/\theta})^{k-j} \,. \tag{24.5.6}$$

Substituting the MLE $\widehat{\theta}$ in (24.5.6) we obtain the MLE of $R_{\mathrm{sys}}(t; \theta)$. The formula for the Fisher information function $I(\theta; t_0)$ is given in Zacks (1986). It is shown that $\widehat{\theta}$ is consistent and asymptotically normal. From these results one can derive the asymptotic variance of the MLE $\widehat{R}_{\mathrm{sys}}(t; \theta)$. This asymptotic variance can be approximated, as shown earlier, by bootstrapping, if the available sample is large.

24.6 Estimating The Time of Shift Into The Wear-Out Phase

Consider a system having several components, whose life distributions depend on the state of the system. Each component is replaced immediately after failure (corrective maintenance), or at scheduled replacement epochs (preventive maintenance). The optimal scheduling of the replacement epochs depends on various economic considerations, and on the particular life distribution of the component. This in turn depends on the state of the whole system. We recognize two phases of the system. Phase I is the phase in which the components have a constant failure rate, λ. Phase II is the phase in which the failure rate is greater than λ and increasing. Phase I is a mature stable phase of the system, while Phase II is a wear-out phase. It is important to detect early the shift of the system from Phase I to Phase II. Optimal replacement scheduling and inventory management during the wear-out phase of the system are different from those appropriate to the constant failure rate phase. In the present section we present an adaptive procedure for detecting the shift epoch to the wear-out phase. For more details see Zacks (1984).

24.6.1 The Wear-Out Life Distribution

The wear-out life distribution is based on the following nondecreasing failure-rate function

$$\rho(t;\lambda,\alpha,\tau)=\begin{cases}\lambda, & \text{if } 0\le t\le \tau_+\\ \lambda+\lambda^{\alpha}\alpha(t-\tau_+)^{\alpha-1}, & \text{if } \tau_+<t,\end{cases} \tag{24.6.1}$$

where λ is the constant failure-rate in Phase I, $0<\lambda<\infty$; α is a shape parameter, $1\le\alpha<\infty$, and τ is the change point parameter $\tau_+=\max(0,\tau)$. If $\tau\le 0$ the system is already in Phase II. The present model represents a superposition of a Weibull failure-rate function on a constant failure-rate function for $t\ge\tau_+$. Notice also that the limiting case of $\alpha=1$ corresponds to a shift at τ from an exponential life distribution with MTBF $\theta_1=1/\lambda$ to an exponential life distribution with an MTBF $\theta_2=1/2\lambda$, i.e., for all $t>\tau$, $P(X\ge t\mid x\ge\tau)=\exp\{-2\lambda(t-\tau)\}$. From this failure-rate model we obtain a life distribution, called the *wear-out life distribution*, with a distribution function

$$\begin{aligned}F(x;\lambda,\alpha,\tau) &= 1-\exp\Big\{-\lambda\int_0^x(1+\alpha(\lambda(t-\tau_+))^{\alpha-1})dt\Big\}\\ &= 1-\exp\{-\lambda x-(\lambda(x-\tau_+)_+)^{\alpha}\},\quad x\ge 0\,.\end{aligned} \tag{24.6.2}$$

Obviously, $F(x;\lambda,\alpha,\tau)=0$ for $x\le 0$.

The corresponding density is

$$f(x;\lambda,\tau,\alpha)=\lambda e^{-\lambda x}(1+\alpha(\lambda(x-\tau_+)_+)^{\alpha-1})\exp\{-\lambda^{\alpha}(x-\tau_+)_+^{\alpha}\},\quad x\ge 0\,. \tag{24.6.3}$$

We remark here that, if τ is very large compared to $\theta=1/\lambda$, the wear-out distribution is practically an exponential distribution with mean $\theta=1/\lambda$. This is the case when the system is far from its wear-out phase. We conclude the present section with a presentation of the formula for the moments of the wear-out distribution (24.6.3). Since

$\theta = 1/\lambda$ is a scale parameter, we will present the moment $\mu_r(\alpha, \tau)$ for the case of $\lambda = 1$. If $\lambda \neq 1$ then

$$E[X^r; \lambda, \alpha, \tau] = \frac{1}{\lambda^r}\, \mu_r(\alpha, \lambda\tau).$$

The rth moment in the standard case is for $\tau > 0$,

$$\begin{aligned} \mu_r(\alpha, \tau) &= \int_0^\tau x^r e^{-x} dx + \int_\tau^\infty x^r e^{-x-(x-\tau)^\alpha}(1 + \alpha(x-\tau)^{\alpha-1})dx \\ &= r!(1 - \mathrm{Pos}(r \mid \tau)) + e^{-\tau} \sum_{j=0}^{r} \binom{r}{j} \tau^{r-j} M_j(\alpha)\ , \end{aligned} \tag{24.6.4}$$

where

$$\mathrm{Pos}(k \mid \lambda) = e^{-\lambda} \sum_{j=0}^{k} \frac{\lambda^j}{j!} \tag{24.6.5}$$

is the Poisson distribution function and

$$M_j(\alpha) = \int_0^\infty y^j e^{-y-y^\alpha}(1 + \alpha \cdot y^{\alpha-1})dy, \quad j \geq 0\ . \tag{24.6.6}$$

Integration by parts yields, for every $j \geq 1$,

$$M_j(\alpha) = j \int_0^\infty y^{j-1} e^{-(y+y^\alpha)} dy\ , \tag{24.6.7}$$

which is smaller than $\Gamma(j+1) = j!$ for all $j \geq 1$. Notice that $M_0(\alpha) = 1$ for all $\alpha \geq 1$. The integral in (24.6.7) can be evaluated numerically. In the case of $\alpha = 2$ we obtain, for all $j \geq 1$,

$$\begin{aligned} M_j(2) &= je^{1/4} \int_0^\infty y^{j-1} e^{-(y+1/2)^2} dy \\ &= j\, \frac{e^{1/4}}{\sqrt{2}} \sum_{l=0}^{j-1} (-1)^l \binom{j-1}{l} \left(\frac{1}{2}\right)^{j-1-l/2} B(l; \sqrt{1/2})\ , \end{aligned} \tag{24.6.8}$$

where, for all $l \geq 2$,

$$\begin{aligned} B(l; \xi) &= \int_\xi^\infty z^l e^{-1/2 z^2} dz \\ &= \xi^{l-1} \exp\left\{-\frac{1}{2}\,\xi^2\right\} + (l-1)B(l-2; \xi) \end{aligned} \tag{24.6.9}$$

and

$$\begin{aligned} B(0; \xi) &= \sqrt{2\pi}\,(1 - \Phi(\xi))\ , \\ B(1; \xi) &= \exp\left\{-\frac{1}{2}\xi^2\right\}\ , \end{aligned} \tag{24.6.10}$$

where $\Phi(z)$ is the standard normal distribution function.

24.6.2 Bayes Estimation of the Shift Epoch, λ and α Known

We develop the Bayesian framework for estimating the shift point, τ.

Given a sequence $t_1, t_2, \cdots$ of replacement epochs, where $0 < t_1 < t_2 < \cdots$, we consider the random variables $X_i = t_i - t_{i-1}$ $(i = 1, 2, \cdots)$, where $t_0 \equiv 0$. X_i represents the operational time length of ith sequentially installed component. Since the previous maintenance replaces the component after Δ units of time, $X_i = \min(Y_i, \Delta)$, where $Y_1, Y_2, \cdots$ are independent random variables having the wear-out distribution, with parameters λ, α and $\tau_i = \tau - t_{i-1}$ $(i = 1, 2, \cdots)$. τ_i is the length of time left for the system, after the ith installment of the component till the transition to Phase II. If $\tau_i < 0$, the change to Phase II has taken place before t_{i-1}.

We assume that λ and α are known. Without loss of generality we assume that $\lambda = 1$. If $\lambda \neq 1$ replace X_i in the formulae below by λX_i $(i = 1, \cdots, n)$.

Given n replacement points $t_1 < t_2 < \cdots < t_n$, let $\boldsymbol{X}^{(n)} = (X_1, \cdots, X_n)'$ where $X_i = t_i - t_{i-1}$, and define

$$J_i = \begin{cases} 1, & \text{if } X_i < \Delta \\ 0, & \text{if } X_i = \Delta \,. \end{cases} \tag{24.6.11}$$

The likelihood function of τ, given $\boldsymbol{X}^{(n)}$, is

$$\begin{aligned} L(\tau; \boldsymbol{X}^{(n)}) \;=\;& I(\tau \leq 0) \exp\left\{ -\sum_{i=1}^{n} X_i^{\alpha} \right\} \prod_{i=1}^{n} (1 + \alpha J_i X_i^{\alpha-1}) \\ & + \sum_{j=1}^{n-1} I(t_{j-1} < \tau \leq t_j) \exp\left\{ -\sum_{k=j+1}^{n} X_k^{\alpha} \right\} \prod_{k=j+1}^{n} (1 + \alpha J_k X_k^{\alpha-1}) \\ & \cdot [1 + \alpha J_i (X_j + t_{j-1} - \tau)^{\alpha-1}] \exp\{-(X_j + t_{j-1} - \tau)^{\alpha}\} \\ & + I(t_{n-1} < \tau \leq t_n)[1 + \alpha J_n (X_n + t_{n-1} - \tau)^{\alpha-1}] \\ & \cdot \exp\{-(X_n + t_{n-1} - \tau)^{\alpha}\} + I(\tau > t_n) \,, \end{aligned} \tag{24.6.12}$$

where $I(A)$ is the indicator function of the set A.

Generally, let $\xi(\tau)$ denote the prior density of τ. Then, the posterior density of τ, given $\boldsymbol{X}^{(n)}$, is

$$\xi(\tau \mid \boldsymbol{X}^{(n)}) = \xi(\tau) L(\tau; \boldsymbol{X}^{(n)}) / D(\xi; \boldsymbol{X}^{(n)}) \,, \tag{24.6.13}$$

where

$$D(\xi; \boldsymbol{X}^{(n)}) = \int L(t; \boldsymbol{X}^{(n)}) dP_{\xi}(\tau \leq t) \,. \tag{24.6.14}$$

In the present study we consider a prior distribution with density

$$\xi(\tau) = \begin{cases} p, & \text{if } \tau \leq \tau_0 \\ (1-p)\psi e^{-\psi(\tau - \tau_0)}, & \text{if } \tau > \tau_0 \,, \end{cases} \tag{24.6.15}$$

where $0 < p < 1$ and $0 < \psi < \infty$. τ_0 is a time point such that, with high prior probability, $1 - p$, the true wear-out point, τ, exceeds it. We are interested therefore in time points greater than τ_0. We shall assume accordingly, without loss of generality, that $\tau_0 = 0$. For the prior density (24.6.15) under consideration, the function (24.6.14) assumes the form

$$\begin{aligned} D(p;\psi;\boldsymbol{X}^{(n)}) &= p\exp\left\{-\sum_{i=1}^{n} X_i^{\alpha}\right\}\prod_{i=1}^{n}[1+\alpha J_i X_i^{\alpha-1}] \\ &\quad +(1-p)\sum_{j=1}^{n-1}\exp\left\{-\sum_{k=j+1}^{n} X_k^{\alpha}\right\}\prod_{k=j+1}^{n}[1+\alpha J_k X_k^{\alpha-1}] \\ &\quad \cdot e^{-\psi t_{j-1}}\psi\int_0^{X_j} e^{-\psi u}[1+\alpha J_j(X_j-u)^{\alpha-1}]e^{-(X_j-u)^{\alpha}}du \\ &\quad +(1-p)\psi e^{-\psi t_{n-1}}\int_0^{X_n} e^{-\psi u} \\ &\quad \cdot[1+\alpha J_n(X_n-u)^{\alpha-1}]e^{-(X_n-u)^{\alpha}}du + (1-p)e^{-\psi t_n}\ . \end{aligned} \tag{24.6.16}$$

The posterior probability, after observing $\boldsymbol{X}^{(n)}$, that the transition to Phase II has already taken place (i.e., $\{\tau \le t_n\}$) is $p_n = 1 - q_n$, where

$$q_n = (1-p)e^{-\psi t_n}/D(p,\psi;\boldsymbol{X}^{(n)})\ . \tag{24.6.17}$$

If p_n is large we have high evidence that the shift has already taken place. In addition, after observing $\boldsymbol{X}^{(n)}$ the Bayes estimator of τ, for squared error loss function, is the posterior expectation of τ, given $\boldsymbol{X}^{(n)}$. This estimator is determined by the formula

$$\begin{aligned} \hat{\tau}_n(p,\psi;\boldsymbol{X}^{(n)}) &= E[\tau \mid \boldsymbol{X}^{(n)},p,\psi] \\ &= \left\{(1-p)\psi\sum_{j=1}^{n-1}\prod_{k=j+1}^{n}[1+\alpha J_k X_k^{\alpha-1}]\exp\left\{-\sum_{k=j+1}^{n} X_k^{\alpha}\right\}\right. \\ &\quad \cdot e^{-\psi t_{j-1}}\int_0^{X_j}(t_{j-1}+u)e^{-\psi u}[1+\alpha J_j(X_j-u)^{\alpha-1}]e^{-(X_j-u)^{\alpha}}du \\ &\quad +(1-p)\psi e^{-\psi t_{n-1}}\int_0^{X_n}(t_{n-1}+u)e^{-\psi u} \\ &\quad \cdot[1+\alpha J_n(X_n-u)^{\alpha-1}]e^{-(X_n-u)^{\alpha}}du \\ &\quad \left.+(1-p)\psi e^{-\psi t_{n-1}}\int_0^{\infty}(t_n+u)e^{-\psi u}du\right\} \div D(p,\psi;\boldsymbol{X}^{(n)})\ . \end{aligned} \tag{24.6.18}$$

Notice that if $\lambda \ne 1$ we determine p_n and $\hat{\tau}_n(p,\psi;X^{(n)})$ by substituting in the above formula λX_i for X_i and ψ/λ for ψ.

Thus, we can compute p_n and $\hat{\tau}_n$ adaptively, after each replacement, and decide that the change has occurred at the first time $\hat{\tau}_n \le t_n$ or p_n is sufficiently large. Recursive formulae can be developed for the computation of $\hat{\tau}_n(p,\psi;\boldsymbol{X}^{(n)})$. For details, see Zacks (1984).

Table 24.1: A simulation run with parameters

$\lambda = 1/200$, $\psi = 1/1000$, $p = .2$, $\tau = 750$

n	X_n	J_n	t_n	p_n	$\widehat{\tau}_n$
1	41.49	1	41.	0.288	742.71
2	156.56	1	198.	0.473	646.85
3	108.39	1	306.	0.631	514.63
4	225.00	0	531.	0.437	921.39
5	225.00	0	756.	0.309	1317.66
6	31.19	1	788.	0.385	1237.91
7	23.03	1	811.	0.446	1170.46
8	115.07	1	926.	0.610	1011.46
9	114.04	1	1040.	0.742	866.27
10	20.53	1	1060.	0.779	815.35
11	27.09	1	1087.	0.820	758.42

24.6.3 Numerical Examples

We present here results of some simulations, which illustrate numerically the characteristics of the procedure developed in the previous sections. In order to simulate a random variable Y, having a wear-out distribution with parameters λ, α and τ, proceed as follows. First, simulate a random variable U, having a uniform distribution on $(0, 1)$. Then, solve for Y the equation

$$\exp\{-\lambda Y - \lambda^\alpha (Y - \tau)_+^\alpha\} = U \ . \tag{24.6.19}$$

Finally, determine $X = \min(Y, \Delta)$.

In the present study we restrict attention to the case of $\alpha = 2$. Accordingly,

$$Y = \begin{cases} -\frac{1}{\lambda} \ln U, & \text{if } U \geq e^{-\lambda\tau} \\ \tau - \frac{1}{2\lambda} + \frac{1}{\lambda}(\frac{1}{4} - \ln U - \lambda\tau)^{1/2}, & \text{if } U < e^{-\lambda\tau} \ . \end{cases} \tag{24.6.20}$$

In the following examples we consider a system with $\lambda^{-1} = 200$ [hr], $\Delta = 225$ [hr], $\tau = 750$ [hr], $\alpha = 2$ and the prior probability that $\{\tau \leq 0\}$ is $p = .2$. For the simulation of the random variable X_n, $(n = 1, 2, \cdots)$ we apply a value of $\tau_n = (\tau - t_{n-1})_+$. The values of the posterior probabilities, p_n, and those of the Bayes estimators, $\widehat{\tau}_n$, are computed adaptively after each stage. In Table 24.1 we present the results of such a simulation run.

We see in the above example that the first value of t_n larger than $\widehat{\tau}_n$ is $t_n^* = 1040$ [hr]. This is greater than τ by 290 [hr], which is a little over than 1 MTBF. The corresponding value of p_n is $p_n^* = .742$. The estimate of τ is $\widehat{\tau}_n = 866.27$. We see also that if we defer the decision until $p_n \geq .8$ then $t_n^* - \tau = 337$ [hr] and $\widehat{\tau}_n = 785.42$. Thus, the second decision (stopping) time provided a more accurate estimate of τ. In Table 24.2 we present frequency distributions and the exact means and standard deviations of $W_n^* = t_n^* - \tau$ as obtained by $M = 100$ independent simulation runs, for each one of the prior parameters $\psi = 1/500$, $1/750$ and $1/1000$. The decision time in these simulations is the first value of t_n greater than $\widehat{\tau}_n$, i.e., $t_n^* =$ least $n \geq 1$, such that $t_n \geq \tau_n$.

Table 24.2: Frequency distributions, means and standard deviations of W_n^* in $M = 100$ simulation runs

Mid Point	Frequencies		
Interval	$1/\psi = 500$	$1/\psi = 750$	$1/\psi = 1000$
-500	10	0	0
-400	35	28	15
-300	17	37	22
-200	21	7	19
-100	8	11	6
0	0	0	10
100	6	7	10
200	3	9	5
300	0	1	7
400	0	0	6
Mean	-278.15	-218.44	-110.71
Std. Dev.	167.44	201.88	239.82

We see in Table 24.2 that the decision time t_n^* tends to yield too many early decision points. If $1/\psi = 1,000$ [hr] the results are significantly better than in the case of $1/\psi = 500$ or 750 [hr]. The situation seems to be better in the case of the decision times defined by

$$t_n^{**} = \text{least } n \text{ s.t. } p_n \geq .9.$$

In Table 24.3 we present the frequency distributions of $W_n^{**} = t_n^{**} - \tau$, obtained in independent simulation runs.

It seems that the Bayes decision times t_n^{**} based on the prior distribution (24.6.15), with a proper choice of the ψ value (not too small), provide good results in the problem of detecting a shift to the wear-out phase.

24.7 Determination of Burn-In Time

Systems, while being constructed, or right after assembly, are subjected to "burn-in" in which, it is expected that, weak components will fail and be replaced. Only components which survived the burn-in remain in the operating system. There are various theories concerning the need for burn-in [see Jensen and Petersen (1982)]. One of the important questions is, how long should a burn-in last? If the time till failure of a component is exponentially distributed, a new component is as good as a used one, and there is no need for burn-in. Jensen and Petersen (1982) consider components which, with small probability, α, have an exponential TTF distribution with mean $\theta_1 = \frac{1}{\lambda_1}$ small, and with probability $1 - \alpha$ the TTF is exponential with mean $\theta_0 = \frac{1}{\lambda_0}$ large. In this case of a mixture of exponentials, with $\lambda_1 > \lambda_0$, the component has a decreasing hazard (failure) rate function

$$\rho(t) = \frac{\lambda_1 + \frac{1-\alpha}{\alpha}\lambda_0 e^{(\lambda_1 - \lambda_0)t}}{1 + \frac{1-\alpha}{\alpha} e^{(\lambda_1 - \lambda_0)t}}, \quad 0 < t < \infty \,. \tag{24.7.1}$$

Table 24.3: Frequency distributions, means and standard deviations of W_n^{**} in $M = 100$ simulation runs

Mid Point	Frequencies		
of Intervals	$1/\psi = 500$	$1/\psi = 750$	$1/\psi = 1000$
-300	15	2	2
-200	17	22	19
-100	17	12	13
0	10	16	10
100	6	0	4
200	4	7	7
300	8	5	7
400	2	4	3
500	12	17	6
600	9	13	25
700	0	2	4
Mean	65.60	152.85	202.13
Std. Dev.	311.78	325.40	322.37

The length of the burn-in time can then be determined as a function of the burn-in cost, and the cost of mission failure [see Clarotti and Spizzichino (1990)]. See also Stewart and Johnson (1972).

We present here a sequential burn-in stopping rule, developed by Boukai (1987). The model involves a shift parameter, τ, such that for t smaller than τ, the hazard rate function is decreasing, and for $t \geq \tau$, the hazard rate function is constant. In contrast to the problem of Section 24.6, here the shift is from the phase of "infant mortality" to the stable (mature) phase of the system. More specifically, the hazard rate function is

$$\rho(t;\theta,\alpha,\tau) = \begin{cases} \frac{\alpha}{\theta}(\frac{t}{\theta})^{\alpha-1}, & 0 < t < \tau \\ \frac{\alpha}{\theta}(\frac{\tau}{\theta})^{\alpha-1}, & \tau \leq t \end{cases} \tag{24.7.2}$$

where α is a shape parameter, θ a scale parameter and τ the shift parameter. We assume that $\tau \in [\tau_1, \tau_2]$, where $0 < \tau_1 < \tau_2 < \infty$ are known. From this hazard rate function we obtain the following distribution function of the TTF,

$$F(t;\theta,\alpha,\tau) = \begin{cases} 0, & t \leq 0 \\ 1 - \exp\{-(\frac{t}{\theta})^\alpha\}, & 0 < t < \tau \\ 1 - \exp\{-(\frac{\tau}{\theta})^\alpha - \frac{\alpha}{\theta}(\frac{\tau}{\theta})^{\alpha-1}(t-\tau)\}, & \tau \leq t\,. \end{cases} \tag{24.7.3}$$

Notice that the rth moment of this Weibull-Exponential distribution is $M_r(\theta,\alpha,\tau) = \theta^r M_r(1,\alpha,\frac{\tau}{\theta})$, where

$$\begin{aligned} M_r(1,\alpha,\sigma) &= \exp\{-\sigma^2(1-\alpha)\}(\alpha\sigma^{\alpha-1})^{-r} \\ &\quad \cdot[r! - \Gamma(\alpha\sigma^\alpha, r+1)] + \Gamma\left(\sigma^\alpha, 1+\frac{r}{\alpha}\right)\,, \end{aligned} \tag{24.7.4}$$

where $\sigma = \tau/\theta$ and $\Gamma(x,\xi)$ is the incomplete gamma integral

$$\Gamma(x,\xi) = \int_0^x y^{\xi-1}e^{-y}dy\,. \tag{24.7.5}$$

Boukai (1987) developed the formula for the Bayes estimator of τ, when θ and α are known. Suppose that N components are put on test (burn-in). Let $0 < t_{(1)} < \cdots < t_{(n)}$ be n observed failure times, $\boldsymbol{\theta}^{(n)} = (t_{(1)}, \cdots, t_{(n)})'$.

Let $I(A)$ denote the indicator variable of the set A. Define the auxiliary functions

$$
\begin{aligned}
C_n &= (\alpha/\theta^\alpha)^n \frac{N!}{(N-n)!} , \\
a_j &= \max(t_{(j)}, \tau_1), \quad b_j = \min(t_{(j)}, \tau_2), \quad j = 1, \cdots, n , \\
\boldsymbol{X}_0 &\equiv 1, \quad \boldsymbol{X}_j = \left(\prod_{i=1}^{j} t_{(i)}\right)^{\alpha-1} \exp\left\{-\sum_{i=1}^{j}\left(\frac{\tau_{(i)}}{\theta}\right)^{\alpha}\right\}, \quad j = 1, \cdots, n , \\
Z_n &= \boldsymbol{X}_n \exp\left\{-\left(\frac{t_{(n)}}{\theta}\right)^{\alpha}(N-n)\right\} , \\
\mu_j &= \frac{N-j}{\theta^\alpha}(1-\alpha), \quad j = 0, \cdots, n-1 , \\
q_{j:n} &= \frac{\alpha}{\theta^\alpha}\left(\sum_{i=j+1}^{n} t_{(i)} + (N-n)t_{(n)}\right), \quad j = 0, \cdots, n-1 ,
\end{aligned}
$$

and finally

$$
\Psi_j^k(j, a, b, u, v) = \int_u^v y^{j(\alpha-1)+k} \exp\{-ay^\alpha - by^{\alpha-1}\} g(y) dy,
$$

where $g(y)$ is the prior density of τ. Then

$$
\begin{aligned}
D_g^k(\boldsymbol{\theta}^{(n)}) &= I(t_{(n)} < \tau_2) C_n Z_n \Psi_g^k(0, 0, 0, a_n, \tau_2) \\
&\quad + \sum_{j=0}^{n-1} I(t_{(j)} < \tau_2, t_{(j+1)} > \tau_1) \boldsymbol{X}_j \Psi_g^k(n-j, \mu_j, q_{j:n}, a_j, b_{j+1}) .
\end{aligned}
\tag{24.7.6}
$$

The Bayes estimator of τ, for the squared error loss, given $\boldsymbol{\theta}^{(n)}$, is

$$
\hat{\tau}_n = D_g^1(\boldsymbol{\theta}^{(n)}) / D_g^0(\boldsymbol{\theta}^{(n)}) . \tag{24.7.7}
$$

Boukai investigated two sequential decision rules, for stopping the burn-in. One stopping rule is to stop as soon as the posterior probability that $\{\tau < t_{(n)}\}$ exceeds a specified threshold. The other one is based on costs of stopping too early or stopping too late. Formulae are provided for the posterior risk and for the predictive risk, if the burn-in will continue until the next failure. After each failure these two quantities are compared. Stopping occurs as soon as the posterior risk is smaller than the predictive risk.

References

Ascher, H. and Feingold, H. (1984). *Repairable Systems Reliability Modeling, Inferences, Misconceptions and Their Causes,* Marcel Dekker, New York.

Barlow, R.E. and Proschan, F. (1965). *Mathematical Theory of Reliability*, John Wiley & Sons, New York.

Barlow, R.E. and Proschan, F. (1975). *Statistical Theory of Reliability and Life Testing: Probability Models*, Holt, Reinhart and Winston, New York.

Boukai, B. (1987). Bayes sequential procedure for estimation and for determination of burn-in time in a hazard rate model with an unknown change-point parameter, *Sequential Analysis*, **6**, 37-53.

Clarotti, C.A. and Spizzichino, F. (1990). Bayes burn-in decision procedures, *Probability in Engineering and Informational Sciences*, **4**, 437-445.

Efron, B. (1982). *The Jackknife, the Bootstrap and Other Resampling Plans*, CBMS-NSF Regional Conferences Series in Applied Mathematics, SIAM, Philadelphia.

Gertsbakh, I.B. (1989). *Statistical Reliability Theory*, Marcel Dekker, New York.

Gnedenko, B.V., Belyayev, Yu.K., and Solovyev, A.D. (1969). *Mathematical Methods of Reliability Theory*, Academic Press, New York.

Ireson, W.G. (Ed.) (1982). *Reliability Handbook*, McGraw-Hill, New York.

Jensen, F. and Petersen, N.E. (1982). *Burn-In: An Engineering Approach to The Design and Analysis of Burn-In Experiments*, John Wiley & Sons, New York.

Kalbfleisch, J.D. and Prentice, R.L. (1980). *The Statistical Analysis of Failure Time Data*, John Wiley & Sons, New York.

Stewart, L.T. and Johnson, J.D. (1972). Determining the optimal burn-in and replacement times using Bayesian decision theory, *IEEE Transactions on Reliability*, **R-21**, 168-173.

Zacks, S. (1984). Estimating the shift to wear-out of systems having Exponential-Weibull life distribution, *Operations Research*, **32**, 741-749.

Zacks, S. (1986). Estimating the scale parameter of an exponential distribution from a sample of time-censored r-th order statistics, *Journal of the American Statistical Association*, **81**, 205-210.

Zacks, S. (1992). *Introduction to Reliability Analysis: Probability Models and Statistical Methods*, Springer-Verlag, New York.

Zacks, S. and Even, M. (1966). The efficiencies in small samples of the maximum likelihood and best unbiased estimators of reliability functions, *Journal of the American Statistical Association*, **61**, 1033-1051.

CHAPTER 25

Exponential Regression with Applications

Richard A. Johnson and Rick Chappell[1]

25.1 Introduction

Most often, in studies of life-length, there are additional explanatory variables which may explain some of the variation in life-length. Let X be the life-length of a unit or individual and let $(z_1, \ldots, z_k)$ be the associated values of k explanatory variables, or covariates. The covariates may include categorical variables that specify the treatment group, gender, or other characteristics or they could be the usual quantitative variables.

The parameter θ for the exponential distribution is a scale parameter. Intuitively, it is easier to think about modifying a location parameter. Therefore, we take the natural logarithm, $Y = \log X$, of life-length. Then Y has density function

$$f_Y(y) = e^{y-\log\theta}\, e^{e^{y-\log\theta}}$$

where $\log\theta$ is now a location parameter. Allowing $\log\theta$ to depend on the covariates $\boldsymbol{z}$, we obtain regression models. One simple choice, as in the normal theory case, is to consider the log-linear model.

$$\log\theta(\boldsymbol{z}) = \beta_0 + \beta_1 z_1 + \cdots + \beta_k z_k = \beta_0 + \boldsymbol{z}'\boldsymbol{\beta} \tag{25.1.1}$$

or $\theta(\boldsymbol{z}) = e^{\beta_0 + \boldsymbol{z}'\boldsymbol{\beta}}$. In the case of a single covariate such as stress in an accelerated life test, if the covariate is expressed as the logarithm of stress, $z_1 = \log(S)$, $\alpha = e^{\beta_0}$, and $\beta_1 = -\eta$ then we obtain the widely studied *inverse power law*

$$\theta = \frac{\alpha}{S^\eta}\ .$$

This is a special case of the log-linear or proportional hazard rate models, described below, when the single covariate is expressed on a log scale.

Another approach to modeling the dependence of an exponential life length on the covariates $\boldsymbol{z}$ is to argue directly in terms of the hazard rate. Starting from the usual

[1] University of Wisconsin, Madison, Wisconsin

hazard rate for the exponential distribution

$$\rho(x) = \frac{f(x)}{1 - F(x)} = \frac{1}{\theta} = \text{ constant },$$

we allow the hazard rate to depend on $\boldsymbol{z}$. In particular, generalizing the exponential constant hazard rate in this manner,

$$\rho(x; \boldsymbol{z}) = \rho(\boldsymbol{z}) = \frac{1}{\theta(\boldsymbol{z})}$$

which is still a constant for any fixed set of covariates $\boldsymbol{z}$ but which may be different for units with different values of the covariates. That is, the constant hazard rate will increase or decrease depending on the values of the covariates.

The function $\rho(\boldsymbol{z})$ can be specified or partially specified. It may be possible to model the effect of $\boldsymbol{z}$ as the linear function $\boldsymbol{z}'\boldsymbol{\beta} = \beta_1 z_1 + \beta_2 z_2 + \cdots + \beta_k z_k$ where $\boldsymbol{\beta} = (\beta_1, \beta_2, \ldots, \beta_k)'$ is the vector of regression parameters. Then, for some constant θ_0 and known function $q(\cdot)$,

$$\rho(x; \boldsymbol{z}) = \frac{1}{\theta_0 q(\boldsymbol{z}'\boldsymbol{\beta})} .$$

If we wish to avoid restrictions being imposed on $\boldsymbol{\beta}$ by the condition that the hazard rate be positive, or equivalently $q(\boldsymbol{z}'\boldsymbol{\beta}) > 0$, a simple choice of $q(\cdot)$ is $q(y) = e^{-y}$. With this choice, the hazard rate becomes

$$\rho(x; \boldsymbol{z}) = \frac{1}{\theta_0} e^{-\boldsymbol{z}'\boldsymbol{\beta}} . \tag{25.1.2}$$

By the well known relation between the hazard function and the probability density function, $f(x; \boldsymbol{z}) = \rho(x; \boldsymbol{z}) \exp(-\int_0^x \rho(s; \boldsymbol{z})\, ds)$, the density function is

$$f(x; \boldsymbol{z}) = \frac{1}{\theta_0} e^{-\boldsymbol{z}'\boldsymbol{\beta}} \exp\left(-\frac{1}{\theta_0} x\, e^{-\boldsymbol{z}'\boldsymbol{\beta}}\right) .$$

Consider the transformation to the natural log of life-length, $Y = \log X$. The density of Y is

$$f_Y(y) = e^{y - \log\theta_0 - \boldsymbol{z}'\boldsymbol{\beta}}\, e^{e^{y - \log\theta_0 - \boldsymbol{z}'\boldsymbol{\beta}}} .$$

Consequently, Y can be expressed as

$$\beta_0 + \boldsymbol{z}'\boldsymbol{\beta} + V , \tag{25.1.3}$$

where V has the extreme value distribution with probability density function $e^{v - e^v}$ for $-\infty < v < \infty$ and $\beta_0 = \log(\theta_0)$.

This is a linear model where the error term has a complete specified error distribution with mean $-\gamma$, where $\gamma = 0.5772$ is Euler's constant, and variance $\pi^2/6$. This is consistent with relation (25.1.1) where β_0 is specifically added.

Two special structures are important. First, from the hazard rate, the covariates $\boldsymbol{z}$ act multiplicatively on the (constant) hazard function. Therefore, (25.1.2) is a special case of the *Proportional hazard rate model*

$$\rho(x; \boldsymbol{z}) = \rho_0(x)\, e^{-\boldsymbol{z}'\boldsymbol{\beta}} , \tag{25.1.4}$$

where the *baseline* hazard $\rho_0(x)$ need not be constant unless the life lengths are distributed as exponential variables.

From the linear model relation (25.1.3), the effect of the covariates is linear on the log of life length. This is a special case of the *log-linear model*

$$Y = \boldsymbol{z}'\boldsymbol{\eta} + \sigma V \ , \tag{25.1.5}$$

where the distribution of V is specified above, $\boldsymbol{z}$ may have been augmented to include 1 as its first element if a constant is included in the model, and $\boldsymbol{\eta}$ is a vector of parameters. For exponential life lengths, $\sigma = 1$. Both of these generalizations were described by Cox (1972). There is now a great deal of literature on both large sample theory of estimation and computational procedures [see Cox and Oakes (1984) or McCullagh and Nelder (1989)].

Accelerated life tests were treated by Zelen (1959) in one of the first papers on the subject. He actually expressed θ as a product of terms under a factorial model which is equivalent to the model above for $\log \theta$. Feigl and Zelen (1965) give one of the first applications of exponential regression to human survival analysis, Lawless (1976) derives the exact procedures for the inverse power law. Singpurwalla (1971) and Kahn (1979) give maximum likelihood and least square approaches, respectively.

25.2 Censoring

When there is no censoring, so all complete life lengths are available, it is straightforward to write down the likelihood. The large sample theory of the maximum likelihood estimators and likelihood ratio statistics follow from the standard results for regression. However, in many applications of exponential regression, the observations are censored. If there are multiple runs at a fixed setting z_i of the input variables, it is possible to consider order statistic censoring. When all of the observations are either complete or order statistic censored, exact distribution theory is possible, at least for the inverse power law. In other instances, approximate large sample distribution theory must be applied.

One general type of censoring, called random censoring, assumes that the censoring point for the ith run, C_i, is a random variable that is independent of the life length X_i. Then, the observed time is $\widetilde{X}_i = \min(X_i, C_i)$ and it is known whether X_i or C_i is observed. Consequently,

$$P[\widetilde{X}_i > t] = P[X_i > t]P[C_i > t]$$

and

$$P[C_i > X_i] = \frac{1}{\theta_i}\int_0^\infty P[C_i > t]\, e^{-t/\theta_i} dt = \frac{1}{\theta_i}\, E[\widetilde{X}_i] \ . \tag{25.2.1}$$

Heavy censoring, that is C_i stochastically small, implies that the expected value of the observed quantity $\widetilde{X}_i$ is small. This formulation includes time censoring where the C_i are fixed ahead of time corresponding to degenerate random variables.

25.3 Estimation

Exact inference procedures have only been developed for special cases and those pertain to the inverse power law model. It does include the important case where complete life lengths are available and the case where multiple runs at a fixed value of the input variables are censored at a fixed order statistic.

We first discuss estimation for the single predictor inverse power law model and then we treat the several predictor proportional hazard or log-linear exponential regression model.

25.3.1 Inference in the Inverse Power Law Model

Lawless (1976) obtained confidence intervals based on pivotals in the model where $\theta = \alpha/S^{\eta}$ or $\theta = \exp(\beta_0 + \beta_1 z_1)$ where $z_1 = \log S$, $\beta_1 = -\eta$ and $\beta_0 = \log \alpha$. Suppose that n_i units are simultaneously placed on lifetest at log stress level z_i and the test is terminated at the time of the r_ith failure, for $i = 1, \ldots, N$. At covariate values z_i let the observed lifetimes be $x_{i,1} < x_{i,2} < \cdots < x_{i,r_i}$ for $i = 1, \ldots, N$. This includes the complete data setting when $r_i = n_i$ for $i = 1, \ldots, N$.

When $n_i = 1$ for all i, this is a regression situation but without any repeat runs. The vector of individual total time on test statistics, $(T_1, T_2, \ldots, T_N)$ is then sufficient for (β_0, β_1), where

$$T_i = \sum_{j=1}^{r_i} x_{i,j} + (n_i - r_i)x_{i,r_i}$$

is the total time on test at z_i. From Epstein and Sobel (1954), the joint density is

$$\prod_{i=1}^{n} \frac{t_i^{r_i - 1}}{\theta_i^{r_i}(r_i - 1)!} e^{-\sum_{i=1}^{n} t_i/\theta_i} . \tag{25.3.1}$$

Here $\theta_i = \theta(z_i)$. It is more natural, in this setting, to obtain pivotals from the distribution of the logs of the sufficient statistics $Y_i = \log T_i$. Their joint density is

$$\prod_{i=1}^{n} \frac{\exp[r_i(y_i - \beta_0 - \beta_1 z_i)]}{(r_i - 1)!} \exp\left[-\sum_{i=1}^{n} \exp(y_i - \beta_0 - \beta_1 z_i) \right] .$$

The maximum likelihood estimators $\widehat{\beta}_0$ and $\widehat{\beta}_1$, obtained from this last expression are such that

$$V_0 = \widehat{\beta}_0 - \beta_0 \qquad \text{and} \qquad V_1 = \widehat{\beta}_1 - \beta_1$$

are pivotals. That is, their joint distribution is free of parameters. The estimated residuals $a_i = y_i - \widehat{\beta}_0 - \widehat{\beta}_1 z_i$, for $i = 1, \ldots, N$, are ancillary statistics. Only $N - 2$ of the a_i are functionally independent. Set $\boldsymbol{a} = (a_1, \ldots, a_{N-2})$. Lawless (1976), following R.A. Fisher, suggests conditioning V_0 and V_1 on the ancillary statistic $\boldsymbol{a}$. The resulting conditional joint distribution has density

$$g(v_0, v_1 | \boldsymbol{a}) = k(\boldsymbol{a}, \boldsymbol{z}) \exp \left\{ n v_0 + v_1 \sum_{i=1}^{N} r_i z_i - \exp v_0 \sum_{i=1}^{n} \exp(a_i + v_1 z_i) \right\} , \tag{25.3.2}$$

where $n = \sum_{i=1}^{N} r_i$ and $k(\boldsymbol{a}, \boldsymbol{z})$, the normalizing constant, must be obtained by numerical integration.

A confidence interval for β_1 can be obtained from the marginal conditional distribution of V_1 given $\boldsymbol{a}$

$$g(v_1|\boldsymbol{a}) = \frac{k_1(\boldsymbol{a}, \boldsymbol{z}) \exp(v_1 \sum_{i=1}^{N} r_i z_i)}{(\sum_{i=1}^{N} \exp(a_i + v_1 z_i))^n} , \tag{25.3.3}$$

where $k_1(\boldsymbol{a}, \boldsymbol{z})$, the normalizing constant, must be obtained by numerical integration. For a fixed N and given $(a_1, \ldots, a_N)$ let $\xi_{.025}$ and $\xi_{.975}$ be the lower and upper .025 quantiles, respectively. Then

$$(\widehat{\beta}_1 - \xi_{.975} , \quad \widehat{\beta}_1 + \xi_{.025}) \tag{25.3.4}$$

is the 95% confidence interval for β_1. An example is given in Section 25.5.

A confidence interval for β_0 can be obtained from the marginal conditional distribution of V_0 given $\boldsymbol{a}$. Its cumulative distribution function is

$$\begin{aligned} P[V_0 \le v|\boldsymbol{a}] &= k_0(\boldsymbol{a}, \boldsymbol{z}) \int_{-\infty}^{\infty} \frac{\exp(u \sum_{i=1}^{N} r_i z_i)}{(\sum_{i=1}^{N} \exp(a_i + u z_i))^n} \\ &\quad \cdot I\left(n, \sum_{i=1}^{N} \exp(a_i + u z_i + v)\right) du , \end{aligned} \tag{25.3.5}$$

where $I(n, y) = \int_0^y u^{n-1} e^{-u} du / \Gamma(n)$ is the incomplete gamma function.

Exact inference procedures are also available for the mean $\theta(z)$ and the qth quantile, $-\theta(z) \log(1-q)$, of the life length distribution for a fixed z. On the log scale, the mean is $\log \theta = \beta_0 + \beta_1 z$ and

$$V_2 = (\hat{\beta}_0 + \hat{\beta}_1 z) - (\beta_0 + \beta_1 z) = V_0 + z V_1 \tag{25.3.6}$$

is also a pivotal. Lawless (1976, 1982) shows that

$$P[V_2 \le v|\boldsymbol{a}] = \int_{-\infty}^{\infty} g(v_1|\boldsymbol{a}) I\left(n, e^{v - z v_1} \sum_{i=1}^{N} e^{a_i + v_1 z_i}\right) dv_1 , \tag{25.3.7}$$

where $I(n, y)$ is the incomplete gamma function. This is a one-dimensional numerical integration in which continued fractions can be used to approximate the incomplete gamma function.

It is important when monitoring the strength of existing materials or when developing new materials, to obtain a lower 95% confidence bound for say the lower 10th quantile of the distribution. Let $\xi_{.95}$ be determined numerically from the requirement

$$P[V_0 \le \xi_{.95}|\boldsymbol{a}] = .95 .$$

Then

$$-\log(1 - 0.10)(\widehat{\beta}_0 + \widehat{\beta}_1 z - \xi_{.95}) \tag{25.3.8}$$

is the 95% lower confidence bound for the lower 10th quantile of the log life length at z. The bound for the 10th quantile of life length is obtained by exponentiating.

The 95% confidence interval for the mean $\theta(z)$ is

$$(\widehat{\beta}_0 + \widehat{\beta}_1 z - \xi_{.975},\ \widehat{\beta}_0 + \widehat{\beta}_1 z + \xi_{.025})\ , \tag{25.3.9}$$

where $P[V_0 \leq \xi_{.975}|\boldsymbol{a}] = .975 = 1 - P[V_0 \leq \xi_{.025}|\boldsymbol{a}]$. Lawless (1976) also derives a pivotal that leads to a forecast interval for a future observation with covariate z. He further suggests that the large sample distribution of the maximum likelihood estimator provides a good approximation to the exact conditional distribution when sample size is greater then 25.

Maximum likelihood estimation was suggested by Singpurwalla (1971). If the model is formulated as $\beta_0 + \beta_1(z_i - \bar{z})$ on the log scale, or $\theta_i = \alpha/(S_i/\dot{S})^\eta$ on the original scale, where

$$\dot{S} = \left(\prod_1^k S_i^{r_i}\right)^{1/\sum_{i=1}^k r_i}$$

is the geometric mean of stress, then the two maximum likelihood estimates are asymptotically independent. According to Singpurwalla (1971), the maximum likelihood estimates are obtained first using numerical methods to find the root(s) of

$$\sum_{i=1}^k r_i \widehat{\theta}_i (S_i/\dot{S})^\eta \log(S_i/\dot{S}) = 0\ . \tag{25.3.10}$$

That estimate, $\widehat{\eta}$, can then be used to obtain the maximum likelihood estimate of α

$$\widehat{\alpha} = \frac{\sum_{i=1}^k r_i \widehat{\theta}_i (S_i/\dot{S})^{\widehat{\eta}}}{\sum_{i=1}^k r_i}\ . \tag{25.3.11}$$

If there is more than one root, $\widehat{\eta}$, to the first equation, the likelihood needs to be maximized over α for each case to determine the maximum likelihood estimator. The asymptotic variances are

$$\begin{aligned} \mathrm{Var}(\widehat{\alpha}) &= \frac{\alpha^2}{\sum_{i=1}^k r_i}\ , \\ \mathrm{Var}(\widehat{\eta}) &= \frac{1}{\sum_{i=1}^k r_i (\log(S_i/\dot{S}))^2}\ . \end{aligned} \tag{25.3.12}$$

Since $\mathrm{Cov}(\widehat{\alpha}, \widehat{\eta}) = 0$ for the limiting normal distribution, the two maximum likelihood estimators can be treated as being asymptotically independent.

A joint confidence region can also be obtained from the large sample theory of likelihood ratio tests. A 95% confidence region consists of all parameter values (α, η) where

$$L(\alpha, \eta) \geq L(\widehat{\alpha}, \widehat{\eta}) e^{-\chi^2_{2,0.05}/2} \tag{25.3.13}$$

and $\chi^2_{2,0.05}$ is the upper 95% point of the chi-square distribution having 2 degrees of freedom.

25.3.2 Inference in the Regression Model with Several Predictors

When a constant is included along with $p-1$ predictor variables, the mean response is of the form

$$\log \theta = \beta_0 + \beta_1 z_1 + \cdots + \beta_{p-1} z_{p-1} .$$

Although Kahn (1979) discusses least squares estimation, maximum likelihood estimation is more efficient so we describe this procedure in detail.

We write the likelihood allowing for censoring that includes random and fixed time censoring. Let the ith unit have life length X_i and covariates $\boldsymbol{z}_i = (z_{i1}, \ldots, z_{ip})'$ for $i = 1, \ldots, N$. Here $\boldsymbol{z}_i$ is $p \times 1$ and its first element is 1 if a constant term is included in model.

Let F be the set of indices such that the actual life length is observed. Its complement, F^c is then the set of indices for which only the censoring time is available. Since we observe $\min(X_i, C_i)$, where C_i is the censoring time, $Y_i = \log X_i$ for i belonging to F. In this formulation, each observation has its own independent censoring time. The likelihood below would have to be modified if several runs at the same $\boldsymbol{z}_i$ are censored at a fixed order statistic. The log-likelihood, using life-length on the log scale, has the form

$$\log L(\boldsymbol{\beta}) = \sum_{i \in F} (y_i - \boldsymbol{z}_i'\boldsymbol{\beta}) - \sum_{i=1}^{N} \exp(y_i - \boldsymbol{z}_i'\boldsymbol{\beta}) . \tag{25.3.14}$$

The first derivatives of $\log L$, which can be solved numerically for $\widehat{\boldsymbol{\beta}}$, are

$$\frac{\partial \log L}{\partial \beta_u} = \sum_{i \in F} z_{iu} + \sum_{i=1}^{N} z_{iu} \exp(y_i - \boldsymbol{z}_i \boldsymbol{\beta}) \qquad \text{for } u = 1, \ldots, p . \tag{25.3.15}$$

The second derivatives are

$$\frac{\partial^2 \log L}{\partial \beta_u \partial \beta_v} = -\sum_{i=1}^{N} z_{iu} z_{iv} \exp(y_i - \boldsymbol{z}_i \boldsymbol{\beta}) \qquad \text{for } u, v = 1, \ldots, p \tag{25.3.16}$$

and these can be used to approximate the limiting variances and covariances.

The observed Fisher information matrix $\boldsymbol{I}_0$ consists of the elements that are the negatives of the elements (25.3.16) evaluated at the maximum likelihood estimate $\widehat{\boldsymbol{\beta}}$. Its inverse provides an estimate of the covariance in the large sample normal distribution of $\widehat{\boldsymbol{\beta}} - \boldsymbol{\beta}$. In the case where all life lengths are observed, the expected information matrix has entries

$$I_{uv} = -\sum_{i=1}^{N} z_{ir} z_{iv} .$$

For large samples, the approximate 95% confidence ellipsoid for $\boldsymbol{\beta}$ is the elliptical region

$$(\widehat{\boldsymbol{\beta}} - \boldsymbol{\beta})' \boldsymbol{I}_0^{-1} (\widehat{\boldsymbol{\beta}} - \boldsymbol{\beta}) \leq \chi^2_{p,0.05}$$

centered at the maximum likelihood estimator $\widehat{\boldsymbol{\beta}}$. Here, $\chi^2_{p,0.05}$ is the upper 5% point of the chi-square distribution with p degrees of freedom.

The 95% one-at-a-time confidence intervals for the individual β's are

$$\beta_j : \widehat{\beta}_j \pm 1.96\, I^{jj} ,$$

where I^{jj} is the jth diagonal element of $\boldsymbol{I}_0^{-1}$. A Bonferroni approach could also be considered.

Lawless (1982) has shown that the centered maximum likelihood estimator $\widehat{\boldsymbol{\beta}} - \boldsymbol{\beta}$ is a pivotal and he gives an expression for the exact density conditional on the ancillary statistics. However, he states that this approach is not particularly useful when there is more than two parameters.

Most of the discussion on large sample consistency and normality of the maximum likelihood estimators is quite heuristic. However, Maller (1988) gives explicit conditions and a careful treatment that allows for a quite general type of censoring. In particular, the life length X_i is censored at C_i where the two are independent random variables. It is possible for C_i to be a degenerate random variable, that is, a fixed constant. Let $\boldsymbol{D}_N$ denote the expected information matrix with elements

$$d_{N\,u,v} = \sum_{i=1}^{N} z_{iu} z_{iv} E[\exp Y_i] e^{-\boldsymbol{z}_i'\boldsymbol{\beta}} \qquad \text{for } u, v = 1, \ldots, p .$$

Then, if $\boldsymbol{D}_N^{-1}$ exists for all sufficiently large N and

$$\max_{1 \le i \le N} \boldsymbol{z}_i' \boldsymbol{D}_N^{-1} \boldsymbol{z}_i \to 0 \qquad \text{as} \qquad n \to \infty ,$$

the maximum likelihood estimator $\widehat{\boldsymbol{\beta}}$ is consistent, and asymptotically normal. It is also asymptotically efficient. That is

$$\widehat{\boldsymbol{\beta}} \text{ is approximately } N_p(\widehat{\boldsymbol{\beta}}, \boldsymbol{D}_N^{-1}) \tag{25.3.17}$$

and the observed information $\boldsymbol{I}_0$ can be used in place of $\boldsymbol{D}_N$.

25.4 Testing

The only exact testing procedures correspond to the cases where exact confidence interval procedures exist for the inverse power law and situations in which H_1 corresponds to a set of univariate models; for instance those with a single factor or factors with all interactions. We mention the test for $H_0 : \eta = 0$ versus $H_1 : \eta > 0$ for the power parameter. The .05 level test rejects H_0 in favor of the alternative if

$$\widehat{\beta}_1 > \xi_{.05} ,$$

where $\xi_{.05}$ is the upper 5th percentile for V_1, determined from (25.3.3), so $P[V_1 \le \xi_{.05} | \boldsymbol{a}] = .95$.

For the general model

$$\log \theta = \boldsymbol{z}'\boldsymbol{\beta}$$

with possibly several predictor variables, tests can be based on the maximum likelihood estimators or likelihood ratio tests can be conducted.

Let $\boldsymbol{\beta}$ be partitioned as $\boldsymbol{\eta}' = (\boldsymbol{\beta}'_{(1)}, \boldsymbol{\beta}'_{(2)})$ where $\boldsymbol{\beta}_{(1)}$ is $k \times 1$ $(k < p)$ and $\boldsymbol{\beta}_2$ is $(p-k) \times 1$. If $\boldsymbol{\beta}^0_{(1)}$ is a specified value for $\boldsymbol{\beta}_{(1)}$, then we can test $H_0 : \boldsymbol{\beta}_{(1)} = \boldsymbol{\beta}^0_{(1)}$ against $H_1 : \boldsymbol{\beta}(1) \neq \boldsymbol{\beta}^0_{(1)}$ using the approximate $N(\boldsymbol{\beta}, \boldsymbol{I}_0^{-1})$ distribution for $\widehat{\boldsymbol{\beta}}$. Specifically, we reject H_0, at level .05 if

$$\Lambda_1 = (\widehat{\boldsymbol{\beta}}_{(1)} - \boldsymbol{\beta}^0_{(1)})' \boldsymbol{C}_{11}^{-1} (\widehat{\boldsymbol{\beta}}_{(1)} - \boldsymbol{\beta}^0_{(1)}) \geq \chi^2_{k,0.05} , \tag{25.4.1}$$

where $\chi^2_{k,0.05}$ is the upper 5th percentile of the chi-square distribution with k degrees of freedom and $\boldsymbol{C}_{11}$ is the large sample covariance matrix for $\widehat{\boldsymbol{\beta}}_{(1)}$. That is, $\boldsymbol{C} = \boldsymbol{I}_0^{-1}$ is partitioned as

$$\boldsymbol{C} = \begin{bmatrix} \boldsymbol{C}_{11} & \boldsymbol{C}_{12} \\ \boldsymbol{C}_{21} & \boldsymbol{C}_{22} \end{bmatrix} . \tag{25.4.2}$$

A likelihood ratio test of the hypothesis $H_0 : \boldsymbol{\beta}_{(1)} = \boldsymbol{\beta}^0_{(1)}$ requires the maximum likelihood estimator $\widehat{\boldsymbol{\beta}}_{(20)}$ for $\boldsymbol{\beta}_{(2)}$ obtained under the restriction that $\boldsymbol{\beta}_{(1)} = \boldsymbol{\beta}^0_{(1)}$. That is, the likelihood now has $\boldsymbol{\beta}^0_{(1)}$ fixed and it is maximized by taking the partial derivatives with respect to the entries of $\boldsymbol{\beta}_{(2)}$. Solving $\partial \log L / \partial \beta_u = 0$ for $u = k+1, \ldots, p$, we obtain $\widehat{\boldsymbol{\beta}}_{(20)}$ and form the likelihood ratio

$$\Lambda = -2 \log \left(\frac{L(\boldsymbol{\beta}^0_{(1)}, \widehat{\boldsymbol{\beta}}_{(20)})}{L(\widehat{\boldsymbol{\beta}}_{(1)}, \widehat{\boldsymbol{\beta}}_{(2)})} \right) , \tag{25.4.3}$$

where $\widehat{\boldsymbol{\beta}} = (\widehat{\boldsymbol{\beta}}'_{(1)}, \widehat{\boldsymbol{\beta}}'_{(2)})'$ is the unconstrained maximum likelihood estimator. The null hypothesis is rejected, at the .05 level, if $\Lambda > \chi^2_{k,0.05}$ which is the upper 5th percentile of the chi-square distribution with k degrees of freedom.

The large sample chi-square approximations are generally good if the maximum likelihood estimators are nearly normal.

25.5 Examples

We consider two examples to illustrate a variety of issues in estimation and exploring goodness-of-fit to the exponential regression model. The first example concerns a small experiment described by Lawless (1980) with complete observations from three groups; the second example concerns a large medical study with censored observations, not previously published in the statistical literature.

25.5.1 Breakdown of Insulating Fluid

Nelson (1990) shows time to breakdown in minutes of electrical insulating fluid under several voltage conditions. The results for three conditions, also given as Table 1.1.1 of Lawless (1982), are given in Table 25.1.

The three treatment conditions 28 kv, 30 kv and 32 kv, given to 5, 11 and 15 samples respectively, might each induce exponential breakdown at different rates. These rates, however, may not be linearly related to voltage so that at least initially we consider

Table 25.1: Insulation fluid breakdown times (minutes)

28 kilovolts	68.85	426.07	110.29	108.29	1067.6	
30 kilovolts	17.05	22.66	21.02	175.88	139.07	
	144.12	20.46	43.40	194.90	47.30	7.74
32 kilovolts	0.40	82.85	9.88	89.29	215.10	
	2.75	0.79	15.93	3.91	0.27	
	0.69	100.58	27.80	13.95	53.24	

Table 25.2: Estimates, standard errors, and p-values

Coefficients	Estimate	S.E. Estimate	p-value
$\widehat{\alpha}$	3.718	0.258	
$\widehat{\beta}$	0.610	0.397	0.12
$\widehat{\gamma}$	2.158	0.516	0.0001

voltage as a factor having three levels so that z_i is a vector with two indicator variables for the 30 kv and 28 kv levels.

Under the proportional hazards model,

$$\begin{aligned} \rho(x; \text{voltage } = 32) &= 1/\theta(32) = e^{-\alpha} \ , \\ \rho(x; \text{voltage } = 30) &= 1/\theta(30) = e^{-\alpha-\beta} \ , \\ \rho(x; \text{voltage } = 28) &= 1/\theta(28) = e^{-\alpha-\gamma} \ . \end{aligned} \tag{25.5.1}$$

The estimates and their asymptotic standard errors and p-values for the null hypothesis that the parameter is zero, are presented in Table 25.2. The maximum of the null one-parameter log-likelihood is

$$\ell_{\text{exponential}}(\widehat{\alpha}; \beta = \gamma = 0 \mid \{z_i\}) = -70.381 \ ; \tag{25.5.2}$$

and the maximum of the full likelihood is

$$\ell_{\text{exponential}}(\widehat{\alpha}, \widehat{\beta}, \widehat{\gamma} \mid \{z_i\}) = -61.630 \ . \tag{25.5.3}$$

Taking twice the difference, or 17.50, and comparing it to a chi-square distribution with 2 d.f., we obtain a p-value of .0002 for the hypothesis of no voltage effect. Considering voltage as a continuous variable and fitting the model $\theta = \exp(\alpha_\ell + \beta_\ell)$ rather than a factor produces a likelihood of -62.308, which is not a significant decrease. This is based only on the five observations with voltage $= 28$, however, so we retain the more general model. The estimated linear coefficient of voltage is $\widehat{\beta}_\ell = -1.413$ minutes/kv.

We may examine exponentiality by plotting the log empirical survival functions for each voltage level, and checking for linearity. This follows from the formula

$$S(x; \theta_i) \equiv 1 - F(x; \theta_i) = e^{-x/\theta_i} \ , \tag{25.5.4}$$

so that under exponentiality

$$-\log S(x;\theta_i) = x/\theta_i \ . \tag{25.5.5}$$

We see in Figures 25.1 and 25.2 that exponentiality may hold, though the right tails are a little long. This may be tested by embedding the exponential in a larger family of distributions by fitting the Weibull model

$$\begin{aligned}
\rho(x;\ \text{voltage } = 32) &= e^{-\alpha} \cdot \kappa \cdot (e^{-\alpha x})^{\kappa-1} \ , \\
\rho(x;\ \text{voltage } = 30) &= e^{-\alpha-\beta} \cdot \kappa \cdot (e^{-\alpha-\beta} x)^{\kappa-1} \ , \\
\rho(x;\ \text{voltage } = 28) &= e^{-\alpha-\gamma} \cdot \kappa \cdot (e^{-\alpha-\gamma} x)^{\kappa-1} \ .
\end{aligned} \tag{25.5.6}$$

We check goodness-of-fit by testing $H_0 : \ \kappa = 1$. This corresponds to (25.1.5) where $\sigma = 1/\kappa$. The maximized Weibull log-likelihood is -58.844, and we compare $-2(61.630 - 58.844) = 5.57$ to a chi-square distribution with 1 d.f. The resultant p-value is .018, indicating that the tails may indeed be long.

Pearson residuals [see McCullagh and Nelder (1989)] for the exponential case are defined as

$$r_{p,i} = \frac{x_i - \widehat{\theta}_i}{\widehat{\theta}_i} \ . \tag{25.5.7}$$

Deviance residuals more directly reflect influence on the likelihood:

$$r_{d,i} = \text{sign}(x_i - \widehat{\theta}_i)\sqrt{2\left[\frac{x_i - \widehat{\theta}_i}{\widehat{\theta}_i} - \log\left(\frac{x_i}{\widehat{\theta}_i}\right)\right]} \ . \tag{25.5.8}$$

The Pearson residuals are plotted in Figure 25.4. Figure 25.3 shows deviance residuals. Although the Pearson residuals are more extreme, both plots tell us that the largest 32 kv breakdown value is an outlier.

We can examine the influence of each datum on $\widehat{\alpha}$, $\widehat{\beta}$ and $\widehat{\gamma}$ by "leaving one out" or using the approximate methods of Pregibon (1985). Here, due to the small sample size, the former is feasible. Figure 25.5 shows the influence of each subject on the respective parameter estimates (observations treated with 32 kv also influence $\widehat{\gamma}$ and $\widehat{\beta}$ by the negative of their $\widehat{\alpha}$ influence). Here we see the effect of the largest 32 kv observation - leaving it out would decrease $\widehat{\gamma}$ by about .25. Due to the small number of 28 kv observations, each observation of this group is also influential. We conclude that careful data checking is warranted and, if all observations are correct, perhaps further modeling is required.

Since the sample size is small here, it is worthwhile to compute exact distributions of estimates (possible in the absence of censoring). This is feasible with one continuous covariate z (here, voltage) as shown by Lawless (1976). Conditioning on the ancillary quantities $a_i = \log(x_i) - \widehat{\beta}_0 - \widehat{\beta}_\ell z_i$, he finds the density of $\widehat{\beta}_\ell - \beta_\ell$ to be given in (25.3.3), with $r_i \equiv 1$.

The exact density of $\widehat{\beta}_\ell - \beta_\ell$ is compared with the normal approximation of (25.3.15) and (25.3.17) in Figure 25.6. The exact density and confidence interval are drawn with solid lines, while approximations are dotted. Since $\widehat{\beta}_\ell = -1.413$ minutes/kv, both methods indicated that β_ℓ is quite probably not 0. Lastly, by (25.3.2) exact distributions of the three estimates produced when voltage is considered to be a factor with three levels are given by a scaled chi-square with $2n$ degrees of freedom. To facilitate comparison with the normal approximation, we give densities on the log scale in Figures 25.7-25.9. For the lowest voltage category,

$$\log(\widehat{\theta}/\theta) + \log(2n) = (\widehat{\alpha} + \widehat{\gamma}) - (\alpha + \gamma) + \log(2n) \sim e^{\chi^2_{2n}} \ . \tag{25.5.9}$$

We see that the exact confidence intervals correspond fairly well to the normal approximations even in the smallest-sized group.

25.5.2 Time to Recurrence in Patients with Laryngeal and Pharyngeal Carcinomas

In 1986, the British Institute of Radiology (BIR) began accrual in a large-scale clinical study to assess the effectiveness of different treatment schedules in radiotherapy of carcinomas of the larynx and pharynx [BIR Working Party (1989) and Wiernik et al. (1991)]. Although patients were randomized to two arms (overall treatment time at most 4 weeks vs. greater than 4 weeks), we will not utilize the randomization in the following analysis.

In addition to the original question of difference between the two regimens, the BIR trial has been used as an observational study providing more detailed information on the effects of varying dose and treatment schedules [Chappell, Nondahl, and Fowler (1994a)]. This is feasible because there was, within each regimen, large variability in scheduling and dosing between clinical centers. A theoretical framework based on physical principles, known as the "linear-quadratic model" [Fowler (1989)] provides a reasonable model in this case. The linear-quadratic model dictates that mean time until tumor recurrence (and also, approximately, time to death) is log-linear in three covariates (implying an accelerated-life relation): total radiation dose (D); total dose squared divided by the number of doses given (Df); and total treatment time (T). In addition, other predictors such as the severity of the cancer (Stage) may be considered. See Chappell, Fowler and Nondahl (1993b) for conditions in which death can be exponentially distributed with mean

$$\theta = \beta_0 + \beta_1 \times I(Stage1) + \beta_2 \times I(Stage2) + \beta_3 \times D + \beta_4 \times Df + \beta_5 \times T \ . \tag{25.5.10}$$

There were 480 patients whose disease had not spread beyond the original tumor. Out of these, 9 died before treatment started or otherwise had missing information. Followup time, death status, Stage, D, Df and T for the remaining 471 subjects are given in Appendix 1. Note that there is substantial censoring; 298 patients finished the trial alive.

Table 25.3 gives the estimated coefficients of the linear-quadratic model for the BIR data. Stage 1 and 2 effects are estimates using Stage 3 as a baseline; they show the expected pattern in that patients with lower stages take longer to recur. Dose delays recurrence in a super-linear fashion, as shown by the positive coefficients of D and Df. Longer treatment times (conditional on dose stage and fractionation) accelerate recurrence. These results are in accordance with radiophysical theory.

Table 25.3: Estimated coefficients of the linear-quadratic model for the BIR model

Variable	Estimated coefficient	Estimated s.e.	p-value
Intercept	$\widehat{\beta}_0 = 1.0572$	0.8721	
Stage			$< .0001$
1	$\widehat{\beta}_1 = 1.3013$	0.1884	
2	$\widehat{\beta}_2 = 0.7589$	0.2018	
D	$\widehat{\beta}_3 = 0.04560/Gy$	0.02031	0.025
Df	$\widehat{\beta}_4 = 0.007690/Gy^2$	0.002954	0.009
T	$\widehat{\beta}_5 = -0.017975/\text{day}$	0.015296	0.24

Asymptotic standard errors are computed using the observed information matrix in (25.3.16) and p-values for the continuous covariates calculated by standardizing the estimates accordingly. The impact of the Stage factor is examined using a likelihood-ratio test (25.4.3) with two degrees of freedom.

Although in the following we illustrate several methods for examining the exponentiality of these data, these are by no means the only diagnostics which should be undertaken. For example, although Df is essentially an interaction term between the dose and the dose per fraction, the lower order term corresponding to the latter quantity is not included in the model. This type of action should always be scrutinized - see Chappell, Nondahl, and Fowler (1994b) for details.

The estimated survival function is useful for graphically depicting distributional form here, but only after modification. First, rather than just using one minus the empirical distribution function, account must be taken of the censoring. In the absence of covariates, Kaplan and Meier (1958) proposed the product-limit estimate:

$$\widehat{S}(x) = \prod_{x_i < x} \left[1 - \frac{r_i}{n_i} \right] , \tag{25.5.11}$$

where the x_i are the unique times at which failures occurred, r_i is the number of failures at x_i, and n_i is the number of patients still at risk (that is, not recurred or censored) at x_i.

In the present case, the effects of covariates would make $\widehat{S}(x)$ an estimate of a continuous mixture of exponential distributions. Under the log-linear exponential model, the unconditional $\widehat{S}(x)$ would therefore have unknown form. Unlike the previous example the covariates do not have a limited number of levels, each of which can be examined for exponentiality.

A solution lies in the model of Cox (1972) which keeps only the proportional hazards formulation of Eq. (25.1.4) but requires no functional assumptions on $\rho_0(x)$. Cox's model is a semiparametric one that allows arbitrary hazard and survival functions as long as the covariates influence the former in a proportional fashion. Bailey (1979) described the properties of an extension to the product-limit estimator after adjusting for covariates using the Cox model [see Kalbfleisch and Prentice (1980) for a summary]. In this way, we may get a conditional estimate of $\widehat{S}(x)$, represented as $\widehat{S}(x \mid \boldsymbol{z})$, for a given set of covariate values $\boldsymbol{z}_i$. As with the product limit estimator, it has the form of a product of hazard estimates:

$$\widehat{S}_{pl}(x \mid \mathbf{0}) = \prod_{t_i \leq x} [1 - \widehat{p}^{(i)}] , \tag{25.5.12}$$

where $\widehat{S}_{pl}(x \mid \mathbf{0})$ is the "baseline" survival estimate for $\boldsymbol{z} = \mathbf{0}$. Bailey estimated $p^{(i)}$ by

$$1 - \widehat{p}^{(i)} = \left[\frac{\sum_{j \in R(t_i)} e^{-\widehat{\boldsymbol{\beta}}' \boldsymbol{z}_j} - e^{-\widehat{\boldsymbol{\beta}}' \boldsymbol{z}_i}}{\sum_{j \in R(x_i)} e^{-\widehat{\boldsymbol{\beta}}' \boldsymbol{z}_j}} \right]^{e^{-\widehat{\boldsymbol{\beta}}' \boldsymbol{z}_i}} . \tag{25.5.13}$$

$R(x_i)$ is the risk set at time x_i, that is the set of all indices $\{j\}$ such that the jth followup time is less than x_i, regardless of censoring. As usual, $\boldsymbol{z}_i$ is the vector of covariates for the ith observation and $\widehat{\boldsymbol{\beta}}$ is the vector of coefficient estimates. $\widehat{S}_{pl}(x \mid \mathbf{0})$ may not always represent a function of real interest (for example, if one of the covariates is weight or other positive quantity). It may be generalized via the transformation

$$\widehat{S}(x \mid \boldsymbol{z}) = \left[\widehat{S}_{pl}(x \mid \mathbf{0}) \right]^{e^{\widehat{\boldsymbol{\beta}}' \boldsymbol{z}}} . \tag{25.5.14}$$

The components of $\boldsymbol{z}$ are generally set equal to their mean, in the case of continuous variables, or a particular level, for factors. Note that $\widehat{S}_{pl}(x \mid \mathbf{0})$ and $\widehat{S}(x \mid \boldsymbol{z})$ are equivalent for diagnostic purposes.

For the BIR data we condition on a stage of 3 with D, Df, and T equaling their means ($\boldsymbol{z}' = [0,\ 0,\ \overline{D},\ \overline{Df},\ \overline{T}]$). The resultant conditional survival curve estimate is graphed in Figure 25.10. Figure 25.11 shows the log estimated survival curve which, showing no major departures from linearity, is consistent with data from the exponential distribution.

The log-likelihood ratio for the exponential model is -455.31. For the Weibull model with -455.26. Doubling their difference and comparing to a chi-square distribution with one degree of freedom yields a p-value of .75, indicating the apparent adequacy of an exponential model.

A test of model adequacy can be undertaken by comparing the two estimated covariance matrices in (25.3.16) and sum-squared terms of (25.3.15) with $\boldsymbol{\beta} = \widehat{\boldsymbol{\beta}}$ [White (1982)]. Testing the diagonals only, we have $p > .10$. Thus, the exponential fit is apparently adequate, an unusual result for a medical data set of this size.

The exponential model's fit in this case may be due partly to censoring the largest observed failure times at about ten years - all later recurrences (as well as many potential earlier ones) are censored. Thus although the hazard of tumor recurrence may eventually decrease, reflecting the possibility of a "cure", this is not seen within the first ten years and so does not affect our model. Additionally, censoring prevents any single observation from being too influential.

References

Bailey, K. (1979). *The General Maximum Likelihood Approach to the Cox Model*, Ph.D. Dissertation, Department of Statistics, University of Chicago.

BIR Working Party (1989). First interim progress report on the second British Institute of Radiology fractionation study: short vs. long overall treatment times for radiotherapy of carcinoma of the laryngo-pharynx, *British Journal of Radiology*, **62**, 450-456.

Chappell, R., Nondahl, D.M., and Fowler, J.F. (1994a). Re-analysis of the effect of dose, time and fraction in the first and second BIR fractionation studies, *The International Journal of Radiation Oncology, Biology & Physics* (To appear).

Chappell, R., Nondahl, D.M., and Fowler, J.F. (1994b). Modeling dose and local control in radiotherapy, *Journal of the American Statistical Association* (To appear).

Cox, D.R. (1972). Regression models and life tables (with discussion), *Journal of the Royal Statistical Society, Series B*, **34**, 187-202.

Cox, D.R. and Oakes, D. (1984). *Analysis of Survival Data*, Chapman and Hall, London.

Epstein, B. and Sobel, M. (1954). Some theorems relevant to life testing from an exponential distribution, *Annals of Mathematical Statistics*, **25**, 373-381.

Feigl, P. and Zelen, M. (1965). Estimating and exponential survival probabilities with concomitant information, *Biometrics*, **21**, 826-838.

Fowler, J.F. (1989). The linear-quadratic formula and progress in fractionated radiotherapy, *British Journal of Radiology*, **62**, 679-694.

Kahn, H.D. (1979). Least squares estimation for the inverse power law for accelerated life tests, *Applied Statistics*, **28**, 40-46.

Kalbfleisch, J.D. and Prentice, R.L. (1980). *The Statistical Analysis of Failure Time Data*, John Wiley & Sons, New York.

Kaplan, E.L. and Meier, P. (1959). Nonparametric estimation from incomplete observations, *Journal of the American Statistical Association*, **53**, 457-481.

Lawless, J.F. (1976). Confidence interval estimation in the inverse power law model, *Applied Statistics*, **25**, 128-138.

Lawless, J.F. (1982). *Statistical Models & Methods for Lifetime Data*, John Wiley & Sons, New York.

Maller, R.A. (1988). On the exponential model of survival, *Biometrika*, **75**, 582-586.

McCullagh, P. and Nelder, J.A. (1989). *Generalized Linear Models*, Second edition, Chapman & Hall, London.

Nelson, W. (1990). *Accelerated Testing: Statistical Models, Test Plans and Data Analysis*, John Wiley & Sons, New York.

Pregibon, D. (1985). Link tests, In *Encyclopedia of Statistical Sciences* (Eds., S. Kotz and N.L. Johnson), John Wiley & Sons, New York.

Singpurwalla, N.D. (1971). A problem in accelerated life testing, *Journal of the American Statistical Association*, **66**, 841-845.

White, H. (1982). Maximum likelihood estimation of misspecified models, *Econometrica*, **50**, 1-25.

Wiernik, G. et al. (1991). Final report on the second British Institute of Radiology fractionation study: short vs. long overall treatment times for radiotherapy of carcinoma of the laryngo-pharynx, *British Journal of Radiology*, **64**, 232-241.

Zelen, M. (1959). Factorial experiments in life testing, *Technometrics*, **1**, 269-288.

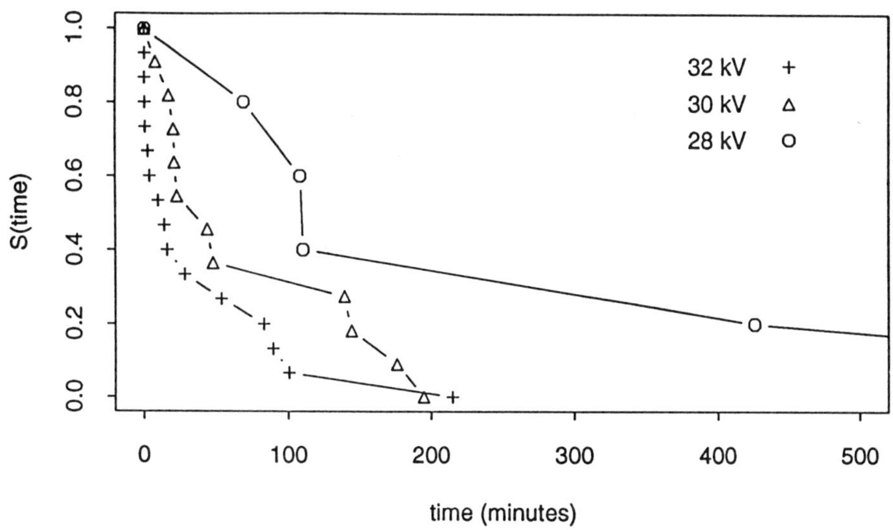

Figure 25.1: Survival curves for insulating fluid data

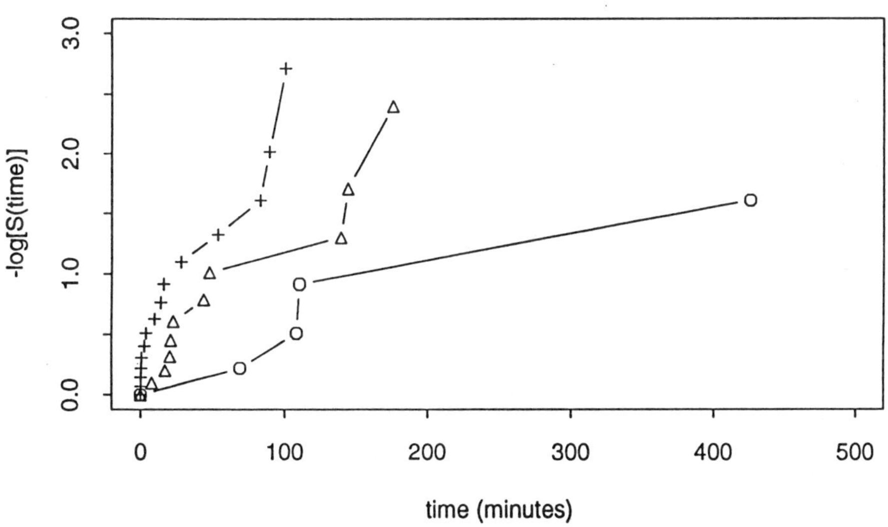

Figure 25.2: -Log survival curves for insulating fluid data

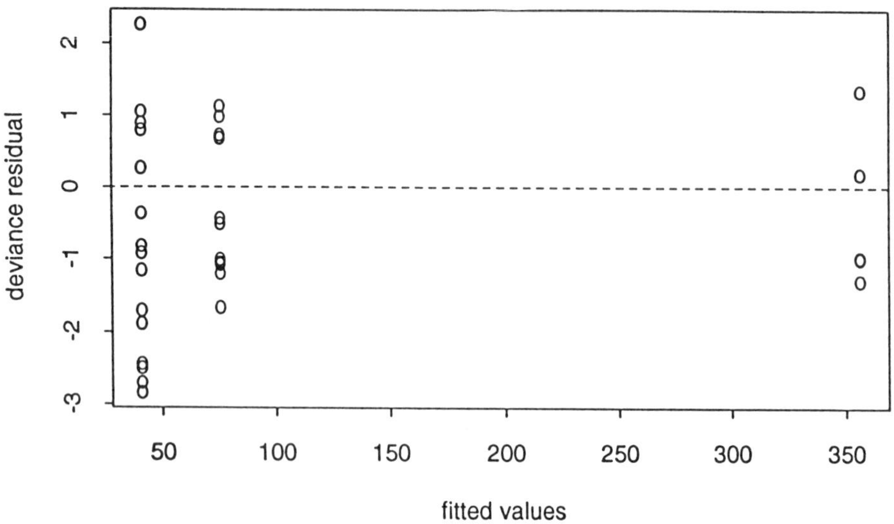

Figure 25.3: Deviance residual plot, insulating fluid example

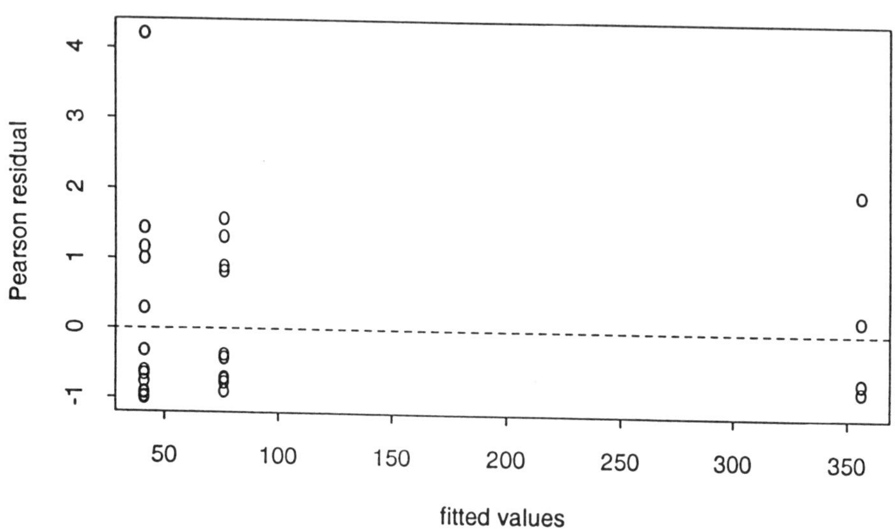

Figure 25.4: Pearson residual plot, insulating fluid example

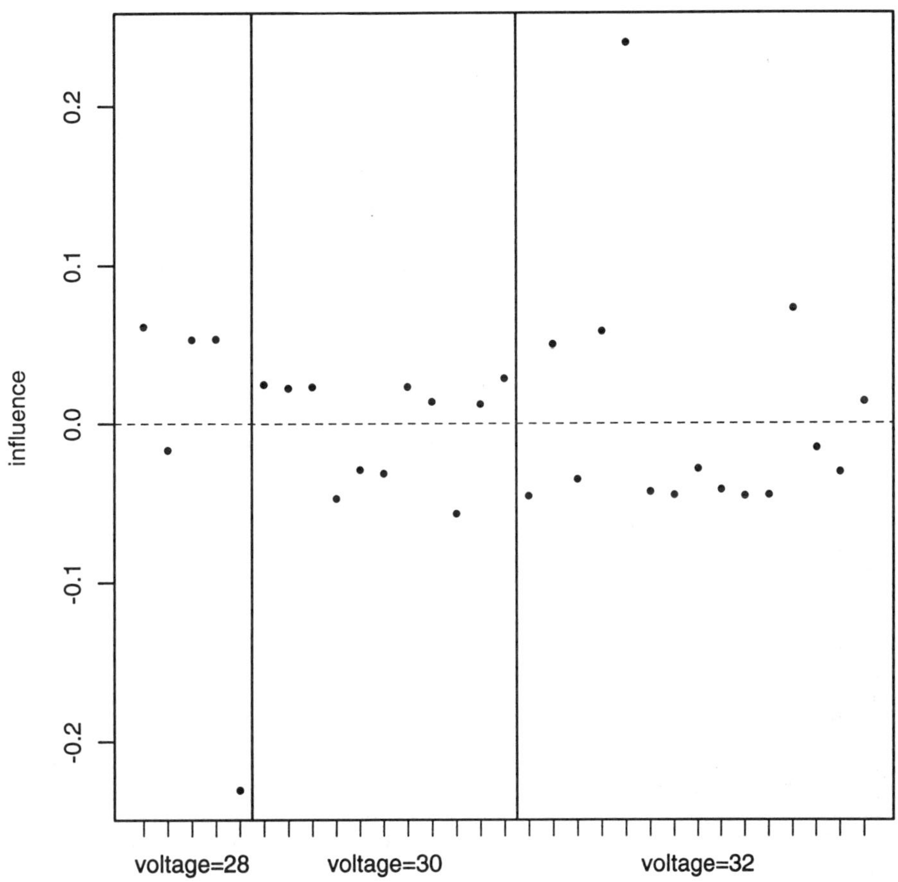

Figure 25.5: Influence plot, insulating fluid example

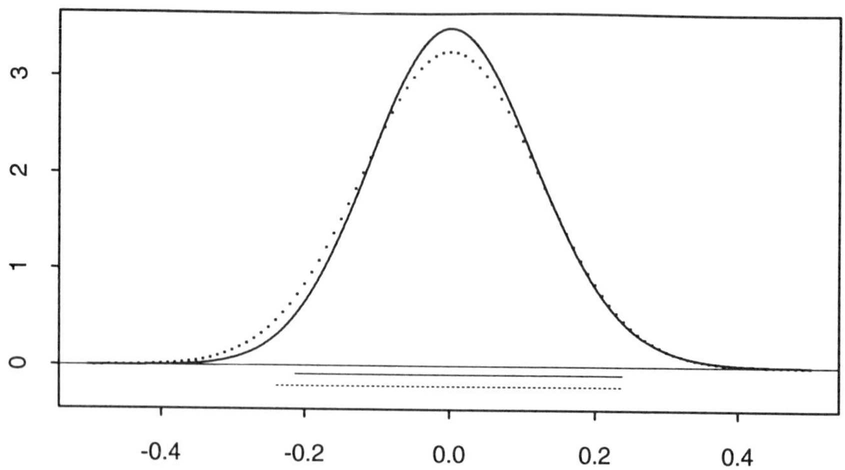

Exact 95% C.I. (-0.213, 0.238); Normal approximation (-0.239, 0.239)

Figure 25.6: Linear coefficient: $\widehat{\boldsymbol{\beta}}_\ell - \boldsymbol{\beta}_\ell$

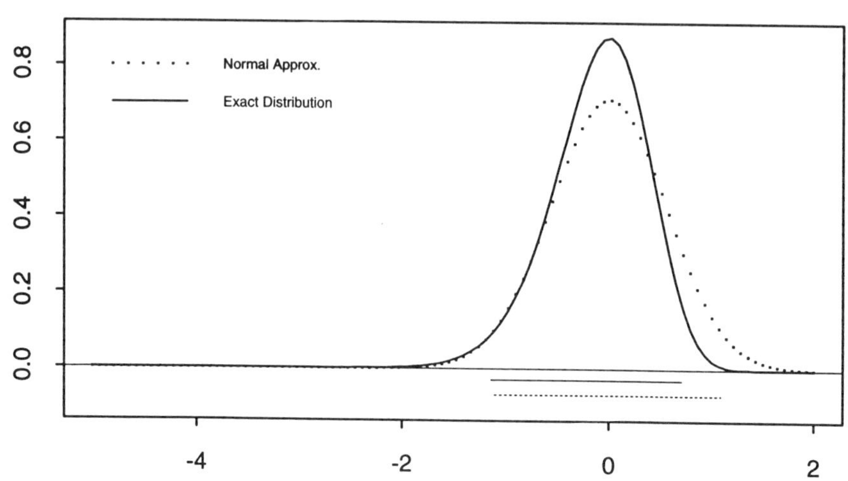

Exact 95% C.I. (-1.129, 0.711); Normal approximation (-1.098, 1.098)

Figure 25.7: $\log(\widehat{\theta}_{28}/\theta_{28})$: $\widehat{\alpha} + \widehat{\gamma} - (\alpha + \gamma)$

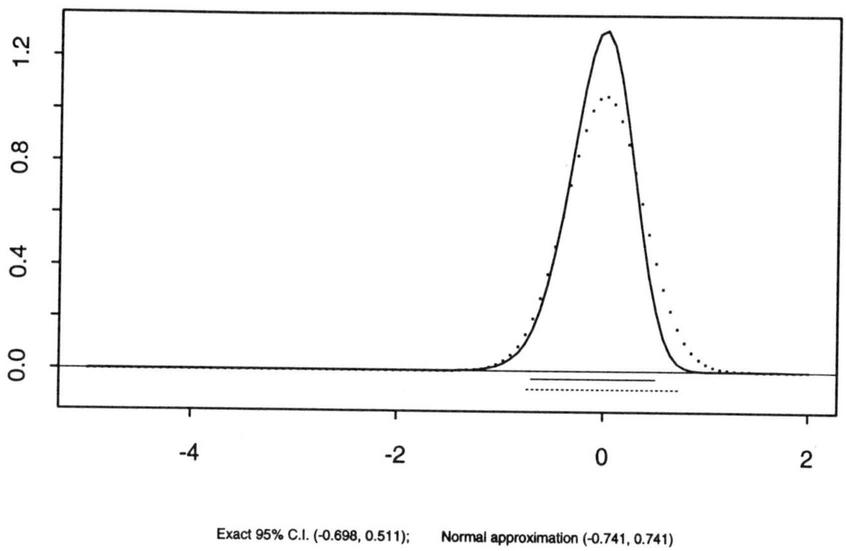

Figure 25.8: $\log(\widehat{\theta}_{30}/\theta_{30})$: $\widehat{\alpha}+\widehat{\beta}-(\alpha+\beta)$

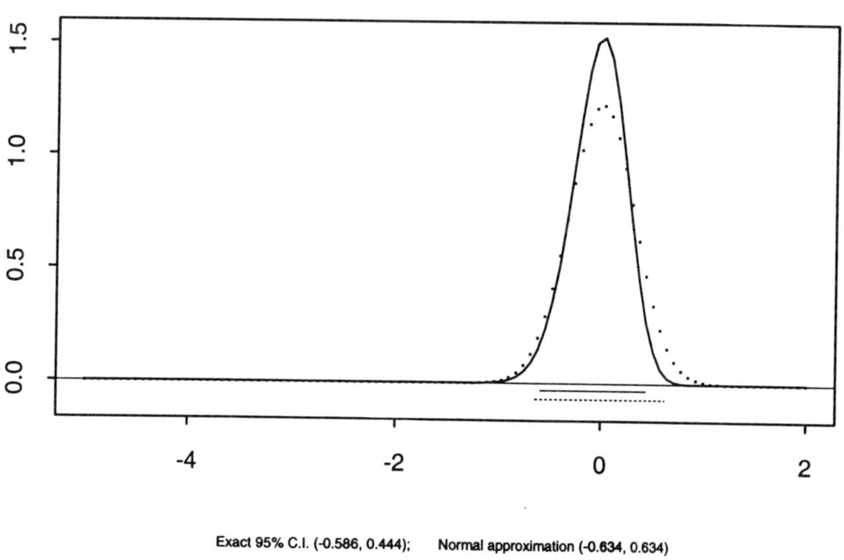

Figure 25.9: $\log(\widehat{\theta}_{32}/\theta_{32})$: $\widehat{\alpha}-\alpha$

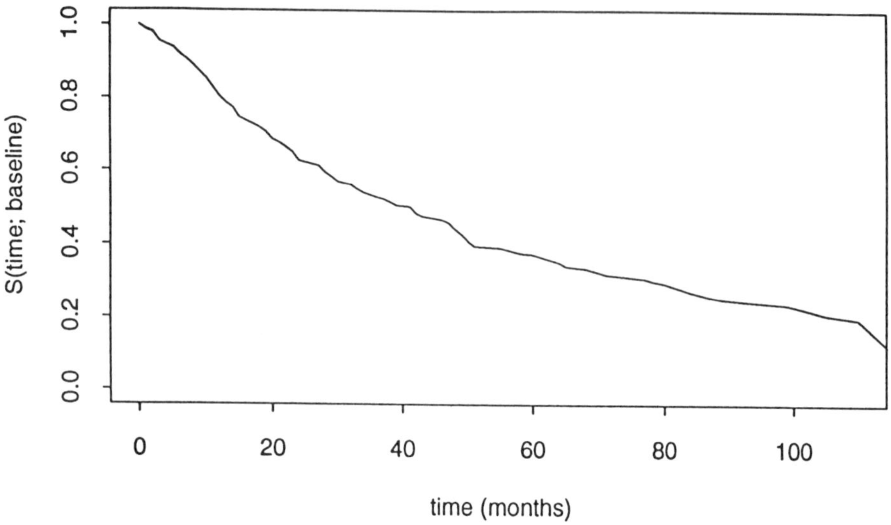

Figure 25.10: Baseline survival curve for BIR data

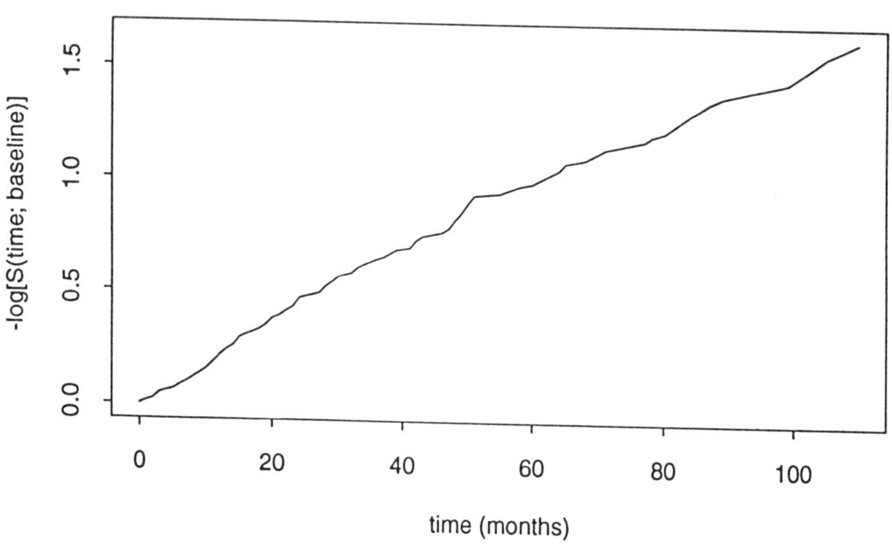

Figure 25.11: -Log baseline survival curve for BIR data

Appendix 1: Outcomes and treatment variables in the BIR study

Followup (F) is in months, censoring (C) is an indicator which equals zero if the patient died or one if censored at the followup time, stage (S) reflects the degree of tumor invasion, dose (D) is in grays, dose × dose per fraction (Df) is in gray^2, and treatment time (T) is in days.

F	C	S	D	DF	T	F	C	S	D	DF	T	F	C	S	D	DF	T
1	0	1	61	122	39	24	0	3	47	171	28	60	1	1	63	128	42
1	0	2	48	174	28	24	1	2	58	177	42	60	1	1	63	128	42
1	1	1	63	209	42	24	1	1	48	174	28	60	1	2	60	120	39
1	0	3	56	174	42	24	0	3	55	252	28	62	0	3	48	174	29
2	0	3	55	189	22	24	1	1	60	120	42	62	0	3	42	176	21
2	1	3	57	173	42	24	1	2	46	135	25	64	0	2	48	177	28
2	1	3	48	174	28	24	1	1	60	120	43	64	0	1	42	176	21
2	1	1	65	132	46	24	1	1	56	174	42	65	0	1	60	120	44
2	1	3	46	135	22	24	1	1	43	182	21	65	0	2	46	135	21
2	1	1	50	156	21	24	1	2	49	150	22	68	0	2	48	144	22
2	0	1	50	156	21	24	1	1	65	132	45	70	0	2	48	174	28
3	0	3	55	189	21	24	1	1	48	147	23	70	0	1	59	180	42
3	0	3	58	177	42	24	1	2	49	150	21	71	0	2	48	177	28
3	1	3	46	166	28	24	1	1	66	137	51	72	1	1	63	133	43
3	1	1	59	181	42	24	1	1	50	195	31	72	1	3	61	125	43
3	1	1	58	177	42	24	0	1	50	156	21	72	1	1	46	135	23
3	1	1	48	174	28	24	1	2	50	156	21	72	1	1	48	174	28
3	0	2	46	135	21	24	1	2	50	156	21	72	1	1	48	174	28
3	1	3	46	135	21	25	0	1	52	131	28	72	1	2	58	177	42
3	0	1	60	120	39	27	0	1	57	170	42	72	1	1	58	177	42
3	0	3	46	135	21	27	0	1	58	177	42	72	1	3	57	174	42
3	1	3	46	135	21	28	0	1	48	174	28	72	1	1	53	146	42
3	1	2	58	186	42	28	0	3	60	120	46	72	1	2	52	184	19
3	0	2	67	137	44	28	0	3	44	195	22	72	1	1	60	189	42
3	1	3	63	128	42	28	0	1	58	186	40	72	1	2	46	135	23
4	0	2	57	172	48	29	0	2	57	173	42	72	1	2	60	120	45
4	0	3	60	120	44	29	0	3	46	135	21	72	1	1	46	135	19
5	0	1	46	135	21	29	0	2	61	125	43	72	1	1	60	120	43
5	0	2	63	128	42	30	0	3	65	137	42	72	1	1	60	120	40
6	1	3	60	120	39	30	0	1	55	161	42	72	1	1	62	144	45
6	1	1	47	170	28	30	0	2	48	174	28	72	1	2	63	128	42
6	1	2	47	169	28	32	0	2	60	120	38	72	1	1	63	128	42
6	1	2	58	177	46	32	0	2	42	176	21	72	1	1	63	128	42
6	0	2	48	174	28	33	0	3	55	189	21	72	1	1	63	128	42
6	1	1	48	174	28	33	0	1	60	120	47	72	1	2	50	156	21
6	1	1	57	174	42	33	0	3	46	135	23	72	1	1	63	128	42
6	1	3	43	141	28	34	0	2	48	174	28	74	0	2	63	128	42
6	0	1	58	177	42	34	0	2	65	132	43	77	0	2	59	180	42
6	1	2	48	174	32	36	0	3	55	189	21	78	0	1	47	169	28
6	1	1	58	167	44	36	0	3	55	189	21	80	0	3	60	120	42
6	1	1	55	252	28	36	1	3	65	141	43	83	0	3	65	141	41
6	1	2	60	120	43	36	1	2	49	152	23	83	0	1	65	137	42
6	0	3	46	135	21	36	1	1	46	135	23	83	0	1	46	135	21
6	1	1	46	135	21	36	1	1	46	135	23	84	1	1	63	133	32
6	1	1	46	135	22	36	1	1	59	181	42	84	1	3	40	102	21
6	1	1	60	120	41	36	1	1	58	177	42	84	0	2	58	177	42
6	1	1	42	176	21	36	0	1	47	170	28	84	1	2	48	174	28
6	1	1	56	174	42	36	1	2	58	177	46	84	1	3	48	174	28
6	1	2	43	189	21	36	1	1	58	180	43	84	1	1	48	174	28
6	1	1	56	174	42	36	1	1	48	144	24	84	1	1	48	144	22
6	1	2	66	137	45	36	1	1	42	176	21	84	1	2	61	148	35
6	1	2	65	132	31	36	1	1	50	138	42	84	1	2	62	156	35
6	1	2	63	128	42	36	1	1	48	115	28	84	1	1	55	252	27
6	1	1	63	128	42	36	1	1	49	150	22	84	1	1	62	201	42
6	1	2	63	128	42	36	1	1	63	128	42	84	1	1	64	213	42
6	0	1	50	156	21	36	1	1	50	156	21	84	1	1	60	120	45
6	1	1	50	156	21	36	1	2	63	128	42	84	1	1	46	135	21

Appendix 1 (Contd)

F	C	S	D	DF	T	F	C	S	D	DF	T	F	C	S	D	DF	T
7	0	3	52	172	21	36	1	3	63	128	42	84	1	1	60	120	43
7	0	3	48	143	27	37	0	3	55	189	21	84	1	2	60	120	42
7	0	3	46	135	22	38	0	1	50	156	21	84	1	1	42	176	21
8	0	3	65	141	45	38	0	3	50	156	21	84	1	1	41	168	21
8	0	1	48	174	28	39	0	3	65	141	42	84	1	1	50	156	21
8	0	1	60	120	43	39	0	1	46	135	21	84	1	1	50	156	21
8	0	1	50	156	21	41	0	1	58	177	42	84	1	2	63	128	42
9	0	3	47	169	30	42	0	2	58	177	42	84	1	2	50	156	21
9	0	1	48	174	28	42	0	2	60	120	41	84	1	1	63	128	42
9	0	2	46	135	21	42	0	3	46	135	20	84	1	1	50	156	22
9	0	3	46	135	26	42	0	3	46	135	21	87	0	3	65	141	42
10	0	3	60	120	41	43	0	2	46	135	21	87	0	1	58	177	42
10	0	1	47	140	21	43	0	2	46	135	21	89	0	1	47	170	28
10	0	1	48	174	28	46	0	2	46	135	22	93	0	3	65	141	41
10	0	3	56	174	40	46	0	2	46	135	23	96	1	3	55	189	21
11	0	3	55	189	21	47	0	2	60	120	43	96	1	3	55	189	21
11	0	1	61	122	39	47	0	2	66	135	45	96	1	3	66	145	39
11	0	2	46	135	21	48	0	3	55	189	21	96	1	1	52	131	28
11	0	3	46	135	21	48	0	3	55	189	21	96	1	1	59	181	48
11	0	2	50	156	21	48	1	3	65	141	47	96	1	1	57	173	42
11	0	1	63	128	42	48	1	3	52	172	22	96	1	1	58	177	42
12	0	3	55	189	21	48	1	3	65	141	42	96	1	2	48	174	28
12	1	3	65	141	42	48	1	2	47	169	28	96	1	1	48	174	28
12	1	3	65	141	42	48	1	1	58	177	42	96	1	2	58	177	42
12	0	3	65	141	42	48	1	1	44	147	30	96	1	2	48	174	28
12	0	3	55	189	21	48	1	1	58	177	42	96	1	1	47	173	28
12	1	3	65	141	42	48	1	1	48	174	28	96	1	1	58	177	42
12	0	3	46	135	21	48	1	1	47	170	28	96	1	3	58	167	44
12	1	1	63	130	39	48	1	3	47	171	28	96	1	2	65	136	45
12	1	2	48	174	28	48	1	1	48	174	30	96	1	1	61	196	42
12	1	1	48	174	28	48	1	1	58	180	42	96	1	1	53	234	28
12	1	2	48	174	28	48	1	1	48	174	28	96	1	1	62	205	42
12	1	1	58	180	42	48	1	1	48	144	21	96	1	1	46	135	21
12	1	1	48	174	28	48	1	2	62	151	34	96	1	2	60	120	44
12	1	1	44	150	29	48	1	1	46	135	19	96	1	1	60	120	45
12	1	2	58	177	42	48	1	1	58	186	38	96	1	1	66	134	44
12	1	1	58	177	42	48	0	2	42	176	21	96	1	1	49	150	21
12	1	1	58	177	42	48	1	1	42	176	21	96	1	1	48	144	21
12	1	1	58	177	48	48	1	2	48	144	21	96	1	1	50	156	21
12	1	2	58	177	42	48	0	3	63	128	42	96	1	1	50	156	21
12	1	3	48	144	22	48	1	1	63	128	42	99	0	3	65	132	44
12	1	2	65	132	44	48	1	1	50	156	21	103	0	1	58	177	42
12	1	1	46	135	21	48	1	1	63	128	43	103	0	1	60	120	41
12	0	3	60	120	42	48	1	1	63	128	42	105	0	1	59	126	42
12	1	2	46	135	21	48	1	2	50	156	21	108	1	3	65	141	42
12	1	2	46	135	21	48	1	1	63	128	42	108	1	3	55	189	21
12	0	1	55	168	39	48	1	1	63	128	42	108	1	3	55	189	21
12	1	1	58	186	38	48	1	2	49	147	21	108	1	3	65	141	42
12	1	1	42	180	21	48	1	1	63	128	42	108	1	3	46	135	21
12	1	1	42	173	21	48	1	2	63	128	42	108	1	2	61	197	42
12	1	1	59	192	42	49	0	3	58	177	42	108	1	1	48	174	28
12	1	1	68	146	47	49	0	1	46	135	21	108	1	1	58	180	42
12	1	1	66	134	45	49	0	1	63	128	42	108	1	1	48	176	28
12	1	2	49	150	25	50	0	3	65	141	41	108	1	1	57	173	42
12	1	2	66	137	45	50	0	3	65	141	42	108	1	2	58	177	42
12	1	1	50	156	21	50	0	2	44	77	34	108	1	1	47	171	28
13	0	3	48	179	28	50	0	1	58	177	42	108	1	2	58	177	47
13	0	2	48	174	34	51	0	3	48	174	28	108	1	1	54	248	28
13	0	1	60	120	43	51	0	1	65	132	44	108	1	1	62	206	42
13	0	1	66	137	45	51	0	2	50	156	21	108	1	1	53	232	28
14	0	3	65	141	41	55	0	1	52	131	28	108	1	1	46	135	26
14	0	3	65	141	42	57	0	1	56	168	42	108	1	2	46	135	21
14	0	1	49	149	21	57	0	1	50	156	21	108	1	2	46	135	22
15	0	3	52	172	21	58	0	2	48	174	28	108	1	1	56	174	42
15	0	3	65	141	42	60	1	3	55	202	21	108	1	1	50	156	21
15	0	2	59	181	42	60	1	3	64	138	46	110	0	3	52	172	21
15	0	3	48	144	21	60	1	3	55	189	26	120	1	3	55	189	22
15	0	3	46	135	21	60	0	2	46	135	23	120	1	3	55	189	21
15	0	1	60	133	42	60	1	1	60	120	39	120	1	3	55	189	21

Appendix 1 (Contd)

F	C	S	D	DF	T	F	C	S	D	DF	T	F	C	S	D	DF	T
16	0	3	65	141	42	60	1	1	46	135	23	120	1	3	55	189	21
16	0	3	65	141	42	60	1	2	60	120	44	120	1	3	65	141	41
17	0	2	48	174	28	60	1	1	46	135	23	120	1	3	65	141	42
17	0	1	48	144	21	60	1	1	47	140	22	120	1	3	55	189	22
18	0	3	48	177	28	60	1	1	48	174	28	120	1	3	65	141	42
18	0	2	60	120	44	60	1	1	58	177	42	120	1	2	58	177	42
19	0	3	65	141	42	60	1	1	58	177	42	120	1	1	58	177	42
19	0	1	58	167	44	60	1	1	48	176	26	120	1	2	47	170	33
19	0	2	62	206	42	60	1	1	65	132	44	120	1	1	58	177	42
20	0	3	55	189	21	60	1	1	64	213	42	120	1	2	56	168	42
20	0	3	55	189	21	60	1	2	55	252	28	120	1	2	58	177	42
20	0	1	60	120	43	60	1	2	46	135	21	120	1	1	48	174	28
20	0	3	60	120	48	60	1	3	46	135	21	120	1	1	60	187	42
20	0	2	50	156	22	60	1	1	60	120	43	120	1	1	58	180	42
21	0	3	65	141	42	60	1	1	46	135	22	120	1	1	57	173	42
21	0	1	60	120	39	60	1	3	46	135	21	120	1	2	48	174	28
22	0	1	47	140	22	60	1	1	46	135	22	120	1	1	58	180	42
22	0	2	60	120	39	60	1	1	60	120	39	120	1	1	48	144	22
22	0	3	65	137	42	60	1	2	46	135	21	120	1	3	65	132	44
23	0	3	47	169	28	60	1	1	42	176	21	120	1	1	65	132	44
23	0	3	60	120	42	60	1	1	49	150	21	120	1	1	63	206	45
23	0	3	60	120	39	60	1	2	48	144	21	120	1	1	54	243	28
24	0	3	65	141	42	60	1	1	48	144	21	120	1	3	62	204	43
24	0	3	52	172	21	60	1	2	50	156	21	120	1	1	55	252	28
24	0	3	47	137	21	60	1	1	50	156	21	120	1	2	60	120	43
24	1	3	61	122	44	60	1	2	63	128	42	120	1	1	46	135	21
24	1	1	58	177	46	60	1	1	50	156	21	120	1	1	60	120	42
24	1	1	57	173	42	60	1	1	50	156	21	120	1	1	46	135	21
24	1	2	58	177	42	60	1	2	63	128	42	120	1	1	45	127	21
24	1	1	47	170	28	60	1	3	63	128	42	120	1	1	65	132	44
24	1	1	56	166	42	60	1	1	50	156	22	122	0	1	65	137	42

Appendix 2: SAS code for parametric and baseline survival analyses for Example 25.5.2

```
APPENDIX 2: SAS code for parametric and baseline survival analyses for
            Example 5.2

data bir (label = 'BIR database of N(0) laryngeal ca.s');
     input OBS FOLLOWUP CENSOR TNM_T D DF T;
* create factor variables from tnm_t;
     kt1 = 0; kt2=0;
     if (tnm_t eq 1) then kt1 = 1;
     if (tnm_t eq 2) then kt2 = 1;
     f = Df/D;
     label D        = 'Total dose (gy)'
           f        = 'Dose per fraction (gy)'
           Df       = 'Dose x Dose per fraction (gy^2)'
           T        = 'Time under treatment (days)'
           followup = 'Time to last followup (months)'
           censor   = 'Indicator variable for death (0)'
           tnm_t    = 'T - stage'
           kt1      = 'T - stage = 1'
           kt2      = 'T - stage = 2';

     drop obs;
     cards;
           1      30        0        3      65.10    136.710     42

                                . . .

           480      60        1        2      60    120.000     39
;

* parametric analyses;
proc lifereg data=bir;
     class tnm_t;
     model followup*censor(1) = tnm_t D Df T / covb corrb d=exponential;
     model followup*censor(1) = tnm_t D Df T / covb corrb d=weibull;

* Set variables at baseline values to calculate estimated S_0(t);
* (PHREG's default baseline estimate won't accomodate factors such as T-stage);
proc means data=bir noprint mean;
     var D Df T;
     output out=bir_base mean=D Df T;

data bir_base; set bir_base; kt1=0; kt2=0;

* SAS version 6.07 or later required for PROC PHREG;
proc phreg data=bir;
     model followup*censor(1) = kt1 kt2 D Df T / ties=exact covb corrb;
     baseline out=bir_surv covariates=bir_base
              survival=survival logsurv=logsurv / nomean;

proc plot data=bir_surv; plot logsurv*followup; plot survival*followup;
```

CHAPTER 26

Two-Stage and Multi-Stage Estimation

N. Mukhopadhyay[1]

26.1 Introduction

We consider the negative exponential distributions having the probability density function (p.d.f.)

$$f(t;\mu,\theta) = \frac{1}{\theta}\, e^{-(t-\mu)/\theta} I(t > \mu) \tag{26.1.1}$$

in this chapter, where $I(\cdot)$ stands for the usual indicator function of $(\cdot)$. We assume that $\mu \in \mathbf{R}$ and $\theta \in \mathbf{R}^+$ are respectively the location and scale parameters of such distributions. Often, in applications, one assumes that $\mu \in \mathbf{R}^+$, but in this chapter we will essentially work under the assumption that $\mu \in \mathbf{R}$, Section 26.3 being the only exception.

Various sequential estimation problems in the case of negative exponential distributions were put together and reviewed in the article of Mukhopadhyay (1988a). These problems and the associated methodologies are briefly included here for completeness. On the other hand, newer methodologies and results have appeared in the literature during the past five years or so. In this chapter, we emphasize the latter aspects, and present the topics as follows:

[1]University of Connecticut, Storrs, Connecticut

In this chapter, one will find extensive coverage of various types of estimation problems. The details and motivations are left out for majority of such topics. In most instances, the original papers plus the review article of Mukhopadhyay (1988a) will however facilitate understanding of the general flow and developments.

Throughout, $I(\cdot)$ continues to stand for the indicator function of $(\cdot)$, and $\langle x \rangle$ stands for the largest integer $< x$.

26.2 Fixed-Width Confidence Intervals for the Location

Let $X_1, X_2, \ldots$ be a sequence of independent and identically distributed random variables (i.i.d. variables) having the density $f(x; \mu, \theta)$ given by (26.1.1). The parameters μ and θ are both assumed unknown. Now, given $d(> 0)$ and $\alpha \in (0,1)$, the goal is to construct a confidence interval I for μ such that (i) the length of I is d, and (ii) $P(\mu \in I)$ is at least $(1-\alpha)$, for all $\mu \in \mathbf{R}$, $\theta \in \mathbf{R}^+$.

Having recorded $X_1, \ldots, X_n$, let

$$T_n = \min\{X_1, \ldots, X_n\}, \qquad U_n = (n-1)^{-1} \sum_{i=1}^{n} (X_i - T_n)$$

for $n \geq 2$. Consider the *fixed-width confidence interval* $I_n = [T_n - d, T_n]$ for μ, and it is easy to see that $P(\mu \in I_n) \geq 1 - \alpha$ provided that n is the smallest integer $\geq a\theta/d = C$, say, where $a = \ell n(1/\alpha)$. One refers to "C" as the *optimal fixed sample size* required had θ been known. However, C is unknown and hence one opts for different types of multi-stage estimation methodologies.

26.2.1 Two-Stage Procedure

Ghurye (1958) proposed the following two-stage procedure along the lines of Stein (1945, 1949). Start with $X_1, \ldots, X_{m_1}$, $m_1 (\geq 2)$, and let

$$N = \max\left\{m_1, \langle b(m_1) U_{m_1} d^{-1} \rangle + 1\right\}, \tag{26.2.1}$$

where $b(m_1)$ is the upper $100\alpha\%$ point of the F-distribution with 2 and $2m_1 - 2$ degrees of freedom. If $N = m_1$, we do not take any more samples in the second stage, but if $N > m_1$, then we sample the difference $(N - m_1)$ in the second stage. Now, based on $X_1, \ldots, X_N$ we propose the confidence interval I_N for μ.

Theorem 26.1 *For the two-stage procedure (26.2.1), we have:*

(a) $P(\mu \in I_N) \geq 1 - \alpha$;

(b) $E[N] > C$;

(c) $P(\mu \in I_N) \to 1 - \alpha$ *as* $d \to 0$;

(d) $E[N/C] \to b(m_1)/a \, (> 1)$ *as* $d \to 0$.

The *consistency* or *exact consistency* property, namely part (a) was proved in Ghurye (1958). The *first-order asymptotic* properties given in parts (c)-(d) can be found in Mukhopadhyay (1982a).

26.2.2 Purely Sequential Procedure

In order to be able to claim the *first-order efficiency* property, that is $E[N/C] \to 1$ as $d \to 0$, a purely sequential procedure was proposed in Mukhopadhyay (1974) along the lines of Chow and Robbins (1965). The concepts introduced in Ghosh and Mukhopadhyay (1981) are particularly relevant here. Let

$$N = \inf\{n \geq m_1 : n \geq aU_n/d\}, \tag{26.2.2}$$

where $m_1 (\geq 2)$ is the starting sample size, followed by one at-a-time sequential sampling. Now, $P(N < \infty) = 1$ and hence, based on $X_1, \ldots, X_N$, we propose the confidence interval I_N for μ.

Theorem 26.2 *For the purely sequential procedure (26.2.2), we have:*

(a) $E[N] \leq C + m_1 + 1$;

(b) $N/C \to 1$ *a.s. as* $d \to 0$;

(c) $E[N/C] \to 1$ *as* $d \to 0$;

(d) $P(\mu \in I_N) \to 1 - \alpha$ *as* $d \to 0$.

These results were obtained in Mukhopadhyay (1974). The properties in parts (c) and (d) respectively show that the purely sequential scheme is both *asymptotically first-order efficient* and *asymptotically consistent.*

Swanepoel and vanWyk (1982) have shown the existence of a nonnegative integer m, independent of α, d, μ and θ, such that $P(\mu \in I_{N+m}) \geq 1 - \alpha$, and obviously $E[N + m]/C \to 1$ as $d \to 0$. However, the magnitude of such a universal integer "m" is yet unknown. In the same paper, these authors also derived *asymptotic second-order expansions* of $E[N]$ and $P(\mu \in I_N)$ via Lombard and Swanepoel's (1978) embedding techniques and the nonlinear renewal theory developed in Woodroofe (1977, 1982). See also Ghosh and Mukhopadhyay (1981) and Section 2.2.2 of Mukhopadhyay (1988a) in this context.

Theorem 26.3 *For the purely sequential procedure (26.2.2), we have as* $d \to 0$ *if* $m_1 \geq 3$:

(a) $E[N] = C + q_1 + o(1)$;

(b) $P(\mu \in I_N) = (1 - \alpha) + d\theta^{-1}\alpha\left\{q_1 - \frac{1}{2}a\right\} + o(d)$;

where q_1 *is an appropriate known real number.*

26.2.3 Modified Two-Stage Procedure

Along the lines of Mukhopadhyay (1980), the two-stage procedure (26.2.1) was modified in Mukhopadhyay (1982a) in such a way that the starting sample size m_1 was allowed to grow, but at a slower rate than C. The net effect being that one could then conclude Theorem 26.1(a) as well as Theorem 26.2(b)-(d). This modification, however, does not lead to *asymptotic second-order efficiency*, in other words one can prove that $E[N] - C$ explodes in the limit in spite of the modification as stated. See also Ghosh and Mukhopadhyay (1981) in this context.

26.2.4 Three-Stage Procedure

Along the lines of Hall (1981), a three-stage procedure was proposed in Mukhopadhyay and Mauromoustakos (1987). A general unified theory for such sampling techniques has appeared in Mukhopadhyay (1990). In this methodology, one first starts with $X_1, \ldots, X_{m_1}$ where $m_1 (\geq 2)$ is such that $m_1 = O\left(C^{1/r}\right)$ for some $r > 1$. Based on these samples, one estimates a fractional part of C by

$$M = \max\left\{m_1, \left\langle \rho a U_{m_1} d^{-1}\right\rangle + 1\right\}, \tag{26.2.3}$$

with fixed prechosen $\rho \in (0,1)$. Extend $X_1, \ldots, X_{m_1}$ to $X_1, \ldots, X_M$ in the second stage and define

$$N = \max\left\{M, \left\langle aU_M d^{-1} + \frac{1}{2}(a+2-\rho)\rho^{-1}\right\rangle + 1\right\}, \tag{26.2.4}$$

and thereby extend $X_1, \ldots, X_M$ to $X_1, \ldots, X_N$ in the third stage. Then, the confidence interval I_N is proposed for μ. The following *second-order asymptotics* were proved in Mukhopadhyay and Mauromoustakos (1987).

Theorem 26.4 *For the three-stage procedure (26.2.3)-(26.2.4), we have as* $d \to 0$:

(a) $E[N] = C + \frac{1}{2}a\rho^{-1} + o(1)$;

(b) $P(\mu \in I_N) = (1-\alpha) + o(d)$.

26.2.5 Accelerated Sequential Procedure

In the three-stage methodology, it consisted of batch sampling in three steps. But, it essentially maintained the forms of asymptotic second-order expansions of $E[N]$ and $P(\mu \in I_N)$ obtained in Theorem 26.3 for the purely sequential scheme. But then, what if one starts out in a sequential fashion by estimating a fraction of C, that is terminating the process early, followed by batch sampling? Such methodologies fall under the category of accelerated sequential procedures, first proposed by Hall (1983) in the context of the corresponding normal problem. Recently, a general unified theory has been proposed in Mukhopadhyay and Solanky (1991).

The sampling techniques can be explained as follows. One first starts with $X_1, \ldots,$ X_{m_1} with $m_1 \geq 2$ and proceeds purely sequentially by taking one sample at-a-time, finally arriving at $X_1, \ldots, X_t$ where (with $U_n^* = n(n-1)^{-1}U_n$)

$$t = \inf\left\{n \geq m_1 : n \geq \rho a U_n^*/d\right\}, \tag{26.2.5}$$

with fixed prechosen $\rho \in (0,1)$. It is clear that t estimates ρC. Now define

$$N = \max\left\{t, \left\langle aU_t^* d^{-1} + q_2\right\rangle + 1\right\}, \tag{26.2.6}$$

with

$$q_2 = \frac{1}{2}(2+a)\rho^{-1}. \tag{26.2.7}$$

One determines N and extends $X_1, \ldots, X_t$ to $X_1, \ldots, X_N$ by means of batch sampling the difference $(N-t)$, if $N > t$. Finally, the interval I_N is proposed for μ. The following *second-order asymptotics* were provided in Mukhopadhyay and Solanky (1991).

Theorem 26.5 *For the accelerated sequential procedure (26.2.5)-(26.2.6), we have as* $d \to 0$:

(a) $q_2 - \rho^{-1} \leq \liminf E[N - C] \leq \limsup E[N - C] \leq q_2 - \rho^{-1} + 1$;

(b) $P(\mu \in I_N) \geq (1 - \alpha) + o(d)$;

where q_2 *had been defined in (26.2.7).*

26.2.6 Piecewise Sequential Procedure

It is well known that the purely sequential stopping variable N has a very complicated and heavy right-tailed distribution. Such features significantly hinder different types of practical applications as well as associated computations. Recently, Mukhopadhyay and Sen (1993) suggested implementing the larger experiment via several identically replicated smaller "pieces". One advantage of this approach is that one can now capitalize frequently on often available system of a parallel network and gain much operational convenience and thus cut cost. The other advantage is that one can easily provide an unbiased estimator of the variance of the final combined stopping variable. Mukhopadhyay and Sen (1993) introduced such techniques in the context of the corresponding normal problem. The following discussion is summarized from the unified general theory paper of Mukhopadhyay and Datta (1994).

In the light of (26.2.2), we now introduce the following modification. Suppose there are $k\ (\geq 2)$ assistants who are independently and simultaneously involved with the sampling methodology. Let $X_{i1}, X_{i2}, \ldots$ be independent sequences of i.i.d. variables having the density $f(x; \mu, \theta)$, $i = 1, \ldots, k$. The ith sequence corresponds to the ith assistant, say, and each investigator will estimate $k^{-1}C$ from the samples obtained from that particular location only. Having observed $X_{i1}, \ldots, X_{in_i}$ for $n_i \geq 2$, let us denote $T_k(\boldsymbol{n}) = \min_{1 \leq i \leq k} T_{in_i}$, $T_{in_i} = \min\{X_{i1}, \ldots, X_{in_i}\}$, $U_{in_i} = (n_i - 1)^{-1} \sum_{j=1}^{n_i} (X_{ij} - T_{in_i})$, $\boldsymbol{n} = (n_1, \ldots, n_k)$, $n = \sum_{i=1}^{k} n_i$, $i = 1, \ldots, k$. We propose $I_{\boldsymbol{n}} = [T_k(\boldsymbol{n}) - d, T_k(\boldsymbol{n})]$ for μ.

Now, each investigator starts with $m_1\ (\geq 2)$ samples and the ith one defines the purely sequential stopping variable

$$N_i = \inf\left\{n \geq m_1 : n \geq aU_{in}\,(kd)^{-1}\right\}, \tag{26.2.8}$$

$i = 1, \ldots, k$. Having observed $X_{i1}, \ldots, X_{iN_i}$, $i = 1, \ldots, k$, we finally combine all the "pieces" and estimate μ by means of the fixed-width confidence interval $I_{\boldsymbol{N}} = [T_k(\boldsymbol{N}) - d, T_k(\boldsymbol{N})]$, where $\boldsymbol{N} = (N_1, \ldots, N_k)$. The total sample size N is given by $\sum_{i=1}^{k} N_i$. Obviously,

$$\Delta = k\,(k-1)^{-1} \sum_{i=1}^{k} \left(N_i - \bar{N}\right)^2, \tag{26.2.9}$$

is an *unbiased estimator of variance* of N, where $\bar{N} = k^{-1}N$.

Theorem 26.6 *For the piecewise sequential procedure (26.2.8), we have as* $d \to 0$ *if* $m_1 \geq 3$:

$$P\left(\mu \in I_{\boldsymbol{N}}\right) = (1 - \alpha) + d\theta^{-1}\alpha\left\{kq_1 - \frac{1}{2}a\right\} + o(d);$$

where q_1 *is the same known real number as in Theorem 26.3.*

26.3 Fixed-Precision Estimation of a Positive Location

Assume that $X_1, X_2, \ldots$ are i.i.d. variables having the density $f(x; \mu, \theta)$ where μ is *known to be positive.* In this situation, fixed-width confidence intervals of the type given by I_n in Section 26.2 will not be appropriate, because the lower confidence limit may turn out to be negative, but we already know that μ is indeed positive. Mukhopadhyay (1988b) reexamined the concept of *fixed-precision* in this situation and proposed a new purely sequential methodology.

Having recorded $X_1, \ldots, X_n$ with $n \geq 2$, we continue using the notations T_n and U_n as in Section 26.2. Given two numbers $0 < d, \alpha < 1$, we propose the confidence interval $I_n^* = [dT_n, T_n]$ for μ. I_n^* is a *fixed-precision* confidence interval for μ in the sense that μ/T_n will then be declared to lie in the fixed interval $[d, 1]$. If d is chosen to be close to unity, then T_n will be quite precise as an estimator of μ. Now, $P(\mu \in I_n^*) \geq 1 - \alpha$, provided that n is chosen to be the smallest integer $\geq a\theta (d^* \mu)^{-1} = C_0^*$, say, where $d^* = (1-d)/d$. But, C_0^* is unknown, and hence Mukhopadhyay (1988b) investigated the following purely sequential procedure.

One starts with $X_1, \ldots, X_{m_1}$ for $m_1 \geq 2$ and proceeds by taking one sample at-a-time according to the stopping rule

$$N = \inf\left\{n \geq m_1 : n \geq aU_n (d^* T_n)^{-1}\right\}. \tag{26.3.1}$$

Finally, one estimates μ by means of the fixed-precision confidence interval $I_N^* = [dT_N, T_N]$. The following results were obtained in Mukhopadhyay (1988b). Asymptotic analysis is carried out as "d" converges to unity from the left.

Theorem 26.7 *For the purely sequential procedure (26.3.1), we have:*

(a) $E[N] \leq C_0^* + m_1 + 1$;

(b) $N/C_0^* \to 1$ *a.s. as* $d \to 1-$;

(c) $E[N/C_0^*] \to 1$ *as* $d \to 1-$;

(d) $P(\mu \in I_N^*) \to 1 - \alpha$ *as* $d \to 1-$.

26.4 Minimum Risk Point Estimation of the Location

Let $X_1, X_2, \ldots$ be a sequence of i.i.d. variables having the density $f(x; \mu, \theta)$ and we continue using notations T_n and U_n as in Section 26.2. Suppose that the loss incurred in estimating μ by means of T_n is given by

$$L_n = A(T_n - \mu)^2 + cn \, , \tag{26.4.1}$$

where A and c are known positive numbers. The risk associated with (26.4.1) is then given by

$$\begin{aligned} R_n(c) &= E[L_n] \\ &= B\theta^2 \left(2n^2\right)^{-1} + cn \, , \end{aligned} \tag{26.4.2}$$

with $B = 4A$. The risk (26.4.2) is minimum for $n = n^*$ where $n^* = \left(B\theta^2/c\right)^{\frac{1}{3}}$ and the corresponding *minimum risk* is given by

$$R^*(c) = R_{n^*}(c) = \frac{3}{2}\, cn^* \; . \tag{26.4.3}$$

The goal is to achieve this minimum risk approximately. However, n^* is indeed unknown and hence one needs to adopt appropriate purely sequential and multi-stage estimation procedures.

Basu (1971) first considered a purely sequential scheme for this problem. Mukhopadhyay (1974) independently put forth a similar purely sequential estimation scheme. We should note that in both these papers, the loss function considered was fairly more general than the one given by (26.4.1). Then, other multi-stage procedures emerged for such problems. In this connection, see Section 3 of Mukhopadhyay (1988a) and more recent articles of Basu (1991) and Mukhopadhyay (1991) on related topics. The procedures considered in Sections 26.4.1-26.4.4 are such that the risk associated with T_N is given by

$$\begin{aligned} R(c) &= E[L_N] \\ &= \frac{1}{2}\, B\theta^2 E[N^{-2}] + cE[N] \, , \end{aligned} \tag{26.4.4}$$

for appropriately defined stopping variables N. Along the lines of Robbins (1959), one defines *risk efficiency* and *regret* of a procedure as

$$e(c) = R(c)/R^*(c) \text{ and } \omega(c) = R(c) - R^*(c) \, , \tag{26.4.5}$$

respectively. After any specific procedure is put forth, one examines asymptotic (as $c \to 0$) behaviors of $e(c)$ and $\omega(c)$.

26.4.1 Purely Sequential Procedure

This procedure is due to Basu (1971) and Mukhopadhyay (1974). One starts with $X_1, \ldots, X_{m_1}$ with $m_1 \geq 2$ and let

$$N = \inf\left\{n \geq m_1 : n \geq \left(BU_n^2/c\right)^{\frac{1}{3}}\right\} , \tag{26.4.6}$$

and finally estimate μ by T_N. Basu (1971) and Mukhopadhyay (1974) proved that $e(c) \to 1$ as $c \to 0$ if $m_1 \geq 3$. Mukhopadhyay (1982b) had shown that $\omega(c) = 0(c)$ if and only if $m_1 \geq 3$. Theorem 10 in Mukhopadhyay (1988a) provides the *second-order* expansion of $\omega(c)$, namely, that

$$\omega(c) = \frac{2}{3}\, c + o(c) \text{ if } m_1 \geq 5 \; . \tag{26.4.7}$$

26.4.2 Two-Stage Procedure

Hilton (1984) had proposed a suitable two-stage procedure for this problem and obtained different types of *first-order asymptotic* results regarding $E[N/n^*]$, $e(c)$ and $(N - n^*)$. One will also find some details in Theorem 11 of Mukhopadhyay (1988a).

26.4.3 Modified Two-Stage Procedure

Along the lines of Mukhopadhyay (1980, 1982a), a modified two-stage procedure was developed in Mukhopadhyay and Hilton (1986). Here, $m_1\ (\geq 2)$ is allowed to grow as $c \to 0$, but at a slower rate than that of n^*, and it was shown that $e(c) \to 1$ as $c \to 0$. The order of $\omega(c)$ was also provided.

26.4.4 Accelerated Sequential Procedure

As in Section 26.2.5, we now work with U_n^* instead of U_n where $U_n^* = n(n-1)^{-1} U_n$. This accelerated sequential procedure was proposed in Mukhopadhyay and Solanky (1991). One starts with $X_1, \ldots, X_{m_1}$ with $m_1 \geq 2$ and proceeds sequentially by taking one sample at-a-time, finally arriving at $X_1, \ldots, X_t$ where

$$t = \inf\left\{n \geq m_1 : n \geq \rho \left(BU_n^{*2}/c\right)^{\frac{1}{3}}\right\}, \tag{26.4.8}$$

with fixed prechosen $\rho \in (0,1)$. In other words, t estimates ρn^*, a fraction of n^*. Now, define

$$N = \max\left\{t, \left\langle \left(BU_t^{*2}/c\right)^{\frac{1}{3}} \right\rangle + 1\right\}, \tag{26.4.9}$$

and sample the difference $(N-t)$ in one batch if $N > t$. Finally, estimate μ by T_N. The asymptotic *second-order* expansion of the corresponding regret $\omega(c)$ can be easily obtained along the lines of Section 4 of Mukhopadhyay and Solanky (1991). Further details are omitted.

26.4.5 Three-Stage Procedure-Weighted Loss

Instead of the loss function in (26.4.1), consider the following weighted version. Let

$$L_n = A\theta (T_n - \mu)^2 + cn \tag{26.4.10}$$

and suppose the goal is to construct an estimator with the approximate *minimum risk*. Along the lines of Mukhopadhyay (1985) and Mukhopadhyay et al. (1987), a three-stage procedure was developed in Hamdy et al. (1988) and they provided the *second-order* asymptotic expansion of the associated regret function.

Remark 26.1 There is significant amount of literature in this context for the corresponding *bounded risk* problems. One should look at Hilton (1984), Mukhopadhyay and Hilton (1986), and Hamdy et al. (1988) for further details.

Remark 26.2 In a one-parameter exponential family, Woodroofe (1987) had discussed a three-stage procedure for estimating the mean in order to achieve asymptotic local minimax regret, a concept defined earlier in Woodroofe (1985).

26.4.6 Improved Purely Sequential Procedure

Here, we go back to the original problem of minimum risk point estimation of μ by means of T_N where N is the purely sequential stopping variable given by (26.4.6) and ask what if we started with some other estimator of μ? Could we then see significant improvement in terms of smaller $E[N]$ and regret? The Ghosh and Mukhopadhyay (1990) paper considered such investigations.

For fixed n and known $\theta, T_n - \theta n^{-1}$ is the best unbiased estimator of μ. Hence, for unknown θ, for $n \geq 2$, Ghosh and Mukhopadhyay (1990) proposed estimating μ by means of $\tilde{T}_n = T_n - (n-1)\, n^{-2} U_n$. Now with the loss function

$$L_n = \left(\tilde{T}_n - \mu\right)^2 + cn \tag{26.4.11}$$

instead of (26.4.1), one can easily verify that the associated risk is minimized when $n = n_0^* = 2^{-\frac{1}{3}} n^* \approx .7937 n^*$. In other words, had θ been known, one could minimize the risk associated with (26.4.11) with approximately 21% fewer samples compared with the fixed sample scenario (with $A = 1$) in terms of (26.4.1). Ghosh and Mukhopadhyay (1990) continued by proposing a purely sequential stopping variable

$$N = \inf\left\{n \geq m_1\, (\geq 2) : n \geq \left(2U_n^2/c\right)^{\frac{1}{3}}\right\}, \tag{26.4.12}$$

finally estimating μ by $\tilde{T}_N$. There is a significant technical difference between the two situations. In the context of (26.4.1) and (26.4.6), it is known that T_n and $I\,(N = n)$ are independent for all $n \geq m_1$. However, in the present situation, in the context of (26.4.11) and (26.4.12), $\tilde{T}_n$ and $I\,(N = n)$ are no longer independent for $n \geq m_1$. Fairly complicated analyses eventually led Ghosh and Mukhopadhyay (1990) to obtain the asymptotic *second-order* expansions of $E[\tilde{T}_N]$ and the regret function associated with $\tilde{T}_N$, for N given by (26.4.12). In conclusion, they noted that, one achieves "significant improvement" in the estimation of μ by means of $\tilde{T}_N$, for N given by (26.4.12).

Remark 26.3 Bose and Boukai (1993) have addressed minimum risk point estimation problems for one of the parameters when the other one is unknown in a two-parameter subfamily of exponential distributions considered by Bar-Lev and Reiser (1982). See also Brown (1986). Bose and Boukai (1993) proposed an appropriate purely sequential estimation methodology and derived the asymptotic *second-order* expansion of the associated regret function. Recently, Bose and Mukhopadhyay (1994 a,b) have proposed both the accelerated and piecewise versions of the original Bose-Boukai (1993) stopping rule and obtained the corresponding asymptotic second-order characteristics.

Remark 26.4 Chaturvedi and Shukla (1990) have also revisited the point estimation problems for the location parameter μ.

26.5 Sequential Point Estimation Procedures for the Scale Parameter, Mean and Percentiles

In this section we summarize various purely sequential procedures for estimating the scale parameter, mean and percentiles of the negative exponential distribution. Throughout, we have available a sequence $X_1, X_2, \ldots$ of i.i.d. variables having the density $f\,(x; \mu, \theta)$. We continue using the notations T_n and U_n as in Section 26.2.

26.5.1 Estimating the Scale Parameter

First, consider the situation when the location parameter μ is known, and it is assumed zero without any loss of generality. Having recorded $X_1, \ldots, X_n$, one estimates θ by $\bar{X}_n = n^{-1} \sum_{i=1}^{n} X_i$. Suppose that the loss function in estimating θ by $\bar{X}_n$ is given by

$$L_n = A\left(\bar{X}_n - \theta\right)^2 + cn \ , \tag{26.5.1}$$

where A and c are known positive numbers. The risk associated with (26.5.1) is given by $R_n(c) = E[L_n] = A\theta^2 n^{-1} + cn$ which is minimized if $n = n^* = (A/c)^{\frac{1}{2}}\,\theta$. But, n^* is unknown. Thus, Starr and Woodroofe (1972) came up with the following one at-a-time purely sequential scheme.

One starts with $X_1, \ldots, X_{m_1}$ where $m_1 \geq 1$. Then, let

$$N = \inf\left\{n \geq m_1 : n \geq (A/c)^{\frac{1}{2}}\,\bar{X}_n\right\}, \tag{26.5.2}$$

and finally one estimates θ by $\bar{X}_N$. The associated risk function is $E[L_N]$ which can no longer be expressed as a function of moments of N alone. But, Starr and Woodroofe (1972) in their remarkable paper were successful in proving that the regret function $\omega(c)$, that is $E[L_N] - R_{n^*}(c)$, is $O(c)$ if $m_1 \geq 2$. Woodroofe (1977) had shown that $\omega(c) = 3c + o(c)$ if $m_1 \geq 3$.

The situation when μ is unknown causes one to rethink about the whole process. Having observed $X_1, \ldots, X_n$, one will now estimate θ by U_n and suppose that the loss function is given by

$$L_n = A\,(U_n - \theta)^2 + cn \ , \tag{26.5.3}$$

where A and c are as before. $E[L_n]$ now becomes $A\theta^2\,(n-1)^{-1} + cn$ and it is minimized when $n = n^* = (A/c)^{\frac{1}{2}}\,\theta + 1$. In the appropriate stopping rule, one plugs in U_n in the place of θ, and finally estimates θ by U_N. The technical difficulties are now many fold. One has to work with the spacings between order statistics and prove various distributional properties of U_N via embedding techniques of Lombard and Swanepoel (1978). In the paper of Mukhopadhyay and Ekwo (1987), they were able to verify that the regret function associated with their purely sequential estimator U_N can be expanded as $3c + o(c)$. A few details are also included in Section 4.1 of Mukhopadhyay (1988a).

Remark 26.5 A few relevant results are implicitly available in Govindarajulu and Sarkar (1991). Govindarajulu (1985) also considered a sequential fixed-width confidence interval problem for θ. Clever generalizations in the presence of random censoring were put forth in Gardiner and Susarla (1983, 1984) as well as in Aras (1987, 1989).

26.5.2 Estimating the Mean

Now, our objective is to describe the point estimation problems for the mean of the distribution, namely $\mu + \theta\,(= \sigma$, say), given by (26.1.1). Having recorded $X_1, \ldots, X_n$ suppose that the loss incurred in estimating σ by $\bar{X}_n \left(= n^{-1} \sum_{i=1}^{n} X_i\right)$ is given by

$$L_n = A\left(\bar{X}_n - \sigma\right)^2 + cn \ , \tag{26.5.4}$$

where A and c are known positive numbers. The associated risk $R_n(c)$ is minimized if $n = n^* = (A/c)^{\frac{1}{2}}\theta$ and the goal is to achieve this minimum risk approximately. Of course, n^* is unknown and hence Mukhopadhyay (1987) proposed the following stopping rule. Let

$$N = \inf\left\{n \geq m_1 (\geq 2) : n \geq (A/c)^{\frac{1}{2}}\left(1-n^{-1}\right)^{-1} U_n\right\}, \tag{26.5.5}$$

and when this purely sequential sampling process stops, one estimates σ by $\bar{X}_N$. The associated risk is given by $R(c) = E[L_N]$ which can no longer be expressed as a function of moments of N alone. Hilton (1984) had considered a slightly different stopping rule and for that estimation rule, he showed that $R(c)/R^*(c) \to 1$ as $c \to 0$ where one writes $R^*(c) = R_{n^*}(c)$, as before. Mukhopadhyay (1987) combined the techniques from Section 26.5.1 with those of Woodroofe (1977, 1982) to prove that the regret function $\omega(c)$, that is $R(c) - R^*(c)$, can be expanded as $3c + o(c)$ as $c \to 0$ if $m_1 \geq 4$. The latter result is a significant improvement over what was claimed in Hilton (1984). The Remark 12 in Mukhopadhyay (1988a) has particular bearing on this problem as well.

26.5.3 Estimating the Percentiles

The percentiles of the negative exponential distribution (26.1.1) are given by $\phi = a\mu + b\theta$ where $a \in [1,\infty)$ and $b \in (0,\infty)$ are known numbers. Ghosh and Mukhopadhyay (1989) considered sequential minimum risk point estimation problems for ϕ when a and b are held fixed. One estimates ϕ by $\hat{\phi}_n = a\left(T_n - n^{-1}U_n\right) + bU_n$ which is unbiased for ϕ when $n \geq 2$. Suppose that the loss incurred in estimating ϕ by $\hat{\phi}_n$ is given by

$$L_n = \left(\hat{\phi}_n - \phi\right)^2 + cn\ , \tag{26.5.6}$$

where c is a known positive number. The associated risk is $R_n(c) = E[L_n] = a^2\theta^2 n^{-2} + b^2\theta^2\left(n^{-1} + n^{-2}\right) - 2ab\theta^2 n^{-2} + cn + o\left(n^{-2}\right)$ which is approximately minimized if $n = n^* = \left(b^2\theta^2/c\right)^{\frac{1}{2}}$. Since n^* is unknown, Ghosh and Mukhopadhyay (1989) considered the following purely sequential stopping variable. Let

$$N = \inf\left\{n \geq m_1 (\geq 2) : n \geq \left(b^2/c\right)^{\frac{1}{2}} U_n\right\}, \tag{26.5.7}$$

and once sampling terminates, one estimates ϕ by means of $\hat{\phi}_N$. The associated risk is given by

$$\begin{aligned} R(c) &= E[L_N] \\ &= a^2\theta^2 E[N^{-2}] + E\left[b^2\left(U_N - \theta\right)^2 - 2ab\left(U_N - \theta\right)^2 N^{-1}\right. \\ &\quad \left. + a^2\left(U_N - \theta\right)^2 N^{-2}\right] + cE[N]\ , \end{aligned} \tag{26.5.8}$$

since T_n and $I(N = n)$ are still independent for all $n \geq m_1$. But, note that $\hat{\phi}_n$ and $I(N = n)$ are not independent and hence (26.5.8) does not simplify to a function of moments of N alone. Again, one combines techniques from Section 26.5.1 with those of Woodroofe (1977) and Mukhopadhyay (1987) in order to derive the asymptotic *second-order* expansion of $R(c)$. The following result appears as Theorem 2.1 in Ghosh and Mukhopadhyay (1989).

Theorem 26.8 *For the purely sequential procedure (26.5.7), we have as* $c \to 0$ *if* $m_1 \geq 4$:

$$R(c) = 2(b\theta)c^{\frac{1}{2}} + c\left(a^2 + 4b^2 - 2ab\right)/b^2 + o(c) \;;$$

where $R(c)$ *has been defined in (26.5.8).*

Remark 26.6 In Theorem 26.8, if one plugs in $a = b = 1$, one obtains the *second order* expansion of the regret, namely $R(c) - 2\theta c^{\frac{1}{2}}$, as $3c + o(c)$ which coincides with the corresponding result in mean estimation problem, earlier described in Section 26.5.2.

Remark 26.7 Isogai and Uno (1993) have recently given an improved sequential estimator of the percentile ϕ. Once the sequential procedure (26.5.7) stops sampling, they propose to estimate ϕ by means of $\hat{\phi}_N + c^{\frac{1}{2}}$ and show that the corresponding regret's *second-order* asymptotic expansion amounts to the cost saving of one observation when compared with that of the estimator $\hat{\phi}_N$ of Ghosh and Mukhopadhyay (1989). Mukhopadhyay's (1994) further extensions of such results are also particularly relevant in this context.

26.6 Multi-Sample Problems

First, we introduce two-sample problems by means of constructing *fixed-width confidence intervals* for the difference of location parameters, both in the case of equal and unknown scale parameter as well as unequal and unknown scale parameters. Then, we move to *minimum risk point estimation* problems, first in the case of comparing two populations, and then dealing with the general multi-sample situation. In the end, we include estimation problems for the largest location parameter among several negative exponential populations having equal but unknown scale parameter. For such restricted parameter estimation problems, we first summarize fixed-width confidence interval procedures, followed by *minimax* and *bounded maximal risk point estimation* procedures.

Let us introduce some of the notations that will be used in this section. Let $X_{i1}, \ldots, X_{in_i}, \ldots$ be independent sequences of i.i.d. variables with the density $f(x; \mu_i, \theta_i)$ given by (26.1.1), $i = 1, \ldots k, k \geq 2$. Let us write $T_{in_i} = \min\{X_{i1}, \ldots, X_{in_i}\}$, $U_{in_i} = (n_i - 1)^{-1} \sum_{j=1}^{n_i} (X_{ij} - T_{in_i})$, $\boldsymbol{n} = (n_1, \ldots, n_k)$, $T(\boldsymbol{n}) = \max_{1 \leq i \leq k} T_{in_i}$, $T^*(\boldsymbol{n}) = \sum_{i=1}^{k} \bar{X}_{in_i}$, $T_*(\boldsymbol{n}) = \bar{X}_{1n_1} - \bar{X}_{2n_2}$, $T_0^*(\boldsymbol{n}) = T_{1n_1} - T_{2n_2}$, $\sigma_i = \mu_i + \theta_i$, $i = 1, \ldots, k$. When $n_1 = \cdots = n_k = n$, say, we write $V_n = k^{-1} \sum_{i=1}^{k} U_{in}$. In the ith population, μ_i and θ_i are respectively estimated by T_{in_i} and U_{in_i}, $i = 1, \ldots, k$. In a k-sample problem, $T(\boldsymbol{n})$ estimates μ^* which is defined as $\max\{\mu_1, \ldots, \mu_k\}$, while $T^*(\boldsymbol{n})$, $T_0^*(\boldsymbol{n})$, and $T_*(\boldsymbol{n})$ respectively estimate $\sum_{i=1}^{k} \sigma_i$, $(\mu_1 - \mu_2)$ and $(\sigma_1 - \sigma_2)$ where σ_i is the mean of the ith population. In the case when $\theta_1 = \cdots = \theta_k (= \theta$, say), V_n is used as the pooled estimator of the common unknown scale parameter θ when sample sizes all equal n.

26.6.1 Fixed-Width Confidence Intervals for the Difference of Locations - Two-Sample Case

Given $d (> 0)$ and $\alpha \in (0, 1)$ we wish to construct a confidence interval for $\delta = \mu_1 - \mu_2$ having width $2d$ and confidence coefficient at least $(1 - \alpha)$.

Equal Scale Parameters

Let $\theta_1 = \theta_2 = \theta$, say and we consider taking equal number of samples n from both populations. $T_0^*(n) = T_{1n} - T_{2n}$ estimates δ and hence one proposes the fixed-width confidence interval $I_n = [T_0^*(n) \pm d]$ for δ. Now, $P(\delta \in I_n) \geq 1 - \alpha$ provided that n is the smallest integer $\geq a\theta/d = C$, say, with $a = \ell n(1/\alpha)$. Of course, C is unknown and hence one will opt for various multi-stage estimation techniques.

Two-Stage Procedure: A two-stage procedure was proposed in Mukhopadhyay and Hamdy (1984). Various exact and asymptotic (as $d \to 0$) results in the spirit of Theorem 26.1 were derived.

Purely Sequential Procedure: A purely sequential procedure was also proposed in Mukhopadhyay and Hamdy (1984). Both *first-* and *second-order* asymptotic results in the spirits of Theorems 26.2 and 26.3 were provided.

Three-Stage Procedure: A suitable three-stage procedure was developed in Mukhopadhyay and Mauromoustakos (1987). Results in the spirit of Theorem 26.4 were also provided.

Unequal Scale Parameters

We assume that θ_1, θ_2 are both unknown and unequal and sample sizes are also unequal. The parameter δ is estimated by $T_0^*(\boldsymbol{n})$ and the proposed fixed-width confidence interval for δ is taken to be $I_{\boldsymbol{n}} = [T_0^*(\boldsymbol{n}) \pm d]$. Now,

$$P(\delta \in I_{\boldsymbol{n}}) = \left[\sum_{i=1}^{2} n_i^{-1}\theta_i \{1 - \exp(-n_i d/\theta_i)\}\right] \left[\sum_{i=1}^{2} n_i^{-1}\theta_i\right]^{-1}, \qquad (26.6.1)$$

and this confidence coefficient will be at least $(1 - \alpha)$ if $n_i \geq a\theta_i/d = C_i$, say, $i = 1, 2$. However, C_1, C_2 are unknown and hence multi-stage estimation techniques have been explored in the literature.

Two-Stage Procedure: A two-stage procedure was proposed in Mukhopadhyay and Hamdy (1984). Various exact and asymptotic results in the spirit of Theorem 26.1 were derived.

Purely Sequential Procedure: A purely sequential procedure was also proposed in Mukhopadhyay and Hamdy (1984). Asymptotic *second-order* properties of such a procedure were provided in the spirit of Theorem 26.3.

Remark 26.8 Further details regarding these procedures can also be found in Sections 5.1 and 5.2 of Mukhopadhyay (1988a).

Three-Stage Procedure: By mimicking the expressions of C_1 and C_2, a three-stage procedure has recently been proposed by Mukhopadhyay and Padmanabhan (1993). One starts with $m_1 (\geq 2)$ samples from both populations where it is assumed that

$m_1 = O\left(d^{-1/r}\right)$ for some $r > 1$. One chooses and fixes two numbers $0 < \rho_1, \rho_2 < 1$ and defines

$$M_i = \max\left\{m_1, \left\langle a\rho_i U_{im_1} d^{-1}\right\rangle + 1\right\}, \tag{26.6.2}$$

$$N_i = \max\left\{M_i, \left\langle aU_{iM_i} d^{-1} + \epsilon_i\right\rangle + 1\right\}, \tag{26.6.3}$$

where ϵ_1, ϵ_2 are known real numbers, $i = 1, 2$. If $M_i > m_1$, $N_i > M_i$, then one samples the differences $(M_i - m_1)$ and $(N_i - M_i)$ in batches respectively in the second and third stages. Finally, the fixed-width confidence interval $I_{\boldsymbol{N}}$ is proposed for the parameter δ. Various types of *second-order* asymptotics have been provided in Mukhopadhyay and Padmanabhan (1993) where the following main result has also been derived.

Theorem 26.9 *For the three-stage procedure (26.6.2)-(26.6.3) with* $\epsilon_i = \frac{1}{2}(a + 3 - \rho_i)/\rho_i$, $i = 1, 2$, *we have as* $d \to 0$:

$$P\left(\delta \in I_{\boldsymbol{N}}\right) = (1 - \alpha) + o(d).$$

26.6.2 Minimum Risk Point Estimation Problems

First, we discuss briefly the two-sample problems. Then, we introduce separately the multi-sample problems because here the approach taken is different from those advocated in earlier published literature.

Two-Sample Problems

Here, what we have to say is mostly derived from the article of Mukhopadhyay and Darmanto (1988). The problems consist of estimating $\delta = \mu_1 - \mu_2$ or $\delta^* (= \sigma_1 - \sigma_2)$ pointwise by approximately achieving the minimum fixed sample risk (had θ's been known) when the loss function is squared error plus linear cost of sampling. In the equal but unknown parameter case, δ is estimated by $T_0^*(n)$, having equal number of samples from both populations. In Section 2 of Mukhopadhyay and Darmanto (1988), one finds a suitable purely sequential procedure for this problem and its associated *first-* and *second-order* asymptotic properties.

In the case when θ_1, θ_2 are both unknown and unequal, δ^* is estimated by $T_*(\boldsymbol{n})$ in the presence of squared error plus linear cost of sampling as the loss function. In Section 3 of Mukhopadhyay and Darmanto (1988), first a sampling scheme was introduced along the lines of Robbins, Simons, and Starr (1967) and Srivastava (1970). The role of the sampling scheme is to decide whether to sample from population 1 or population 2 at any point of time. In conjunction with this sampling scheme, Mukhopadhyay and Darmanto (1988) also developed a purely sequential sampling procedure determining the required sample sizes from each population as N_1, N_2, say. The stopping variables N_1 and N_2 are dependent here because of the involvement of the sampling scheme. The material presented in the next subsection is different because there N_i's are constructed independently and one reaps the benefit of such independence fully. Because of implicit technical difficulties, Mukhopadhyay and Darmanto (1988) had to be satisfied with appropriate *first-order* asymptotics only in the case of unknown and unequal scale parameters.

Generalizations

In this k-sample problem, let us assume throughout that the location parameters are known, which are assumed to be zero without any loss of generality. We now propose estimating $\sum_{i=1}^{k} \sigma_i$ which coincides with $\phi = \sum_{i=1}^{k} \theta_i$ by $T^*(\boldsymbol{n})$ and formulate the problem as in Mukhopadhyay and Chattopadhyay (1991). Suppose that the loss function in estimating ϕ by means of $T^*(\boldsymbol{n})$ is given by

$$L_{\boldsymbol{n}} = A\left[T^*(\boldsymbol{n}) - \phi\right]^2 + c\sum_{i=1}^{k} n_i, \tag{26.6.4}$$

where A and c are known positive numbers. The risk associated with (26.6.4) is given by $R_{\boldsymbol{n}}(c) = A\sum_{i=1}^{k} \theta_i^2 n_i^{-1} + c\sum_{i=1}^{k} n_i$ which is minimized for $\boldsymbol{n}^* = (n_1^*, \ldots, n_k^*)$ where $n_i^* = (A/c)^{\frac{1}{2}}\theta_i$, $i = 1, \ldots, k$. Mukhopadhyay and Chattopadhyay (1991) actually constructed estimation problems for arbitrary linear functions of θ's, that is $\sum_{i=1}^{k} a_i\theta_i$ for known non-zero $a_1, \ldots, a_k$. One cannot necessarily assume $a_1 = \cdots = a_k = 1$, without any loss of generality. In the present summary, however, we focus only on the case when $a_1 = \cdots = a_k = 1$.

The fixed-sample minimum risk is given by $R_{\boldsymbol{n}^*}(c) = 2c\sum_{i=1}^{k} n_i^*$. Note that n_i^* depends only on θ_i and hence one needs to take samples from the ith population only in order to estimate n_i^*, $i = 1, \ldots, k$. In other words, there is no real need for a sampling allocation scheme here and one should proceed with sequential sampling, independently carried out from each population.

The following purely sequential procedure is then in order. One starts with m_1 (≥ 2) samples from the ith population and let

$$N_i = \inf\left\{n \geq m_1 : n \geq (A/c)^{\frac{1}{2}}\bar{X}_{in}\right\}, \tag{26.6.5}$$

$i = 1, \ldots, k$ and define $\boldsymbol{N} = (N_1, \ldots, N_k)$. Once sampling stops from *all* populations, one estimates ϕ by $T^*(\boldsymbol{N})$. The regret is defined as $\omega(c) = E\left[L_{\boldsymbol{N}}\right] - R_{\boldsymbol{n}^*}(c)$. The following Theorem of Mukhopadhyay and Chattopadhyay (1991) provides asymptotic second-order expansion of $\omega(c)$.

Theorem 26.10 *For the purely sequential procedure (26.6.5), we have as* $c \to 0$ *if* $m_1 \geq 3$;

$$\omega(c) = k(2+k)c + o(c)\ .$$

Remark 26.9 One will also find some available interesting results in the area of estimating the common location parameter. Utilizing the concepts of sequential designing from Chernoff (1972) and Zacks (1973), Mukhopadhyay and Narayan (1981) constructed a fixed-width purely sequential confidence interval for the common location parameter in a two-sample situation. Fixed-width confidence intervals were provided both when the scale parameters are (i) equal but unknown, and (ii) unequal and unknown. On the other hand, Costanza et al. (1986) dealt with the same problem purely sequentially in a general k-sample situation, but unfortunately they did not include the important aspects of designing in their formulation. In a problem area like this, the basic idea should be to switch to the population having the smallest scale parameter as early in the sampling process as possible and then, if needed, take all remaining samples from the selected population or implement some type of variation or improvisation of this theme.

Remark 26.10 Various types of simultaneous estimation problems have recently been introduced in Mukhopadhyay (1992). The problems include construction of *minimum risk point estimation* for some of the interesting parameters *simultaneously*.

26.6.3 Estimation of the Largest Location Parameter - Equal and Unknown Scale Parameter Case

Suppose that we have $k\,(\geq 2)$ independent populations, the ith one having the density $f\,(x;\mu_i,\theta)$, $i = 1,\ldots,k$, where all the parameters are assumed unknown. Let us write $\boldsymbol{\mu} = (\mu_1,\ldots,\mu_k) \in R^k$. Now, observe that $P\,(X_{i1} > x_0)$ is maximized for all fixed $x_0 \in R$ provided that the ith population is associated with $\mu^*\,(= \max\{\mu_1,\ldots,\mu_k\})$. Identification problems for such a population falls in the area of selection and ranking. In this section, we are interested in summarizing available results in the context of multi-stage estimation techniques for μ^*.

Fixed-Width Intervals

Given $d\,(> 0)$ and $\alpha \in (0,1)$, and having recorded $X_{i1},\ldots,X_{in}$, $i = 1,\ldots,k$ we consider estimating μ^* by means of the fixed-width confidence interval $I_n = [T\,(n) \pm d]$ where recall that $T(n) = \max_{1\leq i\leq k} T_{in}$. Now, one has

$$\begin{aligned} P\,(\mu^* \in I_n) &= P\,\{T_{in} \leq \mu^* + d \text{ for all } i = 1,\ldots,k\} \\ &= \prod_{i=1}^{k} \left[1 - \exp\{-n\,(d + \mu^* - \mu_i)\,/\theta\}\right] \,, \end{aligned} \tag{26.6.6}$$

and hence it is easy to see that

$$P\,(\mu^* \in I_n) \geq \{1 - \exp\,(-nd/\theta)\}^k \,, \tag{26.6.7}$$

for all $\boldsymbol{\mu}$, while equality in (26.6.7) holds when $\mu_1 = \cdots = \mu_k$, for all fixed values of θ. Thus,

$$\text{Inf}_{\boldsymbol{\mu}} P\,(\mu^* \in I_n) = \{1 - \exp\,(-nd/\theta)\}^k \,. \tag{26.6.8}$$

So, if one wants to have $P\,(\mu^* \in I_n) \geq 1 - \alpha$, then one may equate the expression in (26.6.8) with $(1 - \alpha)$ and conclude that the common sample size n from each population must be the smallest integer $\geq a_k\theta/d = C_k$, say, where $a_k = -\ln\left(1 - \alpha^{\frac{1}{k}}\right)$. Mukhopadhyay and Kuo (1988) proposed the above approach. The analogous theory for normal distributions were put forth in Lal Saxena and Tong (1969) and Tong (1970). But, observe that C_k is unknown, and hence Mukhopadhyay and Kuo (1988) developed various multi-stage estimation methodologies. In particular, they introduced two-stage and modified two-stage procedures and their *first-order* properties. Then, they developed both purely sequential and three-stage estimation procedures and their *second-order* properties. Of course, θ is now estimated by V_n. We just remark that the stopping variables in all these sampling strategies are such that (i) $P\,(N < \infty) = 1$, (ii) $T\,(n)$ and $I\,(N = n)$ are independent for all $n \geq m_1$, and (iii) the distribution of N does not depend upon $\boldsymbol{\mu}$, and hence for the corresponding fixed-width confidence interval I_N, (26.6.8) translates into

$$\text{Inf}_{\boldsymbol{\mu}} P(\mu^* \in I_N) = E\left[\{1 - \exp(-Nd/\theta)\}^k\right], \tag{26.6.9}$$

and the infinum is attained when $\mu_1 = \cdots = \mu_k$. Further details can be obtained from Mukhopadhyay and Kuo (1988).

Minimax and Bounded Maximal Risk Estimation

Now, we summarize what is available in Kuo and Mukhopadhyay (1990a) in terms of the formulation of the problem and results. Suppose that the loss function in estimating μ^* by $T(n)$ is given by

$$L_n = A[T(n) - \mu^*]^s + cn^t, \tag{26.6.10}$$

where A, c, s and t are known positive numbers. Had θ been known, the *minimax* sample size n from each population would be obtained by minimizing the maximum risk. Kuo and Mukhopadhyay (1990a) had shown that $\max_{\boldsymbol{\mu}} E[L_n]$ is attained when $\mu_1 = \cdots = \mu_k$ for all fixed θ, and this *maximal risk* is given by

$$R_n(c) = B\theta^s (sn^s)^{-1} + cn^t, \tag{26.6.11}$$

where $B = Ask \int_0^\infty u^s e^{-u} (1 - e^{-u})^{k-1} du$. The risk in (26.6.11) is minimized when $n = n_k^* = \{B\theta^s/(ct)\}^{1/(t+s)}$ and the *minimax risk* turns out to be $R^*(c) = R_{n_k^*}(c) = c(1 + ts^{-1}) n_k^{*t}$. The goal is to achieve this *minimax risk* approximately. Of course, n_k^* is unknown and hence Kuo and Mukhopadhyay (1990a) proposed a purely sequential methodology and its associated *second-order* asymptotics. The corresponding stopping variable N is such that $\max_{\boldsymbol{\mu}} E[L_N]$ is attained when $\mu_1 = \cdots = \mu_k$ and this maximal sequential risk turns out to be $E[R_N(c)]$. Naturally, in defining N, one uses V_n as the estimator of θ in the boundary condition. Further details are omitted for brevity.

Another way to formulate the problem is to work under the loss function $L_n = A[T(n) - \mu^*]^s$ where A and s are known positive numbers. Given $W(> 0)$, one may wish to achieve *bounded maximal risk*, that is to have $\max_{\boldsymbol{\mu}} E[L_n] \le W$, in the spirit of Hilton (1984) and Mukhopadhyay and Hilton (1986). The analogs are fairly obvious and hence details were not provided in Kuo and Mukhopadhyay (1990a). Remarks 1-3 in that same paper are particularly relevant in this subsection.

Remark 26.11 Problems analogous to those discussed in Section 26.6.3 were developed for normal distributions in Kuo and Mukhopadhyay (1990b).

26.7 Combining Sequential and Multi-Stage Estimation Procedures

Suppose that two divisions or branches of a pharmaceutical company have been individually studying the behavior of the mean response time (or other appropriate parameters) for comparable treatments via separate exponential clinical trials carried out under identical protocols. Each location has been experimenting with just one treatment. The two locations may possibly have implemented *different sequential* or *multi-stage sampling* methodologies, having certain notions of "fixed-precision" tied in with their individual problems of statistical inference. Mukhopadhyay and Chattopadhyay (1992) introduced meta-analytic tools by examining how such *already available* sample resources obtained from independent, but otherwise comparable studies can be fruitfully combined in order to make comparative statements between the treatments. Under two-parameter negative exponential models given by (26.1.1) and different types of sampling strategies, Mukhopadhyay and Chattopadhyay (1992) developed appropriate asymptotic *second-order* characteristics for the final overall treatment comparisons. The normal theory analogs have been developed in Mukhopadhyay and Chattopadhyay (1993).

26.8 Time-Sequential Methodologies

In a controlled clinical trial or reliability study with a large number n of specimens placed on test, usually one continues to monitor each item until a terminal response is obtained. To be specific, suppose that response time is exponentially distributed having the density $f(x;0,\theta)$ given by (26.1.1). On the basis of data gathered in a sequential framework, one may now wish to estimate θ or some other appropriate characteristic of the survival distribution. After obtaining k responses, one may consider an appropriate estimator $\hat{\theta}_{n,k}$ for θ defined in terms of observed samples only. Now, in order to quantify the efficacy or effect due to stopping at this stage and proposing $\hat{\theta}_{n,k}$ as the estimator of θ, one may consider the overall loss function as

$$L_{n,k} = AL\left(\theta, \hat{\theta}_{n,k}\right) + Bn + cV_{n,k}, \tag{26.8.1}$$

where A, B and c are known positive numbers, $L\left(\theta, \hat{\theta}_{n,k}\right)$ is some type of loss function due to estimation such as perhaps $\left(\hat{\theta}_{n,k} - \theta\right)^2$, and $V_{n,k}$ is the total time on test expended by the n items of the sample.

Let $X_1, \ldots, X_n$ be i.i.d. variables having the density $f(x;0,\theta)$, $L(x,y) = (x-y)^2$, $A = 1$. The data recorded up to the kth failure time are $X_{(1)}, \ldots, X_{(k)}$, the first k order statistics. Now, $V_{n,k} = \sum_{i=1}^{k} X_{(i)} + (n-k) X_{(k)}$ and one may wish to estimate θ by means of its maximum likelihood estimator, that is consider $\hat{\theta}_{n,k} = k^{-1}V_{n,k}$ (see Chapter 4). The overall risk, that is $E[L_{n,k}]$ is given by $R_{n,k} = k^{-1}\theta^2 + Bn + ck\theta$. Now, $k = k_n$ minimizes $R_{n,k}$ where

$$k_n = \begin{cases} \min & \{k \le n-1: \quad ck(k+1) \ge \theta\} \\ n & \text{if} \quad cn(n-1) < \theta \,. \end{cases} \tag{26.8.2}$$

Naturally, k_n is unknown and hence Sen (1980) proposed the following purely sequential procedure. Let

$$N = \begin{cases} \min & \{k \le n-1: \quad ck^2(k+1) \ge V_{n,k}\}, \\ n & \text{if no such } k \text{ exists} \,. \end{cases} \tag{26.8.3}$$

Once the sequential process (26.8.3) stops, Sen (1980) proposed considering the estimator $\hat{\theta}_{N,k}$ for θ and put forth various *first-order* asymptotics under appropriate behavioral assumptions involving n, B and c.

Generalizations of this approach to account for random withdrawals of the subjects under study were the focal points in Gardiner, Susarla, and van Ryzin (1986). A thorough overview of this area has appeared in Gardiner and Susarla (1991).

References

Aras, G. (1987). Sequential estimation of the mean exponential survival time under random censoring, *Journal of Statistical Planning and Inference*, **16**, 147-158.

Aras, G. (1989). Second order sequential estimation of the mean exponential survival time under random censoring, *Journal of Statistical Planning and Inference*, **21**, 3-17.

Bar-Lev, S.K. and Reiser, B. (1982). An exponential subfamily which admits UMPU test based on a single test statistic, *Annals of Statistics*, **10**, 979-989.

Basu, A.P. (1971). On a sequential rule for estimating the location parameter of an exponential distribution, *Naval Research Logistics Quarterly*, **18**, 329-337.

Basu, A.P. (1991). Sequential methods in reliability and life testing, In *Handbook of Sequential Analysis*, Chapter 25 (Eds., B.K. Ghosh and P.K. Sen), pp. 581-592, Marcel Dekker, New York.

Bose, A. and Boukai, B. (1993). Sequential estimation results for a two-parameter exponential family of distributions, *Annals of Statistics*, **21**, 484-502.

Bose, A. and Mukhopadhyay, N. (1994a). Sequential estimation by accelerated stopping times in a two-parameter exponential family of distributions, *Statistics & Decisions*, **12**.

Bose, A. and Mukhopadhyay, N. (1994b). Sequential estimation via replicated piecewise stopping number in a two-parameter exponential family of distributions, *Sequential Analysis*, **13**, 1-10.

Brown, L.D. (1986). *Fundamentals of Statistical Exponential Families*, Institute of Mathematical Statistics, Hayward, CA.

Chaturvedi, A. and Shukla, P.S. (1990). Sequential point estimation of location parameter of a negative exponential distribution, *Journal of the Indian Statistical Association*, **28**, 41-50.

Chernoff, H. (1972). *Sequential Analysis and Optimal Design*, SIAM, Philadelphia.

Chow, Y.S. and Robbins, H. (1965). On the asymptotic theory of fixed-width sequential confidence intervals for the mean, *Annals of Mathematical Statistics*, **36**, 457-462.

Costanza, M.C., Hamdy, H.I., and Son, M.S. (1986). Two stage fixed width confidence intervals for the common location parameter of several exponential distributions, *Communications in Statistics - Theory and Methods*, **15**, 2305-2322.

Gardiner, J.C. and Susarla, V. (1983). Sequential estimation of the mean survival time under random censorship, *Sequential Analysis*, **2**, 201-223.

Gardiner, J.C. and Susarla, V. (1984). Risk efficient estimation of the mean exponential survival time under random censoring, *Proceedings of National Academy of Sciences, U.S.A.*, **81**, 5906-5909.

Gardiner, J.C. and Susarla, V. (1991). Time-sequential estimation, In *Handbook of Sequential Analysis*, Chapter 27 (Eds., B.K. Ghosh and P.K. Sen), pp. 613-631, Marcel Dekker, New York.

Gardiner, J.C., Susarla, V., and van Ryzin, J. (1986). Time sequential estimation of the exponential mean under random withdrawals, *Annals of Statistics*, **14**, 607-618.

Ghosh, M. and Mukhopadhyay, N. (1981). Consistency and asymptotic efficiency of two-stage and sequential procedures, *Sankhyā, Series A*, **43**, 220-227.

Ghosh, M. and Mukhopadhyay, N. (1989). Sequential estimation of the percentiles of exponential and normal distributions, *South African Statistical Journal*, **23**, 251-268.

Ghosh, M. and Mukhopadhyay, N. (1990). Sequential estimation of the location parameter of an exponential distribution, *Sankhyā, Series A*, **52**, 303-313.

Ghurye, S.G. (1958). Note on sufficient statistics and two-stage procedures, *Annals of Mathematical Statistics*, **29**, 155-166.

Govindarajulu, Z. (1985). Exact expressions for thc stopping time and confidence coefficient in point and interval estimation of scale parameter of exponential distribution with unknown location, *Technical Report 254*, University of Kentucky, Dept. of Statistics.

Govindarajulu, Z. and Sarkar, S.C. (1991). Sequential estimation of the scale parameter in exponential distribution with unknown location, *Utilitas Mathematica*, **40**, 161-178.

Hall, P. (1981). Asymptotic theory of triple sampling for sequential estimation of a mean, *Annals of Statistics*, **9**, 1229-1238.

Hall, P. (1983). Sequential estimation saving sampling operations, *Journal of the Royal Statistical Society, Series B*, **45**, 219-223.

Hamdy, H.I., Mukhopadhyay, N., Costanza, M.C., and Son, M.S. (1988). Triple stage point estimation for the exponential location parameter, *Annals of the Institute of Statistical Mathematics*, **40**, 785-797.

Hilton, G.F. (1984). Sequential and two-stage point estimation problems for negative exponential distributions, *Ph.D. Dissertation*, Oklahoma State University, Dept. of Statistics.

Isogai, E. and Uno, C. (1993). Sequential estimation of a parameter of an exponential distribution, *Annals of the Institute of Statistical Mathematics*, **45**.

Kuo, L. and Mukhopadhyay, N. (1990a). Point estimation of the largest location of k negative exponential populations, *Sequential Analysis*, **9**, 297-304.

Kuo, L. and Mukhopadhyay, N. (1990b). Multi-stage point and interval estimation of the largest of k normal populations and the associated second-order properties, *Metrika*, **37**, 291-300.

Lal Saxena, K.M. and Tong, Y.L. (1969). Interval estimation of the largest mean of k normal populations with known variances, *Journal of the American Statistical Association*, **64**, 296-299.

Lombard, F. and Swanepoel, J.W.H. (1978). On finite and infinite confidence sequences, *South African Statistical Journal*, **12**, 1-24.

Mukhopadhyay, N. (1974). Sequential estimation of location parameter in exponential distributions, *Calcutta Statistical Association Bulletin*, **23**, 85-95.

Mukhopadhyay, N. (1980). A consistent and asymptotically efficient two-stage procedure to construct fixed-width confidence intervals for the mean, *Metrika*, **27**, 781-784.

Mukhopadhyay, N. (1982a). Stein's two-stage procedure and exact consistency, *Skandinavisk Aktuarietidskrift*, 110-122.

Mukhopadhyay, N. (1982b). A study of the asymptotic regret while estimating the location of an exponential distribution, *Calcutta Statistical Association Bulletin*, **31**, 207-213.

Mukhopadhyay, N. (1985). A note on three-stage and sequential point estimation procedures for a normal mean, *Sequential Analysis*, **4**, 311-320.

Mukhopadhyay, N. (1987). Minimum risk point estimation of the mean of a negative exponential distribution, *Sankhyā, Series A*, **49**, 105-112.

Mukhopadhyay, N. (1988a). Sequential estimation problems for negative exponential populations, *Communications in Statistics - Theory and Methods (Reviews Section)*, **17**, 2471-2506.

Mukhopadhyay, N. (1988b). Fixed precision estimation of a positive location parameter of a negative exponential population, *Calcutta Statistical Association Bulletin*, **37**, 101-104.

Mukhopadhyay, N. (1990). Some properties of a three-stage procedure with applications in sequential analysis, *Sankhyā, Series A*, **52**, 218-231.

Mukhopadhyay, N. (1991). Parametric sequential point estimation, In *Handbook of Sequential Analysis*, Chapter 10 (Eds., B.K. Ghosh and P.K. Sen), pp. 245-267, Marcel Dekker, New York.

Mukhopadhyay, N. (1992). Simultaneous point estimation problems for two-parameter negative exponential populations, *Journal of the Indian Statistical Association*, **30**, 33-41.

Mukhopadhyay, N. (1994). Improved sequential estimation of means of exponential distributions, *Annals of the Institute of Statistical Mathematics*, **46**.

Mukhopadhyay, N. and Chattopadhyay, S. (1991). Sequential methodologies for comparing exponential mean survival times, *Sequential Analysis*, **10**, 139-148.

Mukhopadhyay, N. and Chattopadhyay, S. (1992). Comparing exponential clinical trials by combining individual sequential experiments available from independent studies, *Technical Report 92-09*, University of Connecticut, Dept. of Statistics.

Mukhopadhyay, N. and Chattopadhyay, S. (1993). Comparing means by combining individual sequential estimators available from independent studies, *Sankhyā, Series A*, **54**.

Mukhopadhyay, N. and Darmanto, S. (1988). Sequential estimation of the difference of means of two negative exponential populations, *Sequential Analysis*, **7**, 165-190.

Mukhopadhyay, N. and Datta, S. (1994). Replicated piecewise multistage sampling with applications, *Sequential Analysis*, **13**, 253-276.

Mukhopadhyay, N. and Ekwo, M.E. (1987). A note on minimum risk point estimation of the shape parameter of a Pareto distribution, *Calcutta Statistical Association Bulletin*, **36**, 69-78.

Mukhopadhyay, N. and Hamdy, H.I. (1984). On estimating the difference of location parameters of two negative exponential distributions, *Canadian Journal of Statistics*, **12**, 67-76.

Mukhopadhyay, N., Hamdy, H.I., Al-Mahmeed, M., and Costanza, M.C. (1987). Three-stage point estimation procedures for a normal mean, *Sequential Analysis*, **6**, 21-36.

Mukhopadhyay, N. and Hilton, G.F. (1986). Two-stage and sequential procedures for estimating the location parameter of a negative exponential distribution, *South African Statistical Journal*, **20**, 117-136.

Mukhopadhyay, N. and Kuo, L. (1988). Fixed-width interval estimations of the largest location of k negative exponential populations, *Sequential Analysis*, **7**, 321-332.

Mukhopadhyay, N. and Mauromoustakos, A. (1987). Three-stage estimation for the negative exponential distributions, *Metrika*, **34**, 83-93.

Mukhopadhyay, N. and Narayan, P. (1981). Sequential fixed-width intervals for the common location parameters of two normal or two negative exponential distributions, *Journal of Indian Society of Statistics and Operations Research*, **2**, 1-15.

Mukhopadhyay, N. and Padmanabhan, A.R. (1993). A note on three-stage confidence intervals for the difference of locations - The exponential case, *Metrika*, **40**, 121-128.

Mukhopadhyay, N. and Sen, P.K. (1993). Replicated piecewise stopping numbers and sequential analysis, *Sequential Analysis*, **12**, 179-197.

Mukhopadhyay, N. and Solanky, T.K.S. (1991). Second order properties of accelerated stopping times with applications in sequential estimation, *Sequential Analysis*, **10**, 99-123.

Robbins, H. (1959). Sequential estimation of the mean of a normal population, In *Probability and Statistics*, H. Cramér Volume (Ed., U. Grenander), pp. 235-245, Almquist and Wiksell, Uppsala, Sweden.

Robbins, H., Simons, G., and Starr, N. (1967). A sequential analogue of the Behrens-Fisher problem, *Annals of Mathematical Statistics*, **38**, 1384-1391.

Sen, P.K. (1980). On time-sequential estimation of the mean of an exponential distribution, *Communications in Statistics - Theory and Methods*, **9**, 27-38.

Srivastava, M.S. (1970). On a sequential analogue of the Behrens-Fisher problem, *Journal of the Royal Statistical Society, Series B*, **38**, 219-230.

Starr, N. and Woodroofe, M. (1972). Further remarks on sequential estimation: the exponential case, *Annals of Methematical Statistics*, **43**, 1147-1154.

Stein, C. (1945). A two-sample test for a linear hypothesis whose power is independent of the variance, *Annals of Mathematical Statistics*, **16**, 243-258.

Stein, C. (1949). Some problems in sequential estimation, *Econometrica*, **17**, 77-78.

Swanepoel, J.W.H. and vanWyk, J.W.J. (1982). Fixed-width confidence intervals for the location parameter of an exponential distribution, *Communications in Statistics - Theory and Methods*, **11**, 1279-1289.

Tong, Y.L. (1970). Multi-stage interval estimations of the largest mean of k normal populations, *Journal of the Royal Statistical Society, Series B*, **32**, 272-277.

Woodroofe, M. (1977). Second order approximations for sequential point and interval estimation, *Annals of Statistics*, **5**, 984-995.

Woodroofe, M. (1982). *Nonlinear Renewal Theory in Sequential Analysis*, SIAM, Philadelphia.

Woodroofe, M. (1985). Asymptotic local minimaxity in sequential point estimation, *Annals of Statistics*, **13**, 676-688.

Woodroofe, M. (1987). Asymptotically optimal sequential point estimation in three stages, In *New Perspectives in Theoretical and Applied Statistics* (Ed., M.L. Puri), pp. 397-411, John Wiley & Sons, New York.

Zacks, S. (1973). Sequential design for a fixed-width interval estimation of the common mean of two normal distributions, I. The case of one variance known, *Journal of the American Statistical Association*, **68**, 422-427.

CHAPTER 27

Two-Stage and Multi-Stage Tests of Hypotheses

J.E. Hewett[1]

27.1 Introduction

In this chapter when we make reference to the exponential distribution, we mean a continuous distribution with density given by

$$f(x;\mu,\theta) = \frac{1}{\theta}\, e^{-(x-\mu)/\theta} I(x > \mu)\ ,$$

where $I(\cdot)$ is the usual indicator function of $(\cdot)$, μ could be any real number and $\theta > 0$. For the methods discussed in this chapter, it will usually be the case that $\mu = 0$, and the hypotheses of interest will pertain to θ.

There have been many papers published that contain fixed sample size tests of hypotheses pertaining to the exponential distribution such as Epstein (1954) and Epstein and Sobel (1953). The general format for a fixed sample size test for testing H_0 versus H_1 would be to observe a random sample of size n and compute a statistic, say T. If $T \in A$ for some set A, H_0 is rejected whereas, if $T \in A^c$ (complement of A) then H_0 is not rejected (H_0 is accepted). The general format for a k-stage test would require that the total sample size n be split into k parts denoted by $n_1, n_2, \ldots, n_k$ where $\sum_{i=1}^{k} n_i = n$. We observe a random sample of size n_1 and compute a statistic T_1. If $T_1 \in A_{1r}$ (A_{1A}), then H_0 is rejected (accepted). If $T_1 \in (A_{1r} \cup A_{1A})^c$, then we observe an additional sample of size n_2 and compute a statistic T_2 using all $n_1 + n_2$ observations. If $T_2 \in A_{2r}$ (A_{2A}), then H_0 is rejected (accepted). If $T_2 \in (A_{2r} \cup A_{2A})^c$, we observe an additional sample of size n_3, etc. If H_0 has not been rejected or accepted after the first $k-1$ stages, we then observe an additional sample of size n_k and compute T_k. If $T_k \in A_{kr}$, we reject H_0. If $T_k \in A_{kA}$, we accept H_0 where $A_{kA} = (A_{kr})^c$.

One of the primary objectives of these k-stage tests is to reduce (if possible) the actual sample size needed in order to make a correct decision. The sets A_{ir}, A_{iA}, $i =$

[1]University of Missouri at Columbia, Columbia, Missouri

$1, \ldots, k$ are chosen to simultaneously make the expected sample size, $E[N]$, small and the power function close to that of a fixed sample size test based on all n observations, where

$$\begin{aligned} E[N] &= n_1 P(T_1 \in [A_{1r} \cup A_{1A}]) + \sum_{i=2}^{k-1} \left(\sum_{j=1}^{i} n_j \right) \\ &\cdot P\left(\bigcap_{l=1}^{i-1} \{T_l \in [A_{lr} \cup A_{lA}]^c\} \cap \{T_i \in [A_{ir} \cup A_{iA}]\} \right) \\ &+ \left(\sum_{j=1}^{k} n_j \right) P\left(\bigcap_{l=1}^{k-1} \{T_l \in [A_{lr} \cup A_{lA}]^c\} \right) . \end{aligned}$$

Note that for $k = 2$

$$E[N] = n_1 + n_2 \; P(T_1 \in [A_{1r} \cup A_{1A}]^c) \; .$$

The power function is given by

$$\begin{aligned} \text{Power} &= P(T_1 \in A_{1r}) \\ &+ \sum_{i=2}^{k} P\left(\left[\bigcap_{j=2}^{i} \{T_{j-1} \in [A_{(j-1)r} \cup A_{(j-1)A}]^c\} \right] \cap [T_i \in A_{ir}] \right) . \end{aligned}$$

It is also important to note how much of the significance level α, say α_i, is used at the ith stage where $\alpha = \sum_{i=1}^{k} \alpha_i$. If H_0 is a simple hypothesis, α_1 is given by $\alpha_1 = P(T_1 \in A_{1r} \mid H_0 \text{ true })$ and for $i = 2, \ldots, k$, α_i is given by

$$\alpha_i = P\left(\left[\bigcap_{j=2}^{i} \{T_{j-1} \in [A_{(j-1)r} \cup A_{(j-1)A}]^c \cap [T_i \in A_{ir}]\} \right] \middle| H_0 \text{ true} \right) .$$

If H_0 is a composite hypothesis, a similar statement holds.

The remaining sections will contain discussions of specific methods.

Section 27.2:	One Population Problem
27.2.1:	Two-Stage Methods of Bulgren and Hewett
27.2.2:	Two-Stage Methods of Fairbanks
Section 27.3:	Two Population Problem
27.3.1:	Two-Stage Tests
27.3.2:	Group Sequential Tests
Section 27.4:	Goodness-of-Fit Tests
27.4.1:	Two-Stage Chi-Square Goodness-of-Fit Test
27.4.2:	Two-Stage Test for Exponential Against IFRA Alternatives
Section 27.5:	Distribution Free Tests
27.5.1:	Two-Stage and k-Stage Tests

27.2 One Population Problem

27.2.1 Two-Stage Methods of Bulgren and Hewett

For this section interest is in the scale parameter θ, and μ is assumed to be 0. The methods are discussed in the life testing situation described by Epstein (1954) where n items are placed on test simultaneously, and it is desired to test $H_0 : \theta \geq \theta_0$ versus $H_1 : \theta < \theta_0$. The test is culminated after the first r failures have occurred where r is a preassigned number. Both the replacement and nonreplacement cases were discussed. The test statistic was the total time on test denoted by T_r with H_0 being rejected when $T_r/\theta_0 < c$. When $\theta = \theta_0$, $2T_r/\theta_0$ is distributed as chi-square with $2r$ degrees of freedom.

For the two-stage test of Bulgren and Hewett (1973), n items are placed on test simultaneously and after the r_1th failure has occurred, H_0 is accepted if $2T_{r_1}/\theta_0 \geq d_2$; H_0 is rejected if $2T_{r_1}/\theta_0 \leq d_1$; and if $d_1 < 2T_{r_1}/\theta_0 < d_2$ testing would continue until r_2 additional failures have occurred. If $2T_{r_3}/\theta_0 \leq d_3$, then H_0 is rejected, and if $2T_{r_3}/\theta_0 > d_3$, H_0 is accepted, where $r_3 = r_1 + r_2$.

The operating characteristic function for this two-stage procedure is given by $P_A^*(\lambda) = \int_{d_2/\lambda}^{\infty} g(x)dx + \int_{d_1/\lambda}^{d_2/\lambda} \int_{d_3/\lambda}^{\infty} h(z,y)$ where $g(\cdot)$ is the density of a chi-square distribution with $2r_1$ degrees of freedom and $h(z,y)$ is given by

$$h(z,y) = \begin{cases} k(z-y)^{r_2-1}y^{r_1-1}e^{-z/2} & , \; 0 < y < z, \; 0 < z < \infty \\ 0 & , \; \text{otherwise} \end{cases}$$

where $k = 1/[2^{(r_1+r_2)}\Gamma(r_1)\Gamma(r_2)]$.

The method used to obtain d_1, d_2, d_3 for given r_1 and r_2 was to select r_1 slightly less than or slightly greater than $.5r$ with r_1+r_2 slightly greater than r ($r_1+r_2 = 1.2\,r$). Once r_1 and r_2 are chosen, d_1, d_2, d_3 are obtained by solving the three nonlinear equations $P_A^*(\lambda_1) = p_1$, $P_A^*(\lambda_2) = p_2$, $P_A^*(\lambda_3) = p_3$. Here, λ_1, λ_2, λ_3 are obtained by solving $P_A(\lambda_i) = p_i$ for $i = 1,2,3$ where $P_A(\lambda)$ is the operating characteristic function for the usual procedure based on r failures. p_1, p_2, p_3 usually are chosen such that $p_1 = .95(.90)$, $p_2 = (.05)(.10)$ and $p_3 = .5$. Note that this procedure essentially guarantees that the operating characteristic curve of the two-stage procedure matches that of the usual single stage procedure. Table 1 of Bulgren and Hewett (1973) contains values of d_1, d_2, d_3 for several selections of $\alpha, \beta, \theta_0/\theta$. The potential savings obtained by two-stage procedure is noted by comparing the expected number of failures to r. The graphs of Zeigler and Tietjen (1968) can be used for this purpose. Although this discussion was for the nonreplacement case, similar results hold for the case where failed items are replaced.

27.2.2 Two-Stage Methods of Fairbanks

Fairbanks (1988) considered the same H_0 and H_1 as Bulgren and Hewett, except Fairbanks considers Type I and Type II censoring whereas Bulgren and Hewett only considered Type II censoring. For the Fairbanks test the user places n items on test and has preassigned times t_1^*, t_2^*, t_3^* as well as fixed numbers of failures r_1 and r_2. At the first stage (the first look at the data), the test rejects H_0 if $x_{r_1:n}$, the r_1th failure time is less than t_1^* and accepts H_0 at t_2^* when $x_{r_1:n} \geq t_2^*$. The test continues to the second stage otherwise with the rejection of H_0 at stage two if $x_{r_3:n} < t_3^*$ and the acceptance of H_0

if $x_{r_3:n} \geq t_3^*$. Here, $r_3 = r_1 + r_2$. Figure 1 of Fairbanks (1988) contains an illustration of a set of typical boundaries.

This discussion assumes sampling without replacement; however, analogous results hold for the case where failed items are replaced. Prior to the test, t_3^* is specified and r_1 and r_2 are selected using the guidelines indicated in the Bulgren-Hewett test. Also, as in the Bulgren-Hewett test, the operating characteristic curve of the two-stage test is matched with that of the test of Epstein at three widely spaced values of θ/θ_0. The system of equations

$$P_A(\lambda_0) = 1 - \alpha, \qquad P_A(\lambda_1) = \beta, \qquad P_A(\lambda_2) = .5$$

where λ_0, λ_1 and λ_2 are chosen so that the operating characteristic curve of Epstein has values $1 - \alpha$, β, and .5, respectively, is solved for n, t_1^*, and t_2^*, and

$$\begin{aligned} P_A(\lambda_i) &= \sum_{w=0}^{r_1-1} \mathrm{Bin}(w; n, P_{2i}^*) + \sum_{x=0}^{r_1-1} \sum_{y=r_1-x}^{r_3-x-1} \sum_{z=0}^{r_3-x-y-1} \mathrm{Bin}(x; n, P_{1i}^*) \\ &\quad \cdot \mathrm{Bin}(y; n-x, P_{2i}) \mathrm{Bin}(z; n-x-y, P_{3i}) \,. \end{aligned}$$

Here, $\mathrm{Bin}(x; n, P)$ is the binomial probability of x failures with parameters n and P. Also,

$$\begin{aligned} P_{1i}^* &= P_{\lambda_i}(\text{item fails during } (0, t_1^*]) = 1 - e^{-t_1^*/\theta_i} \,, \\ P_{2i}^* &= P_{\lambda_i}(\text{item fails during } (0, t_2^*]) = 1 - e^{-t_2^*/\theta_i} \,, \\ P_{2i} &= P_{\lambda_i}(\text{item fails during } (t_1^*, t_2^*]) = 1 - e^{-(t_2^*-t_1^*)/\theta_i} \,, \end{aligned}$$

and

$$P_{3i} = P_{\lambda_i}(\text{item fails during } (t_2^*, t_3^*]) = 1 - e^{-(t_3^*-t_2^*)/\theta_i} \,.$$

Fairbanks gives a table where he relates θ_0/θ, θ_0/t_3^*, α, β, r_1, r_2, n, t_1^*, t_2^*, Pr_{A_1,θ_0}, $Pr_{R_1,\theta}$. These latter two probabilities are, respectively, $P(\text{accept } H_0 \text{ in stage } 1 \mid \theta_0)$ and $P(\text{reject } H_0 \text{ in stage } 1 \mid \theta)$. He also gives figures where he compares Bulgren-Hewett, Epstein and his test relative to $E[T]$, the expected time until decision, and the expected number of failures. These figures suggest that considerable savings can be achieved if one uses the new procedure.

27.3 Two Population Problem

27.3.1 Two-Stage Test

We are interesting in developing a two-stage test for testing $H_0 : \theta_1 = \theta_2$ versus $H_A : \theta_1 > \theta_2$ or $H_A^* : \theta_1 \neq \theta_2$ where for $i = 1, 2$, $\theta_i > 0$ and is the unknown parameter of an exponential distribution with density given by $f(x; \theta_i) = (1/\theta_i)\exp[-(x/\theta_i)]$, $x > 0$. The general format for a two-stage test of H_0 versus H_1 will be given for a specific case. The extension to the life testing situation of Epstein (1954) will be obvious.

Let $x_1, \ldots, x_{n_1}$ and $y_1, \ldots, y_{m_1}$ represent random samples from populations one and two, respectively, with $\bar{x}_1$ and $\bar{y}_1$ their respective sample means. The first stage test statistic is denoted by F_1 where $F_1 = \bar{x}_1/\bar{y}_1$. It is well known that if $\theta_1 = \theta_2$, then F_1 has an F-distribution with $2n_1$ and $2m_1$ degrees of freedom. At stage one H_0 is rejected provided $F_1 \geq c_1$ and would be accepted if $F_1 \leq c_2$. If $c_2 < F_1 < c_1$, we take additional observations $x_{n_1+1}, \ldots x_{n_1+n_2}$ and $y_{m_1+1}, \ldots, y_{m_1+m_2}$. Using all $n_1 + n_2$ and $m_1 + m_2$ observations we compute $F_2 = \bar{x}_2/\bar{y}_2$ where $\bar{x}_2 = \frac{1}{n_1+n_2} \sum_{i=1}^{n_1+n_2} x_i$ and $\bar{y}_2 = \frac{1}{m_1+m_2} \sum_{i=1}^{m_1+m_2} y_i$. Note that when $\theta_1 = \theta_2$, F_2 has an F-distribution with $2[n_1 + n_2]$ and $2[m_1 + m_2]$ degrees of freedom. Thus, at stage two we reject H_0 if $F_2 \geq c_3$ and accept H_0 if $F_2 < c_3$. In order to obtain the rejection numbers c_1, c_2, and c_3, Zeigler and Goldman (1972) derived a bivariate density from which the joint density of F_1 and F_2 is easily obtainable. Although they designed a two-stage test for testing equality of two variances of normally distributed populations, their distribution theory is readily adaptable to our problem. They use the concept of matching operating characteristic curves in order to obtain values of c_1, c_2, c_3 in a manner similar to that discussed in Sections 27.2.1 and 27.2.2 here for a specific set of sample sizes. Work currently is ongoing to develop a more extensive set of values of c_1, c_2, and c_3.

The optimality of these tests constructed in this manner would, as before, be evaluated in terms of the expected sample size since the tests are developed to match power to that of the usual one-stage test.

27.3.2 k-Stage Test

Pocock (1977) presented a group sequential design, which has the format of a k-stage test. He presents exact results for two populations which are normally distributed with known variance. He indicated how his methods can be used to obtain an approximate k-stage test for the hypotheses $H_0 : \theta_1 = \theta_2$ and $H_1 : \theta_1 \neq \theta_2$ considered in Section 27.3.1.

For the group sequential procedure n observations are taken from each population at each stage, and the decision at the ith stage is based on $\log(\bar{x}/\bar{y})$ where $\bar{x}$ and $\bar{y}$ are defined as in Section 27.3.1. The hypotheses are restated in terms of $\log(\theta_1/\theta_2)$. The test rejects H_0 at the ith stage if $P_i < \alpha'$ where $P_i = 2\left[1 - \Phi\left\{(in/2)^{1/2}\log(\bar{x}/\bar{y})\right\}\right]$. Values of α' depend on α, the significance level of the overall test and the number of stages denoted by N. Table 1 of Pocock (1977) contains values of α' for $\alpha = .05$, .01 and selected values of N. Table 2 of Pocock contains values of $\Delta = (n/2)^{1/2}\log(\theta_1/\theta_2)$ where power of .5, .75, .9, and .99 are obtained for both $\alpha = .05$ and .01.

27.4 Tests for Exponentiality

27.4.1 Two-Stage Chi-Square Goodness-of-Fit Test

Hewett and Tsutakawa (1972) presented a two-stage chi-square test of homogeneity, which is suitable for testing the null hypothesis H_0 that a random sample comes from a distribution with density given by $f(x;\theta) = (1/\theta)\exp[-x/\theta]I(x > 0)$.

The first-stage test statistic based on a random sample of size n_1 is the usual chi-square statistic Q_1, where under H_0 the limiting distribution of Q_1 is chi-square with $\ell - 2$ degrees of freedom. Here, ℓ represents the number of cells. H_0 is rejected if $Q_1 \geq c_1$ and is accepted if $Q_1 \leq c_2$. If $c_2 < Q_1 < c_1$, then n_2 additional observations are taken. Let Q_2 denote the usual chi-square statistic based on all $n_1 + n_2$ observations. If $Q_2 \geq c_3$, H_0 is rejected; and if $Q_2 < c_3$, H_0 is accepted. Hewett and Tsutakawa (1972) show that the joint limiting distribution of Q_1 and Q_2 is the bivariate chi-square distribution discussed by Kibble (1941) and Jensen (1970). They also investigate the asymptotic power of the two-stage test and compare it with that of the one-stage test. Table 5 of Hewett and Spurrier (1983) contains values of c_1, c_2, c_3 for $\alpha = .05$, $.01$, and different numbers of degrees of freedom.

27.4.2 Two-Stage Test for Exponentiality Against IFRA Alternatives

In this section we discuss the two-stage test of H_0 versus H_1 proposed by Alam and Basu (1989) where

$$\begin{aligned} H_0 &: \quad F(x) = 1 - e^{-\lambda x},\ x \geq 0,\ \lambda > 0 \\ H_1 &: \quad F \text{ is IFRA and not exponential .} \end{aligned}$$

F is IFRA (Increasing Failure Rate in Average) if and only if for $x > 0$, $0 < b < 1$, $\bar{F}(bx) \geq [\bar{F}(x)]^b$ where $\bar{F}(x) = 1 - F(x)$.

For the two-stage test at Stage 1, we observe a random sample $x_1, \ldots, x_{n_1}$ and compute J_{1b} where $J_{1b} = [n_1(n_1-1)]^{-1} \sum_{i,j} h_b(x_i, x_j)$ with $\sum_{i,j}$ denoting summation over $1 \leq i,\ j \leq n_1$ such that $i \neq j$ with $h_b(x_i, x_j) = 1$ if $x_i > bx_j$ and is 0 otherwise. The decision rule at Stage 1 is to reject H_0 if $J_{1b} > c_1$; if $J_{1b} < c_2$, we accept H_0; if $c_2 \leq J_{1b} \leq c_1$ we take n_2 additional observations $x_{n_1+1}, \ldots, x_{n_1+n_2}$. Using all $n_1 + n_2$ observations we then compute J_{2b} where

$$J_{2b} = [(n_1+n_2)(n_1+n_2-1)]^{-1} \sum_{i',j'} h_b(x_{i'}, x_{j'})$$

and $\sum_{i',j'}$ denotes summation over $1 \leq i',\ j' \leq n_1 + n_2$ such that $i' \neq j'$ with $h_b(x_{i'}, x_{j'}) = 1$ if $x_{i'} > bx_{j'}$ and is 0 otherwise. For this discussion b is a constant such that $0 < b < 1$. The decision rule is to reject H_0 at Stage 2 if $J_{2b} > c_3$ and accept H_0 otherwise. In order to execute this two-stage procedure for a given α, it is necessary to obtain c_1, c_2 and c_3. The following theorem [Theorem 3.2 of Alam and Basu (1989)] provides distribution results, which can be used to obtain c_1, c_2, c_3 for a given α.

Theorem 27.1 *If $n_1 \to \infty$ with $n_2 \geq 0$ such that $n_1/(n_1+n_2) \to \rho^2$, then*

$$\begin{pmatrix} n_1^{1/2}\left(J_{1b} - \frac{1}{b+1}\right) \\ (n_1+n_2)^{1/2}\left(J_{2b} - \frac{1}{b+1}\right) \end{pmatrix} \to N\left[\begin{pmatrix} 0 \\ 0 \end{pmatrix}, \sigma^2 \begin{pmatrix} 1 & \rho \\ \rho & 1 \end{pmatrix}\right]$$

where $(n_1+n_2)Cov(J_{1b}, J_{2b}) \to \sigma^2$.

Values of c_1, c_2, c_3, which reduce the expected sample size also reduce power. Hence, it is desirable to find values of c_1, c_2, c_3, which provide good power and also provide a reduction in expected sample size. Alam and Basu provide tables of values of c_1, c_2, c_3 for $\alpha = .05$, $.01$, $b = .5$, $.9$ and also simulation results that indicate the resulting tests have significance levels close to the target values.

27.5 Distribution-Free Tests

27.5.1 Two-Stage and k-Stage Tests

Chase and Hewett (1976) discussed a method of constructing two-stage tests that only requires that the population(s) be of the continuous type and p-values can be computed. The procedure is based on p-values and results from a modification of Fisher's method of combining independent tests. Simulation results indicate that the tests maintain good power and achieve savings in expected sample size.

Madsen (1978) presented a k-stage test procedure using p-values. This procedure is distribution-free in the sense that it is applicable in any situation where a p-value can be calculated. For $k > 2$ the tests are only approximate, but simulation results indicate that the test maintains good power while achieving savings in expected sample size.

References

Alam, M.S. and Basu, A.P. (1989). A two-stage test for exponentiality against IFRA alternatives, In *Recent Developments in Statistics and Their Applications* (Eds., J. Klein and J. Lee), Freedom Academy Publishing Company, Korea.

Bulgren, W.G. and Hewett, J.E. (1973). Double sample tests for hypotheses about the mean of an exponential distribution, *Technometrics*, **15**, 187-190.

Chase, G.R. and Hewett, J.E. (1976). Double sample tests – A distribution free procedure, *Journal of Statistical Computation and Simulation*, **4**, 247-257.

Epstein, B. (1954). Truncated life tests in the exponential case, *Annals of Mathematical Statistics*, **25**, 555-564.

Epstein, B. and Sobel, M. (1953). Life testing, *Journal of the American Statistical Association*, **48**, 485-502.

Fairbanks, K. (1988). A two-stage life test for the exponential parameter, *Technometrics*, **30**, 175-180.

Hewett, J.E. and Spurrier, J.D. (1983). A survey of two-stage tests of hypotheses: Theory and methods, *Communications in Statistics – Theory and Methods*, **12**, 2307-2425.

Hewett, J.E. and Tsutakawa, R.K. (1972). Two-stage chi-square goodness-of-fit test, *Journal of the American Statistical Association*, **67**, 395-401.

Madsen, R.W. (1978). A K-stage test procedure using p-values, *Journal of Statistical Computation and Simulation*, **8**, 117-131.

Pocock, S.J. (1977). Group sequential methods in the design and analysis of clinical trials, *Biometrika*, **64**, 191-199.

Zeigler, R.K. and Goldman, A. (1972). A double sampling plan for comparing two variances, *Journal of the American Statistical Association*, **67**, 698-701.

CHAPTER 28

Sequential Inference

Pranab K. Sen[1]

28.1 Introduction

It is most convenient to introduce a simple exponential model associated with nonnegative random variables through the following life-testing problem. Suppose that n items are put simultaneously to life testing and their failure times are recorded. Let these be denoted by $T_1, \ldots, T_n$, respectively. Note that the T_i are nonnegative and we assume that each of them follow the following simple exponential law:

$$F(x;\theta) = \{1 - \exp(-x/\theta)\}I(x \geq 0) \ , \tag{28.1.1}$$

where θ is an unknown parameter. It is well known that $E[T] = \theta$, so that θ is positive. Statistical inference for θ can be drawn based on the sufficient statistic $S_n = T_1 + \cdots + T_n$, and these have already been discussed in Chapters 4, 5 and 6. If we denote the survival (or reliability) function corresponding to the distribution function $F(\cdot)$ in (28.1.1) by $\overline{F}(\cdot) = 1 - F(\cdot)$, then $\overline{F}(0) = 1$, $\overline{F}(\infty) = 0$ and $\overline{F}(x) = \exp(-x/\theta)$ exponentially converges to 0 as $x \to \infty$. Thus, if we define the hazard function (or failure rate) by $\rho(t) = -(d/dt)\log \overline{F}(t)$, we have the well known characterization of the exponential law that

$$\rho(t) = \theta^{-1}(>0), \qquad \text{for all } t \in \mathbf{R}^+ = [0,\infty) \ . \tag{28.1.2}$$

In various reliability models, often, one may want to test for constant failure rates against an increasing failure rate (IFR) or a decreasing failure rate (DFR). In a life testing model or in the above mentioned reliability model, it is therefore customary to run an experiment with n units (for a prefixed n), record the failure times $T_1, \ldots, T_n$ and then draw statistical conclusions based on the sample outcome. The vintage point in this context is that the observations T_i become available sequentially (in time): The smallest one $T_{1:n}$ occurs first, the second smallest $T_{2:n}$ next, and so on, the largest one $T_{n:n}$ occurs last. Thus, to run the experiment, one needs to take into account the initial

[1]University of North Carolina, Chapel Hill, North Carolina

cost of setting the overall plan plus the cost for the total time on test (TTT) defined as $T_{1:n}+\cdots+T_{n:n}$. The TTT is itself a nonnegative random variable (having a gamma(n,θ) distribution), so that the cost function is typically a stochastic variable. Moreover, the duration of the experiment $[0, T_{n:n}]$, besides being stochastic, may become very large if n is so. For example, suppose that $n = 50$ and $\theta = 1,000$ hours (in a life-testing model for electric lamps). Then, $E[T_{n:n}] = \theta(\log n)$ is about 4,000 hrs compared to the average life length of 1,000 hrs. This may run contrary to the administrative planning, so that a curtailment of the study may be planned in a statistically meaningful way. This may be made for a fixed duration $[0, T]$, for some fixed $T(< \infty)$, resulting in Type I censoring or truncation, or up to a given number (say, $r \leq n$) of failures resulting in Type II censoring. For a Type I censoring scheme, a duration is fixed but the number of failures occurring during the tenure of the study is random, while for a Type II censoring scheme, the number of failures is prefixed but the duration of the study is random (see Chapter 4 for details). In either way, there are some limitations, and hence, a sequential plan may be a better alternative. Since here one may have typically a fixed number (n) of units and the termination is made on a time sequential basis, often, such sequential schemes are referred to as time-sequential schemes. A detailed discussion of time-sequential methods is given in Sen (1981, Chapter 11), while time-sequential estimation theory has been dealt with in some detail in Ghosh, Mukhopadhyay, and Sen (1995, Chapter 12). We shall refer to some of these in a later section. Coming back to the life-testing model, as failures occur, they may be replaced by new units (resulting in *with replacement* schemes) or it may be a *without replacement* scheme with decreasing number of units at successive failures. Since the exponential distributions enjoy the memoryless property in the with replacement scheme, there are certain simplifications which make the Epstein-Sobel (1955) methodology very adoptable in practice. Nevertheless, from a practical consideration, often such replacements may not be possible, and one may have to rely on the without replacement schemes. In a renewal process setup, for exponential inter-arrival times, one encounters a counting process attracted by the classical Poisson process. Therefore, it is quite relevant to discuss some sequential inference procedures relating to such Poisson processes and bring out their relevance to exponential models. Reliability engineers have a very special affection to exponential models, and the classical text of Barlow and Proschan (1975) is an excellent source of such reliability models. In such contexts too, often, sequential analysis may turn out to be a better alternative than the usual fixed-sample size procedures. Limiting availability of a system with a single unit, repair facility, and a spare is a classical example where such sequential procedures arise in a very natural way, and some of these have also been worked out in the literature [see Sen (1986, 1994) and Sen and Bhattacharjee (1986) among others.] Strength of a bundle of parallel filaments is another simple example of this type of reliability model where sequential procedures are available in the literature [see Sen (1974, 1983, 1994) among others.] As such, in this expository article on sequential inference for exponential models, attempts will be made to include this broad scenario, and to report the available methodology with due emphasis on some applications.

Section 28.2 is devoted to the usual Wald (1947) type SPRT for an exponential distribution parameter. Replacement and nonreplacement schemes in life testing are then discussed in Section 28.3. Section 28.4 deals with the sequential point and interval estimation theory. Section 28.5 pertains to the usual random censoring scheme when both the distributions are exponential. The case of two independent exponential distributions is treated in Section 28.6 and relevant sequential analysis is presented. Section 28.7 deals with sequential inference on system availability, and some general remarks are appended in the last section. Methodology and potential applications are presented in a unified manner.

28.2 A Sequential Probability Ratio Test

The theory of sequential probability ratio test (SPRT) due to Wald (1947) is very adoptable for the one-parameter exponential model in (28.1.1). Note that the probability density function corresponding to the exponential distribution in (28.1.1) is

$$f(x;\theta) = \theta^{-1}\exp(-x/\theta)I(x \geq 0) \; . \tag{28.2.1}$$

Suppose that one wants to test a simple null hypothesis $H_0 : \theta = \theta_0$ (specified) against a simple alternative $H_1 : \theta = \theta_1$ (specified), where $\theta_1 > \theta_0 > 0$. Based on n observations $X_1, \ldots, X_n$ from the distribution (28.1.1), in a non-sequential setup, the most powerful test is based on the probability ratio statistic:

$$\lambda_n = \prod_{i=1}^{n} f(X_i;\theta_1)/f(X_i;\theta_0) \; , \tag{28.2.2}$$

rejecting the null hypothesis for large values of λ_n; the Neyman-Pearson fundamental lemma provides the justification and construction of the critical region for any given level of significance α $(0 < \alpha < 1)$. Wald (1947) prescribed the following SPRT: Choose two nonnegative numbers B and A $(B < 1 < A)$, and compute λ_n in (28.2.2), for $n = 1, 2, \ldots$; if

$$\begin{array}{rcl} B < \lambda_n < A & , & \text{continue sampling ,} \\ \lambda_n \leq B & , & \text{terminate sampling along with the acceptance of } H_0 \; ; \\ \lambda_n \geq A & , & \text{terminate sampling along with the acceptance of } H_1 \; . \end{array} \tag{28.2.3}$$

The numbers B and A are to be so determined that the resulting Type I and Type II error probabilities are bounded by some preassigned numbers α and β, where both α and β lie in $(0, 1)$, and in addition Wald (1947) set that $\alpha + \beta < 1$. In this setup the sample size N, termed the ***stopping number***, is a random variable assuming nonnegative integer values, and is defined as

$$N = \inf\{n \geq 0 : \; \lambda_n \notin (B, A)\} \; . \tag{28.2.4}$$

Note that by (28.2.3), for every $n \geq 1$ and $\theta \in \Theta$,

$$P_\theta[N > n] \leq P_\theta[b + n\delta < \Delta S_n < a + n\delta] \; , \tag{28.2.5}$$

where $S_n = X_1 + \cdots + X_n$, and

$$a = \log A, \; b = \log B, \; \delta = \log(\theta_1/\theta_0) \text{ and } \Delta = (\theta_1 - \theta_0)/(\theta_1\theta_0) \; . \tag{28.2.6}$$

Since $n^{-1}S_n \to \theta$ a.s. as $n \to \infty$, and $n^{-1/2}(S_n - n\theta)$ is asymptotically normal, by (28.2.5) and (28.2.6), it follows that for each $\theta \in \Theta$,

$$\lim_{n\to\infty} P_\theta[N > n] = 0, \text{ i.e., } P_\theta[N < \infty] = 1 \; . \tag{28.2.7}$$

Moreover, by (28.2.3) and (28.2.7), we have

$$A \leq (1-\beta)/\alpha \qquad \text{and} \qquad B \geq \beta/(1-\alpha) \; . \tag{28.2.8}$$

Actually, in terms of the partial sum sequence S_n, $n \geq 1$, (28.2.3) can be equivalently expressed as: Continue sampling as long as $b + n\delta < \Delta S_n < a + n\delta$; if for $n = N$, this is vitiated for the first time, accept H_0 or H_1 according as ΔS_N is $\leq b + N\delta$ or $\geq a + N\delta$. The *operating characteristic function* (OC) $L(\theta)$, $\theta \in \Theta$, is defined as the probability of accepting $H_0 : \theta = \theta_0$ when θ is the true parameter point. Then at θ_0, the OC should be $\geq (1-\alpha)$ and at θ_1 it is to be $\leq \beta$. When Δ is small (i.e., θ_1 is close to θ_0), for both the inequalities (for A and B) the inequality signs may be dropped and satisfactory approximations hold for the equality signs. This led Wald (1947) to take $A \simeq (1-\beta)/\alpha$ and $B \simeq \beta/(1-\alpha)$. Moreover, the X_i are i.i.d. random variables with a finite moment generating function, so that for the partial sum process $\{S_n : n \geq 1\}$ the Skorokhod-Strassen embedding of Wiener process holds with an order to approximation as $O(\log n)$. Keeping in mind the plausibility of small values of Δ, we let $\Delta^0 = \theta_0^2\Delta(1+\Delta)$, and

$$\theta = \theta_0 + \phi\Delta^0, \qquad \text{where } \phi \in \Phi = \{\phi : |\phi| \leq K\}, \; K \in (1,\infty) \; . \tag{28.2.9}$$

Then, for small values of Δ, the Wald (1947) approximations yield that

$$L(\theta) \to L^0(\phi) = \begin{cases} (A^{1-2\phi}-1)/(A^{1-2\phi}-B^{1-2\phi}) & , \; \phi \neq 1/2 \; ; \\ a/(a-b) & , \; \phi = 1/2 \; . \end{cases} \tag{28.2.10}$$

We may note further that by (28.2.2), $\log \lambda_n = \sum_{i \leq n} Z_i$, for every $n \geq 1$, where the Z_i are i.i.d. variables; actually, $Z_i = \Delta X_i - \delta$, for every $i \geq 1$. Thus, using the Wald fundamental identity for stopping times, we have on using (28.2.8),

$$E_\theta(N)E_\theta(Z) = E_\theta(\log \lambda_N), \qquad \theta \in \Theta \; , \tag{28.2.11}$$

where $E_\theta(Z) = \Delta\theta - \delta$ and by (28.2.9), we have for small Δ, $E_\theta(Z) = (\phi - 1/2)\Delta^2\theta_0^2 + O(\Delta^3)$. Therefore, defining the *average sample number* (ASN) as $A(\theta)$, we obtain by using (28.2.9), (28.2.10), and (28.2.11) that as $\Delta \to 0$, for $\phi \neq 1/2$,

$$A(\theta) \sim A_\Delta^0(\phi) = \Delta^{-2}\theta_0^{-2}(\phi - 1/2)^{-1}\{L^0(\phi)b + [1 - L^0(\phi)]a\} \; . \tag{28.2.12}$$

Similarly,

$$A((\theta_0+\theta_1)/2) \sim A_\Delta^0(1/2) = \Delta^{-2}\theta_0^{-2}(-ab) \; , \tag{28.2.13}$$

where a and b are defined by (28.2.6) and $b < 0 < a$.

We may note that as (28.1.1) is a single parameter distribution and as both H_0 and H_1 are simple hypotheses, it is possible to construct a fixed-sample size test with a suitably chosen sample size, such that this test is most powerful and has Type I and Type II error probabilities equal to α and β respectively. Since this test is based on λ_n or equivalently on S_n, and S_n is the sum of n i.i.d. variables with finite moment generating function, we have for large n, under θ, $n^{-1/2}(S_n - n\theta)/\theta$ asymptotically normal with 0 mean and unit variance. Based on this normal approximation, for small

Δ, we may verify that the desired sample size, say n_Δ, needed to satisfy the error probability bounds is given by

$$n_\Delta \sim \Delta^{-2}\theta_0^{-2}(\tau_\alpha + \tau_\beta)^2, \qquad \text{where } \Pi(\tau_\varepsilon) = 1 - \varepsilon,\ 0 < \varepsilon < 1 \ , \tag{28.2.14}$$

and $\Pi(\cdot)$ is the distribution function of a standard normal variate. Comparing (28.2.12) and (28.2.13) with (28.2.14), we obtain that the (asymptotic as $\Delta \to 0$) relative efficiency of the SPRT with respect to the fixed-sample size test based on $n = n_\Delta$, is given by

$$e(SPRT/MP) = n_\Delta / A_\Delta^0(\phi), \text{ which depends on } \phi,\ \alpha \text{ and } \beta\ . \tag{28.2.15}$$

Of particular interest is the case of $\phi = 0$ or 1 (i.e., under H_0 or H_1); for $\alpha = \beta$, (28.2.15) assumes the same value at $\alpha = 0$ or 1, and for $\alpha \geq 0.05$, this is greater than 2 with an upper asymptote 3+ when $\alpha \to 0$. For α between 0.05 and 0.20, (28.2.15) lies between 1.62 and 2.02. In any case, this is substantially greater than 1 and indicates the extent of gain due to the use of SPRT over the MP test. For $\phi = 1/2$, the picture is somewhat different. For $\alpha \geq 0.01$, (28.2.15) is greater than 1 (more so for larger values of α), while for $\alpha < 0.01$, (28.2.15) lies below 1 (at $\alpha = 0.0005$, it is equal to 0.94, at 0.0001, it equals to 0.8), and for very small α, it settles to somewhere around 0.55. In practice, generally, α is chosen between 0.01 and 0.10 (depending on other considerations), and in such a case, it can be shown that uniformly in ϕ, (28.2.15) exceeds one.

The above discussion relates to the case where Δ is small, i.e., θ_1/θ_0 is close to 1. On the other hand, if $\mu = \theta_1/\theta_0$ is large, some exact results on the OC and ASN of the SPRT can be obtained along the lines of Raghavachari (1965). In this setup, he assumed that $\mu \geq \max(A, A/B) = A/B$, and in view of (28.2.8), A/B behaves as bounded from above by $(1-\alpha)(1-\beta)/\alpha\beta$, which is generally large for conventional values of α and β. Similar results were also obtained by Weiss (1962). But, from an operational point of view, such a large value of Δ is not of considerable interest in hypothesis testing contexts. For this reason, we shall not go into these details.

As has been mentioned earlier the exponential law in (28.1.1) plays a fundamental role in life testing and reliability theory models. As a nice example, we present here Daniels' (1945) formulation of the bundle strength of parallel filaments, and discuss the relevance of the SPRT discussed above. For n (≥ 1) parallel filaments with individual strengths $X_1, \ldots, X_n$, all following a common distribution F, defined on $\mathbf{R}^+$, the bundle strength B_n is defined as

$$B_n = \max\{(n-k+1)X_{k:n} : \ 1 \leq k \leq n\}\ , \tag{28.2.16}$$

where $X_{1:n} \leq X_{2:n} \leq \cdots \leq X_{n:n}$ are the order statistics corresponding to $X_1, \ldots, X_n$. If we define a parameter (a functional of the distribution function F) $\theta(F)$ by letting

$$\theta(F) = \sup\{[1 - F(x)]x : \ x \in \mathbf{R}^+\}\ , \tag{28.2.17}$$

then, for large values of n, $Z_n = n^{-1}B_n$ is a strongly consistent estimator of $\theta(F)$, and $\theta(F)$ is termed the per unit bundle strength functional. In practice, specifying (mostly from reliability considerations) a minimum load L (usually large) which a bundle can withstand with a very high probability (say, $1-\alpha$, for some α (> 0) close to 0), the problem is to select n large enough so that

$$P_F\{B_n \geq L\} \geq 1 - \alpha\ . \tag{28.2.18}$$

The current literature pertains to a volume of work done in this field, and the asymptotics are reported in Sen (1994). As such, we know that under fairly general regularity conditions, as n increases,

$$n^{-1/2}(B_n - n\theta(F)) \xrightarrow{\mathcal{D}} \mathcal{N}(0, \gamma^2) , \tag{28.2.19}$$

where γ^2 is a functional of F and can also be estimated consistently. In a parametric mold, if F is assumed to be exponential [as in (28.1.1)], then, it is easy to verify that

$$\theta(F) = e^{-1}\theta \qquad \text{and} \qquad \gamma^2 = e^{-2}\theta^2 , \tag{28.2.20}$$

so that for a resolution of (28.2.18), it suffices to reformulate the specifications in terms of the basic parameter $\theta = E[X]$. Of course, this prospect is contingent on the feasibility of measuring the individual X_i, $i \geq 1$ (i.e., without subjecting the whole bundle to a breaking stress), and generally this is not a problem. From cost considerations individual measurements may even be preferable (as the resulting ASN will be smaller than for the bundle testing). In this context, both the sequential estimation problem and hypothesis testing are very relevant. The SPRT may be posed along the lines of the general one considered earlier, while the estimation problem will be considered in a later section.

28.3 Sequential Life Testing Procedures

As in Epstein and Sobel (1955), we consider the following life testing models. Suppose that n (≥ 1) items are drawn at random from a simple exponential population having the distribution F in (28.1.1), so that the corresponding density function is $f(x;\theta) = \theta^{-1}\exp(-x/\theta)I(x \in \mathbf{R}^+)$, where θ ($0 < \theta < \infty$) is the mean of the distribution F. All these n items are put (simultaneously) to life testing. In a replacement scheme as soon as an item fails, it is replaced by a new one which is assumed to have the same density function. Therefore, the total number of items under testing remains equal to n throughout the entire study period. In the context of survival analysis, often, such replacements are not practicable, and typically, we have a nonreplacement scheme where at the successive failure points, the number of operating units is reduced by one. Thus, at the kth failure point, there are $n - k$ surviving units, for $k = 1, \dots, n$. In the replacement scheme the experiment can continue indefinitely, although from practical considerations one needs to set a finite terminal point. In the nonreplacement scheme, the experiment can not go beyond the last failure point, so that an early termination may result in censoring of some of the observations.

We consider the hypothesis testing problem (in a sequential setup) pertaining to both schemes; later on we shall present the related estimation problems. First, in the replacement scheme, the number n of units to start with remains the same throughout the experimental period. We may even take $n = 1$ and reduce the problem to that of a renewal process. The general case of n can be handled in the same way by replacing the parameter θ by $n\theta$. Let t_r be the length of life of the $(r-1)$th renewal item, for $r = 1, \dots$, so that the t_i are i.i.d. variables with exponential density $f(x;\theta) = \theta^{-1}\exp(-x/\theta)I(x \in \mathbf{R}^+)$. Also, corresponding to a time point t (> 0), let N_t be the number of failures occurring in the interval $(0, t]$ (i.e., N_t is the number of replacements in $(0, t]$). Also, let $T_r = t_1 + \cdots + t_r$, for $r \geq 1$ and $T_0 = 0$. Then T_r is the rth renewal time. It is will known that when the inter-arrival times T_i are exponential with mean θ, the counting process $\boldsymbol{N} = \{N_t,\ t > 0\}$ is a homogeneous Poisson process with the intensity function $\lambda(t) = 1/\theta$, for all $t \in \mathbf{R}^+$. Therefore, it suffices to reformulate the hypothesis testing ($H_0 :\ \theta = \theta_0$ vs. $H_1 :\ \theta = \theta_1$, for specified $\theta_1 > \theta_0$) problem with respect to the Poisson process $\boldsymbol{N}$. We write $\lambda = 1/\theta$, and formulate $H_0 :\ \lambda = \lambda_0$ vs. $H_1 :\ \lambda = \lambda_1\ (< \lambda_0)$. Recall that a Poisson process has independent and homogeneous increments, so that using the SPRT in a continuous time parameter setup, we may define a stopping time τ as

$$\tau = \inf\{t > 0 :\ \delta N_t \notin (t\Delta - a,\ t\Delta - b)\}\ , \tag{28.3.1}$$

where $a = \log A,\ b = \log B$ are defined as in (28.2.6) and $\Delta = \lambda_0 - \lambda_1,\ \delta = \log(\lambda_0/\lambda_1)$. Then the (sequential) test is based on $\boldsymbol{N}_\tau$, accepting H_1 or H_0 according as $\delta \boldsymbol{N}_\tau$ is $\leq \tau\Delta - a$ or $\geq \tau\Delta - b$.

Note that the Poisson process $\boldsymbol{N}$ is a continuous time-parameter stochastic process and $N_t - t\lambda,\ t \geq 0$ has the martingale property too. Thus, the classical results of Dvoretzky, Kiefer, and Wolfowitz (1953) on sequential decision problems for processes with continuous time parameter hold for $\boldsymbol{N}$. As such, the excess of N_τ over the boundaries in (28.3.1) can be neglected with probability one. This enabled them to provide some exact formulae for the OC and ASN ($= E[\tau]$) of such an SPRT. For small values of $\Delta = \lambda_0 - \lambda_1$, these exact formulae can be well approximated by (28.2.10), (28.2.11) and (28.2.12) with θ_0^{-2} being replaced by λ_0^2. As such, the asymptotic efficiency results presented in the last section also pertain to this situation.

Since $N_t,\ t \geq 0$ is a jump process with jumps of unit step at the renewal times $T_r,\ r \geq 1$, operationally, it may be convenient to work with the sequence $\{T_r;\ r \geq 1\}$ and apply an SPRT on that sequence instead of $\boldsymbol{N}$. Since for every $n \geq 1$, T_n has the same [gamma(n, θ)] distribution as S_n in Section 28.2 has, we may virtually repeat the procedure in Section 28.2 and derive parallel results. In this case, the excess over the boundaries may not be negligible.

Next, let us consider the nonreplacement scheme and present a parallel testing procedure. We have $n\ (\geq 1)$ units put on life testing, and their ordered failure times are denoted by $T_{1:n} \leq \cdots \leq T_{n:n}$ respectively. Since F is continuous, ties among the $T_{k:n}$ may be neglected with probability one. Conventionally, we let $T_{0:n} = 0$ and $T_{n+1:n} = \infty$ for every $n \geq 1$. Side by side, we introduce a counting process $\boldsymbol{r} = \{r_t :\ t \geq 0\}$ by letting $r_t = \max\{k :\ T_{k:n} \leq t\},\ t \geq 0$. Then, as in the replacement scheme an SPRT may either be based on the sequence $\{T_r,\ r \geq 1\}$ or on the counting process $\boldsymbol{r}$. Note that for the exponential distribution in (28.1.1), $\overline{F}(x+y) = \overline{F}(x)\overline{F}(y)$, for every $x, y \geq 0$, so that this lack of memory property may be used to characterize the following property of the counting process $\boldsymbol{r}$. For every $t > s \geq 0$,

$$P\{r_t = m \mid r_s = \ell,\ s < t\} = \begin{cases} 0 & ,\ \text{if } m < \ell\ , \\ \binom{n-\ell}{m-\ell} \left\{1 - e^{-(t-s)/\theta}\right\}^{m-\ell} e^{-(t-s)(n-m)/\theta} & ,\ \text{if } \ell \leq m \leq n\ . \end{cases} \tag{28.3.2}$$

Thus, $\boldsymbol{r} = \{r_t,\ t \geq 0\}$ is a pure death process with a constant force of mortality $\lambda = 1/\theta$. Notice that by (28.3.2), the time-homogeneity of the process is no longer in tact, and most of the simplicities in the replacement scheme is therefore gone. However, noting r_t has the jump discontinuities at the failure points, we may consider the following process.

The total time on test $V_n(t)$ [up to the point $t > 0$] is defined as

$$V_n(t) = \sum_{i \le r_t} T_{i:n} + t(n - r_t), \qquad t > 0 . \tag{28.3.3}$$

For the accumulated set of observations at time-point t, the likelihood function is

$$L_n(t;\theta) = [n!/(n - r_t)!]\theta^{-r_t} \exp\{-\theta^{-1} V_n(t)\} . \tag{28.3.4}$$

Therefore, the censored (at t) maximum likelihood estimator (MLE) of θ is given by

$$\hat{\theta}_n(t) = V_n(t)/r_t, \qquad \text{whenever } r_t > 0 . \tag{28.3.5}$$

For a $t \in (0, T_{1:n})$, $r_t = 0$, but (28.3.4) is a monotone increasing function of θ, and hence, we take $\hat{\theta}_n(t)$ arbitrarily as $\hat{\theta}_n(T_{1:n})$. Note that in between successive $T_{k:n}$'s, $V_n(t)$ is linear in t, so that $\hat{\theta}_n(t)$, $t > 0$ is piecewise linear with changes in the slope at the failure points $T_{k:n}$, $k \ge 1$. This feature of $\log L_n(t,\theta)$ and $\hat{\theta}_n(t)$, $t > 0$, allows us to concentrate on the discrete time parameter processes

$$L_n(T_{k:n}, \theta) \text{ and } \hat{\theta}_n(T_{k:n}), \qquad \text{for } k = 1, \ldots, n . \tag{28.3.6}$$

Let us introduce the normalized spacings

$$D_{nk} = (n - k + 1)[T_{k:n} - T_{k-1:n}], \qquad \text{for } k = 1, \ldots, n . \tag{28.3.7}$$

Note that as a characterization of the exponential law, we have the D_{nk} independent and identically distributed variables with the same exponential distribution (28.1.1). Moreover,

$$V_n(T_{k:n}) = \sum_{i \le k} D_{ni}, \qquad \text{for } k = 1, \ldots, n . \tag{28.3.8}$$

Note that by (28.3.4) and (28.3.8), for every k $(= 1, \ldots, n)$

$$\log[L_n(T_{k:n};\theta_1)/L_n(T_{k:n};\theta_0)] = \Delta(\sum_{i \le k} D_{ni}) - \delta k, \qquad 1 \le k \le n , \tag{28.3.9}$$

where Δ and δ are defined as in (28.2.6). Therefore, the methodology pertaining to the SPRT in Section 28.2 can as well be adapted here. There is, however, a basic difference between the two situations. Here in (28.3.9), we have a finite sequence (i.e., $1 \le k \le n$) whereas in (28.2.4) the stopping variable N can be larger than n with a positive probability. For this reason, one may need to adapt a truncated SPRT rule for testing $H_0 : \theta = \theta_0$ against $H_1 : \theta = \theta_1$. This truncation is necessary to force a sequential decision before the experiment is terminated due to the failure of all the units. Epstein and Sobel (1955) contains a detailed account of the theory of such a truncated SPRT and also some numerical illustrations. Various approximations for the OC and ASN are also discussed there. In passing, we may enlist a point of clarification in this context. Their $V(t)$ [defined as total life observed up to time t] is nothing but the $V_n(t)$ defined by (28.3.3). Note that for $t \in (T_{r:n}, T_{r+1:n})$, we have by their definition

$$\begin{aligned} V(t) &= \sum_{i \le r}(n - i + 1)(T_{i:n} - T_{i-1:n}) + (n - r)(t - T_{r:n}) \\ &= \sum_{i \le r} T_{i:n} + (n - r)t \ (= V_n(t) \text{ with } r_t = r) . \end{aligned} \tag{28.3.10}$$

But, in their equation (4), the right hand side contains $(n-r)(t-T_{r:n})$ instead of $(n-r)t$, and this may cause some small changes in the results discussed there.

Note that by the characterization in (28.3.7), the independence and homogeneity of the increments in (28.3.8) (over k) are guaranteed. However, the D_{ni} are exponential variables, not normal, and this, in turn, calls for some approximations. It may, however, be noted that for large values of n, weak invariance principles (i.e., weak convergence to a Gaussian process) hold for the partial sum process in (28.3.8). This enables one to approximate the truncated SPRT (for this nonreplacement scheme), for small values of Δ, to that on a Wiener process on the unit interval $[0,1]$. The idea is to set $W_n(t)$, $t \in [0,1]$, by letting $W_n(t) = W_n(k/n)$ for $k/n \leq t < (k+1)/n$, $k = 0, \ldots, n-1$, $W_n(0) = 0$, $W_n(1) = W_n(n/n)$, and

$$W_n(k/n) = [V_n(T_{k:n}) - k\theta]/(\theta\sqrt{n}), \qquad \text{for } k = 1, \ldots, n \ . \tag{28.3.11}$$

$W_n(\cdot)$ belongs to the $D[0,1]$ space, and it converges weakly to a standard Wiener process W (on $[0,1]$), in the Skorokhod J_1-topology. As such, boundary crossing probabilities for W can be used to approximate the same for W_n. In order to justify such asymptotic approximations, it may be necessary to assume Δ so small and n large, such that $n\Delta^2$ is a finite positive number [see Chapter 9 of Sen (1981) for details of this type of asymptotics]. For truncated Wiener processes, boundary crossing probabilities have been nicely established in Anderson (1960), and these may therefore be incorporated in the current context to provide suitable approximations for the OC and ASN functions in the nonreplacement scheme. In this setup, as in Epstein and Sobel (1955), one may prefix an r ($\leq n$) as an upper bound for the number of failures in a nonreplacement scheme; we need to set $r = [np]$, for a prefixed $p \in (0,1)$.

28.4 Sequential Estimation

First, let us consider the interval estimation of the exponential parameter θ in (28.1.1). Note that given n observations $T_1, \ldots, T_n$ from the exponential distribution in (28.1.1), the MLE of θ is $\hat{\theta}_n = (T_1 + \cdots + T_n)/n = n^{-1}S_n$, where S_n/θ has the gamma(n) distribution, which is denoted by $G_n(\cdot)$. Thus,

$$G_n(x) = (\Gamma(n))^{-1} \int_0^x e^{-t} t^{n-1}\, dt, \qquad \text{for } x \in \mathbf{R}^+ \ . \tag{28.4.1}$$

It is therefore possible to choose two numbers, say, $C_{n\alpha}$ and $D_{n\alpha}$, such that

$$G_n(C_{n\alpha}) = \alpha_1 \qquad \text{and} \qquad G_n(D_{n\alpha}) = 1 - \alpha_2;\ \alpha_1 + \alpha_2 = \alpha \ . \tag{28.4.2}$$

where $1-\alpha$ is the desired confidence coefficient (or coverage probability) of the interval we like to construct. Then, from the above, we obtain that

$$P_\theta[C_{n\alpha} \leq S_n/\theta \leq D_{n\alpha}] = 1 - \alpha \ , \tag{28.4.3}$$

and by inversion, we have for every $n \geq 1$,

$$P_\theta[(n\hat{\theta}_n)/D_{n\alpha} \leq \theta \leq (n\hat{\theta}_n)/C_{n\alpha}] = 1 - \alpha \ . \tag{28.4.4}$$

For a given α, α_1 and α_2 may be taken equal to $\alpha/2$, or some other considerations may dominate their choice. Now (28.4.4) provides a confidence interval for θ with confidence coefficient $1-\alpha$ and the width of this confidence interval is $d_n = n\hat{\theta}_n(C_{n\alpha}^{-1} - D_{n\alpha}^{-1})$. Note that d_n depends on the MLE $\hat{\theta}_n$ as well as the percentile points $C_{n\alpha}$ and $D_{n\alpha}$. Therefore, d_n is itself a nonnegative random variable, and we have the following

$$\begin{aligned} \zeta_n &= E[d_n] = \text{ expected width of the confidence interval} \\ &= n\theta(C_{n\alpha}^{-1} - D_{n\alpha}^{-1}) \ . \end{aligned} \tag{28.4.5}$$

In practice, often, one wants to construct a confidence interval for θ such that it has two basic properties: (i) The confidence coefficient is equal to some prefixed number $(1-\alpha)$, and (ii) the width of the interval, d_n is bounded from above by some prefixed positive number $2d$. While (28.4.4) satisfies (i), it may not satisfy (ii) for all θ. As a matter of fact, it is clear from the definition of d_n, that for any fixed sample size n, d_n or even ζ_n can not be smaller than $2d$, simultaneously for every $\theta \in \Theta = \mathbf{R}^+$. For this reason, one may take recourse to a sequential procedure for which a confidence interval based on an appropriate stopping variable satisfies both the requirements (i) and (ii). In passing, we may remark that (28.4.3) may be used to provide a fixed-percentage width confidence interval for θ in a nonsequential setup. To verify this, rewrite (28.4.3) as $P_\theta[n^{-1}C_{n\alpha} \le \hat{\theta}_n/\theta \le n^{-1}D_{n\alpha}] = 1-\alpha$, so that setting n_0 as

$$n_0 = \min\{n \ge 1 : \ n^{-1}C_{n\alpha} \ge (1+\rho)^{-1}, \ n^{-1}D_{n\alpha} \le (1-\rho)^{-1}\} \ , \tag{28.4.6}$$

for some $\rho : \ 0 < \rho < 1$, we obtain that for $n \ge n_0$, $P_\theta\{\theta/\hat{\theta}_n \in [1-\rho, 1+\rho]\} \ge 1-\alpha$; hence, a fixed percentage-width confidence interval exists. In this respect, we need to verify that $n^{-1}C_{n\alpha}$ is $\uparrow$ in n and $n^{-1}D_{n\alpha}$ is $\downarrow$ in n, and this can easily be done by using the right-tiltedness of the gamma(n) distribution with increasing n. In any case, the computation of the $C_{n\alpha}$ and $D_{n\alpha}$ need to be made by reference to the incomplete gamma functions, tables for which are given in most of the statistical tables. If ρ is small, n_0 in (28.4.6) will be large. In such a case, it may be of some advantage to use the normal approximation to gamma distributions. Using the Fisher transformation, we obtain that for large n,

$$2(C_n^{1/2} - n^{1/2}) \sim -\tau_{\alpha/2} \qquad \text{and} \qquad 2(D_n^{1/2} - n^{1/2}) \sim \tau_{\alpha/2} \ , \tag{28.4.7}$$

where τ_ε is defined by (28.2.14), and we choose $\alpha_1 = \alpha_2 = \alpha/2$. With this approximation for the $C_{n\alpha}$ and $D_{n\alpha}$, we may set for small ρ,

$$n_0 = [\tau_{\alpha/2}^2/\{4[1-(1+\rho)^{-1/2}]^2\}] + 1 \ , \tag{28.4.8}$$

so that a neat asymptotic formula holds for the needed sample size.

Coming back to the fixed-width interval estimation problem, we may intuitively consider the following stopping rule. Define $\boldsymbol{N} = N_d, \ d > 0$, by

$$N_d = \inf\{n \ge 1 : \ d_n \le 2d\} \ , \tag{28.4.9}$$

where for each n, d_n is defined as before in (28.4.5). Note that $[d_n \le 2d] <=> [S_n \le 2d(C_{n\alpha}^{-1} - D_{n\alpha}^{-1})^{-1}]$, where S_n being a sum of n i.i.d nonnegative variables is nondecreasing in n with probability one. Thus, we are left with the task of computing the percentile points $C_{n\alpha}$ and $D_{n\alpha}$, for every $n \geq 1$, and this is generally very cumbersome. Moreover, in view of the fact that N_d is itself a positive integer valued variable, the width d_{N_d} being less than or equal to $2d$ does not necessarily imply that the coverage probability of the sequential confidence interval is greater than or equal to $1-\alpha$. Both of these problems can be taken care of nicely if we allow d to be very small, so that the Chow-Robbins (1965) asymptotic theory of bounded-width (sequential) confidence intervals can be adapted to verify the needed technicalities.

First of all, making use of (28.4.5) and (28.4.7), for every $d > 0$, we introduce a positive integer n_d, defined by

$$n_d = \inf\left\{ n \geq 1 : \left(n - \frac{1}{4}\,\tau^2_{\alpha/2} \right)^2 \geq d^{-1}\theta\tau_{\alpha/2}n^{3/2} \right\}, \qquad d > 0 . \qquad (28.4.10)$$

Note that as $d \downarrow 0$, n_d behaves like $d^{-2}\theta^2\tau^2_{\alpha/2}$. Further note that n_d depends on the unknown parameter θ. Finally, note that by (28.4.3), (28.4.4), (28.4.5), and (28.4.10),

$$\lim_{d\downarrow 0} P_\theta[\hat{\theta}_{n_d} - d \leq \theta \leq \hat{\theta}_{n_d} + d] = 1 - \alpha . \qquad (28.4.11)$$

Next, we note that by definition in (28.4.9), for every $d > 0$,

$$S_{N_d} \leq 2d(C^{-1}_{N_d\alpha} - D^{-1}_{N_d\alpha})^{-1} \text{ and } S_{N_d} \geq S_{N_d-1}, \text{ with probability } 1 ; \qquad (28.4.12)$$

$$S_{N_d-1} > 2d(C^{-1}_{N_d-1\alpha} - D^{-1}_{N_d-1\alpha})^{-1} \geq S_{N_d}(C^{-1}_{N_d\alpha} - D^{-1}_{N_d\alpha})(C^{-1}_{N_d-1\alpha} - D^{-1}_{N_d-1\alpha})^{-1} . \qquad (28.4.13)$$

Using (28.4.7), it can be shown that for large n,

$$(C^{-1}_{n\alpha} - D^{-1}_{n\alpha})(C^{-1}_{n-1\alpha} - D^{-1}_{n-1\alpha})^{-1} = 1 - (3/2n) + O(n^{-2}) . \qquad (28.4.14)$$

Therefore, from (28.4.12), (28.4.13) and (28.4.14), we obtain that

$$|S_{N_d}/N_d - S_{N_d-1}/(N_d - 1)| \leq N_d^{-2}S_{N_d} + O(N_d^{-2}), \qquad \text{as } d \downarrow 0 . \qquad (28.4.15)$$

Moreover, by the Khintchine strong law of large numbers,

$$n^{-1}S_n \to \theta \qquad \text{a.s. as } n \to \infty . \qquad (28.4.16)$$

From (28.4.12), (28.4.13) and the monotone convergence theorem, we obtain that

$$N_d \uparrow \infty \text{ a.s. as } d \downarrow 0;\ \hat{\theta}_{N_d} \to \theta \text{ a.s. as } d \downarrow 0 . \qquad (28.4.17)$$

As a result, we have

$$N_d/n_d \to 1 \qquad \text{a.s. as } d \downarrow 0 . \qquad (28.4.18)$$

Since the S_n involve i.i.d. summands with finite moment generating function, we may easily verify the Anscombe (1952) theorem, whereby $n_d^{1/2}(\hat{\theta}_{n_d} - \hat{\theta}_{N_d}) \to 0$, in probability, as $d \downarrow 0$. Combining this with (28.4.10) and (28.4.11), we obtain that

$$\lim_{d\downarrow 0} P_\theta\left[\hat{\theta}_{N_d} - d \leq \theta \leq \hat{\theta}_{N_d} + d\right] = 1 - \alpha \qquad \text{(asymptotic consistency)} \quad (28.4.19)$$

and, (28.4.18) may then be used along with the martingale property of $S_n - n\theta,\ n \geq 1$, to show that

$$\lim_{d \downarrow 0} E_\theta[N_d/n_d] = 1 \qquad \text{(asymptotic efficiency)} . \tag{28.4.20}$$

Therefore, the sequential confidence interval based on the stopping variable N_d in (28.4.9) satisfies both the requirements (i) and (ii) and is asymptotically (as $d \downarrow 0$) efficient.

Let us consider next two sequential point estimation problems related to the parameter θ in (28.1.1). First, the minimum risk estimation of θ incorporating the cost of sampling. If c (> 0) is the cost for drawing a single observation, then for a sample of size n, the variance ($n^{-1}\theta^2$) of the MLE $\hat{\theta}_n$ can be incorporated along with the cost (cn) to formulate a risk function

$$\rho(n;\theta) = cn + an^{-1}\theta^2, \qquad \text{where } a > 0 \text{ and } c > 0 \text{ are given} . \tag{28.4.21}$$

Recall that

$$\rho(n+1;\theta) - \rho(n;\theta) \text{ is } \gtreqless 0 \text{ according as } n(n+1) \text{ is } \gtreqless (a/c)\,\theta^2 . \tag{28.4.22}$$

Thus, the optimal sample size at which the risk is minimized depends on the unknown θ and no fixed sample size solution can be optimal simultaneously for all $\theta \in \Theta$. In this setup too a sequential procedure works out well, and in this context, it is taken for granted that c, the cost per unit sample, is small, while a is held fixed. For simplicity, we take $a = 1$. Then, as $c \downarrow 0$, the optimal sample size n_c^0 behaves as $c^{-1/2}\theta$, and the minimum risk is asymptotically (as $c \downarrow 0$) $2c^{1/2}\theta$. With this motivation, we consider a stopping variable N_c defined by

$$N_c = \inf\{n \geq 1 :\ n(n+1) \geq (a/c)\hat{\theta}_n^2\}, \qquad c > 0 ; \tag{28.4.23}$$

note that by definition, N_c is monotone nonincreasing in c (> 0), and as $c \downarrow 0$, $N_c \to \infty$ a.s.. Mukhopadhyay (1987) studied the properties of the sequential estimator $\hat{\theta}_{N_c}$. Along the same lines as in the interval estimation problem, the following results can be established. Let $\rho^*(c,\theta) = cE[N_c] + aE(\hat{\theta}_{N_c} - \theta)^2$ and $\rho^0(c,\theta) = \rho(n_c^0,\theta),\ c > 0$. Then

$$(i) \qquad N_c/n_c^0 \to 1 \text{ a.s. as } c \downarrow 0;\ E[N_c]/n_c^0 \to 1 \text{ as } c \downarrow 0 , \tag{28.4.24}$$

$$(ii) \qquad \lim_{c \downarrow 0} \rho^*(c,\theta)/\rho^0(c,\theta) = 1, \text{ for every } \theta \in \Theta , \tag{28.4.25}$$

$$\text{and } (iii) \qquad (n_c^0)^{1/2}(\hat{\theta}_{N_c} - \theta)/\theta \xrightarrow{\mathcal{D}} \mathcal{N}(0,1) \text{ as } c \downarrow 0 . \tag{28.4.26}$$

Therefore, the sequential estimator $\hat{\theta}_{N_c}$ has asymptotically (as $c \downarrow 0$) minimum risk and is BAN, with bestness interpreted in terms of the associated risk function. Instead of the mean square error, mean absolute error or some other norm can be used in (28.4.21) and parallel results can be easily established. For a given $c > 0$, the excess of the ratio $\rho^*(c,\theta)/\rho^0(c,\theta)$ over 1 is termed the regret function; asymptotic expressions for such regret functions have also been studied by various workers. Woodroofe (1982) is an excellent source for such studies.

In Section 28.2, we have listed the bundle strength problem in a parametric setup as an important application, and discussed some SPRT type tests. The results on sequential point and interval estimation of the exponential parameter θ remain applicable to the bundle strength problem as well. (28.2.11) provides this link, and hence, we omit details.

We consider now some sequential estimation problems related to the life testing models treated in Section 28.3. In the replacement scheme, by virtue of the equivalence to a Poisson process, the theory follows directly as a corollary to that of sequential estimation of a Poisson parameter, and this runs parallel to the general theory presented earlier in this section. Therefore, we confine ourselves only to the nonreplacement scheme. The MLE of θ based on the accumulated data at time point t (> 0) is given by (28.3.5). Viewed from a slightly different point, at the kth failure point ($T_{k:n}$), the MLE of θ is given by $\hat{\theta}_{n,k} = V_n(T_{k:n})/k = k^{-1}\sum_{i\leq k} D_{ni}$, where the D_{ni} are i.i.d. variables with the same exponential distribution in (28.1.1). Thus $k\hat{\theta}_{n,k}/\theta$ has the gamma(k) distribution, and hence, for the bounded-width (or bounded percentage width) confidence interval for θ based on partial outcome (up to the rth failure point, for some $r \sim np$, $p \in (0,1]$) we may proceed as in (28.4.1) through (28.4.20), with the role of n being replaced by r. There is one major difference in the two setups. Here in order that r is chosen sequentially so that the corresponding confidence interval for θ satisfies both the requirements stated after (28.4.5), we may need that the number of units (n) to start with has to be chosen large adequately. For d sufficiently small and n not so large, if we apply the stopping rule in (28.4.9) (with n replaced by r) then such an r may be larger than n with a positive probability. Thus, with a positive probability, we have insufficient number of failures to estimate θ, and this will make the margin of error larger. To eliminate this problem, we recommend the nonreplacement scheme for the sequential confidence interval estimation problem when n is very large where data set pertaining to the whole experiment may require an enormous amount of time ($\sim \theta \log n$), so that an early truncation (at some r) is desirable, and the sequential scheme would be very appropriate from time and cost saving points. If n is not so large, a replacement scheme eliminates this problem and is a better alternative; but, this may not be universally applicable in survival analysis.

We consider next the sequential point estimation problem relating to the life testing models treated before. Note that in the replacement scheme a failed item is instantaneously replaced by a new item (having the same life distribution). Thus, we may start with an initial cost c_0 (> 0) of setting the experiment, and a cost factor c^* (> 0) for procuring a new item as well as a cost factor c for running the experiment per unit of time/per unit item. Then, noting that there are N_t failures up to a time point t (> 0), we may set a cost function

$$c_n(t) = c_0 + ct + c^* N_t, \qquad t > 0;\ c_0 > 0,\ c > 0,\ c^* > 0\ . \tag{28.4.27}$$

(Note that c may depend on n and so also c_0). Note that $c_n(t)$ is stochastic due to the stochastic element N_t, and as $E[N_t] = nt/\theta$, we may set without any loss of generality $n = 1$. Then the risk function incorporating the cost function $c_n(t)$ in (28.4.27) is given by

$$\rho(t,\theta) = c_0 + ct + c^* t/\theta + t^{-1}\theta^{-1}\ , \tag{28.4.28}$$

where $t^{-1}N_t$ is an optimal estimator of θ^{-1}, and we reparametrize from θ to θ^{-1} for convenience of manipulations. Minimizing the risk in (28.4.28) with respect to t, one obtains an optimal time point t^0 given by

$$t_0 = t_0(c, c^*, c_0, \theta) = (c^* + c\theta)^{-1/2}\ . \tag{28.4.29}$$

Note that t_0 is bounded from above by $(c^*)^{-1/2}$, for all $\theta > 0$, and this additional information should be incorporated in the formulation of a stopping time. Since at a time point t for which $N_t > 0$, t/N_t provides an estimate of θ, motivated by (28.4.29), we may introduce the stopping time

$$\tau^0 = \tau^0(c, c^*) = \inf\{t > 0 : \; N_t \geq k^0 \text{ and } t^2 \geq N_t/(c^* N_t + ct)\} \; , \qquad (28.4.30)$$

where k^0 is a suitable positive integer. A too small value of k^0 may lead to some undesirable feature, and generally, it is chosen to be 2 or more. Then, a sequential estimator $\hat{\theta}_{\tau^0}$ of θ is given by τ^0/N_{τ^0}. Note that it may not be strictly unbiased for θ. Asymptotic (when both c and c^* converge to 0 with c^*/c having a positive finite limit) minimum risk properties of this estimator then follows along the lines of the general results in Section 28.3. We may refer to Chapter 12 of Ghosh, Mukhopadhyay, and Sen (1995) for details.

Let us consider the nonreplacement scheme and present the sequential point estimation theory. The MLE $\hat{\theta}_n(t)$ of θ at time point t is given by $V_n(t)/r_t$ whenever $r_t > 0$, and these notations are all adopted from (28.3.2) - (28.3.10). In this nonreplacement scheme, failed items are not replaced by new ones, so that in (28.4.27), $ct + c^* N_t$ need to be replaced by $cV_n(t)$ alone. Thus, the cost function is taken here as $c_0 + cV_n(t)$, $t > 0$. Although it is possible to work with the continuous time-parameter model, $\hat{\theta}_n(t)$, $t > 0$, as we have observed in Section 28.3, from an operational point of view, we may as well consider the discrete time-parameter model where t is allowed to take on the values $T_{k:n}$, $k = 1, \ldots, n$, leading to the discrete set of estimators $\hat{\theta}_{n,k} = V_n(T_{k:n})/k$, $k = 1, \ldots, n$. The risk of the estimator $\hat{\theta}_{n,k}$ is therefore given by

$$\rho(n, k; \theta) = c_0 + ck\theta + k^{-1}\theta^2, \qquad k = 1, \ldots, n \; . \qquad (28.4.31)$$

In this setup, the problem is to choose k in such a way that for a given n, c_0 and c, $\rho(k, n; \theta)$ is minimized. Note that by definition in (28.4.31),

$$\rho(k+1, n; \theta) - \rho(k, n; \theta) \text{ is } \gtreqless 0 \text{ according as } k(k+1) \text{ is } \gtreqless c^{-1}\theta$$
$$\text{or equivalently, } k \text{ is } \gtreqless (c^{-1}\theta + 1/4)^{1/2} - 1/2 \; . \qquad (28.4.32)$$

Thus, the optimal k not only depends on θ, but also it has one additional restraint that it can not be larger than n. Consequently, n needs to be larger than $c^{-1/2}\theta^{1/2}$. Otherwise, the risk in (28.4.31) is monotone decreasing in k, and the minimum value is attained at $k = n$, i.e., when all the failures have occurred. In the latter case, the situation becomes comparable to the AMRE problem treated in (28.4.21) through (28.4.26), and it was shown there that an optimal n is (asymptotically, as $c \downarrow 0$) $[c^{-1/2}\theta]$. So, from that perspective, there is not much sense in choosing an n less than $c^{-1/2}\theta^{1/2}$, and hence, we may set without any essential loss of generality that

$$n > c^{-1/2}\theta^{1/2}, \qquad \text{as } c \downarrow 0 \; . \qquad (28.4.33)$$

Since θ is unknown, the lower bound in (28.4.32) is also so. However, operationally, we assume that n is very large, but we would like to economize on k, the number of failures by minimizing the associated risk. Looking at (28.4.31) and (28.4.32), we introduce the following stopping number K:

$$\begin{aligned} K &= \inf\{k \geq 1 : \; k(k+1) \geq c^{-1}\hat{\theta}_{n,k}\}; \\ &= n, \text{ if no such } k \; (\leq n) \text{ exists} \; . \end{aligned} \qquad (28.4.34)$$

Having defined K by (28.4.34), we are in a position to incorporate the sequential point estimator $\widehat{\theta}_{n,K}$ in infering on θ, and we denote its risk by

$$\rho^*(c,n;\theta) = c_0 + cE[V_n(T_{K:n})] + E[(\widehat{\theta}_{n,K} - \theta)^2] \, . \qquad (28.4.35)$$

When (28.4.33) does not hold, $K = n$ with probability one, and hence, (28.4.35) reduces to $c_0 + cn\theta + n^{-1}\theta^2$, and the solution is isomorphic to that in (28.4.21) through (28.4.26) with c (there) replaced by $c\theta$. As such, these discussions are omitted. When (28.4.33) holds, making good use of (28.3.7) and (28.3.8), detailed asymptotic analysis of (28.4.35) has been made by Sen (1980), and more recently in Chapter 12 of Ghosh, Mukhopadhyay, and Sen (1995). Avoiding the details, we may summarize the main findings that $\widehat{\theta}_{n,K}$ is asymptotically risk efficient, and when c_0 also converges to 0 with c, (28.4.35) converges to 0 as $c \downarrow 0$ (at a rate $c^{1/2}$). On the other hand, for a fixed c_0 (> 0), the second and third terms on the right hand side of (28.4.35) converge to 0 as $c \downarrow 0$, so that (28.4.35) converges to c_0. In that case, the risk function is asymptotically (as $c \downarrow 0$) flat at c_0, and from a statistical perspective there is not much interest.

28.5 Sequential Inference Under Random Censoring

In reliability theory and survival analysis, censoring is a common phenomenon. A unit may become inoperative for other causes than a failure, or a failure may occur due to other reasons than the one under study. For this reason, often, it is assumed that the failure times (X_i) are coupled with some censoring time (Y_i), and the observable random variables are

$$X_i^0 = \min(X_i, Y_i), \text{ and } \delta_i = I(X_i \le Y_i), \qquad \text{for } i \ge 1 \, . \qquad (28.5.1)$$

Moreover, it is assumed that X_i and Y_i are mutually stochastically independent, and for different i, the vectors (X_i, Y_i) are i.i.d. variables. In the simplest case, in a parametric mold, it is assumed that F, the distribution of X, is exponential [see (28.1.1)] with $\theta - \theta_1$ and, G, the distribution of Y, is also exponential with parameter $\theta = \theta_2$. Note then X_0 has also an exponential distribution with parameter θ_0 where $1/\theta_0 = 1/\theta_1 + 1/\theta_2$. Moreover, $P[\delta = 1] = \int F(x)dG(x) = \theta_2/(\theta_1 + \theta_2) = \theta_0/\theta_1$. Indeed, the joint distribution of (X^0, δ) contains all the pertinent information on θ_1 and θ_2. Our main concern is to draw (sequential) inference on θ_1 treating θ_2 as a nuisance parameter.

Random censoring is typically associated with the nonreplacement scheme, and hence, we discuss the inference procedures for this scheme only. We denote the order statistics associated with X_i^0, $i = 1, \ldots, n$, by $X_{i:n}^0$, $i = 1, \ldots, n$. Note that ties among the observations are negligible with probability 1, and we may set $X_{i:n}^0 = X_{S_i}^0$, $i = 1, \ldots, n$, where $S_1, \ldots, S_n$ stand for the anti-ranks of the observations. We denote by $\delta_{[i]} = \delta_{S_i}$, $i = 1, \ldots, n$. Note that given the order statistics $X_{i:n}^0$, the $\delta_{[i]}$ are conditionally independent, each assuming the 0 or 1 values. Moreover, note that for each i,

$$\begin{aligned} P[\delta_{[i]} = 1 \mid X_{i:n}^0] &= f(X_{i:n})\overline{G}(X_{i:n})/\{f(X_{i:n})\overline{G}(X_{i:n}) + g(X_{i:n})\overline{F}(X_{i:n})\} \\ &= \theta_2/(\theta_1 + \theta_2) \, , \qquad i = 1, \ldots, n \, . \end{aligned} \qquad (28.5.2)$$

Therefore, $\delta_{[i]}$ is independent of $X_{i:n}$, for every $i = 1, \ldots, n$. With this simplification, we may set the likelihood function pertaining to the partial set $\{(X_{i:n}, \delta_{[i]});\ i = 1, \ldots, k\}$ as

$$[n!/(n-k)!] \prod_{i=1}^{k} \left\{ (\theta_2/(\theta_1+\theta_2))^{\delta_{[i]}} (\theta_1/(\theta_1+\theta_2))^{1-\delta_{[i]}} \theta_0^{-1} e^{-X_{i:n}/\theta_0} \right\} e^{-(n-k)X^0_{k:n}/\theta_0} . \tag{28.5.3}$$

Thus, working with the log-likelihood function and noting that $(\partial/\partial\theta_1)\theta_0^{-1} = -\theta_1^{-2}$, we arrive at the following two estimating equations:

$$\theta_1 \sum_{i \le k} \delta_{[i]} = V^0_{nk} \qquad \text{and} \qquad \theta_2 \sum_{i \le k} (1 - \delta_{[i]}) = V^0_{nk} , \tag{28.5.4}$$

where

$$V^0_{nk} = \sum_{i \le k} X^0_{i:n} + (n-k) X^0_{k:n}, \qquad \text{for } k\ 1, \ldots, n . \tag{28.5.5}$$

Note that $E\left[\sum_{i \le k} \delta_{[i]}\right] = k\theta_2/(\theta_1+\theta_2) = k\theta_0/\theta_1$, and hence, the two equations in (28.5.4) are consistent. Moreover, for every k $(= 1, \ldots, n)$, as in Section 28.3,

$$V^0_{nk}/\theta_0 \text{ has the gamma}(k) \text{ distribution} , \tag{28.5.6}$$

and hence, the sequential procedures discussed in Sections 28.3 and 28.4 can be readily modified to suit the present purpose.

First, we present a sequential test for the null hypothesis $H_0 : \theta_1 = \theta_{10}$ against $H_1 : \theta_1 = \theta_{11}$, where $\theta_{10} \neq \theta_{11}$, and both of them are specified, while θ_2 is treated as a nuisance parameter. Instead of the SPRT considered in Section 28.3, we consider a SLRT (sequential likelihood ratio test) wherein an estimator of θ_2 is incorporated in the formulation of the likelihood ratio statistics (see Cox (1963) for some motivation of SLRT's). Using (28.5.3)–(28.5.5), we write the log-likelihood function as

$$\text{const.} - \sum_{i \le k} \delta_{[i]} \log\theta_1 - \sum_{i \le k} (1 - \delta_{[i]}) \log\theta_2 - V^0_{nk}(1/\theta_1 + 1/\theta_2) , \tag{28.5.7}$$

so that letting

$$\Delta = 1/\theta_{10} - 1/\theta_{11} \qquad \text{and} \qquad \delta = \log(\theta_{11}/\theta_{10}) , \tag{28.5.8}$$

we obtain that the log-likelihood ratio statistic is given by

$$\Delta V^0_{nk} - \delta \left(\sum_{i \le k} \delta_{[i]} \right) , \qquad \text{for } k = 1, \ldots, n . \tag{28.5.9}$$

As in (28.3.11), we let

$$W^0_n(k/n) = [V^0_{nk} - k\theta_0] / (n^{1/2}\theta_0), \qquad \text{for } k = 1, \ldots, n;\ W^0_n(0) = 0 , \tag{28.5.10}$$

and by linear interpolation, we complete the definition of $W_n^0(t)$, $t \in [0,1]$. Then W_n^0 converges weakly (in the Skorokhod J_1-topology on $D[0,1]$) to a standard Wiener process on $[0,1]$. Similarly, we let $\pi = \theta_2/(\theta_1+\theta_2) = \theta_0/\theta_1$, and denote by

$$W_n^*(k/n) = \left(\sum_{i\le k} \delta_{[i]} - k\pi\right) \Big/ (n\pi(1-\pi))^{1/2}, \; k=1,\ldots,n; \; W_n^*(0)=0 \;, \quad (28.5.11)$$

and by linear interpolation, we complete the definition of $W_n^*(t)$, $t \in [0,1]$. Then, W_n^* converges weakly to a standard Wiener process on $[0,1]$. Moreover, W_n^0 and W_n^* are stochastically independent of each other. The usual SLRT works out well when H_0 and H_1 are contiguous to each other. In the current setup, this means that the difference $\theta_{11} - \theta_{10}$ is small in the sense that there is some fixed λ, such that

$$H_{1(n)}: \; \theta_1 = \theta_{1(n)} = \theta_{10} + n^{-1/2}\lambda \;. \quad (28.5.12)$$

Though this theory of SLRT has been only motivated by Cox (1963), its justifications for contiguous alternatives have been elaborated in Sen (1981, Chapter 9). As such, in this specific hypothesis testing problem, we replace the pair (H_0, H_1) by the sequence $\{(H_0, H_{1(n)}),\; n=1,\ldots\}$, and using (28.5.8), (28.5.10), and (28.5.11), we obtain that under (28.5.12), (28.5.9) can be equivalently expressed as

$$(\lambda\theta_0/\theta_1^2)\{W_n^0(k/n) - (\theta_1/\theta_2)^{1/2}W_n^*(k/n)\} + o_p(1), \text{ for } k=1,\ldots,n \;. \quad (28.5.13)$$

In view of (28.5.12), we consider values of θ_1 in close proximity to θ_{10}, and as in (28.2.9), we let

$$\theta_1 = \theta_{10} + n^{-1/2}\lambda\phi, \text{ where } \phi \in \Phi = \{\phi: \; |\phi| \le K\}, \; K \in [1,\infty) \;. \quad (28.5.14)$$

Let us also denote by

$$\theta_0^0 = \theta_{10}\theta_2/(\theta_{10}+\theta_2) \qquad \text{and} \qquad \pi_0 = \theta_2/(\theta_{10}+\theta_2) \;. \quad (28.5.15)$$

Then, under (28.5.14), we have

$$n^{1/2}(\theta_0 - \theta_0^0) \to \lambda\phi\,\pi_0^2; \; n^{1/2}(\pi - \pi_0) \to \lambda\phi\,\pi_0(1-\pi_0)/\theta_{10} \;. \quad (28.5.16)$$

Therefore, whenever $k/n \to t \in (0,1)$, under (28.5.14), the drift function for (28.5.12) converges to

$$\begin{aligned} &\lambda\theta_0^0\theta_{10}^{-2}\pi_0 \left\{t\lambda\phi\,\pi_0^2/\theta_0^0 + (1-\pi_0)t\lambda\phi\,\pi_0/\theta_0\right\} \\ &= t\phi\lambda^2/\{\theta_{10}(\theta_{10}+\theta_2)\} = \zeta\, t\phi, \text{ say, for } t \in (0,1) \;. \end{aligned} \quad (28.5.17)$$

On the other hand, the variance function (when $k/n \to t \in (0,1)$) of (28.5.12) is given by

$$(\lambda\theta_0^0/\theta_{10}^2)^2 t\{1 + (1-\pi_0)/\pi_0\} = t\lambda^2\,\pi_0\,\theta_{10}^{-2} = \sigma^2 t, \text{ say} \;. \quad (28.5.18)$$

This shows that the SPRT theory developed in Section 28.2 readily extends to cover the SLRT when n is large. Note that in the uncensored case, $\pi_0 = 1$ (i.e., $\theta_2 = 0$) and in (28.5.17) ζ reduces to λ_2/θ_{10}^2, while for every $\theta_2 > 0$, (28.5.17) is a monotone decreasing function of θ_2. Similarly, in (28.5.18) too, σ^2 is equal to λ^2/θ_{10}^2 when $\pi_0 = 1$ and as π_0 converges to 0, it monotonically converges to 0. In terms of the boundary crossing probabilities, the smaller is the value of π_0, the lesser is the probability of boundary crossing. Hence, a comparatively larger value of n is needed to compensate this loss. Or, in other words, the SLRT for this random censoring scheme, to have the same Type I and Type II error probabilities as in the SPRT for the uncensored case, requires larger sample size. But, this is well anticipated. Note further that (28.5.12) and the weak convergence of W_n^0 and W_n^* also provide the necessary asymptotics on which the sequential estimation theory in Section 28.4 extends directly to the censored case. Gardiner, Susarla, and van Ryzin (1986) contains some useful results in this setup.

28.6 Sequential Comparison of Two Exponential Distributions

In life testing models or survival analysis, we may want to draw sequential inference on two distributions. In a parametric mold when both of these are assumed to be exponential with parameters, say, θ_1 and θ_2, the problem reduces to that of comparing θ_1 and θ_2 in a sequential manner. In this context too, we may have a replacement or a nonreplacement scheme. In a replacement scheme, for n_1 items from distribution 1 and n_2 items from distribution 2, put simultaneously to life testing, we have as in Section 28.3, two counting processes, say $\boldsymbol{N}^{(1)} = \{N^{(1)}(t),\ t > 0\}$ and $\boldsymbol{N}^{(2)} = \{N^{(2)}(t),\ t > 0\}$, where the first one is Poisson with parameter $n_1 t/\theta_1$ while the other one is also Poisson with parameter $n_2 t/\theta_2$, and these two processes are independent. The superposition of these two Poisson processes is therefore a Poisson process with parameter $(n_1/\theta_1 + n_2/\theta_2)t,\ t > 0$. We denote this superimposed process by $\boldsymbol{N}^* = \boldsymbol{N}^{(1)} + \boldsymbol{N}^{(2)}$. Then, given that $\boldsymbol{N}^*(t) = n^*$, $\boldsymbol{N}^{(1)}(t)$ has the binomial law with parameters n^* and $\pi^* = (n_1/\theta_1)/(n_1/\theta_1 + n_2/\theta_2)$. Since $\boldsymbol{N}^*(t)$ is nondecreasing and $\boldsymbol{N}^*(t) \to +\infty$ as $t \to +\infty$, in order to test the null hypothesis $H_0 :\ \theta_1 = \theta_2$ against $H_1 :\ \theta_1 \neq \theta_2$, one may then proceed as in the SPRT for the binomial parameter [Wald (1947)] or may even use the SLRT version along the lines in Section 28.5. Therefore, we omit the details.

The situation is somewhat different in the nonreplacement scheme considered in Section 28.3. For each of the two sets of items under life testing, we may proceed as in (28.3.3) through (28.3.8) [with n replaced by n_1 or n_2] and define the total time on test up to the point $t > 0$ as $V_{n_1}(t)$ and $V_{n_2}(t)$ respectively. If we denote the ordered failures in the jth set by $T_{i:n_j}$, $i = 1, \ldots, n_j$ and let $r_{jt} = \boldsymbol{N}_j(t)$, $j = 1, 2$, then $V_{n_j}(t)$ is given by (28.3.3) with n, $T_{i:n}$ and r_t replaced by n_j, $T_{i:n_j}$ and r_{jt} respectively. Then parallel to (28.3.8), we have here

$$V_{n_j}(T_{k:n_j}) = \sum_{i \le k} D_{n_j,i}, \qquad \text{for } k = 1, \ldots, n_j,\ j = 1, 2\ , \tag{28.6.1}$$

where the $D_{n_j,i}$ are the normalized spacings defined as in (28.3.7), for the jth set, $j = 1, 2$. It is therefore naturally appealing to use a test based on the likelihood ratio process, and a discrete time-parameter version is ideal in this context. However, there is a technical problem. The $T_{i:n_j}$, $i = 1, \ldots, n_j$, $j = 1, 2$, are all distinct with probability one. Hence, at a failure point, say, $T_{k:n_1}$, $V_{n_1}(T_{k:n_1})$ is clearly defined by (28.6.1), while

$V_{n_2}(T_{k:n_1})$ has to be defined by (28.3.3) taking into account the difference of the failure times of the second and the first sets. Thus, if $T_{q:n_2} < T_{k:n_1} < T_{q+1:n_2}$, then

$$V_{n_2}(T_{k:n_1}) = V_{n_2}(T_{q:n_2}) + (n_2 - q)(T_{k:n_1} - T_{q:n_2}) \ . \tag{28.6.2}$$

Note that for each j $(= 1, 2)$ and k $(= 1, \ldots, n_j)$,

$$V_{n_j}(T_{k:n_j})/\theta_j \text{ has the gamma}(k) \text{ distribution} \ , \tag{28.6.3}$$

but in (28.6.2), for a given k, q is random and the second term represents an excess and is also stochastic in nature. This creates some difficulties in an exact test construction. We shall, therefore, take recourse to an asymptotic formulation.

Note that as in (28.3.4), the likelihood function pertaining to the experimental outcome during the interval $[0, t]$ is given by

$$[n_1!n_2!/(n_1 - r_{1t})!(n_2 - r_{2t})!]\theta_1^{-r_{1t}}\theta_2^{-r_{2t}} \exp\{-\theta_1^{-1}V_{n_1}(t) - \theta_2^{-1}V_{n_2}(t)\} \ , \tag{28.6.4}$$

so that setting $r_{0t} = r_{1t} + r_{2t}$ and $V_n(t) = V_{n_1}(t) + V_{n_2}(t)$, $n = n_1 + n_2$, we obtain from (28.6.4) that the MLE of θ_1 and θ_2 are given by

$$\widehat{\theta}_1(t) = V_{n_1}(t)/r_{1t} \qquad \text{and} \qquad \widehat{\theta}_2(t) = V_{n_2}(t)/r_{2t} \ , \tag{28.6.5}$$

whenever $r_{jt} > 0$, $j = 1, 2$. Also, under the null hypothesis $H_0 : \ \theta_1 = \theta_2 = \theta$, the MLE of θ from the same accumulated data is given by

$$\widehat{\theta}(t) = V_n(t)/r_{0t} \ . \tag{28.6.6}$$

As such, if we work with the log-derivative of (28.6.4) with respect to θ_1 evaluated at $\widehat{\theta}(t)$, we obtain the likelihood ratio score statistic

$$\begin{aligned} L_n(t) &= [V_{n_1}(t) - \widehat{\theta}(t)r_{1t}]/\{\widehat{\theta}(t)\}^2 \\ &= r_{1t}r_{2t}r_{0t}^{-1}(\widehat{\theta}_1(t) - \widehat{\theta}_2(t))/\{\widehat{\theta}(t)\}^2 \\ &= r_{1t}r_{2t}r_{0t}(\widehat{\theta}_1(t) - \widehat{\theta}_2(t))/V_n^2(t) \\ &= r_{0t}\{r_{2t}V_{n_1}(t) - r_{1t}V_{n_2}(t)\}/V_n^2(t), \ t > 0 \ . \end{aligned} \tag{28.6.7}$$

Conventionally, we let $L_n(t) = 0$ whenever $r_{0t} = 0$, while (28.6.7) is defined for all t for which r_{0t} is positive. Let us denote the largest failure time in the pooled set by $T_{n:n}^0$. Then note that for $t \geq T_{n:n}^0$, $r_{jt} = n_j$, $r_{0t} = n$, $V_{n_j}(t) = S_{n_j}$ is the sum of n_j i.i.d. variables having exponential (θ_j) distribution, $j = 1, 2$, and under H_0, $V_n(t)$ is the sum of n i.i.d. variables having the exponential (θ) distribution. Hence, we have

$$L_n(t) = n\{n_2S_{n_1} - n_1S_{n_2}\}/(S_{n_1} + S_{n_2})^2, \qquad \text{for } t \geq T_{n:n}^0 \ , \tag{28.6.8}$$

so that its variance function under H_0 may well be approximated by $n_1n_2n^{-1}\theta^{-2}$. Keeping this in mind, we consider the normalized likelihood score process:

$$L_n^*(t) = (n/n_1n_2)^{1/2}r_{1t}r_{2t}\{\widehat{\theta}_1(t) - \widehat{\theta}_2(t)\}/V_n(t), \qquad \text{for } t > 0 \ . \tag{28.6.9}$$

We denote by

$$L_n^*(\mathbf{R}^+) = \{L_n^*(t), \ t \in \mathbf{R}^+\} \ , \tag{28.6.10}$$

and let $W^0 = \{W^0(t),\ t \in \mathbf{R}^+\}$ be a Gaussian function on $\mathbf{R}^+$, with independent increments and variance function $(E[W^0(t)]^2 =)\ F(t) = 1 - e^{-t/\theta},\ t \in \mathbf{R}^+$. Then, we have:

$$L_n^* \text{ converges in law to } W^0, \text{ under the null hypothesis } H_0 \ . \qquad (28.6.11)$$

Note that if we let $W^0(t) = W(F(t))$, for $t \in \mathbf{R}^+$, and let $W = \{W(s),\ s \in [0,1]\}$, then W is a standard Wiener process on $[0,1]$. Thus, W^0 is reducible to a Wiener process on $[0,1]$ by a monotone time-parameter transformation, and this is an advantage in dealing with the distributional problems cropping up in the formulation of sequential tests based on L_n^* in (28.6.9) and (28.6.10). Moreover, as in Sections 28.3 and 28.4, the weak convergence result in (28.6.11) extends directly to any sequence of contiguous alternatives. As in (28.5.12), we consider a sequence of local alternatives

$$H_{1(n)} : \ \theta_1 - \theta_2 = (n_1 n_2/n)^{-1/2}\lambda, \qquad \text{for some } \lambda \in \mathbf{R} \ . \qquad (28.6.12)$$

Under $H_{1(n)}$ in (28.6.12), it is easy to verify that *contiguity* holds and moreover the drift function for $L_n^*(t)$ is asymptotically given by $\lambda F(t)$, so that

$$L_n^* \text{ weakly converges to } \{W^0(t) + \lambda F(t),\ t > 0\} \ , \qquad (28.6.13)$$

while,

$$\{W^0(t) + \lambda F(t),\ t \in \mathbf{R}^+\} <=> \{W(s) + \lambda s,\ s \in [0,1]\} \ . \qquad (28.6.14)$$

The latter result enables us to incorporate the boundary crossing probabilities for Wiener processes in the truncated case, studied by Anderson (1960) and DeLong (1981) among others, to formulate suitable sequential tests for $H_0 :\ \theta_1 = \theta_2$ against a local alternative that $\theta_1 - \theta_2 = \xi$, small when the sample sizes n (and n_1, n_2) are chosen so large and $n\xi^2$ is positive and bounded away from 0. Details of this type of SLRT can be found in Chapter 9 of Sen (1981), and hence, we omit them. The theory of sequential point and interval estimation for the difference $\theta_1 - \theta_2$ also follows along the lines of Section 28.3 (dealing with the single θ), and hence, we do not repeat that part.

28.7 Sequential Inference for Systems Availability Parameter

Consider a system consisting of a single operating unit, a repair facility, and a space. When the operating unit fails, the spare is instantaneously put in its place while the failed unit is transmitted on to the repair shop; upon repair it is sent to the spare box for onward use. Thus, the system fails when at a failure time (other than the first one) of the operating unit there is no spare, which occurs when the repair time of the preceding failed unit exceeds the life time of the currently operating unit. It is thus, quite, conceivable to assume that both the life time and repair time, denoted by X and Y, are random variables having distribution functions F and G respectively, where both F and G are defined on $\mathbf{R}^+$. Exponential forms for these distributions are quite reasonable from reliability theory point of view [see Barlow and Proschan (1975)]. As such, we let

$$\overline{F}(x) = 1 - F(x) = e^{-\theta_1 x} \text{ and } \overline{G}(x) = 1 - G(x) = e^{-\theta_2 x},\ x \in \mathbf{R}^+ \ , \qquad (28.7.1)$$

where θ_1 and θ_2 are positive constants representing the constant failure rates for the two distributions. Let N be the number of operating units failure culminating in a system failure, so that N is ≥ 2, and it has the geometric distribution for which

$$P[N = k + 1] = (1 - \alpha)\alpha^{k-1}, \qquad \text{for } k = 1, 2, \ldots\ , \tag{28.7.2}$$

where

$$\begin{aligned} \alpha = P[X > Y] &= \int_0^\infty \overline{F}(x) dG(x) = \theta_2 \int_0^\infty e^{-(\theta_1+\theta_2)x} dx \\ &= \theta_2/(\theta_1 + \theta_2)\ . \end{aligned} \tag{28.7.3}$$

Note that $E[X] = 1/\theta_1$ and $E[Y] = 1/\theta_2$ and the expected regeneration time is the expected time for the first operating unit failure ($= 1/\theta_1$). Thus, the mean life time of the system measuring from a regeneration point is

$$\tau = (1 - \alpha)^{-1} E[X] = (\theta_1 + \theta_2)\theta_1^{-1} \cdot \theta_1^{-1} = (\theta_1 + \theta_2) \cdot \theta_1^{-2}\ . \tag{28.7.4}$$

The mean system downtime is defined by

$$\begin{aligned} \Delta &= \int_0^\infty \int_0^\infty \{\overline{G}(x + t)/\overline{G}(x)\}\ dF(x) dt \\ &= \int_0^\infty \int_0^\infty e^{-\theta_2 t} \theta_1 e^{-\theta_1 x}\ dx dt \\ &= 1/\theta_2\ . \end{aligned} \tag{28.7.5}$$

Following Barlow and Proschan (1975), we may then define the (limiting) systems availability as

$$\begin{aligned} A(\theta_1, \theta_2) &= \tau/(\tau + \Delta) \\ &= \text{expected proportion of system uptime}\ . \end{aligned} \tag{28.7.6}$$

From (28.7.4), (28.7.5), and (28.7.6), we obtain that

$$\begin{aligned} A(\theta_1, \theta_2) &= (\theta_1 + \theta_2)\theta_1^{-2}/\{(\theta_1 + \theta_2)\theta_1^{-2} + \theta_2^{-1}\} \\ &= (\theta_1 + \theta_2)/\{(\theta_1 + \theta_2) + \theta_1^2/\theta_2\} \\ &= (1 + \rho)/(1 + \rho + \rho^2)\ , \end{aligned} \tag{28.7.7}$$

where

$$\rho = \theta_1/\theta_2 = E[Y]/E[X]\ . \tag{28.7.8}$$

In view of (28.7.7) and (28.7.8), we write $A(\theta_1, \theta_2)$ as $A(\rho)$, and note that

$$(\partial/\partial\rho)A(\rho) = -\rho(2 + \rho)/(1 + \rho + \rho^2)^{-2} < 0, \ \text{for every } \rho > 0\ . \tag{28.7.9}$$

Therefore, $A(\rho)$ is a monotonically decreasing function of ρ; at $\rho = 0$, it is equal to 1 and it converges to 0 as ρ goes to $+\infty$. From reliability point of view, $A(\rho)$ should be close to 1, so that equivalently ρ should be close to 0. Thus, from statistical inference point of view, it suffices to work with the ratio parameter ρ.

Note that a system failure occurs when for the first time, Y_N exceeds X_N where N is defined before in (28.7.2). In that setup the excess of Y_N over X_N may be incorporated in the estimation of $A(\rho)$. But, this will generally lead to loss of information. We may note that at this culmination point, one has already made observations on X_i, Y_i, for $i \leq N$. As such, if these observations are themselves incorporated in this statistical inference problem that would result in a more informative procedure. On the other hand, N is itself a positive integer valued random variable, and hence, the relevance of sequential analysis is quite apparent. In this context, we should be aware of the hidden point: For every $i < N$, Y_i is $\leq X_i$, so that this constraint has to be taken into account in formulating the statistical model. We may bypass this technicality by considering an alternative approach. Let $\mathcal{B}_n = (X_0, \dots, X_n; Y_1, \dots, Y_n)$, $n \geq 1$, and note that $[N = n]$ is $\mathcal{B}_n$-measurable. Therefore, for every $n \geq 1$, $[N \geq n]$ is $\mathcal{B}_{n-1}$-measurable. As such, we have

$$\begin{aligned}
E[X_0 + \cdots + X_N] &= \sum_{n \geq 1} E[(X_0 + \cdots + X_n) I(N = n)] \\
&= \sum_{n \geq 1} E[(X_0 + \cdots + X_n)\{I(N \geq n) - I(N \geq n+1)\}] \\
&= \sum_{n \geq 1} E[X_n I(N \geq n)] + E[X_0] \\
&= E[X] \sum_{n \geq 1} P\{N \geq n\} + E[X_0] \\
&= E[X]E[N] + E[X] = \theta_1^{-1} + (\theta_1 + \theta_2)/\theta_1^2 \\
&= \theta_1^{-1}\{\rho^{-1}(1+\rho) + 1\} \\
&= \theta_1^{-1}(1 + 2\rho)/\rho \,. \qquad (28.7.10)
\end{aligned}$$

As a result,

$$E[X_1 + \cdots + X_N] = \theta_1^{-1}(1+\rho)/\rho \,. \qquad (28.7.11)$$

Similarly,

$$E[Y_1 + \cdots + Y_N] = \theta_1^{-1}(1+\rho) \,. \qquad (28.7.12)$$

Recall that $X_1 + \cdots + X_N$ is the total life time of the system before a failure occurs, and we denote it by L. Similarly, $Y_1 + \cdots + Y_N$ is the total time spent on repair before the system fails; we denote it by R. Now, consider the system operating over a prolonged period of time, and let $N_1, N_2, \dots$ be the successive stopping numbers for the system failures, i.e., the first system failure occurs after N_1 repairs, the next one at $(N_1 + N_2)$th cycle, and so on. We denote the corresponding random vectors of life and repair times by (L_i, R_i), $i \geq 1$. Based on this sequence, we may virtually apply the asymptotic theory of sequential tests and estimators developed in Chapters 9 and 10 of Sen (1981) and carry out suitable sequential procedures for drawing statistical inference on ρ. The only problem is that when $\rho = E[R]/E[L]$ is small, the N_i will be stochastically large, and that would lead to a considerable study period for drawing such statistical conclusions. The alternative plan to be posed now appears to be more adoptable when ρ is small (as is usually the case in practice).

Since we assume that the operating unit is repairable, and upon repairing, the life distribution remains the same, it may be quite reasonable to assume that there exists a sequence $\{X_i,\ i \geq 1\}$ of i.i.d. variables such that their common distribution is exponential with parameter $\theta_1 = 1/E[X]$. Similarly, we assume that there exists a sequence $\{Y_j,\ j \geq 1\}$ of i.i.d. variables whose distribution is exponential with parameter $\lambda_2 = 1/E[Y]$. Based on these two sequences, we may like to consider a SPRT type test for the ratio parameter $\rho = \theta_1/\theta_2$ or to provide a confidence interval for ρ having a prescribed width and coverage probability. Of particular interest is the case where ρ is small. In principle, this runs parallel to the methodology sketched in the preceding section with one exception: There we were interested in the equality of θ_1 and θ_2 (i.e., $\rho = 1$) whereas here the ratio parameter ρ is of interest. Since $\rho = \rho(\theta_1, \theta_2)$ is a smooth and differentiable function of θ_1 and θ_2, starting with the joint likelihood function in (28.6.4), the asymptotic theory of SLRT (for ρ) can be developed in very much the same way. Similarly, the confidence interval theory can be handled in the same setup as in Chapter 10 of Sen (1981). Although these methodologies pertain to the asymptotic case, they fare quite well when compared to the other approach when ρ is small.

28.8 Some General Remarks on Sequential Inference

Exponential models have been extensively used in various applications. Some of the important ones have been treated in the earlier sections. There are some other problems which deserve some discussion. Bivariate or multivariate exponential models have been proposed by a host of workers, and these are discussed in Chapters 20 and 21 of this volume. In the context of competing risk models too, such exponential distributions crop up in a natural manner, and moreover, there is a genuine need for sequential inference (which may have also a time-sequential flavor). To illustrate this point, let us consider the bivariate exponential model due to Marshall and Olkin (1967) [there are scores of other models which are referred to in Chapters 20 and 21 and hence will not be treated in detail; the methodology runs on parallel lines]. The Marshall-Olkin bivariate exponential survival function is of the form

$$P[X > x,\ Y > y] = \exp\{-\lambda_1 x - \lambda_2 y - \lambda_{12} \max(x, y)\},\ \text{ for } x, y \geq 0\ , \qquad (28.8.1)$$

where λ_1, λ_2 and λ_{12} are positive (finite) constants; the marginals for X and Y are exponential. In a competing risk setup, one observes the sequence

$$(Z_i, \delta_i);\ i \geq 1, \text{ where } Z_i = \min(X_i, Y_i) \text{ and } \delta_i = 1 \text{ or } 0 \text{ according as } Z_i = X_i \text{ or } Y_i,\ i \geq 1\ . \qquad (28.8.2)$$

Here also, if we put all the n units (each having two components) in a life testing model, we will observe the order statistics of the Z_i (denoted by $Z_{k:n}$, $k = 1, \ldots, n$) and the corresponding induced order statistics $\delta_{[k]}$, $k = 1, \ldots, n$. The situation is therefore very much comparable to the random censoring scheme presented in Section 28.5. The basic difference between the two is the following: In Section 28.5, we assumed that the failure and censoring distributions are both exponential and are independent. Here both the marginals of (28.8.1) are exponential, but X and Y are not independent of each other.

Let us denote by $\overline{F}_1(x) = P[X > x]$ and $\overline{F}_2(y) = P[Y > y]$, for $x, y \geq 0$. Then, by (28.8.1), we conclude that for $H_i(\cdot) = -\log \overline{F}_i(\cdot)$, $i = 1, 2$,

$$H_1(x)/H_2(x) = (\lambda_1 + \lambda_{12})/(\lambda_2 + \lambda_{12}), \qquad \text{for all } x \geq 0 \ . \tag{28.8.3}$$

Moreover, for the variable Z, defined by (28.8.2), we have the exponential distribution with parameter $\lambda = \lambda_1 + \lambda_2 + \lambda_{12}$. Finally, note that by (28.8.1), (28.8.2), and (28.8.3), we have

$$P[\delta = 1 \mid Z = z] = (\lambda_1 + \lambda_{12})/(\lambda_1 + \lambda_2 + 2\lambda_{12}) \ . \tag{28.8.4}$$

Note that (28.8.4) remains a constant for all z, and hence, the methodology presented in Section 28.5 can be incorporated in the formulation of suitable sequential inference procedures. Towards this, we write the expression on the right hand side of (28.8.4) as π. It is assumed that $\pi \in (0, 1)$. Based on the $Z_{i:n}$, $i \leq k$, we define the V^0_{nk} as in (28.5.5). Then the log-likelihood function of the set $\{(Z_{i:n},\ \delta_{[i]}),\ i \leq k\}$ is given by

$$k \log \lambda - \lambda V^0_{nk} + \sum_{i \leq k} \{\delta_{[i]} \log \pi + (1 - \delta_{[i]}) \log(1 - \pi)\} \ . \tag{28.8.5}$$

This leads to the MLE

$$\widehat{\lambda}_{nk} = k/V^0_{nk} \qquad \text{and} \qquad \widehat{\pi}_{nk} = k^{-1} \sum_{i \leq k} \delta_{[i]} \ . \tag{28.8.6}$$

In the competing risk setup, in a variety of problems, one may want to test the null hypothesis $H_0 : \ \lambda_1 = \lambda_2$, treating λ_{12} and λ as nuisance parameters. In this setup, the reparametrization in (28.8.5) leads to the equivalent null hypothesis $H_0^* \ : \ \pi = 1/2$. If for testing $H_0^* \ : \ \pi = 1/2$ vs. $H_1 \ : \pi \neq 1/2$, we consider the usual SLRT type of test statistic, we have by (28.8.5) and (28.8.6),

$$\begin{aligned} L_{nk} &= \sum_{i \leq k} \{\delta_{[i]} \log \widehat{\pi}_{nk} + (1 - \delta_{[i]}) \log(1 - \widehat{\pi}_{nk})\} - k \log(1/2) \\ &= \widehat{\Delta}_k \log \widehat{\Delta}_k + (k - \widehat{\Delta}_k) \log(k - \widehat{\Delta}_k) - k \log(k/2), \ k = 1, \ldots, n \ , \end{aligned} \tag{28.8.7}$$

where

$$\widehat{\Delta}_k = \sum_{i \leq k} \delta_{[i]}, \qquad \text{for } k = 1, \ldots, n \ . \tag{28.8.8}$$

Let us define a stochastic process $\xi_n = \{\xi_n(t) : \ t \in (0, 1]\}$ by letting

$$\xi_n(t) = 2L_{n[nt]}, \qquad \text{for } t : \ 0 < t \leq 1 \ . \tag{28.8.9}$$

It is known that under H_0^*, for each $t > 0$, $\xi_n(t)$ has asymptotically (as $n \to \infty$) central chi-square distribution with 1 degree of freedom. Moreover, if we consider the process $\xi_n[a, 1] = \{\xi_n(t) : \ a \leq t \leq 1\}$, where $a > 0$, then $\xi_n[a, 1]$ weakly converges to a Bessel (squared) process $\{B^2(t), t \in [a, 1]\}$. The weak convergence result extends directly for contiguous alternatives and we have a noncentral square Bessel process for the limiting case. For such Bessel processes, boundary crossing probabilities have been tabulated by DeLong (1981), and these tables may be used with advantage to construct suitable (time-) sequential tests for H_0^* against alternatives that $\pi \neq 1/2$. Since such tests are similar to the repeated significance tests discussed in Chapter 9 of Sen (1981), we omit their detailed description. Right censoring (on k) is also possible by adjusting the ratio $a : \ell$ to $a : b$, where b is the target proportion of failures allowable in the sequential testing scheme, i.e., the experimentation can be continued at most up to the (nb)th failure ($Z_{nb:n}$). Note that the critical levels reported in DeLong (1981) depend specifically on the ratio b/a (> 1), and hence this does not create any additional problem. In passing, we may remark that unlike the SPRT discussed in Section 28.2, here, the usual emphasis is on the Type I error bounds and 'good' power properties. An analogue of the SPRT (or SLRT) may be worked out more conveniently with the replacement scheme, but in clinical trials and many other biomedical studies, such a scheme may not be that appropriate. For the marginal distribution of Z, the parameter λ is relevant, and, therefore, it may be of some interest to test for a suitable null hypothesis on λ or to estimate λ in a sequential manner. This can be done easily by considering the Z_i alone (i.e., ignoring the $\delta_{[i]}$), and the methodology of Sections 28.2 and 28.4 will hold without any modifications. If, on the other hand, our interest lies in sequential estimation of the parameter π, defined by (28.8.4), we may work with the partial likelihood sequence $\{L_{nk} : 0 \leq k \leq n\}$ in (28.8.7), and in view of the unknown π, we are to replace $k \log(1/2)$ by $\widehat{\Delta}_k \log \pi + (k - \widehat{\Delta}_k) \log(1 - \pi)$, and the resulting estimating equations lead to an invariance principle similar to that in (28.8.9) but relating to a Gaussian process. Therefore, in an asymptotic setup (when n is large), we may use the normal theory sequential estimation plan for the estimation of π. There are some alternative bivariate exponential distributions, among which the later ones due to Block and Basu (1974) and Sarkar (1987) deserve special mention. For such models too, sequential inference on the associated parameters can be drawn (in a competing risk setup) in a very similar manner. We may refer to Wada and Sen (1994) for some of these details. Multivariate generalizations are not conceptually difficult but notationally more complex. We refrain ourselves from their discussions.

Primarily, we have considered sequential estimation and hypothesis testing problems related to a single univariate or bivariate exponential distribution and also to two independent exponential populations. Censoring schemes and competing risk models are also included in this setup. It is therefore quite pertinent to conceive of a more general model where there are one or more exponential distributions in univariate or multivariate models and where censoring may also be a vital factor. In such a case, we would have a vector $\boldsymbol{\lambda} = (\lambda_1, \ldots, \lambda_q)'$ of associated parameters, and one may be specifically interested in formulating a parameter $\theta = \psi(\boldsymbol{\lambda})$, where ψ need not be a ratio type or a linear function. Following the lines of attack of the previous sections, and eliminating the identifiability problems wherever necessary, it may be possible to consider a sequence of estimators, say $\{\widehat{\theta}_n, \; n \geq 1\}$ or $\{\widehat{\theta}_{nk}, \; k \geq 1, \; n \geq 1\}$, such that for such a sequence, an invariance principle holds under appropriate regularity conditions. This will lead us to the access of using appropriate Gaussian functions in simplifying the related distribution theory and the sequential procedures sketched in earlier sections remain adoptable in this general scenario too.

We conclude this section with some remark on sequential inference for the two-parameter exponential model:

$$f(x;\mu,\theta) = \theta^{-1}\exp\{-(x-\mu)/\theta\}I(x \geq \mu),\ \ \mu \text{ real},\ \theta > 0\ , \tag{28.8.10}$$

where based on a sample of size n (with variables $X_1, \ldots, X_n$ associated with it), optimal estimators of μ and θ are respectively

$$X_{1:n} = \min(X_1, \ldots, X_n) \text{ and } \sum_{i \leq n}(X_{i:n} - X_{1:n})/(n-1) = \widehat{\theta}_n, \text{ say}\ . \tag{28.8.11}$$

In a life testing context too, one observes the $X_{i:n}$ in a (time wise) sequential manner so that instead of $\widehat{\theta}_n$ in (28.8.11), one may define at the kth failure time $X_{k:n}$,

$$\widehat{\mu}_{n,k} = X_{1:n} \text{ and } \widehat{\theta}_{n,k} = \left[\sum_{i \leq k}(X_{i:n} - X_{1:n}) + (n-k)(X_{k:n} - X_{1:n})\right]/(k-1)\ . \tag{28.8.12}$$

For drawing sequential inference on θ, the $\widehat{\theta}_{n,k}$ in (28.8.12) can be used in the same manner as before (with the obvious change of k to $(k-1)$). On the other hand, to draw sequential inference on μ, we need to incorporate the $\widehat{\theta}_{n,k}$ as estimates of the nuisance parameter θ. In an asymptotic setup (where n is large), the substitution of θ by the updated $\widehat{\theta}_{n,k}$ makes no real difference, although for $X_{1:n}$ we would have the same exponential density as in (28.8.10) with θ being replaced by θ/n. Recall that we may write

$$n(X_{1:n} - \mu)/\widehat{\theta}_{n,k} = \{n(X_{1:n} - \mu)/\theta\}/(\widehat{\theta}_{n,k}/\theta), \qquad k = 2, \ldots, n\ , \tag{28.8.13}$$

where $n(X_{1:n} - \mu)/\theta$ has the simple exponential law in (28.1.1) with $\theta = 1$. Therefore, if the $\widehat{\theta}_{n,k}$ converge to θ in a stronger mode, we may still take recourse to the Slutsky Theorem, and draw conclusions from the model in Section 28.2. If k is not very large (requiring n to be large too), the convergence result in (28.8.13) may not be that good. However, since for every $k \geq 2$, $\widehat{\theta}_{n,k}$ is independent of $X_{1:n}$, convolution of an exponential and a gamma$(k-1)$ density can be used to provide some closed expression for the distribution and numerically it can be approximated well.

References

Anderson, T.W. (1960). A modification of the sequential probability ratio test to reduce the sample size, *Annals of Mathematical Statistics*, **31**, 165-197.

Anscombe, F.J. (1952). Large sample theory of sequential estimation, *Proceedings of the Cambridge Philosophical Society*, **48**, 600-607.

Barlow, R.E. and Proschan, F. (1975). *Statistical Theory of Reliability and Life Testing: Probability Models*, Holt, Rinehart and Winston, New York.

Block, H.W. and Basu, A.P. (1974). A continuous bivariate exponential distribution, *Journal of the American Statistical Association*, **69**, 1031-1037.

Chow, Y.S. and Robbins, H. (1965). On the asymptotic theory of fixed-width sequential confidence intervals for the mean, *Annals of Mathematical Statistics*, **36**, 457-462.

Daniels, H.A. (1945). The statistical theory of the strength of bundles of threads, *Proceedings of the Royal Society, Series A*, **183**, 405-435.

DeLong, D. (1981). Crossing probabilities for a square root boundary by a Bessel process, *Communications in Statistics - Theory and Methods*, **10**, 2197-2213.

Dvoretzky, A., Kiefer, J., and Wolfowitz, J. (1953). Sequential decision problems for processes with continuous time parameter: Testing hypotheses, *Annals of Mathematical Statistics*, **24**, 254-264.

Epstein, B. and Sobel, M. (1954). Some theorems relevant to life testing from an exponential distribution, *Annals of Mathematical Statistics*, **25**, 373-381.

Epstein, B. and Sobel, M. (1955). Sequential life testing in the exponential case, *Annals of Mathematical Statistics*, **26**, 82-93.

Gardiner, J.C., Susarla, V., and van Ryzin, J. (1986). Time-sequential estimation of the exponential mean under random censoring, *Annals of Statistics*, **14**, 607-618.

Ghosh, M., Mukhopadhyay, N., and Sen, P.K. (1995). *Sequential Estimation*, John Wiley & Sons, New York (To appear).

Marshall, A.W. and Olkin, I. (1967). A multivariate exponential distribution, *Journal of the American Statistical Association*, **62**, 30-44.

Raghavachari, M. (1965). Operating characteristic and expected sample size of a sequential probability ratio test for the simple exponential distribution, *Calcutta Statistical Association Bulletin*, **14**, 65-79.

Sarkar, S.K. (1987). A continuous bivariate exponential distribution, *Journal of the American Statistical Association*, **82**, 667-675.

Sen, P.K. (1973a). An asymptotically optimal test for the bundle strength of filaments, *Journal of Applied Probability*, **10**, 586-596.

Sen, P.K. (1973b). On fixed-size confidence bands for the bundle-strength of filaments, *Annals of Statistics*, **1**, 526-537.

Sen, P.K. (1976). Weak convergence of progressively censored likelihood ratio statistics and its role in asymptotic theory of life testing, *Annals of Statistics*, **4**, 1247-1257.

Sen, P.K. (1980). On time-sequential point estimation of the mean of an exponential distribution, *Communications in Statistics - Theory and Methods*, **9**, 27-38.

Sen, P.K. (1981). *Sequential Nonparametrics: Invariance Principles and Statistical Inference*, John Wiley & Sons, New York.

Sen, P.K. (1986). Nonparametric estimators of availability under provisions of spare and repair, II, In *Reliability and Quality Control* (Ed., A.P. Basu), pp. 297-308, North-Holland, Amsterdam.

Sen, P.K. (1993). Resampling methods for the extrema of certain sample functions, In *Probability and Statistics: Proceedings of the Calcutta Triennial International Conference in Statistics* (December 1991), (Eds., S.K. Basu and B.K. Sinha), pp. 65-79.

Sen, P.K. (1994). Extreme value theory for fibre bundles, In *Extreme Value Theory and Applications*, (Ed., J. Galambos), pp. 77-92, Kluwer, Newell.

Sen, P.K. and Bhattacharjee, M.C. (1986). Nonparametric estimators of availability under provisions of spare and repair, I, In *Reliability and Quality Control* (Ed., A.P. Basu), pp. 281-296, North-Holland, Amsterdam.

Sen, P.K., Bhattacharyya, B.B., and Suh, M.W. (1973). Limiting behavior of the extrema of certain sample functions, *Annals of Statistics*, **1**, 297-311.

Wada, C.Y. and Sen, P.K. (1994). Restricted alternative test in a parametric model with competing risk data, *Journal of Statistical Planning and Inference* (To appear).

Wald, A. (1947). *Sequential Analysis*, John Wiley & Sons, New York.

Weiss, L. (1962). On sequential tests which minimize the maximum expected sample size, *Journal of the American Statistical Association*, **57**, 551-566.

Woodroofe, M. (1982). *Nonlinear Renewal Theory in Sequential Analysis*, SIAM, Philadelphia.

CHAPTER 29

Competing Risks Theory and Identifiability Problems

A.P. Basu[1] and J.K. Ghosh[2]

29.1 Introduction

The problem of competing risks (and complementary risks) occur in a number of different areas. In its simplest form, the problem may be described as follows. Let $X_1, X_2, \ldots, X_p$ be p variables with cumulative distribution functions $F_1(x), F_2(x), \ldots, F_p(x)$, respectively. Let us assume that the X_i's are not observable but $U = \min(X_1, X_2, \ldots, X_p)$ (or $V = \max(X_1, X_2, \ldots, X_p)$) is. This would be the case, for example, for a p-component series system (parallel system) where the component lifetimes $X_1, X_2, \ldots, X_p$ are unknown but the system lifetime U (V) is observable. We would like to uniquely determine the marginal distribution functions $F_1(x), F_2(x), \ldots, F_p(x)$ given the distribution of U (V).

Throughout this chapter, we assume that $X_i \sim e(\lambda_i)$. That is, the X_i's follow the exponential distribution with $F_i(x) \equiv F_i(x_i; \lambda_i) = 1 - \exp(-\lambda_i x)$ and density function $f_i(x) \equiv f_i(x_i; \lambda_i) = \lambda_i \exp(-\lambda_i x)$, $(x > 0,\ i = 1, 2, \ldots, p)$.

Thus, given the distribution of U, we need to uniquely estimate the X_i's.

The problem of competing risks has been extensively studied. For a bibliography, see Basu (1981, 1983), Basu and Ghosh (1978, 1980, 1983), Basu and Klein (1982), David and Moeschberger (1978), Mukhopadhyay and Basu (1993a,b), and the references therein. In this chapter we consider identifiability problems when the underlying distributions follow the exponential distributions. In Section 29.2, the identifiability concepts are discussed. In Section 29.3, inference problems for independent exponential distributions are considered. A Bayesian approach also is mentioned in Section 29.3.

[1] University of Missouri-Columbia, Columbia, Missouri
[2] Indian Statistical Institute, Calcutta, India

29.2 Identifiability

29.2.1 Definition

Before we can consider estimation we need to discuss identifiability problems in the parametric case. The problem, in general terms, can be defined as follows.

Definition 29.1 *Let U be an observable random variable whose distribution function belongs to a family $F = \{F_{\boldsymbol{\theta}} : \theta \in \boldsymbol{\Omega}\}$ of distribution functions indexed by a parameter $\boldsymbol{\theta}$. Here, $\boldsymbol{\Omega}$ is the parameter space, $\boldsymbol{\theta}$ could be scalar or vector valued. We shall say $\boldsymbol{\theta}$ is nonidentifiable by U if there are distinct parameter values, $\boldsymbol{\theta}$ and $\boldsymbol{\theta}'$, such that $F_{\boldsymbol{\theta}}(u) = F_{\boldsymbol{\theta}'}(u)$ for all u. In the contrary case, we shall say $\boldsymbol{\theta}$ is identifiable.*

It may happen that $\boldsymbol{\theta}$ itself is nonidentifiable but a function $g(\boldsymbol{\theta})$ is identifiable in the following sense: for any $\boldsymbol{\theta}$, $\boldsymbol{\eta} \in \boldsymbol{\Omega}$, $F_{\boldsymbol{\theta}}(u) = F_{\boldsymbol{\eta}}(u)$ for all u implies $g(\boldsymbol{\theta}) = g(\boldsymbol{\eta})$. In this case we may say that $\boldsymbol{\theta}$ is partially identifiable.

Example 29.1 Let X_1 and X_2 be independent random variables and let $X_i \sim e(\lambda_i)$, $i = 1, 2$. Then, $U = \min(X_1, X_2) \sim e(\lambda_1 + \lambda_2)$.

If only U is observable, then $\boldsymbol{\theta} = (\lambda_1, \lambda_2)$ is not identifiable. However, $\boldsymbol{\theta}$ is partially identifiable since $g(\boldsymbol{\theta}) = \lambda_1 + \lambda_2$ is identifiable.

In case $\boldsymbol{\theta}$ is not identifiable by U, it may be possible to introduce an additional random variable I so that $\boldsymbol{\theta}$ is identifiable by the augmented random variable (U, I). In this case the original identifiability problem is called rectifiable.

Example 29.2 The problem in Example 29.1 is rectifiable since, as will be shown later, θ is identifiable by (U, I) where $I = i$ if $U = X_i$, $(i = 1, 2)$.

Note (U, I) is called an *identified* minimum if $I = k$ when $U = X_k$. That is, we observe the minimum and know which X_i is the minimum. Similarly, (V, I) is called an *identified* maximum if $I = k$ when $V = X_k$.

29.2.2 Independent Random Variables

If the X_i's are i.i.d. with common distribution function $F(x)$, then $F(x)$ can be trivially obtained from the distribution function of U. Berman (1963) has shown that if the X_i's are independent but not identically distributed and if we have identified minimum, then the unknown distributions functions $F_1(x), F_2(x), \ldots, F_p(x)$ can be uniquely determined in terms of the known monotonic functions $H_k(x) = P(U \leq x,\ I = k)$, $k = 1, 2, \ldots, p$, as

$$F_k(x) = 1 - \exp\left\{ -\int_{-\infty}^{x} \left[1 - \sum_{j=1}^{p} H_j(t) \right]^{-1} dH_k(t) \right\}, \quad k = 1, 2, \ldots, p\,. \tag{29.2.1}$$

Berman's result shows that the problem in Example 29.1 is rectifiable. It is also easy to verify this directly as follows. Let X_1 and X_2 be independent random variables with $X_i \sim e(\lambda_i)$, $i = 1, 2$. Let $Z = \min(X_1, X_2)$. Then

$$f(z) = (\lambda_1 + \lambda_2)\, e^{-(\lambda_1 + \lambda_2) z} \tag{29.2.2}$$

and the joint densities of Z and I are given by

$$f(z, I = 1) = \lambda_1 e^{-(\lambda_1+\lambda_2)z} \tag{29.2.3}$$

and

$$f(z, I = 2) = \lambda_2 e^{-(\lambda_1+\lambda_2)z} \tag{29.2.4}$$

so $(\lambda_1 + \lambda_2)$ can be uniquely estimated given the density (29.2.2) of z. λ_1/λ_2 is identifiable by division of the joint densities (29.2.3) and (29.2.4). So λ_1 and λ_2 are identifiable.

Since $\max(X_1, X_2, \ldots, X_p) = -\min(-X_1, -X_2, \ldots, -X_p)$, we can use previous results to show that the identified maximum (V, I) also uniquely determines that of the X_i's as the identification problem can be restated in terms of the minimum.

In case the extremum (U or V) is not identified, one can still uniquely determine $F_k(x)$ under certain conditions. To this end, Basu and Ghosh (1980) obtained the following.

Theorem 29.1 *Let $\mathcal{J}$ be a family of density functions on R_1 with support (a, b) which are continuous and are positive to the left of some point A and such that if f and g are any two distinct members of $\mathcal{J}$, then $\lim_{x\to a}(f(x)/g(x))$ exists and equals either 0 or ∞. Let $X_1, X_2, \ldots, X_p$ be independent random variables with respective densities $f_1, \ldots, f_p$ in $\mathcal{J}$ and $Y_1, Y_2, \ldots, Y_q$ be independent random variables with respective densities belonging to $\mathcal{J}$. If $\min(X_1, X_2, \ldots, X_p)$ and $\min(Y_1, Y_2, \ldots, Y_q)$ have identical distributions, then $p = q$ and there exists a permutation $(k_1, k_2, \ldots, k_p)$ of $(1, 2, \ldots, p)$ such that the density function of Y_i is f_{k_i} $(i = 1, 2, \ldots, p)$.*

Note a similar theorem for maximum is given by Anderson and Ghurye (1977).

Theorem 29.2 *Let $\mathcal{J}$ be a family of density functions on R_1 which are continuous and positive to the right of some point A and such that if f and g are any two distinct members of $\mathcal{J}$, then $x \to a$ $f(x)/g(x)$ either converges to 0 or diverges to ∞. Let $X_1, X_2, \ldots, X_m$ be independent random variables with respective densities $f_1, \ldots, f_m$ in $\mathcal{J}$ and $Y_1, Y_2, \ldots, Y_n$ be independent random variables with densities in $\mathcal{J}$. If $\max\{X_1, \ldots, X_m\}$ and $\max\{Y_1, \ldots, Y_n\}$ have identical distributions, then $m = n$ and there exists a permutation $\{k_1, k_2, \ldots, k_m\}$ of $\{1, \ldots, m\}$ such that the density function of Y_i is f_{k_i} $(i = 1, 2, \ldots, m)$.*

If $\mathcal{J}$ is the family of negative exponential distributions $f_\lambda(x) = \lambda e^{-\lambda x}$, $x \geq 0$, then the conditions of Theorem 29.1 are not met and, in fact, the distributions are not identifiable. However, if the maximum is observed by Theorem 29.2, the exponential distributions are identifiable.

29.2.3 Dependent Random Variables

Basu and Ghosh (1978) have shown the difficulty in identification based on U or (U, I) unless the class of distributions is restricted to a specific parametric family. This can be seen from the following example. Consider the bivariate exponential distributions of Marshall and Olkin (1967) with survival function

$$\overline{F}(x, y) = P(X > x,\ Y > y) = \exp(-\lambda_1 x - \lambda_2 y - \lambda_{12} \max(x, y)) \,. \tag{29.2.5}$$

Here, $\min(X,Y) \sim e(\lambda_1+\lambda_2+\lambda_{12})$.

Let W_1 and W_2 be independent random variables with $W_1 \sim e(\lambda_1+\lambda_{12})$ and $W_2 \sim e(\lambda_2)$. Then $\min(W_1,W_2) \sim e(\lambda_1+\lambda_2+\lambda_{12})$, also. Thus, the dependent model (X,Y) and the independent model (W_1,W_2) are equivalent, so far as distribution of the minimum is considered.

Basu and Ghosh (1978) have considered the identifiability of a number of bivariate families of exponential distributions useful as models in life testing based on U or (U,I). These include the bivariate exponential distributions of Marshall and Olkin (1967), Block and Basu (1974), and Gumbel (1960). To this end, the following results were obtained.

Theorem 29.3 *Parameters of the Marshall-Olkin distribution are identifiable given the distribution of the identified (U,I).*

For the Block-Basu (1974) and the Freund (1961) distribution, the parameters are not at all identifiable even when the identified minimum is available. Similar results for the identifiability of parameters based on (U,I) have been obtained for the bivariate exponential models proposed by Gumbel (1960). However, if only U is observable, none of the above models, except the following, are identifiable.

Consider the following bivariate distribution of Gumbel:

$$F(x,y) = (1-\exp(-\lambda_1 x))(1-\exp(-\lambda_2 y))(1+\lambda_{12}\exp(-\lambda_1 x-\lambda_2 y)) \, . \tag{29.2.6}$$

Theorem 29.4 *If U is observable, λ_{12} is identifiable and (λ_1,λ_2) is identifiable up to a permutation. That is, if (X_1,X_2) and (X_1',X_2') follow the bivariate exponential distributions in (29.2.6) with parameters $(\lambda_1,\lambda_2,\lambda_{12})$ and $(\lambda_1',\lambda_2',\lambda_{12}')$, respectively, and if $\min(X_1,X_2)$ or $\min(X_1',X_2')$ have the same distribution, then either $(\lambda_1,\lambda_2,\lambda_{12}) = (\lambda_1',\lambda_2',\lambda_{12}')$ or $(\lambda_1,\lambda_2,\lambda_{12}) = (\lambda_2',\lambda_1',\lambda_{12}')$.*

29.3 Estimation of Parameters

29.3.1 Maximum Likelihood Estimates

Estimation of parameters based on (U,I), the identified minimum, has been considered extensively [see David and Moeschberger (1978)]. Let X_{ij} be the jth individual failing from cause C_i ($j=1,2,\ldots,n_i$, $i=1,2,\ldots,p$). Assume the X_{ij}'s to be independent. Let $P_i = P(U=X_i)$ and $X_{ij} \sim f_i(x_{ij})$ with distribution function $F_i(x_{ij})$. Given n observations X_{ij} with $n=\sum_{i=1}^p n_i$,

$$f(n_1,n_2,\ldots,n_p) = \frac{n!}{\prod_{i=1}^p (n_i!)} \prod_{i=1}^p P_i^{n_i} \, .$$

Then the likelihood function, conditioned on the n_i's is given by

$$L = \prod_{i=1}^p \prod_{j=1}^{n_i} \left[\frac{f_i(x_{ij})}{\overline{F}_i(x_{ij})} \prod_{\ell=1}^p \overline{F}_\ell(x_{ij}) \right] \, . \tag{29.3.1}$$

Here, $\overline{F}_\ell(x_{ij})$ is the survival function

$$P(X_\ell > x_{ij}) = 1 - F_\ell(x_{ij}) \, . \tag{29.3.2}$$

In case of exponential distribution $X_{ij} \sim e(\lambda_i)$ and $f_i(x_{ij}) = \lambda_i \exp(-\lambda_i x_{ij})$, and $\overline{F}_i(x_{ij}) = \exp(-\lambda_i x_{ij})$, $i = 1, 2, \ldots, p$, let

$$\lambda = \sum_{i=1}^{p} \lambda_i \quad \text{and} \quad T = \sum_{i=1}^{p} \sum_{j=1}^{n_i} x_{ij} \; ; \tag{29.3.3}$$

then the likelihood function is given by

$$L = \prod_{i=1}^{p} \prod_{j=1}^{n_i} [\lambda_i \exp(-\lambda_i x_{ij})] = \left(\prod_{i=1}^{p} \lambda_i^{n_i}\right) \exp(-\lambda T) \; . \tag{29.3.4}$$

Differentiating natural log of L with respect to λ_i and equating to zero, we obtain

$$\frac{\partial \log L}{\partial \lambda_i} = \frac{n_i}{\lambda_i} - T = 0 \; . \tag{29.3.5}$$

Thus, the maximum likelihood estimate (MLE) of λ_i is given by

$$\widehat{\lambda}_i = \frac{n_i}{T} \, , \qquad i = 1, 2, \ldots, p \; . \tag{29.3.6}$$

Since $\theta_i = \frac{1}{\lambda_i}$, the MLE of $E[X_i]$ is given by

$$\widehat{\theta}_i = \frac{T}{n_i} \, , \qquad i = 1, 2, \ldots, p \; . \tag{29.3.7}$$

Similar results can be obtained when only U (or V) is observed or when the X_i's are dependent. However, the solution requires considerable numerical computation.

Klein and Basu (1981) have considered more general results for the exponential distribution when accelerated testing is considered.

29.3.2 Bayes Estimates

Bayes estimates of λ_i's are also easy to obtain. Mukhopadhyay and Basu (1993) have considered Bayesian analysis when there are competing risks with p independent exponential distributions. The joint prior density of $(\lambda_1, \lambda_2, \ldots, \lambda_p)$ is the Dirichlet-Gamma density of $\mathrm{DG}(\alpha_1, \alpha_2, \ldots, \alpha_p, \alpha, \gamma)$ with parameters $(\alpha_1, \alpha_2, \ldots, \alpha_p, \alpha, \gamma)$. It is given by

$$\pi(\lambda_1, \ldots, \lambda_K) = \left[\frac{\Gamma(\alpha_0)}{\prod_{i=1}^{p} \Gamma(\alpha_i)} \frac{\gamma^\alpha}{\Gamma(\alpha)}\right] \left[\prod_{i=1}^{p} (\lambda_i)^{\alpha_i - 1}\right] \lambda^{\alpha - \alpha_0} \, e^{-\lambda\gamma}, \; \lambda_i > 0 \; . \tag{29.3.8}$$

Here, $\sum_{i=1}^{p} \alpha_i = \alpha_0$, and the parameters $\alpha_1, \ldots, \alpha_p, \alpha, \gamma$ are to be chosen to reflect prior opinions. When the prior distribution is given by (29.3.8), the posterior density is

$$\mathrm{DG}(\alpha_1 + n_1, \ldots, \alpha_p + n_p, \alpha + n, \gamma + T) \; . \tag{29.3.9}$$

Bayesian estimates of λ_i's under squared error loss, are given by the expected values of λ_i's with respect to posterior density (29.3.9). These are denoted by $\widehat{\lambda}_{iB}$ and are given by

$$\widehat{\lambda}_{iB} = \frac{(\alpha_i + n_i)(\alpha + n)}{(\alpha_0 + n)(\gamma + T)} \, , \qquad i = 1, 2, \ldots, p \; . \tag{29.3.10}$$

The Jeffreys' prior corresponds to the special case $\alpha_1 = \alpha_2 = \cdots = \alpha_p = \frac{1}{2}$, and $\alpha = \gamma = 0$. Here, we obtain

$$\widehat{\lambda}_{iB} = \frac{(n_i + \frac{1}{2})}{(n + p/2)} \cdot \frac{n}{T}, \qquad i = 1, 2, \ldots, p . \tag{29.3.11}$$

Note if $n_1 = n_2 = \cdots = n_p$, then MLE's and Bayesian estimates of λ_i's are all equal.

References

Anderson, T.W. and Ghurye, S.G. (1977). Identification of parameters by the distribution of a maximum random variable, *Journal of the Royal Statistical Society, Series B*, **39**, 337-342.

Basu, A.P. (1981). Identifiability problems in the theory of competing and complementary risks - a survey, In *Statistical Distributions in Scientific Work*, Vol. 5 (Eds., C. Taillie, G.P. Patil and B.A. Baldessari), pp. 335-348, Reidel, Dordrecht.

Basu, A.P. (1983). Identifiability, In *Encyclopedia of Statistical Sciences*, Vol. 4 (Eds., S. Kotz and N.L. Johnson), pp. 2-6, John Wiley & Sons, New York.

Basu, A.P. and Ghosh, J.K. (1978). Identifiability of the multinormal distribution under competing risks model, *Journal of Multivariate Analysis*, **8**, 413-429.

Basu, A.P. and Ghosh, J.K. (1980). Identifiability of distributions under competing risks and complementary risks models, *Communications in Statistics - Theory and Methods*, **9**, 1515-1525.

Basu, A.P. and Ghosh, J.K. (1983). Identifiability results for k-out-of-p systems, *Communications in Statistics - Theory and Methods*, **12**, 199-205.

Basu, A.P. and Klein, J.P. (1982). Some recent results in competing risks theory, In *Proceedings on Survival Analysis*, IMS monograph series, **2**, 216-229.

Berman, S.M. (1963). Notes on extreme values, competing risks, and semi-Markov processes, *Annals of Mathematical Statistics*, **34**, 1104-1106.

David, H.A. and Moeschberger, M.L. (1978). *The Theory of Competing Risks*, Griffin, London.

Gumbel, E.J. (1960). Bivariate exponential distributions, *Journal of the American Statistical Association*, **55**, 698-707.

Klein, J.P. and Basu, A.P. (1981). Accelerated life testing under competing exponential failure distributions, *IAPQR Transactions*, **7**, 1-20.

Marshall, A.W. and Olkin, I. (1967). A multivariate exponential distribution, *Journal of the American Statistical Association*, **62**, 30-44.

Mukhopadhyay, C. and Basu, A.P. (1993a). Competing risks with k independent exponentials: a Bayesian analysis, *Technical Report No. 516*, Department of Statistics, Ohio State University, Columbus, Ohio.

Mukhopadhyay, C. and Basu, A.P. (1993b). Maximum likelihood and Bayesian estimation of lifetime parameters from masked system failure data, *Technical Report No. 517*, Department of Statistics, Ohio State University, Columbus, Ohio.

CHAPTER 30

Applications in Survival Analysis

Alan J. Gross[1]

30.1 Introduction

The exponential distribution, being the first distribution to be used in life-testing, has had some use in survival analysis. Unfortunately, since the constant hazard rate rarely holds in human and animal survival studies, its use has been rather limited. That is not to specify the exponential distribution has been without use in survival analysis. Epstein and Sobel (1953) made extensive use of the exponential distribution in their landmark publication. The motivation for this paper as its title indicates was, indeed, life-testing.

Prior to the development of nonparametric methods for analyzing survival data from clinical trials, the exponential distribution played a key role in the analysis of survival data. Some of the early, seminal papers include Harris, Meier, and Tukey (1950), Cox (1953), Bartholomew (1957), Mendenhall and Hader (1958), Cox (1959) and Zelen (1966). Furthermore, as it became apparent that certain patient characteristics were related to survival; e.g. age, smoking status, diet, etc. the need to develop models that account for this relationship became evident. In response to this need, the development of regression models, with an underlying exponential distribution governing survival, took place. The important contributions, in this area, include Cox (1964), Feigl and Zelen (1965), Zippin and Armitage (1966), Glasser (1967), Gehan and Siddiqui (1973), Prentice (1973), and Byar, Huse, and Bailar (1974). Because Cox's proportional hazards' model is nonparametric [Cox (1972)] with respect to the survival distribution, its importance in the analysis of survival data cannot be overstated. The Cox model is not covered in this article because, in the first place, the focus of this paper is the exponential distribution and, secondly, it is well covered by many other authors and is widely used in survival analysis.

Another area, in which the exponential distribution has played an important role in survival analysis, is in the planning aspects of clinical trials. In particular, in determining the sample size, i.e., the number of patients necessary to conduct a clinical

[1]Medical University of South Carolina, Charleston, South Carolina

trial when comparing an intervention therapy to a control therapy, the exponential distribution has been quite useful. Some of the important papers and reports that have used the exponential distribution in sample size determination include Pasternack and Gilbert (1971), Pasternack (1972), Rubinstein, Gail, and Santner (1981), Taulbee and Symons (1983), Morgan (1985), Lachin and Foulkes (1986), Gross et al. (1987) and Craig (1988). A principal reason that the exponential distribution has been useful in sample size determination is given in Gross et al. (1987): "Rubinstein, Gail, and Santner (1981) develop sample size estimates when the log-rank test is used to compare two treatment groups. They allow for loss to follow-up by again employing a competing-cause scenario. Furthermore, they show, by means of computer simulation, that the log-rank test has roughly the same power as either the exponential or Weibull assumption. Morgan (1985) also exploits this method in determining appropriate combinations of accrual and follow-up periods in a clinical trial based on cost minimization."

The final area of survival analysis that is presented in this article is the use of the multivariate exponential distribution in survival analysis. Often, in survival studies, response times are correlated. For example, an individual may have a rash on both of his/her arms and receives two different ointments to relieve the symptoms. The (paired) times to relief are usually correlated. Several bivariate exponential survival distributions have been considered by various authors. The earliest paper of which I am aware is Gumbel (1960). Other articles which followed include: Freund (1961), Marshall and Olkin (1967), Gross, Clark, and Liu (1971), Block and Basu (1974), Gross and Lam (1981), Knapp, Cantor, and Gross (1986), Knapp, Gross, and Cantor (1986), and Lee and Gross (1989, 1991); see also Chapters 20 and 21. Particular emphasis is placed on the Gross-Lam and Knapp et al. papers since they are specifically designed for survival analysis applications and emphasize the exponential distribution.

30.2 Basic Formulation

In most survival analysis studies involving either human beings or experimental animals who are placed on study, the variable of principal interest is the length of time until each responds in some manner. This time to response is generically termed survival time even though the response in question does not have to be death. The vast majority of survival analysis studies is not analyzed by parametric methods, let alone a rather restrictive parametric survival distribution, the exponential. However, when the disease in question has a rapid progression to death such as severe viral hepatitis, the exponential distribution often fits the survival data quite well. Recently, Voit (1992) has described a very general probability distribution, the S-Distribution, that may rival the popular nonparametric techniques for analyzing survival data; see also Johnson, Kotz, and Balakrishnan (1994) for additional details.

To fix ideas, suppose that the ith patient who enters a study is represented by the vector (Y_i, δ_i), where $Y_i = T_i$ the failure time of the ith patient, if $\delta_i = 1$ and $Y_i = C_i$, the censoring time of the ith patient, if $\delta_i = 0$. That is, $Y_i = \min(T_i, C_i)$ and δ_i is the indicator variable that takes the value unity if the ith patient fails on the study or the value zero if the ith patient is censored. It is assumed, for the purpose of this paper, T_i

and C_i are independently distributed. Hence, the pair (Y_i, δ_i) has as its likelihood

$$L(\lambda; t_i, c_i, \delta_i) = \begin{cases} \lambda e^{-\lambda t_i} & , \quad \text{if } \delta_i = 1 , \\ e^{-\lambda c_i} & , \quad \text{if } \delta_i = 0 . \end{cases} \tag{30.2.1}$$

It then follows that if exactly d of the n patients on study fail and $s = n - d$ are censored, the likelihood of the entire sample is, as a function of λ,

$$\begin{aligned} L(\lambda) = \prod_{i=1}^{n} L(\lambda; t_i, c_i, \delta_i) &= \prod_{d} \lambda e^{-\lambda t_i} \prod_{s} e^{-\lambda c_i} , \\ &= \lambda^d e^{-\lambda(\sum_d t_i + \sum_s c_i)} . \end{aligned} \tag{30.2.2}$$

The maximum likelihood estimator $\widehat{\lambda}$ of λ is then

$$\widehat{\lambda} = d \,/\, v , \tag{30.2.3}$$

where, by definition,

$$v = \left(\sum_d t_i + \sum_s c_i \right) .$$

Furthermore,

$$\frac{\partial^2 \ln L}{\partial \lambda^2} = -d \,/\, \lambda^2$$

and so, for large samples,

$$\sqrt{d}(\widehat{\lambda} - \lambda) \,/\, \lambda \sim N(0,1) ,$$

approximately. It is easily shown by the delta method that

$$\sqrt{d}(\ln \widehat{\lambda} - \ln \lambda) \sim N(0,1) ,$$

and since the asymptotic variance of $\ln \widehat{\lambda}$ does not depend on the unknown parameter λ and skewness is reduced, convergence to normality is improved.

It is also of interest to obtain the expected number of failures in a clinical trial when the trial begins at $t = 0$, has a patient recruitment period of length t_0 which runs concurrently with the trial which is of length $t_1 > t_0$. If D is the total number of failures during the trial and if patients enter the trial at random and uniformly during the recruitment period, it is not difficult to show that if n patients enter the trial,

$$E[D] = n \left[1 - \left\{ \frac{e^{-\lambda(t_1 - t_0)} - e^{-\lambda t_1}}{\lambda t_0} \right\} \right] . \tag{30.2.4}$$

Finally, there is a surprisingly not too well known result concerning the confidence interval for $S(t)$, the survival function. That is, if $t_1, \ldots, t_n$ are n distinct times at which survival is measured, and if $S(t_i)$ is the survival function at t_i with $100(1-\alpha)\%$ confidence interval $(\widehat{S}_\ell(t_i),\ \widehat{S}_u(t_i))$, the joint confidence region for all n times; i.e.,

$$\bigcap_{i=1}^{n} (\widehat{S}_\ell(t_i),\ \widehat{S}_u(t_i)) , \tag{30.2.5}$$

is also at level $100(1-\alpha)\%$. To demonstrate this, it is noted that in large samples, for example, a $100(1-\alpha)\%$ confidence for λ, the failure rate is $\widehat{\lambda}_\ell < \lambda < \widehat{\lambda}_u$ where $\widehat{\lambda}_\ell = \widehat{\lambda}\exp(-Z_{\alpha/2} / \sqrt{d})$ and $\widehat{\lambda}_u = \widehat{\lambda}\exp(Z_{\alpha/2} / \sqrt{d})$ and thus, for any t (or set of values $t_1, \ldots, t_n$ for that matter)

$$\widehat{S}_\ell(t) < S(t) < \widehat{S}_u(t)$$

is also a $100(1-\alpha)\%$ confidence interval for $S(t)$. However, subject to the condition that $\widehat{\lambda}_\ell < \lambda < \widehat{\lambda}_u$, $\widehat{S}_\ell(t_i) < S(t) < \widehat{S}_u(t_i)$ for all $i = 1, \ldots, n$ and the result follows.

30.3 The Exponential Survival Distribution in the Presence of Covariables

A seminal paper that looks at the relationship between patient survival (usually in the context of a clinical trial) and covariables that are measured on patients was published by Feigl and Zelen (1965). As with ordinary linear regression, the assumption of an independent, identically distributed sample no longer holds but, rather, it is assumed that in a clinical trial n patients are on study and the failure density function for the ith patient is

$$\begin{aligned} f_i(t_i) &= \lambda_i e^{-\lambda_i t_i} \quad , \ \lambda_i > 0;\ t_i \geq 0 \ , \\ &= 0 \qquad\quad , \ \text{elsewhere} \ , \end{aligned}$$

$i = 1, \ldots, n$. It is assumed, initially, all patients are followed until failure occurs. Furthermore, let x_i be the observed value of a covariable on the ith patient, so that

$$E[T_i] = \lambda_i^{-1} = a + bx_i \ , \tag{30.3.1}$$

$i = 1, \ldots, n$. Thus, each patient's mean failure time is assumed to be linearly related to the covariable. In the example presented by Feigl and Zelen, patient survival times T are recorded for patients with acute myelogenous leukemia and are recorded in weeks from date of diagnosis. The covariable of interest was $\log_{10}$ WBC (white blood count) also recorded at time of diagnosis. The white blood count has long been recognized as a predictor of survival for acute leukemia patients - the higher the initial white blood count the lower the probability of survival for a specified time.

The problem at hand is to estimate the parameters a and b in (30.3.1). To this end and assuming that no censoring occurs, the method of maximum likelihood is used. If $L(a,b)$ is the likelihood function relating to the parameters a and b, then

$$L(a,b) = \prod_{i=1}^{n} (a+bx_i)^{-1} e^{-(a+bx_i)^{-1} t_i} \ . \tag{30.3.2}$$

Using a standard iterative procedure for solving nonlinear equations; e.g., the two dimensional Newton-Raphson method, the maximum likelihood estimates $\widehat{a}$ and $\widehat{b}$ are obtainable in most cases where the data are available in this form.

When censoring occurs $L(a,b)$ can be modified without difficulty. Zippin and Armitage (1966) investigated the censoring problem in a rework of the Feigl-Zelen model to account for censoring. Analogous to (30.2.2)

$$L_m(a,b) = \prod_{d} (a+bx_i)^{-1} e^{-(a+bx_i)^{-1} t_i} \prod_{s} e^{-(a+bx_i)^{-1} c_i} \tag{30.3.3}$$

where $L_m(a,b)$ is the modified likelihood function that reflects the censored observations. As in the uncensored case, standard iterative procedures are used to obtain the maximum likelihood estimates $\widehat{a}$ and $\widehat{b}$.

Concurrently, to the Feigl-Zelen and Zippin-Armitage regression models, Glasser (1967) developed a regression model for which the functional relationship between the hazard rate and the covariates of interest is

$$\lambda_i = \lambda_0 \exp(\beta x_i) \ , \tag{30.3.4}$$

where λ_0 is baseline hazard function for all patients on study and x_i is the observed covariable or covariable vector for the ith patient in the study. The Glasser model allows somewhat more flexibility than the Feigl-Zelen/Zippin-Armitage model since there is no restriction on β as there is on a and b in the prior model to insure a positive hazard function. It then follows, for the Glasser model,

$$L(\lambda_0, \beta) = \lambda_0^d e^{\beta \sum_d x_i} \exp\left[-\lambda_0 \left\{\sum_d e^{\beta x_i t_i} + \sum_s e^{\beta x_i c_i}\right\}\right] . \tag{30.3.5}$$

As is noted from (30.3.5), the Glasser model already accounts for censoring in that $\sum_d$ represents the sum over all deaths and $\sum_s$ represents the sum over all survivors with t_i and c_i representing the times to death and censoring, respectively, in the study. It is noted, just as with the Feigl-Zelen/Zippin-Armitage model, addition of more than a single covariate presents no theoretical difficulties, albeit the computational problems become more complex.

To conclude this section, the problem of analyzing failure data in which the population is heterogeneous consisting of individuals with some serious, perhaps life-threatening disease such as cancer, as well as normal individuals, i.e., individuals with a normal life expectancy. A complete discussion of this problem can be found in Farewell (1977a,b, 1982) and Pian (1987).

Assume that patients suffering cancer have a failure density $\lambda_1 e^{-\lambda_1 t}$ and normal individuals a failure density $\lambda_2 e^{-\lambda_2 t}$; $t \geq 0$; $\lambda_1 > \lambda_2 > 0$. A binary variable Y identifies the group in which any given individual is a member; $Y = 0$ indicates he/she is healthy; i.e., not suffering from cancer and $Y = 1$ indicates the individual has cancer. It is assumed there is a related covariable vector available for everyone in the study. Furthermore, it is assumed that Y is associated with this vector by means of logistic regression. Thus,

$$P[Y = 1] = e^{\boldsymbol{\beta}'\boldsymbol{x}} / (1 + e^{\boldsymbol{\beta}'\boldsymbol{x}}) \ , \tag{30.3.6}$$

where $\boldsymbol{x}'$ is the vector of covariates and $\boldsymbol{\beta}'$ is the corresponding vector of parameters.

Persons under study may then be categorized into three observable groups: Group 1 - death from cancer, Group 2 - death from other causes, and Group 3 - survival at study's end. It follows, the likelihood function $L(\lambda_1, \lambda_2, \boldsymbol{\beta}')$ for the entire sample in terms of the parameters λ_1, λ_2 and $\boldsymbol{\beta}'$, the vector of covariables is

$$\begin{aligned} &L(\lambda_1, \lambda_2, \boldsymbol{\beta}') \\ &= \frac{\prod_1 \lambda_1 e^{-[(\lambda_1+\lambda_2)-\boldsymbol{\beta}'\boldsymbol{x}]} \prod_2 \lambda_2 e^{-\lambda_2 t}(1 + e^{-[\lambda_1 t - \boldsymbol{\beta}'\boldsymbol{x}]}) \prod_3 e^{-\lambda_2 t}(1 + e^{-[\lambda_1 t - \boldsymbol{\beta}'\boldsymbol{x}]})}{\prod(1 + e^{\boldsymbol{\beta}'\boldsymbol{x}})} \end{aligned} \tag{30.3.7}$$

where $\prod_i$ refers to the product over the ith group, $i = 1, 2, 3$ and $\prod$ refers to the product over all three groups. The maximum likelihood estimator $\widehat{\lambda}_2$ of λ_2 is immediate; $\widehat{\lambda}_2 = n_2/\sum t_j$, where n_2 is the number of individuals dying from causes other than cancer and $\sum t_j$ is the total time on study for persons in all groups. The estimators $\widehat{\lambda}_1$ and $\boldsymbol{\beta}'$ must be obtained iteratively. Properties of the estimators can be found in Farewell (1977a,b, 1982) and Pian (1987).

Finally, it is clear that other commonly used distributions in survival analysis; e.g. the Weibull, gamma and lognormal may be used instead of the exponential in modelling this mixture type situation.

30.4 Planning of Clinical Trials: The Use of the Exponential to Obtain Sample Size Estimates

Sample size determination in clinical studies is an extremely important aspect for at least two reasons, which, at first, may seem to be conflicting with one another. First of all, the sample size should be large enough so as to be able to detect any real difference in the measure (parameter) of interest. On the other hand, since cost is a primary factor in the planning of all clinical trials, the sample size cannot be unduly large. While generally a rather complex problem, many authors have found that determination of sample size can be facilitated by using the exponential distribution as the underlying survival distribution even when the conditions necessary for the exponential distribution to prevail are not met. Papers that have exploited this method include: Rubinstein, Gail, and Santner (1981), Taulbee and Symons (1983), Morgan (1985) and Gross et al. (1987).

In this section, the method is discussed in general, albeit sketchy form. The above referenced papers contain the details. The principal factors that are considered in these articles and, indeed, most articles that deal with sample size determination in clinical trials are: (i) the distribution of the response times for patients in the two treatment groups, (ii) specification of Type I and Type II error levels, (iii) consideration of various censoring possibilities, including patient withdrawal and dropout as well as survival of patients beyond the end of the trial and (iv) the asymptotic normality of the estimators. Besides these factors, Morgan (1985) and Gross et al. (1987) consider patient accrual rates, optimal length of the recruitment period and the follow-up period of the trial wherein these periods are so selected to optimize (minimize) the cost of the trial. Taulbee and Symons (1983) include consideration of a covariable in their sample size determination scheme. Specifically, in the article by Gross et al. (1987) the purpose was to examine the optimal length of a clinical trial on the basis of minimizing its cost under the following assumptions:

(i) Patient failure times are exponentially distributed. The exponential assumption turns out to be rather robust as the power of the test for the equality of mean survival times based on the asymptotic theory of maximum likelihood estimators under the exponential assumption is roughly the same as that of the log-rank statistic under the exponential assumption [see Morgan (1985)].

(ii) Patients are recruited for the clinical trial according to a uniform distribution over a fixed period of time. The trial and recruitment times begin simultaneously, however, the trial is necessarily longer and is determined when the minimum cost with respect to both the recruitment and trial times is found.

(iii) Patients are randomized to the two treatment groups in question, i.e., the control group and the intervention group. Each treatment group has the same number of patients.

(iv) Parameter values are specified for the failure rates under the null and alternate hypotheses and, as is the usual case, Type I and Type II error levels are fixed prior to the trial. A linear cost function (linear as a function of the length of the trial) with known slope and intercept is assumed. Finally, dropout rates are assumed known for both treatment groups.

(v) If T_0 is the length of the recruitment period and $2n$ is the total sample size for the two treatment groups, then the minimum recruitment rate is $(2n/T_0)$.

The costs associated with the clinical trial are patient start-up costs, independent of time, patient per-unit-time costs for patients who are still on the trial as well as those off the trial and per-unit-time costs of the clinical trial not directly related to patient costs; e.g. data processing costs.

The cost function as a function of the length of the clinical trial takes into account the fact that if the trial is of a short duration, the requisite sample size is large, usually so large as to be unrealistic. On the other hand, if a longer trial takes place, then while the sample size becomes much smaller, the sheer length of the trial increases the cost beyond a reasonable figure. Thus, heuristically, as well as mathematically, Gross et al. (1987) are able to find a trial length that is optimum in the sense of minimizing the overall trial cost.

While many articles consider the sample size determination problem from many different aspects; e.g. inclusion of covariables, cost consideration, different types of censoring and different recruitment patterns, there does not appear to be any single article that incorporates all these factors. What is important, however, is that use of the exponential distribution in sample size determination is a robust procedure and hence enjoys its well deserved popularity in this role.

30.5 Bivariate and Multivariate Exponential Distributions in Survival Analysis

Bivariate and multivariate exponential models have been studied by many authors. The earliest of whom was Gumbel (1960). His work was followed by Freund (1961), Marshall and Olkin (1967), Downton (1970), Hawkes (1972), and Block and Basu (1974). In survival analysis, however, the Block-Basu model, based on Freund's model has particular appeal since it considers a two component (organ) system wherein if component 1 (2) fails prior to component 2 (1) an additional strain is placed on the remaining component which manifests itself by incurring a higher failure rate (still constant, however) at the time of the first component failure. Another consideration of multivariate survival models is in the realm of competing risks in which an individual, initially, is at risk of death from a number of distinct causes; e.g. heart disease, lung cancer, kidney disease, accidental death, etc., each of which has its own event time. The individual's time to death in only observable to a specific cause. That is, the failure times for all the other causes are censored by the death of the individual due to the specific cause. The important references for competing risks include Chiang (1968), Moeschberger and David (1971) and Prentice et al. (1978).

In this section, the bivariate exponential model, easily extended to a multivariate model, is introduced using conditional independence. The articles that are pertinent to this discussion are Cantor and Knapp (1985), Knapp, Cantor, and Gross (1986) and Knapp, Gross, and Cantor (1986). Assume (X, Y, Z) is a triple of observations for which X is the survival time of the first component, Y is the survival time of the second component and Z is an unobservable "susceptibility" factor common to X and Y that may vary from individual to individual. Furthermore, it is assumed that for a fixed z,

$$f_1(x|z) = (\lambda_1/z)e^{-\lambda_1 x/z} , \quad \lambda_1 > 0, \ x \geq 0, \ z > 0 ,$$
$$f_2(y|z) = (\lambda_2/z)e^{-\lambda_2 y/z} , \quad \lambda_2 > 0, \ y \geq 0, \ z > 0$$

and conditional on z, X and Y are independent. No assumption is made concerning the distribution of Z except that $Z > 0$ and $E[Z^2] < \infty$. Unconditionally,

$$f(x, y) = \int_{z \in D_z} (\lambda_1 \lambda_2/z^2)\, e^{-(\lambda_1 x + \lambda_2 y)/z}\, dH(z) \tag{30.5.1}$$

where $H(z)$ is the cdf of Z and D_z is the domain over which Z is defined. Clearly, the marginal density functions for X and Y are

$$f_1(x) = \int_{z \in D_z} (\lambda_1/z)\, e^{-\lambda_1 x/z}\, dH(z) \tag{30.5.2}$$

and

$$f_2(y) = \int_{z \in D_z} (\lambda_2/z)\, e^{-\lambda_2 y/z}\, dH(z) , \tag{30.5.3}$$

respectively.

Evidently, X and Y are not independent and, in fact, it is not difficult to show that the correlation between X and Y, ρ (say), is given as

$$\rho = \{2 + [cv(Z)]^{-2}\}^{-1} , \tag{30.5.4}$$

where $cv(Z)$ is the coefficient of variation of Z. Thus, $0 < \rho < 0.5$.

Using the joint density of X and Y given by (30.5.1) and performing the transformation $V = Y/X$, the density of V is

$$g(v) = \xi/(1 + \xi v)^2 , \tag{30.5.5}$$

where $v > 0$, and $\xi = \lambda_2/\lambda_1 > 0$. Thus, a test of the hypothesis $H_0 : \lambda_1 = \lambda_2$ against the alternative $H_1 : \lambda_1 \neq \lambda_2$ is equivalent to testing $H_0 : \xi = 1$ versus $H_1 : \xi \neq 1$ in (30.5.5). Knapp, Gross, and Cantor (1986) consider this testing problem under the following scenario: (X, Y) are both uncensored, X is censored but Y is uncensored and Y is censored but X is uncensored. An example of how this testing problem may arise in a clinical setting is to consider the problem of testing time to relief of headache pain for two different analgesics A and B within the same individual. Half the patients in the study receive A prior to B and the other half receive B prior to A. Enough time is allowed between receipt of the analgesics so there is no carry-over effect. If the times to relief are assumed to have the joint exponential density (30.5.1), then an application of (30.5.5) under the scenario described can be made to determine which, if either of the two analgesics, has the shorter relief time. If the sample size is large enough, 25 patients (say), then the likelihood ratio procedure can be used to test $H_0 : \xi = 1$ versus $H_1 : \xi \neq 1$. Again, the details are presented in Knapp, Gross, and Cantor (1986).

To conclude this section, I mention the paired survival model introduced by Gross and Lam (1981). This model was developed using the correlated exponential model by Block and Basu (1974). The idea, as with the Knapp studies, was to develop a test of the hypothesis of equality of the mean survival times for correlated exponential random variables according to the Block-Basu structure which also includes a parameter, actually a nuisance parameter, that describes the correlation. The model put forth by Gross and Lam (1981) was the starting point for the work by Knapp and her colleagues whose results substantially improved the earlier work.

References

Bartholomew, D.J. (1957). A problem in life testing, *Journal of the American Statistical Association*, **52**, 350-355.

Block, H.W. and Basu, A.P. (1974). A continuous bivariate exponential extension, *Journal of the American Statistical Association*, **69**, 1031-1037.

Byar, D.R., Huse, R., and Bailar, J.C., III (1974). An exponential model relating censored survival data and concomitant information for prostatic cancer patients, *Journal of the National Cancer Institute*, **52**, 321-326.

Cantor, A.B. and Knapp, R.G. (1985). A test of the equality of survival distributions based on paired observations from conditionally independent exponential distributions, *IEEE Transactions on Reliability*, **R-34**, 342-346.

Chiang, C.L. (1968). *Introduction to Stochastic Processes in Biostatistics*, John Wiley & Sons, New York.

Cox, D.R. (1953). Some simple tests for Poisson variates, *Biometrika*, **40**, 354-360.

Cox, D.R. (1959). The analysis of exponentially distributed life-times with two types of failure, *Journal of the Royal Statistical Society, Series B*, **21**, 411-421.

Cox, D.R. (1964). Some applications of exponential ordered scores, *Journal of the Royal Statistical Society, Series B*, **26**, 103-110.

Cox, D.R. (1972). Regression models and life tables (with discussion), *Journal of the Royal Statistical Society, Series B*, **34**, 187-220.

Craig, J.B. (1988). Sample size determination in clinical trials considering nonuniform patient entry, loss to follow-up, noncompliance and cost optimization, *Unpublished Ph.D. Dissertation*, Medical University of South Carolina.

Downton, F. (1970). Bivariate exponential distributions in reliability theory, *Journal of the Royal Statistical Society, Series B*, **32** 408-417.

Epstein, B. and Sobel, M. (1953). Life testing, *Journal of the American Statistical Association*, **48**, 486-502.

Farewell, V.T. (1977a). A model for a binary variable with time-censored observations, *Biometrika*, **64**, 43-46.

Farewell, V.T. (1977b). The combined effect of breast cancer risk factors, *Cancer*, **40**, 931-936.

Farewell, V.T. (1982). The use of mixture models for the analysis of survival data with long-term survivors, *Biometrics*, **38**, 1041-1046.

Feigl, P. and Zelen, M. (1965). Estimation of exponential survival probabilities with concomitant information, *Biometrics*, **21**, 826-838.

Freund, J.E. (1961). A bivariate extension of the exponential distribution, *Journal of the American Statistical Association*, **56**, 971-977.

Gehan, E.A. and Siddiqui, M.M. (1973). Simple regression methods for survival time studies, *Journal of the American Statistical Association*, **68**, 848-856.

Glasser, M. (1967). Exponential survival with covariance, *Journal of the American Statistical Association*, **62**, 561-568.

Gross, A.J., Clark, V.A., and Liu, V. (1971). Estimation of survival parameters when one of two organs must function for survival, *Biometrics*, **27**, 369-377.

Gross, A.J., Hunt, H.H., Cantor, A.B., and Clark, B.C. (1987). Sample size determination in clinical trials with an emphasis on exponentially distributed responses, *Biometrics*, **43**, 875-883.

Gross, A.J. and Lam, C.F. (1981). Paired observations from a survival distribution, *Biometrics*, **37**, 505-511.

Gumbel, E.J. (1960). Bivariate exponential distributions, *Journal of the American Statistical Association*, **55**, 698-707.

Harris, T.E., Meier, P., and Tukey, J.W. (1950). The timing of the distribution of events between observations, *Human Biology*, **22**, 249-270.

Hawkes, A.G. (1972). A bivariate exponential distribution with applications to reliability, *Journal of the Royal Statistical Society, Series B*, **34**, 129-131.

Johnson, N.L., Kotz, S., and Balakrishnan, N. (1994). *Continuous Univariate Distributions - Vol. 1*, John Wiley & Sons, New York.

Knapp, R.G., Cantor, A.B., and Gross, A.J. (1986). Estimators of the ratio of means of paired survival data, *Communications in Statistics - Simulation and Computation*, **15**, 85-100.

Knapp, R.G., Gross, A.J., and Cantor, A.G. (1986). A likelihood ratio test of the equality of paired survival data with censoring, *Biometrical Journal*, **28**, 665-672.

Lachin, J.M. and Foulkes, M.A. (1986). Evaluation of sample size and power for analyses of survival with allowance for nonuniform patient entry, losses to follow-up, noncompliance, and stratification, *Biometrics*, **42**, 507-519.

Lee, M-L.T. and Gross, A.J. (1989). Properties of conditionally independent generalized gamma distributions, *Probability in the Engineering and Informational Sciences*, **3**, 289-297.

Lee, M-L.T. and Gross, A.J. (1991). Lifetime distributions under unknown environment, *Journal of Statistical Planning and Inference*, **29**, 137-143.

Marshall, A.W. and Olkin, I. (1967). A multivariate exponential distribution, *Journal of the American Statistical Association*, **62**, 30-44.

Mendenhall, W. and Hader, R.J. (1958). Estimation of parameters of mixed exponentially distributed failure time distributions from censored life test data, *Biometrika*, **45**, 504-520.

Moeschberger, M.L. and David, H.A. (1971). Life tests under competing causes of failure and the theory of competing risks, *Biometrics*, **27**, 909-933.

Morgan, T.M. (1985). Planning the duration of accrual and follow-up for clinical trials, *Journal of Chronic Diseases*, **38**, 1009-1018.

Pasternack, B.S. (1972). Sample sizes for clinical trials designed for patient accrual by cohorts, *Journal of Chronic Diseases*, **25**, 673-681.

Pasternack, B.S. and Gilbert, M.S. (1971). Planning the duration of long-term survival time studies designed for accrual by cohorts, *Journal of Chronic Diseases*, **24**, 681-700.

Pian, L.-P. (1987). Application of a mixed model to survival data with long-term survivors, *Unpublished Ph.D. Dissertation*, Medical University of South Carolina.

Prentice, R.L. (1973). Exponential survivals with censoring and explanatory variables, *Biometrika*, **60**, 279-288.

Prentice, R.L., Kalbfleisch, J.D., Peterson, A.W., Flournoy, N., Farewell, V.T., and Breslow, N.E. (1978). The analysis of failure times in the presence of competing risks, *Biometrics*, **34**, 541-554.

Rubinstein, L.V., Gail, M.H., and Santner, T.J. (1981). Planning the duration of a comparative clinical trial with loss-to-follow-up and a period of continued observation, *Journal of Chronic Diseases*, **20**, 230-239.

Taulbee, J.D. and Symons, M.J. (1983). Sample size and duration for cohort studies of survival time with covariables, *Biometrics*, **39**, 351-360.

Voit, E.O. (1992). The S-distribution: A tool for approximation and classification of univariate, unimodal probability distributions, *Biometrical Journal*, **34**, 855-878.

Zelen, M. (1966). Applications of exponential models to problems in cancer research (with discussion), *Journal of the Royal Statistical Society, Series A*, **129**, 368-398.

Zippin, C. and Armitage, P. (1966). Use of concomitant variables and incomplete survival information in the estimation of an exponential survival parameter, *Biometrics*, **22**, 665-672.

CHAPTER 31

Applications in Queueing Theory

J. George Shanthikumar[1]

31.1 Introduction

Most of the early developments in queueing theory are based on queueing models that assume exponential distributions for the *i.i.d.* inter-arrival times and/or for the *i.i.d.* service times of customers. The purpose of this paper is to discuss the important role that exponential distribution has played and continues to play in the development of queueing theory. If a non-negative random variable X has an exponential distribution, then it is well known that $P\{X - x > t | X > x\} = P\{X > t\}$, $t > 0$; $x > 0$. This is the mostly used memoryless (the big M) property of exponential distribution; see Chapter 1. Thus assuming a Poisson arrival process to a queueing system, one can often formulate a Markov chain model (either in continuous time or in discrete time by appropriately choosing an embedded time sequence) for the queueing system. It also allows one to apply the PASTA (Poisson arrivals sees time average) principle. Combining all these makes the analysis of such queueing systems rather simple. It is therefore no wonder that several hundreds of papers on such queueing systems have been written [for example see Takagi (1991), for a comprehensive treatment of such a class of queueing systems]. It is fruitless to summarize these developments here. Rather, we will demonstrate the use of exponential distributions by considering, as an example, a single server queueing system and illustrating ways in which the queueing theory has evolved as a result of it. Particularly we should mention the development of

(1) the theory of queueing networks (as a consequence of the detailed study of single stage queueing systems with Poisson arrivals and exponentially distributed *i.i.d.* service times [see, for example, Kelly (1979) and Walrand (1988)],

(2) the matrix analytic method for queueing systems with continuous time or embedded discrete time Markov chain models [see, for example, Neuts (1981), (1989)],

and

[1]University of California, Berkeley, California

(3) stochastic bounds using queueing systems with exponential inputs and outputs [see, for example, Buzacott and Shanthikumar (1993), Kingman (1970), and Stoyan (1983)].

We will discuss the use of exponential distributions as

(A) **inputs** (that is the use of exponential distribution for the inter-arrival times and/or for the service times) and as

(B) **outputs** (such as the time spent in the system by an arbitrary customer and inter-departure times).

To be more concrete in summarizing the value of exponential distribution in queueing theory, we will need to first describe the single server queueing system that will serve us as the test bed.

Consider a single sever queueing system $GI/GI/1$ where customers arrive according to a renewal arrival process with inter-arrival times $\{A_n, n = 1, 2, \ldots\}$. Let A be the generic inter-arrival time with mean $1/\lambda$ and squared coefficient of variation C_A^2. The service times $\{S_n, n = 0, 1, 2, 3, \ldots\}$ of the customers form an *i.i.d.* sequence of non-negative random variables independent of the arrival process. Let S be the generic service time with mean $1/\mu$ and squared coefficient of variation C_S^2. Customers are served on a first-come first served (FCFS) basis. We will assume that $\lambda < \mu$ (this is to ensure that the queueing system is stable and that stationary distributions exist for the system performance measures such as the number of customers in the system and the time spent in the system by an arbitrary customer). Let $N(t)$ be the number of customers in the system at time t. In general $\{N(t), t \geq 0\}$ is not a continuous time Markov chain. So in order to analyze $\{N(t), t \geq 0\}$ one uses supplementary variables, e.g., representing the time since the last arrival $(A_a(t))$ and the time since the last service initiation $(A_s(t))$ at time t. Then $\{(N(t), A_a(t), A_s(t)), t \geq 0\}$ is a Markov process. Yet the analysis of this process if often very tedious. Also, all attempts to extend this analysis to meaningful queueing networks have been futile. In Section 31.3 we will see that when the inter-arrival times and/or service times have exponential distributions this analysis can be substantially simplified and that it can be easily extended to queueing networks.

Let W_n be the waiting time of the nth customer in the system before he or she receives service. Then $T_n = W_n + S_n$ is the time spent by the nth customer in the system. $\{W_n, n = 0, 1, 2, \ldots\}$ (and $\{T_n, n = 0, 1, 2, \ldots\}$) is a Markov process. Specifically observe that we have the Lindley's recursion

$$W_n = \max\{W_{n-1} + S_{n-1} - A_n, 0\}, n = 1, 2, \ldots \tag{31.1.1}$$

and

$$T_n = \max\{T_{n-1} - A_n, 0\} + S_n, n = 1, 2, \ldots \tag{31.1.2}$$

Though $\{W_n, n = 0, 1, 2, \ldots\}$ (and $\{T_n, n = 0, 1, 2, \ldots\}$) is a Markov process, it is, in general not easy to obtain the stationary distribution of this process. In Section 31.3 we will see that when the inter-arrival times have an exponential distribution this analysis can be substantially simplified. In the same section we will see that when the service times have an exponential distribution the stationary distribution of the time spent in the system will have an exponential distribution. The rest of this chapter is organized as follows:

In Section 31.2, we present some general principles in queueing theory that play a crucial role in simplifying the analysis of queueing systems. Particularly we describe *Discrete State Level Crossing Analysis, PASTA; Poisson Arrivals Sees Time Averages, Distributional form of Little's* Law and *Time Reversibility of Markov chains.*

In Section 31.3 we demonstrate the use of the exponentiality assumption for the inputs by developing (i) continuous time Markov chain models, or (ii) embedded discrete time Markov chain models for the single server queueing system and apply Level crossing analysis, PASTA, Little's Law and Reversibility to simplify the analysis of these queueing systems.

Also we will show that when the inputs are exponentially distributed, the outputs (performance measures) such as the time spent in the system and inter-departure times may have exponential distributions.

In Section 31.4 we develop bounds for the general case using exponential inputs. Also, we will develop exponential bounds for the outputs in the general case. For these, we will use the external and internal stochastic monotonicity of the associated queueing processes.

Whenever needed we will guide the reader to appropriate references where further extensions of the basic ideas of using exponentiality assumption have been exploited. These extensions may be for complex queueing systems (such as queueing networks) or for queueing systems with generalized exponential distributions (such as phase type and generalized phase type distributions).

31.2 Some Useful General Principles in Queueing Theory

In this section we will present some general principles in queueing theory that plays a crucial role in simplifying the analysis of queueing systems. Particularly we will discuss (1) Discrete state level crossing analysis, (2) (PASTA) Poisson arrivals see time average (3) Distributional from of Little's law and (4) Time reversibility of Markov chains. We shall consider (1), (2) and (3) for a restricted class of queueing systems (even though extensions of these results are valid for more general queueing systems). Specifically, consider a stable queueing system where customers arrive one at a time at times τ_n, $n = 1, 2, 3, \ldots$, with rate λ. They leave also one at a time at times δ_n, $n = 1, 2, 3, \ldots$ There may be other types of customers that may be arriving at these systems and receive service and leave. Their presence or interaction does not matter with respect to the results we are to present below.

Let $N(t)$ be the number of customers in the system at time t and let T_n be the time spent in the system by the nth customer. Define

$$\begin{aligned}
\pi_k &= \lim_{t\to\infty} P\{N(t) = k\},\ k = 0, 1, 2, \ldots \\
a_k &= \lim_{n\to\infty} P\{N(\tau_n^-) = k\},\ k = 0, 1, 2, \ldots \\
d_k &= \lim_{n\to\infty} P\{N(\delta_n^+) = k\},\ k = 0, 1, 2, \ldots \qquad (31.2.1)
\end{aligned}$$

and

$$N \stackrel{d}{=} \lim_{t\to\infty} N(t);\ N^a \stackrel{d}{=} \lim_{n\to\infty} N(\tau_n^-);\ N^d \stackrel{d}{=} \lim_{n\to\infty} N(\delta_n^+);\ T \stackrel{d}{=} \lim_{n\to\infty} T_n\ . \tag{31.2.2}$$

Here $\stackrel{d}{=}$ stands for equality in distribution.

31.2.1 Discrete State Level Crossing Analysis

The rate of up-crossings of the process $\{N(t),\, t \geq 0\}$ from level n is λa_n and the rate of down-crossing to level n is λd_n. Equating these rates one has

$$a_n = d_n,\ n = 0, 1, 2, \dots . \tag{31.2.3}$$

That is $N^a \stackrel{d}{=} N^d$. If the customer departure rate when there are n customers in the system is γ_n then we have by equating the rates of up- and down-crossings,

$$\lambda a_n = \gamma_{n+1}\pi_{n+1}, n = 0, 1, 2, \dots \tag{31.2.4}$$

These and other results related to discrete state level crossings can be found in Shanthikumar and Chandra (1982).

31.2.2 PASTA, Poisson Arrivals See Time Averages

Suppose the customers arrive at the system according to a Poisson process and that the evolution of the queueing system up to any time t is independent of the future arrivals after time t. Then

$$a_n = \pi_n,\ n = 0, 1, 2, 3, \dots \tag{31.2.5}$$

That is $N^a \stackrel{d}{=} N$. This result is true for any state description (such as the workload) of the queueing system [see Wolff (1989)].

31.2.3 Distributional Form of Little's Law

Little's law states that

$$E[N] = \lambda E[T]. \tag{31.2.6}$$

[see, for example, Whitt (1991)]. Suppose the customers arrive at the system according to a Poisson process and that they depart the queueing system in the order they arrived. It is then trivial to observe that the number of customers left behind (in the system) by the nth customer are those arrived while he or she was in the system. If the progress of a customer through the system is unaffected by the future arrivals (of the types of customers that we are concerned with), then

$$N^d \stackrel{d}{=} Z(T)\ , \tag{31.2.7}$$

where $\{Z(t),\, t \geq 0\}$ is a Poisson process with rate λ independent of T. Then with discrete state level crossing analysis (that is with $N^a \stackrel{d}{=} N^d$) and with PASTA (that is with $N \stackrel{d}{=} N^a$), we have

$$N \stackrel{d}{=} Z(T) \tag{31.2.8}$$

This is the distributional form of Little's law [see, for example, Whitt (1991)]. Though this is trivial to derive it can be used to simplify the analysis of queueing systems with Poisson arrival process (see Section 31.3).

31.2.4 Time Reversibility of Markov Chains

Let $\{X(t), t \in \mathcal{R}\}$ be a stationary continuous time Markov chain on $\mathcal{S}$ with transition rates $q_{i,j}$, $i, j \in \mathcal{S}$. Here $\mathcal{R} = (-\infty, \infty)$. Then $\{X(t), t \in \mathcal{R}\}$ is time reversible (that is, $\{X(t), t \in \mathcal{R}\} =^{\ulcorner} \{\mathcal{X}(-\sqcup), \sqcup \in \mathcal{R}\}$) iff $P\{X(t) = i\}q_{i,j} = P\{X(t) = j\}q_{j,i}$, $i, j \in \mathcal{S}$ [for example, see Kelly (1979), Ross (1983), and Walrand (1988)].

31.3 Exponential Inputs: Simplified Analysis and Exponential Outputs

In this section, we consider the single server queueing system described in Section 31.1. We will see how using exponential distribution as inputs will (i) simplify the analysis of the single server queueing system and may (ii) lead to exponential outputs. Particularly we will see that continuous time Markov chain or embedded discrete time Markov chain models can be formulated for the number of customers in the system (that is for $\{N(t), t \geq 0\}$) and can be analyzed easily. We will also see that in some cases, the stationary time spent in the system (T) and the stationary inter-departure time (D) will have exponential distributions.

31.3.1 $M/M/1$ Queueing system

Suppose the generic inter-arrival time A and the service time S have exponential distributions. That is,

$$\begin{aligned} P\{A > t\} &= \exp\{-\lambda t\}, t \geq 0 , \\ P\{S > t\} &= \exp\{-\mu t\}, t \geq 0 . \end{aligned} \tag{31.3.1}$$

Number of customers in the system (and discrete state level crossing analysis): Let $N(t)$ be the number of customers in the system at time t. Then $\{N(t), t \geq 0\}$ is a continuous time Markov chain on $\mathcal{N}_+ := \{\imath, \infty, \in, \ni, \ldots\}$ (indeed it is a spatially homogeneous birth-death process). Then by discrete state level crossing analysis and PASTA we have [see Eqs. (31.2.3) and (31.2.5)],

$$d_n = a_n = \pi_n, \qquad n = 0, 1, 2, 3, \ldots \tag{31.3.2}$$

That is $N^d \stackrel{d}{=} N^a \stackrel{d}{=} N$. Also we have, by discrete state level crossing analysis [Eq. (31.2.4)],

$$\lambda \pi_n = \gamma_{n+1}\pi_{n+1}, \qquad n = 0, 1, 2, 3, \ldots \tag{31.3.3}$$

By the memoryless property of the service time, the (state-dependent) customer departure rate is $\gamma_n = \mu$, $n = 1, 2, 3, \ldots$ Therefore, we have (from the above equation),

$$\lambda \pi_n = \mu \pi_{n+1}, \qquad n = 0, 1, 2, \ldots . \tag{31.3.4}$$

Solving these equations with the boundary condition, $\sum_{n=0}^{\infty} \pi_n = 1$, we have

$$\pi_n = (1-\rho)\rho^n, \qquad n = 0, 1, 2, 3, \ldots \tag{31.3.5}$$

Here $\rho = \lambda/\mu < 1$ is the server utilization. Then

$$E[N] = \frac{\rho}{1-\rho} . \tag{31.3.6}$$

Inter-departure times (and the memoryless property of the inter-arrival times): Let D_n, $n = 1, 2, 3, \ldots$, be the sequence of inter-departure times of the customers from the system and D be the generic stationary inter-departure time. Then one has the following recursion for the inter-departure times:

$$D_n = S_n + I_n, \qquad n = 1, 2, 3, \ldots \tag{31.3.7}$$

where

$$I_n = \max\{A_n - T_{n-1}, 0\}, \qquad n = 1, 2, 3, \ldots \tag{31.3.8}$$

is the idle time of the server between the service completion of the $(n-1)$th customer and the service initiation of the nth customer. The idle time is non-negative iff the $(n-1)$th customer on its departure leaves behind an empty system. Now using the memoryless property of the exponential inter-arrival time distribution, one has [from Eq. (31.3.7)]

$$D_n \stackrel{d}{=} S_n + I\{N_{n-1}^d = 0\}\hat{A}_n, \qquad n = 1, 2, 3, \ldots, \tag{31.3.9}$$

where $\{\hat{A}_n, \; n = 1, 2, 3, \ldots\}$ is a sequence of *i.i.d.* exponential random variables, independent of the queueing process. Then from Eqs. (31.3.2), (31.3.5) and (31.3.9) it is easily verified that

$$\begin{aligned} P\{D > t\} &= \rho P\{S_n > t\} + (1-\rho)P\{S_n + A_n > t\} \\ &= \exp\{-\lambda t\}, \qquad t > 0 . \end{aligned} \tag{31.3.10}$$

That is we see that the stationary inter-departure time from an $M/M/1$ queueing system has an exponential distribution (an exponential output) with mean $1/\lambda$. It is then natural to seek whether the stationary departure process is itself a Poisson process. We shall look into this next.

Departure process (and time reversibility): It is easily seen from (31.3.4) that the stationary version of $\{N(t), \, t \geq 0\}$ is time reversible (see Section 31.2.4). We will use this observation to study the characteristic of the departure process from the $M/M/1$ queueing system. Suppose $\{N(t), \, t \in \mathcal{R}\}$ is a stationary version of the number of customers in the $M/M/1$ queueing system and let $A(t)$ and $D(t)$ be the associated arrival and departure counting processes. That is $A(t)$ is the number of customers arrived and $D(t)$ is the number of customers departed from the system during $(0, t)$ for $t > 0$ [and during $(t, 0)$ for $t < 0$]. Let $\{\hat{N}(t), \, t \in \mathcal{R}\}$ be the time reversed version of the number of customers in the $M/M/1$ queueing system. That is $\hat{N}(t) = N(-t)$, $t \in \mathcal{R}$. Also let $\hat{A}(t)$

and $\hat{D}(t)$ be the associated arrival and departure counting processes of $\{\hat{N}(t),\, t \in \mathcal{R}\}$. Since an arrival in $\{\hat{N}(t),\, t \in \mathcal{R}\}$ corresponds to a departure in $\{N(t),\, t \in \mathcal{R}\}$ it can be seen that

$$\{D(t),\, t \in \mathcal{R}\} = \{\hat{A}(-t),\, t \in \mathcal{R}\}. \tag{31.3.11}$$

Since $\{N(t),\, t \in \mathcal{R}\}$ is time reversible we have

$$\{N(t),\, t \in \mathcal{R}\} \stackrel{d}{=} \{N(-t),\, t \in \mathcal{R}\} = \{\hat{N}(t),\, t \in \mathcal{R}\}. \tag{31.3.12}$$

Hence

$$\{A(t),\, t \in \mathcal{R}\} \stackrel{d}{=} \{\hat{A}(t),\, t \in \mathcal{R}\}. \tag{31.3.13}$$

Now combining Eqs. (31.3.11) and (31.3.13) one finds that

$$\{D(t),\, t \in \mathcal{R}\} \stackrel{d}{=} \{A(-t),\, t \in \mathcal{R}\} \stackrel{d}{=} \{A(t),\, t \in \mathcal{R}\}, \tag{31.3.14}$$

where the last equality follows from the time reversibility of Poisson processes. That is the stationary departure (counting) process from an $M/M/1$ queueing system is a Poisson process. Indeed this is the case even if the single server is replaced by a state dependent service rate. Since the departure process is Poisson, feeding this into another queueing system with exponential server will lead to Poisson departures. It is then easy to see how the exponentiality for the arrival and service times allows one to extend the analysis to queueing networks. For such extensions see, for example, Kelly (1979) and Walrand (1988).

Time spent in the system (and the memoryless property of service times): Now we will consider the time spent in the system. Observe that if a customer on its arrival sees n other customers in the system, then it will have to spend a time equal in distribution to the total of the service times of $n+1$ customers (note that here we are using the fact that the remaining service time of the customer currently in service at the time of our customer's arrival, has the same distribution as the original service time – which is exponential with mean $1/\mu$ – that is, we are using the memoryless property of the exponential distribution). Hence the distribution of the time spent in the system can be obtained by observing that

$$T \stackrel{d}{=} \sum_{k=1}^{N+1} S_k \,. \tag{31.3.15}$$

Here we used PASTA (that is, $N^a \stackrel{d}{=} N$). Simple calculation reveals that [see Eq. (31.3.5)]

$$\begin{aligned} P\{T > t\} &= E\left[P\left\{\sum_{k=1}^{N+1} S_k > t | N\right\}\right] \\ &= E\left[\sum_{k=0}^{N} \exp\{-\mu t\}\frac{(\mu t)^k}{k!}\right] \\ &= \exp\{-(\mu-\lambda)t\}, \qquad t \geq 0\,. \end{aligned} \tag{31.3.16}$$

We see that the total time spent in the system has an exponential distribution (that is an exponential output – later in this section we will see that such a result can be found for the $GI/M/1$ queue as well).

The Markovian nature of the process $\{N(t), t \geq 0\}$ and the above analysis can be extended to queueing systems even when the inter-arrival time distribution is of the phase (PH) type [see Neuts (1981)] and the service time distribution is of the generalized phase (GPH) type [see Shanthikumar (1985)]. That is, the above analysis form the basis for the analysis of the $PH/GPH/1$ queueing systems.

31.3.2 $M/GI/1$ Queueing system

Suppose the generic inter-arrival time A has an exponential distribution and the generic service time has a general distribution function F_S with a Laplace-Stieltjes transform $\tilde{F}_S$. That is,

$$\begin{aligned} P\{A > t\} &= \exp\{-\lambda t\}, \qquad t \geq 0 , \\ P\{S > t\} &= 1 - F_S(t) = \bar{F}(t), \qquad t \geq 0 . \end{aligned} \tag{31.3.17}$$

Number of customers in the system (and embedded Markov chain): In general $\{N(t), t \geq 0\}$ is not a continuous time Markov chain. However if we define $N_n^d = N(\delta_n^+)$ we see that $\{N_n^d, n = 0, 1, 2, 3, \ldots\}$ is an imbedded discrete time Markov chain on $\mathcal{N}_+$. (Recall that δ_n is the departure time of the nth customer). Then by discrete state level crossing analysis and PASTA we have [see Eqs. (31.2.3) and (31.2.5)],

$$d_n = a_n = \pi_n, \qquad n = 0, 1, 2, 3, \ldots \tag{31.3.18}$$

That is $N^d \stackrel{d}{=} N^a \stackrel{d}{=} N$. Let Z_n be the number of customers arrived during the service time S_n of the nth customer. Then $\{Z_n \mid S_n = x\}$ is a Poisson random variable with mean λx. Hence

$$E[z^{Z_n}] = E[E[z^{Z_n}|S_n]] = E[\exp\{-\lambda(1-z)S_n\}] = \tilde{F}_S(\lambda(1-z)) . \tag{31.3.19}$$

With this definition we have,

$$N_n^d = \max\{N_{n-1}^d - 1, 0\} + Z_n, \qquad n = 1, 2, 3, \ldots . \tag{31.3.20}$$

Taking the z-transform on both sides of the above equation and taking the limit as $n \to \infty$ we get [after noting that $d_0 = \pi_0$, Eq. (31.3.18)],

$$E[z^{N^d}] = \{z^{-1}E[z^{N^d}] + (1 - z^{-1})\pi_0\}E[z^{Z_n}] . \tag{31.3.21}$$

Solving for $E[z^{N^d}]$ (with the normalizing condition $E[z^{N^d}]|_{z=1} = 1$) one finds that,

$$E[z^{N^d}] = \frac{(1-\rho)(z-1)\tilde{F}_S(\lambda(1-z))}{z - \tilde{F}_S(\lambda(1-z))} . \tag{31.3.22}$$

Here $\rho = \lambda/\mu < 1$. Then by PASTA [see Eq. (31.3.18)] we have $N \stackrel{d}{=} N^d$. Hence,

$$E[z^N] = \frac{(1-\rho)(z-1)\tilde{F}_S(\lambda(1-z))}{z - \tilde{F}_S(\lambda(1-z))} \tag{31.3.23}$$

from which we find that the server idle probability (π_0) and the mean number of customers in the system ($E[N]$) are

$$\pi_0 = E[z^N]|_{z=0} = 1 - \rho \ . \tag{31.3.24}$$

and

$$\begin{aligned} E[N] &= \frac{d}{dz} E[z^N]|_{z=1} \\ &= \left(\frac{1 + C_S^2}{2}\right)\left(\frac{\rho^2}{1-\rho}\right) + \rho \ . \end{aligned} \tag{31.3.25}$$

Note that the server idleness probability is independent of the service time distribution [compare Eqs. (31.3.5) and (31.3.24)]. When $C_S^2 = 1$ (which is the case for exponentially distributed service times), the result (31.3.25) coincides with (31.3.6).

The embedded discrete time Markovian nature of this process and the above analysis can be extended to queueing systems even when the inter-arrival time distribution is of the phase type [that is for the $PH/GI/1$ queueing systems, see Neuts (1989)]. Because of the simplicity in analysis that arise with the assumption of Poisson arrival process (that is, exponentially distributed inter-arrival times), there has been several hundreds of variants of $M/GI/1$ queues that have been proposed (and continue to be proposed) in the queueing literature; see, for example, Takagi (1991) for an extensive survey of the results available for such variants.

Inter-departure times (and the memoryless property of the inter-arrival times): Let D_n, $n = 1, 2, 3, \ldots$, be the sequence of inter-departure times of the customers from the system and D be the generic stationary inter-departure time. Then as before one has [see Eq. (31.3.9)]

$$D_n \stackrel{d}{=} S_n + I\{N_{n-1}^d = 0\}\hat{A}_n, \ n = 1, 2, 3, \ldots, \tag{31.3.26}$$

where $\{\hat{A}_n, \ n = 1, 2, 3, \ldots\}$ is a sequence of *i.i.d.* exponential random variables, independent of the queueing process. Then from (31.3.18), (31.3.24) and (31.3.26) it is easily verified that

$$\begin{aligned} E[\exp\{-sD\}] &= \rho E[\exp\{-sS_n\}] + (1-\rho)E[\exp\{-s(S_n + \hat{A}_n)\}] \\ &= \left(\rho + (1-\rho)\frac{\lambda}{s+\lambda}\right)\tilde{F}_S(s) \\ &= \left(\frac{\lambda}{s+\lambda}\right)\left(\frac{\tilde{F}_S(s)}{\frac{\mu}{s+\mu}}\right) \ . \end{aligned} \tag{31.3.27}$$

Hence the mean and squared coefficient of variation (C_D^2) of the inter-departure times are

$$\begin{aligned} E[D] &= -\frac{d}{ds} \ E[\exp\{-sD\}]|_{s=0} = 1/\lambda \\ C_D^2 &= \frac{\frac{d^2}{ds^2} E[\exp\{-sD\}]|_{s=0}}{E[D]^2} - 1 = 1 - \rho^2 + \rho^2 C_S^2 \ . \end{aligned} \tag{31.3.28}$$

Now suppose we wish this output to be exponential. That is we wish to have $E[\exp\{-sD\}] = \lambda/(s+\lambda)$. Feeding this in (31.3.28) we find that $\tilde{F}_S(s) = \mu/(s+\mu)$. That is, the stationary inter-departure time in the $M/GI/1$ queueing system has an exponential distribution iff the service time has an exponential distribution. The departure process from the $M/GI/1$ queueing system is in general not even a renewal process!

Time spent in the system (and the distributional form of Little's law): To find the waiting time statistics, observe that we cannot use the argument used for the $M/M/1$ queue (since the service times no longer have the memoryless property). Instead we will utilize the distributional form of Little's law. From it [see Eq. (31.2.8)] we have,

$$E[z^N] = E[z^{Z(T)}] = E[\exp\{-\lambda(1-z)T\}] \ . \tag{31.3.29}$$

Then substituting (31.3.23) in (31.3.29) and replacing z by $1 - s/\lambda$ we get

$$E[\exp\{-sT\}] = \frac{(1-\rho)s\tilde{F}_S(s)}{s - \lambda + \lambda\tilde{F}_S(s)} \tag{31.3.30}$$

from which we find that the mean time spent in the system is

$$\begin{aligned} E[T] &= -\frac{d}{ds}E[\exp\{-sT\}]|_{s=0} \\ &= \left(\frac{1+C_S^2}{2}\right)\left(\frac{\rho^2}{\lambda(1-\rho)}\right) + \frac{1}{\mu} \ . \end{aligned} \tag{31.3.31}$$

We may also use the Lindley's recursion [Eq. (31.1.1)] to obtain the above result directly.

31.3.3 $GI/M/1$ Queueing system

Suppose the generic inter-arrival time A has a general distribution function F_A with a Laplace-Stieltjes transform $\tilde{F}_A$ and the generic service time has an exponential distribution function. That is,

$$\begin{aligned} P\{A > t\} &= 1 - F_A(t) = \bar{F}_A(t), \qquad t \geq 0 \ , \\ P\{S > t\} &= \exp\{-\mu t\}, \qquad t \geq 0 \ . \end{aligned} \tag{31.3.32}$$

Number of customers in the system (and embedded Markov chain) : In general $\{N(t), t \geq 0\}$ is not a continuous time Markov chain. However if we define $N_n^a = N(\tau_n^-)$ we see that $\{N_n^a, n = 0,1,2,3,\ldots\}$ is an embedded discrete time Markov chain on $\mathcal{N}_+$. (Recall that τ_n is the arrival time of the nth customer). By discrete state level crossing analysis we have [see Eqs. (31.2.3) and (31.2.4)],

$$d_n = a_n = \mu\pi_{n+1}, \qquad n = 0,1,2,3,\ldots \tag{31.3.33}$$

That is $N^d \stackrel{d}{=} N^a$. Let Z_n be the number of customers that could have been served during the inter-arrival time A_n. Note that if the server is busy throughout the nth inter-arrival time, then this number will be the same as the number of customers actually served. Otherwise this number will be larger than the number of customers actually served. Observe that $\{Z_n,\ n = 1, 2, 3, \ldots\}$ is a sequence of *i.i.d.* random variables. As before, $\{Z_n | A_n = x\}$ is a Poisson random variable with mean μx. Hence

$$E[z^{Z_n}] = E[E[z^{Z_n}|A_n]] = E[\exp\{-\mu(1-z)A_n\}] = \tilde{F}_A(\mu(1-z)) \ . \qquad (31.3.34)$$

With this definition we have,

$$N_n^a = \max\{N_{n-1}^a + 1 - Z_n, 0\}, \qquad n = 1, 2, 3, \ldots. \qquad (31.3.35)$$

Let $q_j = P\{Z_n = j\}$, $j = 0, 1, 2, \ldots$ Then from (31.3.35) one has

$$a_n = \sum_{j=0}^{\infty} a_{n-1+j} q_j, \qquad n = 1, 2, 3, \ldots \qquad (31.3.36)$$

We may rewrite (31.3.36) as [note that $q_0 = \tilde{F}_A(\mu) > 0$],

$$a_{n-1} = \frac{1}{q_0}(a_n - \sum_{j=0}^{\infty} a_{n+j} q_{j+1}), \qquad n = 1, 2, 3, \ldots \qquad (31.3.37)$$

The structure of Eq. (31.3.37) implies that a_n should be of the geometric form. That is, if we assume that $a_{n+j} = \sigma a_{n+j-1}$, $j = 0, 1, 2, 3, \ldots$, and substitute in (31.3.37) we get $a_{n-1} = \sigma a_{n-2}, n = 2, 3, \ldots$ Hence $a_n = a_0 \sigma^n$, $n = 1, 2, 3, \ldots$ satisfies (31.3.36). By the normalizing condition $\sum_{n=0}^{\infty} a_n = 1$, the unique solution to (31.3.36) is then (for $0 < \sigma < 1$),

$$a_n = (1-\sigma)\sigma^n, \qquad n = 0, 1, 2, 3, \ldots \qquad (31.3.38)$$

Substituting (31.3.38) in (31.3.36) (for $n = 1$) and using (31.3.34), one finds that σ is the unique solution in $(0, 1)$ to

$$\sigma = \tilde{F}_A(\mu(1-\sigma)) \ . \qquad (31.3.39)$$

When A has an exponential distribution, $\sigma = \rho$ as one would expect. Now using (31.3.33) and the normalizing condition $\sum_{n=0}^{\infty} \pi_n = 1$ one finds that

$$\begin{aligned} \pi_0 &= 1 - \rho \\ \pi_n &= \rho(1-\sigma)\sigma^{n-1}, \qquad n = 1, 2, 3, \ldots \end{aligned} \qquad (31.3.40)$$

from which we find that server idle probability (π_0) is the same as that in the $M/M/1$ and $M/GI/1$ queueing systems (as should be the case). The mean number of customers in the system is

$$E[N] = \frac{\rho}{1-\sigma} \ . \qquad (31.3.41)$$

The imbedded discrete time Markovian nature of this process and its geometric solution can be extended to queueing systems even when the service time distribution is of the phase type [that is for the $GI/PH/1$ queueing systems, see Neuts (1981)]. Indeed these analyses easily extend to multiple server queueing systems as well.

Since the inter-arrival times do not possess the memoryless property, it is not very easy to study the inter-departure times of the customers from the system or the departure process (of course, it is possible to study this using supplementary variable techniques!).

Time spent in the system (and the memoryless property of the service times): To find the waiting time statistics, observe that we can use the same argument as that used for the $M/M/1$ queue. Specifically we have,

$$T \stackrel{d}{=} \sum_{k=1}^{N^a+1} S_k \ . \tag{31.3.42}$$

Simple calculation reveals that [see Eq. (31.3.38)]

$$\begin{aligned} P\{T > t\} &= E\left[P\left\{\sum_{k=1}^{N^a+1} S_k > t | N^a\right\}\right] \\ &= E\left[\sum_{k=0}^{N^a} \exp\{-\mu t\}\frac{(\mu t)^k}{k!}\right] \\ &= \exp\{-\mu(1-\sigma)t\}, \qquad t \geq 0 \ . \end{aligned} \tag{31.3.43}$$

We see that the total time spent in the system has an exponential distribution (that is an exponential output). We may also use the Lindley's recursion [Eq. (31.1.2)] to obtain the above result directly.

31.4 Bounds and Approximations: Use of Exponential Inputs and Outputs

In this section we will demonstrate

(1) how the results obtained in Section 31.3 for queueing systems with exponential inputs can be used to develop bounds for the outputs, and

(2) how assuming an exponential output can result in obtaining bounds for the exact outputs.

As we will see, for (1) we will use the external stochastic monotonicity and for (2) we will use the internal stochastic monotonicity of the $GI/GI/1$ queueing system. External monotonicity refers to the monotonicity with different inputs and internal monotonicity refers to the monotonicity in the sequential behavior. Before we can make this statement more precise, we need to give some definitions and results on stochastic orders and non-parametric families of distribution functions (that will be used below). These definitions and results can be found in Shaked and Shanthikumar (1994). A random variable X is larger than a random variable Y in the usual stochastic order if $Ef(X) \geq Ef(Y)$ for all increasing functions f, whenever these expectations exist. We denote this $X \geq_{st} Y$. Observe that

$$X \geq_{st} Y \Leftrightarrow P\{X > t\} \geq P\{Y > t\} \ . \tag{31.4.1}$$

A random variable X is larger than a random variable Y in the increasing convex (decreasing convex) order if $Ef(X) \geq Ef(Y)$ for all increasing convex (decreasing convex) functions f, whenever these expectations exist. We denote this $X \geq_{icx}$ ($\geq_{dcx}$) Y. We have

$$X \geq_{icx} Y \Leftrightarrow \int_{t=x}^{\infty} P\{X > t\}dt \geq \int_{t=x}^{\infty} P\{Y > t\}dt \ , \tag{31.4.2}$$

and

$$X \geq_{dcx} Y \Leftrightarrow \int_{t=0}^{x} P\{X \leq t\}dt \geq \int_{t=0}^{x} P\{Y \leq t\}dt \ . \tag{31.4.3}$$

If $X \geq_{icx} Y$ and $X \geq_{dcx} Y$, then we have $X \geq_{cx} Y$. That is, for all convex functions f we have $Ef(X) \geq Ef(Y)$. In this case $E[X] = E[Y]$. Note that $X \geq_{cx} Y \Leftrightarrow X \geq_{icx} Y$ and $E[X] = E[Y] \Leftrightarrow X \geq_{dcx} Y$ and $E[X] = E[Y]$, $X \geq_{cx} Y \Rightarrow X \geq_{icx} Y$ and $X \geq_{st} Y \Rightarrow X \geq_{icx} Y$. Suppose a random variable X has a harmonic new better [worse] than used in expectation (denoted HNBUE [HNWUE]) distribution function, that is

$$\frac{1}{E[X]} \int_{t=x}^{\infty} P\{X > t\}dt \leq [\geq] \exp\{-x/E[X]\}, \qquad x > 0 \ . \tag{31.4.4}$$

If Y is an exponential random variable with $E[Y] = E[X]$, then

$$X \in HNBUE[HNWUE] \Leftrightarrow Y \geq_{cx} X \, [X \geq_{cx} Y]. \tag{31.4.5}$$

It is worth noting that almost all standard distribution functions fall into either one of these two families. Exponential distribution is the only one that belongs to both these families.

31.4.1 External Monotonicity and Bounds using Exponential Inputs

Let

$$f(t, a, s) = \max\{t - a, 0\} + s, \qquad t, a, s \geq 0 \ . \tag{31.4.6}$$

Then [from the Lindley's recursion, Eq. (31.1.2)],

$$T_n = f(T_{n-1}, A_n, S_n), \qquad n = 1, 2, 3, \ldots \tag{31.4.7}$$

$f(t, a, s)$ is increasing in (t, s), decreasing in a and convex in (t, a, s). Consider a different $GI/GI/1$ queueing systems with inter-arrival times $\{\hat{A}_n, \ n = 1, 2, 3, \ldots\}$ and service times $\{\hat{S}_n, \ n = 1, 2, 3, \ldots\}$. Let $\hat{T}_n$ be the time spent by the nth customer in this system. Then

$$\hat{T}_n = f(\hat{T}_{n-1}, \hat{A}_n, \hat{S}_n), \qquad n = 1, 2, 3, \ldots \tag{31.4.8}$$

Using the properties of the function f one can see from Eqs. (31.4.7) and (31.4.8) that if

$$\hat{A}_n \geq_{dcx} A_n \text{ and } \hat{S}_n \geq_{icx} S_n \text{ then } \hat{T}_0 \geq_{icx} T_0 \Rightarrow \hat{T}_n \geq_{icx} T_n, \ n = 1, 2, 3, \ldots \tag{31.4.9}$$

The above is an external monotonicity property of the queueing process. Let $T_{HNBUE/GI/1}$ be the stationary time spent in the system by an arbitrary customer in a $HNBUE/GI/1$ queueing system with HNBUE inter-arrival time. We will use a similar (but obvious) notation. Then from (31.4.9) and the definition of HNBUE and HNWUE distributions one has

$$\begin{aligned} T_{GI/HNWUE/1} &\geq_{icx} T_{GI/M/1};\, T_{GI/M/1} \geq_{icx} T_{GI/HNBUE/1} \\ T_{HNWUE/GI/1} &\geq_{icx} T_{M/GI/1};\, T_{M/GI/1} \geq_{icx} T_{HNBUE/GI/1}\ . \end{aligned} \qquad (31.4.10)$$

In the above comparisons, the mean inter-arrival times and the mean service times are the same for both queueing systems. Thus we see that the stationary time spent in the $GI/GI/1$ queueing system in many cases can be bounded by the results derived (in Section 31.3) with exponential inputs. Additional bounds may be found in Stoyan (1983).

31.4.2 Internal Monotonicity and Bounds using Exponential Outputs

From the monotonicity of $f(t, a, s)$ defined in (31.4.6) one has

$$T_0 \geq_{st} T_1 \Rightarrow T_{n-1} \geq_{st} T_n, \qquad n = 1, 2, 3, \ldots \qquad (31.4.11)$$

This is an internal monotonicity property of the queueing process. From (31.4.11) we have,

$$T_0 \geq_{st} T_1 \Rightarrow T_0 \geq_{st} T\ . \qquad (31.4.12)$$

Now suppose

$$P\{T_0 > t\} = \exp\{-\eta t\}, \qquad t > 0\ , \qquad (31.4.13)$$

where η ($\eta > 0$) is the largest solution of

$$\tilde{F}_A(\eta)\tilde{F}_S(-\eta) = 1\ . \qquad (31.4.14)$$

Then

$$\begin{aligned} P\{T_1 > t\} &= P\{\max\{T_0 - A_1, 0\} + S_1 > t\} \\ &= E[\exp\{-\eta(A_1 + \max\{t - S_1, 0\})] \\ &= \tilde{F}_A(\eta)E[\exp\{\eta \min\{S_1 - t, 0\}]\ . \end{aligned} \qquad (31.4.15)$$

Therefore

$$\frac{P\{T_1 > t\}}{P\{T_0 > t\}} = \tilde{F}_A(\eta)E[\exp\{\eta \min\{S_1, t\}]\ , \qquad (31.4.16)$$

is increasing in t. From (31.4.14), we note that

$$\lim_{t\to\infty} \frac{P\{T_1 > t\}}{P\{T_0 > t\}} = \lim_{t\to\infty} \tilde{F}_A(\eta)E[\exp\{\eta \min\{S_1, t\}] = \tilde{F}_A(\eta)\tilde{F}_S(-\eta) = 1\ . \qquad (31.4.17)$$

Hence $P\{T_1 > t\} \leq P\{T_0 > t\}$. Therefore from (31.4.11) we have the exponential bound

$$P\{T > t\} \leq \exp\{-\eta t\}, \qquad t > 0\ , \tag{31.4.18}$$

where η is defined in (31.4.14). When we have exponentially distributed service times $\eta = \mu(1-\sigma)$ and the bound is tight. This result was first derived by Kingman (1970). However, not much work has been done after this paper using this idea of internal monotonicity. There exists a lot of potential to apply this idea to other queueing systems.

References

Buzacott, J.A. and Shanthikumar, J.G. (1993). *Stochastic Models of Manufacturing Systems*, Prentice Hall, Englewood Cliffs, New Jersey.

Kelly, F. (1979). *Reversibility and Stochastic Networks*, John Wiley & Sons, New York.

Kingman, J.F.C. (1970). Inequalities in the theory of queues, *Journal of the Royal Statistical Society, Series B,* **32**, 102-110.

Neuts, M. (1981). *Matrix-Geometric Solutions in Stochastic Models*, John Hopkins University Press, Baltimore, Maryland.

Neuts, M. (1989). *Structured Stochastic Matrices of M/G/1 Types and Their Applications*, Marcel Dekker, New York.

Ross, S.M. (1983). *Stochastic Processes*, John Wiley & Sons, New York.

Shaked, M. and Shanthikumar, J.G. (1994). *Stochastic Orders and Their Applications*, Academic Press, San Diego, California.

Shanthikumar, J.G. (1985). Bilateral phase type distributions, *Naval Research Logistics Quarterly,* **32**, 119-136.

Shanthikumar, J.G. and Chandra, M.J. (1982). Application of level crossing analysis to discrete state processes in queueing systems, *Naval Research Logistics Quarterly,* **29**, 593-608.

Stoyan, D. (1983). *Comparison Methods for Queues and Other Stochastic Models*, John Wiley & Sons, New York.

Takagi, H. (1991). *Queueing Analysis, Vol. 1: Vacation and Priority Systems*, North-Holland, Amsterdam.

Walrand, J. (1988). *An Introduction to Queueing Networks*, Prentice Hall, Englewood Cliffs, New Jersey.

Whitt, W. (1991). A review of L=λW and extensions, *Queueing Systems: Theory and Applications*, **9**, 235-269.

Wolff, R. (1989). *Stochastic Modeling and the Theory of Queues*, Prentice Hall, Englewood Cliffs, New Jersey.

CHAPTER 32

Exponential Classification and Applications

N. Balakrishnan[1], S. Panchapakesan,[2] and Q. Zhang[1]

32.1 Introduction

The theory of classification is of considerable interest to practitioners, and it is used to categorize an object (or an individual) on the basis of a profile of its characteristics. Most of the statistical developments in classification theory have primarily been on the normal (univariate or multivariate) populations; see, for example, Rao (1973), Lachenbruch (1974), and Gordon (1981). The classification problem for the exponential distribution has also been treated by some authors including Basu and Gupta (1974, 1977), Giri (1977), Onyeagu (1985), Zhang (1992), and Adegboye (1993). A Bayesian treatment of this problem has already been seen briefly in Chapter 10.

In this chapter, we consider two distinct exponential populations and discuss the classification of a new observation z, which is known to have come from one of these two populations (but exactly which one we do not know), as belonging to one of the two populations; we also study the performance of our classification procedures in terms of error rates. Several different cases are discussed according to whether the parameters are known or unknown, sample is complete or Type-II right censored, one- or two-parameter exponentials are being considered. Some illustrative examples are also presented.

Let the two exponential populations be denoted by A and B. Let z denote the new observation that needs to be classified. There are two possible errors of misclassification: (i) wrongly classifying z into B when in fact it has come from A, denoted by e_{BA}; and (ii) wrongly classifying z into A when in fact it has come from B, denoted by e_{AB}. These two errors are analogous to Type I and Type II errors in a hypothesis testing set-up. Hence, we discuss the effectiveness of a classification procedure by fixing the probability of the error of misclassification e_{BA} and then studying the probability of correct classification of z into B, i.e. $1 - \Pr(e_{AB})$ (referred to as the *power*). Let us denote $\Pr(e_{BA})$ by e_1 and $\Pr(e_{AB})$ by e_2, so the power is $1 - e_2$.

[1]McMaster University, Hamilton, Ontario, Canada
[2]Southern Illinois University, Carbondale, Illinois

32.2 Classification When Parameters are Known

32.2.1 Model

Let us consider $e(\theta_1)$ (A) and $e(\theta_2)$ (B) to be the two exponential populations. Suppose x is an observation from A and y is an independent observation from B. We shall assume that the parameters θ_1 and θ_2 are both known, and that $\theta_1 > \theta_2$ without loss of any generality. Then, we have

$$f(x;\theta_1) = \frac{1}{\theta_1}\, e^{-x/\theta_1}, \qquad x \geq 0,\ \theta_1 > 0\ , \tag{32.2.1}$$

and

$$f(y;\theta_2) = \frac{1}{\theta_2}\, e^{-y/\theta_2}, \qquad y \geq 0,\ \theta_2 > 0\ . \tag{32.2.2}$$

Let z be the new observation that has to be classified into either A or B.

32.2.2 Likelihood-Ratio Classification Procedure

The likelihood-ratio rule is to classify z into A or B according as

$$\frac{f(z;\theta_1)}{f(z;\theta_2)} = \frac{\theta_2}{\theta_1}\, e^{z(\theta_2^{-1}-\theta_1^{-1})} \gtrless C\ , \tag{32.2.3}$$

or equivalently, according as

$$z \gtrless \frac{\theta_1\theta_2}{\theta_1-\theta_2}\ \log(C\theta_1/\theta_2)\ . \tag{32.2.4}$$

In (32.2.4), C is chosen in such a way that e_1 is at a pre-specified level (as mentioned earlier in Section 32.1).

32.2.3 Errors of Misclassification

Once we fix the value of e_1, then C may be obtained from the equation

$$\Pr\left\{ z < \frac{\theta_1\theta_2}{\theta_1-\theta_2}\ \log(C\theta_1/\theta_2) \mid z \in A \right\} = e_1\ , \tag{32.2.5}$$

or equivalently

$$C = \frac{\theta_2}{\theta_1}\,(1-e_1)^{1-(\theta_1/\theta_2)}\ . \tag{32.2.6}$$

Thus, for a fixed e_1, the likelihood-ratio rule is to classify z into A or B according as

$$z \gtrless -\theta_1 \log(1-e_1)\ . \tag{32.2.7}$$

From (32.2.7), we readily obtain

$$e_2 = \Pr\{z > -\theta_1\log(1-e_1) \mid z \in B\} = (1-e_1)^{\theta_1/\theta_2} \tag{32.2.8}$$

so that the power of the classification procedure is

$$\text{Power} = 1 - e_2 = 1 - (1-e_1)^{\theta_1/\theta_2}\ . \tag{32.2.9}$$

Table 32.1: Values of power of the classification procedure

	Ratio θ_1/θ_2								
e_1	2	3	4	5	8	10	12	15	20
0.010	0.020	0.030	0.039	0.049	0.077	0.095	0.114	0.140	0.182
0.050	0.098	0.143	0.185	0.226	0.337	0.401	0.460	0.537	0.642
0.075	0.114	0.209	0.268	0.323	0.464	0.541	0.608	0.689	0.790
0.100	0.190	0.271	0.344	0.410	0.570	0.651	0.718	0.794	0.878

Remark 32.1 From Eq. (32.2.8) we observe that, when the ratio θ_1/θ_2 is fixed, e_1 and e_2 move in opposite directions and that e_2 tends to 1 as e_1 tends to 0. We also note that, when e_1 is fixed, e_2 depends only on the ratio θ_1/θ_2. From Eq. (32.2.9) we observe that, for a fixed value of e_1, the power increases with the ratio θ_1/θ_2.

In Table 32.1, we have presented the power values $(1 - e_2)$ for different choices of the ratio θ_1/θ_2 when e_1 is fixed at 0.01, 0.05, 0.075, 0.10.

32.3 Classification When Parameters are Unknown

In the last section, we presented the likelihood-ratio rule for classification when the parameters θ_1 and θ_2 are both known. In practice, however, the parameters θ_1 and θ_2 will be unknown. In this section, we first derive the likelihood-ratio rule and also propose a plug-in procedure. We next present exact expressions for the errors of this plug-in procedure. Finally, we present an example to illustrate this method of classification.

32.3.1 Model

Let $x_1, x_2, \ldots, x_m$ be a random sample of size m from Population A which is $e(\theta_1)$ where θ_1 is unknown, and $y_1, y_2, \ldots, y_n$ be an independent random sample of size n from Population B which is $e(\theta_2)$ where θ_2 is unknown. Let z be an independent new observation that needs to be classified into either A or B.

As already seen in Chapter 4, the MLEs of θ_1 and θ_2 are the sample means $\bar{x}$ and $\bar{y}$.

32.3.2 Likelihood-Ratio Classification Procedure

When z comes from A, the joint likelihood function is

$$L_1 = \frac{1}{\theta_1^{m+1}}\, e^{-\{\sum_{i=1}^{m} x_i + z\}/\theta_1}\, \frac{1}{\theta_2^n}\, e^{-\sum_{i=1}^{n} y_i/\theta_2} \tag{32.3.1}$$

from which we obtain the MLEs of θ_1 and θ_2 to be

$$\hat{\theta}_1 = \frac{m\bar{x} + z}{m+1} \qquad \text{and} \qquad \hat{\theta}_2 = \bar{y}\ ;$$

upon substituting these expressions in (32.3.1), we obtain

$$\sup_{\theta_1,\theta_2} L_1 = \left(\frac{m+1}{m\bar{x}+z}\right)^{m+1} e^{-(m+n+1)} \bar{y}^{-n} . \tag{32.3.2}$$

Similarly, when z comes from B, we get

$$\sup_{\theta_1,\theta_2} L_2 = \left(\frac{n+1}{n\bar{y}+z}\right)^{n+1} e^{-(m+n+1)} \bar{x}^{-m} . \tag{32.3.3}$$

From Eqs. (32.3.2) and (32.3.3), we derive the likelihood-ratio classification procedure as to classify z into A or B according as

$$W = \frac{\bar{x}^m (n\bar{y}+z)^{n+1}}{\bar{y}^n (m\bar{x}+z)^{m+1}} \gtrless C , \tag{32.3.4}$$

where C is to be determined appropriately after pre-fixing the value of e_1. However, the distribution of W in (32.3.4) is intractable which makes the determination of C impossible.

32.3.3 Plug-in Classification Procedure

By using the MLEs of θ_1 and θ_2 given by $\bar{x}$ and $\bar{y}$, respectively, we propose a simple plug-in procedure [from (32.2.7)] that classifies z into A or B according as

$$z \gtrless -\bar{x}\log(1-e_1) \tag{32.3.5}$$

(with a pre-fixed value of e_1).

Realize here that, unlike in the last section, designation of the two populations as A (with the larger mean) and B (with the smaller mean) is impossible as the means θ_1 and θ_2 are unknown. In order to resolve this difficulty, we will denote the larger sample mean by $\bar{x}$ (and the corresponding population by A) and the smaller sample mean by $\bar{y}$ (and the corresponding population by B). When we use this scheme, we realize that there are now two sources of error: one arises when wrongly classifying an observation from Population $A(B)$ into Population $B(A)$; the other error arises simply when $\bar{x} < \bar{y}$ even though $\theta_1 > \theta_2$. This could happen when either the two sample sizes are small (so that there is a large variation in $\bar{x}$ and $\bar{y}$) or the two populations are close (so that there is a considerable overlap in the distributions of $\bar{x}$ and $\bar{y}$). However, since $2m\bar{x}/\theta_1$ and $2n\bar{y}/\theta_2$ are independently distributed as χ^2_{2m} and χ^2_{2n}, respectively, we have

$$\Pr(\bar{x} > \bar{y}) = \Pr\left(F_{2m,2n} > \frac{\theta_2}{\theta_1}\right) , \tag{32.3.6}$$

where $F_{2m,2n}$ is a F-variable with $(2m, 2n)$ degrees of freedom. In Table 32.2, we have presented the values of $\Pr(\bar{x} > \bar{y})$ [computed from (32.3.6)] for various sample sizes $m = n$ and different choices of the ratio θ_1/θ_2. It is quite clear from Table 32.2 that $\Pr(\bar{x} > \bar{y})$ is almost 1 even for samples of size 20 as long as the ratio θ_1/θ_2 is at least 2.

Table 32.2: Values of $\Pr(\bar{x} > \bar{y})$

	Ratio θ_1/θ_2						
$m = n$	2	3	4	5	6	7	8
10	0.93523	0.98450	0.99594	0.99890	0.99970	0.99992	0.99998
20	0.99110	0.99963	0.99998	1.0	1.0	1.0	1.0
30	0.99842	0.99999	1.0	1.0	1.0	1.0	1.0
50	0.99990	1.0	1.0	1.0	1.0	1.0	1.0
100	1.0	1.0	1.0	1.0	1.0	1.0	1.0

32.3.4 Errors of Misclassification

Without loss of any generality, let us assume that Population A is $e(\theta_1)$ and Population B is $e(\theta_2)$ where $\theta_1 > \theta_2$, and the two parameters are estimated using the MLEs $\bar{x}$ and $\bar{y}$ respectively. We derive here exact expressions for the true e_1 and the power $(1 - e_2)$ of the plug-in procedure in (32.3.5).

When z belongs to Population A, it will get wrongly classified into Population B by the plug-in procedure in the following two ways:

$$\begin{array}{lll} (i) & z < -\bar{x}\log(1-e_1) & \text{when } x > y \\ (ii) & z > -\bar{y}\log(1-e_1) & \text{when } \bar{x} < \bar{y}\,. \end{array} \tag{32.3.7}$$

Here, e_1 is at the pre-fixed level. Let e_1^* denote the true value of e_1, $U = \sum_{i=1}^{m} x_i$, $V = \sum_{i=1}^{n} y_i$, $b_1 = m/n$, $b_2 = -\frac{1}{m}\log(1-e_1)$, and $b_3 = -\frac{1}{n}\log(1-e_1)$. We then have

$$\begin{aligned} &\Pr\{\bar{x} > \bar{y},\ z < -\bar{x}\log(1-e_1) \mid z \in A\} \\ &= \Pr\left\{V < \frac{1}{b_1}U,\ z < b_2 U \middle|\ z \in A\right\} \\ &= \int_0^\infty \int_0^{u/b_1} \int_0^{b_2 u} \frac{1}{\theta_1^m \Gamma(m)} e^{-u/\theta_1} u^{m-1} \cdot \frac{1}{\theta_2^n \Gamma(n)} e^{-v/\theta_2} v^{n-1} \\ &\qquad \times \frac{1}{\theta_1} e^{-z/\theta_1}\, dz\, dv\, du \\ &= 1 - (1+b_2)^{-m} + \frac{1}{\theta_1^m \Gamma(m)} \sum_{i=0}^{n-1} \frac{\Gamma(m+i)}{(b_1\theta_2)^i\, i!} \left\{ \left(\frac{1}{\theta_1} + \frac{1}{b_1\theta_2} + \frac{b_2}{\theta_1}\right)^{-m-i} \right. \\ &\qquad \left. - \left(\frac{1}{\theta_1} + \frac{1}{b_1\theta_2}\right)^{-m-i} \right\}. \end{aligned} \tag{32.3.8}$$

Similarly, we obtain

$$\begin{aligned}
&\Pr\{\bar{x} < \bar{y},\ z > -\bar{y}\log(1-e_1) \mid z \in A\} \\
&= \left(1 + \frac{\theta_2}{\theta_1}\, b_3\right)^{-n} - \frac{1}{\theta_2^n\,\Gamma(n)} \sum_{i=0}^{m-1} \frac{\Gamma(n+i)(b_1/\theta_1)^i}{i!} \left(\frac{1}{\theta_1} + \frac{b_3}{\theta_1} + \frac{b_1}{\theta_1}\right)^{-n-i} .
\end{aligned} \tag{32.3.9}$$

Making use of the expressions in (32.3.8) and (32.3.9), we obtain from (32.3.7)

$$\begin{aligned}
e_1^* &= 1 - (1+b_2)^{-m} + \left(1 + \frac{\theta_2}{\theta_1}\, b_3\right)^{-n} \\
&\quad + \frac{1}{\theta_1^m \Gamma(m)} \sum_{i=0}^{n-1} \frac{\Gamma(m+i)}{(b_1\theta_2)^i\, i!} \left\{ \left(\frac{1}{\theta_1} + \frac{1}{b_1\theta_2} + \frac{b_2}{\theta_1}\right)^{-m-i} \right. \\
&\qquad\qquad \left. - \left(\frac{1}{\theta_1} + \frac{1}{b_1\theta_2}\right)^{-m-i} \right\} \\
&\quad - \frac{1}{\theta_2^n\,\Gamma(n)} \sum_{i=0}^{m-1} \frac{\Gamma(n+i)(b_1/\theta_1)^i}{i!} \left(\frac{1}{\theta_1} + \frac{b_3}{\theta_1} + \frac{b_1}{\theta_1}\right)^{-n-i} .
\end{aligned} \tag{32.3.10}$$

When z belongs to Population B, it will get correctly classified into Population B by the plug-in procedure in the following two ways:

$$\begin{array}{lll}
(i) & z < -\bar{x}\log(1-e_1) & \text{when } \bar{x} > \bar{y} \\
(ii) & z > -\bar{y}\log(1-e_1) & \text{when } \bar{x} < \bar{y}\,.
\end{array}$$

Then, by proceeding exactly on similar lines we obtain the true power to be

$$\begin{aligned}
\text{Power}^* &= 1 - e_2^* \\
&= 1 - \left(1 + \frac{\theta_1}{\theta_2}\, b_2\right)^{-m} + (1+b_3)^{-n} \\
&\quad + \frac{1}{\theta_1^m \Gamma(m)} \sum_{i=0}^{n-1} \frac{\Gamma(m+i)}{(b_1\theta_2)^i\, i!} \left\{ \left(\frac{1}{\theta_1} + \frac{1}{b_1\theta_2} + \frac{b_2}{\theta_2}\right)^{-m-i} \right. \\
&\qquad\qquad \left. - \left(\frac{1}{\theta_1} + \frac{1}{b_1\theta_2}\right)^{-m-i} \right\} \\
&\quad - \frac{1}{\theta_2^n\,\Gamma(n)} \sum_{i=0}^{m-1} \frac{\Gamma(m+i)(b_1/\theta_1)^i}{i!} \left(\frac{1}{\theta_2} + \frac{b_3}{\theta_2} + \frac{b_1}{\theta_1}\right)^{-m-i} .
\end{aligned} \tag{32.3.11}$$

The values of e_1^* and Power* [computed from (32.3.10) and (32.3.11)] are presented in Tables 32.3 and 32.4, respectively, when pre-fixed $e_1 = 0.05, 0.075, 0.10$ and sample sizes $m = n = 10, 20, 30$. Since $\Pr(F_{2m,2n} > \theta_2/\theta_1)$ is almost 1 even for m and n as small as 20 whenever $\theta_1/\theta_2 \geq 2$, we observe from Tables 32.3 and 32.4 that

$$e_1^* \simeq e_1 \text{ and Power}^* \simeq 1 - (1-e_1)^{\theta_1/\theta_2} \tag{32.3.12}$$

Table 32.3:

True values of e_1 when pre-fixed $e_1 = 0.05$

	Ratio θ_1/θ_2						
$m = n$	2	3	4	5	8	10	12
10	0.11043	0.05835	0.05140	0.05022	0.04988	0.04988	0.04988
20	0.06442	0.04997	0.04995	0.04994	0.04994	0.04994	0.04994
30	0.05375	0.04998	0.04996	0.04996	0.04996	0.04996	0.04996

True values of e_1 when pre-fixed $e_1 = 0.075$

	Ratio θ_1/θ_2						
$m = n$	2	3	4	5	8	10	12
10	0.13315	0.08297	0.07621	0.07505	0.07472	0.07472	0.07472
20	0.08881	0.07521	0.07487	0.07486	0.07486	0.07486	0.07486
30	0.07855	0.07493	0.07491	0.07491	0.07491	0.07491	0.07491

True values of e_1 when pre-fixed $e_1 = 0.10$

	Ratio θ_1/θ_2						
$m = n$	2	3	4	5	8	10	12
10	0.15578	0.10753	0.10096	0.09983	0.09950	0.09950	0.09950
20	0.11319	0.10008	0.09976	0.09975	0.09975	0.09975	0.09975
30	0.10334	0.09985	0.09983	0.09983	0.09983	0.09983	0.09983

Table 32.4:

True power values when pre-fixed $e_1 = 0.05$

	Ratio θ_1/θ_2						
$m = n$	2	3	4	5	8	10	12
10	0.15536	0.14964	0.18522	0.22403	0.33112	0.39360	0.44972
20	0.11122	0.14245	0.18465	0.22496	0.33382	0.39738	0.45460
30	0.10099	0.14229	0.18492	0.22538	0.33473	0.39866	0.45626
	0.098	0.143	0.185	0.226	0.337	0.401	0.460

True power values when pre-fixed $e_1 = 0.075$

	Ratio θ_1/θ_2						
$m = n$	2	3	4	5	8	10	12
10	0.19848	0.21400	0.26576	0.31808	0.45393	0.52797	0.59112
20	0.15704	0.20779	0.26615	0.32027	0.45891	0.53458	0.59921
30	0.14748	0.20785	0.26673	0.32111	0.46060	0.53683	0.60198
	0.114	0.209	0.268	0.323	0.464	0.541	0.608

True power values when pre-fixed $e_1 = 0.10$

	Ratio θ_1/θ_2						
$m = n$	2	3	4	5	8	10	12
10	0.24020	0.27458	0.33948	0.40182	0.55480	0.63275	0.69595
20	0.20153	0.26950	0.34103	0.40547	0.56203	0.64185	0.70653
30	0.19265	0.26981	0.34197	0.40680	0.56450	0.64496	0.71016
	0.190	0.271	0.344	0.410	0.570	0.651	0.718

whenever m and n are large (at least 20) and θ_1/θ_2 is at least 2. Realize that the power expression on the right-hand side of (32.3.12) is the true power of the likelihood-ratio procedure when θ_1 and θ_2 are known [see Eq. (32.2.9)]. These values (taken from Table 32.1) are also presented in the bottom rows of Table 32.4 for convenience of comparison.

32.3.5 Illustrative Example

Let us consider the following two samples of size 50 each, representing the lifetimes (in hours) of two brands of components. These are simulated data with $\theta_1 = 40$ and $\theta_2 = 10$.

x_i: 31.99, 45.07, 82.20, 7.02, 18.90, 85.77, 6.45, 30.97, 35.51, 5.89, 85.22, 0.41, 89.79, 4.52, 47.02, 20.79, 23.88, 12.32, 87.51, 31.96, 105.01, 117.18, 7.45, 23.15, 28.31, 45.91, 53.03, 35.91, 11.78, 47.78, 113.15, 82.94, 42.66, 19.26, 46.93, 49.29, 65.73, 55.83, 97.24, 67.89, 21.12, 56.55, 20.64, 26.62, 6.68, 13.35, 51.45, 48.29, 0.94, 79.23

y_i: 1.07, 1.82, 2.53, 1.67, 11.67, 2.24, 0.35, 6.26, 0.23, 2.80, 9.90, 4.18, 10.88, 17.32, 27.22, 7.81, 36.00, 23.53, 3.15, 4.72, 0.44, 4.12, 4.19, 4.38, 10.90, 14.94, 13.81, 2.82, 7.19, 0.38, 25.87, 17.78, 2.72, 3.92, 0.42, 0.42, 3.22, 15.91, 5.16, 20.54, 4.71, 35.61, 8.49, 10.71, 5.11, 17.79, 49.58, 19.92, 4.52, 2.95

Since the first sample mean is larger than the second sample mean, we designate the first sample as x-sample (corresponding population as A) and the second sample as y-sample (corresponding population as B). We observe $\bar{x} = 43.89$ and $\bar{y} = 9.90$. Using the plug-in classification procedure with e_1 fixed as 0.05, and classifying the observations in the two samples one by one into either Population A or Population B, 2 out of 50 from A are wrongly classified into B giving an estimate of e_1 as 0.04 (when it was pre-fixed as 0.05) and that 10 out of 50 from B are correctly classified into itself giving an estimate of the power as 0.20 (when the value of power is 0.185 when θ_1 and θ_2 are known and that $\theta_1/\theta_2 = 4$).

32.4 Classification Under Type-II Right Censoring

In the last section, we discussed the classification problem when the available samples are both complete. However, in many life-testing experiments one may often observe a Type-II right censored sample. Hence, we discuss in this section the classification problem when the available samples are both Type-II right censored. We first derive the likelihood-ratio classification procedure and then propose a plug-in procedure. We next present exact expressions for the errors of the plug-in procedure. Finally, we present an example to illustrate this method of classification.

32.4.1 Model

Let $x_{1:m}, x_{2:m}, \ldots, x_{m-r:m}$ be a Type-II right censored sample from Population A which is $e(\theta_1)$ where θ_1 is unknown, and $y_{1:n}, y_{2:n}, \ldots, y_{n-s:n}$ be an independent Type-II right censored sample from Population B which is $e(\theta_2)$ where θ_2 is unknown. Let z be an independent new observation that needs to be classified into either A or B. As already seen in Chapter 4, the MLEs of θ_1 and θ_2 are

$$\hat{\theta}_1 = \frac{1}{m-r}\left\{\sum_{i=1}^{m-r} x_{i:m} + r x_{m-r:m}\right\}, \quad \hat{\theta}_2 = \frac{1}{n-s}\left\{\sum_{i=1}^{n-s} x_{i:n} + s x_{n-s:n}\right\}. \tag{32.4.1}$$

32.4.2 Likelihood-Ratio Classification Procedure

When z comes from A, the joint likelihood function is

$$L_1 = \frac{m!}{r!\theta_1^{m-r+1}} e^{-\left\{\sum_{i=1}^{m-r} x_{i:m} + r x_{m-r:m} + z\right\}/\theta_1} \frac{n!}{s!\,\theta_2^n} e^{-\left\{\sum_{i=1}^{n-s} x_{i:n} + s x_{n-s:n}\right\}/\theta_2} \tag{32.4.2}$$

from which we obtain the MLEs of θ_1 and θ_2 to be

$$\frac{1}{m-r+1}\left\{\sum_{i=1}^{m-r} x_{i:m} + r x_{m-r:m} + z\right\}, \quad \frac{1}{n-s}\left\{\sum_{i=1}^{n-s} x_{i:n} + s x_{n-s:n}\right\}, \tag{32.4.3}$$

respectively; upon substituting these expressions in (32.4.2), we obtain

$$\sup_{\theta_1,\theta_2} L_1 = \frac{m!n!}{r!s!}\left\{\frac{m-r+1}{\sum_{i=1}^{m-r} x_{i:m} + r x_{m-r:m} + z}\right\}^{m-r+1}\left\{\frac{n-s}{\sum_{i=1}^{n-s} y_{i:n} + s y_{n-s:n}}\right\}^{n-s} \times e^{-(m-r+1+n-s)}. \tag{32.4.4}$$

Similarly, when z comes from B, we get

$$\sup_{\theta_1,\theta_2} L_2 = \frac{m!n!}{r!s!}\left\{\frac{m-r}{\sum_{i=1}^{m-r} x_{i:m} + r x_{m-r:m}}\right\}^{m-r}\left\{\frac{n-s+1}{\sum_{i=1}^{n-s} y_{i:n} + s y_{n-s:n} + z}\right\}^{n-s+1} \times e^{-(m-r+1+n-s)}. \tag{32.4.5}$$

From Eqs. (32.4.4) and (32.4.5), we derive the likelihood-ratio classification procedure as to classify z into A or B according as

$$W' = \frac{\left\{\frac{m-r+1}{\sum_{i=1}^{m-r} x_{i:m}+r x_{m-r:m}+z}\right\}^{m-r+1}\left\{\frac{n-s}{\sum_{i=1}^{n-s} y_{i:n}+s y_{n-s:n}}\right\}^{n-s}}{\left\{\frac{m-r}{\sum_{i=1}^{m-r} x_{i:m}+r x_{m-r:m}}\right\}^{m-r}\left\{\frac{n-s+1}{\sum_{i=1}^{n-s} y_{i:n}+s y_{n-s:n}+z}\right\}^{n-s+1}} \gtrless C, \tag{32.4.6}$$

where C is to be determined so that e_1 is at a pre-specified level. However, the distribution of W' in (32.4.6) is intractable which makes the determination of C impossible.

32.4.3 Plug-in Classification Procedure

Since it is not possible to implement the likelihood-ratio rule for a fixed e_1, we propose a simple plug-in procedure [from (32.2.7)] by using the MLEs of θ_1 and θ_2 in (32.4.1) and classify z into A or B according as

$$z \gtrless -\hat{\theta}_1 \log(1-e_1) \ (= C) \tag{32.4.7}$$

(with a pre-fixed value of e_1).

As in Section 32.3.3, since we do not know the exact values of θ_1 and θ_2, we will denote the larger MLE in (32.4.1) by $\hat{\theta}_1$ (and the corresponding population by A) and the smaller MLE by $\hat{\theta}_2$ (and the corresponding population by B). Since $2(m-r)\hat{\theta}_1/\theta_1$ and $2(n-s)\hat{\theta}_2/\theta_2$ are independently distributed as $\chi^2_{2(m-r)}$ and $\chi^2_{2(n-s)}$, respectively, we have

$$\Pr(\hat{\theta}_1 > \hat{\theta}_2) = \Pr\left(F_{2m-2r,2n-2s} > \frac{\theta_2}{\theta_1}\right), \tag{32.4.8}$$

where $F_{2m-2r,2n-2s}$ is an F-variable with $(2m-2r, 2n-2s)$ degrees of freedom. Thus, these values may be obtained from Table 32.2 itself with m and n replaced by $m-r$ and $n-s$, respectively. For example, if $m = n = 50$ and $r = s = 20$, we have the value of $\Pr(\hat{\theta}_1 > \hat{\theta}_2)$ to be 0.99842 when $\theta_1/\theta_2 = 2$.

32.4.4 Errors of Misclassification

Proceeding exactly on the same lines of Section 32.3.4 and denoting $c_1 = (m-r)/(n-s)$, $c_2 = -\frac{1}{m-r}\log(1-e_1)$ and $c_3 = -\frac{1}{n-s}\log(1-e_1)$, we obtain exact expressions for the true values of e_1 and the power $(1-e_2)$ as follows:

$$\begin{aligned} e_1^* &= 1-(1+c_2)^{-m+r} + \left(1+\frac{\theta_2}{\theta_1}c_3\right)^{-n+s} \\ &\quad + \frac{1}{\theta_1^{m-r}\Gamma(m-r)} \sum_{i=0}^{n-s-1} \frac{\Gamma(m-r+i)}{(c_1\theta_2)^i\, i!} \left\{\left(\frac{1}{\theta_1}+\frac{1}{c_1\theta_2}+\frac{c_2}{\theta_1}\right)^{-m+r-i} \right. \\ &\quad \left. - \left(\frac{1}{\theta_1}+\frac{1}{c_1\theta_2}\right)^{-m+r-i}\right\} \\ &\quad - \frac{1}{\theta_2^{n-s}\Gamma(n-s)} \sum_{i=0}^{m-r-1} \frac{\Gamma(n-s+i)(c_1/\theta_1)^i}{i!} \left(\frac{1}{\theta_1}+\frac{c_3}{\theta_1}+\frac{c_1}{\theta_1}\right)^{-n+s-i}; \end{aligned} \tag{32.4.9}$$

$$\begin{aligned} \text{Power}^* &= 1-\left(1+\frac{\theta_1}{\theta_2}c_2\right)^{-m+r} + (1+c_3)^{-n+s} \\ &\quad + \frac{1}{\theta_1^{m-r}\Gamma(m-r)} \sum_{i=0}^{n-s-1} \frac{\Gamma(m-r+i)}{(c_1\theta_2)^i\, i!} \left\{\left(\frac{1}{\theta_1}+\frac{1}{c_1\theta_2}+\frac{c_2}{\theta_2}\right)^{-m+r-i} \right. \\ &\quad \left. - \left(\frac{1}{\theta_1}+\frac{1}{c_1\theta_2}\right)^{-m+r-i}\right\} \\ &\quad - \frac{1}{\theta_2^{n-s}\Gamma(n-s)} \sum_{i=0}^{m-r-1} \frac{\Gamma(m-r+i)(c_1/\theta_1)^i}{i!} \left(\frac{1}{\theta_2}+\frac{c_3}{\theta_2}+\frac{c_1}{\theta_1}\right)^{-m+r-i}. \end{aligned} \tag{32.4.10}$$

Upon comparing these expressions with those in Section 32.3.4, it is clear that these are exactly the same as (32.3.10) and (32.3.11) with m and n replaced respectively by

$m - r$ and $n - s$. Thus, the true values of e_1 and the power may just be obtained from Tables 32.3 and 32.4 with m and n replaced by $m - r$ and $n - s$.

In order to give a good idea of the effect of censoring, we have presented in Tables 32.5 and 32.6 the true values of e_1 and the power [computed from (32.4.9) and (32.4.10)] when the pre-fixed value of e_1 is 0.05 for the case when $m = m = 10$ and $r = s = 0(1)3$. In Table 32.6, the power values of the likelihood-ratio procedure when the parameters are known (for the case $m = n = 10$), taken from Table 32.1, are also presented in the last row for the sake of convenience of comparison.

From Table 32.5, we observe that the true value of e_1 is much larger than the pre-fixed value of 0.05 since the sample sizes are as small as 10 and increases further as the number of censored observations r and s increase, when the ratio θ_1/θ_2 is small. This happens primarily due to the fact that the third type of error (which arises when $\hat{\theta}_1 > \hat{\theta}_2$ even though $\theta_1 < \theta_2$) occurs with a higher probability in this case. Note, however, that even in the case of small sample sizes and larger levels of censoring, the true values of e_1 are quite close to the pre-fixed value of 0.05 and the true power values are quite close to the power values of the likelihood-ratio procedure when θ_1 and θ_2 are known, whenever the ratio θ_1/θ_2 is large.

32.4.5 Illustrative Example

Let us consider the following two samples of size 50 each with the largest 10 observations censored (i.e., $m = n = 50$, $r = s = 10$), representing the lifetimes (in hours) of two brands of components. These are simulated data with $\theta_1 = 40$ and $\theta_2 = 10$.

$x_{i:50}$: 0.43, 1.22, 2.14, 3.23, 3.45, 3.88, 4.22, 6.07, 6.63, 7.68, 8.85, 8.92, 10.10, 10.34, 10.43, 10.83, 11.30, 11.45, 11.81, 12.19, 13.89, 15.40, 16.20, 18.54, 19.16, 19.46, 19.50, 32.40, 34.32, 34.80, 36.54, 37.94, 45.35, 49.81, 51.42, 52.04, 54.38, 55.78, 57.73, 60.00

$y_{i:50}$: 0.11, 0.25, 0.80, 0.88, 0.92, 1.51, 1.93, 2.01, 2.02, 2.67, 2.89, 3.38, 3.68, 3.82, 3.93, 4.02, 4.61, 5.03, 5.15, 5.15, 5.54, 5.69, 6.05, 6.25, 6.93, 7.14, 7.92, 8.80, 10.59, 10.89, 11.31, 13.04, 13.49, 14.63, 14.79, 15.73, 15.89, 15.91, 16.46, 16.78

Since the MLE based on the first sample is larger than that based on the second sample, we designate the first sample as x-sample (corresponding population as A) and the second sample as y-sample (corresponding population as B). We then observe $\hat{\theta}_1 = 36.74$ and $\hat{\theta}_2 = 11.16$. Using the plug-in classification procedure with e_1 fixed as 0.05, we get the cut-off point C in (32.4.7) as 1.88. Then, upon classifying the observations in two samples one by one into either Population A or Population B, 2 out of 40 from A are wrongly classified into B giving an estimate of e_1 as 0.05 (when it was pre-fixed as 0.05) and that 6 out of 40 from B are correctly classified into itself giving an estimate of the power as 0.15 (when the value of power is 0.185 when θ_1 and θ_2 are known and that $\theta_1/\theta_2 = 4$).

32.5 Classification With Guaranteed Life-time

Suppose the life-times of components in the two populations are represented by two-parameter exponential distributions with a common guaranteed life-time μ. In this

Table 32.5: True values of e_1 when pre-fixed $e_1 = 0.05$ for $m = n = 10$ and $r = s = 0(1)3$

		Ratio θ_1/θ_2						
r	s	2	3	4	5	8	10	12
0	1	0.11329	0.05918	0.05916	0.05027	0.04989	0.04988	0.04988
	2	0.11673	0.06026	0.05188	0.05035	0.04990	0.04988	0.04988
	3	0.12093	0.06167	0.05277	0.05277	0.04991	0.04998	0.04988
1	0	0.11771	0.08103	0.05208	0.05042	0.04988	0.04986	0.04986
	1	0.12044	0.06165	0.05235	0.05050	0.04989	0.04987	0.04986
	2	0.12373	0.06280	0.05269	0.05061	0.04989	0.04987	0.04986
	3	0.12774	0.06431	0.05315	0.05077	0.04990	0.04987	0.04986
2	0	0.12672	0.06408	0.05319	0.05080	0.04989	0.04985	0.04984
	1	0.12929	0.06503	0.05350	0.05092	0.04990	0.04985	0.04984
	2	0.13237	0.06631	0.05392	0.05107	0.04991	0.04986	0.04985
	3	0.13613	0.06790	0.05448	0.05126	0.04992	0.04986	0.04985
3	0	0.13810	0.06890	0.05500	0.05152	0.04995	0.04985	0.04982
	1	0.14044	0.06995	0.05539	0.05168	0.04996	0.04986	0.04983
	2	0.14327	0.07126	0.05590	0.05187	0.04998	0.04996	0.04993
	3	0.14669	0.07295	0.05656	0.05215	0.05001	0.04987	0.04984

Table 32.6: True power values when pre-fixed $e_1 = 0.05$ for $m = n = 10$ and $r = s = 0(1)3$

		Ratio θ_1/θ_2						
r	s	2	3	4	5	8	10	12
0	1	0.18698	0.15595	0.18660	0.22439	0.33116	0.39362	0.44973
	2	0.23563	0.16776	0.18949	0.22518	0.33120	0.39364	0.44974
	3	0.31118	0.19021	0.19580	0.27710	0.33132	0.39368	0.44978
1	0	0.14086	0.14787	0.23556	0.22372	0.33053	0.39277	0.44865
	1	0.16495	0.15266	0.18594	0.22404	0.33055	0.39277	0.44865
	2	0.20263	0.16160	0.18822	0.22470	0.33060	0.39280	0.44867
	3	0.26128	0.17853	0.19308	0.22625	0.33071	0.39284	0.44870
2	0	0.12979	0.14646	0.18444	0.22336	0.32979	0.39174	0.44733
	1	0.14786	0.15019	0.18538	0.22365	0.32980	0.39174	0.44733
	2	0.17638	0.15696	0.18721	0.22423	0.32986	0.39176	0.44733
	3	0.22162	0.16967	0.19099	0.22552	0.32997	0.39180	0.44736
3	0	0.12151	0.14541	0.18406	0.22292	0.32885	0.39044	0.44564
	1	0.13493	0.14833	0.18486	0.22320	0.32885	0.39044	0.44564
	2	0.16512	0.15352	0.18637	0.22302	0.32891	0.39043	0.44564
	3	0.19009	0.16308	0.18940	0.22483	0.32903	0.39047	0.44566
		0.098	0.143	0.185	0.226	0.337	0.401	0.460

section, we present the likelihood-ratio rule when all the parameters are known and also propose a plug-in procedure when the parameters are unknown. Finally, we present an example to illustrate this method of classification.

32.5.1 Model

Let us consider $e(\mu, \theta_1)$ (A) and $e(\mu, \theta_2)$ (B) to be the two exponential populations; suppose x is an observation from A and y is an independent observation from B. We shall assume that the parameters μ, θ_1 and θ_2 are all known, and that $\theta_1 > \theta_2$ without loss of any generality. Then, we have

$$f(x;\mu,\theta_1) = \frac{1}{\theta_1}\, e^{-(x-\mu)/\theta_1}, \qquad x \geq \mu,\ \theta_1 > 0\ , \tag{32.5.1}$$

and

$$f(y;\mu,\theta_2) = \frac{1}{\theta_2}\, e^{-(y-\mu)/\theta_2}, \qquad y \geq \mu,\ \theta_2 > 0\ . \tag{32.5.2}$$

Let z be the new observation that needs to be classified into either A or B.

32.5.2 Likelihood-Ratio Classification Procedure

The likelihood-ratio rule is to classify z into A or B according as

$$\frac{f(z;\mu,\theta_1)}{f(z;\mu,\theta_2)} = \frac{\theta_2}{\theta_1}\, e^{(z-\mu)(\theta_2^{-1}-\theta_1^{-1})} \gtrless C\ , \tag{32.5.3}$$

or equivalently, when

$$z \gtrless \mu + \frac{\theta_1\theta_2}{\theta_1-\theta_2}\, \log(C\theta_1/\theta_2)\ . \tag{32.5.4}$$

In (32.5.4), C is chosen in such a way that e_1 is at a pre-specified level.

Once we fix the value of e_1, we simply obtain the likelihood-rule as to classify z into A or B according as

$$z \gtrless \mu - \theta_1 \log(1-e_1)\ . \tag{32.5.5}$$

Note that the cut-off point on the right-hand side of (32.5.5) is just μ units different from that in (32.2.7). So naturally we have the power of the classification procedure in (32.5.5) to be the same as in (32.2.9), viz.,

$$\text{Power} = 1 - (1-e_1)^{\theta_1/\theta_2}\ . \tag{32.5.6}$$

32.5.3 Plug-in Classification Procedure

Suppose $x_{1:m}, x_{2:m}, \ldots, x_{m-r:m}$ and $y_{1:n}, y_{2:n}, \ldots, y_{n-s:n}$ are the Type-II right censored samples available from Populations A and B, respectively. Then the MLEs of μ, θ_1 and θ_2 are

$$\begin{aligned}
\hat{\mu} &= \min(x_{1:m}, y_{1:n})\ ,\\
\hat{\theta}_1 &= \frac{1}{m-r-1}\left\{\sum_{i=1}^{m-r} x_{i:m} + r x_{m-r:m} - m\hat{\mu}\right\}
\end{aligned}$$

and

$$\hat{\theta}_2 = \frac{1}{n-s-1}\left\{\sum_{i=1}^{n-s} y_{i:n} + s y_{n-s:n} - n\hat{\mu}\right\} . \tag{32.5.7}$$

Then we propose the plug-in classification procedure as to classify z into A or B according as

$$z \gtrless \hat{\mu} - \hat{\theta}_1 \log(1 - e_1) . \tag{32.5.8}$$

In order to examine the performance of this plug-in procedure, we used 5000 Monte Carlo simulations and simulated the true e_1 and the power values for the cases when $m = n = 10, 20, 30, 50, 100$ (with $r = s = 0$) and different choices of the ratio θ_1/θ_2, by taking $\mu = 1$ and 9. These values are presented in Tables 32.7 and 32.8. In Table 32.8, we have also presented the true power values [computed from (32.5.6)] in the last rows for the sake of convenience of comparison.

A comparison of the values in Tables 32.7 and 32.8 with the corresponding entries in Tables 32.3 and 32.4 reveals readily that the estimation of the extra parameter (guaranteed life-time) μ has a negligible effect on the performance of the plug-in procedure with regard to the e_1 and power values.

32.5.4 Illustrative Example

Let us consider the following two samples of size 50 each (simulated with $\mu = 5$, $\theta_1 = 40$ and $\theta_2 = 10$).

x_i: 5.43, 5.90, 6.78, 7.99, 8.99, 9.47, 9.60, 10.21, 11.05, 11.31, 11.62, 12.05, 12.18, 14.09, 14.45, 15.50, 17.86, 18.55, 19.22, 19.27, 21.50, 23.58, 23.76, 24.17, 29.06, 29.17, 29.58, 30.54, 36.70, 37.29, 37.90, 40.46, 40.74, 45.19, 48.49, 48.61, 51.08, 51.80, 55.62, 71.67, 89.40, 94.48, 97.38, 98.70, 103.62, 104.57, 111.56, 113.37, 145.99, 192.23

y_i: 5.04, 5.21, 5.27, 5.55, 5.61, 5.84, 5.87, 6.17, 6.20, 6.33, 6.37, 7.13, 8.49, 8.56, 8.84, 9.63, 9.67, 9.84, 9.89, 9.97, 10.33, 10.35, 10.38, 10.66, 10.87, 11.07, 11.99, 12.43, 12.47, 13.34, 14.42, 14.48, 15.18, 16.03, 16.52, 16.83, 16.87, 17.76, 17.87, 20.13, 20.67, 21.15, 21.19, 23.63, 25.27, 33.30, 34.08, 36.50, 36.76, 58.28

From the above two samples, we find the MLEs of μ, θ_1 and θ_2 to be

$$\hat{\mu} = 5.04, \ \hat{\theta}_1 = 38.35 \text{ and } \hat{\theta}_2 = 9.69 .$$

Now upon using the plug-in procedure and classifying each of the observations in the two samples one by one into Population A or B, 3 of the 50 observations in A is wrongly classified into B giving an estimate of e_1 as 0.06 (when it was pre-fixed as 0.05), and 11 of the 50 observations in B are correctly classified into itself giving an estimate of the power as 0.22.

Table 32.7:

Simulated values of e_1 when pre-fixed $e_1 = 0.05$ and $\mu = 1$

	Ratio θ_1/θ_2								
$m = n$	2	3	4	5	8	10	12	15	20
10	0.123	0.060	0.052	0.047	0.051	0.054	0.060	0.046	0.051
20	0.054	0.056	0.044	0.050	0.058	0.053	0.042	0.049	0.039
30	0.058	0.033	0.063	0.046	0.041	0.057	0.043	0.052	0.048
50	0.054	0.045	0.049	0.062	0.049	0.049	0.048	0.056	0.052
100	0.054	0.053	0.039	0.055	0.040	0.057	0.043	0.052	0.045

Simulated values of e_1 when pre-fixed $e_1 = 0.10$ and $\mu = 1$

	Ratio θ_1/θ_2								
$m = n$	2	3	4	5	8	10	12	15	20
10	0.151	0.114	0.108	0.096	0.105	0.104	0.100	0.101	0.121
20	0.121	0.090	0.096	0.092	0.109	0.086	0.093	0.108	0.097
30	0.108	0.114	0.080	0.107	0.095	0.102	0.112	0.109	0.090
50	0.096	0.096	0.105	0.105	0.098	0.097	0.110	0.095	0.107
100	0.097	0.107	0.114	0.097	0.103	0.102	0.092	0.107	0.084

Simulated values of e_1 when pre-fixed $e_1 = 0.05$ and $\mu = 9$

	Ratio θ_1/θ_2								
$m = n$	2	3	4	5	8	10	12	15	20
10	0.103	0.053	0.057	0.057	0.048	0.046	0.053	0.058	0.045
20	0.045	0.038	0.057	0.041	0.062	0.051	0.052	0.059	0.053
30	0.064	0.042	0.048	0.055	0.044	0.045	0.050	0.039	0.041
50	0.064	0.062	0.053	0.057	0.045	0.053	0.057	0.043	0.062
100	0.053	0.046	0.058	0.048	0.056	0.057	0.042	0.032	0.042

Simulated values of e_1 when pre-fixed $e_1 = 0.10$ and $\mu = 9$

	Ratio θ_1/θ_2								
$m = n$	2	3	4	5	8	10	12	15	20
10	0.132	0.102	0.099	0.088	0.083	0.090	0.102	0.106	0.109
20	0.089	0.098	0.115	0.088	0.095	0.097	0.093	0.092	0.119
30	0.106	0.098	0.093	0.090	0.085	0.084	0.098	0.086	0.081
50	0.099	0.097	0.096	0.092	0.090	0.096	0.092	0.101	0.091
100	0.089	0.105	0.094	0.108	0.104	0.101	0.086	0.092	0.112

Table 32.8:

Simulated power values when pre-fixed $e_1 = 0.05$ and $\mu = 1$

	Ratio θ_1/θ_2								
$m = n$	2	3	4	5	8	10	12	15	20
10	0.165	0.126	0.174	0.203	0.335	0.383	0.446	0.526	0.673
20	0.108	0.149	0.178	0.225	0.328	0.395	0.444	0.537	0.648
30	0.108	0.159	0.182	0.222	0.335	0.426	0.451	0.531	0.648
50	0.104	0.146	0.176	0.202	0.357	0.417	0.469	0.540	0.642
100	0.094	0.145	0.179	0.234	0.349	0.397	0.473	0.505	0.640
	0.098	0.143	0.185	0.226	0.337	0.401	0.460	0.537	0.642

Simulated power values when pre-fixed $e_1 = 0.10$ and $\mu = 1$

	Ratio θ_1/θ_2								
$m = n$	2	3	4	5	8	10	12	15	20
10	0.209	0.284	0.329	0.369	0.537	0.619	0.678	0.775	0.851
20	0.217	0.270	0.330	0.427	0.554	0.627	0.697	0.768	0.882
30	0.201	0.281	0.333	0.398	0.582	0.659	0.700	0.787	0.877
50	0.185	0.266	0.354	0.415	0.553	0.637	0.718	0.800	0.879
100	0.193	0.293	0.340	0.404	0.581	0.679	0.700	0.769	0.871
	0.190	0.271	0.344	0.410	0.570	0.651	0.718	0.794	0.790

Simulated power values when pre-fixed $e_1 = 0.05$ and $\mu = 9$

	Ratio θ_1/θ_2								
$m = n$	2	3	4	5	8	10	12	15	20
10	0.164	0.138	0.186	0.229	0.317	0.436	0.486	0.491	0.629
20	0.093	0.154	0.158	0.228	0.357	0.414	0.463	0.509	0.624
30	0.090	0.160	0.173	0.224	0.337	0.391	0.459	0.509	0.648
50	0.088	0.135	0.177	0.203	0.321	0.389	0.458	0.548	0.630
100	0.097	0.148	0.158	0.212	0.330	0.391	0.437	0.520	0.636
	0.098	0.143	0.185	0.226	0.337	0.401	0.460	0.537	0.642

Simulated power values when pre-fixed $e_1 = 0.10$ and $\mu = 9$

	Ratio θ_1/θ_2								
$m = n$	2	3	4	5	8	10	12	15	20
10	0.220	0.270	0.348	0.397	0.545	0.650	0.690	0.749	0.862
20	0.213	0.256	0.297	0.442	0.555	0.622	0.688	0.794	0.869
30	0.189	0.253	0.249	0.396	0.595	0.658	0.702	0.788	0.874
50	0.193	0.264	0.339	0.430	0.583	0.665	0.696	0.796	0.870
100	0.194	0.276	0.366	0.410	0.545	0.646	0.742	0.779	0.873
	0.190	0.271	0.344	0.410	0.570	0.651	0.718	0.794	0.878

32.6 Classification When Parameters are Unknown but with Known Ordering

In this section, we consider the situation in which the parameters θ_1 and θ_2 of the populations $e(\theta_1)$ and $e(\theta_2)$ are unknown but it is known that $\theta_1 \geq \theta_2$. Such a situation arises, for example, when we know that the items produced by one process are superior (tend to have longer lifetimes) to those produced by another.

32.6.1 Model

Let $x_1, x_2, \ldots, x_m$ be a random sample of size m from Population A which is $e(\theta_1)$ and $y_1, y_2, \ldots, y_n$ be an independent random sample of size n from Population B which is $e(\theta_2)$. Let z be an independent new observation that needs to be classified into either A or B.

32.6.2 Likelihood-Ratio Classification Procedure

When z comes from Population A, the joint likelihood function is

$$L_1 = \frac{1}{\theta_1^{m+1}} e^{-\{\sum_{i=1}^m x_i + z\}/\theta_1} \frac{1}{\theta_2^n} e^{-\sum_{i=1}^n y_i/\theta_2} \tag{32.6.1}$$

from which we find the MLEs of θ_1 and θ_2 subject to the restriction that $\theta_1 \geq \theta_2$ (see Barlow, Bartholomew, Bremner and Brunk, 1972, p. 45) to be isotonic estimators

$$(\hat{\theta}_1, \hat{\theta}_2) = \begin{cases} \left(\frac{m\bar{x}+z}{m+1}, \bar{y}\right) & \text{if } \frac{m\bar{x}+z}{m+1} \geq \bar{y}, \\ \left(\frac{m\bar{x}+n\bar{y}+z}{m+n+1}, \frac{m\bar{x}+n\bar{y}+z}{m+n+1}\right) & \text{otherwise.} \end{cases} \tag{32.6.2}$$

Upon substituting these estimates in (32.6.1), we obtain

$$\sup_{\theta_1 \geq \theta_2} L_1 = \begin{cases} \left(\frac{m+1}{m\bar{x}+z}\right)^{m+1} \bar{y}^{-n} e^{-(m+n+1)} & \text{if } m\bar{x} + z \geq (m+1)\bar{y}, \\ \left(\frac{m+n+1}{m\bar{x}+n\bar{y}+z}\right)^{m+n+1} e^{-(m+n+1)} & \text{otherwise.} \end{cases} \tag{32.6.3}$$

Similarly, when z comes from Population B, we get

$$\sup_{\theta_1 \geq \theta_2} L_2 = \begin{cases} \left(\frac{n+1}{n\bar{y}+z}\right)^{n+1} \bar{x}^{-m} e^{-(m+n+1)} & \text{if } (n+1)\bar{x} \geq n\bar{y} + z, \\ \left(\frac{m+n+1}{m\bar{x}+n\bar{y}+z}\right)^{m+n+1} e^{-(m+n+1)} & \text{otherwise.} \end{cases} \tag{32.6.4}$$

Let $W = \sup_{\theta_1 \geq \theta_2} L_1 / \sup_{\theta_1 \geq \theta_2} L_2$. Then the likelihood-ratio classification procedure classifies z into A or B according as

$$W \gtrless C \tag{32.6.5}$$

where C is to be determined appropriately after pre-fixing the value of e_1. However, the distribution of W defined in terms of (32.6.3) and (32.6.4) is intractable which makes the determination of C impossible.

32.6.3 Plug-in Classification Procedure

When the parameters θ_1 and θ_2 ($\theta_1 > \theta_2$) are known, the likelihood-ratio classification procedure is given by (32.2.7). So when $\theta_1 > \theta_2$ are unknown, we propose a simple plug-in procedure by using the MLEs of θ_1 and θ_2 under the restriction $\theta_1 \geq \theta_2$ in the procedure defined in (32.2.7). This procedure classifies z into A or B according as

$$z > -\hat{\theta}_1 \log(1 - e_1) , \tag{32.6.6}$$

where

$$\hat{\theta}_1 = \begin{cases} \bar{x} & \text{if } \bar{x} \geq \bar{y} , \\ \frac{m\bar{x}+n\bar{y}}{m+n} & \text{otherwise,} \end{cases} \tag{32.6.7}$$

and e_1 is the pre-fixed probability of misclassifying z when it really belongs to A.

32.6.4 Errors of Misclassification

We derive here exact expressions for the error of misclassification (denoted by e_1^*) and the power $(1-e_2^*)$ of the plug-in procedure in (32.6.6). Let $U = \sum_{i=1}^m x_i$, $V = \sum_{i=1}^n y_i$, $b_1 = m/n$, $b_2 = -\frac{1}{m}\log(1-e_1)$, $b_3 = -\frac{1}{n}\log(1-e_1)$, and $b_4 = b_2 m/(m+n)$.

When z belongs to Population A, it will get wrongly classified into Population B by the plug-in procedure in the following two ways:

$$\begin{array}{llll} (i) & z < -\bar{x}\log(1-e_1) & \text{when} & \bar{x} \geq \bar{y} \\ (ii) & z < -\frac{m\bar{x}+n\bar{y}}{m+n}\log(1-e_1) & \text{when} & \bar{x} < \bar{y} . \end{array} \tag{32.6.8}$$

First,

$$\begin{aligned} \Pr\{\bar{x} \geq \bar{y},\ z &< -\bar{x}\log(1-e_1) \mid z \in A\} \\ &= \Pr\{U \geq b_1 V,\ z < b_2 U \mid z \in A\} \\ &= \text{the expression in (32.3.8).} \end{aligned} \tag{32.6.9}$$

Next,

$$\begin{aligned} &\Pr\left\{\bar{x} < \bar{y},\ z < -\frac{m\bar{x}+n\bar{y}}{m+n}\log(1-e_1) \mid z \in A\right\} \\ &= \Pr\{U < b_1 V,\ z < b_4(U+V) \mid z \in A\} \\ &= \int_0^\infty \int_{u/b_1}^\infty \int_0^{b_4(u+v)} \frac{1}{\theta_1^m \Gamma(m)} e^{-u/\theta_1} u^{m-1} \cdot \frac{1}{\theta_2^n \Gamma(n)} e^{-v/\theta_2} v^{n-1} \\ &\qquad \times \frac{1}{\theta_1} e^{-z/\theta_1}\, dz\, dv\, du \\ &= \mathrm{I}_1 - \mathrm{I}_2 , \end{aligned} \tag{32.6.10}$$

where

$$\mathrm{I}_1 = \int_0^\infty \int_{u/b_1}^\infty \frac{u^{m-1}}{\theta_1^m \Gamma(m)} e^{-u/\theta_1} \cdot \frac{v^{n-1}}{\theta_2^n \Gamma(n)} e^{-v/\theta_2}\, dv\, du$$

and

$$\mathrm{I}_2 = \int_0^\infty \int_{u/b_1}^\infty \frac{u^{m-1}}{\theta_1^m \Gamma(m)} e^{-u(1+b_4)/\theta_1} \cdot \frac{v^{n-1}}{\theta_2^n \Gamma(n)} e^{-v(\theta_1+b_4\theta_2)/(\theta_1\theta_2)}\, dv\, du .$$

Straightforward integration of the above integrals yields:

$$\mathrm{I}_1 = \frac{1}{\Gamma(m)}\left(\frac{b_1}{\theta_1}\right)^m \sum_{i=0}^{n-1} \frac{\Gamma(m+i)}{i!\theta_2^i}\left(\frac{1}{\theta_2}+\frac{b_1}{\theta_1}\right)^{-m-i},$$

$$\mathrm{I}_2 = \frac{1}{\Gamma(m)\theta_2^n}\sum_{i=0}^{n-1}\frac{\Gamma(m+i)}{i!}\left(\frac{\theta_1}{b_1}\right)^i\left(\frac{1}{\theta_2}+\frac{b_4}{\theta_1}\right)^{-n+i}\left(1+b_2+\frac{\theta_1}{b_1\theta_2}\right)^{-m-i}. \qquad (32.6.11)$$

Letting I_3 denote the expression in (32.6.9), which is the same as the one in (32.3.8), the error of misclassification e_1^* is given by

$$e_1^* = \mathrm{I}_3 + \mathrm{I}_1 - \mathrm{I}_2 . \qquad (32.6.12)$$

When z belongs to Population B, it will get correctly classified into Population B by the plug-in procedure in the two ways given in (32.6.8).

Then, by proceeding exactly as in the case of e_1^*, we obtain

$$\begin{aligned}\text{Power}^* = {} & 1-\left(1+\frac{b_2\theta_1}{\theta_2}\right)^{-m} + \frac{1}{\Gamma(m)}\sum_{i=0}^{n-1}\frac{\Gamma(m+i)}{i!}\left(\frac{\theta_1}{b_1\theta_2}\right)^i \\ & \times\left(1+\frac{b_2\theta_1}{\theta_2}+\frac{\theta_1}{b_1\theta_2}\right)^{-m-i} \\ & -\frac{(1+b_4)^{-(m+n)}}{\Gamma(m)}\sum_{i=0}^{n-1}\frac{\Gamma(m+i)}{i!}\left(\frac{\theta_1}{b_1\theta_2}\right)^i \\ & \times\left(\frac{1}{1+b_4}+\frac{\theta_1}{(1+b_4)\theta_2}+\frac{\theta_1}{b_1(1+b_4)\theta_2}\right)^{-m-i}.\end{aligned} \qquad (32.6.13)$$

In the data for the illustrative example in Section 32.3.5, suppose that the x-sample (y-sample) came from Population A (Population B). We can then evaluate e_1^* and Power* for different choices of e_1 and θ_1/θ_2. It will also be interesting to compare these with earlier results (when the ordering of θ_1 and θ_2 is not known) to evaluate the gain due to the knowledge of the ordering.

References

Adegboye, O.S. (1993). The optimal classification rule for exponential populations, *Australian Journal of Statistics*, **35**, 185-194.

Barlow, R.E., Bartholomew, D.J., Bremner, J.M., and Brunk, H.D. (1972). *Statistical Inference under Order Restrictions*, John Wiley & Sons, New York.

Basu, A.P. and Gupta, A.K. (1974). Classification results for exponential populations, In *Reliability and Biometry* (Eds., F. Proschan and R.J. Serfling), pp. 637-650, SIAM, Philadelphia.

Basu, A.P. and Gupta, A.K. (1977). Classification rules for exponential populations: Two parameter case, In *Theory and Applications of Reliability* (Eds., C.P. Tsokos and I.N. Shimi), Vol. 1, 507-525, Academic Press, New York.

Giri, N.C. (1977). *Multivariate Statistical Inference*, Academic Press, New York.

Gordon, A.D. (1981). *Classification: Methods for the Exploratory Analysis of Multivariate Data*, Chapman and Hall, London.

Lachenbruch, P.A. (1974). *Discriminant Analysis*, Hafner Press, New York.

Onyeagu, S.I. (1985). A review of estimation of error rates in discriminant analysis, *M.Sc. Dissertation*, A.B.U. Zaria, Nigeria.

Rao, C.R. (1973). *Linear Statistical Inference and Its Applications*, John Wiley & Sons, New York.

Zhang, Q. (1992). Two-way exponential classification, *M.Sc. Project*, McMaster University, Hamilton, Ontario, Canada.

CHAPTER 33

Computer Simulations

Pandu R. Tadikamalla[1] **and N. Balakrishnan**[2]

33.1 Introduction

Recall (from Chapter 1), that the random variable Y has an exponential distribution if it has the cdf of form:

$$F_Y(y) = 1 - \exp[-(y-\mu)/\theta], \qquad (y > \mu;\ \theta > 0) \tag{33.1.1}$$

and the pdf is

$$f_Y(y) = \exp[-(y-\mu)/\theta]/\theta, \qquad (y > \mu;\ \theta > 0)\ . \tag{33.1.2}$$

The percentile function (inverse cdf) is

$$y = \mu + \theta[-\log(1 - F_Y(y))]\ . \tag{33.1.3}$$

If we set

$$X = (Y-\mu)/\theta\ , \tag{33.1.4}$$

then the random variable X has the standard exponential distribution with cdf

$$F_X(x) = 1 - e^{-x}, \qquad x \geq 0 \tag{33.1.5}$$

and the pdf is

$$f_X(x) = e^{-x}, \qquad x \geq 0\ . \tag{33.1.6}$$

The percentile function (inverse cdf) is

$$x = -\log[1 - F_X(x)]\ . \tag{33.1.7}$$

If $\mu = 0$ in (33.1.1)-(33.1.3), we have the most commonly used form of the *one-parameter* exponential distribution.

[1]University of Pittsburgh, Pittsburgh, Pennsylvania
[2]McMaster University, Hamilton, Ontario, Canada

33.2 Simulation

In this section, we briefly discuss (i) the *general* methods for generating random variates from continuous distributions, (ii) the criteria for random variate generation algorithm evaluation and comparisons. Then, we present the state of the art algorithms for exponential variate generation.

33.2.1 General Methods

Most of the random variate generation algorithms fall into one of the following techniques [see Tadikamalla (1975, 1984)]: (i) inverse transformation method, (ii) special transformation methods, (iii) acceptance/rejection method and (iv) the composition or the mixture method. We briefly describe each of these techniques in this section.

The Inverse Transformation Method is based on the following well-known result [see Halton (1970)]: If X has distribution function $F(x)$, then $U = F(X)$ is distributed with uniform density on $(0, 1)$; conversely, if U is uniform in $(0, 1)$, then $G(U) = \inf\{x : U \leq F(x)\}$ has distribution function $F(x)$. This method, yields simple algorithms for variates having $G = F^{-1}$ in simple close form. Examples include the Cauchy, the exponential and the Weibull variates.

The Transformation Method: This method cannot be generalized like the inverse transformation method and yields unique algorithms based on specific properties of a random variable. Examples include (i) the generation of k-Erlang variates by adding k i.i.d. standard exponential variates, ii) the Box-Muller (1958) transformation of a pair of independent uniform (0,1) random numbers (u_1, u_2) into two independent standard normal variates (x_1, x_2) by

$$\begin{aligned} x_1 &= (-2\log u_1)^{1/2}\cos(2\pi u_2) \\ x_2 &= (-2\log u_1)^{1/2}\sin(2\pi u_2)\ . \end{aligned}$$

The Acceptance/Rejection Technique: The acceptance/rejection technique, due to von Neumann (1951) has been presented by various authors in different ways. A generalized version of the rejection technique is given in Tadikamalla (1978) and Tadikamalla and Johnson (1981). In a simple form, the rejection technique is as follows:

Let $f(x)$, $x \in I$, be the probability density function (pdf) from which the samples are required with $f(x) < \infty$, $x \in I$. Let $h(x)$, $x \in J$, $(I \subset J)$, be another pdf, such that $f(x) < Kh(x)$ for all $x \in J$. Note that $K = \text{Max}\{f(x/h(x))\}$ and $K \geq 1$. Also the pdf $h(\cdot)$ is referred to as an enveloping distribution. Then, sampling from $f(x)$ can be carried out as follows:

Step 1. Generate a random variate x from $h(x)$.

Step 2. Generate a uniform $(0, 1)$ random number u.

Step 3. If $u \leq T(x) = f(x)/Kh(x)$, accept $X = x$ as a random variate from the required $f(x)$. Otherwise return to Step 1.

For this rejection procedure to be of practical value, the following criteria must be used in selecting $h(x)$:

(i) It should be easy to generate random variates from $h(x)$.

(ii) The efficiency of the procedure $(1/K)$ should be as high as possible.

Examples of this method include most of the algorithms for gamma variate generation [see Tadikamalla and Johnson (1981)] and beta variate generation [see Schmeiser and Lal (1980)]. Also note that the ratio of uniforms method [Kinderman and Monahan (1977)] is essentially an acceptance/rejection method.

The Composition or Mixture Method: The composition/mixture method [Butler (1956) and Butcher (1961)] is based upon representing the required f as $f(x) = p_1 f_1 + p_2 f_2 + \cdots + p_n f_n$, where $\sum p_i = 1$ and each of the f_i's are pdf's. The rule of thumb for developing successful algorithms using the mixture method is (i) to select the f_i's so that f_1 is the simplest and the fastest f_i to generate variates from, (ii) p_1 is large (close to 1) and (iii) the other f_i's are not unduly difficult to generate. Then, the corresponding composition/mixture algorithm would be simply to generate variates from each f_i with probability p_i. If the required density has no parameters such as in the case of the standard normal or the standard exponential, efficient mixture algorithms can be developed by carefully choosing the f_i's. In fact, some of the efficient algorithms for the normal and the exponential distributions are mixture algorithms [see Kinderman and Ramage (1976) and Marsaglia (1961)]. In the case of other distributions such as the gamma and the beta distributions, the mixture method has not received the same attention as it has for the normal and the exponential distributions. This is primarily due to the awkward problem that a different mixture has to be used for different parameters.

These four different methods are not mutually exclusive in the sense that an algorithm for generating random variates from a specified distribution may use more than one method. For example, in the rejection method, random variates from $h(\cdot)$ may be obtained using the inverse transformation method. Similarly, most of the algorithms based on the mixture method use either the inverse transformation method or rejection method to sample from f_i's.

33.2.2 Criteria for Algorithm Evaluations

The important criteria used in judging random variate generation algorithms are *speed, simplicity, accuracy and generality of the technique. Speed* can be measured in terms of the marginal CPU time required for generating one variate from a distribution for a given set of parameters. The set-up time required when a change in the parameters occurs should be taken into consideration in comparing algorithms. It is difficult to measure *simplicity*, but it can be viewed as a combination of ease of implementation (which in turn can be considered in terms of program length, portability, and special functions or subprograms required, if any) and the core storage required for the algorithm.

Random variate generation algorithms can be approximate or exact, theoretically. We will limit the discussion to exact methods. These methods are exact, subject only to the *accuracy* of the machine and of the standard machine functions logarithm, exponential, and square root. Thus, for comparison purposes only the speed and simplicity become relevant criteria.

33.2.3 Algorithms and Comparisons

If X is the standard exponential variate with c.d.f. (33.1.5), then the random variate $Y = \mu + \theta X$ will have the general exponential distribution given by (33.1.1). Thus, it is only necessary to discuss the simulation of standard exponential variates, X from (33.1.5).

Almost all the algorithms for generating non-uniform random variates require the use of a uniform $(0, 1)$ random number generator. Not all random number generators are good and the reader should beware. He or she should understand the properties of good random number generators and should have access to some good random number generators. Most of the random number generators available on computers are in some form or other, linear congruential generators, initially proposed by Lehmer (1954). One of the popular generators (available on DEC and VAX computers in FORTRAN language and also in the IMSL package) that has been extensively tested by Learmonth and Lewis (1973) and Lewis, Goodman, and Miller (1969) is a multiplicative generator and belongs to this class. We assume that a good uniform random number generator is available. We now describe a few exponential variate generation algorithms [see Devroye (1986)].

A. **A Simple algorithm**, based on the inverse transformation method, can be obtained for sampling from the exponential distribution. Equating $F_X(x)$ in (33.1.5) to a uniform random number value u and solving for x, we obtain

$$x = -\log(1 - u) . \qquad (33.2.1)$$

Note that x is monotonically increasing with u. We can replace $(1 - u)$ in (33.2.1) by u, since if $(1 - u)$ is uniform $(0, 1)$ so is u. Although doing so reduces one subtraction operation, the use of (33.2.1) has some advantages for some variance reduction techniques such as antithetic sampling.

Generating exponential variates, using (33.2.1), probably is the easiest way although it may not be the fastest way. The logarithm function in (33.2.1) is slow and takes relatively a lot of computer time. This has given incentive to several authors to develop faster algorithms.

B. **Fast Algorithms:** Here, we discuss a few algorithms that are significantly faster than the one in (33.2.1). All these algorithms are considerably lengthy and if implemented in machine languages (such as assembler) may yield faster algorithms that may be worth the effort.

Two of the fastest algorithms are due to Marsaglia (1961) and Marsaglia, MacLaren, and Bray (1964). Marsaglia's algorithm is based on the following result. If N is a truncated Poisson random variate with pmf

$$P(N = i) = \frac{1}{e^{\lambda} - 1} \cdot \frac{\lambda^i}{i!} \qquad (i \geq 1)$$

where $\lambda > 0$ is a constant and if M is a geometric random variate with pmf

$$P(M = i) = (1 - e^{-\lambda})e^{-\lambda i} \qquad (i \geq 0) ,$$

then $x = \lambda(M + \min(u_1, u_2, \ldots, u_N))$ is exponentially distributed where, $u_1, u_2, \ldots$ are independent uniform $(0, 1)$ random variables.

Faster algorithms for generating Poisson and geometric variates and the range for the optimal values of the parameter λ are discussed in Marsaglia (1961).

Another fast method, based on partial table look up called the *rectangle-wedge-tail method* is given by Marsaglia, MacLaren, and Bray (1964). As mentioned before, it is ideally suited for machine language implementations and is available under the name *superduper* package from McGill University [Marsaglia, Ananthanarayanan, and Paul (1973)]. The *Super Duper* package also has fast uniform random number and normal variate generators.

Ahrens and Dieter (1972) developed a fast algorithm based on Marsaglia's method. They took the special case $\lambda = \log 2$ which allows one to generate the desired geometric random variate by manipulating the random bits in a uniform $(0,1)$ random number.

C. Von Neumann-Forsythe Generators

von Neumann (1951) presented a very interesting method for generating exponential variates requiring only comparisons and a uniform $(0,1)$ random number generator. Forsythe (1972) later generalized the technique to other distributions and the resulting algorithms are not simple since they require more than simple comparisons. However, most of the algorithms based on von Neumann-Forsythe method have been dominated by other algorithms based on the alias method [Kronmal and Peterson (1979)] and acceptance-rejection methods [Kronmal and Peterson (1981)]. Later, Monahan (1979) generalized the von Neumann-Forsythe method for generating random variates.

D. Other Methods

D.1 The Ratio of uniforms method was developed by Kinderman and Monahan (1977) and is suitable for densities whose tails decrease at least as fast as x^{-2}. The algorithm based on the ratio of uniforms are reasonably fast and are moderately short. Random variate generation algorithms for several distributions, including the exponential distribution, are given in Kinderman and Monahan (1977).

D.2 Devroye (1980, 1981) developed the **series methods** (including series method, convergent series method and alternating series method) for generating random variates from arbitrary distributions. These series methods are essentially acceptance-rejection methods where the density of the required random variable is bounded below and above by a series of functions. If these functions are simple to evaluate, the resulting algorithm is fast due to quick acceptance or rejection.

D.3 Uniform spacings method (Exponential Variates in Batches).

A batch of i.i.d. exponential variates can be generated as follows:

1. Generate an ordered sample from the uniform $(0,1)$ distribution, $u_{1:n-1} \leq u_{2:n-1} \leq \cdots \leq u_{n-1:n-1}$.
2. Generate a gamma variate G_n with shape parameter, n (an n-Erlang variate).
3. Then $E_1 = G_n\, u_{1:n-1}$, $E_2 = G_n\,(u_{2:n-1} - u_{1:n-1}), \ldots, E_n = G_n\,(1 - u_{n-1:n-1})$ are i.i.d. exponential variates. With fast sorting algorithms for obtaining $u_{1:n-1}$, $u_{2:n-1}$, $\ldots, u_{n-1:n-1}$ (such as a bucket sort with a large number of buckets) and a fast

gamma variate generator such as the one in Schmeiser and Lal (1980), we can obtain an efficient exponential variate generator. An interesting and useful special case is that UG_2 and $(1-U)G_2$ are i.i.d. exponential variates (U is a uniform $(0,1)$ and G_2 is a 2-Erlang variate).

33.3 Computer Simulation of Order Statistics

Recall (from Chapter 3) that if $X_1, X_2, \ldots, X_n$ denote a random sample from a distribution with distribution function (df) $F(x)$ and $X_{1:n} \leq X_{2:n} \leq \cdots \leq X_{n:n}$ are the ordered X's, then $X_{1:n}, X_{2:n}, \ldots, X_{n:n}$ are called the *order statistics* corresponding to $X_1, X_2, \ldots, X_n$. As pointed out in Chapter 3, order statistics play an important role in statistical inference and in simulation studies for a number of reasons. For example, some of the properties of order statistics do not depend upon the distribution from which the random sample is obtained. In some simulation studies, only a portion of the order statistics is of interest.

Most of the algorithms for computer simulation of exponential order statistics mainly fall into two categories: based on exponential variate generation (discussed in the previous section) and based on uniform $(0,1)$ order statistics.

A. Sorting Method: This method simply requires generating n exponential variates $E_1, E_2, \ldots, E_n$ and sorting them to obtain $E_{1:n} \leq E_{2:n} \leq \cdots \leq E_{n:n}$. A naive approach to sorting based on pairwise comparisons of the numbers could be very time consuming. The worst-case and expected times taken by these algorithms are called omega $(n \cdot \log n)$ [Knuth (1973)]. Quick sort algorithms have expected time $0(n \cdot \log n)$ and worst-case times of $0(n^2)$.

Lurie and Hartley (1972), Schucany (1972) and Lurie and Mason (1973) developed bucket sort/bucket-search algorithms that are $0(n)$ in the worst-case and in expected time $0(n)$.

B. Via Exponential Distribution: Using the memoryless property of the exponential distribution, we can generate the exponential order statistics as follows:

1. Generate n independent exponential variates $E_1, E_2, \ldots, E_n$.
2. Then $E_{i:n} = E_{i-1:n} + \frac{E_i}{n-i+1}$ are the required order statistics (Note $E_{0:n} = 0$.)

This algorithm takes time $0(n)$.

C. Based on Uniform Order Statistics: Since the inverse transformation method can be used to generate exponential variates, there is a direct correspondence between the order statistics of the exponential distribution $(X_{1:n}, X_{2:n}, \ldots, X_{n:n})$ and the order statistics of the associated uniform sample $(U_{1:n}, U_{2:n}, \ldots, U_{n:n})$. Since F^{-1} is a monotonic function, $X_{i:n} = F^{-1}(U_{i:n})$, $i = 1, 2, \ldots, n$, represent the order statistics from the distribution $F(\cdot)$. Thus, the problem of generating exponential order statistics reduces to the problem of generation of order statistics from the uniform $(0,1)$ distribution.

C.1 A *straightforward approach* for generating the uniform $(0,1)$ order statistics is to generate a sample of size n and then sort the sample to obtain the desired order statistics. For large values of n this sorting procedure can be very time consuming relative to a direct method for the generation of observations from the joint distribution of the extreme values. Now we discuss the generation of certain sets of order statistics from the uniform $(0,1)$ distribution.

Balakrishnan and Cohen (1991), David (1981), and Arnold, Balakrishnan and Nagaraja (1992) (also see Chapter 3) discuss the marginal and joint probability densities of order statistics from a continuous distribution. These results take a simple form in the case of the uniform $(0,1)$ distribution and yield simple methods for sequential generation of order statistics.

C.2 Lurie and Hartley (1972) gave a direct method for generating a complete set of n uniform order statistics in *ascending* order, as follows: Let $v_1, v_2, v_3 \ldots$ be observations from the uniform $(0,1)$ distribution. Set

$$\begin{aligned} u_{1:n} &= 1 - v_1^{1/n} \\ u_{2:n} &= 1 - (1 - u_{1:n})\, v_2^{1/(n-1)} \\ &\vdots \\ u_{i+1:n} &= 1 - (1 - u_{i:n})\, v_{i+1}^{1/(n-i)} . \end{aligned}$$

Then $u_{1:n}, u_{2:n}, \ldots, u_{n:n}$ are order statistics from the uniform $(0,1)$ distribution.

C.3 Schucany (1972) used a similar approach to generate the order statistics in *descending* order, without sort, as follows:

$$\begin{aligned} u_{n:n} &= v_1^{1/n} \\ &\vdots \\ u_{n-i:n} &= u_{n-i+1:n} v_{i+1}^{1/(n-i)} \quad (i = 1, 2, \ldots, n-1) . \end{aligned}$$

Reeder (1972) suggested that generating order statistics in *descending* order is faster than that in ascending order, and Lurie and Mason (1973) confirmed this claim.

C.4 Suppose that one is interested in generating a subset of k consecutive uniform $(0,1)$ order statistics from a sample of size n. If the required k order statistics are at the upper extreme of the sample n, they can be generated using Schucany's descending method. If the required subset is at the lower extreme, they can be generated using Lurie and Hartley's (1972) ascending method.

Neither of these methods is convenient if the required subset falls near the *middle* of the sample size n. For example, if we are interested in generating the middle two order statistics from a sample of size $n = 40$, then both of the previously described methods require the generation of twenty one order statistics. A more efficient approach may be to generate the largest required order statistic (21st in our example) directly; then generate the remaining set conditionally using the descending method. Ramberg and Tadikamalla (1978) suggest generating $u_{i:n}$ as a beta variate (with parameters i and $n-i+1$) and generating $u_{i-1:n}, u_{i-2:n}, \ldots$ using the above recursion algorithm suggested by Schucany (1972). If fast algorithms for generating beta variates are used [Schmeiser and Babu (1980)], this procedure may be worth the effort. This approach will probably be faster in generating a subset of order statistics when the required subset is somewhere in the middle of the sample of size n.

Devroye (1980) considers the case of $U_{n:n}$ when n is very large and evaluation of $U_{n:n} = V_1^{1/n}$ may cause numerical problems. Gerondtidis and Smith (1982) review the inversion method and the grouping method for generating order statistics from a general distribution. The *grouping method* can briefly be described as follows. Given an integer k (a suggested value is $n/4$), divide the range of the distribution into k equal probability intervals. Next generate a multinomial vector $(m_1, m_2, \ldots, m_k)$ corresponding to the division of n objects independently among k equally likely cells [see Fishman (1978) and Ho, Gentle, and Kennedy (1979)]. Then draw m_j variables (from the specified general distribution) from the jth interval for $1 \leq j \leq k$, and sort each group of m_j variables directly and put the k groups together to obtain a complete ordered sample. Gerontidis and Smith (1982) recommend the inversion method if F^{-1} exists in simple closed form as is in the case of the exponential distribution. Schmeiser (1978) provides a general survey of simulation of order statistics.

33.4 Computer Simulation of Progressive Type-II Censored Samples

As in the case of the simulation of order statistics, efficient algorithms may be developed for the computer generation of progressive Type-II censored samples from the exponential distribution.

A progressive Type-II censoring scheme is one in which n identical units are placed on a life-test; after the first failure, R_1 surviving items are removed at random from further observation; after the next failure, R_2 surviving items are removed at random, and so on; finally, after the mth failure, R_m remaining items are withdrawn. Thus, under this scheme, we observe in all m failures and $R_1 + R_2 + \cdots + R_m$ items are progressively censored so that $n = m + R_1 + R_2 + \cdots + R_m$. In Chapters 3, 4 and 5, we have already discussed some properties of such progressively censored samples and inference based on them.

As in Chapter 3, let us use $X_{i:m:n}$ to denote the ith observation in such a progressively censored sample. Then, as shown in Eq. (3.9.4),

$$X_{i:m:n} \stackrel{D}{=} \sum_{r=1}^{i} E_i \bigg/ \left(n - \sum_{j=1}^{r-1} (R_j + 1) \right)$$

where E_i's are independent standard exponential variables. Consequently, analogous to the algorithm in B for the order statistics, we have in this case

$$X_{i:m:n} = X_{i-1:m:n} + E_i \bigg/ \left(n - \sum_{j=1}^{i-1} (R_j + 1) \right)$$

with $X_{0:m:n} = 0$.

Balakrishnan and Sandhu (1995) have recently proposed an alternative method of simulation of progressive Type-II censored samples based on uniform (0,1) distribution.

This is analogous to the algorithm in C.2. Specifically, they have proposed the following algorithm:

Step 1 Generate m independent Uniform(0, 1) observations $U_1, U_2, \ldots, U_m$.

Step 2 Set $V_i = U_i^{1/(i+R_m+R_{m-1}+\cdots+R_{m-i+1})}$, $i = 1, 2, \ldots, m$.

Step 3 Set $W_i = 1 - V_m V_{m-1} \cdots V_{m-i+1}$, $i = 1, 2, \ldots, m$.

Then, $W_1 = U_{1:m:n}$, $W_2 = U_{2:m:n}, \ldots, W_m = U_{m:m:n}$ is the required progressive Type-II censored sample from the Uniform(0, 1) distribution. Next, by using the inverse distribution function $F^{-1}(w) = -\log(1-w)$, one can simulate the required progressive Type-II censored sample from the exponential distribution.

References

Ahrens, J.H. and Dieter, U. (1972). Computer methods for sampling from the exponential and the normal distributions, *Communications of the ACM*, **15**, 873-882.

Arnold, B.C., Balakrishnan, N., and Nagaraja, H.N. (1992). *A First Course in Order Statistics*, John Wiley & Sons, New York.

Balakrishnan, N. and Cohen, A.C. (1991). *Order Statistics and Inference: Estimation Methods*, Academic Press, San Diego.

Balakrishnan, N. and Sandhu, R.A. (1995). A simple simulational algorithm for generating progressive Type-II censored samples, *The American Statistician* (To appear).

Box, G.E.P. and Muller, M.E. (1958). A note on the generation of random normal deviates, *Annals of Mathematical Statistics*, **29**, 610-611.

Butcher, J.C. (1961). Random sampling from the normal distribution, *Computer Journal*, **3**, 251-253.

Butler, J.W. (1956). Machine sampling from given probability distributions, In *Symposium on Monte Carlo Methods* (Ed. H.A. Meyer), pp. 249-264, John Wiley & Sons, New York.

David, H.A. (1981). *Order Statistics*, Second edition, John Wiley & Sons, New York.

Devroye, L. (1980). Generating the maximum of independent identically distributed random variables, *Computers and Mathematics with Applications*, **6**, 305-315.

Devroye, L. (1981). The series method for random variate generation and its application to the Kolmogorov-Smirnov distribution, *American Journal of Mathematical and Management Sciences*, **1**, 359-379.

Devroye, L. (1986). *Non-uniform Random Variate Generation*, Springer-Verlag, New York.

Dudewicz, E.J. and Ralley, T.G. (1981). *The Handbook of Random Number Generation and Testing with TESTRAND Computer Code*, American Sciences Press, Columbus, Ohio, U.S.A.

Fishman, G.S. (1978). Sampling from the multinomial distribution on a computer, *Technical Report #78-5*, Curriculum in Operations Research and Systems Analysis, University of North Carolina, Chapel Hill.

Forsythe, G.E. (1972). von Neumann's comparison method for random sampling from the normal and other distributions, *Mathematics of Computation*, **26**, 817-826.

Gerontidis, I. and Smith, R.L. (1982). Monte Carlo generation of order statistics from general distributions, *Applied Statistics*, **31**, 238-243.

Halton, J.H. (1970). A retrospective and prospective survey of the Monte Carlo method, *SIAM Review*, **8**, 5-15.

Ho, F.C.M., Gentle, J.E., and Kennedy, W.J. (1979). Generation of random variates from the multinomial distribution, *Proceedings of the Statistical Computing Section, American Statistical Association*, 336-339.

Kennedy, Jr., W.J. and Gentle, J.E. (1980). *Statistical Computing*, Marcel Dekker, New York.

Kinderman, A.J. and Monahan, J.F. (1977). Computer generation of random variables using the ratio of uniform deviates, *Association for Computing Machinery, Monograph Transactions, Mathematical Software*, **3**, 257-260.

Kinderman, A.J. and Ramage, J.G. (1976). Computer generation of normal random variables, *Journal of the American Statistical Association*, **71**, 893-896.

Knuth, D.E. (1973). *The Art of Computer Programming, Vol. 3: Searching and Sorting*, Addison-Wesley, Reading, Massachusetts.

Kronmal, R.A. and Peterson, Jr., A.V. (1979). On the alias method for generating random variables from a discrete distribution, *The American Statistician*, **33**, 214-218.

Kronmal, R.A. and Peterson, Jr., A.V. (1981). A variant of the acceptance-rejection method for computer generation of random variables, *Journal of the American Statistical Association*, **76**, 446-451.

Learmonth, G.P. and Lewis, P.A.W. (1973). Naval Postgraduate School random number generator package LLRANDOM, *Naval Postgraduate School Technical Report*, Monterey, California.

Lehmer, D.H. (1954). Random Number Generation on the BRL High Speed Computing Machines, *Proceedings of the Second Symposium on Large-Scale Digital Computing Machinery*, Harvard University Press, Cambridge, Massachusetts.

Lewis, P.A.W., Goodman, A.S., and Miller, J.M. (1969). A pseudo-random number generator for the system/360, *IBM Systems Journal*, **8**, 136-145.

Lurie, D. and Hartley, H.O. (1972). Machine generation of order statistics for Monte Carlo computations, *The American Statistician*, **26**, 26-27.

Lurie, D. and Mason, R.L. (1973). Empirical investigation of general techniques for computer generation of order statistics, *Communications in Statistics*, **2**, 363-371.

Marsaglia, G. (1961). Generating exponential random variables, *Annals of Mathematical Statistics*, *32*, 899-902.

Marsaglia, G. (1962). Random variables and computers, *Transactions of the Third Prague Conference* (Ed., J. Kozesnik), pp. 499-510, Czechoslovak Academy of Sciences, Prague.

Marsaglia, G., Ananthanarayanan, K., and Paul, N. (1973). How to use the McGill random number package, *McGill University Report*, Montreal, Canada.

Marsaglia, G., MacLaren, M.D., and Bray, T.A. (1964). A fast procedure for generating exponential random variables, *Communications of the ACM*, **7**, 298-300.

Monahan, J.F. (1979). Extensions of von Neumann's method for generating random variables, *Mathematics of Computation*, **33**, 1065-1069.

Neumann, J. von (1951). Various techniques in connection with random digits, Monte Carlo methods, *National Bureau of Standards*, Applied Mathematics Series, **12**, 36-38, U.S. Government Printing Office, Washington, D.C.

Ramberg, J.S. and Tadikamalla, P.R. (1978). On generation of subsets of order statistics, *Journal of Statistical Computation and Simulation*, **6**, 239-241.

Reeder, H.A. (1972). Machine generation of order statistics, *The American Statistician*, **26**, 56-57.

Schmeiser, B.W. (1978). Order statistics in digital computer simulation. A survey, *Proceedings of the 1978 Winter Simulation Conference*, 136-140.

Schmeiser, B.W. and Babu, A.J.G. (1980). Beta variate generation via exponential majorizing functions, *Operations Research*, **28**, 917-926.

Schmeiser, B.W. and Lal, R. (1980). Squeeze methods for generating gamma variates, *Journal of the American Statistical Association*, **75**, 679-682.

Schucany, W.R. (1972). Order statistics in simulation, *Journal of Statistical Computation and Simulation*, **1**, 281-286.

Tadikamalla, P.R. (1975). Modeling and generating stochastic methods for simulation studies, *Ph.D. Thesis*, University of Iowa, Iowa City.

Tadikamalla, P.R. (1978). Computer generation of gamma random variables, *Communications of the ACM*, **21**, 419-422.

Tadikamalla, P.R. (1984). Modeling and generating stochastic inputs for simulation studies, *American Journal of Mathematics and Management Sciences*, **3** & **4**, 203-223.

Tadikamalla, P.R. and Johnson, M.E. (1981). A complete guide to gamma variate generation, *American Journal of Mathematical and Management Sciences*, **1**, 213-236.

Bibliography

Abu-Salih, M.S., Ali Khan, M.S., and Husein, A. (1987). Prediction intervals of order statistics for the mixture of two exponential distributions, *Aligarh Journal of Statistics*, **7**, 11-22.

Aczél, J. (1966). *Lectures on Functional Equations and Their Applications*, Academic Press, New York.

Adegboye, O.S. (1993). The optimal classification rule for exponential populations, *Australian Journal of Statistics*, **35**, 185-194.

Ahrens, J.H. and Dieter, U. (1972). Computer methods for sampling from the exponential and the normal distributions, *Communications of the ACM*, **15**, 873-882.

Ahsanullah, M. (1975). A characterization of the exponential distribution, In *Statistical Distributions in Scientific Work* (Eds., G.P. Patil, S. Kotz, and J.K. Ord), Vol. **3**, pp. 131-135, D. Reidel, Dordrecht.

Ahsanullah, M. (1977). A characteristic property of the exponential distribution, *Annals of Statistics*, **5**, 580-582.

Ahsanullah, M. (1979). Characterization of the exponential distribution by record values, *Sankhyā, Series B*, **41**, 116-121.

Ahsanullah, M. (1980). Linear prediction of record values for the two parameter exponential distribution, *Annals of the Institute of Statistical Mathematics*, **32**, 363-368.

Ahsanullah, M. (1981a). Record values of the exponentially distributed random variables, *Statistische Hefte*, **2**, 121-127.

Ahsanullah, M. (1981b). On a characterization of the exponential distribution by weak homoscedasticity of record values, *Biometrical Journal*, **23**, 715-717.

Ahsanullah, M. (1982). Characterization of the exponential distribution by some properties of record values, *Statistische Hefte*, **23**, 326-332.

Ahsanullah, M. (1987a). Two characterizations of the exponential distribution, *Communications Statistics - Theory and Methods*, **16**, 375-381.

Ahsanullah, M. (1987b). Record statistics and the exponential distribution, *Pakistan Journal of Statistics, Series A*, **3**, 17-40.

Ahsanullah, M. (1988). *Introduction to Record Statistics*, Ginn Press, Needham Heights, Massachusetts.

Ahsanullah, M. (1991). Some characteristic properties of the record values from the exponential distribution, *Sankhyā, Series B*, **53**, 403-408.

Ahsanullah, M. and Kirmani, S.N.U.A. (1991). Characterizations of the exponential distribution by lower record values, *Communications in Statistics - Theory and Methods*, **20**, 1293-1299.

Aitchison, J. and Dunsmore, I.R. (1975). *Statistical Prediction Analysis*, Cambridge University Press, Cambridge, England.

Aitchison, J. and Silvey, S.D. (1960). Maximum-likelihood estimation procedures and associated tests of significance, *Journal of the Royal Statistical Society, Series B*, **22**, 154-171.

Aitkin, M., Anderson, D., and Hinde, J. (1981). Statistical modelling of data on teaching styles (with discussion), *Journal of the Royal Statistical Society, Series A*, **144**, 414-461.

Alam, M.S. and Basu, A.P. (1989). A two-stage test for exponentiality against IFRA alternatives, In *Recent Developments in Statistics and Their Applications* (Eds., J. Klein and J. Lee), Freedom Academy Publishing Company, Korea.

Ali, M. Masoom, Umbach, D., and Hassanein, K.M. (1981). Estimation of quantiles of exponential and double exponential distributions based on two order statistics, *Communications in Statistics - Theory and Methods*, **10**, 1921-1932.

Ali, M. Masoom, Umbach, D., and Saleh, A.K.Md.E. (1982). Small sample quantile estimation of the exponential distribution using optimal spacings, *Sankhyā, Series B*, **44**, 135-142.

Ali, M. Masoom, Umbach, D., and Saleh, A.K.Md.E. (1985). Tests of significance for the exponential distribution based on selected quantiles, *Sankhyā, Series B*, **47**, 310-318.

Al-Saadi, S.D. and Young, D.H. (1980). Estimators for the correlation coefficient in a bivariate exponential distribution, *Journal of Statistical Computation and Simulation*, **11**, 13-20.

Al-Saadi, S.D. and Young, D.H. (1982). A test for independence in a multivariate exponential distribution with equal correlation coefficients, *Journal of Statistical Computation and Simulation*, **14**, 219-227.

Al-Saadi, S.D., Scrimshae, D.F., and Young, D.H. (1979). Tests for independence of exponential variables, *Journal of Statistical Computation and Simulation*, **9**, 217-233.

Alzaid, A.A. and Al-Osh, M.A. (1992). Characterization of probability distributions based on the relation $X \stackrel{d}{=} U(X_1 + X_2)$, *Sankhyā, Series B*, **53**, 188-190.

Anderson, T.W. (1958). *An Introduction to Multivariate Statistical Analysis*, John Wiley & Sons, New York.

Anderson, T.W. (1960). A modification of the sequential probability ratio test to reduce the sample size, *Annals of Mathematical Statistics*, **31**, 165-197.

Anderson, T.W. and Ghurye, S.G. (1977). Identification of parameters by the distribution of a maximum random variable, *Journal of the Royal Statistical Society, Series B*, **39**, 337-342.

Angus, J.E. (1982). Goodness-of-fit tests for exponentiality based on a loss-of memory type functional equation, *Journal of Statistical Planning and Inference*, **6**, 241-251.

Angus, J.E. (1989). A note on the use of the Shapiro-Wilk test of exponentiality for complete samples, *Communications in Statistics - Theory and Methods*, **18**, 1819-1830.

Anscombe, F.J. (1952). Large sample theory of sequential estimation, *Proceedings of the Cambridge Philosophical Society*, **48**, 600-607.

Anscombe, F.J. (1960). Rejection of outliers, *Technometrics*, **2**, 123-147.

Anscombe, F.J. (1961). Estimating a mixed-exponential response law, *Journal of the American Statistical Association*, **56**, 493-502.

Aras, G. (1987). Sequential estimation of the mean exponential survival time under random censoring, *Journal of Statistical Planning and Inference*, **16**, 147-158.

Aras, G. (1989). Second order sequential estimation of the mean exponential survival time under random censoring, *Journal of Statistical Planning and Inference*, **21**, 3-17.

Arnold, B.C. (1968). Parameter estimation for a multivariate exponential distribution, *Journal of the American Statistical Association*, **63**, 848-852.

Arnold, B.C. (1971). Two characterizations of the exponential distribution, *Unpublished manuscript*, Iowa State University, Ames, Iowa.

Arnold, B.C. (1973). Some characterizations of the exponential distribution by geometric compounding, *SIAM Journal of Applied Mathematics*, **24**, 242-244.

Arnold, B.C. (1975a). Multivariate exponential distributions based on hierarchical successive damage, *Journal of Applied Probability*, **12**, 142-147.

Arnold, B.C. (1975b). A characterization of the exponential distribution by multivariate geometric compounding, *Sankhyā, Series A*, **37**, 164-173.

Arnold, B.C. (1987). *Majorization and the Lorenz Order: A Brief Introduction*, Lecture Notes in Statistics No. 43, Springer-Verlag, New York.

Arnold, B.C. and Balakrishnan, N. (1989). *Relations, Bounds and Approximations for Order Statistics*, Lecture Notes in Statistics No. 53, Springer-Verlag, New York.

Arnold, B.C., Balakrishnan, N., and Nagaraja, H.N. (1992). *A First Course in Order Statistics*, John Wiley & Sons, New York.

Arnold, B.C. and Isaacson, D. (1976). On solutions to $\min(X,Y) \stackrel{d}{=} aX$ and $\min(X,Y) \stackrel{d}{=} aX \stackrel{d}{=} bY$, *Zeitschrift Wahrscheinlichkeitstheorie verwandte Gebiete*, **35**, 115-119.

Arnold, B.C. and Nagaraja, H.N. (1991). Lorenz ordering of exponential order statistics, *Statistics & Probability Letters*, **11**, 485-490.

Arnold, B.C. and Strauss, D. (1988). Bivariate distributions with exponential conditionals, *Journal of the American Statistical Association*, **83**, 522-527.

Arnold, B.C. and Strauss, D. (1991). Bivariate distributions with conditionals in prescribed exponential families, *Journal of the Royal Statistical Society, Series B*, **53**, 365-375.

Arnold, B.C. and Villaseñor, J.A. (1991). Lorenz ordering of order statistics, In *Stochastic Order and Decisions under Risk*, IMS Lecture Notes - Monograph Series, **19**, 38-47.

Ascher, H. and Feingold, H. (1984). *Repairable Systems Reliability Modeling, Inferences, Misconceptions and Their Causes*, Marcel Dekker, New York.

Ashour, S.K. and Shoukry, S.E. (1981). The Bayesian predictive distribution of order statistics of a future sample from a mixed exponential distribution, *Egyptian Statistical Journal*, **25**, 94-102.

Azlarov, T.A. and Volodin, N.A. (1986). *Characterization Problems Associated with the Exponential Distribution*, Springer-Verlag, New York.

Bahadur, R.R. (1971). *Some Limit Theorems in Statistics*, SIAM, Philadelphia.

Bahadur, R.R. and Raghavachari, M. (1971). Some asymptotic properties of likelihood ratios on general sample spaces, In *Proceedings of the Sixth Berkeley Symposium*, Vol. 1, pp. 129-152.

Bailey, K. (1979). *The General Maximum Likelihood Approach to the Cox Model*, Ph.D. Dissertation, Department of Statistics, University of Chicago.

Bain, L.J. (1974). Analysis for the linear failure rate distribution, *Technometrics*, **16**, 551-559.

Bain, L.J. (1978). *Statistical Analysis of Reliability and Life-Testing Models: Theory and Methods*, Marcel Dekker, New York.

Bain, L.J. and Engelhardt, M. (1991). *Statistical Analysis of Reliability and Life-Testing Models*, Second edition, Marcel Dekker, New York.

Bain, L.J. and Engelhardt, M. (1992). *Introduction to Probability and Mathematical Statistics*, Second edition, PWS-Kent, Boston.

Bain, L.J., Engelhardt, M., and Wright, F.T. (1977). Inferential procedures for the truncated exponential distribution, *Communications in Statistics*, **A2**, 103-112.

Bain, L.J. and Weeks, D.L. (1964). A note on the truncated exponential distribution, *Annals of Mathematical Statistics*, **35**, 1366-1367.

Bain, P.T. and Patel, J.K. (1991). Factors for calculating prediction intervals for samples from a one-parameter exponential distribution, *Journal of Quality Technology*, **23**, 48-52.

Balakrishnan, N. (1983). Empirical power study of a multi-sample test of exponentiality based on spacings, *Journal of Statistical Computation and Simulation*, **18**, 265-271.

Balakrishnan, N. (1989). Recurrence relations among moments of order statistics from two related sets of independent and non-identically distributed random variables, *Annals of the Institute of Statistical Mathematics*, **41**, 323-329.

Balakrishnan, N. (1990). On the maximum likelihood estimation of the location and scale parameters of exponential distribution based on multiply Type II censored samples, *Journal of Applied Statistics*, **17**, 55-61.

Balakrishnan, N. (1994a). Order statistics from non-identical exponential random variables and some applications (with discussion), *Computational Statistics & Data Analysis*, **18**, 203-253.

Balakrishnan, N. (1994b). On order statistics from non-identical right-truncated exponential random variables and some applications, *Communications in Statistics - Theory and Methods*, **23**, 3373-3393.

Balakrishnan, N. and Ahsanullah, M. (1994). Relations for single and product moments of record values from exponential distribution, *Journal of Applied Statistical Science* (To appear).

Balakrishnan, N. and Ambagaspitiya, R.S. (1988). Relationships among moments of order statistics from two related outlier models and some applications, *Communications in Statistics - Theory and Methods*, **17**, 2327-2341.

Balakrishnan, N., Balasubramanian, K., and Panchapakesan, S. (1994). δ-exceedance records, *Submitted for publication.*

Balakrishnan, N. and Barnett, V. (1994). Outlier-robust estimation of the mean of an exponential distribution, *Submitted for publication.*

Balakrishnan, N. and Cohen, A.C. (1991). *Order Statistics and Inference: Estimation Methods*, Academic Press, San Diego.

Balakrishnan, N., Govindarajulu, Z., and Balasubramanian, K. (1993). Relationships between moments of two related sets of order statistics and some extensions, *Annals of the Institute of Statistical Mathematics*, **45**, 243-247.

Balakrishnan, N. and Gupta, S.S. (1992). Higher order moments of order statistics from exponential and right-truncated exponential distributions and applications to life-testing problems, *Technical Report No. 92-07C*, Department of Statistics, Purdue University, West Lafayette, IN.

Balakrishnan, N. and Malik, H.J. (1986). Order statistics from the linear-exponential distribution, Part I: Increasing hazard rate case, *Communications in Statistics - Theory and methods*, **15**, 179-203.

Balakrishnan, N. and Joshi, P.C. (1984). Product moments of order statistics from doubly truncated exponential distribution, *Naval Research Logistics Quarterly*, **31**, 27-31.

Balakrishnan, N. and Sandhu, R.A. (1995a). A simple simulational algorithm for generating progressive Type-II censored samples, *The American Statistician* (To appear).

Balakrishnan, N. and Sandhu, R.A. (1995b). Best linear unbiased and maximum likelihood estimation for exponential distributions under general progressive Type-II censored samples, *Sankhyā, Series B* (To appear).

Balasooriya, U. (1989). Detection of outliers in the exponential distribution based on prediction, *Communications in Statistics - Theory and Methods*, **18**, 711-720.

Balasubramanian, K. and Balakrishnan, N. (1992). Estimation for one- and two-parameter exponential distributions under multiple type-II censoring, *Statistische Hefte*, **33**, 203-216.

Bancroft, G.A. and Dunsmore, I.R. (1976). Predictive distributions in life tests under competing causes of failure, *Biometrika*, **63**, 195-198.

Bancroft, G.A. and Dunsmore, I.R. (1978). Predictive sequential life testing, *Biometrika*, **65**, 609-614.

Bandyopadhyay, D. and Basu, A.P. (1990). On a generalization of a model by Lindley and Singpurwalla, *Advances in Applied Probability*, **22**, 498-500.

Bar-Lev, S.K. and Reiser, B. (1982). An exponential subfamily which admits UMPU test based on a single test statistic, *Annals of Statistics*, **10**, 979-989.

Barlow, R.E. (1968). Likelihood ratio tests for restricted families of probability distributions, *Annals of Mathematical Statistics*, **39**, 547-560.

Barlow, R.E. (1979). Geometry of the total time on test transform, *Naval Research Logistics Quarterly*, **26**, 393-402.

Barlow, R.E., Bartholomew, D.J., Bremner, J.M. and Brunk, H.D. (1972). *Statistical Inference under Order Restrictions*, John Wiley & Sons, New York.

Barlow, R.E. and Campos, R. (1975). Total time on test processes and applications to failure data analysis, In *Reliability and Fault Tree Analysis* (Eds., R.E. Barlow, J. Fussel and N. Singpurwalla), pp. 451-481, SIAM, Philadelphia.

Barlow, R.E., Madansky, A., Proschan, F., and Schever, F. (1968). Statistical estimation procedures for the "burn-in" process, *Technometrics*, **10**, 51-62.

Barlow, R.E. and Proschan, F. (1965). *Mathematical Theory of Reliability*, John Wiley & Sons, New York.

Barlow, R.E. and Proschan, F. (1975). *Statistical Theory of Reliability and Life Testing: Probability Models*, Holt, Reinhart and Winston, New York.

Barlow, R.E. and Proschan, F. (1981). *Statistical Theory of Reliability and Life Testing: Probability Models*, Second edition, To Begin With, Silver Spring, Maryland.

Barnard, G.A. (1963). Contribution to the discussion of paper by M.S. Bartlett, *Journal of the Royal Statistical Society, Series B*, **25**, 294.

Barnard, G.A. (1976). Conditional inference is not inefficient, *Scandinavian Journal of Statistics*, **3**, 132-134.

Barnett, V. (1985). The bivariate exponential distribution: A review and some new results, *Statistica Neerlandica*, **39**, 343-356.

Barnett, V. and Lewis, T. (1978). *Outliers in Statistical Data*, First edition, John Wiley & Sons, Chichester, England.

Barnett, V. and Lewis, T. (1984). *Outliers in Statistical Data*, Second edition, John Wiley & Sons, Chichester, England.

Barnett, V. and Lewis, T. (1994). *Outliers in Statistical Data*, Third edition, John Wiley & Sons, Chichester, England.

Bartholomew, D.J. (1957). A problem in life testing, *Journal of the American Statistical Association*, **52**, 350-355.

Bartholomew, D.J. (1969). Sufficient conditions for a mixture of exponentials to be a probability density function, *Annals of Mathematical Statistics*, **40**, 2183-2188.

Barton, D.E. and Mallows, C.L. (1965). Some aspects of the random sequence, *Annals of Mathematical Statistics*, **36**, 236-260.

Basu, A.P. (1968). On a generalized Savage statistic with applications to life testing, *Annals of Mathematical Statistics*, **39**, 1591-1604.

Basu, A.P. (1971). On a sequential rule for estimating the location parameter of an exponential distribution, *Naval Research Logistics Quarterly*, **18**, 329-337.

Basu, A.P. (1981). Identifiability problems in the theory of competing and complementary risks - a survey, In *Statistical Distributions in Scientific Work*, Vol. 5 (Eds., C. Taillie, G.P. Patil and B.A. Baldessari), pp. 335-348, Reidel, Dordrecht.

Basu, A.P. (1983). Identifiability, In *Encyclopedia of Statistical Sciences*, Vol. 4 (Eds., S. Kotz and N.L. Johnson), pp. 2-6, John Wiley & Sons, New York.

Basu, A.P. (1988). Multivariate exponential distributions and their applications in reliability, In *Handbook of Statistics* (Eds., P.R. Krishnaiah and C.R. Rao), **7**, Elsevier Science Publishers, New York.

Basu, A.P. (1991). Sequential methods in reliability and life testing, In *Handbook of Sequential Analysis*, Chapter 25 (Eds., B.K. Ghosh and P.K. Sen), pp. 581-592, Marcel Dekker, New York.

Basu, A.P. and Ebrahimi, N. (1984). Testing whether survival function is bivariate new better than used, *Communications in Statistics - Theory and Methods*, **13**, 1839-1849.

Basu, A.P. and Ebrahimi, N. (1987). On a bivariate accelerated life test, *Journal of Statistical Planning and Inference*, **16**, 297-304.

Basu, A.P. and Ebrahimi, N. (1991). Bayesian approach to life testing and reliability estimation using asymmetric loss function, *Journal of Statistical Planning and Inference*, **29**, 21-31.

Basu, A.P. and Ebrahimi, N. (1992). Bayesian approach to some problems in life testing and reliability estimation, In *Bayesian Analysis in Statistics and Econometrics* (Eds., P.K. Goel and N.S. Iyenger), pp. 257-266, Springer-Verlag, New York.

Basu, A.P. and Ghosh, J.K. (1978). Identifiability of the multinormal distribution under competing risks model, *Journal of Multivariate Analysis*, **8**, 413-429.

Basu, A.P. and Ghosh, J.K. (1980). Identifiability of distributions under competing risks and complementary risks models, *Communications in Statistics - Theory and Methods*, **9**, 1515-1525.

Basu, A.P. and Ghosh, J.K. (1983). Identifiability results for k-out-of-p systems, *Communications in Statistics - Theory and Methods*, **12**, 199-205.

Basu, A.P. and Gupta, A.K. (1974). Classification results for exponential populations, In *Reliability and Biometry* (Eds., F. Proschan and R.J. Serfling), pp. 637-650, SIAM, Philadelphia.

Basu, A.P. and Gupta, A.K. (1977). Classification rules for exponential populations: Two parameter case, In *Theory and Applications of Reliability* (Eds., C.P. Tsokos and I.N. Shimi), Vol. 1, pp. 507-525, Academic Press, New York.

Basu, A.P. and Habibullah, A. (1987). A test for bivariate exponentiality against BIFRA alternative, *Calcutta Statistical Association Bulletin*, **36**, 79-84.

Basu, A.P. and Klein, J.P. (1982). Some recent results in competing risks theory, In *Proceedings on Survival Analysis*, IMS monograph series, **2**, 216-229.

Basu, A.P. and Klein, J.P. (1985). A model for life testing and for estimating safe dose levels, In *Statistical Theory and Data Analysis* (Ed., Matusita), North-Holland, Amsterdam.

Basu, A.P. and Tarmast, G. (1987). Reliability of a complex system from Bayesian viewpoint, In *Probability and Bayesian Statistics* (Ed., R. Viertt), pp. 31-38, Plenum, New York.

Basu, A.P. and Thompson, R.D. (1992). Life testing and reliability estimation under asymmetric loss, In *Survival Analysis* (Eds., P.K. Goel and J.P. Klein), pp. 3-10, Kluwer Academic, Hingham, Massachusetts.

Basu, D. (1964). Recovery of ancillary information, *Sankhyā, Series A*, **26**, 3-16.

Bechhofer, R.E. (1954). A single-sample multiple decision procedure for ranking means of normal populations with known variances, *Annals of Mathematical Statistics*, **25**, 16-39.

Bechhofer, R.E., Dunnett, C.W., and Sobel, M. (1954). A two sample multiple decision procedure for ranking means of normal populations with a common unknown variance, *Biometrika*, **41**, 170-176.

Bechhofer, R.E., Kiefer, J., and Sobel, M. (1968). *Sequential Identification and Ranking Procedures (with Special Reference to Koopman-Darmois Populations)*, University of Chicago Press, Chicago.

Bechhofer, R.E. and Sobel, M. (1954). A single-sample multiple-decision procedure for ranking variances of normal populations, *Annals of Mathematical Statistics*, **25**, 273-289.

Bemis, B.M., Bain, L.J., and Higgins, J.J. (1972). Estimation and hypothesis testing for the bivariate exponential distribution, *Journal of the American Statistical Association*, **67**, 927-929.

Berger, J.O. (1985). *Statistical Decision Theory and Bayesian Analysis*, Vol. 2, Springer-Verlag, New York.

Berger, R.L. and Kim, J.S. (1985). Ranking and subset selection procedures for exponential populations with type-I and type-II censored data, In *Frontiers of Modern Statistical Inference Procedures* (Ed., E.J. Dudewicz), pp. 425-455 (with discussion), American Sciences Press, Columbus, Ohio.

Berkson, J. and Gage, R.P. (1952). Survival curve for cancer patients following treatment, *Journal of the American Statistical Association*, **47**, 501-515.

Berman, S.M. (1963). Notes on extreme values, competing risks, and semi-Markov processes, *Annals of Mathematical Statistics*, **34**, 1104-1106.

Berndt, E.R., Hall, B.H., Hall, R.E., and Hausman, J.A. (1974). Estimation and inference in nonlinear structural models, *Annals of Economic and Social Measurement*, **3**, 653-665.

Bhattacharyya, G.K. and Johnson, R.A. (1971). Maximum likelihood estimation and hypothesis testing in the bivariate exponential model of Marshall and Olkin, *Technical Report 276*, Department of Statistics, University of Wisconsin, Madison.

Bhattacharyya, G.K. and Johnson, R.A. (1973). On a test of independence in a bivariate exponential distribution, *Journal of the American Statistical Association*, **68**, 704-706.

Bhattacharyya, G.K. and Mehrotra, K.G. (1981). On testing equality of two exponential distributions under combined Type II censoring, *Journal of the American Statistical Association*, **76**, 886-894.

Bibby, J. and Toutenburg, H. (1977). *Prediction and Improved Estimation in Linear Models*, John Wiley & Sons, Chichester, England.

Bickel, P.J. and Doksum, K.A. (1969). Tests for monotone failure rate based on normalized spacings, *Annals of Mathematical Statistics*, **40**, 1216-1235.

Bilodeau, M. and Kariya, T. (1993). LBI tests of independence in bivariate exponential distributions, Presented at the *Seventh International Conference on Multivariate Analysis*, New Delhi.

BIR Working Party (1989). First interim progress report on the second British Institute of Radiology fractionation study: short vs. long overall treatment times for radiotherapy of carcinoma of the laryngo-pharynx, *British Journal of Radiology*, **62**, 450-456.

Black, C.M., Durham, S.D., Lynch, J.D., and Padgett, W.J. (1989). A new probability distribution for the strength of brittle fibers, *Fiber-Tex 1989*, The Third Conference on Advanced Engineering Fibers and Textile Structures for Composites, *NASA Conference Publication 3082*, 363-374.

Black, C.M., Durham, S.D., and Padgett, W.J. (1990). Parameter estimation for a new distribution for the strength of brittle fibers: A simulation study, *Communications in Statistics - Simulation and Computation*, **19**, 809-825.

Block, H.W. (1975). Continuous multivariate exponential extensions, In *Reliability and Fault Tree Analysis* (Eds., R.E. Barlow, J.B. Fussell, and N.D. Singpurwalla), SIAM, Philadelphia.

Block, H.W. and Basu, A.P. (1974). A continuous bivariate exponential extension, *Journal of the American Statistical Association*, **69**, 1031-1037.

Block, H.W. and Savits, T.H. (1981). Multivariate distributions in reliability theory and life testing, In *Statistical Distributions in Scientific Work*, Vol. 5 (Eds., C. Taillie, G.P. Patil, and B.A. Baldessari), pp. 271-288, Reidel, Dordrecht.

Boag, J.Q. (1949). Maximum likelihood estimates of the proportion of patients cured by cancer therapy, *Journal of the Royal Statistical Society, Series B*, **11**, 15-53.

Bofinger, E. (1991). Selecting "demonstrably best" or "demonstrably worst" exponential populations, *Australian Journal of Statistics*, **33**, 183-190.

Bofinger, E. (1992). Comparisons and selection of two-parameter exponential populations, *Australian Journal of Statistics*, **34**, 65-75.

Bondesson, L. (1979). A general result on infinite divisibility, *Annals of Probability*, **7**, 965-979.

Bondesson, L. (1992). *Generalized Gamma Convolutions and Related Classes of Distributions and Densities*, Springer-Verlag, New York.

Bose, A. and Boukai, B. (1993). Sequential estimation results for a two-parameter exponential family of distributions, *Annals of Statistics*, **21**, 484-502.

Bose, A. and Mukhopadhyay, N. (1994a). Sequential estimation by accelerated stopping times in a two-parameter exponential family of distributions, *Statistics & Decisions*, **12**.

Bose, A. and Mukhopadhyay, N. (1994b). Sequential estimation via replicated piecewise stopping number in a two-parameter exponential family of distributions, *Sequential Analysis*, **13**, 1-10.

Boukai, B. (1987). Bayes sequential procedure for estimation and for determination of burn-in time in a hazard rate model with an unknown change-point parameter, *Sequential Analysis*, **6**, 37-53.

Bowman, K.O. and Shenton, L.R. (1988). *Properties of Estimators for the Gamma Distribution*, Marcel Dekker, New York.

Box, G.E.P. (1949). A general distribution theory for a class of likelihood criteria, *Biometrika*, **36**, 317-346.

Box, G.E.P. (1954). Some theorems on quadratic forms applied in the study of analysis of variance problems, I. Effect of inequality of variance in the one-way classification, *Annals of Mathematical Statistics*, **25**, 290-302.

Box, G.E.P. and Muller, M.E. (1958). A note on the generation of random normal deviates, *Annals of Mathematical Statistics*, **29**, 610-611.

Brain, C.W. and Shapiro, S.S. (1983). A regression test for exponentiality: Censored and complete samples, *Technometrics*, **25**, 69-76.

Bristol, D.R., Chen, H.J., and Mithongtae, J.S. (1992). Comparing exponential guarantee times with a control, In *Frontiers of Modern Statistical Inference Procedures - II* (Eds., E.J. Dudewicz et al.), pp. 95-151 (with discussion), American Sciences Press, Syracuse, New York.

Bristol, D.R. and Desu, M.M. (1985). Selection procedures for comparing exponential guarantee times with a standard, In *Frontiers of Modern Statistical Inference Procedures* (Ed., E.J. Dudewicz), pp. 307-324 (with discussion), American Sciences Press, Columbus, Ohio.

Bristol, D.R. and Desu, M.M. (1990). Comparison of two exponential distribution, *Biometrical Journal*, **32**, 267-276.

Broadbent, S. (1958). Simple mortality rates, *Applied Statistics*, **7**, 86-95.

Brockmeyer, F., Halstrom, H.L., and Jensen, A. (1948). The life and works of A.K. Erlang, *Transactions of the Danish Academy of Science*, No. 2.

Brown, L.D. (1986). *Fundamentals of Statistical Exponential Families*, Institute of Mathematical Statistics, Hayward, CA.

Buehler, W.J. and Puri, P.S. (1966). On optimal asymptotic tests of composite hypotheses with several constraints, *Z. Wahrsch. Verw. Gebiete*, **5**, 71-88.

Bulgren, W.G. and Hewett, J.E. (1973). Double sample tests for hypotheses about the mean of an exponential distribution, *Technometrics*, **15**, 187-190.

Büringer, H., Martin, H., and Schriever, K.-H. (1980). *Nonparametric Sequential Procedures*, Birkhauser, Boston.

Butcher, J.C. (1961). Random sampling from the normal distribution, *Computer Journal*, **3**, 251-253.

Butler, J.W. (1956). Machine sampling from given probability distributions, In *Symposium on Monte Carlo Methods* (Ed. H.A. Meyer), pp. 249-264, John Wiley & Sons, New York.

Buzacott, J.A. and Shanthikumar, J.G. (1993). *Stochastic Models of Manufacturing Systems*, Prentice Hall, Englewood Cliffs, New Jersey.

Byar, D.R., Huse, R., and Bailar, J.C., III (1974). An exponential model relating censored survival data and concomitant information for prostatic cancer patients, *Journal of the National Cancer Institute*, **52**, 321-326.

Cantor, A.B. and Knapp, R.G. (1985). A test of the equality of survival distributions based on paired observations from conditionally independent exponential distributions, *IEEE Transactions on Reliability*, **R-34**, 342-346.

Carbone, P.O., Kellerhouse, L.E., and Gehan, E.A. (1967). Plasmacytic myeloma: A study of the relationship of survival to various clinical manifestations and anomalous protein type in 112 patients, *American Journal of Medicine*, **42**, 937-948.

Cauchy, A.L. (1821). *Cours d'anayse de l'École Polytechnique*, Vol. I, Analyse algébrique, V. Paris.

Chan, L.K. (1967). On a characterization of distributions by expected values of extreme order statistics, *American Mathematical Monthly*, **74**, 950-951.

Chan, L.K. and Cheng, S.W. (1988). Linear estimation of the location and scale parameters based on selected order statistics, *Communications in Statistics - Theory and Methods*, **17**, 2259-2278.

Chandler, K.M. (1952). The distribution and frequency of record values, *Journal of the Royal Statistical Society, Series B*, **14**, 220-228.

Chappell, R., Nondahl, D.M., and Fowler, J.F. (1994a). Re-analysis of the effect of dose, time and fraction in the first and second BIR fractionation studies, *The International Journal of Radiation Oncology, Biology & Physics* (To appear).

Chappell, R., Nondahl, D.M., and Fowler, J.F. (1994b). Modeling dose and local control in radiotherapy, *Journal of the American Statistical Association* (To appear).

Chase, G.R. and Hewett, J.E. (1976). Double sample tests – A distribution free procedure, *Journal of Statistical Computation and Simulation*, **4**, 247-257.

Chaturvedi, A. and Shukla, P.S. (1990). Sequential point estimation of location parameter of a negative exponential distribution, *Journal of the Indian Statistical Association*, **28**, 41-50.

Chen, H.J. (1982). A new range statistic for comparison of several exponential location parameters, *Biometrika*, **69**, 257-260.

Chen, H.J. and Dudewicz, E.J. (1984). A new subset selection theory for all best populations and its applications, In *Developments in Statistics and Its Applications* (Eds., A.M. Abouammah et al.), pp. 63-87, King Saud University Press, Riyadh, Saudi Arabia.

Chen, H.J. and Mithongtae, J. (1986). Selection of the best exponential distribution with a preliminary test, *American Journal of Mathematics and Management Sciences*, **6**, 219-249.

Chen, H.J. and Vanichbuncha, K. (1991). Multiple comparisons with the best exponential distribution, In *The Frontiers of Statistical Scientific Theory & Industrial Applications - Vol. II* (Eds., A. Öztürk et al.), pp. 53-76, American Sciences Press, Columbus, Ohio.

Chen, S.-M. and Bhattacharyya, G.K. (1988). Exact confidence bounds for an exponential parameter under hybrid censoring, *Communications in Statistics - Theory and Methods*, **17**, 1857-1870.

Chernoff, H. (1972). *Sequential Analysis and Optimal Design*, SIAM, Philadelphia.

Chiang, C.L. (1968). *Introduction to Stochastic Processes in Biostatistics*, John Wiley & Sons, New York.

Chikkagoudar, M.S. and Kunchur, S.H. (1980). Estimation of the mean of an exponential distribution in the presence of an outlier, *Canadian Journal of Statistics*, **8**, 59-63.

Chikkagoudar, M.S. and Kunchur, S.H. (1983). Distributions of test statistics for multiple outliers in exponential samples, *Communications in Statistics - Theory and Methods*, **12**, 2127-2142.

Childs, A. and Balakrishnan, N. (1995). Some extensions in the robust estimation of parameters of exponential and double exponential distributions in the presence of multiple outliers, In *Handbook of Statistics - 15: Robust Methods* (Eds., C.R. Rao and G.S. Maddala), North-Holland, Amsterdam (To appear).

Chiou, P. (1990). Estimation of scale parameters of two exponential distributions, *IEEE Transactions on Reliability*, **39**, 106-109.

Choi, S.C. (1979). Two-sample tests for compound distributions for homogeneity of mixing proportions, *Technometrics*, **21**, 361-365.

Chou, Y.-M. (1989). One-sided simultaneous prediction intervals for the order statistics of l future samples from an exponential distribution, *Communications in Statistics - Theory and Methods*, **17**, 3995-4003.

Chou, Y.-M. and Owen, D.B. (1989). Simultaneous one-sided prediction intervals for a two-parameter exponential distribution using complete or Type II censored data, *Metrika*, **36**, 279-290.

Chow, Y.S. and Robbins, H. (1965). On the asymptotic theory of fixed-width sequential confidence intervals for the mean, *Annals of Mathematical Statistics*, **36**, 457-462.

Cinlar, E. (1975). *Introduction to Stochastic Processes*, Prentice-Hall, New Jersey.

Cinlar, E. and Jagers, P. (1973). Two mean values which characterize the Poisson process, *Journal of Applied Probability*, **10**, 678-681.

Clarotti, C.A. and Spizzichino, F. (1990). Bayes burn-in decision procedures, *Probability in Engineering and Informational Sciences*, **4**, 437-445.

Clayton, D. and Cuzick, J. (1985). Multivariate generalizations of the proportional hazards model (with discussion), *Journal of the Royal Statistical Society, Series A*, **148**, 82-117.

Cochran, W.G. (1941). The distribution of the largest of a set of estimated variances as a fraction of their total, *Annals of Eugenics*, **11**, 47-52.

Cohen, A.C. (1963). Progressively censored samples in life testing, *Technometrics*, **5**, 327-339.

Cohen, A.C. (1966). Life testing and early failure, *Technometrics*, **8**, 539-549.

Cohen, A.C. (1991). *Truncated and Censored Samples*, Marcel Dekker, New York.

Cohen, A.C. and Helm, R. (1973). Estimation in the exponential distribution, *Technometrics*, **14**, 841-846.

Cohen, A.C. and Whitten, B.J. (1982). Modified moment and maximum likelihood estimators for parameters of the three-parameter gamma distribution, *Communications in Statistics - Simulation and Computation*, **11**, 197-216.

Cohen, A.C. and Whitten, B.J. (1988). *Parameter Estimation In Reliability and Life Span Models*, Marcel Dekker, New York.

Cohen, A.C., Whitten, B.J., and Ding, Y. (1984). Modified moment estimation for the three-parameter Weibull distribution, *Journal of Quality Technology*, **16**, 159-167.

Cook, R.D. and Johnson, M.E. (1981). A family of distributions for modelling nonelliptically symmetric multivariate data, *Journal of the Royal Statistical Society, Series B*, **43**, 210-218.

Costanza, M.C., Hamdy, H.I., and Son, M.S. (1986). Two stage fixed width confidence intervals for the common location parameter of several exponential distributions, *Communications in Statistics - Theory and Methods*, **15**, 2305-2322.

Costigan, T. and Klein, J.P. (1993). Multivariate survival analysis based on frailty models, In *Advances in Reliability* (Ed., A.P. Basu), North-Holland, Amsterdam.

Cox, D.R. (1953). Some simple tests for Poisson variates, *Biometrika*, **40**, 354-360.

Cox, D.R. (1959). The analysis of exponentially distributed life-times with two types of failure, *Journal of the Royal Statistical Society, Series B*, **21**, 411-421.

Cox, D.R. (1964). Some applications of exponential ordered scores, *Journal of the Royal Statistical Society, Series B*, **26**, 103-110.

Cox, D.R. (1972). Regression models and life tables (with discussion), *Journal of the Royal Statistical Society, Series B*, **34**, 187-220.

Cox, D.R. and Hinkley, D.V. (1974). *Theoretical Statistics,* Chapman and Hall, London.

Cox, D.R. and Lewis, P.A. (1966). *The Statistical Analysis of a Series of Events,* Methuen, London.

Cox, D.R. and Oakes, D. (1984). *Analysis of Survival Data*, Chapman & Hall, London.

Craig, J.B. (1988). Sample size determination in clinical trials considering nonuniform patient entry, loss to follow-up, noncompliance and cost optimization, *Unpublished Ph.D. Dissertation*, Medical University of South Carolina.

Crawford, G.B. (1966). Characterization of geometric and exponential distributions, *Annals of Mathematical Statistics*, **37**, 1790-1795.

Crowford, M. (1990). On some nonregular tests for a modified Weibull model, *Biometrika*, **77**, 499-506.

Csörgö, S. and Welsh, A.H. (1989). Testing for exponential and Marshall-Olkin distributions, *Journal of Statistical Planning and Inference*, **23**, 287-300.

D'Agostino, R.B. and Stephens, M.A. (Eds.) (1986). *Goodness-of-Fit Techniques*, Marcel Dekker, New York.

Daniels, H.A. (1945). The statistical theory of the strength of bundles of threads, *Proceedings of the Royal Society, Series A*, **183**, 405-435.

David, H.A. (1979). Robust estimation in the presence of outliers, In *Robustness in Statistics* (Eds., R.L. Launer and G.N. Wilkinson), pp. 61-74, Academic Press, New York.

David, H.A. (1981). *Order Statistics*, Second edition, John Wiley & Sons, New York.

David, H.A. and Moeschberger, M.L. (1978). *The Theory of Competing Risks*, Griffin, London.

Davies, P.L. and Gather, U. (1993). The identification of multiple outliers (with discussion), *Journal of the American Statistical Association*, **88**, 782-792.

Davis, D.J. (1952). An analysis of some failure data, *Journal of the American Statistical Association*, **47**, 113-150.

Day, N.E. (1969). Estimating the components of a mixture of normal distributions, *Biometrika*, **56**, 463-474.

Deemer, W.L. and Votaw, D.F. (1955). Estimation of parameters of truncated or censored exponential distributions, *Annals of Mathematical Statistics*, **26**, 498-504.

DeLong, D. (1981). Crossing probabilities for a square root boundary by a Bessel process, *Communications in Statistics - Theory and Methods*, **10**, 2197-2213.

Dempster, A.P., Laird, N.M., and Rubin, D.B. (1977). Maximum likelihood from incomplete data via the EM algorithm (with discussion), *Journal of the Royal Statistical Society, Series B*, **39**, 1-38.

Desu, M.M. (1971). A characterization of the exponential distribution by order statistics, *Annals of Mathematical Statistics*, **42**, 837-838.

Desu, M.M., Narula, S.C., and Villareal, B. (1977). A two-stage procedure for selecting the best of k exponential distributions, *Communications in Statistics - Theory and Methods*, **6**, 1233-1230.

Desu, M.M. and Sobel, M. (1968). A fixed-subset size approach to a selection problem, *Biometrika*, **55**, 401-410. Corrections and amendments: **63** (1976), 685.

Devroye, L. (1980). Generating the maximum of independent identically distributed random variables, *Computers and Mathematics with Applications*, **6**, 305-315.

Devroye, L. (1981). The series method for random variate generation and its application to the Kolmogorov-Smirnov distribution, *American Journal of Mathematical and Management Sciences*, **1**, 359-379.

Devroye, L. (1986). *Non-uniform Random Variate Generation*, Springer-Verlag, New York.

Doornbos, R. (1976). *Slippage Tests*, Second edition, Mathematical Centre Tracts No. 15, Mathematisch Centrum, Amsterdam.

Downton, F. (1970). Bivariate exponential distributions in reliability theory, *Journal of the Royal Statistical Society, Series B*, **32** 408-417.

Dudewicz, E.J. and Koo, J.O. (1982). The complete categorized guide to statistical selection and ranking procedures, *Series in Mathematical and Management Sciences*, Vol. 6, American Sciences Press, Columbus, Ohio.

Dudewicz, E.J. and Ralley, T.G. (1981). *The Handbook of Random Number Generation and Testing with TESTRAND Computer Code*, American Sciences Press, Columbus, Ohio, U.S.A.

Dunsmore, I.R. (1974). The Bayesian predictive distribution in life testing models, *Technometrics*, **16**, 455-460.

Dunsmore, I.R. (1976). Asymptotic prediction analysis, *Biometrika*, **63**, 627-630.

Dunsmore, I.R. (1983). The future occurrence of records, *Annals of the Institute of Statistical Mathematics*, **35**, 267-277.

Durairajan, T.M. and Kale, B.K. (1982). Locally most powerful similar test for mixing proportion, *Sankhyā, Series A*, **44**, 153-161.

Durbin, J. (1975). Kolmogorov-Smirnov tests when parameters are estimated with application to tests of exponentiality and tests on spacings, *Biometrika*, **62**, 5-22.

Dvoretzky, A., Kiefer, J., and Wolfowitz, J. (1953). Sequential decision problems for processes with continuous time parameter: Testing hypotheses, *Annals of Mathematical Statistics*, **24**, 254-264.

Ebrahimi, N. (1987). Analysis of bivariate accelerated life test data from the bivariate exponential of Marshall and Olkin, *American Journal of Mathematical and Management Sciences*, **6**, 175-190.

Ebrahimi, N. (1993). How to measure uncertainty in the residual life time distribution, *Unpublished report.*

Efron, B. (1979). Bootstrap methods: Another look at the jackknife, *Annals of Statistics*, **7**, 1-26.

Efron, B. (1982). *The Jackknife, the Bootstrap and Other Resampling Plans*, SIAM, Philadelphia.

Efron, B. and Tibshirani, R.J. (1993). *An Introduction to the Bootstrap*, Chapman & Hall, London.

Efron, B. and Hinkley, D.V. (1978). Assessing the accuracy of the maximum likelihood estimator: Observed versus expected Fisher information (with discussion), *Biometrika*, **65**, 457-487.

Engelhardt, M. and Bain, L.J. (1978a). Construction of optimal unbiased inference procedures for the parameters of the gamma distribution, *Technometrics*, **20**, 485-489.

Engelhardt, M. and Bain, L.J. (1978b). Tolerance limits and confidence limits on reliability for the two-parameter exponential distribution, *Technometrics*, **20**, 37-39.

Epstein, B. (1954). Truncated life tests in the exponential case, *Annals of Mathematical Statistics*, **25**, 555-564.

Epstein, B. (1956). Simple estimators of the parameters of exponential distributions when samples are censored, *Annals of the Institute of Statistical Mathematics*, **8**, 15-26.

Epstein, B. (1960a). Estimation from life test data, *Technometrics*, **2**, 447-454.

Epstein, B. (1960b). Tests for the validity of the assumption that the underlying distribution of life is exponential, *Technometrics*, **2**, 83-101 and 167-183.

Epstein, B. (1962). Simple estimates of the parameters of exponential distributions, In *Contributions to Order Statistics* (Eds., A.E. Sarhan and B.G. Greenberg), pp. 361-371, John Wiley & Sons, New York.

Epstein, B. and Sobel, M. (1953). Life testing, *Journal of the American Statistical Association*, **48**, 485-502.

Epstein, B. and Sobel, M. (1954). Some theorems relevant to life testing from an exponential distribution, *Annals of Mathematical Statistics*, **25**, 373-381.

Epstein, B. and Sobel, M. (1955). Sequential life testing in the exponential case, *Annals of Mathematical Statistics*, **26**, 82-93.

Epstein, B. and Tsao, C.K. (1953). Some tests based on ordered observations from two exponential populations, *Annals of Mathematical Statistics*, **24**, 456-466.

Evans, I.G. and Nigm, A.H.M. (1980). Bayesian prediction for the left truncated exponential distribution, *Technometrics*, **22**, 201-204.

Everitt, B.S. and Hand, D.J. (1981). *Finite Mixture Distributions*, Chapman & Hall, London.

Fairbanks, K. (1988). A two-stage life test for the exponential parameter, *Technometrics*, **30**, 175-180.

Farewell, V.T. (1977a). A model for a binary variable with time-censored observations, *Biometrika*, **64**, 43-46.

Farewell, V.T. (1977b). The combined effect of breast cancer risk factors, *Cancer*, **40**, 931-936.

Farewell, V.T. (1982). The use of mixture models for the analysis of survival data with long-term survivors, *Biometrics*, **38**, 1041-1046.

Farewell, V.T. and Sprott, D. (1988). The use of a mixture model in the analysis of count data, *Biometrics*, **44**, 1191-1194.

Faulkenberry, G.D. (1973). A method of obtaining prediction intervals, *Journal of the American Statistical Association*, **68**, 433-435.

Feigl, P. and Zelen, M. (1965). Estimating exponential survival probabilities with concomitant information, *Biometrics*, **21**, 826-838.

Feller, W. (1966). *An Introduction to Probability Theory and Its Applications*, Vol. II, John Wiley & Sons, New York.

Fercho, W.W. and Ringer, L.J. (1972). Small sample power of some tests of constant failure rate, *Technometrics*, **14**, 713-724.

Ferguson, T.S. (1961). On the rejection of outliers, *Proceedings of the Fourth Berkeley Symposium on Mathematical Statistics and Probability*, Vol. 1, pp. 253-287, University of California Press, Berkeley.

Ferguson, T.S. (1964). A characterization of the exponential distribution, *Annals of Mathematical Statistics*, **35**, 1199-1207.

Ferguson, T.S. (1965). A characterization of the geometric distribution, *American Mathematical Monthly*, **72**, 256-260.

Ferguson, T.S. (1967). On characterizing distributions by properties of order statistics, *Sankhyā, Series A*, **29**, 265-278.

Fieller, N.R.J. (1976). Some problems related to the rejection of outlying observations, *Ph.D. Thesis*, University of Hull, U.K.

Filliben, J.J. (1975). The probability plot correlation coefficient test for normality, *Technometrics*, **17**, 111-117.

Finkelstein, J.M. and Schafer, R.E. (1971). Improved goodness of fit tests, *Biometrika*, **58**, 641-645.

Fisher, R.A. (1934). Two new properties of mathematical likelihood, *Proceedings of the Royal Society, Series A*, **144**, 285-305.

Fisher, R.A. (1936). Uncertain inference, *Proceedings of the American Academy of Arts and Sciences*, **71**, 245-258.

Fishman, G.S. (1978). Sampling from the multinomial distribution on a computer, *Technical Report #78-5*, Curriculum in Operations Research and Systems Analysis, University of North Carolina, Chapel Hill.

Fisz, M. (1958). Characterization of some probability distributions, *Skandinavisk Aktuarietidskrift*, **41**, 65-70.

Forsythe, G.E. (1972). von Neumann's comparison method for random sampling from the normal and other distributions, *Mathematics of Computation*, **26**, 817-826.

Fowler, J.F. (1989). The linear-quadratic formula and progress in fractionated radiotherapy, *British Journal of Radiology*, **62**, 679-694.

Frank, M.J. (1979). On the simultaneous associativity of $F(x,y)$ and $x+y-F(x,y)$, *Aequationes Math.*, **19**, 194-226.

Fraser, D.A.S. (1979). *Inference and Linear Models*, McGraw-Hill, New York.

Freireich, E.J., et al. (1963). The effect of 6-mercaptoporine on the duration of steroid-induced remission in acute leukemia, *Blood*, **21**, 699-716.

Freund, J.E. (1961). A bivariate extension of the exponential distribution, *Journal of the American Statistical Association*, **56**, 971-977.

Friday, D.S. and Patil, G.P. (1977). A bivariate exponential model with applications to reliability and computer generation of random variables, In *Theory and Applications of Reliability, Vol. I.* (Eds., C.P. Tsokos and I. Shimi), pp. 527-549, Academic Press, New York.

From, S.G. (1991). Mean squre error efficient estimation of an exponential mean under an exchangeable single outlier model, *Communications in Statistics - Simulation and Computation*, **20**, 1073-1084.

Galambos, J. (1982). Exponential distributions, In *Encyclopedia of Statistical Sciences*, Vol. 2 (Eds., S. Kotz and N.L. Johnson), pp. 582-587, John Wiley & Sons, New York.

Galambos, J. (1987). *The Asymptotic Theory of Extreme Order Statistics*, Second edition, Krieger, Malabar, Florida.

Galambos, J. and Kotz, S. (1978). *Characterizations of Probability Distributions*, Lecture Notes in Mathematics No. 675, Springer-Verlag, New York.

Galambos, J. and Seneta, E. (1975). Record times, *Proceedings of the American Mathematical Society*, **50**, 383-387.

Gan, F.F. (1985). Goodness-of-fit statistics for location-scale distributions, *Unpublished Ph.D. dissertation*, Iowa State University, Department of Statistics.

Gan, F.F. and Koehler, K.J. (1990). Goodness-of-fit tests based on P-P probability plots, *Technometrics*, **32**, 289-303.

Gardiner, J.C. and Susarla, V. (1983). Sequential estimation of the mean survival time under random censorship, *Sequential Analysis*, **2**, 201-223.

Gardiner, J.C. and Susarla, V. (1984). Risk efficient estimation of the mean exponential survival time under random censoring, *Proceedings of National Academy of Sciences, U.S.A.*, **81**, 5906-5909.

Gardiner, J.C. and Susarla, V. (1991). Time-sequential estimation, In *Handbook of Sequential Analysis*, Chapter 27 (Eds., B.K. Ghosh and P.K. Sen), pp. 613-631, Marcel Dekker, New York.

Gardiner, J.C., Susarla, V., and van Ryzin, J. (1986). Time sequential estimation of the exponential mean under random censoring, *Annals of Statistics*, **14**, 607-618.

Gather, U. (1979). Über Ausreissertests und Ausreisseranfälligkeit von Wahrscheinlichkeitsverteilungen, *Doctoral Dissertation*, Aachen Technical University, Germany.

Gather, U. (1984). Tests and estimators in outlier model (in german), *Habilitation Thesis*, Aachen Technical University, Germany.

Gather, U. (1986a). Robust estimation of the mean of the exponential distribution in outlier situations, *Communications in Statistics - Theory and Methods*, **15**, 2323-2345.

Gather, U. (1986b). Estimation of the mean of the exponential distribution under the labelled outlier-model, *Methods in Operations Research*, **53**, 535-546.

Gather, U. (1989a). Testing for multisource contamination in location/scale families, *Communications in Statistics - Theory and Methods*, **18**, 1-34.

Gather, U. (1989b). On a characterization of the exponential distribution by properties of order statistics, *Statistics & Probability Letters*, **7**, 93-96.

Gather, U. (1990). Modelling the occurence of multiple outliers, *Allg. Statist. Archiv*, **74**, 413-428.

Gather, U. and Benda, N. (1989). Adaptive estimation of expected lifetimes when outliers are present, *Manuscript*, FB Statistik, University of Dortmund, Germany.

Gather, U. and Helmers, M. (1983). A locally most powerful test for outliers in samples from the exponential distribution, *Methods in Operations Research*, **47**, 39-47.

Gather, U. and Kale, B.K. (1981). UMP Tests for r upper outliers in samples from exponential families, *Proceedings of the Golden Jubilee Conference, Indian Statistical Institute*, 270-278.

Gather, U. and Kale, B.K. (1988). Maximum likelihood estimation in the presence of outliers, *Communications in Statistics - Theory and Methods*, **17**, 3767-3784.

Gather, U. and Kale, B.K. (1992). Outlier generating models - a review, In *Contributions to Stochastics* (Ed., N. Venugopal), pp. 57-85, Wiley (Eastern), New Delhi, India.

Gather, U. and Pigeot, I. (1994). Identifikation von Ausreissern als multiples Testproblem, In *Medizinische Informatik: Ein integrierender Teil arztunterstützender Technologien* (38. Jahrestagung der GMDS, Lübeck, September 1993), (Eds., Pöppl, S.J., Lipinski, H.-G., Mansky, T.), MMV Medizin Verlag, München, Germany, 474-477.

Gather, U. and Rauhut, B.O. (1990). The outlier behaviour of probability distributions, *Journal of Statistical Planning and Inference*, **26**, 237-252.

Gehan, E.A. (1990). *Unpublished Notes On Survivability Theory*, Univ. of Texas, M.D. Anderson Hospital and Tumor Institute, Houston, Texas.

Gehan, E.A. and Siddiqui, M.M. (1973). Simple regression methods for survival time studies, *Journal of the American Statistical Association*, **68**, 848-856.

Geisser, S. (1984). Predicting Pareto and exponential observables, *Canadian Journal of Statistics*, **12**, 143-152.

Geisser, S. (1985). Interval prediction for Pareto and exponential observables, *Journal of Econometrics*, **29**, 173-185.

Geisser, S. (1986). Predictive analysis, In *Encyclopedia of Statistical Sciences*, Vol. 7 (Eds., S. Kotz, N.L. Johnson and C.B. Read), pp. 158-170, John Wiley & Sons, New York.

Geisser, S. (1990). On hierarchical Bayes procedures for predicting simple exponential survival, *Biometrics*, **46**, 225-230.

Gerontidis, I. and Smith, R.L. (1982). Monte Carlo generation of order statistics from general distributions, *Applied Statistics*, **31**, 238-243.

Gertsbakh, I.B. (1989). *Statistical Reliability Theory*, Marcel Dekker, New York.

Ghitany, M.E. (1993). On the information matrix of exponential mixture models with long-term survivors, *Biometrical Journal*, **35**, 15-27.

Ghosh, M. and Mukhopadhyay, N. (1981). Consistency and asymptotic efficiency of two-stage and sequential procedures, *Sankhyā, Series A*, **43**, 220-227.

Ghosh, M. and Mukhopadhyay, N. (1989). Sequential estimation of the percentiles of exponential and normal distributions, *South African Statistical Journal*, **23**, 251-268.

Ghosh, M. and Mukhopadhyay, N. (1990). Sequential estimation of the location parameter of an exponential distribution, *Sankhyā, Series A*, **52**, 303-313.

Ghosh, M., Mukhopadhyay, N., and Sen, P.K. (1995). *Sequential Estimation*, John Wiley & Sons, New York (To appear).

Ghurye, S.G. (1958). Note on sufficient statistics and two-stage procedures, *Annals of Mathematical Statistics*, **29**, 155-166.

Ghurye, S.G. (1960). Characterization of some location and scale parameter families of distributions, In *Contributions to Probability and Statistics*, pp. 202-215, Stanford University Press, Stanford, California.

Gibbons, J.D., Olkin, I., and Sobel, M. (1977). *Selecting and Ordering Populations: A New Statistical Methodology*, John Wiley & Sons, New York.

Gill, A.N. and Sharma, S.K. (1993). A selection procedure for selecting good exponential populations, *Biometrical Journal*, **35**, 361-369.

Giri, N.C. (1977). *Multivariate Statistical Inference*, Academic Press, New York.

Glasser, M. (1967). Exponential survival with covariance, *Journal of the American Statistical Association*, **62**, 561-568.

Glick, N. (1978). Breaking records and breaking boards, *American Mathematical Monthly*, **85**, 2-26.

Gnedenko, B.V. (1943). Sur la distribution limite du terme maximum d'une serie aletoise, *Annals of Mathematics*, **44**, 423-453.

Gnedenko, B.V., Belyayev, Y.K., and Solovyev, A.D. (1969). *Mathematical Methods of Reliability*, Academic Press, New York.

Goldberger, A.S. (1962). Best linear unbiased prediction in the generalized linear regression model, *Journal of the American Statistical Association*, **57**, 369-375.

Goldman, A.I. (1984). Survivorship analysis when cure is a possibility: a Monte Carlo study, *Statistics in Medicine*, **3**, 153-163.

Gordon, A.D. (1981). *Classification: Methods for the Exploratory Analysis of Multivariate Data*, Chapman and Hall, London.

Gordon, N.H. (1990). Application of the theory of finite mixtures for the estimation of 'cure' rates of treated cancer patients, *Statistics in Medicine*, **9**, 397-407.

Govindarajulu, Z. (1963). Relationships among moments of order statistics in samples from two related populations, *Technometrics*, **5**, 514-518.

Govindarajulu, Z. (1966a). Best linear estimates under symmetric censoring of the parameters of a double exponential population, *Journal of the American Statistical Association*, **61**, 248-258.

Govindarajulu, Z. (1966b). Characterization of the exponential and power distributions, *Skandinavisk Aktuarietidskrift*, **49**, 132-136.

Govindarajulu, Z. (1978). Correction to "Characterization of the exponential and power distributions," *Scandinavian Actuarial Journal*, 175-176.

Govindarajulu, Z. (1985). Exact expressions for the stopping time and confidence coefficient in point and interval estimation of scale parameter of exponential distribution with unknown location, *Technical Report 254*, University of Kentucky, Dept. of Statistics.

Govindarajulu, Z., Huang, J.S., and Saleh, A.K.Md.E. (1975). Expected value of the spacings between order statistics, In *A Modern Course on Statistical Distributions in Scientific Work* (Eds., G.P. Patil, S. Kotz, and J.K. Ord), pp. 143-147, D. Reidel, Dordrecht.

Govindarajulu, Z. and Sarkar, S.C. (1991). Sequential estimation of the scale parameter in exponential distribution with unknown location, *Utilitas Mathematica*, **40**, 161-178.

Greenberg, B.G. and Sarhan, A.E. (1958). Applications of order statistics to health data, *American Journal of Public Health*, **48**, 1388-1394.

Gross, A.J. and Clark, V.A. (1975). *Survival Distributions: Reliability Applications in the Biomedical Sciences*, John Wiley & Sons, New York.

Gross, A.J., Clark, V.A., and Liu, V. (1971). Estimation of survival parameters when one of two organs must function for survival, *Biometrics*, **27**, 369-377.

Gross, A.J., Hunt, H.H., Cantor, A.B., and Clark, B.C. (1987). Sample size determination in clinical trials with an emphasis on exponentially distributed responses, *Biometrics*, **43**, 875-883.

Gross, A.J. and Lam, C.F. (1981). Paired observations from a survival distribution, *Biometrics*, **37**, 505-511.

Grosswald, E. and Kotz, S. (1981). An integrated lack of memory property of the exponential distribution, *Annals of the Institute of Statistical Mathematics*, **33**, 205-214.

Grubbs, F.E. (1971). Fiducial bounds on reliability for the two-parameter negative exponential distribution, *Technometrics*, **13**, 873-876.

Guenther, W.C. (1971). On the use of best tests to obtain best β-content tolerance intervals, *Statistica Neerlandica*, **25**, 191-202.

Guenther, W.C., Patil, S.A., and Uppuluri, V.R.R. (1976). One-sided β-content tolerance factors for the two parameter exponential distribution, *Technometrics*, **18**, 333-340.

Gumbel, E.J. (1960). Bivariate exponential distributions, *Journal of the American Statistical Association*, **55**, 698-707.

Gupta, R.C. (1973). A characteristic property of the exponential distribution, *Sankhyā, Series B*, **35**, 365-366.

Gupta, R.C. (1984). Relationships between order statistics and record values and some characterization results, *Journal of Applied Probability*, **21**, 425-430.

Gupta, R.C. and Langford, E.S. (1984). On the determination of a distribution by its median residual life function: A functional equation, *Journal of Applied Probability*, **21**, 120-128.

Gupta, R.C., Mehrotra, K.G., and Michalek, J.E. (1984). A small sample test for an absolutely continuous bivariate exponential model, *Communications in Statistics - Theory and Methods*, **13**, 1735-1740.

Gupta, R.C., Mehrotra, K.G., and Michalek, J.E. (1984). A small sample test for an absolutely continuous bivariate exponential model, *Communications in Statistics - Theory and Methods*, **13**, 1735-1740.

Gupta, S.S. (1956). On a decision rule for a problem in ranking means, *Mimeograph Series No. 150*, Institute of Statistics, University of North Carolina, Chapel Hill, North Carolina.

Gupta, S.S. (1963). On a selection and ranking procedure for gamma populations, *Annals of the Institute of Statistical Mathematics*, **14**, 199-216.

Gupta, S.S. and Huang, D.-Y. (1981). Multiple decision theory: Recent developments. *Lecture Notes in Statistics*, Vol. 6, Springer-Verlag, New York.

Gupta, S.S. and Kim, W.-C. (1984). A two-stage elimination type procedure for selecting the largest of several normal means with a common unknown variance, In *Design of Experiments: Ranking and Selection* (Eds., T.J. Santner and A.C. Tamhane), pp. 77-93, Marcel Dekker, New York.

Gupta, S.S. and Leu, L.-Y. (1986). Isotonic procedures for selecting populations better than a standard: two-parameter exponential distributions, In *Reliability and Quality Control* (Ed., A.P. Basu), pp. 167-183, Elsevier, Amsterdam.

Gupta, S.S. and Liang, T. (1993). Selecting the best exponential population based on type-I censored data: A Bayesian approach, In *Advances in Reliability* (Ed., A.P. Basu), pp. 171-180, Elsevier, Amsterdam.

Gupta, S.S. and Panchapakesan, S. (1979). *Multiple Decision Procedures: Theory and Methodology of Selecting and Ranking Populations.* John Wiley & Sons, New York.

Gupta, S.S. and Sobel, M. (1962a). On selecting a subset containing the population with the smallest variance, *Biometrika*, **49**, 495-507.

Gupta, S.S. and Sobel, M. (1962b). On the smallest of several correlated F-statistics, *Biometrika*, **49**, 509-523.

Guttman, I. (1970). *Statistical Tolerance Regions: Classical and Bayesian*, Griffin, London.

Guttman, I. (1988). Tolerance regions, statistical, In *Encyclopedia of Statistical Sciences* (Eds., S. Kotz, N.L. Johnson and C.B. Read), pp. 272-287, John Wiley & Sons, New York.

Hahn, G.J. (1975). A simultaneous prediction limit on the means of future samples from an exponential distribution, *Technometrics*, **17**, 341-346.

Hahn, G.J. and Meeker, W.Q., Jr. (1991). *Statistical Intervals - A Guide to Practitioners*, John Wiley & Sons, New York.

Hahn, G.J. and Nelson, W. (1973). A survey of prediction intervals and their applications, *Journal of Quality Technology*, **5**, 178-188.

Hahn, G.J. and Shapiro, S.S. (1967). *Statistical Models in Engineering*, John Wiley & Sons, New York.

Hald, A. (1967). Asymptotic properties of Bayesian single sampling plans, *Journal of the Royal Statistical Society, Series B*, **29**, 162-173.

Hald, A. (1981). *Statistical Theory of Sampling Inspection by Attributes*, Academic Press, New York.

Hall, I.J. and Prairie, R. (1973). One-sided prediction intervals to contain at least m out of k future observations, *Technometrics*, **15**, 897-914.

Hall, P. (1927). Distribution of the mean of samples from a rectangular population, *Biometrika*, **19**, 240.

Hall, P. (1981). Asymptotic theory of triple sampling for sequential estimation of a mean, *Annals of Statistics*, **9**, 1229-1238.

Hall, P. (1983). Sequential estimation saving sampling operations, *Journal of the Royal Statistical Society, Series B*, **45**, 219-223.

Hall, P. and Titterington, D.M. (1989). The effect of simulation order on level accuracy and power of Monte Carlo tests, *Journal of the Royal Statistical Society, Series B*, **51**, 459-467.

Hall, W.J. and Mathiason, D.J. (1990). On large sample estimation and testing in parametric models, *International Statistical Review*, **58**, 77-97.

Hall, W.J. and Wellner, J.A. (1981). Mean Residual Life, In *Statistics and Related Topics* (Eds., M. Csörgö, D. Dawson, J.N.K. Rao, and A.K.Md.E. Saleh), pp. 169-184, North-Holland, Amsterdam.

Halton, J.H. (1970). A retrospective and prospective survey of the Monte Carlo method, *SIAM Review*, **8**, 5-15.

Hamdy, H.I., Mukhopadhyay, N., Costanza, M.C., and Son, M.S. (1988). Triple stage point estimation for the exponential location parameter, *Annals of the Institute of Statistical Mathematics*, **40**, 785-797.

Hamel, G. (1905). Eine Basis aller Zahlen und die unstetigen Lösungen der Funktionalgleichung $f(x+y) = f(x) + f(y)$, *Mathematische Annalen*, **60**, 459-462.

Hampel, F.R., Ronchetti, E.M., Rousseeuw, P.J., and Stahel, W.A. (1986). *Robust Statistics: The Approach Based on Influence Functions*, John Wiley & Sons, New York.

Hanagal, D.D. (1991). Large sample test of independence and symmetry in the multivariate exponential distribution, *Journal of the Indian Statistical Association*, **29**, 89-93.

Hanagal, D.D. (1992). Some inference results in bivariate exponential distributions based on censored samples, *Communications in Statistics - Theory and Methods*, **21**, 1273-1295.

Hanagal, D.D. (1993). Some inference results in an absolutely continuous multivariate exponential model of Block, *Statistics & Probability Letters* (To appear).

Hanagal, D.D. and Kale, B.K. (1991a). Large sample tests of independence for an absolutely continuous bivariate exponential model, *Communications in Statistics - Theory and Methods*, **20**, 1301-1313.

Hanagal, D.D. and Kale, B.K. (1991b). Large sample tests of l_3 in the bivariate exponential distribution, *Statistics & Probability Letters*, **12**, 311-313.

Hanagal, D.D. and Kale, B.K. (1992). Large sample tests for testing symmetry and independence in some bivariate exponential models, *Communications in Statistics - Theory and Methods*, **21**, 2625-2643.

Harris, C.M. (1976). A note on testing for exponentiality, *Naval Research Logistic Quarterly*, **23**, 169-175.

Harris, C.M., Marchal, W.G., and Botta, R.F. (1992). A note on generalized hyperexponential distributions, *Communications in Statistics - Stochastic Models*, **8**, 179-191.

Harris, T.E., Meier, P., and Tukey, J.W. (1950). The timing of the distribution of events between observations, *Human Biology*, **22**, 249-270.

Harter, H.L. (1961). Estimating the parameters of negative exponential populations from one or two order statistics, *Annals of Mathematical Statistics*, **32**, 1078-1090.

Harter, H.L. (1970). *Order Statistics and Their Use in Testing and Estimation, Vol. 2*, U.S. Government Printing Office, Washington, D.C.

Hashino, M. (1985). Formulation of the joint return period of two hydrologic variates associated with a Poisson process, *Journal of Hydroscience and Hydraulic Engineering*, **3**, 73-84.

Hawkes, A.G. (1972). A bivariate exponential distribution with applications to reliability, *Journal of the Royal Statistical Society, Series B*, **34**, 129-131.

Hawkins, D.M. (1980). *Identification of outliers*, Chapman and Hall, London.

Heckman, J.J., Robb, R., and Walker, J.R. (1990). Testing the mixture of exponentials hypothesis and estimating the mixing distribution by the method of moments, *Journal of the American Statistical Association*, **85**, 582-589.

Herd, G.R. (1956). *Estimation of the Parameters From a Multicensored Sample*, Ph.D. dissertation, Iowa State College, Iowa.

Hewett, J.E. (1968). A note on prediction intervals based on partial observations in certain life test experiments, *Technometrics*, **10**, 850-853.

Hewett, J.E. and Bulgren, W.G. (1971). Inequalities for some multivariate F-distributions with applications, *Technometrics*, **13**, 397-402.

Hewett, J.E. and Spurrier, J.D. (1983). A survey of two-stage tests of hypotheses: Theory and methods, *Communications in Statistics - Theory and Methods*, **12**, 2307-2425.

Hewett, J.E. and Tsutakawa, R.K. (1972). Two-stage chi-square goodness-of-fit test, *Journal of the American Statistical Association*, **67**, 395-401.

Hilton, G.F. (1984). Sequential and two-stage point estimation problems for negative exponential distributions, *Ph.D. Dissertation*, Oklahoma State University, Dept. of Statistics.

Hinkley, D.V. (1979). Predictive likelihood, *Annals of Statistics*, **7**, 718-728.

Ho, F.C.M., Gentle, J.E., and Kennedy, W.J. (1979). Generation of random variates from the multinomial distribution, *Proceedings of the Statistical Computing Section, American Statistical Association*, 336-339.

Hoaglin, D.C. (1985). Using quantiles to study shape, In *Exploring Data Tables, Trends and Shapes* (Eds., D.C. Hoaglin, F. Mosteller, and J.W. Tukey), pp. 417-460, John Wiley & Sons, New York.

Hoeffding, W. (1953). On the distribution of the expected values of the order statistics, *Annals of Mathematical Statistics*, **24**, 93-100.

Hogg, R.V. and Craig, A.T. (1978). *Introduction to Mathematical Statistics*, Macmillan, New York.

Hogg, R.V. and Tanis, E.A. (1963). An iterated procedure for testing the equality of several exponential distributions, *Journal of the American Statistical Association*, **58**, 435-443.

Hollander, M. and Proschan, F. (1972). Testing whether new is better than used, *Annals of Mathematical Statistics*, **43**, 1136-1146.

Hope, A.C.A. (1968). A simplified Monte Carlo significance test procedure, *Journal of the Royal Statistical Society, Series B*, **30**, 582-598.

Houchens, R.L. (1984). Record value theory and inference, *Ph.D. Thesis*, University of California, Riverside.

Hougaard, P. (1986). A class of multivariate failure time distributions, *Biometrika*, **73**, 671-678.

Hougaard, P., Harvard, B., and Holm, N. (1992). Assessment of dependence in the life times of twins, In *Survival Analysis: State of the Art* (Eds., J.P. Klein and P.K. Goel), pp. 77-98, Kluwer Academic Publishers, Boston.

Hsieh, H.K. (1979). On asymptotic optimality of likelihood ratio tests for multivariate normal distributions, *Annals of Statistics*, **7**, 592-598.

Hsieh, H.K. (1981). On testing the equality of two exponential distributions, *Technometrics*, **23**, 265-269.

Hsieh, H.K. (1986). An exact test for comparing location parameters of k exponential distributions with unequal scales based on Type II censored data, *Technometrics*, **28**, 157-164.

Huang, J.S. (1974a). On a theorem of Ahsanullah and Rahman, *Journal of Applied Probability*, **11**, 216-218.

Huang, J.S. (1974b). Characterization of the exponential distribution by order statistics, *Journal of Applied Probability*, **11**, 605-608.

Huang, J.S. (1981). On a 'Lack of memory' property, *Annals of the Institute of Statistical Mathematics*, **33**, 131-134.

Huang, J.S., Arnold, B.C., and Ghosh, M. (1979). On characterizations of the uniform distribution based on identically distributed spacings, *Sankhyā, Series B*, **41**, 109-115.

Huang, W.-T. and Huang, K.-C. (1980). Subset selections of exponential populations based on censored data, *Proceedings of the Conference on Recent Developments in Statistical Methods and Applications*, pp. 237-254, Directorate-General of Budget, Accounting and Statistics, Executive Yuan, Taipei, Taiwan, Republic of China.

Hyakutake, H. (1990). Statistical inferences on location parameters of bivariate exponential distributions, *Hiroshima Mathematical Journal*, **20**, 525-547.

IMSL (1991). STAT/LIBRARY, Version 2.0. IMSL, Inc. Houston.

Ireson, W.G. (Ed.) (1982). *Reliability Handbook*, McGraw-Hill, New York.

Isham, V., Shanbhag, D.N., and Westcott, M. (1975). A characterization of the Poisson process using forward recurrence times, *Mathematical Proceedings of the Cambridge Philosophical Society*, **78**, 513-516.

Isogai, E. and Uno, C. (1993). Sequential estimation of a parameter of an exponential distribution, *Annals of the Institute of Statistical Mathematics*, **45**.

Jain, R.B. and Pingel, L.A. (1981). A procedure for estimating the number of outliers, *Communications in Statistics - Theory and Methods*, **10**, 1029-1041.

Jensen, F. and Petersen, N.E. (1982). *Burn-In: An Engineering Approach to The Design and Analysis of Burn-In Experiments*, John Wiley & Sons, New York.

Jewell, N.P. (1982). Mixtures of exponential distributions, *Annals of Statistics*, **10**, 479-484.

Jeyaratnam, S. and Panchapakesan, S. (1986). Estimation after subset selection from exponential populations, *Communications in Statistics - Theory and Methods*, **10**, 1869-1878.

Joe, H. (1985). Characterizations of life distributions from percentile residual lifetimes, *Annals of the Institute of Statistical Mathematics*, **37**, 165-172.

John, S. (1971). Some optimal multivariate tests, *Biometrika*, **58**, 123-127.

Johnson, N.L. and Kotz, S. (1977). *Distributions in Statistics: Continuous Multivariate Distributions,* John Wiley & Sons, New York.

Johnson, N.L., Kotz, S., and Balakrishnan, N. (1994). *Continuous Univariate Distributions - I*, Second edition, John Wiley & Sons, New York.

Johnson, N.L., Kotz, S., and Balakrishnan, N. (1995). *Continuous Univariate Distributions - Volume 2*, Second edition, John Wiley & Sons, New York (to appear).

Johnson, R.A. and Mehrotra, K.G. (1972). Locally most powerful rank tests for the two-sample problem with censored data, *Annals of Mathematical Statistics*, **43**, 823-831.

Jones, P.N. and McLachlan, G.J. (1992). Improving the convergence rate of the EM algorithm for a mixture model fitted to grouped truncated data, *Journal of Statistical Computation and Simulation*, **43**, 31-44.

Joshi, P.C. (1972). Efficient estimation of a mean of an exponential distribution when an outlier is present, *Technometrics*, **14**, 137-144.

Joshi, P.C. (1978). Recurrence relations between moments of order statistics from exponential and truncated exponential distributions, *Sankhyā, Series B*, **39**, 362-371.

Joshi, P.C. (1979). A note on the moments of order statistics from doubly truncated exponential distribution, *Annals of the Institute of Statistical Mathematics*, **31**, 321-324.

Joshi, P.C. (1982). A note on the mixed moments of order statistics from exponential and truncated exponential distributions, *Journal of Statistical Planning and Inference*, **6**, 13-16.

Joshi, P.C. (1988). Estimation and testing under an exchangeable exponential model with a single outlier, *Communications in Statistics - Theory and Methods*, **17**, 2315-2326.

Joshi, P.C. and Balakrishnan, N. (1984). Distribution of range and quasi-range from doubly truncated exponential distribution, *Trabajos de Estadistica y de Investigaciones Operationes*, **35**, 231-236.

Kahn, H.D. (1979). Least squares estimation for the inverse power law for accelerated life tests, *Applied Statistics*, **28**, 40-46.

Kakosyan, A.V., Klebanov, L.B., and Melamed, J.A. (1984). *Characterization of Distribution by the Method of Intensively Monotone Operators*, Lecture Notes in Mathematics No. 1088, Springer-Verlag, New York.

Kalbfleisch, J.D. (1971). Likelihood methods of prediction, In *Foundations of Statistical Inference - A Symposium* (Eds., V.P. Godambe and D.A. Sprott), Holt, Rinehart, and Winston, Toronto, Canada.

Kalbfleisch, J.D. (1982). Ancillary statistics, In *Encyclopedia of Statistical Sciences, Volume 1* (Eds., S. Kotz and N.L. Johnson), pp. 77-81, John Wiley & Sons, New York.

Kalbfleisch, J.D. and Prentice, R.L. (1980). *The Statistical Analysis of Failure Time Data*, John Wiley & Sons, New York.

Kale, B.K. (1975). Trimmed means and the method of maximum likelihood when spurious observations are present, In *Applied Statistics* (Ed., R.P. Gupta), North-Holland, Amsterdam.

Kale, B.K. and Sinha, S.K. (1971). Estimation of expected life in the presence of an outlier observation, *Technometrics*, **13**, 755-759.

Kambo, N.S. (1978). Maximum likelihood estimators of the location and scale parameters of the exponential distribution from a censored sample, *Communications in Statistics - Theory and Methods*, **7**, 1129-1132.

Kambo, N.S. and Awad, A.M. (1985). Testing equality of location parameters of k exponential distributions, *Communications in Statistics - Theory and Methods*, **14**, 567-583.

Kaminsky, K.S. (1972). Confidence intervals for the exponential scale parameter using optimally selected order statistics, *Technometrics*, **14**, 371-383.

Kaminsky, K.S. (1973). Comparison of approximate confidence intervals for the exponential scale parameter from sample quantiles, *Technometrics*, **15**, 483-487.

Kaminsky, K.S. (1974). Confidence intervals and tests for two exponential scale parameters based on order statistics in compressed samples, *Technometrics*, **16**, 251-254.

Kaminsky, K.S. (1977a). Best prediction of exponential failure times when items may be replaced, *Australian Journal of Statistics*, **19**, 61-62.

Kaminsky, K.S. (1977b). Comparison of prediction intervals for failure times when life is exponential, *Technometrics*, **19**, 83-86.

Kaminsky, K.S., Mann, N.R., and Nelson, P.I. (1975). Best and simplified linear invariant prediction of order statistics in location and scale families, *Biometrika*, **62**, 525-526.

Kaminsky, K.S. and Nelson, P.I. (1974). Prediction intervals for the exponential distribution using subsets of the data, *Technometrics*, **16**, 57-59.

Kaminsky, K.S. and Nelson, P.I. (1975a). Best linear unbiased prediction of order statistic in location and scale families, *Journal of the American Statistical Association*, **70**, 145-150.

Kaminsky, K.S. and Nelson, P.I. (1975b). Characterization of distributions by the form of predictors of order statistics, In *Statistical Distributions in Scientific Work* (Eds., G.P. Patil, S. Kotz, and J.K. Ord), Vol. 3, pp. 113-116, D. Reidel, Dordrecht.

Kaminsky, K.S. and Rhodin, L.S. (1978). The prediction information in the latest failure, *Journal of the American Statistical Association*, **73**, 863-866.

Kaminsky, K.S. and Rhodin, L.S. (1985). Maximum likelihood prediction, *Annals of the Institute of Statistical Mathematics*, **37**, 507-517.

Kaplan, E.L. and Meier, P. (1959). Nonparametric estimation from incomplete observations, *Journal of the American Statistical Association*, **53**, 457-481.

Karlin, S. and Taylor, H.M. (1975). *A First Course in Stochastic Processes*, Second edition, Academic Press, New York.

Keating, J.P., Mason, R.L., Rao, C.R., and Sen, P.K. (1991). Pitman's measure of closeness, *Communications in Statistics - Theory and Methods, Special Issue*, **20**(11).

Kelly, F. (1979). *Reversibility and Stochastic Networks*, John Wiley & Sons, New York.

Kendall, M.G. and Stuart, A. (1958). *The Advanced Theory of Statistics*, Hafner, New York.

Kennedy, Jr., W.J. and Gentle, J.E. (1980). *Statistical Computing*, Marcel Dekker, New York.

Khatri, C.G. (1974). On testing the equality of location parameters in k censored exponential distributions, *Australian Journal of Statistics*, **16**, 1-10.

Khatri, C.G. (1981). Power of a test for location parameters of two exponential distributions, *Aligarh Journal of Statistics*, **1**, 8-12.

Kiefer, J. (1982). Conditional inference, In *Encyclopedia of Statistical Sciences, Volume 2* (Eds., S. Kotz and N.L. Johnson), pp. 103-109, John Wiley & Sons, New York.

Killeen, T.J., Hettmansperger, T.P., and Sievers, G.Z. (1972). An elementary theorem on probability of large deviations, *Annals of Mathematical Statistics*, **43**, 181-192. Correction, *Annals of Statistics* (1974), 1357.

Kim, J.S. (1988). A subset selection procedure for exponential populations under random censoring, *Communications in Statistics - Theory and Methods*, **17**, 183-206.

Kim, W.-C. and Lee, S.-H. (1985). An elimination type two-stage selection procedure for exponential distributions, *Communications in Statistics - Theory and Methods*, **14**, 2563-2571.

Kimber, A.C. (1979). Tests for a single outlier in a gamma sample with unknown shape and scale parameters, *Applied Statistics*, **28**, 243-250.

Kimber, A.C. (1982). Tests for many outliers in an exponential sample, *Applied Statistics*, **32**, 304-310.

Kimber, A.C. (1983). Comparison of some robust estimators of scale in gamma samples with known shape, *Journal of Statistical Computation and Simulation*, **18**, 273-286.

Kinderman, A.J. and Monahan, J.F. (1977). Computer generation of random variables using the ratio of uniform deviates, *Association for Computing Machinery, Monograph Transactions, Mathematical Software*, **3**, 257-260.

Kinderman, A.J. and Ramage, J.G. (1976). Computer generation of normal random variables, *Journal of the American Statistical Association*, **71**, 893-896.

Kingman, J.F.C. (1970). Inequalities in the theory of queues, *Journal of the Royal Statistical Society, Series B*, **32**, 102-110.

Kirmani, S.N.U.A. and Beg, M.I. (1984). On characterization of distributions by expected records, *Sankhyā, Series A*, **46**, 463-465.

Kitagawa, G. (1979). On the use of AIC for the detection of outliers, *Technometrics*, **21**, 193-199.

Klein, J.P. and Basu, A.P. (1981). Accelerated life testing under competing exponential failure distributions, *IAPQR Transactions*, **7**, 1-20.

Klein, J.P. and Basu, A.P. (1982). Accelerated life testing under competing exponential failure distributions, *IAPQR Transactions*, **7**, 1-20.

Klein, J.P. and Basu, A.P. (1982). Accelerated life tests under competing Weibull causes of failure, *Communications in Statistics - Theory and Methods*, **11**, 2271-2287.

Klein, J.P. and Basu, A.P. (1985). On estimating reliability for bivariate exponential distributions, *Sankhyā, Series B*, **47**, 346-353.

Klein, J.P., Moeschberger, M.L., Li, Y.I., and Wang, S.T. (1992). Estimating random effects in the Framingham heart study, In *Survival Analysis: State of the Art* (Eds., J.P. Klein and P.K. Goel), pp. 99-120, Kluwer Academic Publishers, Boston.

Knapp, R.G., Cantor, A.B., and Gross, A.J. (1986). Estimators of the ratio of means of paired survival data, *Communications in Statistics - Simulation and Computation*, **15**, 85-100.

Knapp, R.G., Gross, A.J., and Cantor, A.G. (1986). A likelihood ratio test of the equality of paired survival data with censoring, *Biometrical Journal*, **28**, 665-672.

Knuth, D.E. (1973). *The Art of Computer Programming, Vol. 3: Searching and Sorting*, Addison-Wesley, Reading, Massachusetts.

Kocherlakota, S. and Balakrishnan, N. (1986). One- and two-sided sampling plans based on the exponential distribution, *Naval Research Logistics Quarterly*, **33**, 513-522.

Kondo, T. (1931). Theory of sampling distribution of standard deviations, *Biometrika*, **22**, 31-64.

Konheim, A.G. (1971). A note on order statistics, *American Mathematical Monthly*, **78**, 524.

Kourouklis, S. (1988). Asymptotic optimality of likelihood ratio tests for exponential distributions under Type II censoring, *Australian Journal of Statistics*, **30**, 111-114.

Krishnaji, N. (1970). Characterization of the Pareto distribution through a model of under-reported income, *Econometrica*, **38**, 251-255.

Krishnaji, N. (1971). Note on a characterizing property of the exponential distribution, *Annals of Mathematical Statistics*, **42**, 361-362.

Krishnamoorthy, A.S. and Parthasarathy, M. (1951). A multivariate gamma type distribution, *Annals of Mathematical Statistics*, **22**, 549-557.

Kronmal, R.A. and Peterson, Jr., A.V. (1979). On the alias method for generating random variables from a discrete distribution, *The American Statistician*, **33**, 214-218.

Kronmal, R.A. and Peterson, Jr., A.V. (1981). A variant of the acceptance-rejection method for computer generation of random variables, *Journal of the American Statistical Association*, **76**, 446-451.

Kudo, A. (1956). On the invariant multiple decision procedures, *Bulletin of Mathematical Statistics*, **6**, 57-68.

Kuk, A.Y.C. (1992). A semiparametric mixture model for the analysis of competing risks data, *Australian Journal of Statistics*, **34**, 169-180.

Kuk, A.Y.C. and Chen, C.-H. (1992). A mixture model combining logistic regression with proportional hazards regression, *Biometrika*, **79**, 531-541.

Kulldorff, G. (1963). Estimation of one or two parameters of the exponential distribution on the basis of suitably chosen order statistics, *Annals of Mathematical Statistics*, **34**, 1419-1431.

Kumar, S. and Patel, H.I. (1971). A test for the comparison of two exponential distributions, *Technometrics*, **13**, 183-189.

Kuo, L. and Mukhopadhyay, N. (1990a). Point estimation of the largest location of k negative exponential populations, *Sequential Analysis*, **9**, 297-304.

Kuo, L. and Mukhopadhyay, N. (1990b). Multi-stage point and interval estimation of the largest of k normal populations and the associated second-order properties, *Metrika*, **37**, 291-300.

Lachenbruch, P.A. (1974). *Discriminant Analysis*, Hafner Press, New York.

Lachin, J.M. and Foulkes, M.A. (1986). Evaluation of sample size and power for analyses of survival with allowance for nonuniform patient entry, losses to follow-up, noncompliance, and stratification, *Biometrics*, **42**, 507-519.

Lal Saxena, K.M. and Tong, Y.L. (1969). Interval estimation of the largest mean of k normal populations with known variances, *Journal of the American Statistical Association*, **64**, 296-299.

Lam, K. and Ng, C.K. (1990). Two-stage procedures for comparing several exponential populations with a control when the scale parameters are unknown and unequal, *Sequential Analysis*, **9**, 151-164.

Lam, Y. (1988a). A decision theory approach to variable sampling plans, *Scientia Sinica, Series A*, **31**, 120-140.

Lam, Y. (1988b). Bayesian approach to single variable sampling plans, *Biometrika*, **75**, 387-391.

Lam, Y. (1990). An optimal single variable sampling plan with censoring, *The Statistician*, **39**, 53-67.

Lam, Y. (1994). Bayesian variable sampling plans for the exponential distribution with Type I censoring, *Annals of Statistics*, **22** (to appear).

Lam, Y. and Choy, S.T.B. (1994). Bayesian variable sampling plans for the exponential distribution with random censoring, *Journal of Statistical Planning and Inference* (revised).

Lam, Y. and Lau, L.C. (1993). Optimal single variable sampling plans, *Communications in Statistics - Simulation and Computation*, **22**, 371-386.

Larsen, R.J. and Marx, M.L. (1981). *An Introduction to Mathematical Statistics and Its Applications*, Prentice-Hall, Engelwood Cliffs, New Jersey.

Lau, K.S. and Rao, C.R. (1982). Integrated Cauchy functional equation and characterizations of the exponential law, *Sankhyā, Series A*, **44**, 72-90.

Lawless, J.F. (1971). A prediction problem concerning samples from the exponential distribution, with application to life testing, *Technometrics*, **13**, 725-730.

Lawless, J.F. (1972). On prediction intervals for samples from the exponential distribution and prediction limits for system survival, *Sankhyā, Series B*, **34**, 1-14.

Lawless, J.F. (1976). Confidence interval estimation in the inverse power law model, *Applied Statistics*, **25**, 128-138.

Lawless, J.F. (1977). Prediction intervals for the two parameter exponential distribution, *Technometrics*, **19**, 469-472.

Lawless, J.F. (1982). *Statistical Models & Methods for Lifetime Data*, John Wiley & Sons, New York.

Learmonth, G.P. and Lewis, P.A.W. (1973). Naval Postgraduate School random number generator package LLRANDOM, *Naval Postgraduate School Technical Report*, Monterey, California.

Lee, M.-L.T. and Gross, A.J. (1989). Properties of conditionally independent generalized gamma distributions, *Probability in the Engineering and Informational Sciences*, **3**, 289-297.

Lee, M.-L.T. and Gross, A.J. (1991). Lifetime distributions under unknown environment, *Journal of Statistical Planning and Inference*, **29**, 137-143.

Lehmann, E.L. (1959). *Testing of Statistical Hypotheses*, John Wiley & Sons, New York.

Lehmann, E.L. (1983). *Theory of Point Estimation*, John Wiley & Sons, New York.

Lehmer, D.H. (1954). Random Number Generation on the BRL High Speed Computing Machines, *Proceedings of the Second Symposium on Large-Scale Digital Computing Machinery*, Harvard University Press, Cambridge, Massachusetts.

Leong, C.Y. and Ling, K.D. (1976). Bayesian predictive distributions for future observations from exponential variate in a life testing model, *Nanta Mathematica*, **9**, 171-177.

Leu, L.-Y. and Liang, T. (1990). Selection of the best with a preliminary test for two-parameter exponential distributions, *Communications in Statistics - Theory and Methods*, **19**, 1443-1455.

Leurgans, S., Tsai, T., Wei, Y., and Crowley, J. (1982). Freund's bivariate exponential distribution and censoring, In *Survival Analysis* (Eds., R.A. Johnson and J. Crowley), pp. 230-242, IMS Lecture Notes.

Lewis, P.A.W., Goodman, A.S., and Miller, J.M. (1969). A pseudo-random number generator for the system/360, *IBM Systems Journal*, **8**, 136-145.

Li, Y.I., Klein, J.P., and Moeschberger, M.L. (1993). Semi-parametric estimation of covariate effects using the inverse Gaussian frailty model, *Technical Report*, The Ohio State University, Columbus.

Liang, T. and Panchapakesan, S. (1992). A two-stage procedure for selecting the δ^*-optimal guaranteed lifetimes in the two-parameter exponential model, In *Multiple Comparisons, Selection, and Applications in Biometry, A Festschrift in Honor of Charles W. Dunnett* (Ed., F.M. Hoppe), pp. 353-365, Marcel Dekker, New York.

Likeš, J. (1974). Prediction of s-th ordered observation for the two-parameter exponential distribution, *Technometrics*, **16**, 241-244.

Likeš, J. and Nedělka, S. (1973). Note on studentized range in samples from an exponential distribution, *Biometrical Journal*, **15**, 545-555.

Lilliefors, H.W. (1969). On the Kolmogorov-Smirnov test for the exponential distribution with mean unknown, *Journal of the American Statistical Association*, **64**, 387-389.

Lin, C.C. and Mudholkar, G.S. (1980). A test for exponentiality based on the bivariate F distribution, *Technometrics*, **22**, 79-82.

Lindley, D.V. (1980). Approximate Bayesian methods, In *Bayesian Statistics* (Eds., J.M. Bernardo, M. De Groot, D.V. Lindley, and A.F.M. Smith), pp. 223-245, Valencia Press, Spain.

Ling, K.D. (1975). On structural prediction distribution for samples from exponential distribution, *Nanta Mathematica*, **8**, 47-52.

Ling, K.D. and Lee, G.C. (1977). Bayesian predictive distributions for samples from exponential distribution based on partially grouped samples: I, *Nanta Mathematica*, **10**, 166-173.

Ling, K.D. and Lee, G.C. (1978). Bayesian predictive distributions for samples from exponential distribution based on partially grouped samples: II, *Nanta Mathematica*, **11**, 55-62.

Ling, K.D. and Leong, C.Y. (1977). Bayesian predictive distributions for samples from exponential distribution: I, *Tamkang Journal of Mathematics*, **8**, 11-16.

Ling, K.D. and Siaw, V.C. (1975). On prediction intervals for samples from exponential distribution, *Nanyang University Journal, Part III*, **9**, 41-46.

Ling, K.D. and Tan, C.K. (1979). On predictive distributions based on samples with missing observations – Exponential distribution, *Nanta Mathematica*, **12**, 173-181.

Lingappaiah, G.S. (1973). Prediction in exponential life testing, *Canadian Journal of Statistics*, **11**, 113-117.

Lingappaiah, G.S. (1979a). Bayesian approach to prediction and the spacings in the exponential distribution, *Annals of the Institute of Statistical Mathematics*, **31**, 391-401.

Lingappaiah, G.S. (1979b). Bayesian approach to the prediction of the restricted range in the censored samples from the exponential population, *Biometrical Journal*, **21**, 361-366.

Lingappaiah, G.S. (1979c). Bayesian approach to the prediction problem in complete and censored samples from the gamma and exponential populations, *Communications in Statistics - Theory and Methods*, **8**, 1403-1424.

Lingappaiah, G.S. (1980). Intermittant life testing and Bayesian approach to prediction with spacings in the exponential model, *Statistica*, **40**, 477-490.

Lingappaiah, G.S. (1981). Mixture of exponential populations and prediction by Bayesian approach, *Revue Roumaine de Mathematiques Pures et Appliquees*, **26**, 753-760.

Lingappaiah, G.S. (1984). Bayesian prediction regions for the extreme order statistics, *Biometrical Journal*, **26**, 49-56.

Lingappaiah, G.S. (1986). Bayes prediction in exponential life-testing when sample size is a random variable, *IEEE Transactions on Reliability*, **35**, 106-110.

Lingappaiah, G.S. (1989). Bayes prediction of maxima and minima in exponential life tests in the presence of outliers, *Industrial Mathematics*, **39**, 169-182.

Lingappaiah, G.S. (1990). Inference with samples from an exponential population in the presence of an outlier, *Publications of the Institute of Statistics, University of Paris*, **35**, 43-54.

Lingappaiah, G.S. (1991). Prediction in exponential life tests where average lives are successively increasing, *Pakistan Journal of Statistics*, **7**, 33-39.

Lloyd, E.H. (1952). Least-squares estimation of location and scale parameters using order statistics, *Biometrika*, **39**, 88-95.

Lobachevskii, N.L. (1829). On the foundation of geometry, III, Section 12, *Kasaner Bote*, **27**, 227-243 (in Russian).

Lombard, F. and Swanepoel, J.W.H. (1978). On finite and infinite confidence sequences, *South African Statistical Journal*, **12**, 1-24.

Louis, T.A. (1982). Finding the observed information matrix when using the EM algorithm, *Journal of the Royal Statistical Society, Series B*, **44**, 226-233.

Lu, J. and Bhattacharyya, G.K. (1991a). Inference procedures for bivariate exponential model of Gumbel, *Statistics& Probability Letters*, **12**, 37-50.

Lu, J. and Bhattacharyya, G.K. (1991b). Inference procedures for a bivariate exponential model of Gumbel based on life test of component and system, *Journal of Statistical Planning and Inference*, **27**, 383-396.

Lurie, D. and Hartley, H.O. (1972). Machine generation of order statistics for Monte Carlo computations, *The American Statistician*, **26**, 26-27.

Lurie, D. and Mason, R.L. (1973). Empirical investigation of general techniques for computer generation of order statistics, *Communications in Statistics*, **2**, 363-371.

Madi, M. and Tsui, K.-W. (1990a). Estimation of the ratio of the scale parameters of two exponential distributions with unknown location parameters, *Annals of the Institute of Statistical Mathematics*, **42**, 77-87.

Madi, M. and Tsui, K.-W. (1990b). Estimation of the common scale of several exponential distributions with unknown locations, *Communications in Statistics - Theory and Methods*, **19**, 2295-2313.

Madsen, R.W. (1978). A K-stage test procedure using p-values, *Journal of Statistical Computation and Simulation*, **8**, 117-131.

Malik, H.J. (1970). A characterization of the Pareto distribution, *Skandinavisk Aktuarietidskrift*, 115-117.

Maller, R.A. (1988). On the exponential model of survival, *Biometrika*, **75**, 582-586.

Malmquist, S. (1950). On a property of order statistics from a rectangular distribution, *Skandinavisk Aktuarietidskrift*, **33**, 214-222.

Mann, N.R. (1972). Design of over-stress life-test experiments when failure times have a two-parameter Weibull distribution, *Technometrics,* **14,** 437-451.

Mann, N.R. and Grubbs, F.E. (1974). Chi-square approximations for exponential parameters, prediction intervals and beta percentiles, *Journal of the American Statistical Association*, **69**, 654-661.

Mann, N.R., Schafer, R.E., and Singpurwalla, N.D. (1974). *Methods for Statistical Analysis of Reliability and Life Data*, John Wiley & Sons, New York.

Marasinghe, M.G. (1985). A multistage procedure for detecting several outliers in linear regression, *Technometrics*, **27**, 395-399.

Marsaglia, G. (1961). Generating exponential random variables, *Annals of Mathematical Statistics*, *32*, 899-902.

Marsaglia, G. (1962). Random variables and computers, *Transactions of the Third Prague Conference* (Ed., J. Kozesnik), pp. 499-510, Czechoslovak Academy of Sciences, Prague.

Marsaglia, G., Ananthanarayanan, K., and Paul, N. (1973). How to use the McGill random number package, *McGill University Report*, Montreal, Canada.

Marsaglia, G., MacLaren, M.D., and Bray, T.A. (1964). A fast procedure for generating exponential random variables, *Communications of the ACM*, **7**, 298-300.

Marsaglia, G. and Tubilla, A. (1975). A note on the 'Lack of memory' property of the exponential distribution, *Annals of Probability*, **3**, 353-354.

Marshall, A.W. and Olkin, I. (1967). A multivariate exponential distribution, *Journal of the American Statistical Association*, **62**, 30-44.

Marshall, A.W. and Olkin, I. (1979). *Inequalities: Theory of Majorization and Its Applications*, Academic Press, New York.

McCool, J.I. (1974). Inferential techniques for Weibull populations, *Aerospace Research Laboratories Report ARL TR 74-0180*, Wright-Patterson AFB, Ohio.

McCullagh, P. and Nelder, J.A. (1989). *Generalized Linear Models*, Second edition, Chapman & Hall, London.

McGiffin, D.C., O'Brien, M.F., Galbraith, A.J., McLachlan, G.J., Stafford, E.G., Gardiner, M.A.H., Pohlner, P.G., Early, L., and Kear, L. (1994a). An analysis of risk factors for death and mode-specific death following aortic valve replacement using allograft, xenograft and mechanical valves, *Journal of Cardiac and Thoracic Surgery*, **106**, 895-911.

McGiffin, D.C., O'Brien, M.F., Galbraith, A.J., McLachlan, G.J., Stafford, E.G., Gardiner, M.A.H., Pohlner, P.G., Early, L., and Kear, L. (1994b). An analysis of risk factors for reoperation following aortic valve replacement using allograft, xenograft and mechanical valves, *Journal of Cardiac and Thoracic Surgery* (To appear).

McLachlan, G.J. (1987). On bootstrapping the likelihood ratio test statistic for the number of components in a normal mixture, *Applied Statistics*, **36**, 318-324.

McLachlan, G.J. (1992). *Discriminant Analysis and Statistical Pattern Recognition*, John Wiley & Sons, New York.

McLachlan, G.J., Adams, P., McGiffin, D.C., and Galbraith, A.J. (1993). Fitting mixtures of Gompertz distributions to censored survival data, *Submitted for publication*.

McLachlan, G.J. and Basford, K.E. (1988). *Mixture Models: Inference and Applications to Clustering*, Marcel Dekker, New York.

McLachlan, G.J., Lawoko, C.R.O., and Ganesalingam, S. (1982). On the likelihood ratio test for compound distributions for homogeneity of mixing proportions, *Technometrics*, **24**, 331-334.

Meeker, W.Q. (1977). Limited failure population life tests: Application to integrated circuit reliability, *Technometrics*, **29**, 51-65.

Meeker, W.Q. and Hahn, G.J. (1980). Prediction intervals for the ratio of normal distribution sample variances and exponential distribution sample means, *Technometrics*, **22**, 357-366.

Mehrotra, K.G. and Bhattacharyya, G.K. (1982). Confidence intervals with jointly Type-II censored samples from two exponential distributions, *Journal of the American Statistical Association*, **77**, 441-446.

Mehrotra, K.G. and Michalek, J.E. (1976). Estimation of parameters and tests of independence in a continuous bivariate exponential distribution, *Technical Report*, Syracuse University, Syracuse.

Meilijson, I. (1972). Limiting properties of the mean residual life function, *Annals of Mathematical Statistics*, **43**, 354-357.

Meilijson, I. (1989). A fast improvement to the EM algorithm on its own terms, *Journal of the Royal Statistical Society, Series B*, **51**, 127-138.

Mendenhall, W. and Hader, R.J. (1958). Estimation of parameters of mixed exponentially distributed failure time distributions from censored life-test data, *Biometrika*, **45**, 504-519.

Meng, X.L. and Rubin, D.B. (1991). Using EM to obtain asymptotic variance-covariance matrices: The SEM algorithm, *Journal of the American Statistical Association*, **86**, 899-909.

Miller, D.S. and Mukhopadhyay, N. (1990a). Likelihood based sequential procedures for selecting the best exponential population, *Metron*, **48**, 255-282.

Miller, D.S. and Mukhopadhyay, N. (1990b). Sequential likelihood rules for selecting the negative exponential population having the smallest scale parameter, *Calcutta Statistical Association Bulletin*, **39**, 31-44.

Moeschberger, M.L. and David, H.A. (1971). Life tests under competing causes of failure and the theory of competing risks, *Biometrics*, **27**, 909-933.

Monahan, J.F. (1979). Extensions of von Neumann's method for generating random variables, *Mathematics of Computation*, **33**, 1065-1069.

Moran, P.A.P. (1967). Testing for correlation between non-negative variates, *Biometrika*, **54**, 385-394.

Morgan, T.M. (1985). Planning the duration of accrual and follow-up for clinical trials, *Journal of Chronic Diseases*, **38**, 1009-1018.

Mosteller, F. (1946). On some useful "inefficient" statistics, *Annals of Mathematical Statistics*, **17**, 377-408.

Mount, K.S. and Kale, B.K. (1973). On selecting a spurious observation, *Canadian Mathematical Bulletin*, **16**, 75-78.

Mukherjee, S.P. and Roy, D. (1985). Some characterizations of the exponential and related life distributions, *Calcutta Statistical Association Bulletin*, 189-197.

Mukhopadhyay, C. and Basu, A.P. (1993a). Competing risks with k independent exponentials: a Bayesian analysis, *Technical Report No. 516*, Department of Statistics, Ohio State University, Columbus, Ohio.

Mukhopadhyay, C. and Basu, A.P. (1993b). Maximum likelihood and Bayesian estimation of lifetime parameters from masked system failure data, *Technical Report No. 517*, Department of Statistics, Ohio State University, Columbus, Ohio.

Mukhopadhyay, N. (1974). Sequential estimation of location parameter in exponential distributions, *Calcutta Statistical Association Bulletin*, **23**, 85-95.

Mukhopadhyay, N. (1980). A consistent and asymptotically efficient two-stage procedure to construct fixed-width confidence intervals for the mean, *Metrika*, **27**, 781-784.

Mukhopadhyay, N. (1982a). Stein's two-stage procedure and exact consistency, *Skandinavisk Aktuarietidskrift*, 110-122.

Mukhopadhyay, N. (1982b). A study of the asymptotic regret while estimating the location of an exponential distribution, *Calcutta Statistical Association Bulletin*, **31**, 207-213.

Mukhopadhyay, N. (1984). Sequential and two-stage procedures for selecting the better exponential population covering the case of scale parameters being unknown and unequal, *Journal of Statistical Planning and Inference*, **9**, 33-44.

Mukhopadhyay, N. (1985). A note on three-stage and sequential point estimation procedures for a normal mean, *Sequential Analysis*, **4**, 311-320.

Mukhopadhyay, N. (1986). On selecting the best exponential population, *Journal of Indian Statistical Association*, **24**, 31-41.

Mukhopadhyay, N. (1987a). Three-stage procedures for selecting the best exponential population, *Journal of Statistical Planning and Inference*, **16**, 345-352.

Mukhopadhyay, N. (1987b). Minimum risk point estimation of the mean of a negative exponential distribution, *Sankhyā, Series A*, **49**, 105-112.

Mukhopadhyay, N. (1988a). Sequential estimation problems for negative exponential populations, *Communications in Statistics - Theory and Methods (Reviews Section)*, **17**, 2471-2506.

Mukhopadhyay, N. (1988b). Fixed precision estimation of a positive location parameter of a negative exponential population, *Calcutta Statistical Association Bulletin*, **37**, 101-104.

Mukhopadhyay, N. (1990). Some properties of a three-stage procedure with applications in sequential analysis, *Sankhyā, Series A*, **52**, 218-231.

Mukhopadhyay, N. (1991). Parametric sequential point estimation, In *Handbook of Sequential Analysis*, Chapter 10 (Eds., B.K. Ghosh and P.K. Sen), pp. 245-267, Marcel Dekker, New York.

Mukhopadhyay, N. (1992). Simultaneous point estimation problems for two-parameter negative exponential populations, *Journal of the Indian Statistical Association*, **30**, 33-41.

Mukhopadhyay, N. (1994). Improved sequential estimation of means of exponential distributions, *Annals of the Institute of Statistical Mathematics*, **46**.

Mukhopadhyay, N. and Chattopadhyay, S. (1991). Sequential methodologies for comparing exponential mean survival times, *Sequential Analysis*, **10**, 139-148.

Mukhopadhyay, N. and Chattopadhyay, S. (1992). Comparing exponential clinical trials by combining individual sequential experiments available from independent studies, *Technical Report 92-09*, University of Connecticut, Dept. of Statistics.

Mukhopadhyay, N. and Chattopadhyay, S. (1993). Comparing means by combining individual sequential estimators available from independent studies, *Sankhyā, Series A*, **54**.

Mukhopadhyay, N. and Darmanto, S. (1988). Sequential estimation of the difference of means of two negative exponential populations, *Sequential Analysis*, **7**, 165-190.

Mukhopadhyay, N. and Datta, S. (1994). Replicated piecewise multistage sampling with applications, *Sequential Analysis*, **13**, 253-276.

Mukhopadhyay, N. and Ekwo, M.E. (1987). A note on minimum risk point estimation of the shape parameter of a Pareto distribution, *Calcutta Statistical Association Bulletin*, **36**, 69-78.

Mukhopadhyay, N. and Hamdy, H.I. (1984a). On estimating the difference of location parameters of two negative exponential distributions, *Canadian Journal of Statistics*, **12**, 67-76.

Mukhopadhyay, N. and Hamdy, H.I. (1984b). Two-stage procedures for selecting the best exponential population when the scale parameters are unknown and unequal, *Sequential Analysis*, **3**, 51-74.

Mukhopadhyay, N., Hamdy, H.I., Al-Mahmeed, M., and Costanza, M.C. (1987). Three-stage point estimation procedures for a normal mean, *Sequential Analysis*, **6**, 21-36.

Mukhopadhyay, N., Hamdy, H.I., and Darmanto, S. (1988). Simultaneous estimation after selection and ranking and other procedures, *Metrika*, **35**, 275-286.

Mukhopadhyay, N. and Hilton, G.F. (1986). Two-stage and sequential procedures for estimating the location parameter of a negative exponential distribution, *South African Statistical Journal*, **20**, 117-136.

Mukhopadhyay, N. and Kuo, L. (1988). Fixed-width interval estimations of the largest location of k negative exponential populations, *Sequential Analysis*, **7**, 321-332.

Mukhopadhyay, N. and Mauromoustakos, A. (1987). Three-stage estimation for the negative exponential distributions, *Metrika*, **34**, 83-93.

Mukhopadhyay, N. and Narayan, P. (1981). Sequential fixed-width intervals for the common location parameters of two normal or two negative exponential distributions, *Journal of Indian Society of Statistics and Operations Research*, **2**, 1-15.

Mukhopadhyay, N. and Padmanabhan, A.R. (1993). A note on three-stage confidence intervals for the difference of locations - The exponential case, *Metrika*, **40**, 121-128.

Mukhopadhyay, N. and Sen, P.K. (1993). Replicated piecewise stopping numbers and sequential analysis, *Sequential Analysis*, **12**, 179-197.

Mukhopadhyay, N. and Solanky, T.K.S. (1991). Second order properties of accelerated stopping times with applications in sequential estimation, *Sequential Analysis*, **10**, 99-123.

Mukhopadhyay, N. and Solanky, T.K.S. (1992). Accelerated sequential procedure for selecting the best exponential population, *Journal of Statistical Planning and Inference*, **32**, 347-361.

Mukhopadhyay, N. and Solanky, T.K.S. (1993). Estimation after selection and ranking, *Technical Report No. 93-32*, The University of Connecticut, Storrs, CT 06269.

Mukhopadhyay, N. and Solanky, T.K.S. (1994). *Multistage Selection and Ranking Procedures: Second-Order Asymptotics*, Marcel Dekker, New York.

Muntz, C.H. (1914). Uber den approximationssatz von Weierstrass, Schwarz-Festschrift.

Murphy, R.B. (1951). On tests for outlying observations, *Ph.D. Thesis*, Princeton University.

Nagao, M. and Kadoya, M. (1971). Two-variate exponential distribution and its numerical table for engineering applications, *Bulletin of Disaster Prevention Research Institute*, **20**, 183-215.

Nagaraja, H.N. (1977). On a characterization based on record values, *Australian Journal of Statistics*, **19**, 70-73.

Nagaraja, H.N. (1986). Comparison of estimators and predictors from two-parameter exponential distribution, *Sankhyā, Series B*, **48**, 11-18.

Nagaraja, H.N. (1988a). Some characterizations of continuous distributions based on regressions of adjacent order statistics and record values, *Sankhyā, Series A*, **50**, 70-73.

Nagaraja, H.N. (1988b). Record values and related statistics - a review, *Communications in Statistics - Theory and Methods*, **17**, 2223-2238.

Nagarsenkar, B.N. and Nagarsenkar, P.B. (1988). Non-null distribution of a modified likelihood ratio criterion associated with exponential distribution, *ASA Proceedings of Statistical Computing Section*, 288-290.

Nagarsenkar, P.B. (1980). On a test of equality of several exponential survival distributions, *Biometrika*, **67**, 475-478.

Nagarsenkar, P.B. and Nagarsenkar, B.N. (1988). Asymptotic distribution of a test of equality of exponential distributions, *ASA Proceedings of Statistical Computing Section*, 284-287.

Nayak, S.S. (1981). Characterization based on record values, *Journal of the Indian Statistical Association*, **19**, 123-127.

Nelson, W. (1968). A method for statistical hazard plotting of incomplete failure data that are arbitrarily censored, *TIS Report 68-C-007*, General Electric Research and Development Center, Schenectady, New York.

Nelson, W. (1969). Hazard plotting for incomplete failure data, *Journal of Quality Technology*, **1**, 27-52.

Nelson, W. (1970). A statistical prediction interval for availability, *IEEE Transactions on Reliability*, **19**, 179-182.

Nelson, W. (1972a). Theory and applications of hazard plotting for censored failure data, *Technometrics*, **14**, 945-966.

Nelson, W. (1972b). Graphical analysis of accelerated life test data with the inverse power law model, *IEEE Transactions on Reliability*, **21**, 2-11.

Nelson, W. (1982). *Applied Life Data Analysis*, John Wiley & Sons, New York.

Nelson, W. (1986). *How to Analyse Data with Simple Plots*, Volume 1, ASQC Basic References in Quality Control: Statistical Techniques, American Society for Quality Control, Milwaukee.

Nelson, W. (1990). *Accelerated Testing: Statistical Models, Test Plans and Data Analysis*, John Wiley & Sons, New York.

Neumann, J. von (1951). Various techniques in connection with random digits, Monte Carlo methods, *National Bureau of Standards*, Applied Mathematics Series, **12**, 36-38, U.S. Government Printing Office, Washington, D.C.

Neuts, M. (1981). *Matrix-Geometric Solutions in Stochastic Models*, John Hopkins University Press, Baltimore, Maryland.

Neuts, M. (1989). *Structured Stochastic Matrices of M/G/1 Types and Their Applications*, Marcel Dekker, New York.

Nevzorov, V.B. (1988). Records, *Theory of Probability and Its Applications*, **32**, 201-228.

Neyman, J. (1959). Optimal asymptotic tests of composite statistical hypotheses, In *Probability and Statistics* (Ed., W. Grenander), pp. 213-234, John Wiley & Sons, New York.

Ng, V.-M. (1984). Prediction intervals for the 2-parameter exponential distribution using incomplete data, *IEEE Transactions on Reliability*, **33**, 188-191.

Oakes, D. (1989). Bivariate survival models induced by frailties, *Journal of the American Statistical Association*, **84**, 487-493.

Oakes, D. and Desu, M.M. (1990). A note on residual life, *Biometrika*, **77**, 409-410.

Ofosu, J.B. (1974). On selection procedures for exponential distributions, *Bulletin of Mathematical Statistics*, **16**, 1-9.

Ogawa, J. (1951). Contributions to the theory of systematic statistics, I, *Osaka Mathematical Journal*, **3**, 175-213.

Ogawa, J. (1952). Contributions to the theory of systematic statistics, II, *Osaka Mathematical Journal*, **4**, 41-61.

O'Hagan, A. (1994). *Bayesian Inference, Kendall's Advanced Theory of Statistics*, Vol. 2B, Edward Arnold, London.

O'Neill, T.J. (1985). Testing for symmetry and independence in a bivariate exponential distribution, *Statistics & Probability Letters*, **3**, 269-274.

Onyeagu, S.I. (1985). A review of estimation of error rates in discriminant analysis, *M.Sc. Dissertation*, A.B.U. Zaria, Nigeria.

Pasternack, B.S. (1972). Sample sizes for clinical trials designed for patient accrual by cohorts, *Journal of Chronic Diseases*, **25**, 673-681.

Pasternack, B.S. and Gilbert, M.S. (1971). Planning the duration of long-term survival time studies designed for accrual by cohorts, *Journal of Chronic Diseases*, **24**, 681-700.

Patel, J.K. (1989). Prediction intervals - A review, *Communications in Statistics - Theory and Methods*, **18**, 2393-2465.

Patnaik, P.B. (1949). The non central χ^2 and F distributions and their applications, *Biometrika*, **36**, 202-232.

Paulson, A.S. (1974). A characterization of the exponential distribution and a bivariate exponential distribution, *Sankhyā, Series A,* **35,** 69-78.

Paulson, E. (1952). A optimum solution to the k-sample slippage problem for the normal distribution, *Annals of Mathematical Statistics,* **23**, 610-616.

Pearson, E.S. and Chandra Sekar, C. (1936). The efficiency of statistical tools and a criterion for the rejection of outlying observations, *Biometrika*, **28**, 308-320.

Pearson, K. (1895). Contributions to the mathematical theory of evolution. II. Skew variations in homogeneous material, *Philosophical Transactions of the Royal Society of London, Series A*, **186**, 343-414.

Pečarić, J.E., Proschan, F., and Tong, Y.L. (1992). *Convex Functions, Partial Orderings, and Statistical Applications*, Academic Press, San Diego.

Perng, S.K. (1978). A test for the equality of two exponential distributions, *Statistica Neerlandica*, **32**, 93-102.

Pettitt, A.N. (1977). Tests for the exponential distribution with censored data using Cramer-von Mises statistics, *Biometrika*, **64**, 629-632.

Pexider, J.V. (1903). Notiz uber funktional theoreme, *Monatsh. Math. Phys.*, **14**, 293-301.

Pian, L.-P. (1987). Application of a mixed model to survival data with long-term survivors, *Unpublished Ph.D. Dissertation*, Medical University of South Carolina.

Pitman, E.J.G. (1937). The closest estimate of statistical parameters, *Proceedings of the Cambridge Philosophical Society*, **33**, 212-222.

Pocock, S.J. (1977). Group sequential methods in the design and analysis of clinical trials, *Biometrika*, **64**, 191-199.

Pollak, M. (1973). On equal distributions, *Annals of Statistics*, **1**, 180-182.

Pregibon, D. (1985). Link tests, In *Encyclopedia of Statistical Sciences* (Eds., S. Kotz and N.L. Johnson), John Wiley & Sons, New York.

Prentice, R.L. (1973). Exponential survivals with censoring and explanatory variables, *Biometrika*, **60**, 279-288.

Prentice, R.L., Kalbfleisch, J.D., Peterson, A.W., Flournoy, N., Farewell, V.T., and Breslow, N.E. (1978). The analysis of failure times in the presence of competing risks, *Biometrics*, **34**, 541-554.

Press, S.J. (1972). *Applied Multivariate Analysis*, Holt, Rinehart, and Winston, New York.

Press, S.J. (1989). *Bayesian Statistics: Principles, Models, and Applications*, John Wiley & Sons, New York.

Proschan, F. and Pike, K. (1967). Tests for monotone failure rate, In *Proceedings of the 5th Berkeley Symposium in Mathematics, Statistics and Probability*, Vol. 3, 293-312, University of California Press, Berkeley.

Proschan, F. and Sullo, P. (1974). Estimating the parameters of a bivariate exponential distribution in several sampling situations, In *Reliability and Biometry* (Eds., F. Proschan and R.J. Serfling), pp. 423-440, SIAM, Philadelphia.

Proschan, F. and Sullo, P. (1976). Estimating the parameters of a multivariate exponential distribution, *Journal of the American Statistical Association*, **71**, 465-472.

Puri, P.S. and Rubin, H. (1970). A characterization based on the absolute difference of two I.I.D. random variables, *Annals of Mathematical Statistics*, **41**, 2113-2122.

Quinn, B.G., McLachlan, G.J., and Hjort, N.L. (1987). A note on the Aitkin-Rubin approach to hypothesis testing in mixture models, *Journal of the Royal Statistical Society, Series B*, **49**, 311-314.

Rachev, S.T. and SenGupta, A. (1992). Geometric stable distributions and Laplace-Weibull mixtures, *Statistics and Decision*, **10**, 251-271.

Raftery, A.E. (1984). A continuous multivariate exponential distribution, *Communications in Statistics - Theory and Methods,* **13,** 947-965.

Raghavachari, M. (1965). Operating characteristic and expected sample size of a sequential probability ratio test for the simple exponential distribution, *Calcutta Statistical Association Bulletin*, **14**, 65-79.

Raghavachari, M. and Starr, N. (1970). Selection problems for some terminal distributions, *Metron*, **28**, 185-197.

Ramachandran, B. (1979). On the 'Strong memorylessness property' of the exponential and geometric probability laws, *Sankhyā, Series A*, **41**, 244-251.

Ramberg, J.S. and Tadikamalla, P.R. (1978). On generation of subsets of order statistics, *Journal of Statistical Computation and Simulation*, **6**, 239-241.

Rao, C.R. (1973). *Linear Statistical Inference and Its Applications*, John Wiley & Sons, New York.

Rao, C.R. and Shanbhag, D.N. (1994). *Choquet-Deny Type Functional Equations with Applications to Stochastic Models*, John Wiley & Sons, Chichester, England.

Rao, V., Savage, I.R., and Sobel, M. (1960). Contribution to the theory of rank order statistics: The two sample censored case, *Annals of Mathematical Statistics*, **31**, 415-426.

Raqab, M.Z. (1992). *Predictors of Future Order Statistics from Type II Censored Samples*, Ph.D. Dissertation, The Ohio State University, Department of Statistics.

Redner, R.A. and Walker, H.F. (1984). Mixture densities, maximum likelihood and the EM algorithm, *SIAM Review*, **26**, 195-239.

Reeder, H.A. (1972). Machine generation of order statistics, *The American Statistician*, **26**, 56-57.

Rényi, A. (1953). On the theory of order statistics, *Acta Mathematica Academiae Scientarium Hungaricae*, **4**, 191-232.

Rényi, A. (1956). A characterization of the Poisson process, *Magyar Tud. Akad. Mat. Kutato Int. Kozl.*, **1**, 519-527 (in Hungarian). Translated into English in *Selected Papers of Alfréd Rényi*, Vol. 1, Akademiai Kiadó, Budapest, 1976.

Robbins, H. (1959). Sequential estimation of the mean of a normal population, In *Probability and Statistics*, H. Cramér Volume (Ed., U. Grenander), pp. 235-245, Almquist and Wiksell, Uppsala, Sweden.

Robbins, H. (1980). Estimation and prediction for mixtures of the exponential distribution, *Proceedings of the National Academy of Science*, **77**, 2382-2383.

Robbins, H., Simons, G., and Starr, N. (1967). A sequential analogue of the Behrens-Fisher problem, *Annals of Mathematical Statistics*, **38**, 1384-1391.

Robinson, G.K. (1991). That BLUP is a good thing: The estimation of random effect, *Statistical Science*, **6**, 15-51.

Rogers, G.S. (1963). An alternative proof of the characterization of the density Ax^B, *American Mathematical Monthly*, **70**, 857-858.

Ross, S.M. (1983). *Stochastic Processes*, John Wiley & Sons, New York.

Rossberg, H.J. (1960). Über die Verteilungsfunktionen der Differenzen und Quotienten von Ranggrössen, *Mathematische Nachrichten*, **21**, 37-79.

Rossberg, H.-J. (1972a). Characterization of the exponential and the Pareto distributions by means of some properties of the distributions which the differences and quotients of order statistics are subject to, *Mathematisch Operationsforschung und Statistik, Series Statistics*, **3**, 207-215.

Rossberg, H.-J. (1972b). Characterization of distribution functions by the independence of certain functions of order statistics, *Sankhyā, Series A*, **34**, 111-120.

Rubinstein, L.V., Gail, M.H., and Santner, T.J. (1981). Planning the duration of a comparative clinical trial with loss-to-follow-up and a period of continued observation, *Journal of Chronic Diseases*, **20**, 230-239.

Ryu, K. (1993). An extension of Marshall and Olkin's bivariate exponential distribution, *Journal of the American Statistical Association*, **88**, 1458-1465.

Sackrowitz, H.B. and Samuel-Cahn, E. (1984). Estimation of the mean of a selected negative exponential population, *Journal of the Royal Statistical Society, Series B*, **46**, 242-249.

Saleh, A.K.Md.E. (1966). Estimation of the parameters of the exponential distribution based on order statistics in censored samples, *Annals of Mathematical Statistics*, **37**, 1717-1735.

Saleh, A.K.Md.E. (1967). Determination of the exact optimum order statistics for estimating the parameters of the exponential distribution from censored samples, *Technometrics*, **9**, 279-292.

Saleh, A.K.Md.E. (1981). Estimating quantiles of exponential distribution, In *Statistics and Related Topics* (Eds., M. Csörgö, D.A. Dawson, J.N.K. Rao and A.K.Md.E. Saleh), pp. 279-283, North-Holland, Amsterdam.

Saleh, A.K.Md.E. and Ali, M.M. (1966). Asymptotic optimum quantiles for the estimation of the parameters of the negative exponential distribution, *Annals of Mathematical Statistics*, **37**, 143-151.

Saleh, A.K.Md.E., Scott, C., and Junkins, D.B. (1975). Exact first and second order moments of order statistics from the truncated exponential distribution, *Naval Research Logistics Quarterly*, **22**, 65-77.

Samanta, M. (1986). On asymptotic optimality of some tests for exponential distribution, *Australian Journal of Statistics*, **28**, 164-172.

Santner, T.J. (1976). A two-stage procedure for selecting δ^*-optimal means in the normal model, *Communications in Statistics - Theory and Methods*, **5**, 283-292.

Sarhan, A.E. (1954). Estimation of the mean and standard deviation by order statistics, *Annals of Mathematical Statistics*, **25**, 317-328.

Sarhan, A.E. (1955). Estimation of the mean and standard deviation by order statistics, Part III, *Annals of Mathematical Statistics*, **26**, 576-592.

Sarhan, A.E. and Greenberg, B.G. (1957). Tables for best linear estimates by order statistics of the parameters of single exponential distributions from singly and doubly censored samples, *Journal of the American Statistical Association*, **52**, 58-87.

Sarhan, A.E. and Greenberg, B.G. (Eds.) (1962). *Contributions to Order Statistics*, John Wiley & Sons, New York.

Sarkadi, K. (1975). The consistency of the Shapiro-Francia test, *Biometrika*, **62**, 445-450.

Sarkar, S.K. (1987). A continuous bivariate exponential distribution, *Journal of the American Statistical Association*, **82**, 667-675.

Sathe, Y.S. and Varde, S.D. (1969). Minimum variance unbiased estimation of reliability for the truncated exponential distribution, *Technometrics*, **11**, 609-612.

Satterthwaite, F.E. (1941). Synthesis of variance, *Psychometrika*, **6**, 309-316.

Schmeiser, B.W. (1978). Order statistics in digital computer simulation. A survey, *Proceedings of the 1978 Winter Simulation Conference*, 136-140.

Schmeiser, B.W. and Babu, A.J.G. (1980). Beta variate generation via exponential majorizing functions, *Operations Research*, **28**, 917-926.

Schmeiser, B.W. and Lal, R. (1980). Squeeze methods for generating gamma variates, *Journal of the American Statistical Association*, **75**, 679-682.

Schucany, W.R. (1972). Order statistics in simulation, *Journal of Statistical Computation and Simulation*, **1**, 281-286.

Sen, A. and Bhattacharyya, G.K. (1995). Inference procedures for the linear failure rate model, *Journal of Statistical Planning and Inference* (to appear).

Sen, K. and Jain, M.B. (1990). A test for bivariate exponentiality against BHNBUE alternative, *Communications in Statistics - Theory and Methods*, **19**, 1827-1835.

Sen, P.K. (1973a). An asymptotically optimal test for the bundle strength of filaments, *Journal of Applied Probability*, **10**, 586-596.

Sen, P.K. (1973b). On fixed-size confidence bands for the bundle-strength of filaments, *Annals of Statistics*, **1**, 526-537.

Sen, P.K. (1976). Weak convergence of progressively censored likelihood ratio statistics and its role in asymptotic theory of life testing, *Annals of Statistics*, **4**, 1247-1257.

Sen, P.K. (1980). On time-sequential estimation of the mean of an exponential distribution, *Communications in Statistics - Theory and Methods*, **9**, 27-38.

Sen, P.K. (1981). *Sequential Nonparametrics: Invariance Principles and Statistical Inference*, John Wiley & Sons, New York.

Sen, P.K. (1986). Nonparametric estimators of availability under provisions of spare and repair, II, In *Reliability and Quality Control* (Ed., A.P. Basu), pp. 297-308, North-Holland, Amsterdam.

Sen, P.K. (1993). Resampling methods for the extrema of certain sample functions, In *Probability and Statistics: Proceedings of the Calcutta Triennial International Conference in Statistics* (December 1991), (Eds., S.K. Basu and B.K. Sinha), pp. 65-79.

Sen, P.K. (1994). Extreme value theory for fibre bundles, In *Extreme Value Theory and Applications*, (Ed., J. Galambos), pp. 77-92, Kluwer, Newell.

Sen, P.K. and Bhattacharjee, M.C. (1986). Nonparametric estimators of availability under provisions of spare and repair, I, In *Reliability and Quality Control* (Ed., A.P. Basu), pp. 281-296, North-Holland, Amsterdam.

Sen, P.K., Bhattacharyya, B.B., and Suh, M.W. (1973). Limiting behavior of the extrema of certain sample functions, *Annals of Statistics*, **1**, 297-311.

SenGupta, A. (1991). A review of optimality of multivariate tests, In "Special Issue on Multivariate Optimality and Related Topics" (Eds., S.R. Jammalamadaka and A. SenGupta), *Statistics & Probability Letters*, **12**, 527-535.

SenGupta, A. (1993). On construction of optimal tests in several multivariate exponential distributions, *Submitted for publication.*

SenGupta, A. and Pal, C. (1991). Locally optimal tests for no contamination in standard symmetric multivariate normal mixtures, In "Special Issue on Reliability Theory" (Ed., A.P. Basu), *Journal of Statistical Planning and Inference*, **29**, 145-155.

SenGupta, A. and Pal, C. (1993). Optimal tests for no contamination in symmetric multivariate normal mixtures, *Annals of the Institute of Statistical Mathematics*, **45**, 137-146.

SenGupta, A. and Vermeire, L. (1986). Locally optimal tests for multiparameter hypotheses, *Journal of the American Statistical Association*, **81**, 819-825.

Seshadri, V., Csörgö, M., and Stephens, M.A. (1969). Tests for the exponential distribution using Kolmogorov-type statistics, *Journal of the Royal Statistical Society, Series B*, **31**, 499-509.

Sethuraman, J. (1965). On a characterization of the three limiting types of the extreme, *Sankhyā, Series A*, 357-364.

Shah, S.M. and Rathod, V.R. (1978). Prediction intervals for future order statistics in two-parameter exponential distributions, *Journal of the Indian Statistical Association*, **16**, 113-122.

Shaked, M. and Shanthikumar, J.G. (1994). *Stochastic Orders and Their Applications*, Academic Press, San Diego, California.

Shannon, C.E. (1948). The mathematical theory of communication, *Bell System Technical Journal*, 279-423, 623-656.

Shanthikumar, J.G. (1985). Bilateral phase type distributions, *Naval Research Logistics Quarterly*, **32**, 119-136.

Shanthikumar, J.G. and Chandra, M.J. (1982). Application of level crossing analysis to discrete state processes in queueing systems, *Naval Research Logistics Quarterly*, **29**, 593-608.

Shapiro, S.S. (1990). *How To Test Normality and Other Distributional Assumptions*, Volume 3, ASQC Basic References in Quality Control: Statistical Techniques, American Society for Quality Control, Milwaukee.

Shapiro, S.S. and Balakrishnan, N. (1994). Testing for exponentiality with progressively censored data, *Submitted for publication.*

Shapiro, S.S. and Francia, R.S. (1972). Approximate analysis of variance test for normality, *Journal of the American Statistical Association*, **67**, 215-225.

Shapiro, S.S. and Wilk, M.B. (1972). An analysis of variance test for the exponential distribution (complete samples), *Technometrics*, **14**, 355-370.

Shayib, M.A. and Awad, A.M. (1990). Prediction interval for the difference between two sample means from exponential population: A Bayesian treatment, *Pakistan Journal of Statistics*, **6**, 1-23.

Shayib, M.A., Awad, A.M., and Dawagreh, A.M. (1986). Large sample prediction intervals for a future sample mean: A comparative study, *Journal of Statistical Computation and Simulation*, **24**, 255-270.

Shetty, B.N. and Joshi, P.C. (1986). Testing equality of location parameters of two exponential distributions, *Aligarh Journal of Statistics*, **6**, 11-25.

Shetty, B.N. and Joshi, P.C. (1987). Estimation of parameters of k exponential distributions in doubly censored samples, *Communications in Statistics - Theory and Methods*, **16**, 2115-2123.

Shetty, B.N. and Joshi, P.C. (1989). Likelihood ratio test for testing equality of location parameters of two exponential distributions from doubly censored samples, *Communications in Statistics - Theory and Methods*, **18**, 2063-2072.

Shimizu, R. (1979). On a lack of memory property of the exponential distribution, *Annals of the Institute of Statistical Mathematics*, **31**, 309-313.

Shorack, G.R. (1972). The best test of exponentiality against gamma alternatives, *Journal of the American Statistical Association*, **67**, 213-214.

Singh, N. (1983). The likelihood ratio test for the equality of location parameters of k (≥ 2) exponential populations based on Type II censored samples, *Technometrics*, **25**, 193-195.

Singh, N. (1985). A simple and asymptotically optimal test for the equality of k (≥ 2) exponential distributions based on Type II censored samples, *Communications in Statistics - Theory and Methods*, **14**, 1615-1625.

Singh, N. and Narayan, P. (1983). The likelihood ratio test for the equality of k (≥ 3) two-parameter exponential distributions based on Type II censored samples, *Journal of Statistical Computation and Simulation*, **18**, 287-297.

Singpurwalla, N.D. (1971). A problem in accelerated life testing, *Journal of the American Statistical Association*, **66**, 841-845.

Sinha, B.K. and Sinha, B.K. (1993). An application of bivariate exponential models an related inference, *Journal of Statistical Planning and Inference* (To appear).

Sinha, S.K. (1986). *Reliability and Life Testing*, John Wiley & Sons, New York.

Sloan, J.A. and Sinha, S.K. (1991). Bayesian predictive intervals for a mixture of exponential failure-time distributions with censored samples, *Statistics & Probability Letters*, **11**, 537-545.

Smith, R.M. and Bain, L.J. (1976). Correlation type goodness of fit statistics with censored data, *Communications in Statistics - Theory and Methods*, **5**, 119-132.

Spinelli, J.J. and Stephens, M.A. (1987). Tests for exponentiality when origin and scale parameters are unknown, *Technometrics*, **29**, 471-476.

Spurrier, J.D. (1984). An overview of tests for exponentiality, *Communications in Statistics - Theory and Methods*, **13**, 1635-1654.

Spurrier, J. and Wei, L.J. (1980). A test for the parameter of the exponential distribution in the Type I censoring case, *Journal of the American Statistical Association*, **75**, 405-409.

Srivastava, M.S. (1970). On a sequential analogue of the Behrens-Fisher problem, *Journal of the Royal Statistical Society, Series B*, **38**, 219-230.

Stadje, W. (1994). A characterization of the exponential distribution involving absolute differences of i.i.d. random variables, *Proceedings of the American Mathematical Society* (To appear).

Starr, N. and Woodroofe, M. (1972). Further remarks on sequential estimation: the exponential case, *Annals of Methematical Statistics*, **43**, 1147-1154.

Steffensen, J.F. (1930). *Some Recent Research in the Theory of Statistics and Actuarial Science*, Cambridge University Press, Cambridge, England.

Stein, C. (1945). A two-sample test for a linear hypothesis whose power is independent of the variance, *Annals of Mathematical Statistics*, **16**, 243-258.

Stein, C. (1949). Some problems in sequential estimation, *Econometrica*, **17**, 77-78.

Stephens, M.A. (1974). EDF statistics for goodness of fit and some comparisons, *Journal of the American Statistical Association*, **69**, 730-737.

Stephens, M.A. (1976). Asymptotic results for goodness-of-fit statistics with unknown parameters, *Annals of Statistics*, **4**, 357-368.

Stephens, M.A. (1978). On the W test for exponentiality with origin known, *Technometrics*, **20**, 33-35.

Steutel, F.W. (1967). Note on the infinite divisibility of exponential mixtures, *Annals of Mathematical Statistics*, **38**, 1303-1305.

Steutel, F.W. and Thiemann, J.G.F. (1989). On the independence of integer and fractional parts, *Statistica Neerlandica*, **43**, 53-59.

Stewart, L.T. and Johnson, J.D. (1972). Determining the optimal burn-in and replacement times using Bayesian decision theory, *IEEE Transactions on Reliability*, **R-21**, 168-173.

Stoyan, D. (1983). *Comparison Methods for Queues and Other Stochastic Models*, John Wiley & Sons, New York.

Sukhatme, P.V. (1936). On the analysis of k samples from exponential populations with especial reference to the problem of random intervals, *Statistical Research Memoirs*, **1**, 94-112.

Sukhatme, P.V. (1937). Tests of significance for samples of the χ^2 population with two degrees of freedom, *Annals of Eugenics*, **8**, 52-56.

Swanepoel, J.W.H. and vanWyk, J.W.J. (1982). Fixed-width confidence intervals for the location parameter of an exponential distribution, *Communications in Statistics - Theory and Methods*, **11**, 1279-1289.

Sweeting, T.J. (1983). Independent scale-free spacings for the exponential and uniform distribution, *Statistics & Probability Letters*, **1**, 115-119.

Sweeting, T.J. (1986). Asymptotically independent scale-free spacings with applications to discordancy testing, *Annals of Statistics*, **14**, 1485-1496.

Tadikamalla, P.R. (1975). Modeling and generating stochastic methods for simulation studies, *Ph.D. Thesis*, University of Iowa, Iowa City.

Tadikamalla, P.R. (1978). Computer generation of gamma random variables, *Communications of the ACM*, **21**, 419-422.

Tadikamalla, P.R. (1984). Modeling and generating stochastic inputs for simulation studies, *American Journal of Mathematics and Management Sciences*, **3** & **4**, 203-223.

Tadikamalla, P.R. and Johnson, M.E. (1981). A complete guide to gamma variate generation, *American Journal of Mathematical and Management Sciences*, **1**, 213-236.

Takada, Y. (1979). The shortest invariant prediction interval for the largest observation from the exponential distribution, *Journal of the Japan Statistical Society*, **9**, 87-91.

Takada, Y. (1981). Relation of the best invariant predictor and the best unbiased predictor in location and scale families, *Annals of Statistics*, **9**, 917-921.

Takada, Y. (1985). Prediction limit for observation from the exponential distribution, *Canadian Journal of Statistics*, **13**, 325-330.

Takada, Y. (1991). Median unbiasedness in an invariant prediction problem, *Statistics & Probability Letters*, **12**, 281-283.

Takagi, H. (1991). *Queueing Analysis, Vol. 1: Vacation and Priority Systems*, North-Holland, Amsterdam.

Tanis, E.A. (1964). Linear forms in the order statistics from an exponential distribution, *Annals of Mathematical Statistics*, **35**, 270-276.

Taulbee, J.D. and Symons, M.J. (1983). Sample size and duration for cohort studies of survival time with covariables, *Biometrics*, **39**, 351-360.

Teicher, H. (1961). Identifiability of mixtures, *Annals of Mathematical Statistics*, **32**, 244-248.

Teicher, H. (1961). Maximum likelihood characterization of distributions, *Annals of Mathematical Statistics*, **32**, 1214-1222.

Teissier, G. (1934). Recherches sur le vicillissement et sur les lois de mortalité, *Annals of Physics, Biology and Physical Chemistry*, **10**, 237-264.

Thall, P.F. (1979). Huber-sense robust M-estimation of a scale parameter, with application to the exponential distribution, *Journal of the American Statistical Association*, **74**, 147-152.

Thall, P.F. (1979). Huber-sense robust M-estimation of a scale parameter, with application to the exponential distribution, *Journal of the American Statistical Association*, **74**, 147-152.

Thiagaraja, K. and Paul, S.R. (1990). Testing for the equality of scale parameters of K (≥ 2) exponential populations based on complete and Type II censored samples, *Communications in Statistics - Simulation and Computation*, **19**, 891-902.

Thoman, D.R., Bain, L.J., and Antle, C.E. (1969). Inference on the parameters of the Weibull distribution, *Technometrics*, **11**, 445-460.

Thompson, R.D. and Basu, A.P. (1993). Bayesian reliability of stress-strength systems, In *Advances in Reliability* (Ed., A.P. Basu), pp. 411-421, North-Holland, Amsterdam.

Tierney, L. and Kadane, J. (1986). Accurate approximations for posterior moments and marginals, *Journal of the American Statistical Association*, **81**, 82-86.

Tietjen, G.L. and Moore, R.H. (1972). Some Grubbs-type statistics for the detection of several outliers, *Technometrics*, **14**, 583-597.

Tiku, M.L. (1975). A new statistic for testing suspected outliers, *Communications in Statistics, Series A*, **4**, 737-752.

Tiku, M.L. (1980). Goodness of fit statistics based on the spacings of complete or censored samples, *Australian Journal of Statistics*, **22**, 260-275.

Tiku, M.L. (1981). Testing equality of location parameters of two exponential distributions, *Aligarh Journal of Statistics*, **1**, 1-7.

Tiku, M.L., Tan, W.Y., and Balakrishnan, N. (1986). *Robust Inference*, Marcel Dekker, New York.

Titterington, D.M., Smith, A.F.M., and Makov, U.E. (1985). *Statistical Analysis of Finite Mixture Distributions*, John Wiley & Sons, New York.

Tong, Y.L. (1970). Multi-stage interval estimations of the largest mean of k normal populations, *Journal of the Royal Statistical Society, Series B*, **32**, 272-277.

Tosch, T.J. and Holmes, P.T. (1980). A bivariate failure model, *Journal of the American Statistical Association,* **75,** 415-417.

Truax, D.R. (1953). An optimum slippage test for the variances of k, normal distributions, *Annals of Mathematical Statistics*, **24**, 669-674.

Tukey, J.W. (1960). A survey of sampling from contaminated distributions, In *Contributions to Probability and Statistics* (Ed., I. Olkin), Stanford University Press, Stanford, California.

Umbach, D., Ali, M. Masoom, and Hassanein, K.M. (1981). Small sample estimation of exponential quantiles with two order statistics, *Aligarh Journal of Statistics*, **1**, 113-120.

Upadhyay, S.K. and Pandey, M. (1989). Prediction limits for an exponential distribution: A Bayes predictive distribution approach, *IEEE Transactions on Reliability*, **38**, 599-602.

Uthoff, V.A. (1970). An optimum test property of two well known statistics, *Journal of the American Statistical Association*, **65**, 1597-1600.

van Soest, J. (1969). Some goodness of fit tests for the exponential distribution, *Statistica Neerlandica*, **23**, 41-51.

Varde, S.D. (1969). Life testing and reliability estimation for the two-parameter exponential distribution, *Journal of the American Statistical Association*, **64**, 621-631.

Veale, J.R. (1975). Improved estimation of expected life when one identified spurious observation may be present, *Journal of the American Statistical Association*, **70**, 398-401.

Vellaisamy, P. and Sharma, D. (1988). Estimation of the mean of the selected gamma population, *Communications in Statistics - Theory and Methods*, **17**, 2797-2817.

Vellaisamy, P. and Sharma, D. (1989). A note on the estimation of the mean of the selected gamma population, *Communications in Statistics - Theory and Methods*, **18**, 555-560.

Verhagen, A.M.W. (1961). The estimation of regression and error-scale parameters when the joint distribution of the errors is of any continuous form and known apart from a scale parameter, *Biometrika*, **48**, 125-132.

Viveros, R. and Balakrishnan, N. (1994). Interval estimation of parameters of life from progressively censored data, *Technometrics*, **36**, 84-91.

Voit, E.O. (1992). The S-distribution: A tool for approximation and classification of univariate, unimodal probability distributions, *Biometrical Journal*, **34**, 855-878.

Wada, C.Y. and Sen, P.K. (1994). Restricted alternative test in a parametric model with competing risk data, *Journal of Statistical Planning and Inference* (To appear).

Wadsworth, H.M. (1990). *Handbook of Statistical Methods for Engineers and Scientists*, McGraw-Hill, New York.

Wald, A. (1947). *Sequential Analysis*, John Wiley & Sons, New York.

Walrand, J. (1988). *An Introduction to Queueing Networks*, Prentice Hall, Englewood Cliffs, New Jersey.

Wang, S.T., Klein, J.P., and Moeschberger, M.L. (1993). Semi-parametric estimation of covariate effects using the positive stable frailty model, *Technical Report*, The Ohio State University, Columbus.

Wang, Y.H. and Srivastava, R.C. (1980). A characterization of the exponential and related distributions by linear regression, *Annals of Statistics*, **8**, 217-220.

Weibull, W. (1939). The phenomenon of rupture in solids, *Ingen. Vetensk. Akad. Handl.*, **153**.

Weibull, W. (1951). A statistical distribution function of wide applicability, *Journal of Applied Mechanics*, **18**, 293-297.

Weier, D.R. (1981). Bayes estimation for bivariate survival models based on the exponential distribution, *Communications in Statistics - Theory and Methods*, **10**, 1415-1427.

Weier, D.R. and Basu, A.P. (1980). Testing for independence in multivariate exponential distributions, *Australian Journal of Statistics*, **22**, 276-288.

Weier, D.R. and Basu, A.P. (1981). On tests of independence under bivariate exponential models, In *Statistical Distributions in Scientific Work - Vol. 5* (Eds., C.P. Taillie, G.P. Patil, and B. Baldessari), pp. 169-180, D. Reidel, Dordrecht.

Weinman, D.G., Dugger, G., Franck, W.E., and Hewett, J.E. (1973). On a test for the equality of two exponential distributions, *Technometrics*, **15**, 177-182.

Weiss, L. (1962). On sequential tests which mimize the maximum expected sample size, *Journal of the American Statistical Association*, **57**, 551-566.

Welch, B.L. (1939). On confidence limits and sufficiency, with special reference to parameters of location, *Annals of Mathematical Statistics*, **10**, 58-69.

Wetherill, G.B. (1977). *Sampling Inspection and Quality Control*, Second edition, Chapman and Hall, London.

White, H. (1982). Maximum likelihood estimation of misspecified models, *Econometrica*, **50**, 1-25.

Whitmore, G.A. and Lee, M.-L.T. (1991). A multivariate survival distribution generated by an inverse Gaussian mixture of exponentials, *Technometrics*, **33**, 39-50.

Whitt, W. (1991). A review of $L=\lambda W$ and extensions, *Queueing Systems: Theory and Applications*, **9**, 235-269.

Wiernik, G. et al. (1991). Final report on the second British Institute of Radiology fractionation study: short vs. long overall treatment times for radiotherapy of carcinoma of the laryngo-pharynx, *British Journal of Radiology*, **64**, 232-241.

Wijsman, R.A. (1967). Cross-section of orbits and appliations to densities of maximal invariants, In *Proceedings of the Fifth Berkeley Symposium*, Vol. 1, pp. 389-400.

Wilson, E.B. (1899). Note on the function satisfying the functional relation $\phi(u)\phi(v) = \phi(u+v)$, *Annals of Mathematics*, **1**, 47-48.

Wolff, R. (1989). *Stochastic Modeling and the Theory of Queues*, Prentice Hall, Englewood Cliffs, New Jersey.

Woodroofe, M. (1977). Second order approximations for sequential point and interval estimation, *Annals of Statistics*, **5**, 984-995.

Woodroofe, M. (1982). *Nonlinear Renewal Theory in Sequential Analysis*, SIAM, Philadelphia.

Woodroofe, M. (1985). Asymptotic local minimaxity in sequential point estimation, *Annals of Statistics*, **13**, 676-688.

Woodroofe, M. (1987). Asymptotically optimal sequential point estimation in three stages, In *New Perspectives in Theoretical and Applied Statistics* (Ed., M.L. Puri), pp. 397-411, John Wiley & Sons, New York.

Wright, F.T., Engelhardt, M., and Bain, L.J. (1978). Inferences for the two-parameter exponential distribution under type I censored sampling, *Journal of the American Statistical Association*, **73**, 650-655.

Zacks, S. (1973). Sequential design for a fixed-width interval estimation of the common mean of two normal distributions, I. The case of one variance known, *Journal of the American Statistical Association*, **68**, 422-427.

Zacks, S. (1984). Estimating the shift to wear-out systems having exponential-Weibull life distribution, *Operations Research*, **32**, 741-749.

Zacks, S. (1986). Estimating the scale parameter of an exponential distribution from a sample of time-censored r-th order statistics, *Journal of the American Statistical Association*, **81**, 205-210.

Zacks, S. (1992). *Introduction to Reliability Analysis: Probability Models and Statistical Methods*, Springer-Verlag, New York.

Zacks, S. and Even, M. (1966). The efficiencies in small samples of the maximum likelihood and best unbiased estimators of reliability functions, *Journal of the American Statistical Association*, **61**, 1033-1051.

Zeigler, R.K. and Goldman, A. (1972). A double sampling plan for comparing two variances, *Journal of the American Statistical Association*, **67**, 698-701.

Zelen, M. (1959). Factorial experiments in life testing, *Technometrics*, **1**, 269-288.

Zelen, M. (1966). Applications of exponential models to problems in cancer research (with discussion), *Journal of the Royal Statistical Society, Series A*, **129**, 368-398.

Zelen, M. and Dannemiller, M.C. (1961). The robustness of life testing procedures derived from the exponential distribution, *Technometrics*, **3**, 29-49.

Zhang, Q. (1992). Two-way exponential classification, *M.Sc. Project*, McMaster University, Hamilton, Ontario, Canada.

Zippin, C. and Armitage, P. (1966). Use of concomitant variables and incomplete survival information in the estimation of an exponential survival parameter, *Biometrics*, **22**, 665-672.

Author Index

Subject Index